Electrochemical Sensors, Biosensors and Their Biomedical Applications

Electrochemical Sensors, Biosensors and Their Biomedical Applications

Edited by

Xueji Zhang
World Precision Instruments, Inc., Sarasota, Florida, USA

Huangxian Ju
Nanjing University, Nanjing, P.R. China

Joseph Wang
Arizona State University, Tempe, Arizona, USA

ELSEVIER

AMSTERDAM • BOSTON • HEIDELBERG • LONDON • NEW YORK • OXFORD
PARIS • SAN DIEGO • SAN FRANCISCO • SINGAPORE • SYDNEY • TOKYO
Academic Press is an imprint of Elsevier

Academic Press is an imprint of Elsevier
525 B Street, Suite 1900, San Diego, CA 92101-4495, USA
360 Park Avenue South, New York, NY 10010-1710
84 Theobald's Road, London WC1X 8RR, UK
30 Corporate Drive, Suite 400, Burlington, MA 01803, USA

First edition 2008

Library of Congress Cataloging in Publication Data
A catalog record for this book is available from the Library of Congress

British Library Cataloguing in Publication Data
A catalogue record for this book is available from the British Library

ISBN: 978-0-12-373738-0

For information on all Academic Press publications
visit our web site at http://books.elsevier.com

Typeset by Charon Tec Ltd (A Macmillan Company), Chennai, India
www.charontec.com

Printed and bound by CPI Group (UK) Ltd, Croydon, CR0 4YY

Transferred to Digital Print 2011

CONTENTS

Chapter 2 Biosensors for pesticides

Huangxian Ju and Vivek Babu Kandimalla

Chapter 3 Electrochemical glucose biosensors

Joseph Wang

Chapter 8 Electrochemical sensors for the determination of hydrogen sulfide production in biological samples

David W. Kraus, Jeannette E. Doeller, and Xueji Zhang

Chapter 13 Chemical and biological sensors based on electroactive inorganic polycrystals

Arkady Karyakin

Chapter 14 Nanoparticles-based biosensors and bioassays

Guodong Liu, Jun Wang, Yuehe Lin, and Joseph Wang

Chapter 15 Electrochemical sensors based on carbon nanotubes

Manliang Feng, Heyou Han, Jingdong Zhang, and Hiroyasu Tachikawa

LIST OF CONTRIBUTORS

Eric Bakker, Nanochemistry Research Institute, Department of Applied Chemistry, Curtin University of Technology, Perth, WA 6845, Australia

Jeannette E. Doeller, University of Alabama at Birmingham, Department of Environmental Health Sciences 1665 University Boulevard, Ryals 530B Birmingham AL 35294-0022, USA

Hua Dong, School of Chemical and Biomedical Engineering, 70 Nanyang Drive, Nanyang Technological University, Singapore 637457

Manliang Feng, Department of Chemistry, Jackson State University, 1400 J.R. Lynch Street, Jackson, MS 39217, USA

Jeremy M. Fowler, Department of Chemistry and Biomolecular Sciences, Macquarie University, Sydney, NSW 2109, Australia

Kai H. Goh, School of Chemical and Biomedical Engineering, 70 Nanyang Drive, Nanyang Technological University, Singapore 637457

H. Brian Halsall, Department of Chemistry, University of Cincinnati, P.O. Box 210172, Cincinnati, OH 45221-0172, USA

Heyou Han, Department of Chemistry, Huazhong Agricultural University, Wuhan 430070, China

William R. Heineman, Department of Chemistry, University of Cincinnati, P.O. Box 210172, Cincinnati, OH 45221-0172, USA

Shengshui Hu, Department of Chemistry, Wuhan University, Wuhan 430072, P.R. China

Huangxian Ju, Key Laboratory of Analytical Chemistry for Life Science (Education Ministry of China), Department of Chemistry, Nanjing University, Nanjing 210093, China

Vivek Babu Kandimalla, Key Laboratory of Analytical Chemistry for Life Science (Education Ministry of China), Department of Chemistry, Nanjing University, Nanjing 210093, China

Arkady Karyakin, Chemistry Faculty of M.V. Lomonosov Moscow State University, Moscow, Russia

David W. Kraus, University of Alabama at Birmingham, Departments of Biology and Environmental Health Sciences 1665 University Boulevard, Ryals 530B Birmingham AL 35294-0022, USA

Dónal Leech, Department of Chemistry, National University of Ireland, Galway, Ireland Marie Pellissier and Frédéric Barrière Université de Rennes I, UMR CNRS 6226, 34052 Rennes, France

Chang Ming Li, School of Chemical and Biomedical Engineering, 70 Nanyang Drive, Nanyang Technological University, Singapore 637457

Yuehe Lin, Pacific Northwest National Laboratory, Richland, WA, 99352, USA

Guodong Liu, Pacific Northwest National Laboratory, Richland, WA, 99352, USA

Qing Lu, Department of Chemistry, Wuhan University, Wuhan 430072, P.R. China

Sergey Makarychev-Mikhailov, Department of Chemistry, Purdue University, West Lafayette, IN 47907, USA

Lanqun Mao, Center for Molecular Science, Institute of Chemistry, Chinese Academy of Sciences, Beijing 100080, P.R China

Takeo Ohsaka, Department of Electronic Chemistry, Interdisciplinary Graduate School of Science and Engineering, Tokyo Institute of Technology, 4259 Nagatsuta, Midori-ku, Yokohama 226-8502, Japan

Arshak Poghossian, Aachen University of Applied Sciences, Jülich Campus, Institute of Nano- and Biotechnologies, Ginsterweg 1, D-52428 Jülich, Germany; and Research Centre Jülich, Institute of Bio- and Nanosystems, D-52425 Jülich, Germany

Michael J. Schöning, Aachen University of Applied Sciences, Jülich Campus, Institute of Nano- and Biotechnologies, Ginsterweg 1, D-52428 Jülich, Germany; and Research Centre Jülich, Institute of Bio- and Nanosystems, D-52425 Jülich, Germany

Guoli Shen, State Key Laboratory for Chemo/Biosensing and Chemometrics, College of Chemistry and Chemical Engineering, Hunan University, Changsha, Hunan 410082, P.R. China

Alexey Shvarev, Department of Chemistry, Oregon State University, Corvallis, OR 97331, USA

Hiroyasu Tachikawa, Department of Chemistry, Jackson State University, 1400 J.R. Lynch Street, Jackson, MS 39217, USA

Yang Tian, Department of Chemistry, Tongji University, Siping Road 1239, Shanghai 200092, P.R China

Vijay Shyam Tripathi, Key Laboratory of Analytical Chemistry for Life Science (Education Ministry of China), Department of Chemistry, Nanjing University, Nanjing 210093, China

Hua Wang, State Key Laboratory for Chemo/Biosensing and Chemometrics, College of Chemistry and Chemical Engineering, Hunan University, Changsha, Hunan 410082, P.R. China

Joseph Wang, Biodesign Institute, Center for Bioelectronics and Biosensors, Departments of Chemical and Materials Engineering and Biochemistry, Box 875801, Arizona State University, Tempe, AZ 85387-5801, USA

Jun Wang, Pacific Northwest National Laboratory, Richland, WA, 99352, USA

Danny K.Y. Wong, Department of Chemistry and Biomolecular Sciences, Macquarie University, Sydney, NSW 2109, Australia

Ruqin Yu, State Key Laboratory for Chemo/Biosensing and Chemometrics, College of Chemistry and Chemical Engineering, Hunan University, Changsha, Hunan 410082, P.R. China

Jingdong Zhang, College of Environmental Science and Engineering Huazhong University of Science and Technology, Wuhan 430074, China

Xueji Zhang, Department of Chemistry, World Precision Instruments Inc., 175 Sarasota Center Boulevard Sarasota, FL 34240-9258, USA

David Daomin Zhou, Second Sight Medical Products, Inc., Sylmar, California, USA

Qin Zhou, School of Chemical and Biomedical Engineering, 70 Nanyang Drive, Nanyang Technological University, Singapore 637457

Yanxia Xu, Department of Chemistry, Wuhan University, Wuhan 430072, P.R. China

PREFACE

The development of chemical and biological sensors is currently one of the most active areas of analytical research. Sensors are small devices that incorporate a recognition element with a signal transducer. Such devices can be used for direct measurement of the analyte in the sample matrix. There are a variety of combinations of recognition elements and signal transducers. Electrochemical sensors, in which an electrode is used as the transduction element, represent an important subclass of chemical sensors. Such devices hold a leading position among sensors presently available and have found a vast range of important applications in the fields of clinical, industrial, environmental, and agricultural analyses. The field of sensors, in general, and electrochemical sensors, is interdisciplinary and future advances are likely to occur from progress in several disciplines.

Up to the 1960s, the pH glass electrode could be considered as the only widely used chemical sensor. The modern concept of biosensors owes much to L. Clark Jr. who introduced the amperometric glucose enzyme electrode in 1962. Over the past four decades we have witnessed the evolution of sophisticated sensing devices based on different transduction principles and recognition elements. With the start of the 21st century, such devices are routinely being used for a wide range of clinical, environmental, industrial and security applications. Future progress in sensor development would require extensive multidisciplinary efforts for meeting emerging needs ranging from early detection of disease biomarkers or minimally invasive continuous monitoring of glucose and lactate, to early detection of biological warfare agents. Research into electrochemical sensors and their biomedical applications is proceeding in a number of exciting directions, as reflected by the content of this book.

The goal of this book is to cover the full scope of electrochemical sensors and biosensors. It offers a survey of the principles, design and biomedical applications of the most popular types of electrochemical devices in use today. The book is aimed at all scientists and engineers who are interested in developing and using chemical sensors and biosensors. By discussing recent advances, it is hoped to bridge the common gap between research literature and standard textbooks.

The material is presented in 17 chapters, covering topics such as trends in ion selective electrodes, advances in electrochemical immunosensors, modern glucose biosensors for diabetes management, biosensors based on nanomaterials (e.g. nanotubes or nanocrystals), biosensors for nitric oxide and superoxide, or biosensors for pesticides.

We are fortunate to have assembled contributions from leading experts in the field, and we are grateful to all contributors to this book. Last but not least, we warmly acknowledge the gracious support of our families.

Xueji Zhang, Sarasota, FL, USA
Huangxian Ju, Nanjing, P.R. China
Joseph Wang, Tempe, AZ, USA

CHAPTER 1

Nitric oxide (NO) electrochemical sensors

Xueji Zhang

1.1 INTRODUCTION

1.1.1 Significance of nitric oxide in life science

Nitric oxide (NO) is reported to have been first prepared by the Belgian scientist Jan Baptist van Helmont in about 1620 [1]. The chemical properties of NO were first characterized by Joseph Priestly in 1772. However, until the mid-1980s, NO was regarded as an atmospheric pollutant and bacterial metabolite. Nitric oxide (NO) is a hydrophobic, highly labile free radical that is catalytically produced in biological systems from the reduction of L-arginine by nitric oxide synthase (NOS) to form L-citrulline, which produces NO in the process. In biological systems NO has long been known to play various roles in physiology, pathology and pharmacology [2]. In 1987 NO was identified as being responsible for the physiological actions of endothelium-derived relaxing factor (EDRF) [3]. Since that discovery, NO has been shown to be involved in numerous biological processes such as: vasodilatation and molecular messaging [3]; penile erection [4]; neurotransmission [5, 6]; inhibition of platelet aggregation [7]; blood pressure regulation [8]; immune response [9]; and as a mediator in a wide range of both anti-tumor and anti-microbial activities [10, 11]. In addition, NO has been implicated in a number of diseases including diabetes [12], and Parkinson's and Alzheimer's diseases [13]. The importance of NO was confirmed in 1992 when *Science* magazine declared NO the "Molecule of the Year" and in 1998, F. Furchgott, Louis J. Ignarro, and Ferid Murad were awarded the Nobel Prize in Physiology and Medicine for unraveling the complex nature of this simple molecule. Despite the obvious importance of NO in so many biological processes, less than 10% of the thousands of scientific publications over the last decade dedicated to the field of NO research involve its direct measurement.

1.1.2 Methods of measurement of nitric oxide in physiology

As stated above, NO plays a significant role in a variety of biological processes where its spatial and temporal concentration is of extreme importance. However, the measurement of NO is quite difficult due to its short half-life ($\approx$5 sec) and high reactivity with other biological components such as superoxide, oxygen, thiols, and others. To date, several techniques have been developed for the measurement of NO including: chemiluminescence [14, 15]; Griess method [16]; paramagnetic resonance spectrometry [17]; paramagnetic resonance imaging; spectrophotometry [18]; and bioassay [19]. Each of these techniques has certain benefits associated with it but suffer from poor sensitivity and the need for complex and often expensive experimental apparatus. In addition, the above NO sensing techniques are limited when it comes to continuous monitoring of NO concentration in real time and most importantly *in vivo*.

1.1.3 Advantages of electrochemical sensors for determination of NO

To date, electrochemical (amperometric) detection of NO is the only available technique sensitive enough to detect relevant concentrations of NO in real time and *in vivo* and suffers minimally from potential interfering species such as nitrite, nitrate, dopamine, ascorbate, and L-arginine. Also, because electrodes can be made on the micro- and nano-scale these techniques also have the benefit of being able to measure NO concentrations in living systems without any significant effects from electrode insertion.

The first amperometric NO electrode used for direct measurement was described in 1990 [20]. In 1992, the first commercial NO sensor system was developed. Over subsequent years a range of highly specialized and sensitive NO electrodes have been developed offering detection limits for NO ranging from below 1 nM up to 100 μM [21]. Most recently, a unique range of high sensitivity NO sensors based on a membrane coated activated carbon microelectrode with diameters ranging from 200 μm down to 100 nm have been developed by this lab. These electrodes exhibit superior performance during NO measurement and feature a detection limit of less than 0.1 nM NO.

1.2 PRINCIPLES OF DETERMINATION OF NO BY ELECTROCHEMICAL SENSORS

NO can be oxidized or reduced on an electrode surface. Since the reduction potential of NO is close to that of oxygen which causes huge interference NO measurement, therefore, usually oxidation of NO is used for measurement of NO. NO oxidation on solid electrodes proceeds via an "EC mechanism" electrochemical reaction [22] followed by chemical reaction [23]. First, one-electron transfer from the NO molecule to the electrode occurred and resulted in the formation of a cation:

$$NO - e^- \rightarrow NO^+ \tag{1}$$

NO^+ is immediately, irreversibly converted into nitrite in the presence of OH^-, since it is a relatively strong Lewis acid:

$$NO^+ + OH^- \rightarrow HNO_2 \tag{2}$$

According to equation 2, the rate of the chemical reaction increases with the increase of pH.

Among the several electrochemical techniques that have been shown to be useful for the measurement of NO, the most popular is amperometry. This technique uses the model set forth by Clark and Lyons in 1962 for continuous gas monitoring during cardiovascular surgery [24]. Generally, this technique involves applying a fixed (poise) voltage potential to a working electrode, versus a reference electrode, and monitoring the very low redox current produced (e.g. pAs) by the oxidation of NO. This technique has proven to be very useful for NO detection due to its fast response time, which is less than a few seconds, and its high sensitivity. As a result it is possible to monitor changes in NO concentration on biologically relevant time scales and concentrations, which are typically in the nm range. A multitude of other electrochemical techniques have been used to detect NO including differential pulse voltammetry (DPV), differential normal pulse voltammetry (DNPV), linear scanning voltammetry (LSV), square wave voltammetry (SWV), and fast scan voltammetry (FSV). These methods typically employ a classical three-electrode configuration consisting of a working electrode, reference electrode, and a counter electrode. Scanning techniques, with the exception of fast scanning voltammetry, require approximately ten seconds for the voltammogram to be recorded, which precludes its use in most NO research applications. Moreover, since scanning voltammetry-based NO instrumentation is not commercially available, NO researchers typically prefer to use the two-electrode amperometric technique. Since the amperometric method is so widely used the discussion below will focus on this technique.

1.3 FABRICATION OF ELECTRODES FOR NO DETERMINATION

1.3.1 Clark type NO electrodes

The first described electrochemical NO sensor was based on a classical Clark electrode design, where NO was directly oxidized on the working electrode surface [25]. The NO sensor was composed of a fine platinum wire and a separate silver wire, which were then inserted into a glass micropipette. The micropipette was then filled with 30 mM NaCl and 0.3 mM HCl and sealed at the tip with a chloroprene rubber membrane. The platinum (working) electrode was positioned close to the surface of the membrane. The silver wire was then used as the reference/counter electrode. Although such electrodes could be used to measure NO, their inherent low sensitivity, narrow linear concentration measurement range and fragility rendered them unsuitable for most research applications. In 1992, utilizing the Clark type design, WPI produced the first commercial electrochemical NO sensor (ISO-NOP) for use with their NO detection meter (ISO-NO). The ISO-NOP sensor consisted of a platinum wire disk working electrode and an Ag/AgCl reference electrode. Both electrodes were encased within a protective Faraday-shielded stainless steel sleeve. The tip of the sleeve was covered with a NO-selective membrane and the sleeve itself contained an electrolyte. The rugged design of this sensor made it extremely convenient in many research applications and the sensor became widely used and established in numerous NO measurement research applications. The basic design of this type of NO sensor is illustrated in Fig. 1.1 [26].

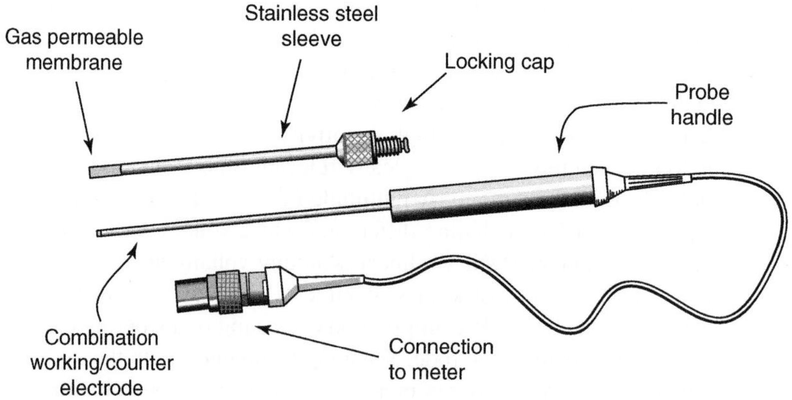

FIGURE 1.1 Illustration of WPI's ISO-NOP NO sensor. (Reprinted with permission from *Frontiers in Bioscience* [26].)

In principle the ISO-NOP sensor works as follows. The sensor is immersed in a solution containing NO and a positive potential of ~860 mV (vs Ag/AgCl reference electrode) is applied. NO diffuses across the gas permeable/NO-selective membrane and is oxidized at the working electrode surface producing a redox current. This oxidation proceeds via an electrochemical reaction followed by a chemical reaction. The electrochemical reaction is a one-electron transfer from the NO molecule to the electrode, resulting in the formation of the nitrosonium cation:

$$NO - e^- \rightarrow NO^+$$

NO^+ is a relatively strong Lewis acid and in the presence of OH^-, it is converted into nitrite (NO_2^-). Hence:

$$NO^+ + OH^- \rightarrow HNO_2 \rightarrow H^+ + NO_2^-$$

Nitrite can then be further oxidized into nitrate. The amount of NO oxidized is thus proportional to the current flow between the working and reference electrodes, which is measured by an NO meter.

The amount of redox current that is typically generated by the oxidation of NO in biological systems is extremely small, typically on the order of 1–10 pA. As a result of these extremely small currents the design of an amperometric-based electrode NO detection system requires very sensitive electronics and ultra low noise amplification circuitry. These measurement limitations were overcome by this lab with the development of the low noise and isolated circuit electronic devices ISO-NO and ISO-NO Mark II NO meters. These meters employed a unique electrically isolated low noise circuit that permitted measurement of redox currents as small as 0.1 pA. The design of the instrument also allowed measurement of NO to be performed without the need for special electrical screening, such as a Faraday cage. Recently, the world's only fully integrated multiple channel electrochemical nitric oxide/free radical detection system, the Apollo-4000, was developed at WPI. The detection system is an optically isolated multiple channel nitric oxide/free radical analyzer designed specially for the detection of NO and other free radicals such as oxygen, hydrogen peroxide, hydrogen sulfide, and superoxide [22]. Although the Clark type electrode has enjoyed such great success for NO detection there have been other significant advances for the detection of NO that deserve mention.

1.3.2 Modified carbon fiber NO microelectrodes

Surface modified NO sensors incorporate an electrode surface that has been modified or treated in some way so as to increase the selectivity of the sensor for NO and promote catalytic oxidation of NO. An early example of such a sensor was presented by Malinski and Taha in 1992 [27]. In this publication an ≈500 nm diameter carbon fiber electrode was coated with tetrakis(3-methoxy-4-hydroxyphenyl)porphyrin, via oxidative polymerization, and Nafion. This electrode was shown to have a detection limit of ~10 nM for NO and great selectivity against common interferences. However, recently it has been shown that this electrode suffers severe interference from H_2O_2 [28].

Following this publication, Schuhmann showed that pyrrole functionalized porphyrins, containing metals such as Ni, Pd and Mn, can be immobilized on carbon microelectrode surfaces via oxidative polymerization and be used for NO detection [29, 30]. Other researchers have shown that carbon fibers coated with a variety of porphyrins such as iron porphyrin [31] and cobalt porphyrin [23, 27, 32–34] are also effective for NO detection. Although metal porphyrin coated electrodes were successfully used to some extent for various applications [35–37], subsequent studies have shown that carbon fibers modified with unmetallated porphyrins as well as bare carbon fibers can detect NO with similar sensitivity [38, 39]. The sensitivity and selectivity of porphyrinic NO sensors vary significantly from electrode to electrode and depended not only on the potential at which NO oxidizes, but also on the effects of axial ligation to the central metal in the porphyrin, modification/treatment procedure and other experimental variations. Furthermore, because the surface of the electrode remained in direct contact with the measurement medium a variety of biological species were shown to interfere (i.e. give false responses) with the measurement of NO. Adding a Nafion layer to the porphyrin coated fibers could minimize these interferences. Other practical problems such as easy porphyrin removal and degradation have limited their usefulness for most applications [40]. Phthalocyanines, with a similar structure to porphyrins, containing metals such as Fe, Ni and Co have also been used to modify electrode surfaces for NO sensing [41, 42]. Phthalocyanine modified electrodes have comparable detection limits and selectivity to porphyrin modified electrodes with the added benefit of being more stable to degradation.

1.3.3 Integrated NO microelectrodes

During the mid- to late 1990s a new range of combination NO sensors with tip diameters between 7 μm to 200 μm were developed by this lab [43]. These sensors combine a carbon fiber working electrode with a separate integrated Ag/AgCl reference electrode. The resulting combination sensor was then coated with a proprietary gas permeable/NO-selective membrane. A high performance Faraday-shielded layer was then added to the sensor outer to minimize susceptibility to environmental noise. This electrode is then operated exactly as outlined above for the Clark type NO sensors. The use of these proprietary diffusion membranes and the novel design allows for NO measurement in small volumes and confined spaces with great selectivity against a wide range of interferences such as ascorbic acid, nitrite, and dopamine. This sensor design was elaborated upon by creating L-shaped sensors designed specifically for tissue bath studies [22]. The above design was further elaborated upon by our lab by creating flexible, virtually unbreakable NO sensors designed specifically for use in measuring NO concentrations in arteries and microvessels. This electrode combines a Pt/Ir wire with a separate integrated Ag/AgCl reference electrode [44]. The resulting combination sensor was again coated with a proprietary gas permeable/NO-selective membrane and a high performance Faraday-shielded layer and operated as described above. Recently, we developed a novel combination NO nanosensor, which had a tip diameter of just 100 nm [45]. The design of this sensor can be seen in Fig. 1.2. This

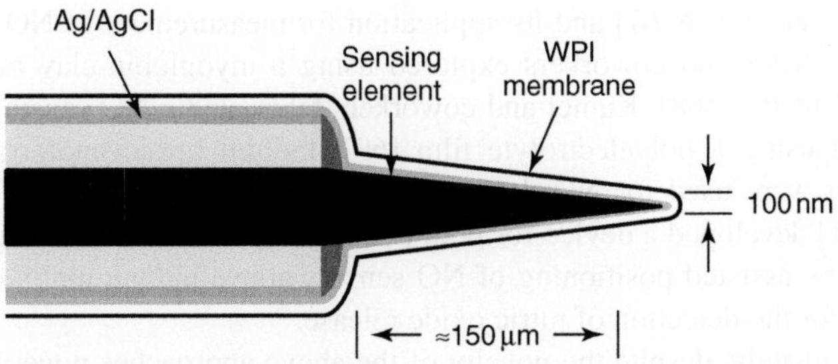

FIGURE 1.2 Illustration of WPI's ISO-NOPNM 100 nm combination NO nanosensor. (Reprinted with permission from Elsevier Publishing [45].)

sensor was constructed using a 7 μm carbon fiber that was etched with an Ar ion beam to result in fibers with diameters in the 100 nm range. These 100 nm fibers were then used to construct combination electrodes, as described above. These sensors are capable of making NO measurements on the cellular level. Later in 2000, our group developed a unique "microchip" combination sensor [46]. This sensor is quite unique in its design but its principle of operation is the same as for all the other combination electrodes presented in this section. Briefly, a 5000 Å thick carbon film was deposited on an Si wafer by RF sputtering. This carbon layer was then covered with a resist layer. A number of 2 μm diameter carbon electrodes were then exposed by a dry etching process in a 50 × 50 array on the wafer surface, where each carbon electrode is 20 μm away from its nearest neighbor. An Ag/AgCl reference electrode and a Pt counter electrode were added to the wafer surface. The surface was then modified with WPI's proprietary gas permeable, NO-selective membrane and made ready for use. The resulting sensor (ISO-NOP-MC) exhibited an extremely low NO detection limit on the order of 300 pM, great selectivity against ascorbic acid, nitrite and dopamine and a superior response time.

1.3.4 Other NO electrodes

Various other types of carbon fiber NO sensors, which utilize a variety of different coatings, have been described. Coatings used for these sensors include: conducting and non-conducting polymers [47–51]; multiple membranes [52, 53]; ruthenium [54]; iridium and palladium [55]; heated-denatured cytochrome c [56]; Nafion-Co (II)-1,10-phenanthroline[57]; ferrioxamine [58]; a microcoaxial microelectrode was reported for *in vivo* nitric oxide measurement [59]; siloxane polymer [60]; Nafion and cellulose [61]; Hb/phosphatidylcholine films [62]; hemoglobin-DNA film [63] and ionic polymers; and α-cyclodextrin [64]. Recently, Mizutani's group reported the coating of a 10 μm Pt disc electrode with a cross-linked sol-gel Langmuir-Blodgett film of siloxane polymer to render the electrode permselective for NO [51]. Schoenfisch's group recently reported the addition of a sol-gel film to a Pt electrode for NO detection [65]. Meyerhoff's group described an improved planar amperometric NO sensor based on a platinized anode [66, 67] and its application for measurement of NO release from NO donors. Scheler and coworkers explored using a myoglobin-clay modified electrode for NO detection [68]. Kamei and coworkers fabricated a NO sensing device for drug screening using a polyelectrolyte film [69]. Indium hexacyanoferrate film-modified electrodes were used for NO detection by Casero and coworkers [70]. Schuhmann's group [71] developed a device for both *in-situ* formation and scanning electrochemical microscopy assisted positioning of NO sensors above human umbilical vein endothelial cells for the detection of nitric oxide release.

Unfortunately, despite the novelty of the above approaches none of the sensors has stood the test of time, mostly due to various practical difficulties and/or poor sensitivity/selectivity. Furthermore, the lack of any published data describing the use of these sensors in any biological research applications limits any conclusion that can be made on their individual performance.

1.4 CALIBRATION OF NO ELECTRODES

Routine calibration of an NO sensor is essential in order to ensure accurate experimental results. One of three calibration techniques is generally used, depending on the sensor type, and will be described in the following section. Each of these methods has already been the subject of several reviews [23, 72–74] and will therefore only be summarized here. NO sensors are typically sensitive to temperature. Therefore, calibration is usually best performed at the temperature at which the measurements will be made.

1.4.1 Calibration using an NO standard solution

This technique involves the production of an NO stock solution using a supply of compressed NO gas. One advantage of this method is that it allows an NO sensor to be calibrated in a similar environment in which the experimental measurements will be made. However, the major drawback is that it requires a source of compressed NO gas and since NO gas is toxic, the whole procedure must be performed in a fume hood.

This method can be summarized as follows. A vacutainer is first filled with 10 mL deionized water and agitated ultrasonically for 10 min. Purified argon is then passed through an alkaline pyrogallol solution (5% w/v) to scavenge any traces of oxygen before being purged through a deionized water solution for 30 min. NO stock solution is prepared by bubbling compressed NO gas through the argon-treated water for 30 min. The NO gas is first purified by passing it through 5% pyrogallol solution in saturated potassium hydroxide (to remove oxygen) and then 10% (w/v) potassium hydroxide to remove all other nitrogen oxides. The resultant concentration of saturated NO in the water is 2 mM at 22°C [75]. This can be confirmed further by a photometric method based on the conversion of oxyhemoglobin to methemoglobin in the presence of NO [76]. NO standard solutions can then be freshly prepared by serial dilution of saturated NO solution with oxygen-free deionized water prior to each experiment.

1.4.2 Calibration based on decomposition of SNAP

For this method S-nitroso-N-acetyl-D, L-penicillamine (SNAP, FW = 220.3) is decomposed to NO in solution in the presence of a Cu (I) catalyst [77]. The resultant NO generated can then be used to calibrate the sensor. The reaction proceeds in accordance to the following reaction:

$$2\ \text{RSNO} \rightarrow 2\ \text{NO} + \text{RS-SR}$$

The stoichiometry of the reaction dictates that the final generated NO concentration will be equal to the concentration of SNAP in the solution. The method can be summarized as follows. Saturated cuprous chloride solution is first prepared by adding 150 mg CuCl to 500 mL distilled water. This solution is then deoxygenated by purging with pure nitrogen or argon gas for 15 min. The final, saturated CuCl solution will have a concentration of approximately 2.4 mM at room temperature. The solution is light sensitive and must therefore be kept in the dark prior to use.

The SNAP solution is then prepared separately as follows. EDTA (5 mg) is dissolved in 250 mL of pure water (HPLC grade, Sigma) and then adjusted to pH 9.0 using 0.1 M NaOH. The solution is then deoxygenated using the method described above. A quantity of 5.6 mg of SNAP is then added to the solution to result in a SNAP concentration of ~0.1 mM. SNAP solution is also extremely sensitive to light and temperature and must therefore be stored refrigerated and in the dark until required. Under these conditions, and in the presence of the cheating reagent (EDTA), the decomposition of SNAP occurs extremely slowly. This allows the solution to be used to calibrate NO probes throughout the day. In practice, calibration is performed by placing an NO sensor into a vial containing a measured amount of the CuCl solution and known volumes of the SNAP stock solution are then injected into the vial and the final concentration of NO can be calculated using dilution factors.

The concentration of SNAP in the stock solution is calculated as follows:

$$[C] = [A \times W/(M \times V)] \times 1000$$

where C = concentration of SNAP (μM); A = purity of SNAP; W = weight of SNAP (mg); M = MW of SNAP (mg/mmole); and V = volume of the solution in liters (l).

If SNAP purity, for example, is 98.5% then the concentration of SNAP is calculated as:

$$[C] = [98.5\% \times 5.6/(220.3 \times 0.25)] \times 1000 = 100.1\ \mu\text{M}$$

Figure 1.3 shows a typical calibration curve generated using an NO microsensor and the SNAP method described.

1.4.3 Calibration based on chemical generation of NO

This method of calibration generates known concentrations of NO based on the reaction of nitrite with iodide in acid according to the following equation:

$$2\ KNO_2 + 2\ KI + 2\ H_2SO_4 \rightarrow 2\ NO + I_2 + 2\ H_2O + 2\ K_2SO_4$$

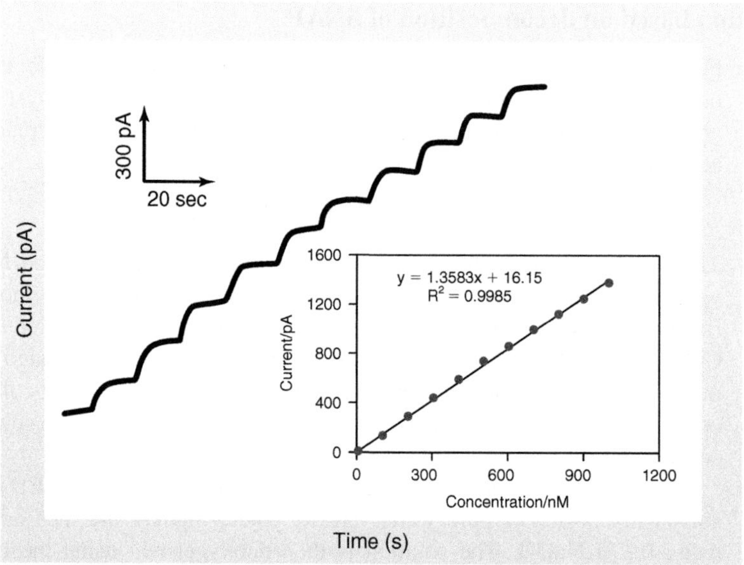

FIGURE 1.3 Response of WPI's ISO-NOP007 NO sensor to increasing concentration of NO produced by introduction of SNAP to a solution of CuCl. (Reprinted with permission from Wiley Publishing [77].)

The NO generated from the reaction is then used to calibrate the sensor. Since the conversion of NO_2^- to NO is stoichiometric (and KI and H_2SO_4 are present in excess) the final concentration of NO generated is equal to the concentration of KNO_2 in the solution. Hence, the concentration of NO can be easily calculated by simple dilution factors. Experiments have demonstrated that NO generated from this reaction will persist sufficiently long enough to calibrate an NO sensor. However, since this technique involves the use of a strong acid, which can damage the delicate selective membranes of most NO microsensors, it is only suitable for use with Clark type stainless steel encased NO sensors such as the ISO-NOP. Figure 1.4 illustrates the amperometric response of a 2.0 mm ISO-NOP sensor following exposure to increasing concentrations of NO. The sensor responds rapidly to NO and reaches steady state current within a few seconds. The data generated from Fig. 1.4 is then used to construct a final calibration curve (Fig. 1.4, inset). The calibration curve illustrates the good linearity that exists between NO concentration and the current produced by its oxidation.

1.5 CHARACTERIZATION OF NO ELECTRODES

NO sensors can be characterized in terms of sensitivity, detection limit, selectivity, response time, stability, linear range, lifetime, reproducibility, and biocompatibility. Sensor stability is important especially when measuring low NO concentrations. For example, when measuring low NO concentrations it must be the case that the noise

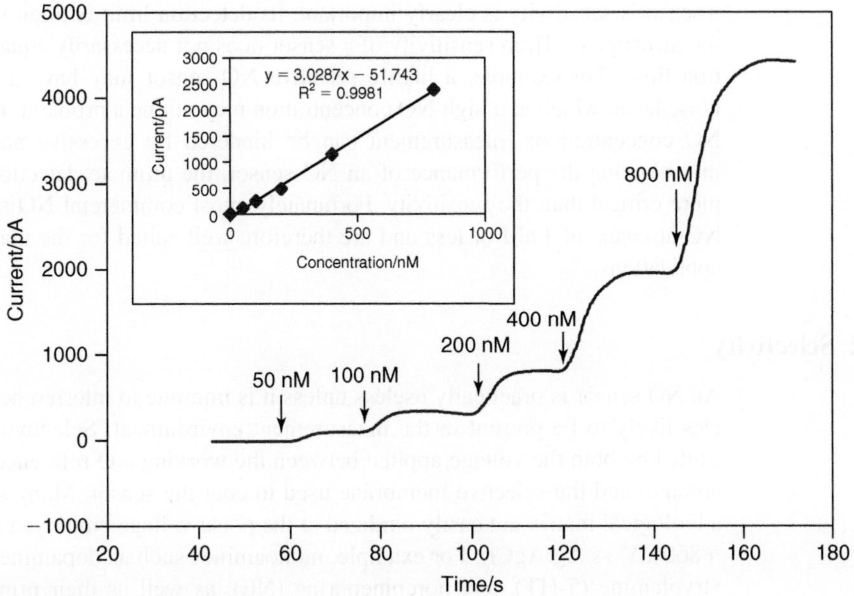

FIGURE 1.4 Response of WPI's ISO-NOP NO sensor to increasing concentration of chemically generated NO, inset shows the resulting calibration curve. (Reprinted with permission from *Frontiers in Bioscience* [26].)

is lower than the anticipated current change upon NO addition. The linear range of an NO sensor is important to know before performing any measurement. For example, the concentration of NO that is being measured must be in the linear range of the sensor in order for the measurement to be accurate. Typical commercially available NO sensors from WPI have a wide linear range of 1 nM–10 µM. Sensor lifetime and reproducibility are important to know when using a sensor and can be determined by frequent calibration, as described above. Biocompatibility is also extremely important when making NO measurements *in vivo*. Frequently, NO sensors are impaled into living tissue so the reactivity of the tissue toward the NO-selective membrane must be minimal and cause a minimal amount of irritation in the tissue. The NO-selective membranes used by this lab fulfill the requirements of being unreactive toward biomolecules and do not contribute to tissue inflammation. In most applications detection limit, sensitivity, selectivity and response time are usually the most important requirements and will be described in further detail.

1.5.1 Sensitivity and detection limit

The sensitivity of an NO sensor depends largely on the reactive surface area of the sensor and the electrode materials used in the design and can range from 0.03 pA/nM to 100 pA/nM NO. Generally speaking, this sensitivity is directly proportional to the electrode size and surface status where an electrode with a small surface area will generally have a lower sensitivity compared to one with a larger surface area. Although a sensor's sensitivity is clearly important, its detection limit is often more important to the investigator. High sensitivity of a sensor does not necessarily equate to a low detection limit. For example, a highly sensitive NO sensor may have a high background noise level, which at a high NO concentration may not be a problem. However, at lower NO concentrations, measurement can be hindered by excessive noise. Accordingly, in evaluating the performance of an NO sensor, the ultimate detection limit is usually more critical than the sensitivity. Fortunately, most commercial NO sensors can detect NO at levels of 1 nM or less and are therefore well suited for the majority of research applications.

1.5.2 Selectivity

An NO sensor is practically useless unless it is immune to interference from other species likely to be present in the measurement environment. Selectivity is usually controlled by both the voltage applied between the working and reference electrode (poise voltage) and the selective membrane used to coat the sensor. Many species present in a biological matrix are easily oxidized at the poise voltage employed to detect NO (i.e. +860 mV vs Ag/AgCl). For example, monoamines such as dopamine (DA), 5-hydroxytryptamine (5-HT), and norepinephrine (NE), as well as their primary metabolites, can be oxidized at 0.3 V (and higher) vs Ag/AgCl. Ascorbic acid can be oxidized at 0.4 V (and higher). A Clark type NO sensor (e.g. ISO-NOP) is covered with a gas permeable membrane, hence the selectivity of such sensors in biological samples is extremely good. With other types of NO sensors selectivity is usually achieved by coating the sensor surface with Nafion and other gas permeable membranes. Nafion is widely used to eliminate interference caused by anions, such as ascorbate and nitrite, during measurement of catecholamine species. When used for NO detection the negatively charged Nafion layer can stabilize NO^+ formed upon the oxidation of NO and prevent a complicated pattern of reactions that could lead to the formation of nitrite and nitrate. However, the main drawback with Nafion is that it does not eliminate interference from cationic molecules such as dopamine, serotonin, epinephrine and other catecholamines. Consequently, selectivity of the conventional Nafion coated NO sensors is very poor. Nafion coated NO sensors also exhibit other undesirable characteristics including unstable background current, continuous drift in the base line, and extended polarization requirements. These problems significantly limit the use of Nafion coated carbon fiber electrodes for measurement of NO. During the late 1990s this lab developed a unique multi-layered proprietary membrane configuration. NO sensors coated with this membrane exhibited increased selectivity and sensitivity for NO, and moreover were shown to be immune from interference caused from a wide range of potentially interfering species [43].

1.5.3 Response time

Response and recovery times of NO sensors are extremely important for their use *in vivo*. Theoretically, since the rate of mass transport at a microelectrode is very

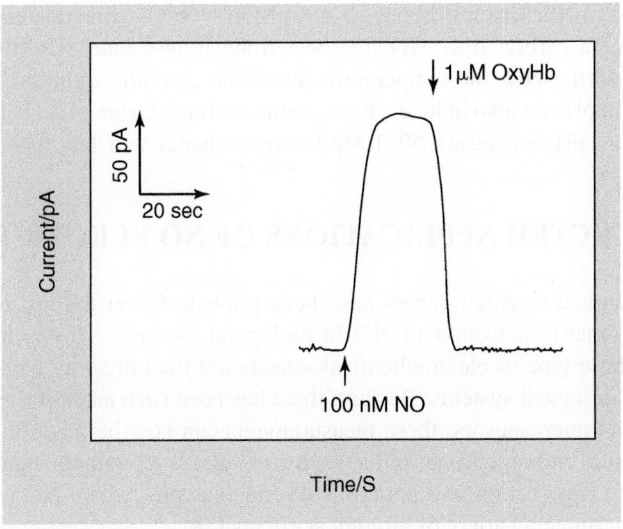

FIGURE 1.5 Amperometric response of an NO nanosensor upon addition of 100 nM NO and 1 μM oxyhemoglobin to a stirred 0.1 M phosphate buffer solution (pH 7.4).

high it should have a response time on the order of μsec; however, the addition of NO-selective membranes to the electrode surface decreases this response time significantly. With a membrane present the response time is now dependent on the diffusion rate of NO across the membrane, which is highly dependent on the nature of the membrane as well as its thickness. The response time not only depends on the electrode being used but also on the electronics being used to read out the current. For example, since the current being read out is typically on the order of pAs an electronic filter is usually applied, which also slows the system response. Since the half-life of NO is from a few seconds to minutes in a biological system, a sensor response on the order of 3–4 sec will work fine. Figure 1.5 is a typical response of an NO microsensor to the addition of NO in PBS solution. It shows the 90% response of the sensor to a step from 0 to 100 nM NO and subsequently injection of NO scavenger 1 μM of oxyhemoglobin. As can be seen, the response times to both additions of NO and oxyhemoglobin are less than 4 seconds. This indicates that even with a heavy filter, the response is within a few seconds, which is within the time frame required to measure NO. To decrease the response time of NO sensors, fewer filters can be applied, which will sacrifice the detection limit of NO sensors.

1.5.4 Effect of temperature and pH on NO electrodes

Background currents of all NO electrodes are sensitive to changes of temperature and pH. Depending on type of electrodes, the effect may be more or less. Clark type NO electrodes are very sensitive to temperature change. The temperature induced response of this type of NO sensor is about 50–100 nM of NO/°C, while the temperature induced response of a carbon fiber NO microelectrode is less than 10 nM of NO/°C. Thus, when measuring NO, the temperature should be carefully monitored simultaneously. A change in pH can also induce a background current change of NO electrodes. Usually a change of 1 pH can cause a 50–100 pA current change on Clark type NO sensors.

1.6 SELECTED APPLICATIONS OF NO ELECTRODES

Several hundred research papers have been published over the last decade describing the amperometric detection of NO in biological systems. This is mainly due to the fact that these type of electrochemical sensors are the only way NO can be measured *in vivo* in biological systems. Because there has been such an explosion in the development of NO microsensors, these measurements can now be made in a variety of biological tissues and organs, as well as on the cellular level without significant damage to the system. This section will point out several examples where NO microsensors were used to determine a variety of biological effects [78].

Determinations of NO in a variety of biological systems have been made. For example, measurement of NO has been made in eyes [79–81], gastrointestinal tract [82, 83], brain tissue [47, 50, 84–87], kidney and kidney tubule fluid [88–93], rat and guinea pig isolated and intact hearts [94, 95], rat spinal cord [96], human monocyte cells [97], human endothelial cells [98], mitochondria [99, 100], rat penis corpus cavernosum [101], granulocytes [102], invertebrate ganglia and immunocytes [103], choroidal endothelial cells [104], cancer cells [105, 106], peripheral blood [107], human blood [108], human leukocytes [109], platelets [110–112], ears [113, 114], plants [115–118], and pteropod mollusk [119].

Levine and coworkers first reported on the real-time profiling of kidney tubular fluid nitric oxide concentration *in vivo* [89, 91]. In the 2001 publication, a modified version of a combination NO electrode (WPI, ISONOP007) was successfully used to measure NO concentration profiles along the length of a single nephron of a rat kidney tubular segment. Since it was shown that the electrode is sensitive to NO in the rat tubule it was used to detect NO concentration differences in rat kidney tubules before and after 5/6 nephrectomy. The results clearly showed that the NO concentration was much higher in nephrectomized rats vs unnephrectomized rats.

In a recent publication, investigators used a specially customized ISO-NOP sensor to monitor, in real time, NO production in the stomach and esophagus of human patients [82]. In this method a patient first swallowed two NO electrodes (see Fig. 1.6), which were then withdrawn slowly at 1 cm increments every 2 min. The investigators were then able to establish a profile of NO concentration in the upper gastrointestinal tract.

Simonsen's group has performed some elegant work over the years on NO release characteristics from rat superior mesenteric artery. Initially, Simonsen's group simultaneously monitored artery relaxation and NO concentration in the artery using a NO microsensor in response to various drugs [120]. NO concentration was monitored via an ISONOP30 electrode, purchased from WPI and inserted into the artery lumen using

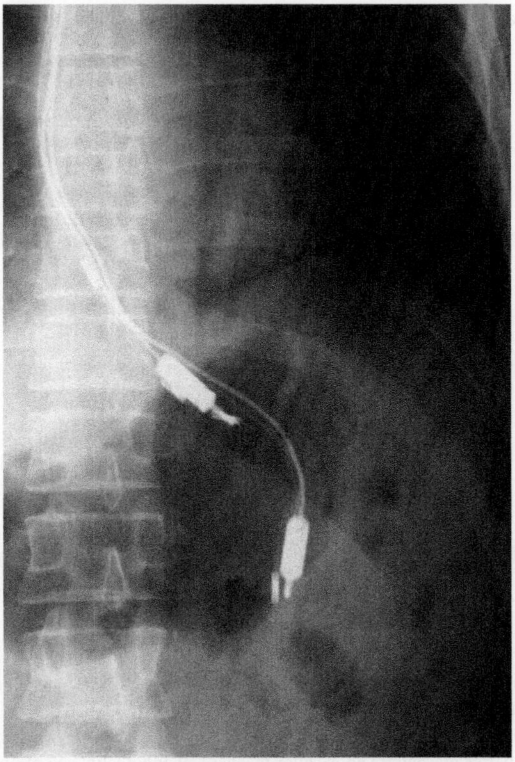

FIGURE 1.6 Abdominal X-ray showing the apparatus consisting of two nitric oxide sensors, a 4-channel pH catheter, and a Teflon nasogastric tube. (Reprinted with permission from the American Gastroenterological Association [82].)

a micromanipulator. The results of this work are shown in Fig. 1.7. The figure shows the force (upper traces) and NO concentration (lower traces) in an endothelium intact (+E) segment of the rat superior mesenteric artery and the same segment after mechanical endothelial cell removal (−E). As can be seen from the traces, if endothelial cells are present the artery is capable of relaxation due to the endothelial cells releasing NO in response to acetylcholine but if the endothelial cells are removed the artery is insensitive to acetylcholine injection but relaxes upon the introduction of the NO releasing molecule SNAP. In a subsequent publication, Simonsen's group again monitored artery relaxation and NO concentration in the rat superior mesenteric artery to monitor its hyporeactivity to various endotoxins [121]. In this study an ISONOP30 electrode (WPI) was inserted into the lumen of the artery and the NO concentration, as well as artery relaxation, was measured as a function of endotoxin introduction to determine what effects lipopolysaccharide had on the artery function. The results nicely showed that lipopolysaccharide resulted in induction of iNOS and SOD associated with endotoxin and NO concentration increased only in response to L-arginine. Simonsen's

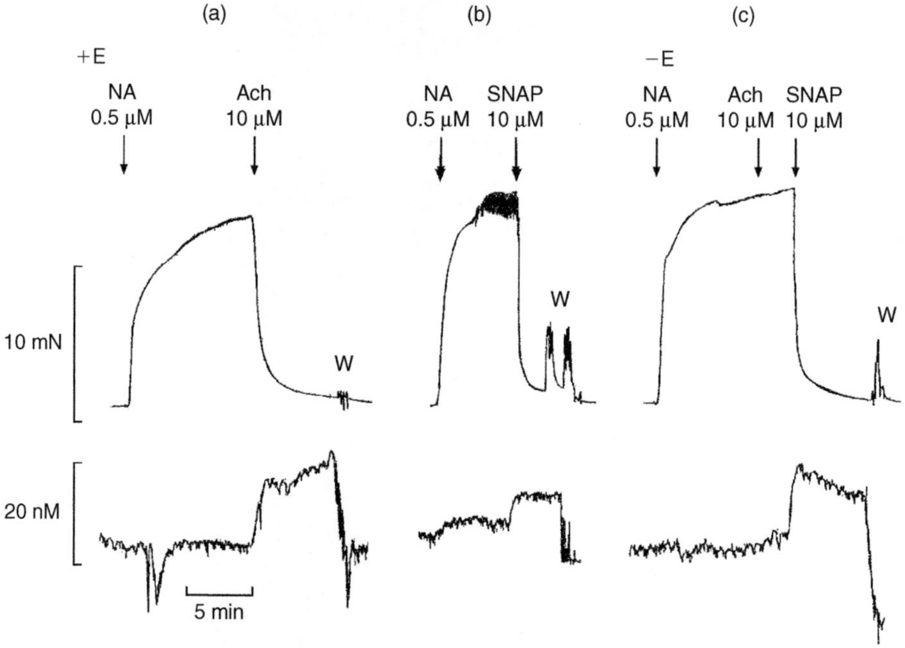

FIGURE 1.7 Simultaneous measurements of force (upper traces) and NO concentration (lower traces) in an endothelium intact (+E) segment of rat superior mesenteric artery contracted with 0.5μM noradrenaline (NA) and relaxed with either 10μM acetylcholine (ACh) (a), or 10μM SNAP (b). Panel C shows a similar measurement in the rat superior mesenteric artery after mechanical endothelial cell removal. As can be seen in C, ACh addition does not cause NO production from the artery but shows an NO increase upon SNAP addition causing artery relaxation. W = washout. (Reprinted with permission from Blackwell Publishing [120].)

group has also experimented with measuring NO concentration in the artery of a hypertensive rat [122] and in isolated human small arteries [123]. Schuhmann's group has recently published some very interesting results describing the measurement of NO release from human umbilical vein endothelial cells (HUVEC) using a unique array of microelectrodes [124]. Figure 1.8 (top) shows an SEM image of the microelectrode array. Following microelectrode array construction the surfaces of the electrodes were modified with nickel tetrasulfonate phthalocyanine tetrasodium salt using electrochemically induced deposition. Following this deposition the HUVEC cells were allowed to grow in the interstitial spaces between the electrodes (Fig. 1.8, bottom) and NO release from the cells was monitored as a function of growth and stimulation from bradykinin. This work is unique because the cells are actually being grown on the Si_3N_4 insulating layer so the cells are not being affected by the microelectrode working potential, which typically causes cell death, allowing for NO concentration to be monitored.

A recent publication by Millar has shown that NO concentrations can be measured in bovine eyes' trabecular meshwork *in situ* using an ISONOP200 electrode from WPI [79]. For this study, the tip of the 200μm diameter electrode was inserted

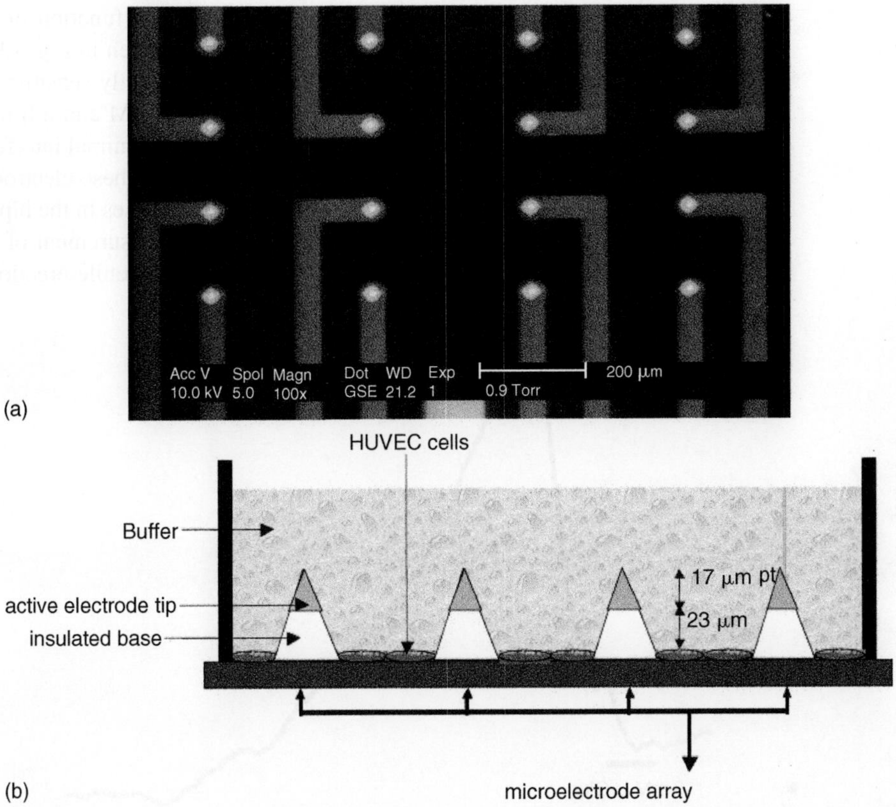

FIGURE 1.8 (a) A representative SEM image of the microelectrode array and (b) a schematic representation of the experimental setup. (Reprinted with permission from Elsevier Publishing [124].)

into the region of the trabecular meshwork and NO monitored as a function of epinephrine concentration. This study found that NO generation increased as a function of epinephrine addition resulting in a reduction of intraocular pressure. This finding is important for shedding light on treatments for patients with primary open-angle glaucoma. Researchers such as Akeo and Amaki have used a Pt/Ir electrode, coated with NO selective membranes, for the determination of what effects L-DOPA has on eyes [80, 81].

Because NO is a diffusible messenger molecule in the brain the measurement of NO concentration *in vitro* is quite important to understand its action. In order to accomplish this general goal, Gerhardt's group recently developed an 8 μm diameter carbon fiber electrode coated with Nafion and *o*-phenylenediamine [50]. Nafion was added to the electrode surface by dipping the electrode into a 5% solution followed by drying for 10 min at 170°C followed by *o*-phenylenediamine electropolymerization on the electrode by holding the potential at +0.9 V vs Ag/AgCl reference electrode. These electrodes were then inserted into the CA1 region of the hippocampus, from a Wistar rat, and the NO concentration monitored as a function of L-glutamate and *N*-methyl-D-aspartate introduction. The results can be seen in Fig. 1.9. The main finding of this study was that these electrodes were sufficiently sensitive to monitor brain NO concentrations with a sensitivity of $954 \pm 271\,pA/\mu M$ and a limit of detection of $6 \pm 2\,nM$. These measurements can also be made with minimal interference from common interferences such as ascorbate, nitrite, and H_2O_2. These electrodes can be useful for unraveling pathways for memory and learning processes in the hippocampus.

The Mas group recently published results on the measurement of NO release from the corpus cavernosum of the penis and its relation to penile erection [101]. For this

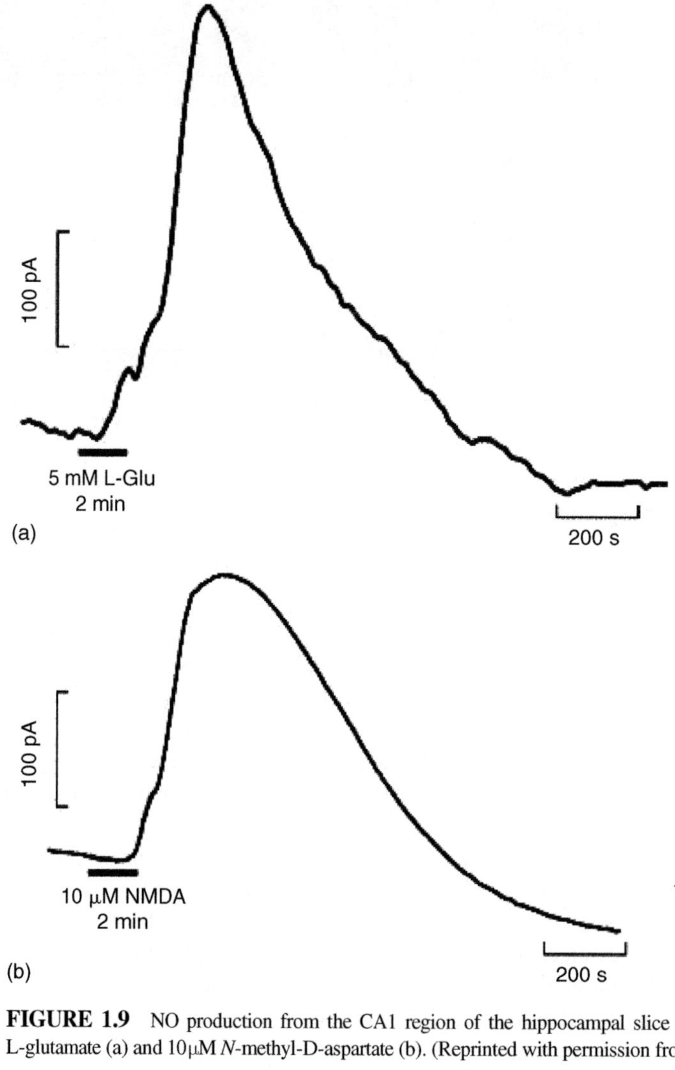

FIGURE 1.9 NO production from the CA1 region of the hippocampal slice following addition of 5 mM L-glutamate (a) and $10\,\mu M$ *N*-methyl-D-aspartate (b). (Reprinted with permission from Elsevier Publishing [50].)

study a 30 μm diameter carbon fiber was coated with nickel tetrakis(3-methoxy-4-hydroxyphenyl)porphyrin, via electrodeposition using differential pulse voltammetry, and Nafion. This electrode was then inserted into the cavernous bodies of a rat penis. The results showed an increase in NO concentration, and intracavernous pressure, upon cavernous nerve stimulation and a subsequent decrease upon the introduction of NO synthase isoenzymes. This study was important to facilitate further measurements of NO concentrations *in vivo* in the penis.

NO measurements have also been made in the ears of a guinea pig. The first of these studies was performed by Nuttall and coworkers in 2002 [113]. For this investigation an ISONOP30 carbon fiber electrode from WPI was inserted into the perilymph of the basal turn of the guinea pig ear to measure changes in NO concentration as a result of noise stimulation. This study showed that guinea pigs exposed to broadband noise for 3 h/day at 120 dBA, for 3 days, exhibited an increase in NO concentration in the perilymph. This result is important in order to understand the role that NO plays in hearing loss. A subsequent publication by Nuttall and coworkers used the ISONOP200 from WPI to measure NO concentration in the spiral modiolar artery (SMA) of a guinea pig [114]. This study is important to gain a better understanding of how NO potentially regulates cochlear blood flow to set a benchmark for pharmacological and pathological evaluation. To perform this study a 3 mm section of the SMA was added to a bath solution and the complete tip of the 200 μm electrode was inserted into the bath, parallel to the SMA, to measure basal NO concentration as well as drug induced NO release and how that relates to cochlear blood flow regulation. The key findings of this study were that the SMA continuously releases NO and that a blockage of this release by L-NAME causes a decrease in NO production and a vasoconstriction. These findings are important in order to understand and interpret future findings.

An interesting study was performed by Kashiwagi and coworkers studying the role that NO plays in tumor vessel morphogenesis and maturation [105]. For this study, B16 tumor cells were injected into mice and tumor tissue removed from the mice when it reached ~8 mm in diameter. The tip of a Nafion polymer coated Au microelectrode was subsequently inserted into the tumor to monitor NO production. The results of this study showed that NO mediates mural cell coverage and vessel branching/longitudinal extension but does not play a part in the growth of tumor blood vessels. The investigators also used an NOS inhibitor to show that NOS from endothelial cells in tumors is the primary source of NO and mediates tumor growth.

Kishi and coworkers used a commercial NO-selective microelectrode to monitor the effect that exercise has on platelet-derived NO [110]. This study used 23 healthy male non-smokers who underwent treadmill exercise. Blood samples were taken from the subjects before and directly following exercise and blood platelets isolated. The study showed that NO concentration and platelet levels were increased following exercise. This increase in NO concentration is thought to play a role in the prevention of exercise induced platelet activation in humans.

Kellogg and coworkers recently reported on the measurement of NO under the human skin in response to heat stress. For this study a flexible 200 μm NO microelectrode from

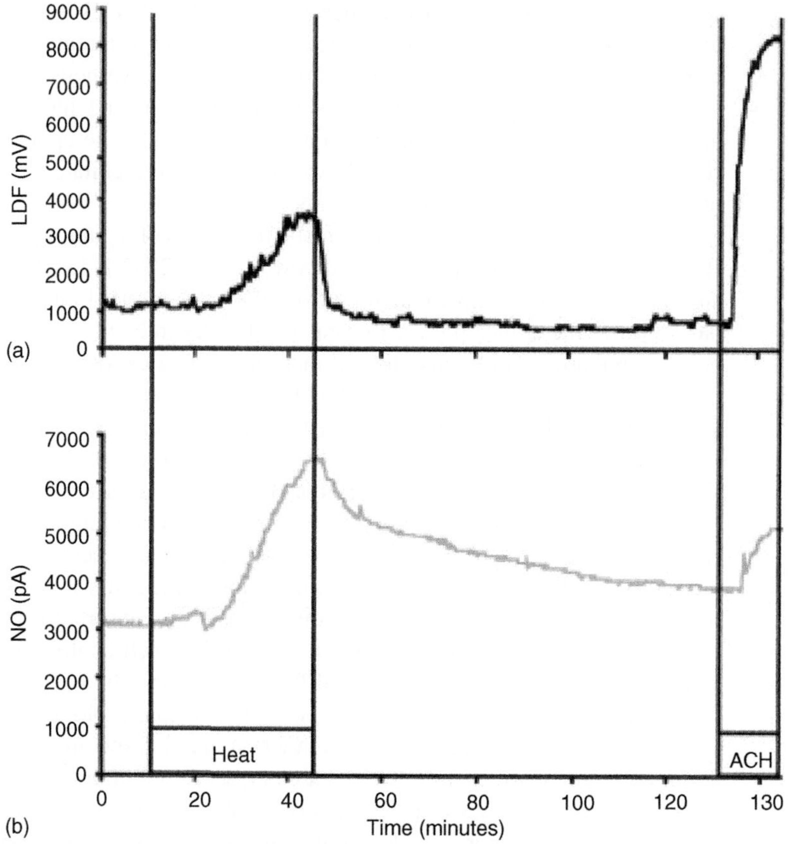

FIGURE 1.10 Laser-doppler flowmetry (a) and NO measurement (b) from one subject. Upon heating the subject to 39°C, at ~10 min, NO production and skin blood flow increased, which instantly returned to normal upon cooling the subject at ~45 min. After heat stress and cooling, ACh was administered by intradermal microdialysis to confirm the ability of the microelectrode to measure NO concentrations. (Reprinted with permission from the American Physiological Society [125].)

WPI was inserted into the cutaneous interstitial space of the forearm of nine human patients to measure NO concentration while the subjects were at low (34°C) and high (39°C) temperature. Laser-Doppler flowmetry (LDF) was used to monitor skin blood flow (SkBF) [125]. This publication demonstrated that NO concentration, as well as SkBF increased in the cutaneous interstitial space during heat stress in humans. Figure 1.10 shows the key results for these experiments. As the temperature is increased at ~10 min the data show that the blood flow (top plot) and NO concentration (bottom plot) increased as a function of temperature. Also, as the temperature was decreased again, at ~45 min, the blood flow and NO concentration returned to its original value. At ~130 min, while the subjects were at low temperature, the investigators injected acetylcholine to show that

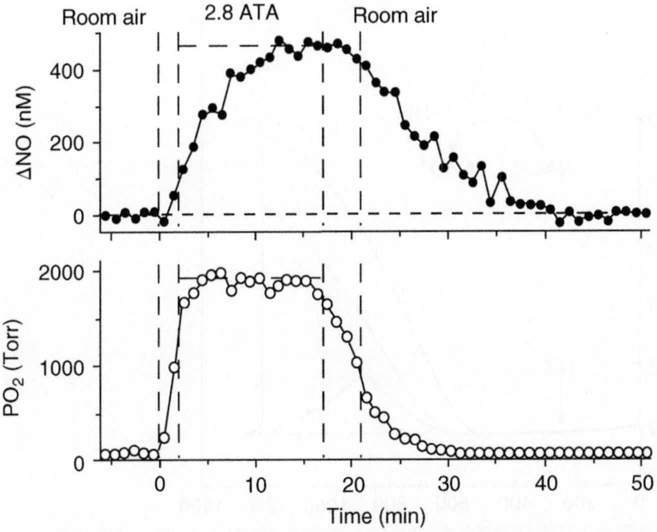

FIGURE 1.11 NO concentration (top) and O_2 concentration (bottom) as a function of O_2 pressure. As can be seen, NO and O_2 concentrations increase significantly when a rat is exposed to pressurized O_2 atmospheres (2.8 atmospheres absolute, ATA). (Reprinted with permission from the American Physiological Society [127].)

NO was indeed being detected in the subjects. The same group used a similar experimental design to monitor NO concentration, as well as SkBF, during reactive hyperemia under the human skin [126].

Thom and coworkers published results on the stimulation of perivascular NO synthesis by oxygen [127]. To perform this study a 200 μm diameter electrode (ISONOP200) was placed between the aorta and vena cava of anesthetized rats and mice (rodents) and then the rodents were placed inside a hyperbaric chamber. Inside the hyperbaric chamber the partial pressure of O_2 was regulated/changed as NO concentration was monitored. Figure 1.11 shows that NO concentration increased as a function of O_2 partial pressure. This experiment is important for understanding how NO synthesis, by NOS, is altered and regulated by O_2.

Studies have also been conducted on the NO release from plants and the effects of a plant diet on atherosclerosis and endothelial cell dysfunction. Visioli and coworkers studied the effects that a diet of wild artichoke and thyme had on the release of NO from porcine aortic endothelial cells and cerebral cell membranes [118]. For this study a rat brain was homogenized and cell membranes isolated by ultracentrifugation and NO release monitored using a 2 mm nitric oxide sensor (ISONOP). The cell membranes were then exposed to wild artichoke and thyme extracts. This study showed that NO release was significantly increased following wild artichoke and thyme extract addition. These results suggest that eating a diet rich in phenolic compounds, such as wild artichoke and thyme, contributes to maintenance of a healthy cardiovascular system. Yamasaki and coworkers used an ISONOP from WPI to study the ability of plant nitrate reductase to produce NO [117]. Figures 1.12 and 1.13 show that plant nitrate reductase produces NO

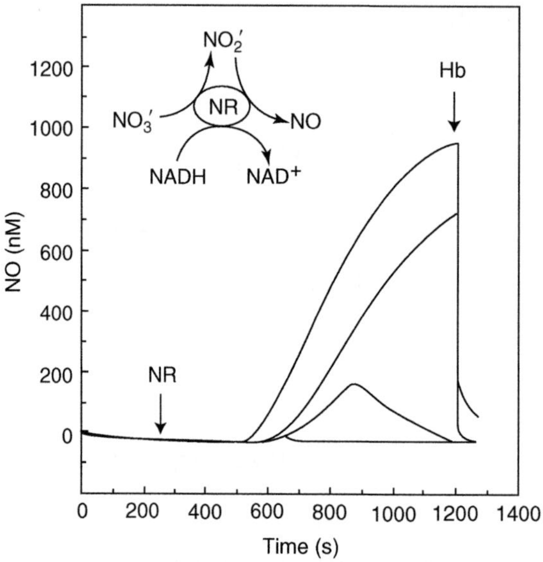

FIGURE 1.12 NO production from the addition of nitrate reductase (NR = 15 mU/mL) to a solution containing 50 μM sodium nitrate and various concentrations of NADH. The curves from top to bottom were obtained in solutions containing 100, 50, 40, and 0 μM NADH. Hemoglobin (Hb) was introduced to quench the production of NO. (Reprinted with permission from Elsevier Publishing [117].)

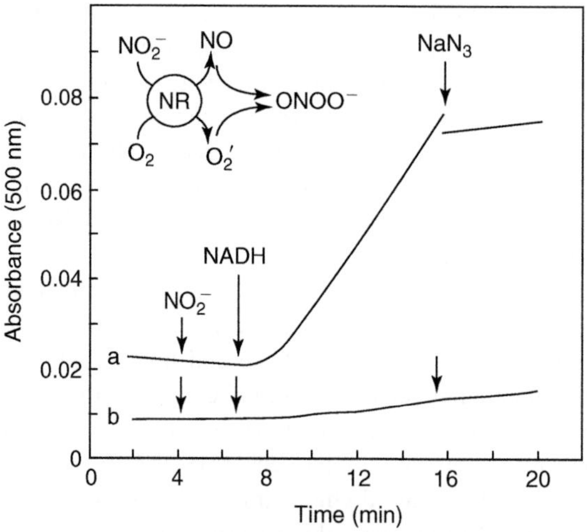

FIGURE 1.13 Absorbance measurement of 2′,7′-dichlorodihydrofluorescein (DCDHF), a peroxynitrite-sensitive dye, as a function of nitrite (1 mM) and NADH (1 mM) introduction to a solution containing 100 μM DCDHF and 30 mU/mL NR under ambient (a) and nitrogen saturated conditions (b). As can be seen, the absorbance of DCDHF increases upon nitrite and NADH introduction only under an oxygen atmosphere, indicative of peroxynitrite production. (Reprinted with permission from Elsevier Publishing [117].)

in vitro as well as its toxic derivative peroxynitrite. Vinik *et al.* reported that pioglitazone treatment improves nitrosative stress in type 2 diabetes. In this research, nitric oxide was measured *in vivo* using an electrochemical NO microsensor inserted directly into the skin [128]. They found the NO production was significantly decreased in the pioglitazone treated group in the basal condition.

1.7 CONCLUDING REMARKS AND OTHER DIRECTIONS

The use of the electrochemical NO microsensor provides an elegant and convenient way to detect NO in real time and in biological samples. Currently they provide the only means by which to measure NO continuously, accurately and directly within living tissue without significant damage. The increasing acceptance of such sensors and their diversity of use in many NO research applications will help to further the current understanding of the various clinical roles of this interesting and ubiquitous molecule. Continual improvements being made to NO microsensor design and technology will facilitate these studies in the foreseeable future.

1.8 ACKNOWLEDGMENTS

This research was supported by NIH grants (1 R43 GM62077-01, 2R44 GM62077-02, 5R GM62077-3) to XZ and a WPI R&D priority research fund.

1.9 REFERENCES

1. *Encyclopaedia Britannica*, 15th edition, vol. 8, p. 726.
2. S. Moncada, R.M.J. Palmer, and E.A. Higgs, Nitric-oxide – physiology, pathophysiology, and pharmacology. *Pharmacol. Rev.* **43**, 109–142 (1991).
3. L.J. Ignarro, G.M. Buga, K.S. Wood, R.E. Byrns, and G. Chaudhuri, Endothelium-derived relaxing factor produced and released from artery and vein is nitric-oxide. *Proc. Natl. Acad. Sci. U.S.A.* **84**, 9265–9269 (1987).
4. L.J. Ignarro, Nitric-Oxide – A novel signal transduction mechanism for transcellular communication. *Hypertension* **16**, 477–483 (1990).
5. D.S. Bredt, P.M. Hwang, C.E. Glatt, C. Lowenstein, R.R. Reed, and S.H. Snyder, Cloned and expressed nitric-oxide synthase structurally resembles cytochrome-P-450 reductase. *Nature* **351**, 714–718 (1991).
6. P.L. Feldman, O.W. Griffith, and D.J. Stuehr, The surprising life of nitric-oxide. *Chem. Eng. News* **71**, 26–37 (1993).
7. M.W. Radomski, R.M.J. Palmer, and S. Moncada, An L-arginine nitric-oxide pathway present in human platelets regulates aggregation. *Proc. Natl. Acad. Sci. U.S.A.* **87**, 5193–5197 (1990).
8. S. Moncada Moncada, M.W. Radomski, and R.M.J. Palmer, Endothelium-derived relaxing factor – identification as nitric-oxide and role in the control of vascular tone and platelet-function. *Biochem. Pharmacol.* **37**, 2495–2051 (1988).
9. T. Akaike, M. Yoshida,Y. Miyamoto, K. Sato, M. Kohno, K. Sasamoto, K. Miyazaki, S. Ueda, and H. Maeda, Antagonistic action of imidazolineoxyl N-oxides against endothelium-derived relaxing factor. NO through a radical reaction. *Biochemistry* **32**, 827–832 (1993).

10. J.B. Hibbs, R.R. Taintor, Z. Vavrin, and E.M. Rachlin, Nitric-oxide – a cyto-toxic activated macrophage effector molecule. *Biochem. Biophys. Res. Commun.* **157**, 87–94 (1988).
11. M. Anbar, Hyperthermia of the cancerous breast – analysis of mechanism. *Cancer Lett.* **84**, 23–29 (1994).
12. H. Schmidt, T.D. Warner, K. Ishii, H. Sheng, and F. Murad, Insulin-secretion from pancreatic B-cells caused by L-arginine derived nitrogen-oxides. *Science* **255**, 721–723 (1992).
13. S. Moncada, R.M.J. Palmer, and E.A. Higgs, Biosynthesis of nitric-oxide from L-arginine – a pathway for the regulation of cell-function and communication. *Biochem. Pharmacol.* **38**, 1709–1715 (1989).
14. J.S. Beckman and K.A. Congert, Direct measurement of dilute nitric oxide in solution with an ozone chemiluminescent detector. *Methods* **7**, 35–38 (1995).
15. J.K. Robinson, M.J. Bollinger, and J.W. Birks, Luminol/H2O_2 chemiluminescence detector for the analysis of nitric oxide in exhaled breath. *Anal. Chem.* **71**, 5131–5136 (1999).
16. L.C. Green, D.A. Wagner, J. Glogowski, P.L. Skipper, J.S. Wishnok, and S.R. Tannenbaum, Analysis of nitrate, nitrite, and N-15-labeled nitrate in biological-fluids. *Anal. Biochem.* **126**, 131–138 (1982).
17. A. Wennmalm, B. Lanne, and A.S. Petersson, Detection of endothelial-derived relaxing factor in human plasma in the basal state and following ischemia using electron-paramagnetic resonance spectrometry. *Anal. Biochem.* **187**, 359–363 (1990).
18. D.S. Bredt and S.H. Snyder, Nitric-oxide mediates glutamate-linked enhancement of cgmp levels in the cerebellum. *Proc. Natl. Acad. Sci. U.S.A.* **86**, 9030–9033 (1989).
19. J.L. Wallace and R.C. Woodman, Detection of nitric oxide by bioassay. *Methods* **7**, 55 (1995).
20. *World Precision Instruments Catalog*, 42–49 (2005).
21. X. Zhang and M. Broderick, Amperometric detection of nitric oxide. *Mod. Asp. Immunobiol.* **1**, 160 (2000).
22. S. Trevin, F. Bedioui, and F. Devynck, Electrochemical and spectrophotometric study of the behavior of electropolymerized nickel prophyrin films in the determination of nitric oxide in solution. *Talanta* **43**, 303–311 (1996).
23. T. Malinski, Z.Taha, S. Grunfeld, A. Burewicz, P. Tomboulian, and F. Kiechle, Measurement of nitric oxide in biological materials using a porphyrinnic microsensor. *Anal. Chim. Acta* **279**, 135–140 (1993).
24. L.C. Clark and C. Lyons, Electrode systems for continuous monitoring in cardiovascular surgery. *Ann. NY Acad. Sci.* **102**, 29 (1962).
25. K. Shibuki, An Electrochemical microprobe for detecting nitric-oxide release in brain-tissue. *Neurosci. Res.* **9**, 69–76 (1990).
26. X. Zhang, Real time and in vivo monitoring of nitric oxide by electrochemical sensors – from dream to reality. *Front. Biosci.* **9**, 3434–3446 (2004).
27. T. Malinski and Z. Taha, Nitric-oxide release from a single cell measured in situ by a porphyrinic-based microsensor. *Nature* **358**, 676–678 (1992).
28. S. Nagase, N. Ohkoshi, A. Ueda, K. Aoyagi, and A. Koyama, Hydrogen peroxide interferes with detection of nitric oxide by an electrochemical method. *Clin. Chem.* **43**, 1246 (1997).
29. N. Diab and W. Schuhmann, Electropolymerized manganese porphyrin/polypyrrole films as catalytic surfaces for the oxidation of nitric oxide. *Electrochim. Acta* **47**, 265–273 (2001).
30. N. Diab, J. Oni, A. Schulte, I. Radtke, A. Blochl, and W. Schuhmann, Pyrrole functionalised metalloporphyrins as electrocatalysts for the oxidation of nitric oxide. *Talanta* **61**, 43–51 (2003).
31. J. Hayon, D. Ozer, J. Rishpon, and A. Bettelheim, Spectroscopic and electrochemical response to nitrogen monoxide of a cationic iron porphyrin immobilized in nafion-coated electrodes or membranes. *J. Chem. Soc.-Chem. Commun.* 619–620 (1994).
32. A.V. Kashevskii, J. Lei, A.Y. Safronov, and O. Ikeda, Electrocatalytic properties of meso-tetraphenylporphyrin cobalt for nitric oxide oxidation in methanolic solution and in Nafion® film. *J. Electroanal. Chem.* **531**, 71–79 (2002).
33. A. Brunet, C. Privat, O. Stepien, M. David-Dufilho, J. Devynck, and M.A. Devynck, Advantages and limits of the electrochemical method using Nafion and Ni-porphyrin-coated microelectrode to monitor NO release from cultured vascular cells. *Analusis* **28**, 469 (2000).
34. F. Bedioui, S. Trevin, and J. Devynck, Chemically modified microelectrodes designed for the electrochemical determination of nitric oxide in biological systems. *Electroanalysis* **8**, 1085–1091 (1996).

35. T. Malinski, Z. Taha, S. Grunfeld, A. Burewicz, P. Tomboulian, and F. Kiechle, Measurements of nitric-oxide in biological-materials using a porphyrinic microsensor. *Anal. Chim. Acta* **279**, 135 (1993).
36. T. Malinski, F. Bailey, Z.G. Zhang, and M. Chopp, Nitric-oxide measured by a porphyrinic microsensor in rat-brain after transient middle cerebral-artery occlusion. *J. Cereb. Blood Flow Metab.* **13**, 355–358 (1993).
37. Z.G. Zhang, M. Chopp, F. Bailey, and T. Malinski, Nitric-oxide changes in the rat-brain after transient middle cerebral-artery occlusion. *J. Neurol. Sci.* **128**, 22–27 (1995).
38. F. Lantoine, S. Trevin, F. Bedioui, and J. Devynck, Selective and sensitive electrochemical measurement of nitric-oxide in aqueous-solution – discussion and new results. *J. Electroanal. Chem.* **392**, 85–89 (1995).
39. F. Bedioui, S. Trevin, and J. Devynck, The use of gold electrodes in the electrochemical detection of nitric-oxide in aqueous-solution. *J. Electroanal. Chem.* **377**, 295 (1994).
40. H. Yokoyama, N. Mori, N. Kasai, T. Matsue, I. Uchida, N. Kobayashi, N. Tsuchihashi, T. Yoshimura, M. Hiramatsu, and S.I. Niwa, Direct and continuous monitoring of intrahippocampal nitric oxide (NO) by an NO sensor in freely moving rat after N-methyl-D-aspartic acid injection. *Denki Kagaku* **63**, 1167–1170 (1995).
41. K.I. Kim, H.Y. Chung, G.S. Oh, H.O. Bae, S.H. Kim, and H.J. Chun, Integrated gold-disk microelectrode modified with iron(II)-phthalocyanine for nitric oxide detection in macrophages. *Microchem J.* **80**, 219–226 (2005).
42. S.L Vilakazi and T. Nyokong, Voltammetric determination of nitric oxide on cobalt phthalocyanine modified microelectrodes. *J. Electroanal. Chem.* **512**, 56–63 (2001).
43. X.J. Zhang, L. Cardosa, M. Broderick, H. Fein, and J. Lin, An integrated nitric oxide sensor based on carbon fiber coated with selective membranes. *Electroanalysis* **12**, 1113–1117 (2001).
44. A. Dickson, J. Lin, J. Sun, M. Broderick, H. Fein and X.J. Zhang, Construction and characterization of a new flexible and nonbreakable nitric oxide microsensor. *Electroanalysis* **16**, 640–643 (2004).
45. X.J. Zhang, Y. Kislyak, J. Lin, A. Dickson, L. Cardosa, M. Broderick, and H. Fein, Nanometer size electrode for nitric oxide and S-nitrosothiols measurement. *Electrochem. Commun.* **4**, 11–16 (2002).
46. X.J. Zhang, J. Lin, L. Cardoso, M. Broderick, and V. Darley-Usmar, A novel microchip nitric oxide sensor with sub-nM detection limit. *Electroanalysis* **14**, 697–703 (2002).
47. B. Fabre, S. Burlet, R. Cespuglio, and G. Bidan, Voltammetric detection of NO in the rat brain with an electronic conducting polymer and Nafion®bilayer-coated carbon fibre electrode. *J. Electroanal. Chem.* **426**, 75–83 (1997).
48. M.N. Friedemann, S.W. Robinson, and G.A. Gerhardt, o-phenylenediamine-modified carbon fiber electrodes for the detection of nitric oxide. *Anal. Chem.* **68**, 2621–2628 (1996).
49. J.K. Park, P.H. Tran, J.K.T. Chao, R. Ghodadra, R. Rangarajan, and N.V. Thakor, In vivo nitric oxide sensor using non-conducting polymer-modified carbon fiber. *Biosens. Bioelectron.* **13**, 1187–1195 (1998).
50. N.R. Ferreira, A. Ledo, J.G. Frade, G.A. Gerhardt, J. Laranjinha, and R.M. Barbosa, Electrochemical measurement of endogenously produced nitric oxide in brain slices using Nafion/o-phenylenediamine modified carbon fiber microelectrodes. *Anal. Chim. Acta* **535**, 1–7 (2005).
51. D. Kato, M. Kunitake, M. Nishizawa, T. Matsue, and F. Mizutani, Amperometric nitric oxide microsensor using two-dimensional cross-linked Langmuir-Blodgett films of polysiloxane copolymer. *Sens. Actuator B-Chem.* **108**, 384–388 (2005).
52. K. Ichimori, H. Ishida, M. Fukahori, H. Nakazawa, and E. Murakami, Practical nitric-oxide measurement employing a nitric oxide-selective electrode. *Rev. Sci. Instrum.* **65**, 2714–2718 (1994).
53. M. Pontie, F. Bedioui, and J. Devynck, New composite modified carbon microfibers for sensitive and selective determination of physiologically relevant concentrations of nitric oxide in solution. *Electroanalysis* **11**, 845–850 (1999).
54. B.W. Allen, C.A. Piantadosi, and L.A. Coury, Electrode materials for nitric oxide detection. *Nitric Oxide-Biol. Chem.* **4**, 75–84 (2000).
55. Y.Z. Xian, W.L. Sun, J.A. Xue, M. Luo, and L.T. Jin, Iridium oxide and palladium modified nitric oxide microsensor. *Anal. Chim. Acta* **381**, 191–196 (1999).
56. T. Haruyama, S. Shiino, Y. Yanagida, E. Kobatake, and M. Aizawa, Two types of electrochemical nitric oxide (NO) sensing systems with heat-denatured Cyt C and radical scavenger PTIO. *Biosens. Bioelectron.* **13**, 763–769 (1998).

57. X.C. He and J.Y. Mo, Electrocatalytic oxidation of NO at electrode modified with Nafion-Co-II-1, 10-phenanthroline film and its application to NO detection. *Analyst* **125**, 793–795 (2000).
58. S.R. Smith and H.H. Thorp, Application of the electrocatalytic reduction of nitric oxide mediated by ferrioxamine B to the determination of nitric oxide concentrations in solution. *Inorg. Chim. Acta* **273**, 316–319 (1998).
59. Y. Kitamura, T. Uzawa, K. Oka, Y. Komai., H. Ogawa, N. Takizawa, H. Kobayashi, and K. Tanishita, Microcoaxial electrode for in vivo nitric oxide measurement. *Anal. Chem.* **72**, 2957–2962 (2000).
60. F. Mizutani, S. Yabuki, T. Sawaguchi, Y. Hirata,Y. Sato, and S. Iijima, Use of a siloxane polymer for the preparation of amperometric sensors: O-2 and NO sensors and enzyme sensors. *Sens. Actuator B-Chem.* **76**, 489–493 (2001).
61. J. Katrlik and P. Zalesakova, Nitric oxide determination by amperometric carbon fiber microelectrode. *Bioelectrochemistry* **56**, 73–76 (2002).
62. C.H. Fan, J.T. Pang, P.P. Shen, G.X. Li, and D.X. Zhu, Nitric oxide biosensors based on Hb/phosphatidylcholine films. *Anal. Sci.* **18**, 129–132 (2002).
63. C.H. Fan, G.X. Li, J.Q. Zhu, and D.X. Zhu, A reagentless nitric oxide biosensor based on hemoglobin-DNA films. *Anal. Chim. Acta* **423**, 95–100 (2000).
64. A. Kitajima, T. Teranishi, and M. Miyake, Detection of nitric oxide on carbon electrode modified with ionic polymers and alpha-cyclodextrin. *Electrochemistry* **69**, 16–20 (2001).
65. J.H. Shin, S.W. Weinman, and M.H. Schoenfisch, Sol-gel derived amperometric nitric oxide microsensor. *Anal. Chem.* **77**, 3494–3501 (2005).
66. Y. Lee, B.K. Oh, and M.E. Meyerhoff, Improved planar amperometric nitric oxide sensor based on platinized platinum anode. 1. Experimental results and theory when applied for monitoring NO release from diazeniumdiolate-doped polymeric films. *Anal. Chem.* **76**, 536–544 (2004).
67. Y. Lee, J. Yang, S.M. Rudich, R.J. Schreiner, and M.E. Meyerhoff, Improved planar amperometric nitric oxide sensor based on platinized platinum anode. 2. Direct real-time measurement of NO generated from porcine kidney slices in the presence of L-arginine, L-arginine polymers, and protamine. *Anal. Chem.* **76**, 545–551 (2004).
68. S. Kroning, F.W. Scheller, U. Wollenberger, and F. Lisdat, Myoglobin-clay electrode for nitric oxide (NO) detection in solution. *Electroanalysis* **16**, 253–259 (2004).
69. K. Kamei, T. Haruyama, M. Mie, Y. Yanagida, M. Aizawa, and E. Kobatake, The construction of endothelial cellular biosensing system for the control of blood pressure drugs. *Biosens. Bioelectron.* **19**, 1121–1124 (2004).
70. E. Casero, F. Pariente, and E. Lorenzo, Electrocatalytic oxidation of nitric oxide at indium hexacyanoferrate film-modified electrodes. *Anal. Bioanal. Chem.* **375**, 294–299 (2003).
71. A. Pailleret, J. Oni, S. Reiter, S. Isik, M. Etienne, F. Bedioui, and W. Schuhmann, In situ formation and scanning electrochemical microscopy assisted positioning of NO-sensors above human umbilical vein endothelial cells for the detection of nitric oxide release. *Electrochem. Commun.* **5**, 847–852 (2003).
72. H.I. Magazine, Detection of endothelial cell-derived nitric oxide: current trends and future directions. *Adv. Neuroimmunol.* **5**, 479–490 (1995).
73. F.L. Kiechle and T. Malinski, Nitric-oxide – biochemistry, pathophysiology, and detection. *Am. J. Clin. Pathol.* **100**, 567–575 (1993).
74. S. Archer, Measurement of nitric-oxide in biological models. *Faseb J.* **7**, 349–360 (1993).
75. *Handbook of Chemistry and Physics*; 76th ed.; CRC Press: Boca Raton, FL, 1995.
76. M. Kelm and J. Schrader, Control of coronary vascular tone by nitric-oxide. *Circ. Res.* **66**, 1561–1575 (1990).
77. X.J. Zhang, L. Cardosa, M. Broderick, H. Fein, and I.R. Davies, Novel calibration method for nitric oxide microsensors by stoichiometrical generation of nitric oxide from SNAP. *Electroanalysis* **12**, 425–428 (2000).
78. M. Tristani-Firouzi, E.G. DeMaster, B.J. Quast, D.P. Nelson, and S.L. Archer, Utility of a nitric oxide electrode for monitoring the administration of nitric oxide in biologic systems. *J. Lab. Clin. Med.* **131**, 281–285 (1998).
79. J.C. Millar, Real-time direct measurement of nitric oxide in bovine perfused eye trabecular meshwork using a Clark-type electrode. *J. Ocular Pharmacol. Ther.* **19**, 299–313 (2003).

80. K. Akeo, S. Amaki, T. Suzuki, and T. Hiramitsu, Melanin granules prevent the cytotoxic effects of L-DOPA on retinal pigment epithelial cells in vitro by regulation of NO and superoxide radicals. *Pigm. Cell. Res.* **13**, 80–88 (2000).
81. S.K. Amaki,Y. Oguchi, T. Ogata, T. Suzuki, K. Akeo, and T. Hiramitsu, L-DOPA produced nitric oxide in the vitreous and caused greater vasodilation in the choroid and the ciliary body of melanotic rats than in those of amelanotic rats. *Pigm. Cell. Res.* **14**, 256–263 (2001).
82. K. Iijima, E. Henry, A. Moriya, A. Wirz, A.W. Kelman, and K.E.L. McColl, Dietary nitrate generates potentially mutagenic concentrations of nitric oxide at the gastroesophageal junction. *Gastroenterology* **122**, 1248–1257 (2002).
83. G.B. Stefano, W. Zhu, P. Cadet, T.V. Bilfinger, and K. Mantione, Morphine enhances nitric oxide release in the mammalian gastrointestinal tract via the mu 3 opiate receptor subtype: a hormonal role for endogenous morphine. *J. Physiol. Pharmacol.* **55**, 279–288 (2004).
84. L. Cherian, J.C. Goodman, and C.S. Robertson, Brain nitric oxide changes after controlled cortical impact injury in rats. *J. Neurophysiol.* **83**, 2171–2178 (2000).
85. D.G. Buerk, B.M. Ances, J.H. Greenberg, and J.A. Detre, Temporal dynamics of brain tissue nitric oxide during functional forepaw stimulation in rats. *Neuroimage* **18**, 1–9 (2003).
86. G. Rocchitta, R. Migheli, M.P. Mura, G. Esposito, M.S. Desole, E. Miele, M. Miele, and P.A. Serra, Signalling pathways in the nitric oxide donor-induced dopamine release in the striatum of freely moving rats: evidence that exogenous nitric oxide promotes Ca2+ entry through store-operated channels. *Brain Res.* **1023**, 243–252 (2004).
87. A. Meulemans, A brain nitric oxide synthase study in the rat: production of a nitroso-corn pound NA and absence of nitric oxide synthesis. *Neurosci. Lett.* **321**, 115–119 (2002).
88. M. Saitoand and I. Miyagawa, Real-time monitoring of nitric oxide in ischemia-reperfusion rat kidney. *Urol. Res.* **28**, 141–146 (2000).
89. D.Z. Levine, M. Iacovitti, K.D. Burns, and X.J. Zhang, Real-time profiling of kidney tubular fluid nitric oxide concentrations in vivo. *Am. J. Physiol.-Renal Physiol.* **281**, F189–194 (2001).
90. D.Z. Levine and M. Iacovitti, Real time microelectrode measurement of nitric oxide in kidney tubular fluid in vivo. *Sensors* **3**, 314 (2003).
91. D.Z. Levine, K.D. Burns, J. Jaffey, and M. Iacovitti, Short-term modulation of distal tubule fluid nitric oxide in vivo by loop NaCl reabsorption. *Kidney Int.* **65**, 184–189 (2004).
92. C. Thorup, M. Kornfeld, J.M. Winaver, M.S. Goligorsky, and L.C. Moore, Angiotensin-II stimulates nitric oxide release in isolated perfused renal resistance arteries. *Pflugers Arch.* **435**, 432–434 (1998).
93. B. Arregui, B. Lopez, M.G. Salom, F. Valero, C. Navarro, and F.J. Fenoy, Acute renal hemodynamic effects of dimanganese decacarbonyl and cobalt protoporphyrin. *Kidney Int.* **65**, 564–574 (2004).
94. S. Fujita, D.L. Roerig, Z.J. Bosnjak, and D.F. Stowe, Effects of vasodilators and perfusion pressure on coronary flow and simultaneous release of nitric oxide from guinea pig isolated hearts. *Cardiovasc. Res.* **38**, 655–667 (1998).
95. E. Novalija, S. Fujita, J.P. Kampine and D.F. Stowe, Sevoflurane mimics ischemic preconditioning effects on coronary flow and nitric oxide release in isolated hearts. *Anesthesiology* **91**, 701–712 (1999).
96. D. Schulte and J. Millar, The effects of high- and low-intensity percutaneous stimulation on nitric oxide levels and spike activity in the superficial laminae of the spinal cord. *Pain* **103**, 139–150 (2003).
97. G.B. Stefano, V. Prevot, J.C. Beauvillain, C. Fimiani, I. Welters, P. Cadet, C. Breton, J. Pestel, M. Salzet, and T.V. Bilfinger, Estradiol coupling to human monocyte nitric oxide release is dependent on intracellular calcium transients: evidence for an estrogen surface receptor. *J. Immunol.* **163**, 3758–3763 (1999).
98. G.B. Stefano, V. Prevot, J.C. Beauvillain, P. Cadet, C. Fimiani, I. Welters, G.L. Fricchione, C. Breton, P. Lassalle, M. Salzet, and T.V. Bilfinger, Cell-surface estrogen receptors mediate calcium-dependent nitric oxide release in human endothelia. *Circulation* **101**, 1594–1597 (2000).
99. B. Beltran, M. Quintero, E. Garcia-Zaragoza, E. O'Connor, J.V. Esplugues, and S. Moncada, Inhibition of mitochondrial respiration by endogenous nitric oxide: a critical step in Fas signaling. *Proc. Natl. Acad. Sci. U.S.A.* **99**, 8892 (2002).
100. S. Shiva, P.S. Brookes, P. Patel, P.G. Anderson, and V.M. Darley-Usmar, Nitric oxide partitioning into mitochondrial membranes and the control of respiration at cytochrome c oxidase. *Proc. Natl. Acad. Sci. U.S.A.* **98**, 7212–7217 (2001).

101. M. Mas, A. Escrig and J.L. Gonzalez-Mora, In vivo electrochemical measurement of nitric oxide in corpus cavernosum penis. *J. Neurosci. Methods* **119**, 143–150 (2002).
102. K.Z. Kedziora-Kornatowska, M. Luciak, J. Blaszczyk, and W. Pawlak, Effect of aminoguanidine on the generation of superoxide anion and nitric oxide by peripheral blood granulocytes of rats with streptozotocin-induced diabetes. *Clin. Chim. Acta* **278**, 45–53 (1998).
103. G.B. Stefano, P. Cadet, C. Breton, Y. Goumon, V. Prevot, J. P. Dessaint, J.C. Beauvillain, A.S. Roumier, I. Welters, and M. Salzet, Estradiol-stimulated nitric oxide release in human granulocytes is dependent on intracellular calcium transients: evidence of a cell surface estrogen receptor. *Blood* **95**, 3951–3958 (2000).
104. S. Uhlmann, U. Friedrichs,W. Eichler, S. Hoffmann, and P. Wiedemann, Direct measurement of VEGF-induced nitric oxide production by choroidal endothelial cells. *Microvasc. Res.* **62**, 179–189 (2001).
105. S. Kashiwagi, Y. Izumi, T. Gohongi, Z.N. Demou, L. Xu, P.L. Huang, D.G. Buerk, L.L. Munn, R K. Jain, and D. Fukumura, NO mediates mural cell recruitment and vessel morphogenesis in murine melanomas and tissue-engineered blood vessels. *J. Clin. Invest.* **115**, 1816–1827 (2005).
106. M. Tsatmali, A. Graham, D. Szatkowski, J. Ancans, P. Manning, C.J. McNeil, A.M. Graham, and A.J. Thody, Alpha-melanocyte-stimulating hormone modulates nitric oxide production in melanocytes. *J. Invest. Dermatol.* **114**, 520–526 (2000).
107. J. Rysz, M. Luciak, J. Kedziora, J. Blaszczyk, and E. Sibinska, Nitric oxide release in the peripheral blood during hemodialysis. *Kidney Int.* **51**, 294–300 (1997).
108. M. Rievaj, J. Lietava, and D. Bustin, Electrochemical determination of nitric oxide in blood samples. *Chem. Pap.-Chem. Zvesti* **58**, 306–310 (2004).
109. G. Larfars, F. Lantoine, M.A. Devynck, and H. Gyllenhammar, Electrochemical detection of nitric oxide production in human polymorphonuclear neutrophil leukocytes. *Scand. J. Clin. Lab. Invest.* **59**, 361–368 (1999).
110. N. Kasuya, Y. Kishi, S. Sakita, F. Numano, and M. Isobe, Acute vigorous exercised primes enhanced NO release in human platelets. *Atherosclerosis* **161**, 225–232 (2002).
111. J.P. de la Cruz, J.A. Gonzalez-Correa, A. Guerrero, E. Marquez, F. Martos and F.S. de la Cuesta, Differences in the effects of extended-release aspirin and plain-formulated aspirin on prostanoids and nitric oxide in healthy volunteers. *Fundam. Clin. Pharmacol.* **17**, 363–372 (2003).
112. J.E. Freedman, J. Loscalzo, M.R.M.R. Barnard, C. Alpert, J.F. Keaney, and A.D. Michelson, Nitric oxide released from activated platelets inhibits platelet recruitment. *J. Clin. Invest.* **100**, 350–356 (1997).
113. X.R. Shi, T.Y. Ren, and A.L. Nuttall, The electrochemical and fluorescence detection of nitric oxide in the cochlea and its increase following loud sound. *Hear. Res.* **164**, 49–58 (2002).
114. Z.G. Jiang, X.R. Shi, H. Zhao, J.Q. Si, and A.L. Nuttall, Basal nitric oxide production contributes to membrane potential and vasotone regulation of guinea pig in vitro spiral modiolar artery. *Hear. Res.* **189**, 92–100 (2004).
115. Y. Sakihama, S. Nakamura, and H. Yamasaki, Nitric oxide production mediated by nitrate reductase in the green alga Chlamydomonas reinhardtii: an alternative NO production pathway in photosynthetic organisms. *Plant Cell Physiol.* **43**, 290–297 (2002).
116. H. Yamasaki, Y. Sakihama, and S. Takahashi, An alternative pathway for nitric oxide production in plants: new features of an old enzyme. *Trends Plant Sci.* **4**, 128–129 (1999).
117. H. Yamasaki and Y. Sakihama, Simultaneous production of nitric oxide and peroxynitrite by plant nitrate reductase: in vitro evidence for the NR-dependent formation of active nitrogen species. *FEBS Lett.* **468**, 89–92 (2000).
118. S. Grande, P. Bogani, A. de Saizieu, G. Schueler, C. Galli, and F. Visioli, Vasomodulating potential of Mediterranean wild plant extracts. *J. Agric. Food Chem.* **52**, 5021–5026 (2004).
119. L.L. Moroz, T.P. Norekian, T.J. Pirtle, K.J. Robertson, and R.A. Satterlie, Distribution of NADPH-diaphorase reactivity and effects of nitric oxide on feeding and locomotory circuitry in the pteropod mollusc, Clione limacina. *J. Comp. Neurol.* **427**, 274–284 (2000).
120. U. Simonsen, R.W. Wadsworth, N.H. Buus, and M.J. Mulvany, In vitro simultaneous measurements of relaxation and nitric oxide concentration in rat superior mesenteric artery. *J. Physiol.-London* **516**, 271–282 (1999).

121. R. Hernanz, M.J. Alonso, H. Zibrandtsen, Y. Alvarez, M. Salaices, and U. Simonsen, Measurements of nitric oxide concentration and hyporeactivity in rat superior mesenteric artery exposed to endotoxin. *Cardiovasc. Res.* **62**, 202–211 (2004).
122. E. Stankevicius, A.C. Martinez, M.J. Mulvany, and U. Simonsen, Blunted acetylcholine relaxation and nitric oxide release in arteries from renal hypertensive rats. *J. Hypertens.* **20**, 1571–1579 (2002).
123. N.H. Buus, U. Simonsen, H.K. Pilegaard, and N.J. Mulvaney, Nitric oxide, prostanoid and non-NO, non-prostanoid involvement in acetylcholine relaxation of isolated human small arteries. *Br. J. Pharmacol.* **129**, 184–192 (2002).
124. S. Isik, L. Berdondini, J. Oni, A. Blochl, M. Koudelka-Hep, and W. Schuhmann, Cell-compatible array of three-dimensional tip electrodes for the detection of nitric oxide release. *Biosens. Bioelectron.* **20**, 1566–1572 (2005).
125. D.L. Kellogg, J.L. Zhao, C. Friel and L.J. Roman, Nitric oxide concentration increases in the cutaneous interstitial space during heat stress in humans. *J. Appl. Physiol.* **94**, 1971–1977 (2003).
126. J.L. Zhao, P.E. Pergola, L.J. Roman, and D.L. Kellogg, Bioactive nitric oxide concentration does not increase during reactive hyperemia in human skin. *J. Appl. Physiol.* **96**, 628–632 (2004).
127. S.R. Thom, D. Fisher, J. Zhang, V.M. Bhopale, S.T. Ohnishi, Y. Kotake, T. Ohnishi, and D.G. Buerk, Stimulation of perivascular nitric oxide synthesis by oxygen. *Am. J. Physiol.-Heart Circul. Physiol.* **284**, H1230–1239 (2003).
128. A.I. Vinik, P.M.M. Barlow, J. Ullal, C.M. Casellini, H.K. Parson, Pioglitazone treatment improves nitrosative stress in type 2 diabetes. *Diabetes Care* **29**, 869–876 (2006).

CHAPTER 2

Biosensors for pesticides

Huangxian Ju and Vivek Babu Kandimalla

2.1 INTRODUCTION

2.1.1 Need for pesticide biosensors

In agriculture, farmers randomly use several organic toxic compounds, i.e. pesticides, herbicides, insecticides, to control diseases for obtaining high yields. Pesticide is a term used in a broad sense for chemicals, synthetic or natural, that are used for the control of insects, fungi, bacteria, weeds, nematodes, rodents and other pests [1, 2]. The undegraded pesticide residues may enter into the food chains through air, water and soil and cause several health problems to ecosystems, birds, animals and human beings. Pesticides can be carcinogenic or citogenic. They can produce bone marrow diseases, infertility, nerve disorders and immunological and respiratory diseases. To circumvent these problems the USA and European governments have imposed new legislation. Enforcement of this legislation necessitates reliable monitoring of the environment for the presence of compounds, which may adversely affect human health and local ecosystems. The highest permissible levels of different pesticides in water for human use go from 0.3 to 400 $\mu g\ l^{-1}$ [3]. In the last two decades organochlorine insecticides (e.g. DDT, aldrin and lindane) have been progressively replaced by organophosphorus (e.g. parathion and malathion) and derivatives of carbamic acid (e.g. carbaryl and aldicarb) insecticides that show low persistence in the environment but represent a serious risk due to their high acute toxicity.

The commonly used analytical methods for pesticide analysis include high pressure liquid chromatography, gas chromatography or coupled techniques of GC-MS and ELISA [4–6]. These conventional methods are sensitive, reliable and precise. In spite of their advantages, these techniques need expensive instrumentation, require skilled technicians, and are time consuming, laborious and not easily adoptable for field analysis. The simple and advantageous alternative is the use of biosensors. A biosensor is a probe that integrates a biological component (e.g. an enzyme, whole cell, antibody, etc.) with an electronic component to yield a measurable signal. Even though biosensors do not compete with the above potent chromatographic techniques, they do provide fast and reliable analysis. These devices are highly useful for preliminary screening before applying more costly techniques. In the last three decades several quick and low cost pesticide biosensors have also been developed using enzymes, antibodies, and whole cells. This chapter comprehensively discusses various aspects of the pesticide biosensors and tries to incorporate all-important aspects; however, addressing all reports in one chapter is very difficult considering the size of the available literature.

2.1.2 Developments in pesticide biosensors

Acetylcholinesterase (AChE) isolated from various organisms has been used in the majority of pesticide biosensors. In the early 1950s potentiometric detection was adopted for pesticide detection. In the middle of the 1980s it was used for the construction of the first integrated biosensors for detection of pesticides based on inhibition of AChE. Later rapid changes in science and technology introduced novel genetically modified AChE, and other enzymes were also employed in construction of biosensors. The main drawback of AChE amperometric biosensors is that the enzymatic product, thiocholine oxidation, needs a high applied potential, which reduces the stability of the biosensor. To solve this problem mediators such as 7, 7′, 8, 8′-tetracyanoquinodimethane (TCNQ) [7] and highly conductive materials such as carbon nanotubes [8] have been employed. To avoid enzyme purification and reduce the cost, whole cells have also been directly interfaced with transducer surface. The major limitation with enzyme inhibition-based pesticide determination is non-selectivity. To improve the selectivity of the pesticide biosensors, immunosensors have been introduced, where antibodies act as recognition receptors. Several detection methods have been interfaced with biosensors/immunosensors such as potentiometric, amperometric, differential pulse voltammetry (DPV), chemiluminescence, piezoelectric, surface plasmon resonance, etc. for the improvement of the biosensor analytical performance [9]. The sensitivity of the biosensor and stability of the biocatalyst are mainly governed by the immobilization method adopted for the interfacing of the biocatalyst with the transducer surface. Several immobilization methods such as covalent linking, adsorption and entrapment have been developed. Furthermore the exploitation of advanced microfabrication technology allows the design of portable, easy-to-use, inexpensive and environmental friendly biosensor chips and strips for pesticide monitoring in real samples. Although pesticide biosensor technology has been progressing dramatically, commercialization is still in its infancy and needs more research.

2.1.3 Thrust areas for pesticide biosensors

Pesticide biosensors are highly useful in agriculture, the food industry and the medical sector. In the export and import of food materials portable biosensors or kits with quick responses are highly convenient for fast detection. Similarly if the biosensor has the capability to detect the pesticide/toxicant selectively present in clinical samples, it would be highly advantageous for diagnosis and treatment in the medical sector. Miniaturized sensor strips or kits are highly advantageous and applicable for the domestic applications to check the water quality for daily use.

2.2 BIOCATALYSTS USED IN PESTICIDE BIOSENSORS

The selectivity and sensitivity of biosensors are governed by the biocatalyst. Commonly used biocatalysts in biosensors are enzymes [10, 11], antibodies [12, 13], whole cells [14], and artificial receptors such as molecularly imprinted polymers [15, 16].

2.2.1 Enzymes used in pesticide biosensors and their features

Most of the pesticide biosensors are designed based on the inhibitory property of enzymes. AChE and butyrylcholinesterase (BChE) are widely used in the development of pesticide biosensors [17, 18]. Inhibition leads to a decrease in activity, which

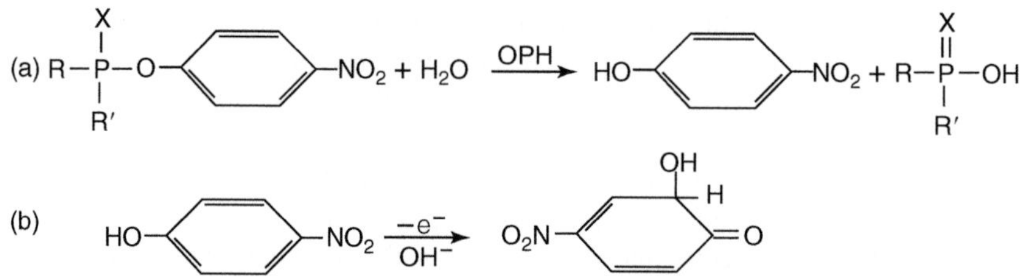

FIGURE 2.1 Reaction scheme for OPH hydrolysis of methyl parathion and paraoxon (a) followed by the electrochemical oxidation of p-phenol (b). R and R′ are ethoxy and methoxy and X is O and S in paraoxon and methyl parathion, respectively.

is indirectly proportional to the amount of inhibitor or pesticide in the sample. Mainly AChE-based biosensors are superior, due to the strong enzyme inhibition by organophosphorus pesticides, enabling a higher sensitivity. The other often-employed enzymes in pesticide biosensors are acetolactate synthase [19], acid phosphatase [20], alkaline phosphatase [21], tyrosinase [22, 23], ascorbate oxidase [14], etc. Pesticides can also inhibit the activity of luciferase, which is a major enzyme in bioluminescence reactions. By employing firefly, luciferase pesticide concentrations have been determined, where the pesticide concentration is indirectly proportional to the bioluminescence [24]. The enzyme organophosphorus hydrolase (OPH) has been used as an alternative recognition component. OPH can hydrolyze an organophosphate molecule, the resulting products can be monitored either spectrophotometrically or electrochemically (Fig. 2.1). Because organophosphate is the substrate for OPH, this scheme leads to a direct determination of analyte, as the rate of signal generation is directly proportional to the concentration of organophosphate [10, 25–28]. The above enzymes are used singly; sometimes bienzymes such as AChE and choline oxidase are used [29, 30]. In the case of bienzymatic systems a number of parameters need to be optimized and they are more complex than single enzyme systems [8].

2.2.2 Immobilization methods used in pesticide biosensors design

All the enzymes are highly selective to the substrates and sensitive to the environmental factors such as pH, temperature and inhibitors, denaturing and chelating agents. Enzyme activity and stability on the transducer surface is governed by the procedure followed for the immobilization and chemical nature of used matrices. The adopted immobilization method should be sufficiently strong to provide good mechanical stability of biosensor, and sufficiently soft to provide optimal conformation and freedom of the enzymes, which is crucial for reaching sufficient enzymatic activity [31]. Several approaches have been followed for immobilization of enzyme on the transducer/electrode surface, such as adsorption [32], cross-linking with bifunctional chemical reagents [33], binding with dendrimer layers [34], entrapment in different matrices [35], including layer of cross-linked bovine serum albumin [36, 37] and electropolymerization [38], and more recently bioaffinity attachment using concanavalin A [39], etc.

In view of the conductive and electrocatalytic features of carbon nanotubes (CNTs), AChE and choline oxidases (COx) have been covalently coimmobilized on multiwall carbon nanotubes (MWNTs) for the preparation of an organophosphorus pesticide (OP) biosensor [40, 41]. Another OP biosensor has also been constructed by adsorption of AChE on MWNTs modified thick film [8]. More recently AChE has been covalently linked with MWNTs doped glutaraldehyde cross-linked chitosan composite film [11], in which biopolymer chitosan provides biocompatible nature to the enzyme and MWNTs improve the conductive nature of chitosan. Even though these enzyme immobilization techniques have been reported in the last three decades, no method can be commonly used for all the enzymes by retaining their complete activity.

2.3 ENZYME-BASED BIOSENSORS CONSTRUCTION

2.3.1 Pesticides measuring principles

Mainly, two principles are used in electrochemical pesticide biosensor design, either enzyme inhibition or hydrolysis of pesticide. Among these two approaches inhibition-based biosensors have been widely employed in analysis due to the simplicity and wide availability of the enzymes. The direct enzymatic hydrolysis of pesticide is also extremely attractive for biosensing, because the catalytic reaction is superior and faster than the inhibition [27].

2.3.2 Inhibition-based biosensors

Organophosphate and carbamate pesticides are potent inhibitors of the enzyme cholinesterase. The inhibition of cholinesterase activity by the pesticide leads to the formation of stable covalent intermediates such as phosphoryl–enzyme complexes, which makes the hydrolysis of the substrate very slow. Both organophosphorus and carbamate pesticides can react with AChE in the same manner because the acetylation of the serine residue at the catalytic center is analogous to phosphorylation and carbamylation. Carbamated enzyme can restore its catalytic activity more rapidly than phosphorylated enzyme [17, 42]. Kok and Hasirci [43] reported that the total anti-cholinesterase activity of binary pesticide mixtures was lower than the sum of the individual inhibition values.

In AChE-based biosensors acetylthiocholine is commonly used as a substrate. The thiocholine produced during the catalytic reaction can be monitored using spectrometric, amperometric [44] (Fig. 2.2) or potentiometric methods. The enzyme activity is indirectly proportional to the pesticide concentration. La Rosa *et al.* [45] used 4-aminophenyl acetate as the enzyme substrate for a cholinesterase sensor for pesticide determination. This system allowed the determination of esterase activities via oxidation of the enzymatic product 4-aminophenol rather than the typical thiocholine. Sulfonylureas are reversible inhibitors of acetolactate synthase (ALS). By taking advantage of this inhibition mechanism ALS has been entrapped in photo cured polymer of polyvinyl alcohol bearing styrylpyridinium groups (PVA-SbQ) to prepare an amperometric biosensor for

5′–Mercapto –2 –nitrobenzoic acid (yellow)

↑ 5,5′–dithio–2–nitrobenzoic acid (Chromogen)

$$\text{Acetylcholine} + H_2O \xrightarrow{\text{AChE}} \text{Thiocholine} + \text{Acetic acid}$$

$$2\ \text{Thiocholine} \xrightarrow{\text{Oxidation}} \text{Dithio–bis–choline} + 2H^+ + 2e^-$$

FIGURE 2.2 Enzymatic formation of thiocholine and detection principles for pesticide.

$$\text{Acetylcholine} + H_2O \xrightarrow{\text{AChE}} \text{Choline} + \text{Acetic acid}$$

$$\text{Choline} + O_2 + H_2O \xrightarrow{\text{COx}} \text{Betaine} + H_2O_2$$

FIGURE 2.3 Bienzymatic reaction for pesticide detection.

the detection of sulfometuron methyl (herbicide) [19, 46]. Immobilized ALS is more stable than the native enzyme, but the inhibition potency is higher for the free enzyme. Polyphenol oxidase (tyrosinase) can also be irreversibly inhibited by atrazin, diazinon, dichlorvos, etc. Besombes *et al.* [47] used an electropolymerized pyrrole/tyrosinase coating for the detection of carbamate pesticides and reported a detection limit of 2 μM for chloroisopropylphenyl carbamate. In another study a tyrosinase electrode was constructed by direct adsorption of the enzyme on the surface of a graphite disc electrode. Three carbamate pesticides, ziram, diram, and zinc diethyldithiocarbamate, inhibited the tyrosinase in a medium of reversed micella and the detection limits of this amperometric biosensor were 0.074, 1.3 and 1.7 μM, respectively [22]. To improve the stability and sensitivity of a single enzyme-based biosensor the modified enzymes have been employed instead of the wild type. A genetically modified *Drosophila melanogaster* AChE (DmAChE) allows its oriented immobilization on electrodes [7].

Another stable and versatile method for pesticide detection is the use of the second enzyme, such as AChE and choline oxidase (COx). Multienzyme-based biosensors use cholinesterase in conjunction with choline oxidase to measure hydrogen peroxide production or oxygen consumption. COx oxidizes choline to betaine and hydrogen peroxide (Fig. 2.3). These methods couple an enzyme that is inhibited by the pesticide in conjunction with another enzyme, which uses the product of the first enzymatic reaction as the substrate. Another bienzymatic amperometric biosensor has been reported for the detection of dithiocarbamate fungicides [48, 49]. An aldehyde dehydrogenase and diaphorase were coimmobilized in a PVA-SbQ layer and mounted on a screen-printed electrode (SPE) with Pt sputtered carbon paste. The detection limits of bienzymatic amperometric biosensor for nabam [48] and zineb [49] were 8 and 9 ng mL^{-1}, respectively. Organophosphorus and carbamate pesticides can reversibly inhibit the catalytic activity of acid phosphatase (AP); hence AP can be reused with normal washing with buffer. But amperometric detection of AP inhibition requires a bienzymatic system (Fig. 2.4). Mazzei *et al.* [20] immobilized AP and GOD on a separate dialytic membrane. The bienzyme modified membrane was placed on Pt electrode using an

$$\text{Glucose} - 6 - \text{phosphate} + H_2O \xrightarrow{\text{Acid phosphatase}} \text{Glucose} + \text{Inorganic phosphate}$$

$$\text{Glucose} + O_2 + H_2O \xrightarrow{\text{GOD}} \text{Gluconolactone} + H_2O_2$$

FIGURE 2.4 Acid phosphatase and GOD bioenzymatic reaction for pesticide detection.

"O" ring. At 20 min inhibition time the detection limits for malathion, parathion methyl and paraoxon were 3, 0.5 and 5 $\mu g\ l^{-1}$, respectively. Although these bienzymatic systems look simple, it is difficult to provide optimal conditions for both enzymes. In general the optimum pH, temperature and buffer molarity for different enzymes are different. The experimental conditions are at the levels below the optimum capacity of both enzymes [14]. This disadvantage can be minimized by use of a single enzyme system, which is readily inhibited by the pesticide.

Usually the inhibition results in the decay of the enzyme activity so that the number of consecutive measurements with the same biosensor is limited. To overcome the above limitations and determine the pesticides faster and at a cheaper cost disposable sensors have been developed [7, 39, 40, 50]. Screen-printed electrodes have been widely used in the design of disposable sensors using AChE as a biocatalyst [7, 50] and doped TCNQ in carbon paste to decrease the working potential and to avoid the interfering influence of electroactive impurities by reducing the applied potential. Using thick film technology, a biosensor strip was reported by integrating photolithographic conducting copper tracks, graphite–epoxy composite applied by screen printing and enzyme immobilized by cross-linking with the bifunctional reagent glutaraldehyde [51]. The detection limit of the disposable strip was on the order of 10^{-9} to 10^{-11} M for paraoxon and carbofuran. Wang *et al.* [52] reported a tyrosinase-based screen-printed biosensor for the determination of carbamate pesticides with fast response time and without the preincubation period. Recently a disposable carbon nanotube modified screen-printed biosensor has been reported using the bienzymatic system (AChE/CHO) [40]. After inhibition by methyl parathion, the bienzymatic amperometric response shows a wide dynamic linear range (up to 200 μM) and a detection limit of 0.05 μM (Fig. 2.5). These characteristics are attributed to the catalytic activity of carbon nanotubes to promote the redox reaction of the H_2O_2 produced from the AChE/CHO enzymatic reaction with their substrates and a large surface of the carbon nanotube materials.

2.3.3 Catalysis-based biosensors

Although the inhibition-based biosensors are sensitive, they are poor in selectivity and are rather slow and tedious since the analysis involves multiple steps of reaction such as measuring initial enzyme activity, incubation with inhibitor, measurement of residual activity, and regeneration and washing. Biosensors based on direct pesticide hydrolysis are more straightforward. The OPH hydrolyzes ester in a number of organophosphorus pesticides (OPPs) and insecticides (e.g. paraoxon, parathion, coumaphos, diazinon) and chemical warfare agents (e.g. sarin) [53]. For example, OP parathion hydrolyzes by the OPH to form *p*-nitrophenol, which can be measured by anodic oxidation. Rainina

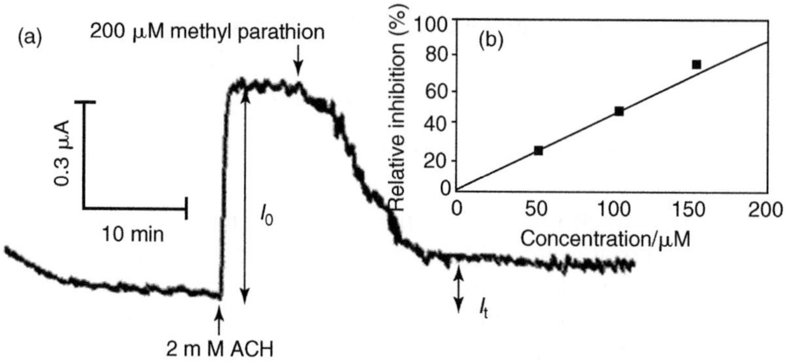

FIGURE 2.5 (a) Amperometric response of CNT modified screen-printed biosensor for methyl parathion in 0.1 M PBS containing 0.1 M NaCl (pH 7.4) at applied potential of 0.50 V and (b) the relative inhibition of CNT AChE/CHO biosensor as a function of methyl parathion concentration (adapted from [40]).

et al. [54] used a recombinant, OPH-expressing *Eschericia coli* (*E. coli*) immobilized on a pH electrode, to develop a microbial biosensor for the direct detection of OP. Mulchandani *et al.* [55] purified OPH from a recombinant *E. coli* and immobilized it on a pH electrode to develop a potentiometric biosensor by catalyzing the hydrolysis of organophosphorus pesticides (parathion, paraoxon, methyl parathion) to release protons, the concentration of which was proportional to the amount of hydrolyzed substrate. Mulchandani *et al.* [56] reported another amperometric OP biosensor using thick film strip electrode and genetically engineered *E. coli* OPH. The detection limit of this sensor was 7×10^{-8} M.

2.3.4 Flow injection biosensors

Flow injection analysis (FIA) is a highly convenient method for fast detection and automatization. Several reports exist on the FIA of pesticides using enzyme-based biosensors with a wide linear range of detection. Kindervator *et al.* [57] proposed an exchangeable enzyme reactor by immobilizing AChE and COx on magnetic particles using glutaraldehyde as a cross-linker. The main advantages of this method were that the magnetic particles could be easily held inside the reactor under the magnetic field and the completely inactivated enzyme could be replaced. An FIA biosensor for paraoxon by immobilizing AChE in PVA-SbQ polymer and integrating with a continuous flow system exhibited a detection limit of 1.0 nM (0.3 ng mL^{-1}) (inhibition %=10). The analytical time for one sample including inhibition reactivation using 1.0 mM 2-pyridinealdoxime methiodide was 1 h [44]. Recently Kandimalla and Ju [11] reported an FIA system for the detection of the organophosphorus insecticide suflotep by modifying the GCE with cross-linked chitosan-multiwall carbon nanotube (MWNT) composite and covalently linking AChE to the chitosan via glutaraldehyde. This membrane was crack free and showed a homogeneous porous structure and distribution of MWNTs (Fig. 2.6). Both biocompatibility of chitosan and the inherent conductive properties of

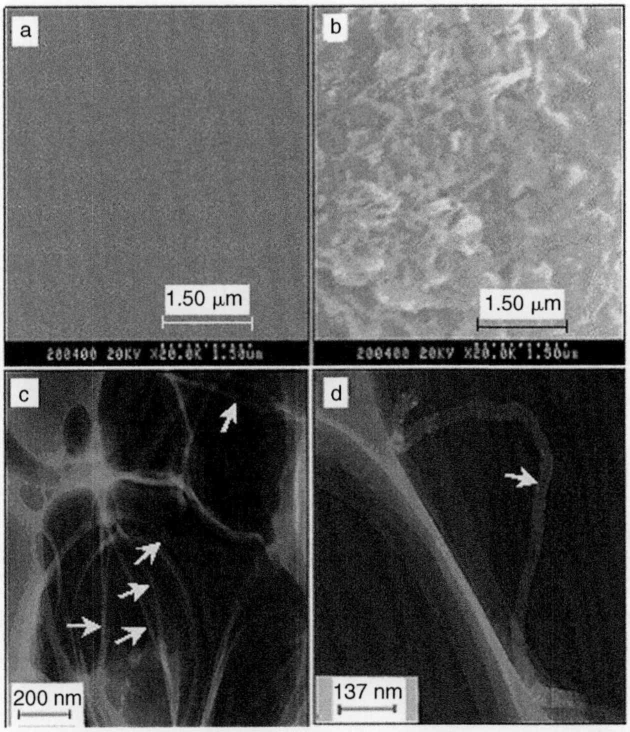

FIGURE 2.6 SEM images of cross-linked chitosan (a) and CMC (b) membranes and TEM images of CMC (c) and single CNT (d) (adapted from [11]).

MWNTs favored the detection of insecticide from 1.5 to 80 nM with a detection limit of 1 nM at an inhibition of 10% (Fig. 2.7). Bucur *et al.* [58] employed two kinds of AChE, wild type *Drosophila melanogaster* and a mutant E69W, for the pesticide detection using flow injection analysis. Mutant AChE showed lower detection limit (1×10^{-7} M) than the wild type (1×10^{-6} M) for omethoate. An amperometric FIA biosensor was reported by immobilizing OPH on aminopropyl control pore glass beads [27]. The amperometric response of the biosensor was linear up to 120 and 140 μM for paraoxon and methyl-parathion, respectively, with a detection limit of 20 nM (for both the pesticides). Neufeld *et al.* [59] reported a sensitive, rapid, small, and inexpensive amperometric microflow injection electrochemical biosensor for the identification and quantification of dimethyl 2,2′-dichlorovinyl phosphate (DDVP) on the spot. The electrochemical cell was made up of a screen-printed electrode covered with an enzymatic membrane and combined with a flow cell and computer-controlled potentiostat. Potassium hexacyanoferrate (III) was used as mediator to generate very sharp, rapid, and reproducible electric signals. Other reports on pesticide biosensors could be found in review [17].

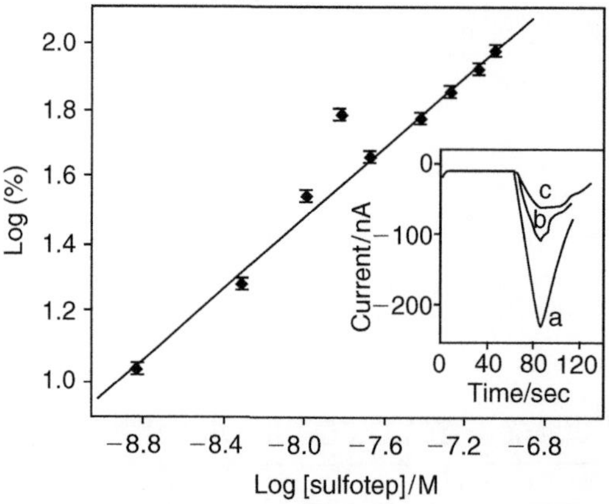

FIGURE 2.7 Logarithmic plot of *I*% vs sulfotep concentration for flow-injection detection. Inset: amperometric responses at (a) 0, (b) 20, and (c) 50 nM sulfotep [11].

Phosphorylated enzyme   Oxime

Enzyme—O—P(=O)(R_1)(−OR_2) + O—N=CHR

Reactivated   Phosphorylated oxime

⟶ Enzyme + CHR=N—O—P(=O)(R_1)(−OR_2)

FIGURE 2.8 Mechanism for inhibited enzyme reactivation using oxime.

2.3.5 Enzyme reactivation

The reactivation of inhibited enzyme is highly desirable for the repeated use of FIA. Wilson and his colleague [60] reported a classic study and demonstrated that strong nucleophiles such as oximes could reactivate organophosphatecholiesterase complex to give rise to free enzyme. Various oximes have been used as reactivators for cholinesterase inhibition [61, 62]. Figure 2.8 shows the reactivation mechanism. One of the limitations in use of oximes is the relatively slow rate of reactivation. Of concern to pesticide biosensors, 2-pyridinealdoxime methiodide and 2-pyridinealdoxime methochloride have been widely employed as reactivators for continuous monitoring of the OPPs [11, 61]. Dioximes such as 1,1′-trimethylene bis 4-formylpyridinium bromide dioxime (TMB-4) have been also employed as reactivators, particularly under conditions of exposure of the enzyme to higher pesticide concentrations. TMB-4 performs a more effective reactivation of the immobilized AChE inhibited by the paraoxon than 2-pyridinealdoxime methiodide [62]. The reactivation % and stability of enzyme depend on the degree of inhibition and type of pesticide.

2.4 PESTICIDE IMMUNOSENSORS

An antibody can recognize an antigen in particular even in the presence of structurally close or related compounds. Hence antibodies have been used as potential recognition

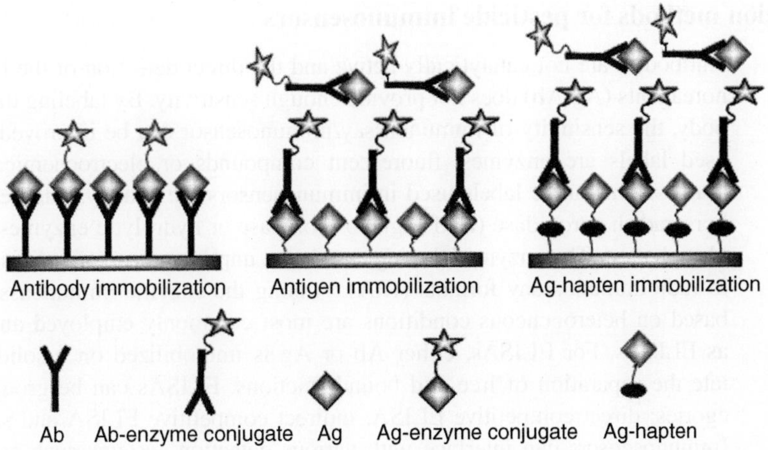

FIGURE 2.9 Methods for modification of transducer surface in immunosensor development.

agents in a number of immunoassays and immunosensors. Immunoassays have been widely used in clinic. Its main disadvantage is the need for a long incubation time, which implies several hours from sample collection to results generation, so it is difficult to use this technique in alarm station or process control. Immunosensors are regarded as devices able to partially circumvent these drawbacks. Immunosensors consist of a bioactive surface (usually an antibody immobilized on a sensing surface) and a transducer system able to generate a physical signal when the immunochemical reaction takes place [63]. Two competitive enzyme-linked immunosorbent assay (ELISA) approaches have been followed in the development of immunosensors by immobilizing either antibody or antigen/hapten. The first is the most common approach (Fig. 2.9) [64]. The main advantages of this option are the economy of not using expensive antibodies and the reduction of assay steps. The technique of immobilization of the antigen or hapten, even though less employed, has the advantage that the regeneration process can be performed without loss of activity of the immobilized reagent. Another important issue to be considered in immunosensor development is the choice of immobilization support/immobilization method [65]. The immobilization method/matrices used in immunosensor development are discussed in later sections.

The determination of an analyte at $\mu g\ l^{-1}$ levels requires the use of extraction, cleanup, and preconcentration processes [66]. This is usually accomplished using organic solvents or combining them with solid-phase extraction. Most official methods of residue analysis are based on these methodologies employing pure organic solvents such as methanol, acetone, acetonitrile, or hexane [67, 68]. Hence the biosensor/immunosensor should be stable to perform analysis even in trace amounts of organic solvents. The flow immunosystems have proven their stability and compatibility in high concentrations of organic solvents and also observed improved antigen–antibody binding [69, 70]. Penalva *et al.* [70] reported that the performance of monoclonal antibodies is more stable than polyclonal antibodies in organic solvents.

2.4.1 Detection methods for pesticide immunosensors

Antibodies are not catalytically active and the direct detection of the binding of immunoreagents (Ag-Ab) does not provide enough sensitivity. By labeling the antigen or antibody, the sensitivity of immunoassay/immunosensor can be improved. The commonly used labels are enzymes, fluorescent compounds or electrochemically active compounds. Enzymatic labels used in immunosensors are usually oxidoreductases such as horseradish peroxidase (HRP), glucose oxidase or hydrolytic enzymes such as alkaline phosphatase. The enzyme labeling avoids the unpleasant use of radioisotopes employed in radioimmunoassay formats (RIA). Among the enzyme immunoassays (EIA), those based on heterogeneous conditions are most commonly employed and are referred to as ELISAs. For ELISAs, either Ab or Ag is immobilized on a solid phase to facilitate the separation of free and bound fractions. ELISAs can be grouped in three categories: direct competitive ELISA, indirect competitive ELISA and sandwich ELISA. Immunosensors can interface with various detection systems such as electrochemical analysis, piezoelectric detection, surface plasmon resonance analysis, etc.

2.4.2 Immunosensors for pesticides

2.4.2.1 Piezoelectric immunosensors

In piezoelectric immunosensors, the sensor surface is usually coated with an antibody or an antigen (hapten) and the mass variation induced by the antigen–antibody binding is correlated to the concentration of the antigen [9]. During the initial studies on pesticides, piezoelectric immunosensor response was monitored after dying sensor surface [71]. Later pesticide ligands were immobilized on sensor surface and indirect competitive assay was carried out using high molecular antibody as a tracer [72]. Horáček and Skládal [73] carried out piezoelectric measurements under wet conditions. In this study the piezoelectric crystals were modified with 2,4-dichlorophenoxyacetic acid (2,4-D) using the self-assembled monolayer of thio compounds on the gold surface of the crystals. The biosensor was placed in a flow-through cell and the affinity binding of monoclonal antibodies against 2,4-D on the modified piezoelectric crystals was studied. One measurement could be completed within 25 min with a detection limit of 0.24 ng mL^{-1}. In another piezoelectric immunosensor, bovine serum albumin-linked atrazine was covalently immobilized on silanized piezoelectric crystals activated using glutaraldehyde [74]. The modified piezoelectric crystal was placed in a flow cell and competitive immunoassay was performed directly in a flowing solution. These crystals could be easily regenerated using 100 mM NaOH within 5 min. Recently another piezoelectric biosensor has been reported for the determination of atrazine employing both direct and indirect immunoassay methods [75]. The detection limit for atrazine was 0.025 and 1.5 ng mL^{-1} with assay times less than 10 and 25 min for indirect and direct immunoassays, respectively.

2.4.2.2 Optical immunosensors

Optical immunosensors are based on the measurement of the absorption or emission of light induced by the immunoreactants [9]. They can also be based on evanescent wave transducers [9]. Minunni and Mascini [76] used the commercial SPR apparatus BIAcore™ from Pharmacia (Uppsala, Sweden) to detect the herbicide atrazine. This non-labeled competitive device allowed achieving a detection limit of 0.05 ng mL^{-1}. Mouvel *et al.* [77] reported an immunosensor based on waveguide surface plasmon resonance (WSPR). The analyte derivatives were immobilized on modified gold film; surface change caused by immunocomplexes was detected by silicon photodetectors. The detection limit was 0.2 ng mL^{-1}. A label-free and highly sensitive immunosensor using optical waveguide lightmode spectroscopy (OWLS) has been reported for the detection of the herbicide trifluralin [78]. The competitive assay allowed the detection of trifluralin in the concentration range of 2×10^{-7} to 3×10^{-5} ng mL^{-1}. A photometric flow injection immunoassay has been reported for the determination of atrazine, simazine, and 2,4-D by immobilizing the respective antibody on aminopropyl glass particles by means of avidin/biotin conjugation [79]. Microcantilevers have also been employed in the detection of pesticides. DDT-BSA conjugate has been immobilized on a gold coated side of cantilever by using thiol self-assembled monolayers. The cantilever deflection (shift of the resonant frequencies) occurred though indirect competitive immunoassay was measured for the quantification of DDT without any labeling [80].

In view of the fast detection and possibility of automatization, several flow injection analysis biosensors have been reported. González-Martínez *et al.* [81] reported an FIA immunosensor for carbaryl with special attention to assay sensitivity and sensor regeneration properties. This immunoassay used mouse monoclonal anti-carbaryl antibodies either in solution (indirect format) or surface (direct format). In both formats, the same enzyme label (HRP) and fluorometric detection method was employed. A sensitive detection (26 ng l^{-1}) and fast response (11 min/assay) were obtained using direct immunoassay, while in indirect format the detection limit was 284 ng l^{-1} and the response time was 17 min/assay. However, the useful life of the sensor was 60–70 cycles and 200 cycles for direct and indirect assays, respectively. During the regeneration steps the antibody lost activity in direct competitive assay, whereas the pesticide underwent regeneration in indirect assay, hence the sensor could be reused for longer. An ng mL^{-1} level detection of isoproturon has been reported using an optical immunosensor based on solid phase fluorescence immunoassay [82]. In this approach, analyte derivative was immobilized on glass slides in agreement with the highest stability of the immobilization structure. Detection limits for Milli-Q and river water samples were 0.01 and 0.14 ng mL^{-1}, respectively. The major degradation product of the insecticides chlorpyrifos is 3,5,6-trichloro-2-pyridinol (TCP). The presence of TCP in human urine is considered as a biomarker of exposure to chlorpyrifos insecticides. A flow injection immunosensor has been reported for the detection of TCP, based on the immunocomplex formed in solution by an antibody-binder protein A/G derivatized polymeric gel [83]. An optical immunosensor coupled to an FIA system has been reported to monitor chlorotriazine pesticides in river water samples [84]. In this study a solid-phase fluorescence immunoassay was employed with immobilized analyte derivative and free, fluorescence-labeled anti-atrazine or anti-simazine. The limits of detection for atrazine and simazine varied between 0.06 and 0.2 ng mL^{-1}, depending on Milli-Q water or river water samples used.

Multianalyte immunosensor is highly advantageous for field applications. It allows several compounds in a single run to be determined, irrespective to the nature of the

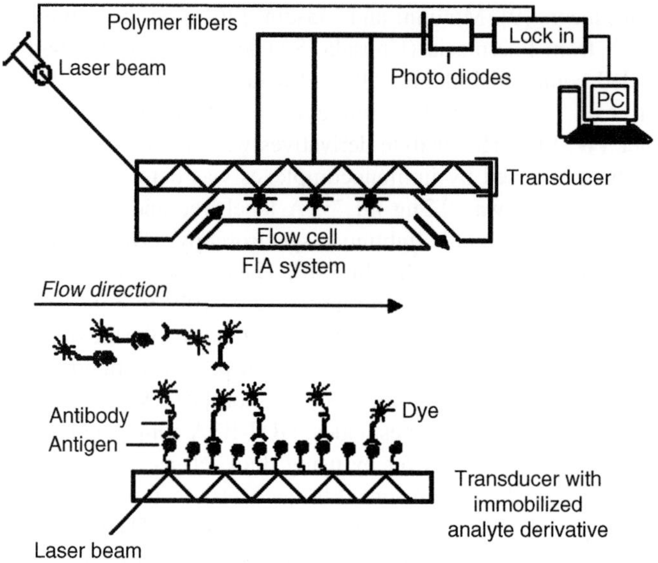

FIGURE 2.10 Schematic diagram of the RIANA (adapted from [86]).

target molecules [85]. Barzen *et al.* [86] reported a prototype optical immunosensor called RIver ANAlyser (RIANA). RIANA is based on a solid-phase fluoroimmunoassay that takes place at an optical transducer chip. The transducer surface was chemically modified with three analyte derivatives placed in different discrete locations. The transducer was mounted in a flow cell and passed a mixture of analyte and fluorescent dye (Cy5.5 dye) conjugated antibody after preincubation. The excitation of bound fluorescently labeled antibodies was accomplished using a collimated He-Ne laser that was directly launched into the transducer. The collected light was subsequently filtered to avoid any collected pump radiation, and detected with photodiodes (Fig. 2.10). The RIANA could be used for simultaneous detection of various pesticides in one sample. Two sets of three different analytes (atrazine, alachlor, pentachlorophenol and 2, 4-D, simazine, isoproturon) were detected successfully and the limit of detection was comparable with single analyte immunoassay [86]. In another study RIANA has also been employed for the simultaneous determination of three different contaminants, atrazine, isoproturon, and estrone, present in real samples [87]. The limits of detection for atrazine, isoproturon and estrone were 0.155, 0.046 and 0.084 ng mL^{-1}, respectively. Propanil, a herbicide present in water, has also been quantified up to the subnanogram level (0.6 ng l^{-1}) using RIANA, without any preconcentration [88]. Mastichiadis *et al.* [89] reported a four band disposable optical capillary immunosensor, which was capable of detecting pesticides such as mesotrione, paraquat, diquat and hexaconazole with detection limits of 0.04, 0.06, 0.09 and 0.10 ng mL^{-1} and dynamic ranges up to 9, 6, 12 and 15 ng mL^{-1}, respectively.

2.4.2.3 Electrochemical immunosensors

Although optical and piezoelectric immunosensor systems are capable of achieving the required detection limits for pesticides and herbicides, these are at the expense of experimental complexity and instrumental portability. Most potentiometric biosensors for detection of environmental pollutants use enzyme to catalyze the consumption or production of protons. However, the reports on potentiometric immunosensors for pesticide determination are less. Dzantiev *et al.* [90] reported a detection of 50 ng mL^{-1} of the herbicide 2,4,5-trichlorophenoxyacetic acid. This method was based on the competitive binding of free pesticide and pesticide–peroxidase conjugate with antibodies immobilized on a graphite electrode. Another potentiometric immunosensor for simazine has been reported by using HRP as tracer [91]. The antibodies immobilized on gold planar electrodes were stable and could be easily regenerated with 0.04 M HCl. Among the electrochemical methods a widely used method was amperometric detection. Amperometric biosensors typically rely on an enzyme system. The enzyme can convert catalytically electrochemically non-active analytes into products that can be oxidized or reduced at a working electrode. The current produced is directly related to the concentration of the electroactive species, which in turn is proportional to the non-electroactive enzyme substrate. Recently simazine antibodies and recombinant HRP with histidine tag were coimmobilized on gold electrode surface to construct an amperometric immunosensor, which was used in a flow immunoassay [79]. The detection was performed by passing analyte and GOD conjugated analyte through the immobilized antibody surface in the presence of glucose and dissolved oxygen at a potential of −50 mV vs Ag/AgCl. The detection limit was 0.1 ng mL^{-1}.

Gascón *et al.* [92] reported a flow injection immunoassay (FIIA) method for the direct and accurate determination of atrazine, without purification or preconcentration. This FIIA method showed an IC_{50} of 2 nM (0.47 ng mL^{-1}) and a detection limit of 0.35 nM (0.075 ng mL^{-1}). López *et al.* [93] reported an immunosensor by modifying a GCE with redox polymer $PVPOs(bpy)_2Cl_2$, where PVP was poly(4-vinylpyridine), and the specific antibody. Atrazine could be detected up to 1 ng mL^{-1} using this immunosensor. A label-free electrochemical immunosensor has been reported by immobilizing the antibody on gold nanoparticles [94]. Gold nanoparticle labeled antibody was fixed on GCE surface by Nafion. After the formation of paraoxon immunocomplex, analyte could be directly quantified at an applied potential of −0.03 mV.

Screen-printed electrodes (SPEs) have also been used in immunosensor design [95, 96]. Kaláb and Skládal [97] reported a disposable screen-printed immunosensor for the detection of 2,4-D. 2,4-D was covalently immobilized on the surface of SPE. Through the indirect competitive immunoassay the detection limit for 2,4-D was close to 0.1 ng mL^{-1}. Keay and McNeil [98] reported a separation-free and disposable electrochemical screen-printed atrazine immunosensor. The SPE was prepared with carbon ink incorporating HRP, and an atrazine antibody immobilized Biodyne C membrane was placed over the electrode surface. The assay was carried out based on the competition between available Ab binding sites and atrazine and atrazine–GOD conjugate. In the presence of glucose, H_2O_2 formed by the conjugated GOD was reduced by enzyme

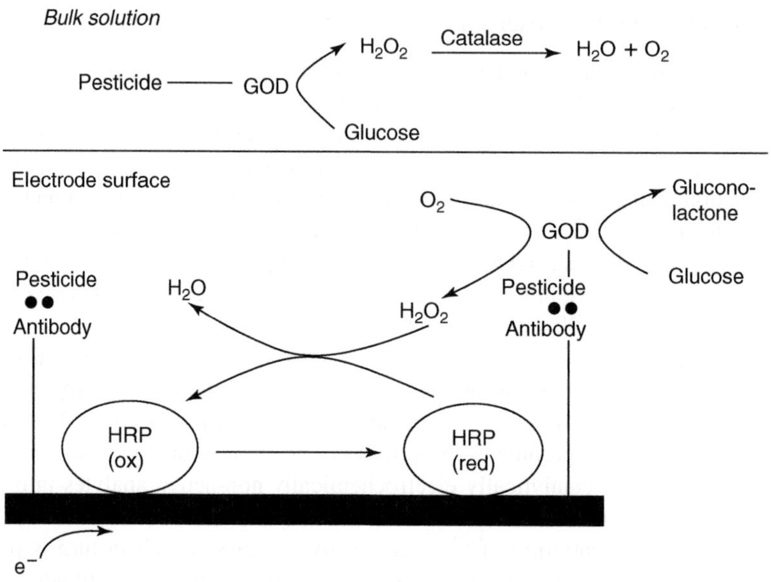

FIGURE 2.11 Schematic diagram of the separation-free immunosensor principle for pesticides (H_2O_2 detected at +50 mV vs Ag/AgCl via direct enzymatic reduction of HRP) (adapted from [98]).

channeling via the HRP electrode. The linear detection range was 0.01–1.0 ng mL^{-1} (Fig. 2.11). By employing the above working principle, another amperometric immunosensor for chlorosulfuron has been reported [99]. The linear detection of chlorosulfuron was 0.01–1 ng mL^{-1}. The recombinant single-chain antibody (scAb) fragments were also used as a recognition receptor in the development of the atrazine immunosensor using SPE [96]. This real-time and separation-free FIA system was capable of measuring atrazine with a detection limit of 0.1 ng mL^{-1}. Some other recent reports on pesticide biosensors could be found in a review [9] and Table 2.1.

2.4.3 Regeneration of pesticide immunosensors

Immunoaffinity-based biosensor performance for repeated pesticide detection is dependent on the binding efficiency of the immobilized antibody. The mild and effective dissociation method can retain the antibody affinity for longer. For the regeneration either highly acidic (pH 1.8 to 3.8) or basic (pH 9.5–11) buffers or buffers along with organic solvents or enzymes such as pepsin (2 mg mL^{-1}, pH 1.9 solution) was used. In a report by Fránek *et al.* [118] 0.04 M NaOH, 0.1 M glycine/HCl and 0.1 M NaOH were suitable dissociation buffers for effective regeneration of atrazine, 2,4-D, and simazine, respectively. One percent DMSO in pH 2.3 Gly–HCl buffer showed 97% dissociation and the immobilized antibody retained binding affinity to carry out 14 reproducible assays for ethyl parathion [13]. Some of the reports used diluted HCl for the dissociation of immunocomplexes [91]. 0.04 M HCl showed efficient dissociation

TABLE 2.1
Some of the recent reports on pesticide biosensors

Pesticide	Mode of detection	Receptor molecule	Detection limit	Ref.
Carbofuran	Amp/FIA	AChE	10^{-7} M	[100]
Dichlorvos Parathion Azinphos	Amp/FIA	AChE	Dichlorvos: 1×10^{-17} M Parathion: 1×10^{-16} M Azinphos: 1×10^{-16} M	[101]
Dichlorvos	Amp	AChE	10^{-10} M	[102]
Methamidophos	DPV/Chronoamperometry	AChE	1 ng mL^{-1}	[103]
Fenitrothion	Fluorescence	Antibody	0.8 nM	[104]
Carbofuran Carbaryl Paraoxon Dichlorvos	Fluorescence/FIA	AChE	Carbofuran: 3.5 ng mL^{-1} Carbaryl: 50 ng mL^{-1} Paraoxon: 12 ng mL^{-1} Dichlorvos: 25 ng mL^{-1}	[105]
Methyl parathion	Fluorescence polarization	Antibody	15 ng mL^{-1}	[106]
Trichlorfon Coumaphos Methiocarb	Potentiometric	AChE	Trichlorfon: 1.5×10^{-7} M Coumaphos: 5×10^{-9} M Methiocarb: 8×10^{-7} M Coumaphos: 5×10^{-8} M	[107]
Coumaphos Carbofuran Chloropyrifos methyl	Amp/SPE	AChE	Carbofuran: 8×10^{-9} M Chloropyrifos methyl: 2×10^{-8} M	[108]
Carbofuran Eserine	Amp	AChE	Carbofuran: 0.01 ng mL^{-1} Eserine: 0.03 ng mL^{-1}	[34]
Methyl-parathion	Fluorescence polarization	Antibody	15 ppb	[109]
Paraoxon carbofuran	Amp	AChE	Both pesticides: 10^{-10} M	[110]
Dipterex Paraoxon	Amp	AChE	Dipterex: 5 μM Paraoxon: 0.4 μM	[111]
Trichlorfon	Potentiometric	BChE	10^{-7} M	[112]
Aldicarb Carbaryl	Potentiometric	AChE	12 ppb Carbaryl: 108 μg l^{-1}	[113]
Dichlorvos	Fiber optic	AChE	Dichlorvos: 5.2 μg l^{-1}	[114]
Chlorpyrifos Paraoxon	Amp/SPE	AChE	0.1 mg kg^{-1} (wheat/barley) Paraoxon: 0.32 ng mL^{-1}	[115]
Carbofuran Triazophos Oxydemeton methyl	Spectrometric/FIA	AChE	Carbofuran: 3.8 ng mL^{-1} Triazophos: 90 ng mL^{-1} Oxydemeton methyl: 200 ng mL^{-1}	[116]
Trichlorfon Coumaphos	Amp/SPE	BChE	Trichlorfon: 3.5×10^{-7} M Coumaphos: 1.5×10^{-7} M	[50]
Methsulfuron methyl	Fiber optic	Antibody	0.1 ng mL^{-1}	[117]

(95%) of simazine immunocomplex within 5 min. However, at lower concentrations of HCl dissociation was poor, and at higher concentrations antibody was destroyed drastically, though the dissociation was fast. The enzyme pepsin has also been employed in the regeneration of sensing surface. Pepsin acts on bound antibodies and breaks up the peptide chains into less active and inactive fragments, which could be more easily removed in acidic conditions (pH 1.9) [82]. Rodriguez-Mozaz *et al.* [87] employed SDS solution (pH 1.9, 5 mg mL^{-1}) for the regeneration of the immunosensor surface.

2.5 WHOLE CELL AND TISSUE-BASED PESTICIDE BIOSENSORS

In whole cell biosensors the enzyme is stable, as it is present in natural biological surroundings and can reduce the cost of enzyme purification. Potato slices are rich in polyphenol oxidase (PPO), and PPO activity can be inhibited by atrazine. Thus, a PPO whole cell biosensor has been reported [119]. A thin slice of potato tissue was clamped onto the surface of an O_2 selective Clark electrode and catechol was used as substrate for PPO. The linear detection range and detection limit of this low cost whole cell-based biosensor are 20–130 μM and 10 μM, respectively. A cucumber tissue was sandwiched between Teflon and nylon membranes at the surface of an oxygen electrode [14]. An amperometric biosensor has been designed for the detection of ethyl paraoxon based on the inhibition activity of ascorbate oxidase. The detection limit is 1 μg mL^{-1} and analysis time for one sample is 10 min. Other plant materials such as tissues [120] and thylakoid membranes [121] were also employed in pesticide determination. A novel whole cell biosensing method has been reported for the determination of herbicide glyphosate by assessing electrophysiological interactions with cell components (*Johnsongrass*) immobilized in an agarose gel matrix that preserves cell "physiological" functions. In this biorecognition assay (BERA) cellular response to the herbicide was evaluated by measuring the cellular electric potential. This preliminary work showed linear response in potentiometric measurements from 0.1 to 1 mM of glyphosate [122].

Like plant cells, microbial cells have also been used in whole cell biosensors. Recombinant *E. coli* cells, which are capable of hydrolyzing a wide spectrum of OPs and chemical warfare agents, were entrapped in PVA to prepare a packed column. The packed column was interfaced with a pH electrode to construct a whole cell biosensor [54]. The proton released through the hydrolysis of OP pesticides by organophosphate hydrolase was correlated with the concentration of pesticide. The linear detection range of paraoxon was from 0.25 to 250 μg mL^{-1} (0.001–1.00 mM) and the cryoimmobilized organophosphate hydrolase exhibited stability over 2 months in phosphate buffer at 4°C. Luciferease is another enzyme whose activity can be inhibited by the herbicides. It presents in bioluminescent cyanobacteria *Synechocystis* sp. (rec. PCC6803) cells and has been used in the detection of atrazine, propazine, simazine, paraquat, and glyphosate by measuring the luminescence after pesticide inhibition [123]. Interfacing of this type of luminescent cells with optical transducers may be used to develop low cost sensors. Herbicides can reversibly inhibit the photosynthetic electron transport (PET) chain. The PET includes two photosystems, photosystem-I, and photosystem-II. The main components of the reaction center photosystem-II (PSII) are D1 and D2 proteins, where herbicidal inhibition can occur. The inhibition at the D1 protein has been monitored through the mediated photocurrent by employing diaminodurene as a mediator. With the above inhibition principle a photosynthetic cyanobacteria-based biosensor has been developed. The detection limit for atrazine was 1 ng mL^{-1} [124].

2.6 MAJOR INTERFERING COMPOUNDS AND SAMPLE PRETREATMENT

Real samples contain several interfering compounds along with the target analyte. The most common ways to reduce matrix effects are (a) selective extraction ("cleanup") and (b) dilution to bring the interfering substances below a concentration [109]. According to the literature AchE activity can be inhibited not only by insecticides but also by other substances like heavy metals and hypochloride (e.g. sodium hypochloride) [125]. The heavy metals such as Cu^{2+}, Cd^{2+} (pH 4.5), Fe^{3+} (pH 2) [126] and Hg^{2+} (pH 7) [127] can produce a non-competitive AChE inhibition. In immunoassays the main interfering compounds are anions such as azide and cations such as Ca^{2+}. They inhibit the enzyme used as a label. The humic substances (humic acids) present in water or soil extracts may bind non-specifically to the Ab and thereby interfere with the specific binding of the analyte. Matrix effects in food samples frequently occur owing to colored extracts or to the content of lipids, proteins or polyphenols, which may be coextracted during sample preparation [128]. Usually, the pesticides are extracted from soil with organic solvents such as acetone, ethyl acetate or methanol. Immunoassay is tolerant to a variety of solvents up to a certain degree [6, 129, 130]. In some cases a cleanup step is introduced such as passing through C18 columns or immunoaffinity columns, in which the analyte is separated from the matrix [131–133]. Another approach is the supercritical fluid extraction (SFE) prior to immunoanalysis [67, 114].

2.7 CONCLUSIONS

The current pesticide biosensors/immunosensors are competing with other, fairly well-established field analytical methods such as chemical sensors, immunoassays, and chemical test kits. Although biosensors have the potential to replace chromatographic methods in the identification of the pesticides in simple and less expensive ways, accurate validation methods are still very much necessary. More genetically modified sensitive biocatalysts and highly specific antibodies are necessary for the development of stable and robust biosensors/immunosensors. This can be achieved by adopting genetically modified receptor molecules. New immobilization methods/matrices should be explored to improve the sensor stability and fast signal transfer to the transducer surface. Even though a few of the methods do not require any pretreatment of sample, sensor stability is another factor. Hence for the on-site applications fewer pretreatment sample preparation or direct analysis methods should be identified. Utilizing the advances in microfabrication technologies the development of miniaturized sensors or disposable kits is highly beneficial for household purposes and real-time analysis in various sectors such as industry, medicine, etc. Although efforts have been made in the development of multi-pesticide biosensors, in the future more emphasis is required to design a highly selective multi-pesticide flow injection biosensor for the fast and automatic analysis of real samples.

2.8 ACKNOWLEDGMENTS

We gratefully acknowledge the financial support of the National Science Funds for Distinguished Young Scholars (20325518) and Creative Research Groups (20521503), the Key (20535010) and General (20275017) Programs from the National Natural Science Foundation of China, the Specialized Research Fund for Excellent Young Teachers from the Ministry of Education of China and the Science Foundation of Jiangsu (BS2006006, BS2006074). KVB is highly thankful to the postdoctoral fellowship from Nanjing University.

2.9 REFERENCES

1. A.M. Jiménez and M.J. Navas, Crit. Chemiluminescent methods in agro-chemical analysis. *Rev. Anal. Chem.* **27**, 291–305 (1997).
2. E.M. Garrido, C. Delerue-Matos, J.L.F.C. Lima, and A.M. Brett, Electrochemical methods in pesticides control. *Anal. Lett.* **37**, 1755–1791 (2004).
3. H. Kidd and D. Hartley, UK pesticides for farmers and growers, in *The Royal Society of Chemistry*, Nottingham (1987).
4. J. Sherma, in *Analytical Methods for Pesticides and Plant Growth Regulators*, Academic Press, San Diego (1990).
5. D.G. Burek, *Biosensors*, Chapter 4, Technomic Publishing, Pennsylvania (1993).
6. J.H. Skerritt and B.E.A. Rani, in *Residue Analysis in Food Safety: Applications of Immunoassay Methods* (R.C. Berier and L.H. Stanker, eds), ACS Symposium Series, Washington (1996).
7. S. Andreescu, V. Magearu, A. Lougarre, D. Fournier, and J.L. Marty, Immobilization of enzymes on screen-printed sensors via a histidine tail. Application to the detection of pesticides using modified cholinesterase. *Anal. Lett.* **34**, 529–540 (2001).
8. K.A. Joshi, J. Tang, R. Haddon, J. Wang, W. Chen, and A. Mulchnadani, A disposable biosensor for organophosphorus nerve agents based on carbon nanotubes modified thick film strip electrode. *Electroanalysis* **17**, 54–58 (2005).
9. J.L. Marty, B. Leca, and T. Noguer, Biosensors for the detection of pesticides. *Analysis Magazine* **26**, M144–M148 (1998).
10. D.P. Dumas, H.D. Durst, W.G. Landis, F.M. Raushel, and J.R. Wild, Inactivation of organophosphorus nerve agents by the phosphotriesterase from Pseudomonas diminuta. *Arch. Biochem. Biophys.* **277**, 155–159 (1990).
11. V.B. Kandimalla and H.X. Ju, Binding of acetylcholinesterase to multiwall carbon nanotube-cross-linked chitosan composite for flow-injection amperometric detection of an organophosphorous insecticide. *Chem. Eur. J.* **12**, 1074–1080 (2006).

12. A.J. Killard, L. Micheli, K. Grennan, M. Franek, V. Kolar, D. Moscone, I. Palchetti, and M.R. Smyth, Amperometric separation-free immunosensor for real-time environmental monitoring. *Anal. Chim. Acta* **427**, 173–180 (2001).
13. V.B. Kandimalla, N.S. Neeta, N.G. Karanth, M.S. Thakur, K.R. Roshini, B.E.A. Rani, A. Pasha, and N.G.K. Karanth, Regeneration of ethyl parathion antibodies for repeated use in immunosensor: a study on dissociation of antigens from antibodies. *Biosens. Bioelectron.* **20**, 903–906 (2004).
14. K. Rekha, M.D. Gouda, M.S. Thakur, and N.G. Karanth, Ascorbate oxidase based amperometric biosensor for organophosphorous pesticide monitoring. *Biosens. Bioelectron.* **15**, 499–520 (2000).
15. A.L. Jenkins, R. Yin, and J.L. Jensen, Molecularly imprinted polymer sensors for pesticide and insecticide detection in water. *Analyst* **126**, 798–802 (2001).
16. V.B. Kandimalla and H.X. Ju, Molecular imprinting: a dynamic technique for diverse applications in analytical chemistry. *Anal. Bioanal. Chem.* **380**, 587–605 (2004).
17. M. Trojanowicz, Determination of pesticides using electrochemical enzymatic biosensors. *Electroanalysis* **14**, 1311–1328 (2002).
18. H. Schulze, S. Vorlova, F. Villatte, T.T. Bachmann, and R.D. Schmid, Design of acetylcholinesterases for biosensor applications. *Biosens. Bioelectron.* **18**, 201–209 (2003).
19. J.L. Marty, N. Mionetto, T. Noguer, F. Ortega, and C. Roux, Enzyme sensors for the detection of pesticides. *Biosens. Bioelectron.* **8**, 273–280 (1993).
20. F. Mazzei, F. Botre, and C. Botre, Acid phosphatase/glucose oxidase-based biosensors for the determination of pesticides. *Anal. Chim. Acta* **336**, 67–75 (1996).
21. M. Ayyagari, S. Kametkar, R. Pande, K.A. Marx, J. Kumar, S.K. Tripathy, J. Akkara, and D.L. Kaplan, Chemiluminescence-based inhibition kinetics of alkaline phosphatase in the development of a pesticide biosensor. *Biotechnol. Prog.* **11**, 699–703 (1995).
22. M.T. Pérez Pita, A.J. Reviejo, F.J.M. de Villena, and J.M. Pingarron, Amperometric selective biosensing of dimethyl- and diethyldithiocarbamates based on inhibition processes in a medium of reversed micelles. *Anal. Chim. Acta* **340**, 89–97 (1997).
23. W.R. Everett and G.A. Rechnitz, Mediated bioelectrocatalytic determination of organophosphorus pesticides with a tyrosinase-based oxygen biosensor. *Anal. Chem.* **70**, 807–810 (1998).
24. S. Trajkovska, K. Tosheska, J.J. Aaron, F. Spirovski, and Z. Zdravkovski, Bioluminescence determination of enzyme activity of firefly luciferase in the presence of pesticides. *Luminescence* **20**, 192–196 (2005).
25. K.I. Dave, C.E. Miller, and J.R. Wild, Characterization of. organophosphorous hydrolases and the genetic manipulation of. the phosphotriesterase from pseudomonas diminuta. *Chem. Biol. Intract.* **87**, 55–68 (1993).
26. P. Mulchandani, A. Mulchandani, I. Kaneva, and W. Chen, Biosensor for direct determination of organophosphate nerve agents. 1. Potentiometric enzyme electrode. *Biosens. Bioelectron.* **14**, 77–85 (1999).
27. P. Mulchandani, W. Chen, and A. Mulchandani, Flow injection amperometric enzyme biosensor for direct determination of organophosphate nerve agents. *Environ. Sci. Technol.* **35**, 2562–2565 (2001).
28. A. Mulchandani, W. Chen, P. Mulchandani, J. Wang, and K.R. Rogers, Biosensors for direct determination of organophosphate pesticides. *Biosens. Bioelectron.* **16**, 225–230 (2001).
29. M. Bernabai, C. Cremisini, M. Mascini, and G. Palleschi, Determination of organophosphorus and carbamic pesticides with a choline and acetylcholine electrochemical biosensor. *Anal. Lett.* **24**, 1317–1331 (1991).
30. C. Cremisini, A.D. Sario, J. Mela, R. Pilloton, and G. Paleshci, Evaluation of the use of free and immobilised acetylcholinesterase for paraoxon detection with an amperometric choline oxidase based biosensor. *Anal. Chim. Acta* **311**, 273–280 (1995).
31. T. Montensinos, S.P. Munguia, F. Valdez, and J.L. Marty, Disposable cholinesterase biosensor for the detection of pesticides in water-miscible organic solvents. *Anal. Chim. Acta* **431**, 231–237 (2001).
32. I. Palchetti, A. Cagnini, M. del Carlo, C. Coppi, M. Mascini, and A.P.F. Turner, Determination of anticholinesterase pesticides in real samples using a disposable biosensor. *Anal. Chim. Acta* **337**, 315–321 (1997).
33. T.T. Bachmann and R.D. Schmid, A disposable multielectrode biosensor for rapid simultaneous detection of the insecticides paraoxon and carbofuran at high resolution. *Anal. Chim. Acta* **401**, 95–103 (1999).

34. M. Snejdakova, L. Svobodova, D.P. Nikolelis, J. Wang, and D. Hianik, Acetylcholine biosensor based on dendrimer layers for pesticides detection. *Electroanalysis* **15**, 1185–1191 (2003).
35. S. Andreescu, L. Barthelmebs, and J.L. Marty, Immobilization of acetylcholinesterase on screen-printed electrodes: comparative study between three immobilization methods and applications to the detection of organophosphorus insecticides. *Anal. Chim. Acta* **464**, 171–180 (2002).
36. M.N. Hendji, N. Jaffrezic-Renault, C. Martelet, P. Clechet, A.A. Shulga, V.I. Strikha, L.I. Netchipruk, A.P. Soldatkin, and W.B. Wlodarski, Sensitive detection of pesticides using a differential ISFET-based system with immobilized cholinesterases. *Anal. Chim. Acta* **281**, 3–11 (1993).
37. R.E. Gyurcsanyi, Z. Vagfoldi, K. Toth, and G. Nagy, Fast response potentiometric acetylcholine biosensor. *Electroanalysis* **11**, 712–718 (1999).
38. A. Cagnini, I. Palchetti, I. Lioni, M. Mascini and A.P.F. Turner, Disposable ruthenized screen-printed biosensors for pesticides monitoring. *Sens. Actuat. B* **24**, 85–89 (1995).
39. B. Bucur, A.F. Danet, and J.L. Marty, Versatile method of cholinesterase immobilisation via affinity bonds using concanavalin A applied to the construction of a screen-printed biosensor. *Biosens. Bioelectron.* **20**, 217–225 (2004).
40. Y. Lin, F. Lu, and J. Wang, Disposable carbon nanotube modified screen-printed biosensor for amperometric detection of organophosphorus pesticides and nerve agents. *Electroanalysis* **16**, 145–149 (2004).
41. M. Musameh, N.S. Lawrence, and J. Wang, Electrochemical activation of carbon nanotubes. *Electrochem. Commun.* **7**, 14–18 (2005).
42. M. Lotti, Cholinesterase inhibition: complexities in interpretation. *Chin. Chem.* **41**, 1814–1818 (1995).
43. F.N. Kok and V. Hasirci, Determination of binary pesticide mixtures by an acetylcholinesterase–choline oxidase biosensor. *Biosens. Bioelectron.* **19**, 661–665 (2004).
44. G. Jeanty and J.L. Marty, Detection of paraoxon by continuous flow system based enzyme sensor. *Biosens. Bioelectron.* **13**, 213–218 (1998).
45. C. la Rosa, F. Pariente, L. Hernandez, and E. Lorenzo, Determination of organophosphorus and carbamic pesticides with an acetylcholinesterase amperometric biosensor using 4-aminophenyl acetate as substrate. *Anal. Chim. Acta* **295**, 273–282 (1993).
46. A. Seki, F. Ortega, and J.L. Marty, Enzyme sensor for the detection of herbicides inhibiting acetolactate synthase. *Anal. Lett.* **29**, 1259–1271 (1996).
47. J. Besombes, S. Cosnier, P. Labbe, and G. Reverdy, A biosensor as warning device for the detection of cyanide, chlorophenols, atrazine and carbamate pesticides. *Anal. Chim. Acta* **311**, 255–263 (1995).
48. T. Noguer, A. Gradinaru, A. Cincu, and J.L. Marty, A new disposable biosensor for the accurate and sensitive detection of ethylenebis(dithiocarbamate) fungicides. *Anal. Lett.* **32**, 1723–1738 (1999).
49. T. Noguer, B. Leca, G. Jeanty, and J.L. Marty, Biosensors based on enzyme inhibition: detection of organophosphorus and carbamate insecticides and dithiocarbamate fungicides. *Field Anal. Chem. Tech.* **3**, 171–178 (1999).
50. E.V. Gogol, G.A. Evtugyn, J.L. Marty, H.C. Budnikov, and V.G. Winter, Amperometric biosensors based on Nafion coated screen-printed electrodes for the determination of cholinesterase inhibitors. *Talanta* **53**, 379–389 (2000).
51. M.A. Sirvent, A. Merkoci, and S. Alegret, Pesticide determination in tap water and juice samples using disposable amperometric biosensors made using thick-film technology. *Anal. Chim. Acta* **442**, 35–44 (2001).
52. J. Wang, V.B. Nascimento, S.A. Kane, K. Rogers, M.R. Smyth, and L. Angnes, Screen-printed tyrosinase-containing electrodes for the biosensing of enzyme inhibitors. *Talanta* **43**, 1903–1907 (1996).
53. J.S. Karns, M.T. Muldoon, W.W. Mulbury, M. Derbyshire, and P.C. Kearn, Use of microorganisms and microbial systems in the degradation of pesticide. *ACS Symp. Ser.* **334**, 157–170 (1987).
54. E.I. Rainina, E.N. Efremenco, S.D. Varfolomeyev, A.L. Simonian, and J.R. Wild, The development of a new biosensor based on recombinant *E. coli* for the direct detection of organophosphorus neurotoxins. *Biosens. Bioeletron.* **11**, 991–1000 (1996).
55. A. Mulchandani, P. Mulchnadani, W. Chen, J. Wang, and L. Chen, Amperometric thick-film strip electrodes for monitoring organophosphate nerve agents based on immobilized organophosphorus hydrolase. *Anal. Chem.* **71**, 2246–2249 (1999).

56. A. Mulchandani, I. Kaneva, and W. Chen, Biosensor for direct determination of organophosphate nerve agents using recombinant *Escherichia coli* with surface-expressed organophosphorus hydrolase. 2. Fiber-optic microbial biosensor. *Anal. Chem.* **70**, 5042–5046 (1998).
57. R. Kindervater, W. Künnecke, and R.D. Schimid, Exchangeable immobilized enzyme reactor for enzyme inhibition tests in flow-injection analysis using a magnetic device. Determination of pesticides in drinking water. *Anal. Chim. Acta* **234**, 113–117 (1990).
58. B. Bucur, M. Dondoi, A. Danet, and J.L. Marty, Insecticide identification using a flow injection analysis system with biosensors based on various cholinesterases. *Anal. Chim. Acta* **539**, 195–201 (2005).
59. T. Neufeld, I. Eshkenazi, E. Cohen, and J. Rishpon, A micro flow injection electrochemical biosensor for organophosphorus pesticides. *Biosens. Bioelectron.* **15**, 323–329 (2000).
60. H.C. Frode and I.B. Wilson, in *The Enzymes* (P.D. Boyer, ed.), Academic Press, New York and London (1971).
61. S. Ozaki, H. Nakagawa, K. Fukuda, S. Asakura, H. Kiuchi, T. Shigemori, and S. Takahashi, Re-activation of an amperometric organophosphate pesticide biosensor by 2-pyridinealdoxime methochloride. *Sens. Actuat. B* **66**, 131–134 (2000).
62. K.C. Gulla, M.D. Gouda, M.S. Thakur, and N.G. Karanth, Reactivation of immobilized acetyl cholinesterase in an amperometric biosensor for organophosphorus pesticide. *Biochim. Biophys. Acta* **1597**, 133–139 (2002).
63. E.P. Meulenberg, W.H. Mulder, and P.G. Stocks, Immunoassays for pesticides. *Environ. Sci. Tech.* **29**, 553–561 (1995).
64. R. Puchades and A. Maquieira, Recent developments in flow injection immunoanalysis. *Crit. Rev. Anal. Chem.* **26**, 195–218 (1996).
65. R.Q. Thompson, H. Kim, and C.E. Miller, Comparison of immobilized enzyme reactors for flow-injection systems. *Anal. Chim. Acta* **198**, 165–172 (1987).
66. L.D. Sawyer, B.M. McMahon, W.M. Newsome, and G.A. Parker, in *Pesticide and Industrial Chemical Residues*, 15th ed., AOAC: Champaign, Illinois (1990).
67. V. Lopez-Avila, C. Charan, and J. van Emon, Supercritical fluid extraction–enzyme-linked immunosorbent assay applications for determination of pesticides in soil and food, in *Immunoassays for Residue Analysis: Food Safety* (R.C. Beier and L.H. Stanker eds), ACS Symposium Series 621, American Chemical Society, Washington (1996).
68. A. Dankwardt, Immunochemical assays in pesticide analysis, in *Encyclopedia of Analytical Chemistry* (R.A. Meyers, ed.), John Wiley & Sons Ltd, Chichester (1997).
69. S. Kröger, S.J. Setford, and A.P.F. Turner, Immunosensor for 2,4-dichlorophenoxyacetic acid in aqueous/organic solvent soil extracts. *Anal. Chem.* **70**, 5047–5053 (1998).
70. J. Penalva, R. Puchades, and A. Maquieira, Analytical properties of immunosensors working in organic media. *Anal. Chem.* **71**, 3862–3872 (1999).
71. G.G. Guilbault, B. Hock, and R. Schmid, A piezoelectric immunobiosensor for atrazine in drinking water. *Biosens. Bioelectron.* **7**, 411–419 (1992).
72. M. Minunni, P. Skladal, and M. Mascini, A piezoelectric quartz crystal biosensor for atrazine. *Life. Chem. Repts* **11**, 391–398 (1994).
73. J. Horáček and P. Skládal, Improved direct piezoelectric biosensors operating in liquid solution for the competitive label-free immunoassay of 2,4-dichlorophenoxyacetic acid. *Anal. Chim. Acta* **347**, 43–50 (1997).
74. C. Steegborn and P. Skládal, Construction and characterization of the direct piezoelectric immunosensor for atrazine operating in solution. *Biosens. Bioelectron.* **12**, 19–27 (1997).
75. J. Přibyl, M. Hepel, J. Halámeka, and P. Skládal, Development of piezoelectric immunosensors for competitive and direct determination of atrazine. *Sens. Actuat. B* **91**, 333–341 (2003).
76. M. Minunni and M. Mascini, Detection of pesticide in drinking water using real-time biospecific intraction analysis (BIA). *Anal. Lett.* **26**, 1441–1460 (1993).
77. C. Mouvel, R.D. Harris, C. Maciag, B.J. Luff, J.S. Wolknson, J. Piehler, A. Brecht, G. Gauglitaz, R. Abuknesha, and G. Isamil, Determination of simazine in water samples by waveguide surface plasmon resonance. *Anal. Chim. Acta* **338**, 109–117 (1997).

78. A. Székács, N. Trummerb, N. Adányi, M. Váradi, and I. Szendrö, Development of a non-labeled immunosensor for the herbicide trifluralin via optical waveguide lightmode spectroscopic detection. *Anal. Chim. Acta* **487**, 31–42 (2003).
79. J. Zeravik, T. Ruzgas, and M. Fránek, A highly sensitive flow-through amperometric immunosensor based on the peroxidase chip and enzyme-channeling principle. *Biosens. Bioelectron.* **18**, 1321–1327 (2003).
80. M. Alvarez, A. Calle, J. Tamayo, L.M. Lechuga, A. Abad, and A. Montoya, Development of nanomechanical biosensors for detection of the pesticide DDT. *Biosens. Bioelectron.* **18**, 649–653 (2003).
81. M.A. González-Martínez, S. Morais, R. Puchades, A. Maquieira, A. Abad, and A. Montoya, Monoclonal antibody-based flow-through immunosensor for analysis of carbaryl. *Anal. Chem.* **69**, 2812–2818 (1997).
82. E. Mallat, C. Barzen, R. Abuknesha, G. Gauglitz, and D. Barceló, Part per trillion level determination of isoproturon in certified and estuarine water samples with a direct optical immunosensor. *Anal. Chim. Acta* **426**, 209–216 (2001).
83. M.A. González-Martínez, R. Puchades, A. Maquieira, J.J. Manclús and A. Montoya, Automated immunosensing system for 3,5,6-trichloro-2-pyridinol: application to surface water samples. *Anal. Chim. Acta* **392**, 113–123 (1999).
84. E. Mallat, C. Barzeb, A. Klotz, A. Brecht, G. Gauglits, and D. Barceló, River Analyzer for chlorotriazines with a direct optical immunosensor. *Environ. Sci. Technol.* **33**, 965–971 (1999).
85. M.A. González-Martínez, R. Puchades, and A. Maquieira, Comparison of multianalyte immunosensor formats for on-line determination of organic compounds. *Anal. Chem.* **73**, 4326–4332 (2001).
86. C. Barzen, A. Brecht, and G. Gauglitz, Optical multiple-analyte immunosensor for water pollution control. *Biosens. Bioelectron.* **17**, 289–-295 (2002).
87. S. Rodriguez-Mozaz, S. Reder, M. Lopez de Alda, G. Gauglitz, and D. Barceló, Simultaneous multianalyte determination of estrone, isoproturon and atrazine in natural waters by the RIver ANAlyser (RIANA), an optical immunosensor. *Biosens. Bioelectron.* **19**, 633–640 (2004).
88. J. Tschmelak, G. Proll, and G. Gauglitz, Ultra-sensitive fully automated immunoassay for detection of propanil in aqueous samples: steps of progress toward sub-nanogram per liter detection. *Anal. Bioanal. Chem.* **379**, 1004–1012 (2004).
89. C. Mastichiadis, S.E. Kakabakos, I. Christofidis, M.A. Koupparis, C. Willetts, and K. Misiakos, Simultaneous determination of pesticides using a four-band disposable optical capillary immunosensor. *Anal. Chem.* **74**, 6064–6072 (2002).
90. B.B. Dzantiev, A.V. Zherdev, M.F. Yulaev, R.A. Sitdikov, N.M. Dmitrieva, and I.Y. Moreva, Electrochemical immunosensors for determination of the pesticides 2,4-dichlorophenoxyacetic and 2,4,5-tricholorophenoxyacetic acids. *Biosens. Bioelectron.* **11**, 179–185 (1996).
91. M.F. Yulaev, R.A. Sitdikov, N.M. Dmitrieva, E.V. Yazynina, A.V. Zherdev and B.B. Dzantiev, Development of a potentiometric immunosensor for herbicide simazine and its application for food testing. *Sens. Actuat. B* **75**, 129–135 (2001).
92. J. Gascón, A. Oubiña, B. Ballesteros, D. Barceló, F. Camps, M.P. Marco, M.A. González-Martínez, S. Morais, R. Puchades, and A. Maquieira, Development of a highly sensitive enzyme-linked immunosorbent assay for atrazine. Performance evaluation by flow injection immunoassay. *Anal. Chim. Acta* **347**, 149–162 (1997).
93. M.A. López, F. Ortega, E. Domínguez, and I. Katakis, Electrochemical immunosensor for the detection of atrazine. *J. Mol. Recog.* **11**, 178–181 (1998).
94. S.Q. Hu, J.W. Xie, Q.H. Xu, K.T. Rong, G.L. Shen, and R.Q. Yu, A label-free electrochemical immunosensor based on gold nanoparticles for detection of paraoxon. *Talanta* **61**, 769–777 (2003).
95. P. Skládal and T. Kaláb, A multichannel immunochemical sensor for determination of 2,4-dichlorophenoxyacetic acid. *Anal. Chim. Acta* **316**, 73–78 (1993).
96. K. Grennan, G. Strachan, A.J. Porter, A.J. Killard, and M.R. Smyth, Atrazine analysis using an amperometric immunosensor based on single-chain antibody fragments and regeneration-free multi-calibrant measurement. *Anal. Chim. Acta* **500**, 287–298 (2003).
97. T. Kaláb and P. Skládal, A disposable amperometric immunosensor for 2,4-dichlorophenoxyacetic acid. *Anal. Chim. Acta* **304**, 361–368 (1995).
98. R.W. Keay and C.J. McNeil, Separation-free electrochemical immunosensor for rapid determination of atrazine. *Biosens. Bioelectron.* **13**, 963–970 (1998).

99. B.B. Dzantiev, E.V. Yazynina, A.V. Zherdev, Y.V. Plekhanova, A.N. Reshetilov, S.C. Chang, and C.J. McNeil, Determination of the herbicide chlorsulfuron by amperometric sensor based on separation-free bienzyme immunoassay. *Sens. Actuat. B* **98**, 254–261 (2004).
100. D.P. Nikolelisa, M.G. Simantirakia, C.G. Siontoroua, and K. Tothb, Flow injection analysis of carbofuran in foods using air stable lipid film based acetylcholinesterase biosensor. *Anal. Chim. Acta* **537**, 169–177 (2005).
101. K.A. Law and S.P.J. Higson, Sonochemically fabricated acetylcholinesterase micro-electrode arrays within a flow injection analyser for the determination of organophosphate pesticides. *Biosens. Bioelectron.* **20**, 1914–1924 (2005).
102. A. Vakurov, C.E. Simpson, C.L. Daly, T.D. Gibson, and P.A. Millner, Acetylcholinesterase-based biosensor electrodes for organophosphate pesticide detection: I. Modification of carbon surface for immobilization of acetylcholinesterase. *Biosens. Bioelectron.* **20**, 1118–1125 (2004).
103. P.R.B. de O Marques, G.S. Nunes, T.C.R. dos Santos, S. Andreescu, and J.L. Marty, Comparative investigation between acetylcholinesterase obtained from commercial sources and genetically modified *Drosophila melanogaster*: application in amperometric biosensors for methamidophos pesticide detection. *Biosens. Bioelectron.* **20**, 825–832 (2004).
104. T.R. Glass, H. Saiki, T. Joh, Y. Taemi, N. Ohmura, and S.J. Lackie, Evaluation of a compact bench top immunoassay analyzer for automatic and near continuous monitoring of a sample for environmental contaminants. *Biosens. Bioelectron.* **20**, 397–403 (2004).
105. S. Jin, Z. Xu, J. Chen, X. Liang, Y. Wu, and X. Qian, Determination of organophosphate and carbamate pesticides based on enzyme inhibition using a pH-sensitive fluorescence probe. *Anal. Chim. Acta* **523**, 117–123 (2004).
106. A.Y. Kolosova, J.H. Park, S.A. Eremin, S.J. Park, S.J. Kanga, W.B. Shima, H.S. Lee, Y.T. Lee, and D.H. Chunga, Comparative study of three immunoassays based on monoclonal antibodies for detection of the pesticide parathion-methyl in real samples. *Anal. Chim. Acta* **511**, 323–331 (2004).
107. A. Ivanov, G. Evtugyn, L.V. Lukachova, E.E. Karyakina, HC. Budnikov, S.G. Kiseleva, A.V. Orlov, G.P. Karpacheva, and A.A. Karyakin, Cholinesterase potentiometric sensor based on graphite screen-printed electrode modified with processed polyaniline. *IEEE Sensors J.* **3**, 333–340 (2003).
108. A. Ivanov, G. Evtugyn, H. Budnikov, F. Ricci, D. Moscone, and G. Palleschi, Cholinesterase sensors based on screen-printed electrodes for detection of organophosphorus and carbamic pesticides. *Anal. Bioanal. Chem.* **377**, 624–631 (2003).
109. A.Y. Kolosova, J.H. Park, S.A. Eremin, S.J. Kang and D.H. Chung, Fluorescence polarization immunoassay based on a monoclonal antibody for the detection of the organophosphorus pesticide parathion-methyl. *J. Agric. Food Chem.* **51**, 1107–1114 (2003).
110. A.A. Ciucu, C. Negulescu, and R.P. Baldwin, Detection of pesticides using an amperometric biosensor based on ferophthalocyanine chemically modified carbon paste electrode and immobilized bienzymatic system. *Biosens. Bioelectron.* **18**, 303–310 (2003).
111. G.L. Turdean, I.C. Popescu, L. Oniciu, and D.R. Thevenot, Sensitive detection of organophosphorus… Thiocholine electrochemistry and immobilised enzyme inhibition. *J. Enzyme Inhi. Medicinal Chem.* **17**, 107–115 (2002).
112. K. Reybier, S. Zairi, N. Jaffrezic-Renault, and B. Fahys, The use of polyethyleneimine for fabrication of potentiometric cholinesterase biosensors. *Talanta* **56**, 1015–1020 (2002).
113. F.N. Kok, F. Bozoglu, and V. Hasirci, Construction of an acetylcholinesterase–choline oxidase biosensor for aldicarb determination. *Biosens. Bioelectron.* **17**, 531–539 (2002).
114. V.G. Andreou and Y.D. Clonis, A portable fiber-optic pesticide biosensor based on immobilized cholinesterase and sol-gel entrapped bromcresol purple for in-field use. *Biosens. Bioelectron.* **17**, 61–69 (2002).
115. E.V. Gogol, G.A. Evtyugin, E.V. Suprun, G.K. Budnikov, and V.G. Vinter, Determination of residual pesticide in plant materials using planar cholinesterase sensors modified with nafion. *J. Anal. Chem.* **56**, 963–970 (2001).
116. L. Pogacnik and M. Franko, Optimisation of FIA system for detection of organophosphorus and carbamate pesticides based on cholinesterase inhibition. *Talanta* **54**, 631–641 (2001).
117. W.L. Xing, L.R. Ma, Z.H. Jiang, F.H. Cao, and M.H. Jia, Portable fiber-optic immunosensor for detection of methsulfuron methyl. *Talanta* **52**, 879–883 (2000).

118. M. Fránek, A. Deng, and V. Kolář, Performance characteristics for flow injection immunoassay using monoclonal antibodies against s-triazine and 2,4-D herbicides. *Anal. Chim. Acta* **412**, 19–27 (2000).
119. F. Massei, F. Botrè, G. Lorenti, G. Simonetti, F. Porcelli, G. Scibona, and C. Botrè, Plant tissue electrode for the determination of atrazine. *Anal. Chim. Acta* **316**, 79–82 (1995).
120. M.D. Gouda, M.S. Thakur, and N.G. Karanth, Stability studies on immobilized glucose oxidase using an amperometric biosensor. *Biotech. Techniq.* **11**, 653–655 (1997).
121. R. Rouillon, M. Sole, T. Carpentier, and J.L. Marty, Immobilization of thylakoids in polyvinylalcohol for the detection of herbicides. *Sens. Actuat. B* **27**, 477–479 (1995).
122. S. Kintzios, E. Pistola, P. Panagiotopoulos, M. Bomsel, N. Alexandropoulos, F. Bem, G. Ekonomou, J. Biselis, and R. Levin, Bioelectric recognition assay (BERA). *Biosens. Bioelectron.* **16**, 325–336 (2001).
123. C.Y. Shao, C.J. Howe, A.J.R. Porter, and L.A. Glover, Novel cyanobacterial biosensor for detection of herbicides. *Appl. Environ. Microbiol.* **68**, 5026–5033 (2002).
124. M. Preuss and A.H. Halt, Mediated herbicide inhibition in a PET biosensor. *Anal. Chem.* **67**, 1940–1949 (1995).
125. A.P. Soldatkin, D.V. Gorchkoh, C. Martelet, and N. Jaffrezic-Renault, New enzyme potentiometric sensor for hypochlorite species detection. *Sens. Actuat. B* **43**, 99–104 (1997).
126. M. Stoytcheva, Electrochemical evaluation of the kinetic parameters of a heterogeneous enzyme reaction in presence of metal ions. *Electroanalysis* **14**, 923–927 (2002).
127. M. Stoytcheva and V. Sharkova, Kinetics of the inhibition of immobilized acetylcholinesterase with Hg(II). *Electroanalysis* **14**, 1007–1010 (2002).
128. P. Nugent, in *Emerging Strategies for Pesticide Analysis* (T. Cairns and J. Sherma, eds), CRC Press, Boca Raton, Florida (1992).
129. A.S. Hill, J.V. Mei, C. Yin, B.S. Ferguson, and J.H. Skerritt, Determination of the insect growth regulator methoprene in wheat grain and milling fractions using an enzyme immunoassay. *J. Agric. Food Chem.* **39**, 1882–1886 (1991).
130. J.W. King and K.S. Nam, in *Residue Analysis in Food Safety* (R.C. Beier and L.H. Stanker, eds), ACS Symposium Series 621, American Chemical Society, Washington (1996).
131. D.S. Aga, E.M. Thurman, and M.L. Pomes, Determination of alachlor and its sulfonic acid metabolite in water by solid-phase extraction and enzyme-linked immunosorbent assay. *Anal. Chem.* **66**, 1495–1499 (1994).
132. A. Marx, T. Giersch, and B. Hock, Immunoaffinity chromatography of s-triazines. *Anal. Lett.* **28**, 267–278 (1995).
133. V. Pichon, L. Chen, and M.C. Hennion, On-line preconcentration and liquid chromatographic analysis of phenylurea pesticides in environmental water using a silica-based immunosorbent. *Anal. Chim. Acta* **311**, 429–436 (1995).

CHAPTER 3

Electrochemical glucose biosensors

Joseph Wang

3.1 INTRODUCTION

Diabetes mellitus is a worldwide public health problem. It is one of the leading causes of death and disability in the world. Such metabolic disorder results from insulin deficiency and hyperglycemia. The diagnosis and management of diabetes mellitus requires a tight monitoring of blood glucose levels. Millions of diabetics thus test their blood glucose levels daily making glucose the most commonly tested analyte. The challenge of providing such reliable and tight glycemic control remains the subject of a considerable amount of research [1, 2]. Electrochemical biosensors for glucose play a leading role in this direction. Amperometric enzyme electrodes, based on glucose oxidase (GOx) bound to electrode transducers, have thus been the subject of substantial research [2].

Since Clark and Lyons first proposed the initial concept of glucose enzyme electrodes in 1962 [3] we have witnessed tremendous activity towards the development of reliable devices for diabetes control. A variety of approaches have been explored in the operation of glucose enzyme electrodes. In addition to diabetes control, such devices offer great promise for other important applications, ranging from food analysis to bioprocess monitoring. The great importance of glucose has generated an enormous number of publications, the flow of which shows no sign of diminishing. Yet, despite impressive advances in glucose biosensors, there are still many challenges related to the achievement of clinically accurate tight glycemic monitoring.

The goal of this chapter is to examine the history and current status of electrochemical glucose biosensors, and discuss their principles of operation along with future challenges.

3.2 FORTY YEARS OF PROGRESS

The history of glucose enzyme electrodes starts with the first device developed in 1962 by Clark and Lyons from the Children Hospital in Cincinnati [3]. Their first device relied on a thin layer of GOx entrapped over an oxygen electrode (via a semipermeable dialysis membrane), and monitoring the oxygen consumed by the enzyme-catalyzed reaction:

$$\text{Glucose} + \text{Oxygen} \xrightarrow{\text{Gox}} \text{Gluconic acid} + \text{Hydrogen peroxide} \qquad (1)$$

Clark's original patent [4] covers the use of one or more enzymes for converting electroinactive substrates to electroactive products. The effect of interferences was corrected by using two electrodes (one covered with GOx) and measuring the differential current. Clark's technology was subsequently transferred to Yellow Spring Instrument Company, which launched in 1975 the first dedicated glucose analyzer (the Model 23 YSI analyzer) for the direct measurement of glucose in 25 μL samples of whole blood. Updike and Hicks [5] developed further this principle by using two oxygen working electrodes (one covered with the enzyme) and measuring the differential current for correcting the oxygen background variation in samples. In 1973, Guilbault and Lubrano [6] described an enzyme electrode for the determination of blood glucose based on amperometric (anodic) monitoring of the liberated hydrogen peroxide:

$$H_2O_2 \rightarrow O_2 + 2H^+ + 2e^- \qquad (2)$$

Good accuracy and precision were obtained in connection to 100 μL blood samples. A wide range of amperometric enzyme electrodes, differing in the electrode design or material, membrane composition, or immobilization approach have since been described.

During the 1980s biosensors became a "hot" topic, reflecting the growing emphasis on biotechnology. Considerable efforts during this decade focused on the development of mediator-based "second-generation" glucose biosensors [7, 8], the introduction of commercial strips for self-monitoring of blood glucose [9, 10], and the use of modified electrodes for enhancing the sensor performance [11]. In the 1990s we witnessed intense

TABLE 3.1
Historical landmarks in the development of electrochemical glucose biosensors

Date	Event	Ref.
1962	First glucose enzyme electrode	[3]
1973	Glucose enzyme electrode based on peroxide detection	[6]
1975	Launch of the first commercial glucose sensor system	YSI Inc.
1982	Demonstration of *in-vivo* glucose monitoring	[36]
1984	Development of ferrocene mediators	[7]
1987	Launch of the first personal glucose meter	Medisense Inc.
1987	Electrical wiring of enzymes	[12]
1999	Launch of a commercial *in-vivo* glucose sensor	Minimed Inc.
2000	Introduction of a wearable non-invasive glucose monitor	Cygness Inc.

activity towards the establishment of electrical communication between the redox center of GOx and the electrode surface [12, 13], and the development of minimally invasive subcutaneously implantable devices [14, 15]. Table 3.1 summarizes major historical landmarks in the development of electrochemical glucose biosensors.

3.3 FIRST-GENERATION GLUCOSE BIOSENSORS

First-generation devices have relied on the use of the natural oxygen cosubstrate, and the production and detection of hydrogen peroxide [Eqs (1)–(2)]. Such measurements of peroxide formation have the advantage of being simpler, especially when miniaturized sensors are concerned. A very common configuration is the YSI probe, involving the entrapment of GOx between an inner anti-interference cellulose acetate membrane and an outer diffusion-limiting/biocompatible polycarbonate one (Fig. 3.1).

3.3.1 Redox interferences

The amperometric measurement of hydrogen peroxide requires application of a potential at which endogenous reducing species, such as ascorbic and uric acids and some drugs (e.g. acetaminophen), are also electroactive. The anodic contributions of these and other oxidizable constituents of biological fluids can compromise the selectivity and hence the overall accuracy. Extensive efforts during the 1980s were devoted to minimizing the error of electroactive interferences in glucose electrodes. One useful strategy is to employ a permselective coating that minimizes access of such constituents to the transducer surface. Different polymers, multilayers and mixed layers, with transport properties based on size, charge or polarity, have thus been used for discriminating against coexisting electroactive compounds [16, 17]. Such films also exclude surface-active

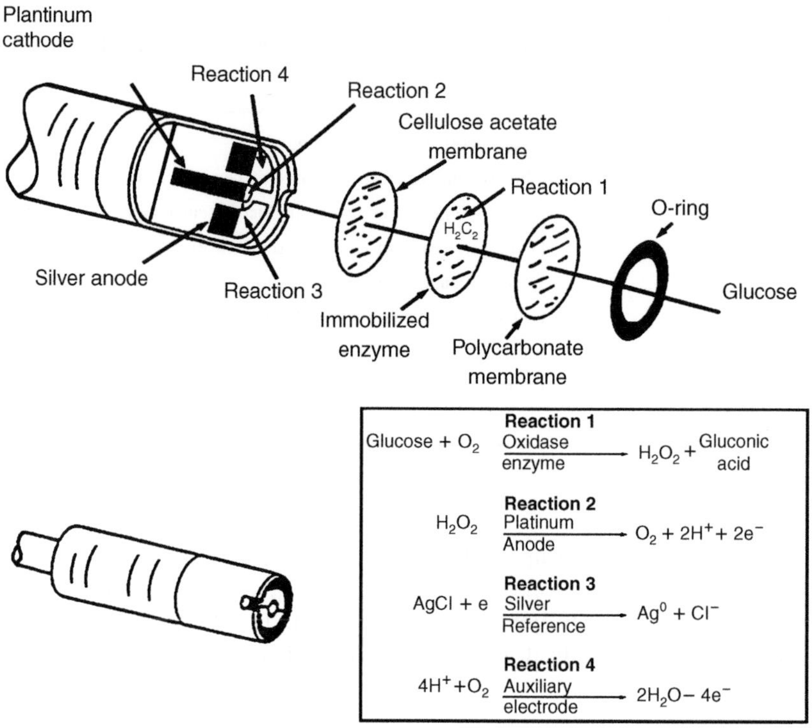

FIGURE 3.1 Schematic of a "first-generation" glucose biosensor (based on a probe manufactured by YSI Inc.).

macromolecules, hence imparting higher stability. Electropolymerized films, particularly poly (phenelendiamine) and overoxidized polypyrrole, have been shown particularly useful for imparting high selectivity (based on size exclusion) while confining the GOx onto the surface [17, 18]. Other widely used coatings include size-exclusion cellulose acetate films, the negatively charged (sulfonated) Nafion or Kodak AQ ionomers, and hydrophobic alkanethiol or lipid layers. The use of overlaid multilayers, combining the properties of different films, offers additional advantages. For example, alternate deposition of Nafion and cellulose acetate was used for eliminating the interference of the neutral acetaminophen and negatively charged ascorbic and uric acids, respectively [19].

Efforts during the 1990s focused on the preferential electrocatalytic detection of the liberated hydrogen peroxide [20, 21, 22, 23]. This has allowed tuning of the detection potential to the optimal region ($\sim$0.0 to -0.20V vs Ag/AgCl) where most unwanted background reactions are negligible. The remarkably high selectivity thus obtained was coupled to a fast and sensitive response. Metallized (Ru,Rh,)-carbon [20, 21] and metal-hexacyanoferrate [22, 23] based transducers have been particularly useful for enhancing the selectivity towards the target glucose substrate. Additional improvements can be achieved by combining this preferential catalytic activity with a discriminative layer, e.g. by dispersing rhodium particles within a Nafion film [24].

3.3.2 Oxygen dependence

Since oxidase-based devices rely on the use of oxygen as the physiological electron acceptor, they are subject to errors accrued from fluctuations in the oxygen tension and the stoichiometric limitation of oxygen. These include fluctuations in the sensor response and a reduced upper limit of linearity. Such limitation (known as the "oxygen deficit") reflects the fact that normal oxygen concentrations are about an order of magnitude lower than the physiological level of glucose.

Several avenues have been proposed for addressing this oxygen limitation. One approach relies on the use of mass-transport limiting films (such as polyurethane or polycarbonate) for tailoring the flux of glucose and oxygen, i.e. increasing the oxygen/glucose permeability ratio [1, 25]. A two-dimensional cylindrical electrode, designed by Gough's group [25, 26], has been particularly attractive for addressing the oxygen deficit by allowing oxygen to diffuse into the enzyme region of the sensor from both directions and glucose diffusion only from one direction. We addressed the oxygen limitation of glucose biosensors by designing an oxygen-rich carbon paste enzyme electrode [27]. Such a biosensor is based on a fluorocarbon (Kel-F oil) pasting liquid, which has very high oxygen solubility, allowing it to act as an internal source of oxygen. The internal flux of oxygen can thus support the enzymatic reaction even in oxygen-free glucose solutions. It is possible also to circumvent the oxygen demand issue by replacing the GOx with glucose dehydrogenase (GDH) that does not require an oxygen cofactor [28].

3.4 SECOND-GENERATION GLUCOSE BIOSENSORS

3.4.1 Electron transfer between GOx and electrode surfaces

Further improvements (and attention to the above errors) can be achieved by replacing the oxygen with a non-physiological (synthetic) electron acceptor, which is able to shuttle electrons from the redox center of the enzyme to the surface of the electrode. Glucose oxidase does not directly transfer electrons to conventional electrodes because a thick protein layer surrounds its flavin redox center. Such a thick protein shell introduces a spatial separation of the electron donor-acceptor pair, and hence an intrinsic barrier to direct electron transfer, in accordance with the distance dependence of the electron transfer rate [29]:

$$K_{\text{et}} = 10^{13}\text{e}^{-0.91(d-3)}\text{e}^{[-(\Delta G+\lambda)/4RT\lambda]} \quad (3)$$

where ΔG and λ correspond to the free and reorganization energies accompanying the electron transfer, respectively, and d the actual electron transfer distance. The minimization of the electron transfer distance (between the immobilized GOx and the electrode surface) is thus crucial for ensuring optimal performance. Accordingly, different innovative strategies have been suggested for establishing and tailoring the electrical contact between the redox center of GOx and electrode surfaces.

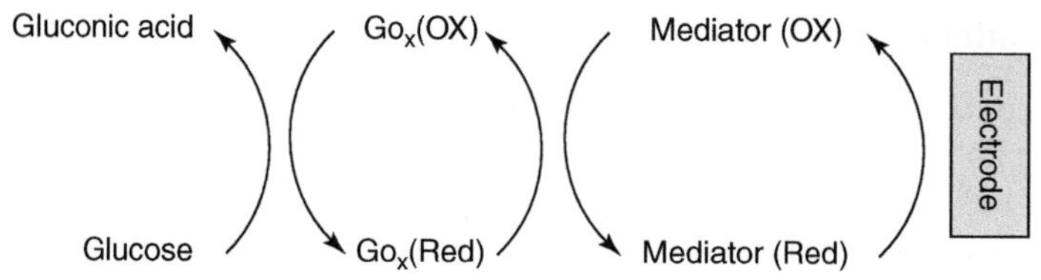

FIGURE 3.2 Sequence of events that occur in "second-generation" (mediator-based) glucose biosensor mediated systems.

3.4.2 Use of artificial mediators

Particularly useful have been the artificial mediators that shuttle electrons between the FAD center and the surface by the following scheme:

$$\text{Glucose} + \text{GOx}_{(\text{ox})} \rightarrow \text{Gluconic acid} + \text{GOx}_{(\text{red})} \quad (4)$$

$$\text{GOx}_{(\text{red})} + 2M_{(\text{ox})} \rightarrow \text{GOx}_{(\text{ox})} + 2M_{(\text{red})} + 2\text{H}^+ \quad (5)$$

$$2M_{(\text{red})} \rightarrow 2M_{(\text{ox})} + 2\text{e}^- \quad (6)$$

where $M_{(\text{ox})}$ and $M_{(\text{red})}$ are the oxidized and reduced forms of the mediator. Such a mediation cycle produces a current dependent on the glucose concentration (Fig. 3.2). Diffusional electron mediators, such as ferrocene derivatives, ferricyanide, conducting organic salts (particularly tetrathiafulvalene-tetracyanoquinodimethane, TTF-TCNQ), phenothiazine and phenoxazine compounds, or quinone compounds have thus been widely used to electrically contact GOx [7, 8]. As a result of using these electron-carrying mediators, measurements become largely independent of oxygen partial pressure and can be carried out at lower potentials that do not provoke interfering reactions from coexisting electroactive species [Eq. (6)]. In order to function effectively, the mediator should react rapidly with the reduced enzyme (to minimize competition with oxygen), possess good electrochemical properties (such as a low redox potential), have low solubility in aqueous medium, and must be non-toxic and chemically stable (in both reduced and oxidized forms). Commercial blood glucose self-testing meters, described in the following section, commonly rely on the use of ferricyanide or ferrocene mediators. Most *in-vivo* devices, however, are mediatorless due to potential leaching and toxicity of the mediator.

3.4.3 Attachment of electron-transfer relays

Heller's group [12] developed an elegant non-diffusional route for establishing a communication link between GOx and electrodes based on "wiring" the enzyme to the surface with a long flexible poly-pyridine polymer having a dense array of osmium-complex electron relays (Fig. 3.3). The resulting three-dimensional redox-polymer/enzyme networks offer high current outputs and stabilize the mediator to electrode surfaces.

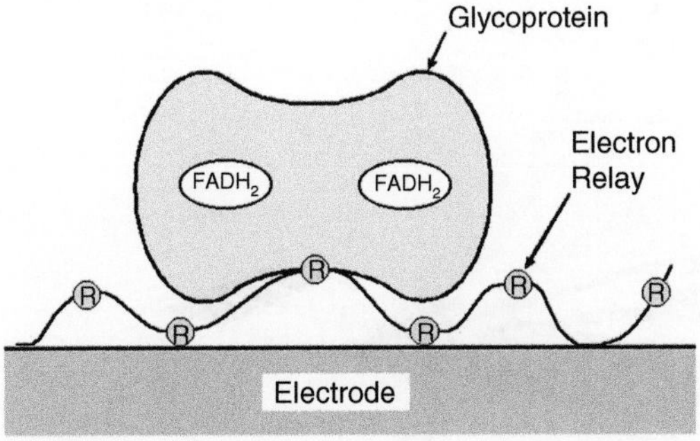

FIGURE 3.3 Electrical contacting of a flavoenzyme by its reconstitution with a relay-FAD semisynthetic cofactor.

Chemical modification of GOx with electron-relay groups represents another novel avenue for facilitating the electron transfer between its redox center and the electrode surface. Willner and coworkers [13] reported on an elegant approach for modifying GOx with electron relays. For this purpose, the FAD active center of the enzyme was removed to allow positioning of an electron-mediating ferrocene unit prior to the reconstitution of the enzyme. The attachment of electron-transfer relays at the enzyme periphery has also been considered for yielding short electron-transfer distances [30]. More sophisticated bioelectronic systems for enhancing the electrical response, based on patterned monolayer or multilayer assemblies and organized enzyme networks on solid electrodes, have been developed for contacting GOx with the electrode support [30]. Functionalized alkanethiol modified gold surfaces have been particularly attractive for such a layer-by-layer (LBL) creation of GOx/mediator networks.

3.5 *IN-VITRO* GLUCOSE TESTING

Electrochemical biosensors are well suited for satisfying the needs of personal (home) glucose testing. The majority of personal blood glucose meters are based on disposable (screen-printed) enzyme electrode test strips [31, 32]. Such single-use disposable electrode strips are mass produced by the thick film (screen-printing) microfabrication technology [21, 33]. The screen-printing technology relies on printing patterns of conductor and insulators onto the surface of planar (plastic or ceramic) substrates. Each strip contains the printed working and reference electrodes (Fig. 3.4, see Plate 1 for color version), with the working one coated with the necessary reagents (i.e. enzyme, mediator, stabilizer, linking and binding agents). Such reagents are commonly being dispensed by inkjet printing technology. A counter and an additional ("baseline") working electrode may also be included. Such single-use devices obviate problems of carryover, contamination or drift. Various membranes (mesh) are incorporated into the test strips, and along with surfactants are used to provide a uniform sample coverage.

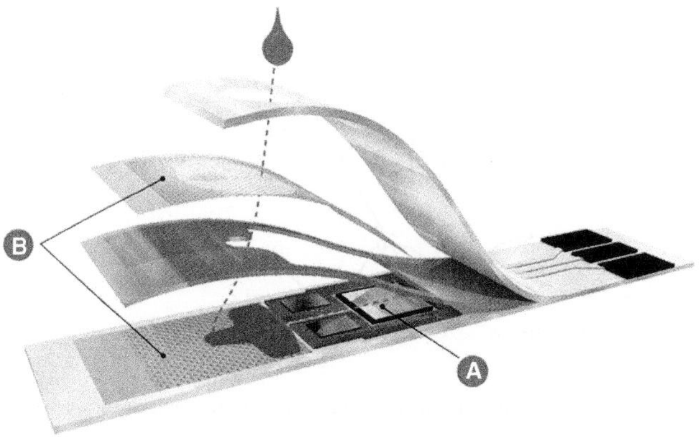

FIGURE 3.4 A typical commercial strip for self-testing of blood glucose (based on a biosensor manufactured by Abbott Inc.) (see Plate 1 for the color version).

TABLE 3.2
Major commercial electrochemical systems for self-monitoring of blood glucose

Source	Trade name	Enzyme	Mediator
Abbott	Precision	GOx	Ferrocene
	FreeStyle	GOx	Osmium "wire"
Bayer	Elite	GOx	Ferricyanide
LifeScan	SureStep	GOx	Ferricyanide
	One Touch Ultra	GOx	
Roche-Diagnostics	Accu-Check	GOx	Ferricyanide

The control meter is typically light and small (pocket size), battery operated, and relies on a potential-step (chronoamperometric) operation, in connection with a short incubation (reaction) step. Such devices offer considerable promise for obtaining the desired clinical information in a simpler ("user-friendly"), faster, and cheaper manner compared to traditional assays. The first product was a pen-style device (the Exactech™), launched by Medisense Inc. in 1987, that relied on the use of a ferrocene-derivative mediator. Since then, over 40 different commercial strips and pocket-sized test meters, for self-monitoring of blood glucose – based on the use of ferricyanide or ferrocene mediators – have been introduced [32]. Yet, over 90% of the market is being shared by Life Scan, Roche Diagnostics, Abbott, and Bayer (Table 3.2). In all cases, the diabetic patient pricks the finger, places the small blood droplet on the sensor strip, and obtains the blood glucose concentration (on an LC display) within 5–30 sec. In addition to fast response and small size, such modern personal glucose meters have features such as extended memory capacity and computer downloading capabilities.

3.6 CONTINUOUS REAL-TIME *IN-VIVO* MONITORING

Although self-testing is considered a major advance in glucose monitoring it is limited by the number of tests per day. The inconvenience associated with standard finger-stick sampling deters patients from frequent monitoring. Such testing neglects night-time variations and may result in poor approximation of blood glucose variations. Tighter glycemic control, through more frequent measurements or continuous monitoring, is desired for detecting sharp changes in the glucose level and triggering proper alarm in cases of hypo- and hyperglycemia, essential for making valid therapeutic decisions [34, 35]. Glucose biosensors are thus key components of closed-loop glycemic control systems. A wide range of possible *in-vivo* glucose biosensors has thus been studied for maintaining glucose levels close to normal. The first application of such devices for *in-vivo* glucose monitoring was demonstrated first by Shichiri *et al.* in 1982 [36]. His needle-type glucose sensor relied on a platinum anode held at +0.6 V (vs silver cathode) for monitoring the enzymatically produced hydrogen peroxide. The enzyme (glucose oxidase) entrapment was accomplished in connection with a cellulose-diacetate/heparin/polyurethane coating. Continuous *ex-vivo* monitoring of blood glucose was proposed already in 1974 [37]. The majority of glucose sensors used for *in-vivo* applications are based on the GOx-catalyzed oxidation of glucose by oxygen.

3.6.1 Requirements

The major requirements of clinically accurate *in-vivo* glucose sensors have been discussed in various review articles [1, 34, 35]. These include proper attention to the issues of biocompatibility (rejection of the sensor by the body), miniaturization, long-term stability of the enzyme and transducer, oxygen deficit, short stabilization times, *in-vivo* calibration, baseline drift, safety, and convenience. The sensor must be of a size and shape that can be easily implanted and cause minimal discomfort. Under biocompatibility one must consider the effect of the sensor upon the *in-vivo* environment as well as the environmental effect upon the sensor performance. Problems with biocompatibility have proved to be the major barriers to the development of reliable implantable devices. Most glucose biosensors lack the biocompatibility necessary for a reliable prolonged operation in whole blood. Alternative sensing sites, particularly the subcutaneous tissue, have thus received growing attention. While the above issues represent a major challenge, significant progress has been made towards the continuous monitoring of glucose.

3.6.2 Subcutaneous monitoring

Most of the recent attention has been given to the development of subcutaneously implantable needle-type electrodes [14, 15, 34, 38]. Such devices track blood glucose levels by measuring the glucose concentration in the interstitial fluid of the subcutaneous tissue (assuming the ratio of the blood/tissue levels is constant). Subcutaneously implantable devices are commonly designed to operate for a few days and be replaced by the patient. Success in this direction has reached the level of short-term human implantation;

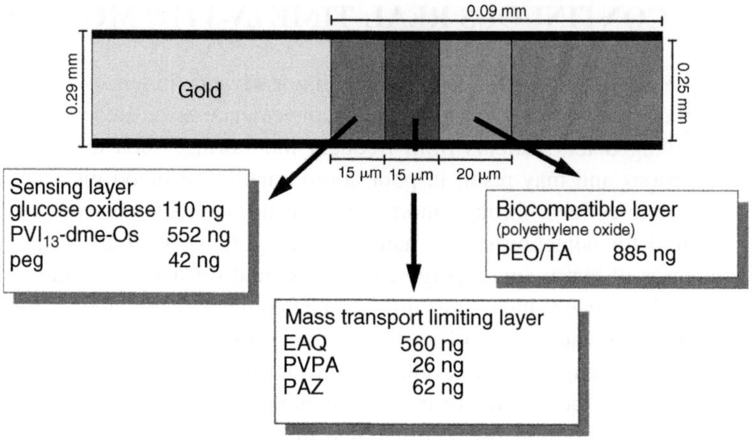

FIGURE 3.5 Design of an implantable three-layered glucose biosensor for subcutaneous monitoring (based on [15]).

continuously functioning devices, possessing adequate (>1 week) stability, are expected in the near future. Such devices would enable a swift and appropriate corrective action (through a closed-loop insulin delivery system, i.e. an artificial pancreas). Algorithms correcting for the transient difference (short time lag) between blood and tissue glucose concentrations have been developed [38]. The CGMS unit of Medtronic Minimed Inc. (Sylmar, CA) offers a 72h period of such subcutaneous monitoring, with measurement of tissue glucose every 5min and data storage in the monitor's memory [39]. After 72h, the sensor is removed, and the information is transferred to a computer for identifying patterns of glucose variations. In addition to easily removable short-term implants, efforts are continuing towards chronically implanted devices (aimed at functioning reliably for 6–12 months). A similar system is currently being developed by Abbott Inc. [40]. This system is based on the wired enzyme technology of Heller's group (Fig. 3.5), an insertion of a short needle into the skin, to yield a reading every minute. Both the Abbott and Minimed devices include a limited range transmitter that relays the sensor data to a pager-like device that provides the necessary warnings and stores the data. Additional devices based on patch-like sensors, nano-needles or microdialysis sampling are currently being developed by different companies.

3.6.3 Towards non-invasive glucose monitoring

Non-invasive glucose sensing is the ultimate goal of glucose monitoring. Such a non-invasive route for continuous glucose monitoring is expected to obviate the challenges of implantable devices. In particular, Cygnus Inc. has developed a wearable glucose monitor, based on the coupling of reverse iontophoretic collection of glucose and biosensor functions [41]. The GlucoWatch biographer provides up to three glucose readings per hour for up to 12h (i.e. 36 readings within a 12h period). The system has been shown to be capable of measuring the electroosmotically extracted glucose with a clinically acceptable level of accuracy. An alarm capability is included to alert the individual of very low or high glucose levels. Other routes for "collecting" the glucose through the skin and for non-invasive glucose testing are currently being examined by various groups and companies.

3.7 CONCLUSIONS AND OUTLOOK

Over the past 40 years we have witnessed a tremendous progress towards the development of electrochemical glucose biosensors. Elegant research on new sensing concepts, coupled with numerous technological innovations, has opened the door to widespread applications of electrochemical glucose biosensors. Such devices account for nearly 85% of the world market of biosensors. Major advances have been made for enhancing the capabilities and improving the reliability of glucose measuring devices. Such intensive activity has been attributed to tremendous economic prospects and fascinating research opportunities. The success of glucose blood meters has stimulated considerable interest in *in-vitro* and *in-vivo* devices for monitoring other physiologically important compounds. Despite a tremendous progress in glucose biosensors, there are still many challenges related to the achievement of tight, stable, and reliable glycemic monitoring. The development of new and improved glucose biosensors thus remains the prime focus of many researchers.

As this field enters the fifth decade of intense research we expect significant efforts coupling fundamental sciences with technological advances. Such stretching of the ingenuity of researchers will result in greatly improved electrical contact between the redox center of GOx and electrode surfaces, enhanced "genetically engineered" GOx, new "painless" *in-vitro* testing, artificial (biomimetic) receptors for glucose, advanced biocompatible membrane materials, the coupling of minimally invasive monitoring with a compact insulin delivery system, new innovative approaches for non-invasive monitoring, and miniaturized long-term implants. These, and similar developments, will greatly improve the control and management of diabetes.

3.8 REFERENCES

1. G. Reach and G.S. Wilson, Can continuous glucose monitoring be used for the treatment of diabetes? *Anal. Chem.* **64**, 381A–389A (1992).
2. J. Wang, Glucose biosensors: 40 yrs of advances and challenges. *Electroanalysis* **12**, 983–988 (2001).
3. L.C. Clark and C. Lyons, Electrode systems for monitoring in cardiovascular surgery. *Ann. NY Acad. Sci.* **102**, 29–45 (1962).
4. L. Clark, Jr., US Patent 33, 539, 455 (1970).
5. S. Updike and G. Hicks, Enzyme electrode. *Nature* **214**, 986–988 (1967).
6. G. Guilbault and G. Lubrano, An enzyme electrode for the amperometric determination of glucose. *Anal. Chim. Acta* **64**, 439–455 (1973).
7. A. Cass, G. Davis, G. Francis, H.A. Hill, W. Aston, J. Higgins, E. Plotkin, L. Scott, and A.P. Turner, Ferrocene-mediated enzyme electrode for amperometric determination of glucose. *Anal. Chem.* **56**, 667–671 (1984).

8. J. Frew and H.A. Hill, Electrochemical biosensors. *Anal. Chem.* **59**, 933A–939A (1987).
9. P. Hilditch and M. Green, Disposable electrochemical biosensors. *Analyst* **116**, 1217–1220 (1991).
10. D. Matthews, R. Holman, E. Brown, J. Streemson, A. Watson, and S. Hughes, Pen-sized digital 30-second blood glucose meter. *Lancet* **2**, 778–779 (1987).
11. R.W. Murray, A. Ewing, and R. Durst, Chemically modified electrodes – molecular design for Electroanalysis. *Anal. Chem.* **59**, 379A–384A (1987).
12. Y. Degani and A. Heller, Direct electrical communication between chemically modified enzymes and metal electrodes. I. Electron transfer from glucose oxidase to metal electrodes via electron relays, bound covalently to the enzyme. *J. Phys. Chem.* **91**, 1285–1289 (1987).
13. I. Willner, V. Heleg-Shabtai, R. Blonder, E. Katz, and G. Tao, Electrical wiring of glucose oxidase by reconstitution of FAD-modified monolayers assembled onto Au-electrodes. *J. Am. Chem. Soc.* **118**, 10321–10322 (1996).
14. D. Bindra, Y. Zhang, G. Wilson, R. Sternberg, D.Trevenot, G. Reach, and D. Moatti, Design and in vitro studies of a needle-type glucose sensor for subcutaneous monitoring. *Anal. Chem.* **63**, 1692–1696 (1991).
15. E. Csoregi, D.W. Schmidtke, and A. Heller, Design and optimzation of a selective subcutaneously implantable gluocse electrode based on wired glucose oxidase. *Anal. Chem.* **67**, 1240–1244 (1995).
16. S. Emr and A. Yacynych, Use of polymer films in amperometric biosensors. *Electroanalysis* **7**, 913–923 (1995).
17. C. Malitesta, F. Palmisano, L. Torsi, and P. Zambonin, Glucose fast-response amperometric sensor based on glucose oxidase immobilized in an electropolymerized poly(o-phenylenediamine) film. *Anal. Chem.* **62**, 2735–2740 (1990).
18. S. Sasso, R. Pierce, R. Walla, and A. Yacynych, Electropolymerized 1,2-diaminobenzene as a means to prevent interferences and fouling and to stabilize immobilized enzyme in electrochemical biosensors. *Anal. Chem.* **62**, 1111–1117 (1990).
19. Y. Zhang, Y. Hu, G.S. Wilson, D. Moatti-Sirat, V. Poitout, and G. Reach, Elimination of the acetaminophen interference in an implantable glucose sensor. *Anal. Chem.* **66**, 1183 (1994).
20. J. Wang, J. Liu, L. Chen, and F. Lu, Highly selective membrane-free, mediator-free glucose biosensor. *Anal. Chem.* **66**, 3600–3603 (1994).
21. J. Newman, S. White, I. Tothill, and A.P. Turner, Catalytic materials, membranes, and fabrication technologies suitable for the construction of amperometric biosensors. *Anal. Chem.* **67**, 4594–4599 (1995).
22. A. Karaykin, O. Gitelmacher, and E. Karaykina, Prussian blue-based first-generation biosensor. A sensitive amperometric electrode for glucose. *Anal. Chem.* **67**, 2419–2423 (1995).
23. Q. Chi and S. Dong, Amperometric biosensors based on the immobilization of oxidases in a Prussian blue film by electrochemical codeposition. *Anal. Chim. Acta* **310**, 429–436 (1995).
24. J. Wang and H. Wu, Highly selective biosensing of glucose utilizing a glucose oxidase + rhodium + Nafion® biocatalytic-electrocatalytic-permselective surface microstructure. *J. Electroanal. Chem.* **395**, 287–291 (1995).
25. D. Gough, J. Lucisano, and P. Tse, Two-dimensional enzyme electrode sensor for glucose. *Anal. Chem.* **57**, 2351–2357 (1985).
26. J. Armour, J. Lucisano, and D. Gough, Application of chronic intravascular blood glucose sensor in dogs. *Diabetes* **39**, 1519–1526 (1990).
27. J. Wang and F. Lu, Oxygen-rich oxidase enzyme electrodes for operation in oxygen-free solutions. *J. Am. Chem. Soc.* **120**, 1048–1050 (1998).
28. E. D'Costa, J. Higgins, and A.P. Turner, Quinoprotein glucose dehydrogenase and its application in an amperometric glucose sensor. *Biosensors* **2**, 71–87 (1986).
29. R.A. Marcus and N. Sutin, Electron transfers in chemistry and biology. *Biochim. Biophys. Acta* **811**, 265–322 (1985).
30. I. Willner and E. Katz, Integration of layered redox proteins and conductive supports for bioelectronic applications. *Angew Chem. Int. Ed.* **39**, 1180–1218 (2000).
31. A.P.F. Turner, B. Chen, and S.A. Piletsky, In vitro diagnostics in diabetes: meeting the challenge. *Clinical Chemistry* **45**, 1596–1601 (1999).

32. J.D. Newman and A.P.F. Turner, Home blood glucose biosensors: a commercial prospective. *Biosensors and Bioelectronics* **20**, 2388–2403 (2005).
33. S. Wring and J. Hart, Chemically modified screen-printed carbon electrodes. *Analyst* **117**, 1281–1286 (1992).
34. C. Henry, Getting under the skin. *Anal. Chem.* **70**, 594A–598A (1998).
35. G.S. Wilson and R. Gifford, Biosensors for real-time in-vivo measurements. *Biosensors and Bioelectronics* **20**, 2388–2403 (2005).
36. M. Shichiri, Y. Yamasaki, N. Hakui, and H. Abe, Wearable artificial endocrine pancreas with needle-type glucose sensor. *Lancet* **2**, 1129–1131 (1982).
37. A. Albisser, B. Lebel, G. Ewart, Z. Davidovac, C. Botz, and W. Zingg, Clinical control of diabetes by artificial pancreas. *Diabetes* **23**, 397–404 (1974).
38. D. Schmidtke, A. Freeland, A. Heller, and R. Bonnecaze, Measurement and modeling of the transient difference between blood and subcutaneous glucose concentrations in the rat after injection of insulin. *Proc. Natl. Acad. Sci. U.S.A.* **95**, 294–299 (1998).
39. T.M. Gross, Efficacy and reliability of the continuous glucose monitoring system. *Diabetes Technol. Ther.* **2**, S19–S26 (2000).
40. B. Feldman, R. Brazg, S. Schwartz, and R. Weinstein, A continuous glucose sensor based on wired enzyme technology. *Diabetes Tech. Ther.* **5**, 769–779 (2003).
41. M. Tierney, H. Kim, J. Tamada, and R. Potts, Electroanalysis of glucose in transcutaneously extracted samples. *Electroanalysis* **12**, 666–671 (2000).

CHAPTER 4

New trends in ion-selective electrodes

Sergey Makarychev-Mikhailov, Alexey Shvarev, and Eric Bakker

4.1 INTRODUCTION

4.1.1 State-of-the-art

The history of ion-selective electrodes (ISEs) [1] starts from the discovery of the pH response of thin film glass membranes by Cremer in 1906, thus making ISEs the oldest class of chemical sensors. They still are superior over other sensor types in a variety of applications, including the biomedical, industrial, and environmental fields. The glass pH electrode is the most widespread sensor, being present in an arsenal of virtually any laboratory. Although the performance of the best glass and crystalline membrane sensors remains unsurpassed, the chemical versatility of these materials is limited, which imposes restrictions on the range of available analytes. During the last decades research and development of potentiometric sensors has shifted primarily towards the more versatile and tunable solvent polymeric membrane ISEs [2]. These sensors originated in the early 1960s, completely replaced older analytical methods in various biomedical applications, and gained a foothold in clinical chemistry. Most of the advantages of solid state ISEs, such as robustness and cost effectiveness, are still present in their polymeric counterparts, but the unique versatile matrix allows a significant increase in the number of available analytes. Indeed, the list of detectible analytes is magnificent, today approaching 100 species, a number hardly accessible with numerous other analytical techniques [3, 4]. The extraordinary measuring range of ISEs (in some cases exceeding eight orders of magnitude in concentration) is also an important intrinsic property of this analytical method. Note that all these advantages are complemented by their simplicity, a distinct characteristic of excellence for ISEs. The modern directions of solvent polymeric sensor research are the focus of the present chapter.

The great success of ISEs achieved during the past decades has not reverted the field into a stagnating mature discipline. There are several dynamically developing research directions that continue to tempest the analytical chemistry community, opening new horizons. Apart from the classical potentiometric response, new transduction schemes for the solvent polymeric membranes have been proposed, making the detection of new analytes accessible, allowing comprehensive instrumental control over sensing characteristics and introducing new detection principles [5, 6]. At the same time, work on a detailed theoretical description of the sensors has kept pace with the experimental progress, being a prerequisite for further development. The phase boundary potential (PBP) model was shown to be a very powerful tool [7] that not only allows one to quantify important sensor properties, but also inspires innovations in the field. The extremely low detection limits (DL), originally predicted by a detailed theoretical analysis of the underlying membrane processes [8], were discovered for the solvent polymeric ISEs [9] and revolutionized the field. More than ten analytes (and this number continues to grow) were reported so far to be detectable at the nanomolar concentration level and some of them even down to picomolar activities [10], which puts potentiometric sensors along with the most sensitive analytical methods.

Examination of the membranes with a variety of physicochemical techniques, from related electrochemical approaches (as electrochemical impedance spectroscopy (EIS), voltammetry and chronoamperometry) to more sophisticated characterization methods (spectroscopy and microscopy), actually serves the same end as the theory and leads to a deeper understanding of the chemistry behind the functioning of these sensors [5, 6].

Improvement and optimization of the characteristics of the existing sensors is an important work that is constantly been addressed by many research groups, and is briefly reviewed here. The rational molecular design principles become important to the search for new sensor materials to be more selective and sensitive, possess better detection limits, and to be more stable and have improved biocompatibility [11]. A small number of the most successful ionophores used in the polymeric ion-selective electrodes are depicted in Fig. 4.1. Significant progress has been achieved in the development of solid contact sensors, especially on those based on conducting polymers [12]. Such sensors can be easily miniaturized, which significantly simplifies the sensor fabrication process, decreases sample volume, enables *in-vivo* measurements and provides compatibility with microchip technologies and other advances [13]. Microsensors can also be incorporated into sensor arrays, capable of multianalyte detection and pattern recognition [14].

4.1.2 Most important biomedical applications of ion-selective electrodes

Clinical chemistry, particularly the determination of the biologically relevant electrolytes in physiological fluids, remains the key area of ISEs application [15], as billions of routine measurements with ISEs are performed each year all over the world [16]. The concentration ranges for the most important physiological ions detectable in blood fluids with polymeric ISEs are shown in Table 4.1. Sensors for pH and for ionized calcium, potassium and sodium are approved by the International Federation of Clinical Chemistry (IFCC) and implemented into commercially available clinical analyzers [17]. Moreover, magnesium, lithium, and chloride ions are also widely detected by corresponding ISEs in blood liquids, urine, hemodialysis solutions, and elsewhere. Sensors for the determination of physiologically relevant polyions (heparin and protamine), dissolved carbon dioxide, phosphates, and other blood analytes, intensively studied over the years, are on their way to replace less reliable and/or awkward analytical procedures for blood analysis (see below).

In contrast to other analytical methods, ion-selective electrodes respond to an ion activity, not concentration, which makes them especially attractive for clinical applications as health disorders are usually correlated to ion activity. While most ISEs are used *in vitro*, the possibility to perform measurements *in vivo* and continuously with implanted sensors could arm a physician with a valuable diagnostic tool. *In-vivo* detection is still a challenge, as sensors must meet two strict requirements: first, minimally perturb the *in-vivo* environment, which could be problematic due to injuries and inflammation often created by an implanted sensor and also due to leaching of sensing materials; second, the sensor must not be susceptible to this environment, and effects of protein adsorption, cell adhesion, and extraction of lipophilic species on a sensor response must be diminished [13]. Nevertheless, direct electrolyte measurements *in situ* in rabbit muscles and in a porcine beating heart were successfully performed with microfabricated sensor arrays [18].

The relative simplicity of the sensor setup allows them to be implemented into portable automated devices or bed-side analyzers (Fig. 4.2), which are easily installed at patient beds, eliminating the time-consuming laboratory analyses. On the other hand, modern high throughput clinical analyzers may process more than 1000 samples per hour and simultaneously determine dozens of analytes, using a handful of analytical methods. Blood electrolyte analysis, however, remains one of the most important in

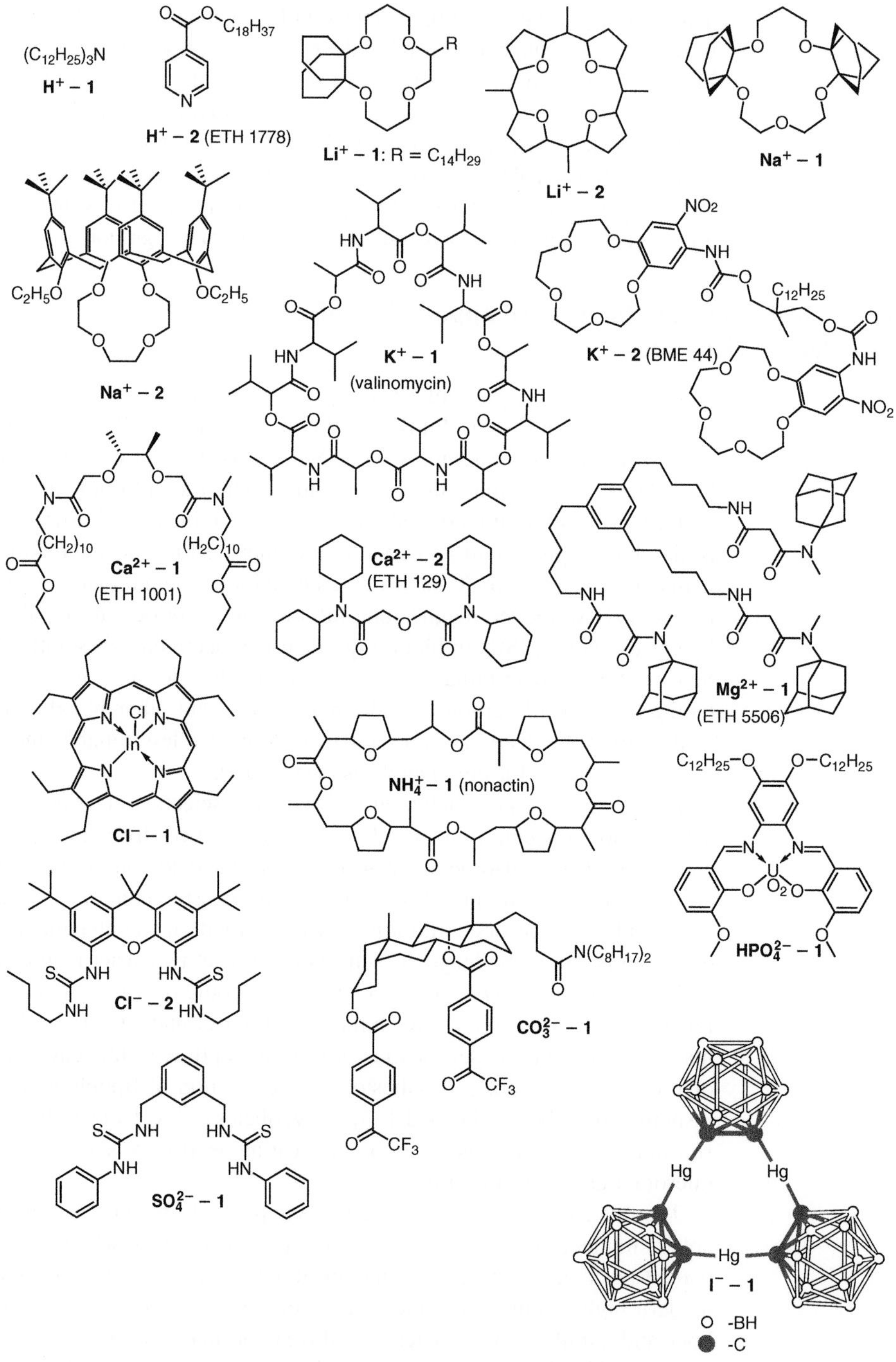

FIGURE 4.1 Structures of common ionophores used in ion-selective membranes.

TABLE 4.1
Concentration range of the most important ions in blood and compositions of the corresponding polymeric membrane sensors and their selectivities

Ion I	Concentration range, mM	Membrane composition	Selectivity, log K^{pot}_{IJ}
H+	pH: 7.35–7.45	Tri-*n*-dodecylamine, KTpClPB, PVC/DOS	Na^+ −10.4 K^+: −9.8 Ca^{2+}: <−11.1
Li^+	0.5–1.5[a]	7-tetradecyl-2,6,9,13-tetraoxatricyclo[12.4.4.0$^{1.14}$] docosane, KTpClPB, PVC/DBPA	Na^+: −3.1 K^+: −3.6 Ca^{2+}: <−5.0
Na^+	136–145	Calix[4]arenecrown-4 ionophore, KTpClPB, PVC/NPOE	Li^+: −2.8 K^+: −5.0 Mg^{2+}: −4.5 Ca^{2+}: −4.4
K^+	3.5–5.0	Valinomycin, NaTFPB, PVC/DOS	Na^+: −4.5 Mg^{2+}: −7.5 Ca^{2+}: −6.9
Ca^{2+}	2.2–2.6	N,N,N′,N′-tetracyclohexyl-3-oxapentanediamide (ETH 129), KTpClPB, PVC/NPOE	Na^+: −8.3 K^+: −10.1 Mg^{2+}: −9.3
Mg^{2+}	0.8–1.3	Tetraamide ionophore, KTpClPB, PVC/NPOE	Li^+: −3.7 Na^+: −3.2 K^+: −1.4 Ca^{2+}: −2.5
Cl^-	98–106	2,7-di-*tert*-butyl-9,9-dimethyl-4,5-xanthenediamine, TDDMACl, PVC/NPOE	Sal^-: +1.8 NO_3^-: +0.7 HCO_3^-: −2.6
CO_3^{2-}	35–40[b]	Tweezer-type carbonate ionophore, TDMACl, PVC/DOA	Cl^-: −6.0 Sal^-: −0.8 NO_3^-: −3.4
Phosphates	0.7–1.4	Uranyl salophene, PVC/NPOE	Cl^-: −2.5 NO_3^-: −1.7
Salicylate	1–2[c]	Sn(II) phthalocyanine	Cl^-: −4.8 OAc^-: −3.4

[a] For patients under treatment with lithium salts.
[b] Total carbon dioxide concentration.
[c] For patients under treatment with aspirin.

clinical chemistry and ion-selective electrodes are implemented in all commercially available clinical analyzers, large and small (Fig. 4.2). As the blood electrolyte activity usually varies in a narrow range, a 10–100 microvolt precision of the potential measurements is required, which is achieved by temperature-controlled flow-through cells and frequent automated recalibrations. The use of disposable probes or cartridges with precalibrated sensors is another option, simplifying analysis for an end-user.

Pharmaceutical analysis is another important area for ISEs [19]. A large number of drugs were reported to be detectable by ISEs in pharmaceutical formulations and during manufacturing processes. The concentration of drugs and their metabolites can be measured in real biological fluids. Although ISEs are currently not widely used in pharmaceutical chemistry, they have a great potential and a number of ISE applications have been developed in the last years [20]. Most drug-selective electrodes are based on ion exchangers and exploit commonly high lipophilicities of drugs and metabolites.

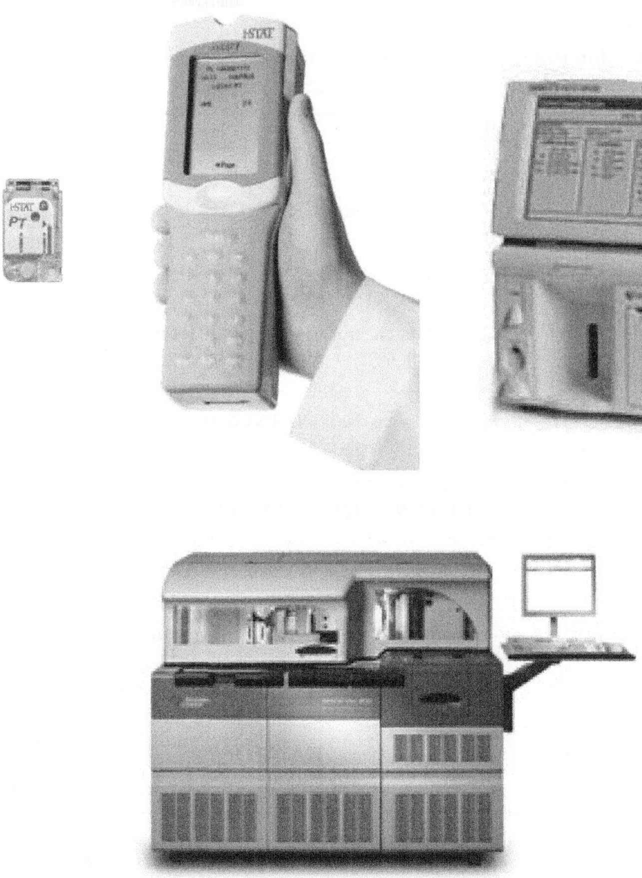

FIGURE 4.2 From top left, clockwise: hand held i-STAT clinical analyzer and cartridge from Abbott Laboratories, bench top GEM 3000 clinical analyzer from IL, Inc., and large mainframe clinical analyzer UniCel DxC800 from Beckman Coulter, Inc. All these systems utilize ion-selective electrodes to determine clinically relevant electrolytes.

Receptor-based sensors are still quite rare and much work is needed to implement ISEs into the general practice of pharmaceutical analysis.

Miniature sensors and microsensor arrays are capable of intracellular measurements, which are demanded by physiologists and biologists for the understanding of the processes taking place in live organisms [21]. The development of micro total analysis systems (μTAS) and lab-on-a-chip devices with integrated microsensors is a logical step for *in-vitro* measurement automation and miniaturization, which is especially important for biomedical applications [22]. Moreover, microfabricated sensors (Fig. 4.3) are highly demanded for *in-vivo* measurements in order to minimize potential injuries [13] and significant advances have been achieved in the field, still requiring further research and development.

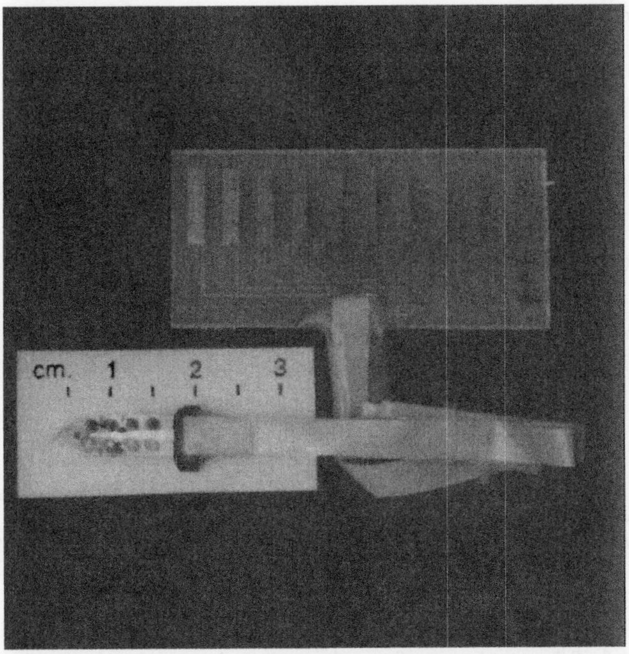

FIGURE 4.3 Miniaturized planar nine-site H^+/K^+ selective sensor array applied to *in-vivo* measurements in porcine cardiovascular system. (From [107].)

4.2 CLASSICAL ION-SELECTIVE ELECTRODES

4.2.1 Understanding of the operational principles

An established setup for measurements with ion-selective electrodes is depicted in Fig. 4.4. An ion-sensitive membrane is placed between two aqueous phases, i.e. a sample and an internal filling solution. A reference Ag/AgCl electrode is placed into the inner filling solution, which contains ions this electrode is responsive to. The external reference electrode is placed in the sample and usually has a salt bridge in order to prevent sample contamination. The membrane is an essential part of the sensor and is made of glass (oxide or chalcogenide), crystalline material (monocrystal or polycrystalline) or water-immiscible liquids (highly plasticized polymers, solvent impregnated porous films, etc.).

Potentiometric measurements are usually performed under zero-current conditions in a galvanic cell of the type:

Ag | AgCl | KCl_{sat} | 3 M KCl || sample solution || membrane || inner filling solution | AgCl | Ag

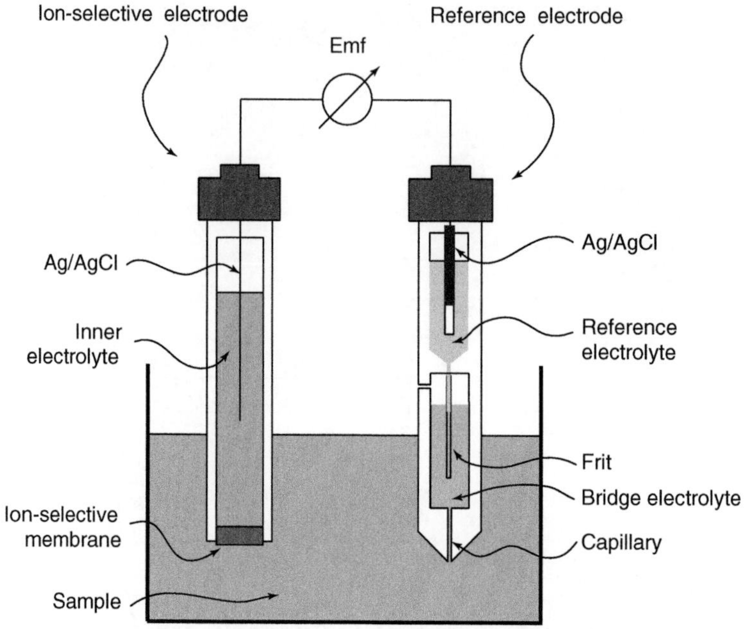

FIGURE 4.4 Schematic diagram of a membrane electrode measuring circuit and cell assembly.

The electromotive force (emf) across the cell is a sum of several potentials; many of them are sample independent:

$$\text{emf} = E_{\text{const}} + E_{\text{M}} + E_{\text{J}} \tag{1}$$

where E_{M} is the membrane potential and E_{J} is the liquid junction potential, formed at the sample–salt bridge interface. The latter is often considered constant and neglected, or it may be estimated according to the Henderson formalism.

Solvent polymeric membranes, conventionally prepared from a polymer that is highly plasticized with lipophilic organic esters or ethers, are the scope of the present chapter. Such membranes commonly contain various constituents such as an ionophore (or ion carrier), a highly selective complexing agent, and ionic additives (ion exchangers and lipophilic salts). The variety and chemical versatility of the available membrane components allow one to tune the membrane properties, ensuring the desired analytical characteristics.

The theory of ion-selective electrode response is well developed, due to the works of Eisenman, Buck and others [23]. Three models used for the description of the ISE response through the years, namely kinetic, membrane surface (or space charge) and phase boundary potential (PBP) models, although being seemingly contradictory, give similar results in most cases [7]. The first two sophisticated models are out of the scope of the present chapter, as the PBP model, despite its simplicity, satisfactorily explains most of the experimental results and thus has become widely applicable. The

PBP model considers the membrane potential as a sum of the potentials formed at the membrane–solution interfaces (phase boundary potentials), and generally neglects any diffusion potential within the membrane:

$$E_M = E_{const} + E_{PB} \tag{2}$$

where E_{PB} is the phase boundary potential at the membrane–sample interface, which can be derived from basic thermodynamic considerations:

$$E_{PB} = E^0 + \frac{RT}{zF} \ln \frac{a_I(aq)}{a_I(org)} \tag{3}$$

where R, T and F are, correspondingly, the universal gas constant, absolute temperature, and the Faraday constant, z is the ion charge and a_I(aq) and a_I(org) are the activities of the ion I in the aqueous and organic (membrane) phases. The standard potential, E^0, is a function of the standard chemical potentials in respective phases:

$$E^0 = \frac{\mu^0(aq) - \mu^0(org)}{zF} \tag{4}$$

If a_I(org) does not depend on a_I(aq) and is therefore constant, Eq. (3) reduces to the well-known Nernst equation:

$$E_I = E_I^0 + \frac{RT}{zF} \ln a_I(aq) \tag{5}$$

and ISE demonstrates a linear response to the logarithmic ion activity with the Nernstian slope of 59.2 mV decade^{-1} for $z = 1$ at 298 K (see Fig. 4.5). For Eq. (5) to hold, the respective ion activity in the membrane phase must be constant, which is usually ensured by the lipophilic ion exchanger present in the membrane. For neutral carrier-based membranes, the ion exchanger carries a charge opposite to the analyte ion, while for charged carrier sensors the same charge is normally required (Fig. 4.6) [24]. Oddly, in the early stages of polymeric sensors research, no ion exchangers were used and it was later found that intrinsic ionic impurities played the role of the ion exchanger in such membranes [25]. Nowadays added ion exchangers are strongly recommended, as they are also beneficial in terms of selectivity optimization and reduction of the membrane electrical resistance.

While ionophore-free membranes based on classical ion exchangers are still in use for the determination of lipophilic ions, such sensors often suffer from insufficient selectivity, as it is governed solely by the lipophilicity pattern of ions, also known for anions as the Hofmeister sequence. This pattern for cations is $Cs^+ > Ag^+ > K^+ > NH_4^+ > Na^+ > Li^+ > Ca^{2+} > Mg^{2+}$; and for anions: $ClO_4^- > SCN^- > I^- > Sal^- > NO_3^- > Br^- > NO_2^- > Cl^- > OAc^- \sim HCO_3^- > SO_4^{2-} > HPO_4^{2-}$. While the ion exchanger fixes the concentration of hydrophilic analyte ions in the membrane on the basis of the electroneutrality condition within the membrane, the second key membrane component is the ionophore that selectively binds to the analyte ions. The selectivity of

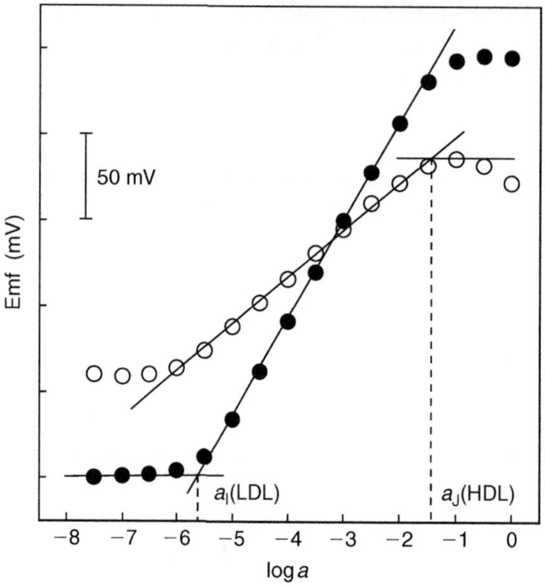

FIGURE 4.5 Typical ISE response curves to monovalent cations (solid circles) with a response slope of 59.2 mV decade^{-1} and to divalent cations (open circles) with a Nernstian slope of 29.6 mV decade^{-1}. Intercepts of the linear ranges of sensor responses define the lower and higher detection limits.

carrier-based sensors, thus, depends on the strength of the ionophore binding with primary ions relative to interfering ones and also on the difference in lipophilicities of the corresponding ions. As the primary and interfering ions may be of different charge and may form complexes of different stoichiometry, the variation of the ionophore–ion exchanger ratio in the membrane allows one to optimize the sensor selectivity to some extent [26]. The selectivity, being an essential sensor characteristic, was traditionally described by the extended Nernst equation, called the Nikolski–Eisenman equation:

$$E = E_I^0 + \frac{RT}{z_I F} \ln\left(a_I(\text{aq}) + K_{IJ}^{\text{pot}} a_J(\text{aq})^{z_I/z_J}\right) \tag{6}$$

where K_{IJ}^{pot} is the potentiometric selectivity coefficient. It was later shown that this semiempirical equation is not valid for $z_I \neq z_J$ in the activity range where both primary I and interfering J ions significantly contribute to the potential. More complex explicit equations were derived for each specific case of z_I and z_J, which are beyond the scope of this chapter [27]. Nonetheless, the Nikolski selectivity coefficient remains an excellent measure of ion interference as it describes the selectivity of the membrane in terms of the equilibrium constant of the exchange reaction of primary and interfering ions between the organic and aqueous phases (Fig. 4.6).

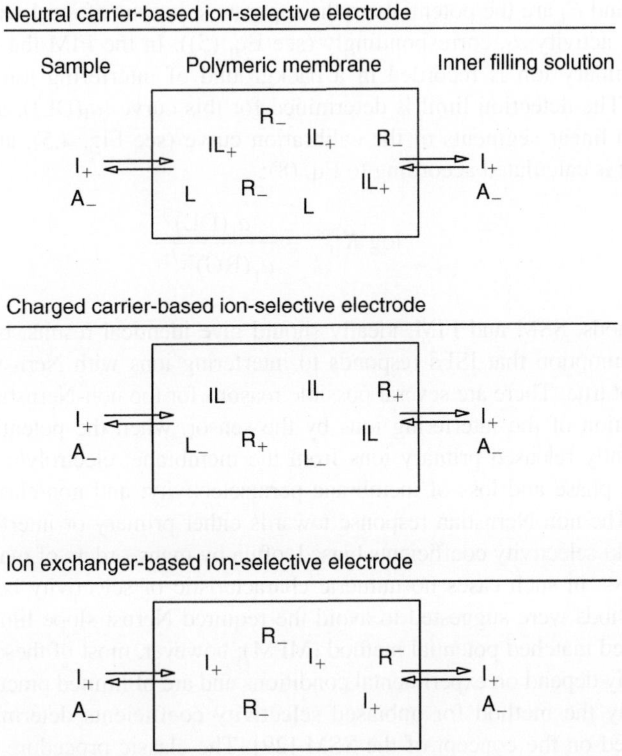

FIGURE 4.6 Schematic view of the equilibria between sample, ion-selective membrane, and inner filling solution for three important classes of solvent polymeric ion-selective membranes. Top: electrically neutral carrier (L) and lipophilic cation exchanger (R^-); center: charged carrier (L^-) and anion exchanger (R^+); and bottom: cation exchanger (R^-).

4.2.2 Response characteristics: selectivity and detection limits

Sensor selectivity improvement is often desired for specific applications. While the progress in the design of new highly selective ionophores for ISEs is discussed below, here we are focused on the selectivity as a sensor characteristic, methods of its determination, and the general ways of its enhancement. Unfortunately, there are large discrepancies in selectivity data, published over decades, mainly due to inappropriate methods used for selectivity determination. IUPAC recommended two procedures for selectivity determination as the separate solution method (SSM) and fixed interference method (FIM) [28]. In the former the potential measurements are performed in two separate solutions, each containing the salt of the primary or interfering ion only. The Nikolski selectivity coefficient is then calculated according to the following equation:

$$\log K_{\mathrm{IJ}}^{\mathrm{pot}} = \frac{z_{\mathrm{I}} F(E_{\mathrm{J}} - E_{\mathrm{I}})}{2.303RT} - \log\left(\frac{a_{\mathrm{I}}}{a_{\mathrm{J}}^{z_{\mathrm{I}}/z_{\mathrm{J}}}}\right) \qquad (7)$$

where E_I and E_J are the potentials in the separate solutions of ion I with activity a_I and ion J with activity a_J, correspondingly (see Eq. (5)). In the FIM the calibration curve for the primary ion is recorded in a background of interfering ion at fixed activity (a_J(BG)). The detection limit is determined for this curve (a_I(DL)), as a cross-section of the two linear segments of the calibration curve (see Fig. 4.5), and the selectivity coefficient is calculated according to Eq. (8):

$$\log K_{IJ}^{pot} = \frac{a_I(DL)}{a_J(BG)^{z_I/z_J}} \tag{8}$$

Both methods, SSM and FIM, ideally should give identical results, but they both rely on the assumption that ISEs responds to interfering ions with Nernstian slope, which often is not true. There are several possible reasons for the non-Nernstian behavior: high discrimination of the interfering ions by the sensor, when the potential is dictated by the constantly released primary ions from the membrane; electrolyte coextraction into membrane phase and loss of membrane permselectivity; and non-classical response of a sensor. The non-Nernstian response towards either primary or interfering ions makes the Nikolski selectivity coefficients biased, often by many orders of magnitude, and thus meaningless; in such cases no numeric characteristic of selectivity becomes available. Other methods were suggested to avoid the required Nernst slope limitation, including the so-called matched potential method (MPM); however, most of these empirical methods strongly depend on experimental conditions and are of limited practical significance.

Recently the method for unbiased selectivity coefficients determination was proposed based on the concept of the SSM [29]. The classic procedure for ISE preparation involves sensor conditioning in a *primary* ion solution prior to use to ensure stable and reproducible behavior. Furthermore, the sensor inner filling solution often contains relatively high concentrations of the *primary* ion as well. These issues may prevent sensors showing Nernstian response slopes to *interfering* ions due to the release of the highly preferred ions into the sample–membrane interfacial layer, where these ions dictate partly or completely the electrode response. However, if fresh sensors, never having had contact with preferred ions, are used for calibrations towards interfering ions, a Nernstian sensitivity can be achieved even for strongly discriminated ions. In order to observe such response slopes the membranes are prepared with the ion exchanger in the form of one of the interfering ions and sensor conditioning and inner filling solutions are also free of the primary ions. Once the Nernstian response curves for a series of discriminated ions are recorded, which fulfills the requirement for the Nikolski selectivity coefficients, the sensor is reconditioned in a primary ion solution and the conventional calibration is performed. The calculation of the unbiased selectivity coefficients is then made according to Eq. (7). The proposed method allows one to significantly widen the applicability of selectivity coefficients and, which is also very important, provides thermodynamically meaningful sensor characteristics. Fortunately, the unbiased true selectivity coefficients often have much more favorable values for the same sensor compositions. This fact calls for the revision and reconsideration of many previously proposed ionophores.

There are two main factors that influence the selectivity of a sensor: limits in discrimination of an interfering ion and upper limits in stability constant of an analyte–ionophore complex. While an ideal ionophore does not form complexes with interfering ions, too strong complexation with the primary ion leads to a massive extraction of analyte into membrane phase coupled with a coextraction of sample counter-ions, known as Donnan exclusion failure. In such cases, at high activities and lipophilicities of sample electrolytes, $a_I(\text{org})$ increases and a breakdown of membrane permselectivity prevents the Nernst equation to hold.

Often, however, the PBP model can be used to describe the influence of key membrane parameters on the selectivity with a simplified equation after making certain assumptions. It is here given for cation-selective electrodes based on neutral carriers [30]:

$$K_{IJ}^{pot} = K_{IJ} \frac{(\beta_{JLnJ})^{z_I/z_J}}{\beta_{ILnI}} \frac{R_T}{z_I\left[L_T - n_I(R_T/z_I)\right]^{n_I}} \left(\frac{z_J\left[L_T - n_J(R_T/z_J)\right]^{n_J}}{R_T} \right)^{z_I/z_J} \tag{9}$$

where K_{IJ} is the equilibrium constant for the ion exchange between uncomplexed primary and interfering ions between the sample and organic phase, β_{ILnI} and β_{JLnJ} are the overall complex formation constants for ions I and J with stoichiometric factors n_I and n_J, L_T and R_T are the total membrane concentrations of ionophore and cation exchanger, correspondingly. A similar equation may be derived for charged-carrier-based ISEs (not shown). As the membrane selectivity depends on the concentrations of the carrier and ionic sites, the modifying effect of the ionophore–ion exchanger ratio can be predicted, which allows one to tune the selectivity of the sensor when the primary and interfering ions are of different charge and/or form complexes with ionophores of different stoichiometry. Although ionophores in some cases may simultaneously form complexes of different stoichiometry, making calculations of the optimal ratio much more complicated, this approach remains useful. For the simplest case where the primary and interfering ions are of the same charge and form complexes of the same stoichiometry, the selectivity coefficient does not depend on ionophore and ion exchanger concentration and the selectivity coefficient reduces to the equilibrium constant, which can be expressed by:

$$K_{IJ}^{pot} = K_{IJ} \frac{\beta_{JLnJ}}{\beta_{ILnI}} \tag{10}$$

The selectivity here is directly proportional to complex formation constants and can be estimated, once the latter are known. Several methods are now available for determination of the complex formation constants and stoichiometry factors in solvent polymeric membranes, and probably the most elegant one is the so-called sandwich membrane method [31]. Two membrane segments of different known compositions are placed into contact, which leads to a concentration polarized sensing membrane, which is measured by means of potentiometry. The power of this method is not limited to complex formation studies, but also allows one to quantify ion pairing, diffusion, and coextraction processes as well as estimation of ionic membrane impurity concentrations.

Aside from the factors accounted for in the simplified Eq. (9), a deeper consideration of the membrane selectivity reveals other influential parameters. The membrane polarity, for example, which depends mainly on the nature of membrane plasticizer, may give a selectivity modifying influence because of the improved solvation of high valence ions by more polar media.

It is the selectivity that generally determines the lower detection limit (DL) of a sensor, as background interferences may influence the electrode potential at low activities of analyte. As the true selectivities of the ISEs were determined to be much better than earlier believed, the detection limits could also be significantly improved. For a long time the sensitivity range of polymeric membrane ISEs was thought to be limited by several orders of magnitude, from about one molar down to micromolar, which is already uniquely wide for an analytical method. As mentioned above, the sensor detection limit is determined by the activity at the cross-section of the two linear segments of the calibration curve (Fig. 4.5). Note, this IUPAC definition for ISEs [28] differs from that of other analytical methods (analyte concentration at which the signal is increased relative to the background level by three times the standard deviation of the noise), which may cause some confusion. Indeed, taking into account the low noise in potentiometric measurements, the traditional analytical DLs of ISEs are orders of magnitude lower [10].

The main limitation of the lower DL was discovered to be caused by the diffusion of the primary ion from the inner filling solution (where the concentration of the ions was traditionally high) through the membrane, which leads to enrichment of the analyte ions at the membrane–sample interface if the sample solution is more dilute. Under such conditions further dilution of the sample does not change the phase boundary potential, determined mostly by the leached primary ions in the aqueous diffusion layer. Control over the ion fluxes in membrane and their minimization made it possible to significantly reduce the sensor DLs, and this led to a real breakthrough for the entire field of ion-selective electrodes. Many sensor parameters were varied in order to minimize the fluxes: using inner-filling solution with primary ions at low activities, buffered by chelating agents or ion exchange resins; reducing ion diffusion coefficients in membrane by loading it with high polymer content or by covalent attachment of the ionophore to the polymeric matrix; using sample stirring or rotating electrodes, or wall-jet techniques to reduce the Nernstian diffusion layer in the aqueous part of the outer interface and so on. The use of a monolithic capillary filled with membrane cocktail as a matrix for ISE is another very recent approach to suppress ionic fluxes in the membrane and to achieve very low DLs (see Fig. 4.7). The response of such sensors does not depend on the composition of the inner filling solution [32]. Complete elimination of the inner filling solution and development of ISEs with a solid inner contact is another successful approach to low DL sensors. Conducting polymers are used as an intermediate layer between an electron-conducting substrate and ion-conducting sensor membrane. Different polymers and sensor manufacture procedures were described to improve the sensor characteristics and although not all experimental challenges have been overcome so far, the concept offers new prospects to highly sensitive electrodes, mainly due to its universality and applicability to different ion sensors.

So far ions as Na^+, K^+, NH_4^+, Ca^{2+}, Ag^+, Cd^{2+}, Cu^{2+}, Pb^{2+}, Vitamin B1, ClO_4^- and I^- are detectable in the range of 10^{-8}–10^{-11} M. One of the challenges of the day is to

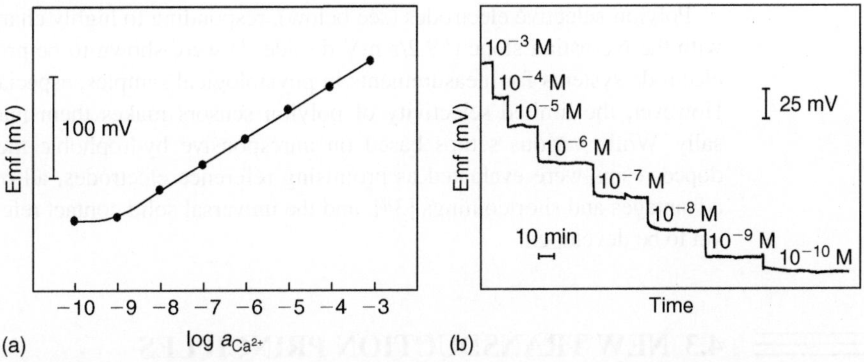

FIGURE 4.7 Calibration curves for Ca^{2+}-selective membrane electrodes made from a monolith of low porosity (i.d., 200 μm, no PVC) with inner filling solution of 0.1 M $CaCl_2$ (a) and time response of the monolithic Ca^{2+}-ISE (b). (Figures adapted from [32].)

extend the list of the available low DL sensors, covering the most important ions, including poisonous metal cations and inorganic anions, which are still poorly detectable by ISEs. Advances of the modern PBP model allow one to predict various influences on the DL of the sensor even in the presence of ion fluxes in the membrane. On the basis of this powerful theory a rational design of the low DL sensors becomes possible. However, a unified and simplified procedure for sensor manufacturing and a universal experimental protocol for low activity measurements still have to be developed. Rapid response times, high long-term stabilities and sufficient chemical ruggedness are key requirements for the low DL sensors. Miniaturization of such sensors in order to perform measurements in small sample volumes is another crucial direction of current research.

4.2.3 Reference electrodes

The proper reference electrode, completing the electrochemical cell, is always required for accurate and reproducible measurements with ISEs. In contrast to tremendous progress in solvent polymeric membrane sensors, much less attention was paid during the last decade to reference electrodes. The reference elements usually contain calomel (Hg/Hg_2Cl_2) or more often silver/silver chloride (Ag/AgCl) electrodes in contact with a concentrated KCl solution. The reference half-cell is connected to the sample via an electrolyte bridge (Fig. 4.4), filled with an equitransferent electrolyte (LiOAc, NH_4NO_3, KCl), which provides a diminished diffusion potential and helps to avoid sample contamination. The liquid junction potential, formed at the sample–bridge interface, can be kept reasonably small under defined conditions or is accessible to calculations according to the Henderson formalism. Unfortunately, the liquid junction, being an essential part of the reference system, is difficult to miniaturize and imposes serious restrictions on microsensor and sensor array development. Although several potentiometric methods eliminate the use of a reference electrode, the problem of solid contact (liquid junction-free), miniature, and robust reference is of a great importance.

Polyion-selective electrodes (see below), responding to highly charged ions ($z \geqslant 60$) with the Nernstian slope ($59.2/z$ mV decade^{-1}) were shown to be promising reference electrode systems for measurements in physiological samples, especially in blood [33]. However, the limited selectivity of polyion sensors makes them inapplicable universally. While various setups based on unresponsive hydrophobic membranes or salt-doped resins were evaluated as promising reference electrodes, all of them have both advantages and shortcomings [34], and the universal solid contact reference electrode is yet to be developed.

4.3 NEW TRANSDUCTION PRINCIPLES

4.3.1 Polyion-selective electrodes

The possibility of detecting polyionic macromolecules added a new thrust to the area of ion-selective electrodes in the past decade. Professors Ma and Meyerhoff from the University of Michigan described in their pioneering work [35] the first polymeric membrane electrodes that respond to the polyanion heparin.

Heparin is a highly sulfated polysaccharide (copolymer of uronic/iduronic acids alternating with sulfated glucosamine residues) with an average charge of -70 and an average molecular mass of 15 000 Da. It is the most commonly used anticoagulant during cardiac or vascular surgery. A typical fragment of heparin molecule is depicted in Fig. 4.8. The anticoagulant activity of heparin is attributed to its interaction with antithrombin III, a serine protease inhibitor, causing the formation of heparin–antithrombin complexes, which drastically increase the inhibition ability of antithrombin on enzymes involved in the coagulation process.

More than 1 trillion units are administered annually to approximately 12 million patients in the United States. Real-time monitoring of heparin concentration in blood prevents the risk of possible bleeding and reduces postoperative complications. The activated clotting time measurement (ACT) is a widespread method for the determination of the heparin concentration in whole blood. Although this method is fast and simple, it is non-specific and indirect, and the results can be affected by many factors, such as hemodilution and hypothermia [36]. Alternative methods such as chromogenic heparin assays based on factor Xa inhibition, although widely used in clinical laboratories, cannot be performed during the surgery with whole blood samples. In contrast to other methods, the heparin-selective electrode was able to detect heparin concentration directly in whole blood or plasma samples.

One may think that the idea of detecting ionic compounds such as heparin using polymeric ion-selective electrodes seems very difficult due to the high charge of polyionic molecules, which makes the slope of the electrode function negligibly small for an analytical application. Indeed, for heparin-selective electrodes the theoretical slope is less than 1 mV decade^{-1} and the potential practically does not depend on heparin concentration, which means that this ISE can be useful as a reference electrode [33]. Nonetheless, Ma and Meyerhoff noticed that the potential of polymeric membrane

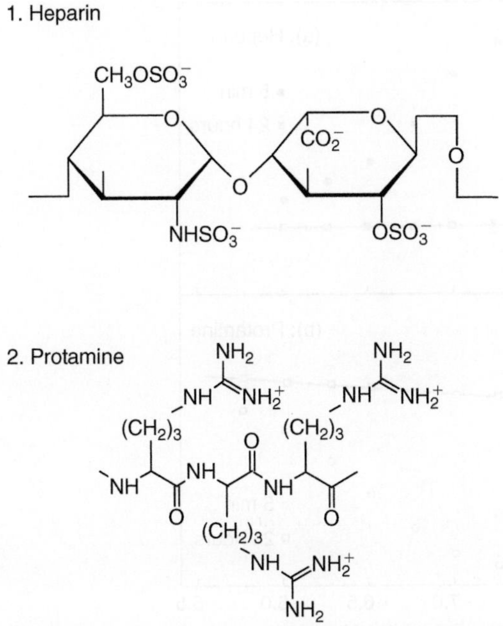

FIGURE 4.8 Structures of representative fragments of heparin (1) and protamine (2) molecules.

electrode doped with TDMACl (tridecylmethylammonium chloride) or Aliquat 336 (tricaprylmethylammonium chloride) exhibited significant potentiometric anion response when heparin was added to 0.15 M NaCl (Fig. 4.9a) [35, 37].

Soon after the initial development of the heparin sensor, an electrode for the detection of the polycation protamine was proposed [38] based on a polymeric membrane doped with the cation exchanger tetrakis-(4-chlorophenyl)borate. Protamine is a polypeptide and usually administered as a heparin antidote. Protamine is a polycation with an average charge of +20 and is rich in arginine (Fig. 4.8). The response function of protamine-selective electrodes is similar to the heparin response function (Fig. 4.9b).

In contrast to conventional ISEs the resulting polyion responses yielded non-linear S-shaped calibration curves. After prolonged exposure to the solutions containing polyions both electrodes lost their response function demonstrating negligible slope close to the calculated according Nernst equation (Fig. 4.9). In order to explain this unusual non-classical response function a theoretical non-equilibrium model was proposed [39]. Briefly, the membrane initially acts as an ion exchanger-based sodium electrode and a Nernstian response slope to sodium ions is expected when protamine is absent. Once a low concentration (ca. 1 μM) of protamine is present in the sample, it is preferably extracted into the membrane owing to the ion exchanger, which stabilizes the polyion charge not unlike that in a reverse micelle. This extraction forces the sodium ions initially present in the membrane to be expelled by ion exchange. The amount of sodium expelled depends directly on the mass transport kinetics of the

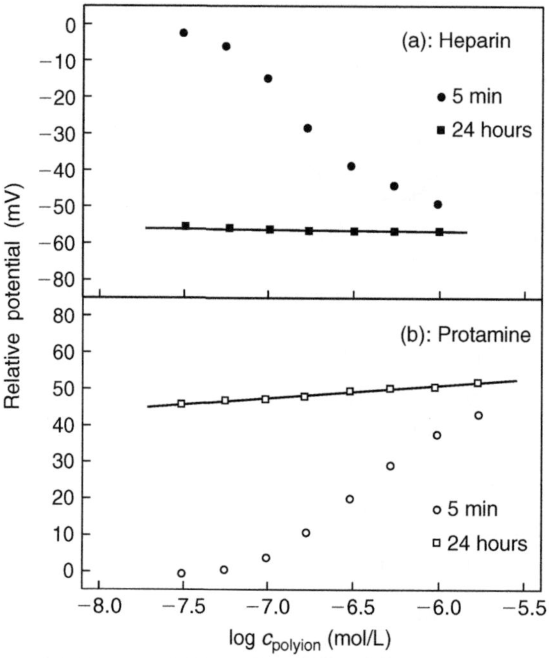

FIGURE 4.9 Potentiometric response of heparin (a) and protamine (b) selective electrodes after different 5 min and 24 h equilibration times, respectively. Sample solutions contained 0.1 M NaCl and 50 mM TRIS (pH 7.40) [39].

protamine to the membrane, and hence on its sample concentration. The approximate equation describing the electrode potential in the presence of protamine can be derived on the basis of a quasi-steady-state flux consideration [39]:

$$E_{PB} = E^0 + \frac{RT}{F} \ln \frac{a_{Na}}{R_T - z_{PA}(D_{aq,PA}\delta_m / D_{m,PA}\delta_{aq})c_{PA,bulk}} \tag{11}$$

where z_{PA} is the charge of the polyion (for protamine, ca. +21), $D_{m,PA}$, $D_{aq,PA}$, δ_m, and δ_{aq} are the diffusion coefficients of protamine in the membrane and aqueous phases and the resulting diffusion layer thicknesses, respectively. Since δ_m increases with time until it reaches the membrane thickness, which may take many hours to accomplish, sensors based on this mechanism do not operate in a reversible manner and are not suited for continuous monitoring purposes. Nonetheless, a range of important applications have emerged with this technology.

Meyerhoff's group developed disposable heparin-selective electrodes based on "coated wire" technology [40]. The electrodes were successfully used as end-point indicators for the detection of heparin concentration via protamine titration in whole blood samples. During clinical trials 44 whole blood specimens were obtained from eight patients undergoing open heart surgery [41]. The results at different stages of surgery were in good agreement with those detected by ACT (Hepcon HMS System).

The protamine ion-selective electrode used as an end-point detector created another possibility for the determination of heparin concentration via titration with protamine, utilizing the specific heparin–protamine interaction. Ramamurthy *et al.* [42] reported on an improved protamine-selective electrode. The polymeric membrane was doped with the ion exchanger dinonylnaphthalenesulfonate (DNNS), which demonstrated better performance in comparison to previously used tetrakis-(4-chlorophenyl)borate [38]. The electrode exhibited significant non-equilibrium potentiometric response to protamine over the concentration range of 0.5–20 mg l^{-1} in undiluted whole blood samples. Whole blood heparin concentrations determined by titration ($n > 157$) correlated well with other methods, including ACT (the Hepcon HMS) ($r = 0.934$) and a previously reported potentiometric heparin sensor-based method [41] ($r = 0.973$). Reasonable correlation was found with a commercial chromogenic anti-Xa heparin assay ($r = 0.891$) with corresponding plasma samples and appropriate correction for hematocrit levels.

Titration with protamine using protamine-selective electrode as an end-point detector was also shown to be a reliable methodology for the detection of low-molecular-weight heparins (LMWH) in whole blood samples at concentrations up to 2 U mL^{-1} [43] (measured heparin activity is in units per milliliter according to the US Pharmacopeia standard). These results were especially important because commercially available LMWH samples are very heterogeneous, since they are prepared using different methods such as chemical and enzymatic hydrolysis of unfractionated heparin (UFH). Results of determination of LMWH biological activity in whole blood are normally not reliable because ACT assays are not always sensitive toward LMWHs [44]. The anti-Xa assay, the most adopted methodology for detection of LMWH, can be performed only with plasma samples and thus cannot be used as a point-of-care method.

Among other polyions Meyerhoff's group demonstrated successful detection of pentosan polysulfate (PPS) [45], an anti-osteoarthritis drug, in plasma and polyanions with high phosphate content, including polyphosphates and nucleic acids in saline solutions [46].

Due to their response mechanism the polyion-selective electrodes are not sensitive to the small fragments of polyionic macromolecules. Thus, if an enzyme cleaves the polyionic molecule these sensors can be used for detection of enzyme activity. Polycation protamine is rich in arginine residues that make it a suitable substrate for protease-sensitive electrochemical assays. Real-time detection of trypsine activity was demonstrated with the protamine-selective electrode as a detector [38].

Protamine is also known as a specific substrate for plasmin, the proteolytic enzyme produced by the trombolytic cascade [47]. It was demonstrated that the protamine-selective electrode can be used as a sensor for the detection of thrombolytic agents such as urokinase in human plasma [48]. Urokinase or plasminogen alone being added to the protamine solution did not produce a change in electrode potential. However, a significant decrease in the potential was observed when both urokinase and plasminogen were added to the protamine solution, suggesting that the conversion of plasminogen to plasmin by urokinase occurred followed by the digestion of protamine by plasmin.

This same approach was used for assaying other protease activities such as chymotrypsin and renin [49]. Chymotrypsin and renin cleave only at specific sites of a polypeptide chain, the carboxylic side of lipophilic amino acids and between sequential Leu-Leu residues, respectively. The researchers used specifically synthesized polyionic oligopeptides containing different di- and tripeptide fragments in order to recognize activities of a specific protease.

Unfortunately, in the presence of detectable polyions in the solution a strong potential drift is normally observed due to the instability of the ion concentration gradients. Moreover, the main disadvantage of polyion-selective potentiometric electrodes lies in the intrinsic irreversibility of the underlying response mechanism. The target polyions eventually displace the counter-ions in the membrane phase and consequently the sensor loses its response.

Extracted polyions may be removed from the membrane phase by reconditioning of the sensor in concentrated (2 M) NaCl, for instance. More recently, a pH cross-sensitive potentiometric heparin sensor was proposed, which contained the ion exchanger and a charged H^+ ionophore [50]. Heparin stripping could now be accomplished by adjusting the pH of the sample. Unfortunately, both methods require prolonged reconditioning of the sensor (15 min) in the stripping solution after each measurement.

Another approach described above, which was employed by Meyerhoff's group [41], requires a set of identical electrodes, in which every electrode is disposed after a single measurement. A typical heparin–protamine titration requires on the order of 10–12 disposable electrodes.

4.3.2 Galvanostatically controlled sensors

The disadvantages described above in terms of the irreversibility of the polyion response stimulated further research efforts in the area of polyion-selective sensors. Recently, a new detection technique was proposed utilizing electrochemically controlled, reversible ion extraction into polymeric membranes in an alternating galvanostatic/potentiostatic mode [51]. The solvent polymeric membrane of this novel class of sensors contained a highly lipophilic electrolyte and, therefore, did not possess ion exchange properties in contrast to potentiometric polyion electrodes. Indeed, the process of ion extraction was here induced electrochemically by applying a constant current pulse.

The experimental setup included a three-electrode electrochemical cell with a liquid contact membrane electrode in which the internal Ag/AgCl electrode acted as a working electrode connected to a potentiostat/galvanostat. The instrument was capable of switching rapidly between potentiostatic and galvanostatic modes [51].

Let us consider, for instance, the response mechanism of a polycation-selective galvanostatically controlled sensor. The polymeric membrane is in contact with a NaCl solution. The membrane of the sensor is formulated with a lipophilic salt, for instance, tetradodecylammonium dinonylnaphthalenesulfonate (TDDA–DNNS), which has a relatively high affinity to protamine. Even though protamine is presented in the sample, spontaneous extraction does not take place due to the high lipophilicity of TDDA–DNNS, thus the initial concentration of protamine or sodium cations in the membrane is close to zero.

The measurement cycle of the sensor begins with a pulse of cathodic current i that induces a net flux J of cations in the direction of the membrane phase. Assuming for simplicity that only sodium and protamine ions may be extracted into the membrane phase, the relationship between current i is a sum of the fluxes of sodium, J_{Na}, and protamine, J_{PA}:

$$i = FA(J_{Na} + z_{PA}J_{PA}) \tag{12}$$

where A is the exposed membrane area. Equations for the sensor response function can be derived on the basis of a steady-state diffusion model similar to that employed for the theoretical description of the potentiometric polyion-selective electrode [52]. If no protamine is present in the solution an equation resembling the Nernst equation can be obtained, but when protamine is present in the sample it will compete with sodium in the extraction process (Eq. (12)). Assuming that the applied current imposes a flux that is always larger than the flux that can be sustained by polycation diffusion alone and that electromigration is neglected, the sensor response function can be written as follows [51]:

$$E_{PB} = E^0 + \frac{RT}{F}\ln\left(\frac{a_{Na}}{\frac{\delta_m}{D_{m,Na}}\left(-\frac{i}{FA} - z_{PA}\frac{D_{aq,PA}}{\delta_{aq}}c_{PA,bulk}\right)}\right) \tag{13}$$

In comparison to Eq. (11) if the activity of sodium ions is constant for a current pulse of fixed duration and magnitude the phase boundary potential is a function of protamine concentration in the sample.

A baseline potential pulse followed each current pulse in order to strip extracted ions from the membrane phase and, therefore, regenerated the membrane, making it ready for the next measurement pulse. This made sure that the potentials are sampled at discrete times within a pulse that correspond to a δ_m that is reproducible from pulse to pulse. This made it possible to yield a reproducible sensor on the basis of a chemically irreversible reaction. It was shown that the duration of the stripping period has to be at least ten times longer than the current pulse [53]. Moreover the value of the baseline (stripping) potential must be equal to the equilibrium open-circuit potential of the membrane electrode, as demonstrated in [52]. This open-circuit potential can be measured prior to the experiment with respect to the reference electrode.

Figure 4.10 shows the applied current (upper plot) and resulting potential (bottom plot) for a protamine-selective sensor [54]. Upon application of current the observed potential decreased, indicating the arising diffusion gradient of ions in the membrane. Sampled potentials that represent the sensor response were obtained as the average value during the last 100 ms of each current pulse. The observed current during the following baseline potential pulse continuously decreased to zero and was indicative of ions diffusing back from the membrane into the sample.

This method was further modified to eliminate any excessive iR drop across the ion-selective membrane when the current was applied. The modified setup (termed pulstrodes) allows one to set the current to zero after the current pulse and to measure the potential, which in this mode does not contain any undesirable iR component [55].

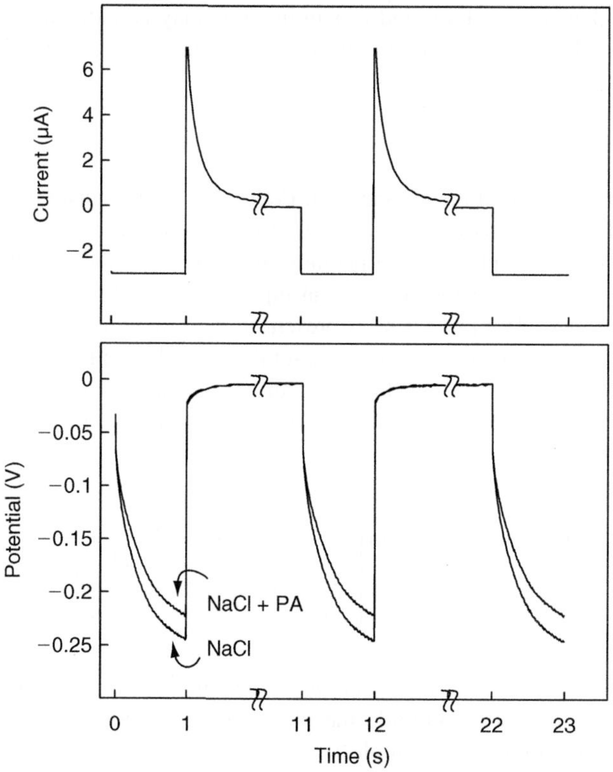

FIGURE 4.10 Current–time traces (top) and potential–time traces (bottom) for the pulsed galvanostatic measurement of 10 mg l^{-1} protamine in aqueous sample containing 0.1 M NaCl and 50 mM TRIS (pH 7.40) [54]. An applied cathodic current of $-3\,\mu A$ leads to the extraction of protamine into the membrane, and the observed potential is significantly different for samples with and without protamine (bottom). The membrane is renewed potentiostatically at 0 V for 10 sec before the next current pulse.

The behavior of potentiometric and pulsed galvanostatic polyion sensors can be directly compared. Figure 4.11 shows the time trace for the resulting protamine calibration curve in 0.1 M NaCl, obtained with this method (a) and with a potentiometric protamine membrane electrode (b) analogous to that described in [42, 43]. Because of the effective renewal of the electrode surface between measuring pulses, the polyion response in (a) is free of any potential drift, and the signal fully returns to baseline after the calibration run. In contrast, the response of the potentiometric protamine electrode (b) exhibits very strong potential drifts.

Galvanostatically pulsed sensors can be employed for heparin determination via titration with protamine using protocol described earlier [42, 43] and initial experiments showed that heparin detection in whole blood samples can be accomplished with this technology.

This method can be applied not only for polyion detection but for the detection of small ions as well. In contrast to potentiometric electrodes the external control of the

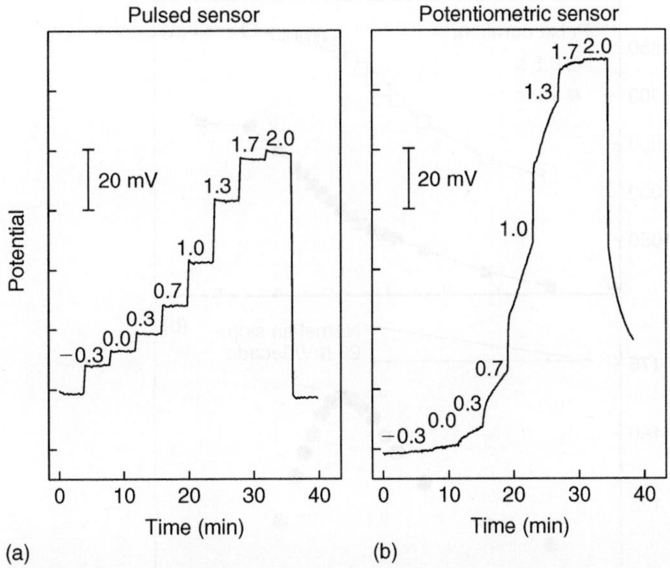

FIGURE 4.11 Amplitude–time behavior of potential during calibration in 0.1 M NaCl solution containing 50 mM TRIS (pH 7.40) with a pulsed galvanostatic sensor (a) and a classical potentiometric protamine-selective electrode (b) [54]. Logarithmic protamine concentrations (mg l^{-1}) are indicated on the traces. The first and last samples did not contain protamine.

applied current gives an additional degree of freedom. The tunability of the selectivity by the magnitude and sign of the applied current and the capability of such sensors to monitor more than one analyte at the same time was demonstrated [51]. Electrochemical control of ion fluxes allows one to detect simultaneously with a single sensor the calcium activity and total concentration of complexed calcium [56].

Perhaps one of the most intriguing applications of this methodology is the detection of ions within the so-called super-Nernstian response region in a defined range of ion activity [55]. The response function of a Ca-selective galvanostatic sensor has a near-Nernstian response slope at relatively high activities of calcium. Indeed, at lower calcium activities a drastic potential change occurs with a slope much higher than Nernstian (Fig. 4.12a). The origin of this behavior lies in a concentration polarization at the phase boundary under an applied constant current pulse, which is caused by the depletion of the aqueous diffusion layer adjacent to the membrane [51]. There is a complete analogy with similar responses of traditional ISEs where the super-Nernstian behavior is due to the uptake of primary ions into membrane phase [57]. However, in traditional ISEs there are a variety of parameters that may influence this response function, including the activities of interfering ions of the sample and inner solution, the concentrations of ionophore and lipophilic ion exchanger in the membrane, the thickness of the Nernstian diffusion layers and the diffusion coefficients in both phases [58]. As full control over these parameters is not possible, a poor reproducibility and significant drift of the sensor

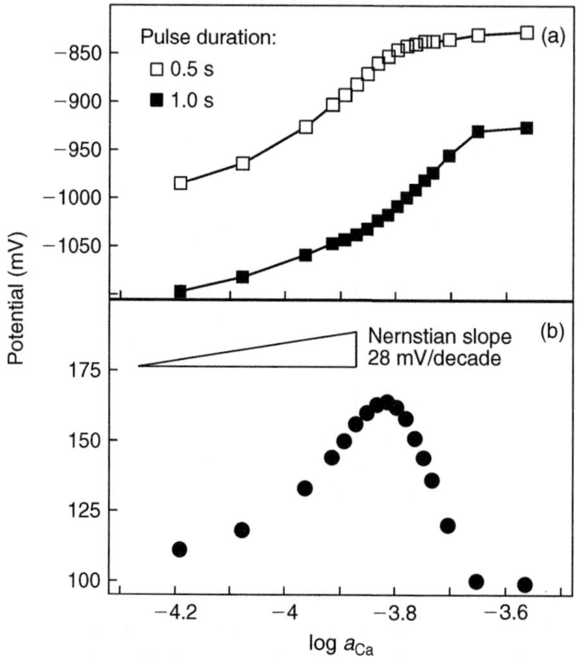

FIGURE 4.12 Pulsed chronopotentiometric (a) and differential (b) response of a calcium pulstrode in artificial ten times diluted blood serum [59]. Cathodic current pulses of 10 μA (125 μA cm^{-1}) and 1 sec duration are followed by a stripping potential of +30 mV applied for 180 sec. Potentials were sampled at 0.5 and 1.0 sec of each pulse and are the averages over preceding 100 msec. The Nernstian response slope of 29 mV $decade^{-1}$ is shown for comparison.

potentials are observed in this super-Nernstian region. In contrast, pulstrodes allow one to obtain robust and fully reproducible potential readings in the super-Nernstian region, as full control of the system is performed by instrumental means.

In the experiment illustrated in Fig. 4.12b differential calcium responses were obtained from a single sensor at different durations of the current pulse in artificial diluted (tenfold) serum [59]. Observed super-Nernstian response was set within the physiological range of calcium activity by choosing the appropriate current. Note that the observed differential response slope in Fig. 4.12b achieved 50 mV per 0.2 logarithmic units of calcium activity, which corresponds to a remarkable eightfold increase in sensitivity in a very narrow activity window compared to the traditional 29 mV $decade^{-1}$, which is the slope of ordinary potentiometric Ca-selective electrodes. The differential response region may be easily tuned by variation of the current pulse parameters such as duration and magnitude of the applied current [55]. Due to the high sensitivity of differential response, pulstrodes are intended for applications requiring detection of a small change in activity of a target ion, which is common in the monitoring of blood electrolytes.

4.3.3 Voltammetric ion-selective electrodes

Controlled potential methods have been successfully applied to ion-selective electrodes. The term "voltammetric ion-selective electrode" (VISE) was suggested by Cammann [60]. Senda and coworkers called electrodes placed under constant potential conditions "amperometric ion-selective electrodes" (AISE) [61, 62]. Similarly to controlled current methods potentiostatic techniques help to overcome two major drawbacks of classic potentiometry. First, ISEs have a logarithmic response function, which makes them less sensitive to the small change in activity of the detected analyte. Second, an increased charge of the detected ions leads to the reduction of the response slope and, therefore, to the loss of sensitivity, especially in the case of large polyionic molecules. Due to the underlying response mechanism voltammetric ISEs yield a linear response function that is not as sensitive to the charge of the ion.

This type of sensor often does not have a membrane; it instead utilizes the properties of a water–oil interface, a boundary between an aqueous and a non-aqueous (organic) phase. Traditionally, sensors based on non-equilibrium ion-selective transport phenomena were distinguished as a separate group and considered as the electrochemistry of the ion transfer between two immiscible electrolyte solutions (ITIES). Here, we will not distinguish polymeric membrane electrodes and ITIES-based electrodes due to the similarity in the theoretical consideration.

Theoretical insight into the interfacial charge transfer at ITIES and detection mechanism of this type of sensor were considered [61–63]. In case of ionophore assisted transport for a cation I the formation of ion–ionophore complexes in the organic (membrane) phase is expected, which can be described with the appropriate complex formation constant, $\beta_{\mathrm{IL}n\mathrm{I}}$.

Using the steady-state diffusion model described in section 4.3.2 one may define the parameter q as:

$$q = D_{\mathrm{aq,I}}\delta_{\mathrm{m}}/D_{\mathrm{m,I}}\delta_{\mathrm{aq}} \tag{14}$$

If q is negligibly small a stagnant diffusion layer in aqueous phase is a rate-limiting step. This case is often the most useful from an analytical point of view. The well-known equation for a reversible polarographic wave can be obtained as:

$$i_{\mathrm{I}} = \frac{i_{\mathrm{I,d}}}{1 + \exp\left(\frac{zF}{RT}(E_{1/2} - E)\right)} \tag{15}$$

where the limiting (diffusion) current and half-wave potential are given by:

$$i_{\mathrm{I,d}} = zFAa_{\mathrm{I}}(\mathrm{aq})\sqrt{D_{\mathrm{m,I}}/t} \tag{16}$$

and

$$E_{1/2} = E^0 + \frac{RT}{zF}\left(\ln\sqrt{\frac{D_{\mathrm{aq,I}}}{D_{\mathrm{m,I}}}} + \ln[\beta_{\mathrm{IL}n\mathrm{I}}a_{\mathrm{L}}(\mathrm{m})^{n_{\mathrm{I}}}]\right) \tag{17}$$

Detection of Li^+ in artificial serum with a voltammetric Li-selective electrode in a flow-through system was demonstrated [64]. Lithium salts such as lithium carbonate have been extensively used for treatment of manic depressive and hyperthyroidism disorders. The therapeutic range of Li concentration is generally accepted to be 0.5–1.5 mM in blood serum. The authors used normal pulse voltammetry in which a stripping potential was applied between pulses in order to renew the membrane surface and expel all of the extracted ions from the membrane, similar to galvanostatically controlled potentiometric sensors described above. Unfortunately, the insufficient selectivity ($K^{pot}_{LiNa} = 2.4 \times 10^{-2}$) of the Li-ionophore dibenzyl-14-crown-4 required the use of an ion exchange column in order to separate lithium in the presence of a high (0.14 M) concentration of sodium. A linear response function in the range of 0.2–2 mM was obtained.

Another important bioanalytical application of voltammetric ISEs is the detection of polyions (see also above). A technique using cyclic voltammetry on micropipette electrodes filled with the organic electrolyte solutions in 1,2-dichloroethane was successfully applied for the detection of protamine [65] in saline solution and heparin in undiluted sheep plasma samples [66]. Protamine transport was facilitated with dinonylnaphthalenesulfonic acid (DNNS). As a heparin-selective component the tetrakis-(4-chlorophenyl)borate salt of trimethyloctadecyl ammonium was used.

Further improvement of the low detection limit was achieved using stripping voltammetry based on facilitated heparin adsorption and desorption [66]. Stripping voltammetry yielded a detection limit of 0.13 U mL^{-1} in sheep blood plasma, which is lower than therapeutic heparin concentrations (>0.2 U mL^{-1}). A linear response function in the range of 0.2–6 U mL^{-1} was observed. The authors also found that blood polypeptides and lipids with a mass above 25 000 significantly interfered with heparin detection, perhaps by hindrance of a charge transfer reaction at the interface.

4.3.4 Light-addressable potentiometric sensors

The concept of light addressable potentiometric sensors (LAPS) was introduced in 1988 [67]. LAPS is a semiconductor-based sensor with either electrolyte–insulator–semiconductor (EIS) or metal–insulator–semiconductor (MIS) structure, respectively. Figure 4.13 illustrates a schematic representation of a typical LAPS with EIS structure. A semiconductor substrate (silicone) is covered with an insulator (SiO_2). A sensing ion-selective layer, for instance, pH-sensitive S_3N_4, is deposited on top of the insulator. The whole assembly is placed in contact with the sample solution.

A constant bias potential is applied across the sensor in order to form a depletion layer at the insulator–semiconductor interface. The depth and capacitance of the depletion layer changes with the surface potential, which is a function of the ion concentration in the electrolytic solution. The variation of the capacitance is read out when the semiconductor substrate is illuminated with a modulated light and the generated photocurrent is measured by means of an external circuit.

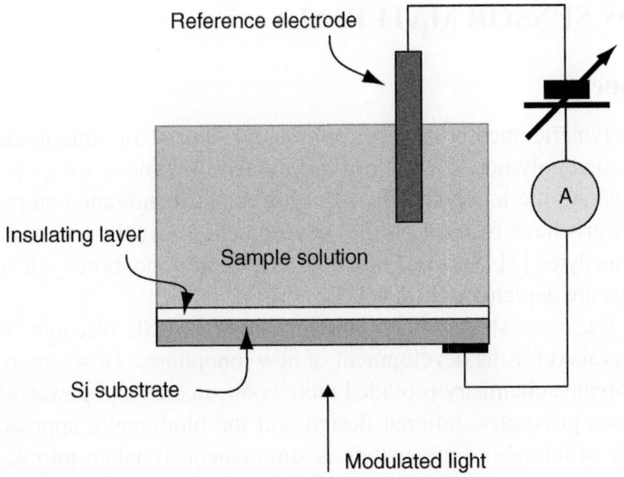

FIGURE 4.13 Schematic representation of a light-addressable potentiometric sensor (LAPS).

The working principle of LAPS resembles that of an ion-selective field effect transistor (ISFET). In both cases the ion concentration affects the surface potential and therefore the properties of the depletion layer. Many of the technologies developed for ISFETs, such as forming of ion-selective layers on the insulator surface, have been applied to LAPS without significant modification.

Electron-hole pairs are only produced in areas illuminated by the light and spatially resolved photocurrents can be achieved using a focused scanning beam. Therefore the main advantage of LAPS is that it can work as a chemical "imaging" sensor: the "light-addressability" of the LAPS allows one to obtain a two-dimensional map of the distribution of the ion concentration.

The first LAPS utilized silicone nitride (S_3N_4) as a pH-sensitive layer [68]. A light-addressable high resolution pH imaging sensor was applied to the detection of spatially resolved metabolic activity of *Escherichia coli* colonies on agar medium [69]. For a silicone substrate thickness of 20 μm the reported spatial resolution was about 10 μm. The observed pH distribution was in good agreement with the results of simulation based on a two-dimensional diffusion model.

Ionophore-based solvent polymeric membranes were used as sensing layers for the development of LAPS selective for lithium [70], potassium and calcium ions [71]. Anion-selective LAPS for the determination of nitrate and sulfate ions were described [72].

LAPS offers a number of advantages in view of multisensor system applications. Each point on the sensing surface can be individually accessed, and therefore multi-sensor functions can be easily accomplished by depositing various sensor materials on different sites of the sensing surface. A number of LAPS-based ion-selective multisensor systems have been reported recently [73, 71, 74].

An enzyme deposited on the LAPS surface allows one to observe the spatial distribution of a specific substrate. In a urea-selective sensor urease was immobilized on a pH-selective LAPS [75].

4.4 NEW SENSOR MATERIALS

4.4.1 Membrane components

Solvent polymeric membranes conventionally consist of ionophore, ion exchanger, plasticizer, and polymer. The majority of modern polymeric ISEs are based on neutral carriers, making the ionophore the most important membrane component. Substantial research efforts have focused on the development of highly selective ionophores for a variety of analytes [3]. Some of the most successful ionophores relevant to biomedical applications are depicted in Fig. 4.1.

During the early stages of polymeric membrane ISE research, the trail-and-error method prevailed for the development of new ionophores. However, recent advances of synthetic organic chemistry provided more sophisticated and powerful strategies, using basically two principles, rational design and the biomimetic approach. In the former the number of ionophore parameters is simultaneously taken into account in order to optimize the molecular structure and to endow an ionophore with a strong binding ability with a target analyte. Computational chemistry methods are used to model binding and stabilizing interactions, including Lewis acid/base chemistry, hydrogen bonding, electrostatic and π-electron interactions. Modification of the binding site setting with various substituents, capable of withdrawing/donating electron density, increasing/decreasing steric hindrance, etc., is also an instrument in the development of superior ionophores. Moreover, the binding sites are to be complemented with a molecular periphery, which ensures sufficient lipophilicity of the ionophore and its compatibility with the sensor membrane matrix. At last, the synthesis of the designed ionophore must not be too complicated. The ionophore optimization is very similar to drug design procedures. Unfortunately, the comprehensive modeling of analyte–ionophore interactions in the membrane still has many shortcomings, mainly due to imperfections of the available theories, which restricts the prediction power of such calculations. Further development of computational chemistry will hopefully allow a better quantification of sensor properties *a priori*.

The biomimetic approach, on the other hand, may help one to avoid some of the complexities of rational design, as it exploits natural binding sites and receptors, fine tuned by millions of years of natural evolution. Valinomycin and nonactin, naturally occurring antibiotics with high selectivity to potassium and ammonium, correspondingly proposed in the mid-1960s, are good examples of such an approach and promoted the development of ISEs for decades. More recent success has been achieved with multiple anion ionophores inspired by nature, e.g. metalloporphyrins, guanidinium salts, and functionalized ureas [11].

Cation-selective ionophores are the most successful in polymeric ISEs and selectivities exceeding ten orders of magnitude became quite common. The cation–ionophore binding occurs dominantly due to Lewis interactions and could be understood in terms of hard and soft acid and bases theory (HSAB). While hard base oxygen atoms originate from ester, ether or carbonyl functionalities, and interact with hard acid alkaline cations, the softer sulfur or nitrogen atoms better bind with transition metal ions. Cation ionophores may be provisionally divided into two main groups, as macrocyclic and acyclic compounds. The first, larger, group includes crowns, cryptands, calixarenes, and cyclic polypeptides, which due to different and variable cavity size may spatially discriminate ions. The high selectivity of macrocyclic ionophores is ensured by the cavity size, but also by high preorganization of ionophore molecules, when its original ion-free structure resembles one after complexation. Under such conditions the conformational change during complex formation is usually of low energy. In preorganized cavities cations may simultaneously interact with several functionalities, which is energetically favorable. The cavity may be ring shaped in crowns, barrel or cylinder shaped in calixarenes and hemispherical in bridged crown ionophores. The best-fit approach, based on cavity size and ionic radius, however, is not the only option, as crowns may form complexes with a 1:2 ion–ionophore stoichiometry, which was used for successful development of bis-crown ionophores, forming intramolecular "sandwiches", where an ion is squeezed between two crowns [76]. Many macrocyclic compounds and especially calixarenes could be easily functionalized with additional chelating moieties to enhance their selectivity and with branched alkyl chains to increase ionophore lipophilicity and to avoid crystallization in the membrane [77].

Acyclic ionophores, although sometimes considered less selective than their macrocyclic counterparts, are often of practical relevance. Open-chained analogs of crown compounds and cryptands, so-called podands, are widely spread. Besides, amide derivatives such as the well-known calcium ionophores ETH 1001 and ETH 129 (Fig. 4.1) exhibit extremely high selectivities and are successfully used for calcium determination in physiological samples [78]. The complexes formed by such ionophores are often of high stoichiometry, so usually two or three ionophore molecules enlace one ion. Stabilization of the resulting complex is achieved due to a chelate effect. Rational design of tweezer type ionophores [79], based on a rigid structure of cholic acid derivatives (Fig. 4.1, $CO_3^{2-}-1$) is another successful approach to acyclic ionophores. Modification at certain and variable positions of cholic acid with functionalities of different types gives a preorganized ionophore that holds the binding sites, spatially oriented as tweezers. Depending on the binding site nature, ionophores may specifically interact with either cations or anions, allowing significant selectivity improvement.

Amid the prosperity of cation-carrier development, advances in anion-selective ionophores so far seem to be modest. Larger size and diversity of anion geometries, higher hydration and less solvation energies for anions than for cations of comparable size, presence of different forms of anions in solution – all these factors hinder the development of anion sensors [80]. The general principles of host–guest and supramolecular chemistries are widely applicable for anion receptor design, but with certain singularities [81, 82]. The development of modern anion ionophores usually exploits two main processes: coordination to a metal center and hydrogen bonding with ion–dipole interactions. The best results are often obtained by combination of these two features in a single receptor.

Organometallic ionophores containing either metal ion complexed with lipophilic organic ligands, or a metal atom covalently bound to organic molecules may serve as charged or neural carriers. Probably the most interesting class of organometallic ionophores consists of the metalloporphyrins [83]. The naturally occurring ionophore Vitamin B_{12}, a derivative of cobalt(III) cobyrinate, was shown to be selective towards chloride and nitrite in the mid-1980s, which boosted further research in this area [84]. A number of metal centers were evaluated in porphyrin structures with a different periphery in order to optimize selectivity. Several highly selective halide and oxo-anion sensors were proposed based on metalloporphyrins with Ga(III), In(III), Mn(III), Zr(IV) metal centers [83]. Other important organometallic receptors include lipophilic uranyl salophenes, which were suggested as phosphate and fluoride ionophores; covalently bound compounds of tin and mercury are used as chloride ionophores. A fruitful "anticrown" concept was proposed, using a macrocycle reciprocal to the original crown structure that holds Lewis acidic hosts interacting with guest anions. The [9]mercuracarborand-3 ionophore (Fig. 4.1, $I^- - 1$) allowed one to achieve nanomolar DLs for iodide, which is exceptionally low for anions [85].

Ionophores based on hydrogen bonding are quite multifarious. Heterocycles with NH groups, capable of interacting with anions in the cycle cavity, were proposed for halides. Sensors for sulfate and phosphate with enhanced selectivity were developed based on bipodal urea and thiourea ionophores. The task is challenging due to a high hydration and low complexation ability of the mentioned anions. Guanidinium ion-based ionophores were suggested for sulfite. Hydrated trifluoroacetophenone derivatives (in the form of a geminal diol) are also believed to be hydrogen bonding ionophores for carbonate. However, there is still discussion on the exact mechanism of the relevant ion–ionophore interaction [86].

Lipophilic ion exchangers traditionally used for polymeric membrane preparation are the anionic tetraphenylborate derivatives and the cationic tetraalkylammonium salts. The charges on both lipophilic ions are localized on a single (boron or nitrogen) atom, but the steric inaccessibility of the charged center, due to bulky substituents, may inhibit ion-pair formation in the membrane and provide, when necessary, non-specific interactions between ionic sites and sample ions.

For a long time tetraphenylborates comprised the only class of cation exchanger used in ISEs, despite of their shortcomings. Simple tetraphenylborates are of low lipophilicity and their leaching from membrane shortens sensor lifetime [87]. Moreover, their decomposition in acidic solutions limits the applicability of this type of cation exchanger. However, incorporation of acceptor substituents into the phenyl rings of tetraphenylborate may to some extent enhance stability and lipophilicity of these compounds. While lipophilic sulfonic acids were tested as ion exchangers, they have not become widespread. Very recently a new class of cation exchangers, named carboranes, was introduced. Perhalogenated *closo*-dodecacarboranes were shown to be not only of higher stability than tetraphenylborates, retaining comparable lipophilicity, but can be with ease covalently attached to the membrane polymeric backbone, leading to the development of plasticizer-free membranes without leachable ion exchangers [88]. Related compounds, metal dicarbollides, could also be incorporated into membrane, providing anionic sites of extremely high chemical stability.

Quaternary ammonium salts (QAS) are anion exchangers used in solvent polymeric membranes. Variation of substituents at the nitrogen atom is an option for tuning QAS properties in terms of steric hindrance and lipophilicity. Control over the ion-pair formation in the membrane, achieved through the QAS structure variation and selection of an appropriate plasticizer, is an additional tool for anion selectivity enhancement, as the latter is often highly desired [89].

The main classes of plasticizers for polymeric ISEs are defined by now and comprise lipophilic esters and ethers [90]. The regular plasticizer content in polymeric membranes is up to 66% and its influence on the membrane properties cannot be neglected. Compatibility with the membrane polymer is an obvious prerequisite, but other plasticizer parameters must be taken into account, with polarity and lipophilicity as the most important ones. The nature of the plasticizer influences sensor selectivity and detection limits, but often the reasons are not straightforward. The specific solvation of ions by the plasticizer may influence the apparent ion–ionophore complex formation constants, as these may vary in different matrices. Ion-pair formation constants also depend on the solvent polarity, but in polymeric membranes such correlations are rather qualitative. Insufficient plasticizer lipophilicity may cause its leaching, which is especially undesired for *in-vivo* measurements, for microelectrodes and sensors working under flow conditions. Extension of plasticizer alkyl chains in order to enhance lipophilicity is only a partial problem solution, as it may lead to membrane component incompatibility. The concept of plasticizer-free membranes with active compounds, covalently attached to the polymer, has been intensively studied in recent years [91].

Dioctyl sebacate (DOS) with relative permittivity ε of 3.9 and 2-nitrophenyl octyl ether (NPOE) with $\varepsilon = 23.9$ are the traditionally used sensor membrane plasticizers. The choice of a plasticizer always depends on a sensor application. Thus, NPOE appears to be more beneficial for divalent ions due to its higher polarity, but for some cases its lipophilicity is insufficient. Furthermore, measurements with NPOE-plasticized sensors in undiluted blood are complicated by precipitation of charged species (mainly proteins) on the sensor surface, which leads to significant potential drifts. Although calcium selectivity against sodium and potassium for NPOE-based membranes is better by two orders of magnitude compared to DOS membranes, the latter are recommended for blood measurements as their lower polarity prevents protein deposition [92].

Although widely employed and comprehensively studied, high molecular weight poly(vinyl chloride) (PVC) as a polymeric matrix is not the only option. While PVC is often preferred due to mechanistic aspects, a number of other polymers have been tested, including functionalized PVC (with carboxyl, amino or hydroxyl groups), polyurethanes, silicon rubbers, polysiloxanes, polystyrenes, etc. with the main objective to develop membranes with better biocompatibility and adhesion properties. While membrane casting is widely used, development of other sensor manufacturing procedures is of a great importance. Photocurable polymers and co-polymers based on various acrylate derivatives not only may produce plasticizer-free membranes, but also allow one to covalently attach other membrane components such as ionophore and ion exchanger to the polymeric backbone. Furthermore, the simplified manufacturing process is also complemented with compatibility with microelectronics technology.

Of the formulated requirements, the glass transition temperature (T_g) of a sensor polymer must be below room temperature, otherwise the polymer should be plasticized (as in case of PVC for which the T_g is 80ºC). The low T_g provides the appropriate mechanical properties, such as elasticity, and ensures sufficient ionic mobilities in sensor membranes. On the other hand, however, some polymers with low T_g may be too soft to be castable into membranes. The polymer itself is often considered an inert matrix; however, it inevitably contains impurities that may influence, sometimes strongly, the membrane properties. Membrane polarity depends on the polymer nature and content. For a traditional PVC to plasticizer ratio of 1:2 (by mass) the relative permittivity of DOS-plasticized membranes is 4.8, which is higher, and for NPOE-based ones the value is 14, which is significantly lower than the respective values for the pure plasticizers.

Non-traditional compounds such as perfluorocarbons are now intensively tested as new matrices for polymeric ISEs [93]. Properties of these species significantly differ from those of other organics. Extremely low polarity and solubility in water, along with very high chemical inertness and decreased affinity to proteins and lipids make them attractive materials for chemical sensors, especially for biomedical applications. However, due to compatibility reasons all sensor membrane components should be fluorinated, which narrows the range of available sensing materials. Moreover, the knowledge of materials on ISEs accumulated over the years may be of limited applicability for perfluorinated membranes because of the widely different matrix properties. For example, ion-pair formation in fluorous phases is many orders of magnitude stronger than in conventional membranes and may compete with the ion–ionophore complex formation, likely co-determining sensor selectivity. These novel membrane materials are very promising but they require purposeful and comprehensive research.

4.4.2 Solid contact

An inner filling solution and internal reference electrode are used in macro ISEs due to a very good stability of the potential at the inner membrane–solution interface in such a setup (see Fig. 4.4). However, the presence of a solution inside a sensor could be a serious limitation for development of microelectrodes and may be undesired for a variety of other reasons, including ionic fluxes in the membrane and limited temperature range of sensor operation. There are several requirements for such an inner contact. First of all, a reversible change of electricity carriers ions–electrons must take place at the membrane–substrate interface. The potential of the electrochemical reaction, ensuring this transfer, has to be constant, stable, and must not depend on the sample composition. At last, the substrate must not influence the membrane analytical performance.

The first and very simple solid contact polymeric sensors were proposed in the early 1970s by Cattrall and Freiser and comprised of a metal wire coated with an ion-selective polymeric membrane [94]. These coated wire electrodes (CWEs) had similar sensitivity and selectivity and even somewhat better DLs than conventional ISEs, but suffered from severe potential drifts, resulting in poor reproducibility. The origin of the CWE potential instabilities is now believed to be the formation of a thin aqueous layer between membrane and metal [95]. The dominating redox process in the layer is likely the reduction of dissolved oxygen, and the potential drift is mainly caused by pH and pO_2 changes in a sample. Additionally, the ionic composition of this layer may vary as a function of the sample composition, leading to additional potential instabilities.

Obviously, a well-defined redox process between metal substrate and membrane is required for optimum sensor stability. In "pseudo" solid contact sensors a hydrogel layer, entrapping a salt such as KCl, is applied between the membrane and the internal reference electrode (Ag/AgCl). When the concentration of chloride ions in the hydrogel layer is kept constant the resulting reference potential remains stable. Importantly, modern highly cross-linked hydrogels exhibit low swelling and prevents sensor membrane mechanical damage and geometry change.

Various redox-active compounds such as salt-doped resins and lipophilic complexes of silver were used to provide a reversible redox reaction at the inner interface. Conducting polymers, due to their unique electrochemical properties are probably the most important solid contact materials [12]. Conducting polymers, such as polypyrrole, polythiophene, polyaniline, and their derivatives, not only provide defined redox couples, but also form layers with mixed ionic/electronic conductivity that are often compatible with the sensor membrane. On the back side of the polymer, however, an aqueous layer still may appear, as in CWEs. The use of lipophilic monolayers assembled on the metal surface could help avoid undesired formation of the aqueous phase. While there is still work required to fully suppress the influences of sample pH, pO_2, pCO_2, and redox species on conducting polymer-based solid contact sensors, conducting polymers are very effective ion-to-electron transducers. Highly stable solid contact ISEs for many ions have been reported, including those for trace level measurements [96]. Furthermore, incorporation of the sensing materials directly into conducting polymer chains, either by covalent attachment or by doping with counter-ions containing desired functionalities, is a very challenging approach to make a sensor and transducer on the same and single molecule [97]. While the redox sensitivity, being an intrinsic property of conducting polymer, may interfere with the sensor response [98], this approach opens new horizons in nanosensor technology.

4.4.3 Biocompatibility improvement

In addition to laboratory blood analyzers and portable point-of-care devices, which require blood collection, continuous monitoring of ion activities in a blood stream via implanted ion-selective electrodes is of great interest. The term "biocompatibility" refers to the ability of a sensor not to cause toxic or injurious effects while being in contact with living tissue. As dealing with any foreign object introduced into the human body, biocompatibility and hemocompatibility particularly are the most important requirements.

The first aspect of biocompatibility is a natural immune response. When a foreign object enters the blood stream, it can be attacked by the body's defense system. The first step is protein adsorption on an object surface. It is believed that the amount and type of protein adsorption is one of the most important steps determining whether the object is tolerated or rejected by the body. The next step is cell adhesion, which may cause aggregation and activation of platelets and triggering of the blood coagulation system with resulting thrombus formation. It may not only lead to sensor failure via surface blocking but directly threatens the patient's health.

In order to improve the biocompatibility of ISEs and reduce adsorption of cells and polypeptides several approaches have been used. Among them are immobilization of anticoagulants such as heparin and the continuous release of biologically active molecules such as nitric oxide [99]. Polymers with better biocompatibility such as polyurethanes have been employed as well. An interesting approach was used by the group of Bachas [100], who prepared a potassium-selective electrode coated with a copolymer containing phosphorylcholine, which mimicked the polar groups on cell surfaces. Decreased adhesion and activation of platelets was demonstrated after soaking the PVC membranes in platelet-rich plasma. At the same time the biocompatible coating did not affect the ISE sensing characteristics.

The second aspect of biocompatibility is a leaching problem. Ion-selective electrode materials, especially components of solvent polymeric membranes, are subject to leaching upon prolonged contact with physiological media. Membrane components such as plasticizers, ion exchangers and ionophores may activate the clotting cascade or stimulate an immune response. Moreover, they can be potentially toxic when released to the blood stream in significant concentrations.

There is a practical limit in the synthesis of ionophores and ion exchangers containing longer alkyl chains that potentially improve component retention. Replacement of solvent polymeric membranes with self-plasticized compositions with all sensing components covalently attached to the polymeric backbone appears to be an elegant solution to the leaching problem. Many self-plasticized polymers have been studied as ion-selective membrane matrices, including polyurethanes, polysiloxanes, and polyacrylates [101]. Among them alkyl acrylate copolymers appear to be very promising, combining several important advantages such as simplicity of variation of physical properties and the possibility of attaching sensing components to the matrix via single-step solution polymerization.

Several examples of ion-selective electrodes with ionophores covalently attached to a self-plasticized polymeric matrix have been reported in the literature. For instance, a Ca-selective electrode with the ionophore attached to a methylmethacrylate-*co*-decyl methacrylate backbone was developed recently [91]. Ion exchangers such as the dodecacarborane anion have been anchored to the polymeric backbone, with grafted dodecarborane showing greatly improved retention in the polymeric phase [88].

4.5 MINIATURIZATION

4.5.1 Miniaturization

Miniaturization of polymeric ISEs has been the focus of efforts of many research groups over decades [21] and still is of high relevance. Biology and medicine greatly demand analytical techniques for intracellular measurements. They provide information on the electrolyte concentration in living cells and give access to fundamental parameters such as cell membrane potential and cellular ion transport and may enable a deeper understanding of the connected physiological processes. Automated sample handling and separation techniques such as various flow injection methods and micro total analysis systems (μTAS) require miniature detectors. Microsensors compatible with existing and well-developed microelectronics technologies may be more easily mass-produced and assembled into sensor arrays for simultaneous multianalyte detection.

There are a variety of methods available for microfabrication of ion-selective electrodes, from relatively simple dip-coating and casting to more sophisticated screen-printing and lamination processes. The first solvent polymeric membrane electrodes proposed for intracellular measurements were made from micropipette tips filled with a liquid membrane solution. This concept is still successfully used with capillary electrode bodies and the reported electrode tip diameters are as small as 0.1 μm. A recently reported microelectrode array platform for intracellular calcium measurements [102] is depicted in Fig. 4.14. During the past decade microelectrode development has been shifting towards planar devices. Modern photolithography and etching techniques are applicable for producing multilayer sensors using thin and thick film technologies [13]. *In-vivo* measurements in the beating mammalian heart were performed with such sensors (Fig. 4.15).

Most of the possible shortcomings of macroelectrodes are even more pronounced in their micro counterparts. For instance, the challenges of a solid contact, discussed above, are of great importance for microsensors. Layers of electrochemically deposited Ag/AgCl or electropolymerized conducting polymers applied on an electron-conducting substrate are used as solid contact layers, which are further covered with an ion-sensitive film. Unfortunately, the application of hydrogels in an inner reference compartment, successfully employed in conventional ISEs, is technically more difficult with microsensors. The compatibility of materials may also be an issue, i.e. poor adhesion of the sensing membrane to a transducer surface could lead to exfoliation and limit the sensor lifetime. Various polymers with better adhesion properties than traditional PVC have been tested as matrices for microelectrodes [103].

Size-related problems may become important for all microsensors. Leakage of sensing materials from a small membrane may lead to rapid deterioration of sensor properties [104]. While the lipophilicity of membrane components cannot be increased infinitely, immobilization of ionophore and ion exchanger in the polymer by covalent attachment or molecular imprinting along with utilization of plasticizer-free membranes could help solve the leakage problem.

4.5.2 Sensor arrays

Our multidimensional and fast-paced world does require robust instruments, capable of tracing the constant changes. Discrete sensors, providing information on a single or few parameters, cannot supply the growing needs. A general and complex approach to gaining information on various systems and processes logically leads to development of multisensor systems or sensor arrays. The signals from potentiometric sensor arrays, comprising electrodes selective to different species in sample solutions, when processed simultaneously using mathematical methods, may reduce the errors of determination of individual discrete sensors. Once the influences of interfering ions are well understood, the mathematical data processing makes it possible to deconvolute the contributions from primary and interfering ions and to introduce automatic correction

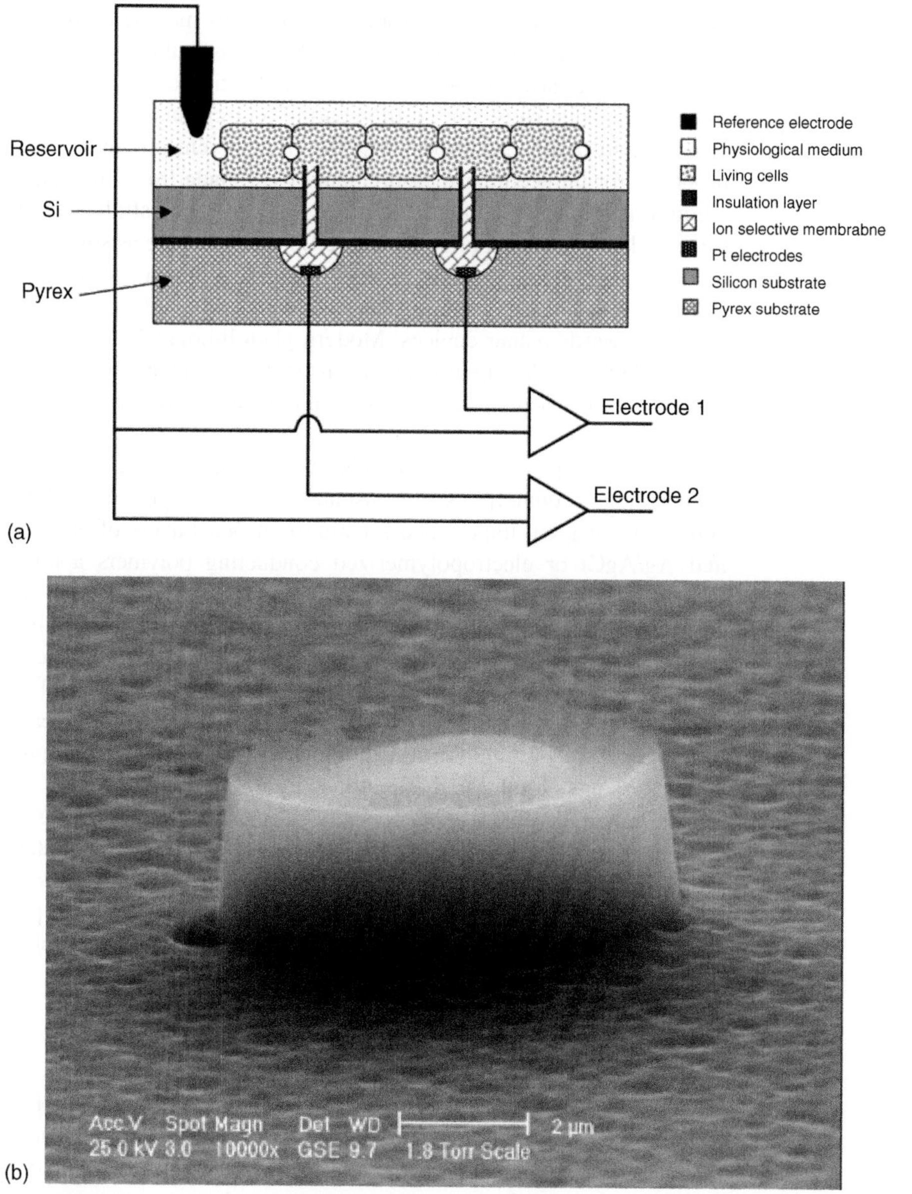

FIGURE 4.14 Schematic cross-section of the platform for *in-vitro* intracellular recording of ion concentration, designed for hepatocyte cell culture (a) and ESEM pictures of a micropipette (tilted at 75°) filled with the membrane cocktail. (From [102].)

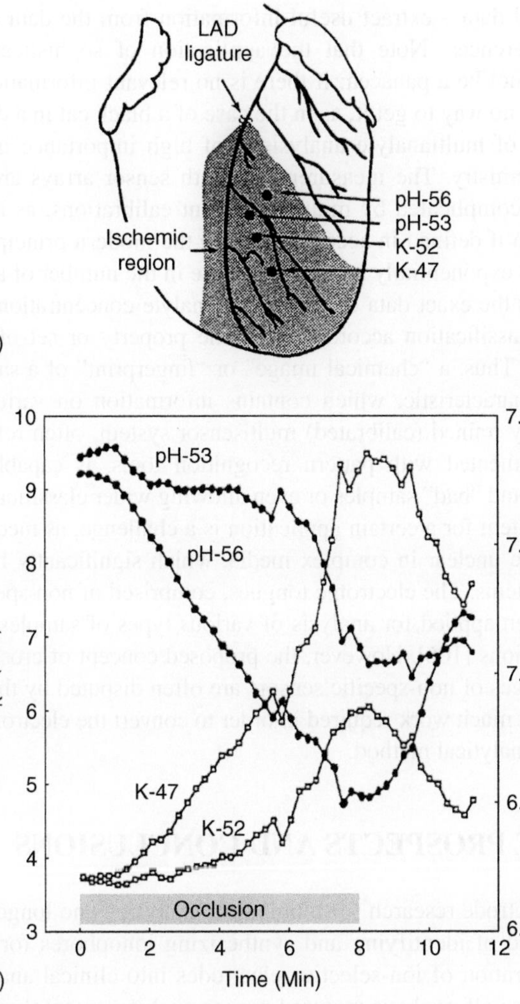

FIGURE 4.15 Two pH and two K^+ planar sensors in the *in-situ* porcine heart preparation at midmyocardial depth (a) and the recorded fall in the pH and increase in K^+ activity, respectively, during the course of coronary artery occlusion (b). (From [18].)

to the analytical interpretation [105]. Furthermore, when no highly selective electrodes are available, this approach allows one to use only sparingly selective electrodes.

A number of mathematical methods for multivariate and multi-way analysis of sensor array complex response are available nowadays, including various regression and pattern recognition techniques. Multivariate regression methods based on classical statistics usually matched against more sophisticated "black box" methods as fuzzy logic and artificial neural networks. In the latter the signal processing is often performed in a non-transparent way, hindering critical evaluation of the results. However, all these methods, while quite diverse, are intended to provide higher signal-to-noise ratio for multidimensional data – extract useful information from the data arrays and suppress unwanted interferences. Note that the application of sophisticated signal processing methods cannot be a panacea: if there is no relevant information in a sensor array response there is no way to get it, as in the case of a black cat in a dark room.

The problem of multianalyte analysis is of high importance in different methods of analytical chemistry. The measurements with sensor arrays are not an exception; they are always complicated by multicomponent calibrations, as the number of these calibrations, even if defined in accordance with the modern principles of experimental design, increases exponentially with the increase in the number of analytes.

In some cases the exact data on numerous analyte concentrations are of less importance. Sample classification according to some property or set of properties is often required instead. Thus, a "chemical image" or "fingerprint" of a sample could be used as an integral characteristic, which contains information on various critical parameters. The properly trained (calibrated) multisensor system, often referred as "electronic tongue", complemented with pattern recognition tools, is capable of distinguishing between "good" and "bad" samples or even allowing wider classification [14]. Design of a multisensor system for a certain application is a challenge, as mechanism(s) of sensor responses may be unclear in complex media, which significantly hinders the development of such systems. The electronic tongues, comprised of non-specific cross-sensitive sensors, have been applied for analysis of various types of samples, including pharmaceutical formulations [106]. However, the proposed concept of cross-sensitivity and the declared advantages of non-specific sensors are often disputed by the scientific community. There is still much work required in order to convert the electronic tongue approach to a recognized analytical method.

4.6 FUTURE PROSPECTS AND CONCLUSIONS

Ion-selective electrode research for biomedical analysis is no longer the relatively narrow, focused field of identifying and synthesizing ionophores for improved selectivity and the integration of ion-selective electrodes into clinical analyzers and portable instruments. These efforts have matured now to such an extent that they can teach valuable lessons to other chemical sensing fields that are just emerging technologies.

Ion-selective electrodes are now well understood in terms of the underlying theory, and this has made it possible for new sensing principles to emerge that make use of the thousands of chemical receptors originally developed for ion-selective electrodes. One is the field of optical sensors, which has not been discussed here because it is outside the focus of this chapter. Such so-called bulk optodes do not require electrical connectivity between the sensing and detection unit and are therefore more easily brought into various shapes and sizes, including particle formats, which suit the need of modern chemical analysis.

Electrochemical sensors, however, currently share one key advantage: an excitation signal may be imposed that can trigger a sensing reaction, and the energy required for an otherwise thermodynamically unfavorable extraction and/or binding process can be instrumentally imposed. While ion-selective electrodes historically have been passive sensing devices where such control is not possible, much of current research deals with non-classical response principles where concentration polarizations take place at the sample–membrane interface. Galvanostatic and voltammetric control of ion-selective membranes is now also possible and offers an exciting path to novel approaches in extraction/complexation-based sensing. The key application discussed above has been the development of reversible sensors for the anticoagulant heparin and its antidote protamine, but many other important sensing principles may be developed since the tools are now in place to fabricate reversible sensors on the basis of otherwise irreversible reactions.

This field is therefore at an exciting stage. Ion-selective electrodes have a proven track record in terms of clinical and biomedical analysis, with a well-developed theory and a solid history of fundamental research and practical applications. With novel directions in achieving extremely low detection limits and instrumental control of the ion extraction process this field has the opportunity to give rise to many new bioanalytical measurement tools that may be truly useful in practical chemical analysis.

4.7 ACKWNOWLEDGMENTS

The authors would like to thank the National Institutes of Health (GM071623 and EB002189) and the National Science Foundation (BIO8-004-00) for financial support of their research.

4.8 REFERENCES

1. R.P. Buck and E. Lindner, Tracing the history of selective ion sensors. *Anal. Chem.* **73**, 88A–97A (2001).
2. E. Bakker, P. Buhlmann, and E. Pretsch, Carrier-based ion-selective electrodes and bulk optodes. 1. General characteristics. *Chem. Rev.* **97**, 3083–3132 (1997).
3. P. Buhlmann, E. Pretsch, and E. Bakker, Carrier-based ion-selective electrodes and bulk optodes. 2. Ionophores for potentiometric and optical sensors. *Chem. Rev.* **98**, 1593–1687 (1998).
4. E. Bakker, P. Buhlmann, and E. Pretsch, Polymer membrane ion-selective electrodes – what are the limits? *Electroanalysis* **11**, 915–933 (1999).
5. E. Bakker and M. Telting-Diaz, Electrochemical sensors. *Anal. Chem.* **74**, 2781–2800 (2002).
6. E. Bakker, P. Buhlmann, and E. Pretsch, The phase-boundary potential model. *Talanta* **63**, 3–20 (2004).
7. S. Mathison and E. Bakker, Effect of transmembrane electrolyte diffusion on the detection limit of carrier-based potentiometric ion sensors. *Anal. Chem.* **70**, 303–309 (1998).
8. T. Sokalski, A. Ceresa, T. Zwickl, and E. Pretsch, Large improvement of the lower detection limit of ion-selective polymer membrane electrodes. *J. Am. Chem. Soc.* **119**, 11 347–11 348 (1997).
9. E. Bakker and E. Pretsch, Potentiometric sensors for trace-level analysis. *TrAC, Trends Anal. Chem.* **24**, 199–207 (2005).
10. R.D. Johnson and L.G. Bachas, Ionophore-based ion-selective potentiometric and optical sensors. *Anal. Bioanal. Chem.* **376**, 328–341 (2003).
11. J. Bobacka, J., A. Ivaska, and A. Lewenstam, Potentiometric ion sensors based on conducting polymers. *Electroanalysis* **15**, 366–374 (2003).
12. E. Lindner and R.P. Buck, Microfabricated potentiometric electrodes and their in vivo applications. *Anal. Chem.* **72**, 336A–345A (2000).

13. A. Legin, A. Rudnitskaya, and Y. Vlasov, Electronic tongues: sensors, systems, applications. *Sensors Update* **10**, 143–188 (2002).
14. E. Bakker and M.E. Meyerhoff, Ion-selective electrodes for measurements in biological fluids, in A.J. Bard, M. Stratmann,. and J.S. Wilson, eds, *Encyclopedia of Electrochemistry,* Wiley (2002).
15. E. Bakker, D. Diamond, A. Lewenstam, and E. Pretsch, Ion sensors: current limits and new trends. *Anal. Chim. Acta* **393**, 11–18 (1999).
16. R.W. Burnett, A.K. Covington, N. Fogh-Andersen, W.R. Kulpmann, A. Lewenstam, A.H.J. Maas, O. Muller-Plathe, A.L. Vankessel, and W.G. Zijlstra, Use of ion-selective electrodes for blood-electrolyte analysis. Recommendations for nomenclature, definitions and conventions. *Clin. Chem. Lab. Med.* **38**, 363–370 (2000).
17. R.P. Buck, V.V. Cosofret, E. Lindner, S. Ufer, M.B. Madaras, T.A. Johnson, R.B. Ash, and M.R. Neuman, Microfabrication technology of flexible membrane-based sensors for in-vivo applications. *Electroanalysis* **7**, 846–851 (1995).
18. V.V. Cosofret and R.P. Buck, Recent advances in pharmaceutical analysis with potentiometric membrane sensors. *Crit. Rev. Anal. Chem.* **24**, 1–58 (1993).
19. S.K. Menon, A. Sathyapalan, and Y.K. Agrawal, Ion selective electrodes in pharmaceutical analysis – a review. *Rev. Anal. Chem.* **16**, 333–353 (1997).
20. D. Ammann, *Ion-Selective Microelectrodes*, Berlin, Springer-Verlag (1986).
21. H. Suzuki, Advances in the microfabrication of electrochemical sensors and systems. *Electroanalysis* **12**, 703–715 (2000).
22. W.E. Morf, *The Principles of Ion-Selective Electrodes and of Membrane Transport,* New York, Elsevier (1981).
23. U. Schaller, E. Bakker, U.E. Spichiger, and E. Pretsch, Ionic additives for ion-selective electrodes based on electrically charged carriers. *Anal. Chem.* **66**, 391–398 (1994).
24. M. Perry, E. Lobel, and R. Bloch, Mechanism of a polymeric valinomycin-based potassium specific electrode. *J. Membr. Sci.* **1**, 223–235 (1976).
25. R. Eugster, P.M. Gehrig, W.E. Morf, U.E. Spichiger, and W. Simon, Selectivity-modifying influence of anionic sites in neutral-carrier-based membrane electrodes. *Anal. Chem.* **63**, 2285–2289 (1991).
26. E. Bakker, R.K. Meruva, E. Pretsch, and M.E. Meyerhoff, Selectivity of polymer membrane-based ion-selective electrodes – self-consistent model describing the potentiometric response in mixed ion solutions of different charge. *Anal. Chem.* **66**, 3021–3030 (1994).
27. R.P. Buck and E. Lindner, Recommendations for nomenclature of ion-selective electrodes (Iupac Recommendations 1994). *Pure Appl. Chem.* **66**, 2527–2536 (1994).
28. E. Bakker, Determination of improved selectivity coefficients of polymer membrane ion-selective electrodes by conditioning with a discriminated ion. *J. Electrochem. Soc.* **43**, L83–L85 (1996).
29. P.C. Meier, W.E. Morf, M. Laubli, and W. Simon, Evaluation of the optimum composition of neutral-carrier membrane electrodes with incorporated cation-exchanger sites. *Anal. Chim. Acta* **156**, 1–8 (1984).
30. M.M. Shultz, O.K. Stefanova, S.S. Mokrov, and K.N. Mikhelson, Potentiometric estimation of the stability constants of ion-ionophore complexes in ion-selective membranes by the sandwich membrane method: theory, advantages, and limitations. *Anal. Chem.* **74**, 510–517 (2002).
31. T. Vigassy, C.G. Huber, R. Wintringer, and E. Pretsch, Monolithic capillary-based ion-selective electrodes. *Anal. Chem.* **77**, 3966–3970 (2005).
32. Y. Mi, S. Mathison, and E. Bakker, Polyion sensors as liquid junction-free reference electrodes. *J. Electrochem. Soc.* **2**, 198–200 (1999).
33. E. Bakker, Hydrophobic membranes as liquid junction-free reference electrodes. *Electroanalysis* **11**, 788–792 (1999).
34. S.-C. Ma, V.C. Yang, and M.E. Meyerhoff, Heparin-responsive electrochemical sensor – a preliminary study. *Anal. Chem.* **64**, 694 (1992).
35. E.J. Cohen, L.J. Camerlengo, and J.P. Dearing, Activated clotting times and cardiopulmonary bypass: I. The effect of hemodilution and hypothermia upon activated clotting time. *J. Extra Corpor. Technol.* **12**, 139–141 (1980).
36. S.-C. Ma, V.C. Yang, B. Fu, and M.E. Meyerhoff, Electrochemical sensor for heparin – further characterization and bioanalytical applications. *Anal. Chem.* **65**, 2078 (1993).

37. J.H. Yun, V.C. Yang, and M.E. Meyerhoff, Protamine-sensitive polymer membrane electrode: characterization and bioanalytical applications. *Anal. Biochem.* **224**, 212–220 (1995).
38. B. Fu, E. Bakker, J.H. Yun, V.C. Yang, and M.E. Meyerhoff, Response mechanism of polymer membrane-based potentiometric polyion sensors. *Anal. Chem.* **66**, 2250–2259 (1994).
39. J.H. Yun, B. Fu, M.E. Meyerhoff, and V.C. Yang, A disposable, coated wire heparin sensor. *ASAIO J.* **40**, M401–M405 (1994).
40. J.H. Yun, L. Lee, J.A. Wahr, V.C. Yang, and M.E. Meyerhoff, Clinical application of disposable heparin sensors. Blood heparin measurements during open heart surgery. *ASAIO J.* **41**, M661–M664 (1995).
41. N. Ramamurthy, N. Baliga, J. Wahr, U. Schaller, V.C. Yang, and M.E. Meyerhoff, Improved polycation-sensitive membrane electrode for monitoring heparin levels in whole blood via protamine titration. *Clin. Chem.*, 606–613 (1998).
42. N. Ramamurthy, N. Baliga, T.W. Wakefield, P.C. Andrews, V.C. Yang, and M.E. Meyerhoff, Determination of low-molecular-weight heparins and their binding to protamine and a protamine analog using polyion-sensitive membrane electrodes. *Anal. Biochem.* **266**, 116–124 (1999).
43. D.P. Thomas, and R.E. Merton, A low-molecular weight heparin compared with unfractionated heparin. *Thromb. Res.* **28**, 343–350 (1982).
44. N. Durust and M.E. Meyerhoff, Determination of pentosan polysulfate and its binding to polycationic species using polyion-sensitive membrane electrodes. *Anal. Chim. Acta* **432**, 253–260 (2001).
45. J.M. Esson and M.E. Meyerhoff, Polyion-sensitive membrane electrodes for detecting phosphate-rich biological polyanions. *Electroanalysis* **9**, 1325–1330 (1997).
46. E.B. Ong and A. Johnson, *Anal. Biochem.*, 568–582 (1976).
47. L.-C. Chang, M.E. Meyerhoff, and V.C. Yang, Electrochemical assay of plasminogen activators in plasma using polyion-sensitive membrane electrode detection. *Anal. Biochem.* **276**, 8–12 (1999).
48. I.S. Han, N. Ramamurthy, J.H. Yun, U. Schaller, M.E. Meyerhoff, and V.C. Yang, Selective monitoring of peptidase activities with synthetic polypeptide substrates and polyion-sensitive membrane electrode detection. *FASEB J.* **10**, 1621–1626 (1996).
49. S. Mathison and E. Bakker, Renewable pH cross-sensitive potentiometric heparin sensors with incorporated electrically charged H^+ ionophores. *Anal. Chem.* **71**, 4614–4621 (1999).
50. A. Shvarev and E. Bakker, Pulsed galvanostatic control of ionophore-based polymeric ion sensors. *Anal. Chem.* **75**, 4541–4550 (2003).
51. A. Shvarev and E. Bakker, Response characteristics of a reversible electrochemical sensor for the polyion protamine. *Anal. Chem.* **77**, 5221–5228 (2005).
52. E. Bakker and A.J. Meir, How do pulsed amperometric ion sensors work? A simple PDE model. *Siam Review* **45**, 327–344 (2003).
53. A. Shvarev and E. Bakker, Reversible electrochemical detection of nonelectroactive polyions. *J. Am. Chem. Soc.* **125,** 11[ts]192–11[ts]193 (2003).
54. S. Makarychev-Mikhailov, A. Shvarev, and E. Bakker, Pulstrodes: triple pulse control of potentiometric sensors. *J. Am. Chem. Soc.* **126**, 10 548–10 549 (2004).
55. A. Shvarev and E. Bakker, Distinguishing free and total calcium with a single pulsed galvanostatic ion-selective electrode. *Talanta* **63**, 195–200 (2004).
56. T. Sokalski, T. Zwickl, E. Bakker, and E. Pretsch, Lowering the detection limit of solvent polymeric ion-selective membrane electrodes. 1. Steady-state ion flux considerations. *Anal. Chem.* **71**, 1204–1209 (1999).
57. T. Zwickl, T. Sokalski, and E. Pretsch, Steady-state model calculations predicting the influence of key parameters on the lower detection limit and ruggedness of solvent polymeric membrane ion-selective electrodes. *Electroanalysis* **11**, 673–680 (1999).
58. S. Makarychev-Mikhailov, A. Shvarev, and E. Bakker, Calcium pulstrodes for measurements in physiological fluids. *Anal. Chem.* (submitted).
59. D. Henn and K. Cammann, Voltammetric ion-selective electrodes (VISE). *Electroanalysis* **12**, 1263–1271 (2000).
60. M. Senda, H. Katano, and M. Yamada, Amperometric ion-selective electrode. Voltammetric theory and analytical applications at high concentration and trace levels. *J. Electroanal. Chem.* **468**, 34–41 (1999).
61. M. Senda, H. Katano, and M. Yamada, Amperometric ion-selective electrode. Voltammetric theory and analytical applications at high concentration and trace levels. *J. Electroanal. Chem.* **475**, 90–98 (1999).

62. Z. Samec, E. Samcova, and H.H. Girault, Ion amperometry at the interface between two immiscible electrolyte solutions in view of realizing the amperometric ion-selective electrode. *Talanta* **63**, 21–32 (2004).
63. S. Sawada, H. Torii, T. Osakai, and T. Kimoto, Pulse amperometric detection of lithium in artificial serum using a flow injection system with a liquid/liquid-type ion-selective electrode. *Anal. Chem.* **70**, 4286–4290 (1998).
64. Y. Yuan and S. Amemiya, Facilitated protamine transfer at polarized water/1,2-dichloroethane interfaces studied by cyclic voltammetry and chronoamperometry at micropipet electrodes. *Anal. Chem.* **76**, 6877–6886 (2004).
65. J.D. Guo, Y. Yuan, and S. Amemiya, Voltammetric detection of heparin at polarized blood plasma/1,2-dichloroethane interfaces. *Anal. Chem.* **77**, 5711–5719 (2005).
66. D.G. Hafeman, J.W. Parce, and H.M. McConnell, Light-addressable potentiometric sensor for biochemical systems. *Science* **240**, 1182–1185 (1988).
67. M. Nakao, T. Yoshinobu, and H. Lwasaki, Scanning-laser-beam semicomductor pH imaging sensor. *Sens. Actuators, B* **20**, 119–123 (1994).
68. E. Bakker, Electrochemical sensors. *Anal. Chem.,* 76, 3285-3298 (2004).
69. M. Nakao, S. Inoue, T. Yoshinobu, and H. Iwasaki, High-resolution pH imaging sensor for microscopic observation of microorganisms. *Sens Actuators, B* **34**, 234–239 (1996).
70. Y. Ermolenko, T. Yoshinobu, Y. Mourzina, K. Furuichi, S. Levichev, Y. Vlasov, M.J. Schoning, and H. Iwasaki, Lithium sensor based on the laser scanning semiconductor transducer. *Anal. Chim. Acta* 459, 1–9 (2002).
71. Y. Ermolenko, T. Yoshinobu, Y. Mourzina, K. Furuichi, S. Levichev, M.J. Schoning, Y. Vlasov, and H. Iwasaki, The double $K^+/Ca2^+$ sensor based on laser scanned silicon transducer (LSST) for multicomponent analysis. *Talanta* **59**, 785–795 (2003).
72. Y. Mourzina, Y. Ermolenko, T. Yoshinobu, Y. Vlasov, H. Iwasaki, and M.J. Schoning, Anion-selective light-addressable potentiometric sensors (LAPS) for the determination of nitrate and sulphate ions. *Sens Actuators, B* **91**, 32–38 (2003).
73. T. Yoshinobu, M.J. Schoning, R. Otto, K. Furuichi, Y. Mourzina, Y. Ermolenko, and H. Iwasaki, Portable light-addressable potentiometric sensor (LAPS) for multisensor applications. *Sens. Actuators, B* **95**, 352–356 (2003).
74. Y. Ermolenko, T. Yoshinobu, Y. Mourzina, Y. Vlasov, M.J. Schoning, and H. Iwasaki, Laser-scanned silicon transducer (LSST) as a multisensor system. *Sens. Actuators, B* **103**, 457–462 (2004).
75. S. Inoue, M. Nakao, T. Yoshinobu, and H. Iwasaki, Chemical-imaging sensor using enzyme. *Sens. Actuators, B* **32**, 23–26 (1996).
76. E. Lindner, K. Toth, M. Horvath, E. Pungor, B. Agai, I. Bitter, L. Toke, and Z. Hell, Bis-crown ether derivatives as ionophores for potassium selective electrodes. *Fres. Z. Anal. Chem.* **322**, 157–163 (1985).
77. R. Ludwig and N.T.K. Dzung, Calixarene-based molecules for cation recognition. *Sensors* **2**, 397–416 (2002).
78. U. Schefer, D. Ammann, E. Pretsch, U. Oesch, and W. Simon, Neutral carrier based $Ca\text{-}2^+$-selective electrode with detection limit in the sub-nanomolar range. *Anal. Chem.* **58**, 2282–2285 (1986).
79. J.H. Shim, I.S. Jeong, M.H. Lee, H.P. Hong, J.H. On, K.S. Kim, H.S. Kim, B.H. Kim, G.S. Cha, and H. Nam, Ion-selective electrodes based on molecular tweezer-type neutral carriers. *Talanta* **63**, 61–71 (2004).
80. M.M.G. Antonisse and D.N. Reinhoudt, Potentiometric anion selective sensors. *Electroanalysis* **11**, 1035–1048 (1999).
81. F.P. Schmidtchen and M. Berger, Artificial organic host molecules for anions. *Chem. Rev.* **97**, 1609–1646 (1997).
82. P.D. Beer and P.A. Gale, Anion recognition and sensing: the state of the art and future perspectives. *Angew. Chem., Int. Ed.* **40**, 487–516 (2001).
83. L. Gorski, E. Malinowska, P. Parzuchowski, W. Zhang, and M.E. Meyerhoff, Recognition of anions using metalloporphyrin-based ion-selective membranes: state-of-the-art. *Electroanalysis* **15**, 1229–1235 (2003).
84. P. Schulthess, D. Ammann, B. Krautler, C. Caderas, R. Stepanek, and W. Simon, Nitrite-selective liquid membrane-electrode. *Anal. Chem.* **57**, 1397–1401 (1985).

85. A. Malon, A. Radu, W. Qin, Y. Qin, A. Ceresa, M. Maj-Zurawska, E. Bakker, and E. Pretsch, Improving the detection limit of anion-selective electrodes: an iodide-selective membrane with a nanomolar detection limit. *Anal. Chem.* **75**, 3865–3871 (2003).
86. S. Makarychev-Mikhailov, A. Legin, J. Mortensen, S. Levitchev, and Y. VLASOV, Potentiometric and theoretical studies of the carbonate sensors based on 3-bromo-4-hexyl-5-nitrotrifluoroacetophenone. *Analyst* **129**, 213–218 (2004).
87. E. Bakker and E. Pretsch, Lipophilicity of tetraphenylborate derivatives as anionic sites in neutral carrier-based solvent polymeric membranes and lifetime of corresponding ion-selective electrochemical and optical sensors. *Anal. Chim. Acta* **309**, 7–17 (1995).
88. Y. Qin and E. Bakker, A copolymerized dodecacarborane anion as covalently attached cation exchanger in ion-selective sensors. *Anal. Chem.* **75**, 6002–6010 (2003).
89. V.V. Egorov, E.M. Rakhman'ko, E.B. Okaev, E.V. Pomelenok, and V.A. Nazarov, Effects of ion association of lipophilic quaternary ammonium salts in ion-exchange and potentiometric selectivity. *Talanta* **63**, 119–130 (2004).
90. R. Eugster, T. Rosatzin, B. Rusterholz, B. Aebersold, U. Pedrazza, D. Ruegg, A. Schmid, U.E. Spichiger, and W. Simon, Plasticizers for liquid polymeric membranes of ion-selective chemical sensors. *Anal. Chim. Acta* **289**, 1–13 (1994).
91. Y. Qin, S. Peper, A. Radu, A. Ceresa, and E. Bakker, Plasticizer-free polymer containing a covalently immobilized $Ca2^{+}$-selective ionophore for potentiometric and optical sensors. *Anal. Chem.* **75**, 3038–3045 (2003).
92. P. Anker, E. Wieland, D. Ammann, R.E. Dohner, R. Asper, and W. Simon, Neutral carrier based ion-selective electrode for the determination of total calcium in blood-serum. *Anal. Chem.* **53**, 1970–1974 (1981).
93. P.G. Boswell and P. Buhlmann, Fluorous bulk membranes for potentiometric sensors with wide selectivity ranges: observation of exceptionally strong ion pair formation. *J. Am. Chem. Soc.* **127**, 8958–8959 (2005).
94. R.W. Cattrall and H. Freiser, Coated wire ion selective electrodes. *Anal. Chem.* **43**, 1905–1906 (1971).
95. M. Fibbioli, W.E. Morf, M. Badertscher, N.F. de Rooij, and E. Pretsch, Potential drifts of solid-contacted ion-selective electrodes due to zero-current ion fluxes through the sensor membrane. *Electroanalysis* **12**, 1286–1292 (2000).
96. K.Y. Chumbimuni-Torres, N. Rubinova, A. Radu, L.T. Kubota, and E. Bakker, A universal recipe for solid contact potentiometric sensors for trace level measurements. (Submitted 2005)
97. M. Vazquez, J. Bobacka, M. Luostarinen, K. Rissanen, A. Lewenstam, and A. Ivaska, Potentiometric sensors based on poly(3,4-ethylenedioxythiophene) (PEDOT) doped with sulfonated calix[4]arene and calix[4]resorcarenes. *J. Solid State Electrochem.* **9**, 312–319 (2005).
98. A. Lewenstam, J. Bobacka, and A. Ivaska, Mechanism of ionic and redox sensitivity of p-type conducting polymers. 1. Theory. *J. Electroanal. Chem.* **368**, 23–31 (1994).
99. M.H. Schoenfisch, K.A. Mowery, M.V. Rader, N. Baliga, J.A. Wahr, and M.E. Meyerhoff, Improving the thromboresistivity of chemical sensors via nitric oxide release: fabrication and in vivo evaluation of NO-releasing oxygen-sensing catheters. *Anal. Chem.* **72**, 1119–1126 (2000).
100. M.J. Berrocal, R.D. Johnson, I.H.A. Badr, M. Liu, D. Gao, and L.G. Bachas, Improving the blood compatibility of ion-selective electrodes by employing poly(MPC-co-BMA), a copolymer containing phosphorylcholine, as a membrane coating. *Anal. Chem.* **74**, 3644–3648 (2002).
101. M.J. Berrocal, I.H.A. Badr, D. Gao, and L.G. Bachas, Reducing the thrombogenicity of ion-selective electrode membranes through the use of a silicone-modified segmented polyurethane. *Anal. Chem.* **73**, 5328–5333 (2001).
102. O.T. Guenat, J.F. Dufour, P.D. van der Wal, W.E. Morf, N.F. de Rooij, and M. Koudelka-Hep, Microfabrication and characterization of an ion-selective microelectrode array platform. *Sens. Actuators, B* **105**, 65–73 (2005).
103. W. Wroblewski, A. Dybko, E. Malinowska, and Z. Brzozka, Towards advanced chemical microsensors – an overview. *Talanta* **63**, 33–39 (2004).
104. B.D. Pendley, R.E. Gyurcsanyi, R.P. Buck, and E. Lindner, A chronoamperometric method to estimate changes in the membrane composition of ion-selective membranes. *Anal. Chem.* **73**, 4599–4606 (2001).

105 D. Diamond (ed.), *Principles of Chemical and Biological Sensors*, New York, Wiley Interscience (1998).
106 A. Legin, A. Rudnitskaya, D. Clapham, B. Seleznev, K. Lord, and Y. Vlasov, Electronic tongue for pharmaceutical analytics: quantification of tastes and masking effects. *Anal. Bioanal. Chem.* **380**, 36–45 (2004).
107 V.V. Cosofret, M. Erdosy, T.A. Johnson, R.P. Buck, R.B. Ash, and M.R. Neuman, Microfabricated sensor arrays sensitive to Ph and K^+ for ionic distribution measurements in the beating heart. *Anal. Chem.* **67**, 1647–1653 (1995).

CHAPTER 5

Recent developments in electrochemical immunoassays and immunosensors

Jeremy M. Fowler, Danny K.Y. Wong, H. Brian Halsall, and William R. Heineman

5.1 INTRODUCTION

There is a continuing demand for fast and simple analytical methods for the determination of many clinical, biochemical and environmental analytes. In this respect, immunoassays and immunosensors that rely on antibody–antigen interactions provide a promising means of analysis owing to their specificity and sensitivity. High specificity is achieved by the molecular recognition of target analytes (usually the antigens) by antibodies to form a stable complex on the surface of an immunoassay system or an immunosensor [1]. On the other hand, sensitivity depends on several factors including the use of high affinity analyte-specific antibody(ies), their orientation after being immobilized on the immunoassay or immunosensor surface and the appropriate detection system for measuring the analytical signal [1]. Electrochemical detection overcomes problems associated with other modes of detection of immunoassays and immunosensors. For example, the short half-life of radioactive agents, concerns of health hazards, and disposal problems are frequently raised in radioimmunoassays, while limited sensitivity in the analysis of colored or turbid samples is achieved in immunoassays coupled with optical detection. In contrast, electrochemical immunoassays and immunosensors enable fast, simple, and economical detection that is free of these problems. Furthermore, electrochemistry is an interfacial process in which the relevant reactions take place at the electrode–solution interface, rather than in bulk solution. Therefore, in conjunction with developments in micro- and nanoelectrochemical sensors, electrochemistry offers an added bonus of detecting analytes in very small volumes [2]. In this chapter, we will regard an electrochemical immunoassay as a solid phase system in which an antibody–antigen reaction takes place but the corresponding electrochemical detection is carried out elsewhere. However, an electrochemical immunosensor is a stand-alone device with the immunoreaction and electrochemical detection occurring within the same device.

There are already several excellent reviews on electrochemical immunoassays available in the literature [2–7]. In this chapter, we will focus our discussions based on work primarily reported post-2002. Many aspects of earlier immunoassays have already been reported in reviews quoted above. We will begin here with some basic background information including antibody structure and the antibody–antigen interaction, which are the crucial components in the development of all immunoassays and immunosensors after adopting a particular immunoassay format. Following this, specific discussions on recent developments in electrochemical immunoassays and immunosensors will be presented so that non-experts can readily gain an understanding and appreciation of the significance of work in the area.

5.2 THE ANTIBODY–ANTIGEN INTERACTION

Immunoassays and immunosensors are analytical systems that both use the remarkable specificity provided by the molecular recognition of an antigen by antibodies. Antibodies are a family of glycoproteins known as immunoglobulins (Ig). There are generally five distinct classes of glycoproteins (IgA, IgG, IgM, IgD, and IgE) with IgG being the most abundant class (approximately 70%) and most often used in immunoanalytical techniques [8]. As depicted in Fig. 5.1, IgG is a "Y"-shaped molecule based upon two distinct types of polypeptide chains. The molecular weight of the smaller (light) chain is approximately 25 000 Da, while that of the larger (heavy) chain is approximately 50 000 Da. In each IgG molecule, there are two light and two heavy chains held together by disulfide linkages. Both heavy and light chains are divided into constant (C) and variable (V) domains based on their amino acid sequence variability. The light chains have a single variable domain (V_L) and a single constant domain (C_L). In comparison, a heavy chain consists of a single variable domain (V_H) and three constant domains (C_H1, C_H2, C_H3). In general, the antibody molecule may be divided into two main fragments, the non-antigen binding fragment (denoted as Fc) and the antigen-binding fragment ($F(ab')_2$), as indicated in Fig. 5.1.

The variable domains in both chain types are the most important regions with regard to the antibody–antigen binding interaction. The specificity of an antibody towards the binding site (or epitope) of its antigen is a function of its amino acid sequence. Within the V_L and V_H domains, there are three distinct subregions of high sequence variability, known as hypervariable regions. There are three on each light chain and three on each heavy chain, forming six hypervariable loops known as complementarity determining regions, which constitute the antigen binding site. It is the diversity in this region that allows antibodies of high affinity to be produced against almost any antigen. It is estimated that 10^8 antibody specificities can be produced from this one basic molecular structure, and an individual antibody will usually recognize only one antigen, although there are possible cross-reactivities [8].

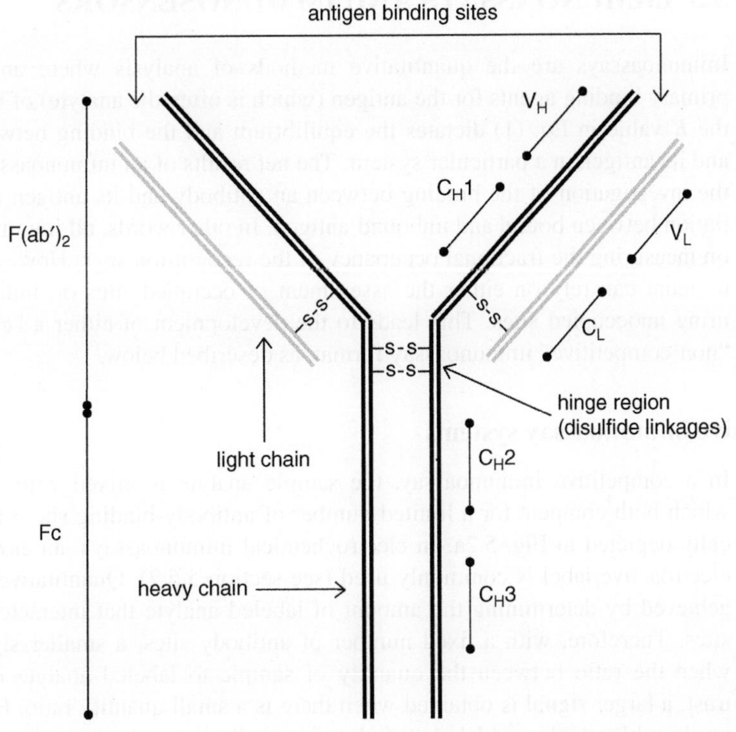

FIGURE 5.1 A schematic illustrating the "Y"-shaped structure of an antibody. The region between the heavy chain and the light chain is where antigen binding occurs. This open arm portion of the "Y" shape is generally denoted as $F(ab')_2$, while the non-antigenic binding site in the base portion is referred to as Fc.

The binding affinity between an antibody (Ab) and an antigen (Ag) can generally be described by the equilibrium expression:

$$K = \frac{[\mathrm{Ab} - \mathrm{Ag}]}{[\mathrm{Ab}][\mathrm{Ag}]} \quad (1)$$

where K is the equilibrium constant for the interaction and Ab—Ag is the immunocomplex formed between Ab and its specific Ag. Typical values of K range from 10^6 to $10^{12}\ \mathrm{L\,mol^{-1}}$. In general, only antibodies yielding a high K value ($\geqslant 10^8\ \mathrm{L\,mol^{-1}}$) will exhibit low cross-reactivity. These features thus make many antibodies ideal biological recognition components in immunoassays and biosensor design. In addition, monoclonal antibodies are particularly suitable for use in immunoassays [8]. These are antibodies produced by a cloned cell line, and thus have the same epitopic specificity and affinity. This homogeneous population can be made in large numbers. Therefore, they offer superior specificity and homogeneity compared to polyclonal antibodies, reducing the need for laborious purification.

5.3 IMMUNOASSAYS AND IMMUNOSENSORS

Immunoassays are the quantitative methods of analysis where antibodies are the primary binding agents for the antigen (which is often the analyte) of interest. Notably, the K value in Eq. (1) dictates the equilibrium and the binding between an antibody and its antigen in a particular system. The net results of an immunoassay are thus often the investigation of the binding between an antibody and its antigen and the differentiation between bound and unbound antigen. In other words, all immunoassays depend on measuring the fractional occupancy of the recognition sites. However, such a measurement can rely on either the assessment of occupied sites or, indirectly, on measuring unoccupied sites. This leads to the development of either a "competitive" or a "non-competitive" immunoassay format, as described below.

5.3.1 Competitive immunoassay systems

In a competitive immunoassay, the sample analyte is mixed with labelled analyte, which both compete for a limited number of antibody-binding sites. This is schematically depicted in Fig. 5.2a. In electrochemical immunoassays, an enzyme label or an electroactive label is commonly used (see section 5.5.2). Quantitative analysis can be achieved by determining the amount of labeled analyte that interacted at the binding sites. Therefore, with a fixed number of antibody sites, a smaller signal is expected when the ratio between the quantity of sample to labeled analyte is large. In contrast, a larger signal is obtained when there is a small quantity ratio. Hence, the signal produced by the bound labeled analyte is usually inversely proportional to the amount of sample analyte. An example involving competitive immunoassays is a voltammetric enzyme immunosensor for the determination of rabbit IgG [9]. In this work, the surface of a screen-printed carbon electrode (SPCE) was modified first with streptavidin. The immobilized streptavidin was used to bind biotinylated anti-rabbit IgG. The modified surface was then exposed to rabbit IgG and rabbit IgG that had been labeled with the enzyme alkaline phosphatase (AP). These two antibodies compete for a limited number of binding sites of the immobilized anti-rabbit IgG and the square wave voltammetric signal produced is due to the oxidation of the product of an enzymatic reaction. A detection limit (based on the analytical signal that is three times greater than the blank signal) of 50 pmol L^{-1} (or 7.0 ng mL^{-1}) for rabbit IgG was achieved by this system.

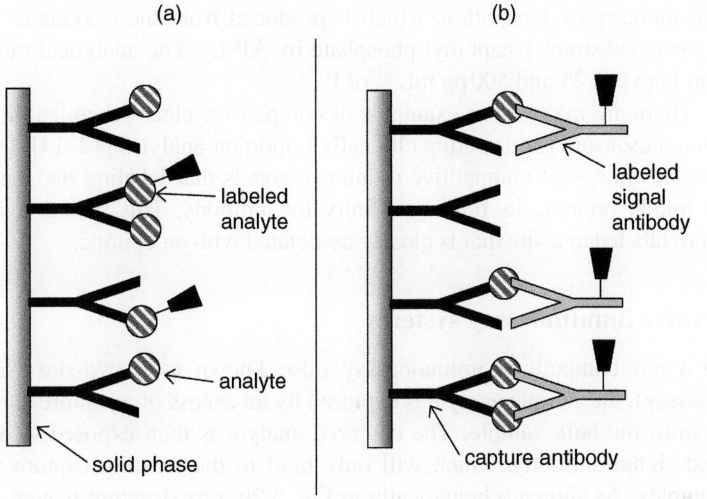

FIGURE 5.2 Schematic representation of (a) competitive and (b) non-competitive immunoassay formats.

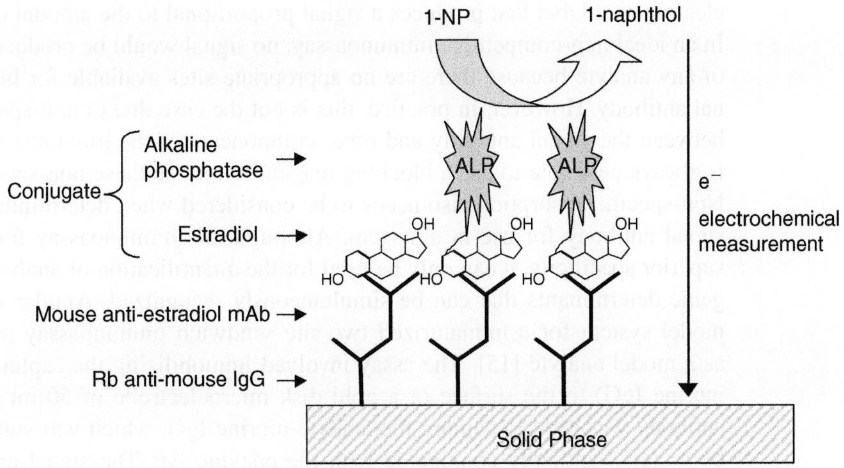

FIGURE 5.3 A schematic illustrating a competitive immunoassay format used for the detection of estradiol. (Reprinted from [11] with permission from Elsevier.)

Competitive immunoassays may also be used to determine small chemical substances [10, 11]. An electrochemical immunosensor based on a competitive immunoassay for the small molecule estradiol has recently been reported [11]. A schematic diagram of this immunoassay is depicted in Fig. 5.3. In this system, anti-mouse IgG was physisorbed onto the surface of an SPCE. This was used to bind monoclonal mouse anti-estradiol antibody. The antibody coated SPCE was then exposed to a standard solution of estradiol (E_2), followed by a solution of AP-labeled estradiol (AP-E_2). The E_2 and AP-E_2 competed for a limited number of antigen binding sites of the immobilized anti-estradiol antibody. Quantitative analysis was based on differential pulse voltammetry of 1-naphthol, which is produced from the enzymatic hydrolysis of the enzyme substrate 1-naphthyl phosphate by AP-E_2. The analytical range of this sensor was between 25 and 500 pg mL^{-1} of E_2.

There are many other examples of competitive electrochemical immunoassays and immunosensors for detecting clinically important analytes [12–14]. Despite simplicity, a disadvantage of competitive immunoassays is that labeling the analyte may reduce, or totally remove, its binding affinity for antibody. This would occur if the analyte were labeled at a site that is closely associated with an epitope.

5.3.2 Non-competitive immunoassay systems

In a non-competitive immunoassay (also known as a two-site "sandwich" immunoassay), the sample analyte is captured by an excess of a capture antibody, separating it from the bulk sample. The captured analyte is then exposed to an excess of second signal antibody, which will only bind to the existing capture antibody–analyte complex. As shown schematically in Fig. 5.2b, this structure is now a classic two-site immunoassay complex in which the analyte is sandwiched between two antibodies. In this system, the signal antibody is often conjugated to either an enzyme label or an electroactive label that produces a signal proportional to the amount of bound analyte. In an ideal non-competitive immunoassay, no signal would be produced in the absence of any analyte because there are no appropriate sites available for binding to the signal antibody. However, in practice, this is not the case due to non-specific interactions between the signal antibody and other components of the immunoassay. Therefore, it is always desirable to use a blocking reagent to reduce these non-specific interactions. Non-specific adsorption also needs to be considered when determining the quantity of signal antibody for use in a system. Although this immunoassay format often offers superior specificity, it can only be used for the quantification of analytes with two antigenic determinants that can be simultaneously recognized. Aguilar *et al.* proposed a model system for a miniaturized two-site sandwich immunoassay using murine IgG as a model analyte [15]. The assay involved immobilizing the capture antibody (anti-murine IgG) to the surface of a gold disk microelectrode of 50 μm in diameter. The antibody was used to capture the analyte murine IgG, which was subsequently bound by a signal antibody conjugated with the enzyme AP. The signal antibody was specific for a different site of the analyte IgG. Upon the catalytic conversion of the substrate, 4-aminophenyl phosphate, by AP to form 4-aminophenol, the magnitude of the oxidation current of 4-aminophenol was used to relate to the quantity of analyte present. Femtogram detection limits for murine IgG were achieved in this electrochemical immunoassay format.

Another example is a non-competitive immunoassay system developed for the detection of pathogenic bacteria in food samples [16]. In this work, capture antibody was immobilized on the surface of highly dispersed carbon particles. The design of the sensing system incorporates a flow-through immunosensor that contains the carbon particles within a disposable centrifugal filtration device at the base of a hollow carbon rod (the working electrode). A hollow Ag|AgCl rod and an additional hollow carbon rod act as the reference and counter electrodes, respectively. A schematic diagram of this flow cell is shown in Fig. 5.4. The design is such that fluid flows through the trapped immunosorbent via the working electrode. The fluid then flows through the counter and reference electrodes before being expelled to waste. After trapping the immunosorbent in the flow cell and an initial washing stage, a solution containing the analyte cells was injected into the system to be captured by the capture antibody. Next, horseradish peroxidase (HRP)-labeled signal antibody was introduced to complete the two-site sandwich immunoassay. After a final washing stage, amperometry was carried out for quantitative determinations by flowing sodium acetate buffer (10 mM, pH 5.6) containing the HRP substrates, hydrogen peroxide and sodium iodide (each 1 mM)

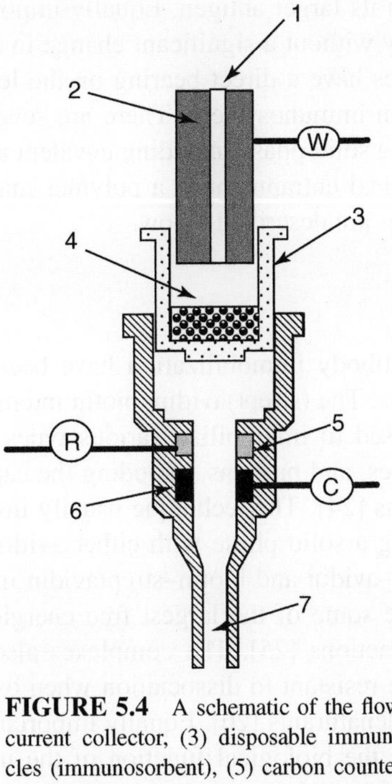

FIGURE 5.4 A schematic of the flow-through immunosensor equipped with (1) fluid inlet, (2) carbon current collector, (3) disposable immuno-column, (4) highly dispersed antibody-modified carbon particles (immunosorbent), (5) carbon counter electrode, (6) Ag|AgCl reference electrode and (7) fluid outlet. (Reprinted from [16] with permission from Elsevier.)

through the system. The potential of the working electrode was set at 105 mV and a steady state current was achieved after 5 min. Three different pathogenic microorganisms were tested in this system, with the best results being for *Listeria monocytogenes*, an organism associated with the cause of the rare and potentially fatal disease Listeriosis. For this organism, the detection limit was 10 cells mL^{-1} and the analytical range was between 10 and 1500 cells mL^{-1}. At cell concentrations greater than 1500 cells mL^{-1}, the amperometric signal decreased due to the "hook effect" [17–20], which is a result of a decrease in the immunoreaction between the capture antibody and the analyte pathogen at high concentrations.

There are also examples of non-competitive assays in the literature for analyzing different clinically important species. For example, an immunosensor for the pathogenic bacterium *Salmonella typhi* and for bacterial toxins from pathogenic *Vibrio cholerae* [21–23].

5.4 MODES OF ANTIBODY IMMOBILIZATION

The manner in which a capture antibody is immobilized on a solid phase is a critical aspect that requires careful consideration in the design of an immunoassay system, whether it is competitive or non-competitive. A desirable feature of the chosen method is that it results in an immobilized capture antibody that is oriented with minimal steric hindrance to interact favorably with its target antigen. Equally important, it is highly desirable to immobilize the antibody without a significant change in its ability to bind its antigen. Clearly, all these features have a direct bearing on the level of sensitivity and dynamic range achievable by an immunosystem. There are several strategies for immobilizing a capture antibody on a solid phase including covalent attachment, physical adsorption or electrostatic/physical entrapment in a polymer matrix. These commonly used immobilization strategies are described below.

5.4.1 Biotin–(strept)avidin interaction

Specific affinity interactions for antibody immobilization have been widely used in immunoassay systems in recent years. The (strept)avidin–biotin interaction is one such example. This technique may be used to immobilize various types of biomolecules such as nucleic acids, polysaccharides, and proteins, including the capture antibody in immunoassay/immunosensor systems [24]. The technique usually involves biotinylating the capture antibody and coating a solid phase with either avidin or streptavidin. The dissociation constants of biotin–avidin and biotin–streptavidin interactions are of the order of 10^{-15} mol L^{-1} and are some of the largest free energies of association yet observed for non-covalent interactions [25]. The complexes also withstand high temperatures, pH variations, and are resistant to dissociation when exposed to chemicals such as detergents and protein denaturants [26]. Equally important, the use of this immobilization technique maintains the biological function of the immobilized antibody [24]. In some cases, neutravidin, which is an almost neutrally charged (pI of 6.3) variation of avidin, is used to minimize any non-specific binding by charged species to maintain high binding affinity for biotin. An electrochemical immunosensor for the detection of *Mycobacterium tuberculosis*, based on the immobilization of a capture antibody using the biotin–streptavidin interaction, has recently been reported [27]. In this system, biotinylated anti-*M. tuberculosis* antibody was immobilized on the surface of a streptavidin-modified SPCE. The electrode was initially oxidized in 0.1 mol L^{-1} H_2SO_4 by applying an anodic current of 25 μA for 2 min. This was to enhance the adsorptive properties of the SPCE as a result of an increased hydrophilicity. Following pre-treatment, streptavidin solution was applied to the electrode surface overnight. Any remaining free sites of the SPCE were blocked with a bovine serum albumin (BSA) solution. Next, biotinylated anti-*M. tuberculosis* was applied to the electrode and the biotin–streptavidin interaction was allowed to proceed for 90 min. Incubation between antigen *M. tuberculosis* and monoclonal mouse anti-*M. tuberculosis* was carried out remotely, then introduced to the sensor surface for capture by the immobilized capture antibody. The immunosensor structure was completed by introducing AP-labeled rabbit anti-mouse antibody (Fig. 5.5a). The substrate 3-indoxyl phosphate was then introduced and converted to its Indigo product by AP. Indigo was converted to hydrosoluble indigo carmine and the analytical signal was produced by either cyclic or square-wave voltammetry. A detection limit of 1 ng mL^{-1} *M. tuberculosis* was achieved by this immunoassay. The results were compared to those of a similar assay, which relied upon the passive adsorption of monoclonal rabbit anti-mouse antibody directly onto

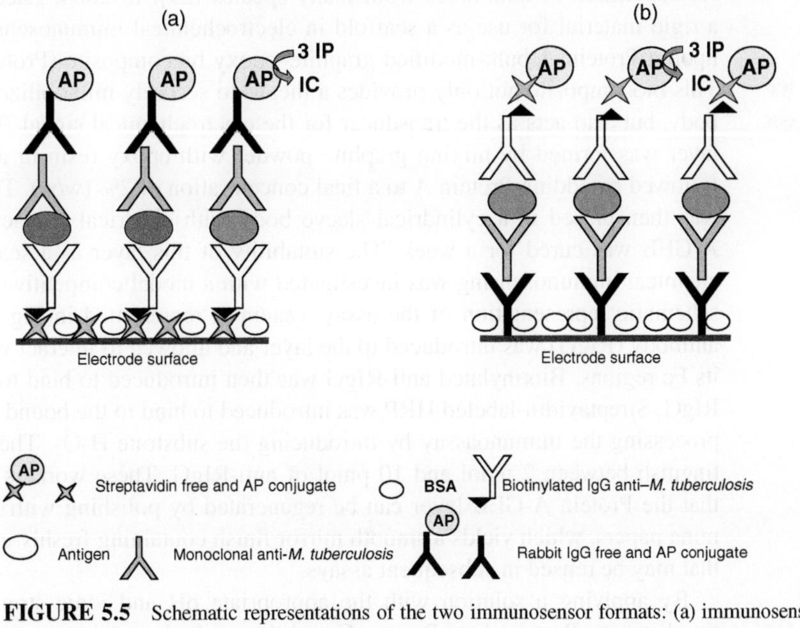

FIGURE 5.5 Schematic representations of the two immunosensor formats: (a) immunosensor based on the biotin–streptavidin interaction and (b) immunosensor based on rabbit IgG-modified SPCEs. (Reprinted from [27] with permission from Elsevier.)

the surface of a pre-treated SPCE (Fig. 5.5b). This assay format yielded a detection limit of 40 ng mL^{-1}, indicating that the biotin–streptavidin interaction used to immobilize capture antibody provides a suitable support for electrochemical immunosensing. Further, it is likely that sufficient capture antibody was immobilized on the surface of the SPCE with appropriate orientation for binding analyte.

5.4.2 Antibody-binding proteins

Another commonly used affinity-based immobilization technique for capture antibodies in immunoassay systems involves a bacterial antibody-binding protein. The two most common of which are Protein A and Protein G. These proteins bind specifically to antibodies through their non-antigenic (Fc) regions, which allow the antigen binding sites of the immobilized antibody to be oriented away from the solid phase and be available to bind the target analyte. As these proteins interact directly with the Fc region of antibodies, there is no need for antibody biotinylation. Protein A has a molecular weight of approximately 42 kDa and was originally isolated from the cell wall of *Staphylococcus aureus* [28]. It is known to contain five Fc binding domains located towards its $-NH_2$ terminal. However, the binding capacity of Protein A is limited to three human IgG subclasses (IgG 1, 2 and 4) [29]. Also, Protein A will not bind to goat and rat IgG, and only weakly to mouse IgG [30]. The second bacterial antibody binding protein, Protein G, is a cell surface protein of group C and G *streptococci* with three Fc binding domains located near its C-terminal, and has specificity for subclasses of antibodies from many species [29]. In 2004, Zacco *et al.* reported a rigid material for use as a scaffold in electrochemical immunosensing that is based upon a Protein A bulk-modified graphite–epoxy biocomposite (Protein A-GEB) [31]. This biocomposite not only provides a means to securely immobilize the capture antibody, but also acts as the transducer for the electrochemical signal. The biocomposite layer was formed by mixing graphite powder with epoxy resin in a 1:4 (w/w) ratio, followed by adding Protein A to a final concentration of 2% (w/w). The resulting paste was then placed in a cylindrical sleeve body with electrical contact and the Protein A-GEB was cured for a week. The suitability of this layer as a scaffold for electrochemical immunosensing was investigated with a model competitive immunoassay. A schematic representation of the assay system is represented in Fig. 5.6. First, rabbit antibody (RIgG) was introduced to the layer and allowed to interact with Protein A via its Fc regions. Biotinylated anti-RIgG was then introduced to bind to the immobilized RIgG. Streptavidin-labeled HRP was introduced to bind to the bound anti-RIgG before processing the immunoassay by introducing the substrate H_2O_2. The assay could distinguish between 2 pmol and 10 pmol of anti-RIgG. These workers have also shown that the Protein A-GEB layer can be regenerated by polishing with abrasive and alumina papers, which yields a smooth mirror finish containing freshly exposed Protein A that may be reused in subsequent assays.

By applying a solution with the appropriate pH and ionic strength, the interaction between Protein A or Protein G and the antibody can be reversed, enabling easy renewal of sensing surfaces [32, 33]. This has been demonstrated by Yakovleva *et al.*

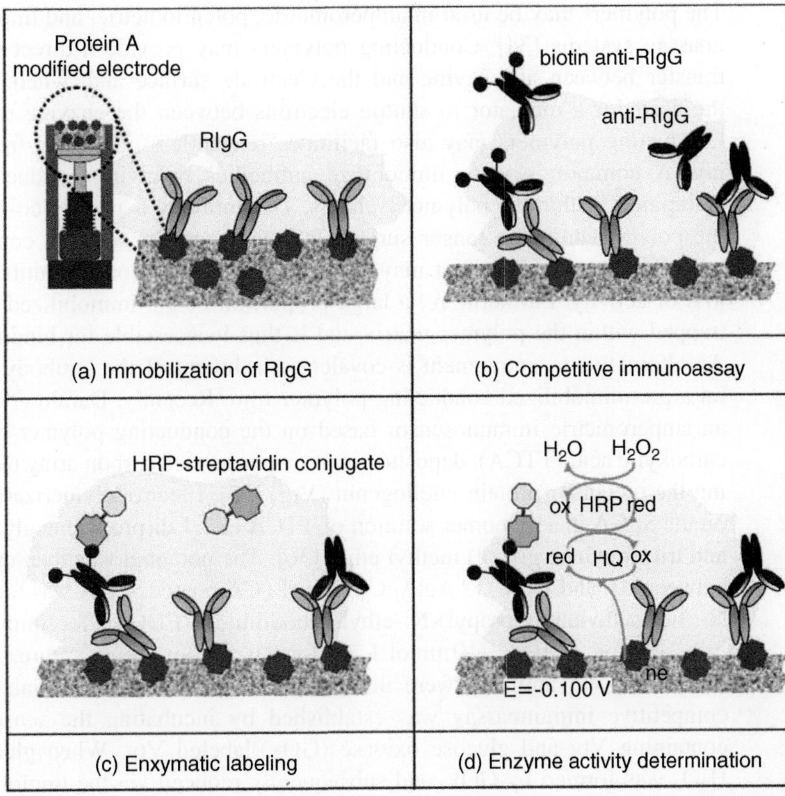

FIGURE 5.6 Schematic representation of the immunosensor based on a Protein A-GEB biocomposite as a transducer. (a) Immobilization of RIgG on the surface via interaction with Protein A, (b) competitive immunoassay using anti-RIgG and biotinylated anti-RIgG, (c) enzyme labeling using HRP-streptavidin and (d) electrochemical enzyme activity determination. (Reprinted from [31] with permission from Elsevier.)

who have developed a renewable microfluidic immunosensor using Protein G as the immobilization aid [33]. In this work, Protein G was covalently bound to a silicon microchip to trap immunocomplexes that were formed off-line in a competitive assay between antibody and either labeled or unlabeled analyte antigen. Following injection of these complexes and processing of the immunoassay, the Protein G bound to the silicon microchip was regenerated by removing the immunocomplexes. This was achieved by injecting glycine-HCl buffer (0.4 mol L^{-1}, pH 2.2) at a flow rate of 50 $\mu L\ min^{-1}$ for 2–3 min. The Protein G chip lost no activity after more than 8 months.

5.4.3 Conducting polymers

The application of conducting polymers such as polyaniline, polypyrrole, and polythiophene for immobilizing capture antibodies in immunoassay systems is widespread.

The polymers may be used in amperometric, potentiometric, and impedimetric immunoassay systems [34]. Conducting polymers may provide a direct route of electron transfer between an enzyme and the electrode surface and, where required, negate the need for a mediator to shuttle electrons between the enzyme and the electrode. Conducting polymers may also facilitate "reagentless" or "label-free" immunosensing. A common way to immobilize antibodies involving conducting polymers is entrapment within the polymeric chains. The antibody is usually co-immobilized with the polymer onto the sensor surface from a monomer solution containing the antibody. However, entrapment may result in denaturation of the antibody, leading to a loss of activity. Furthermore, a large proportion of the immobilized antibody will be trapped within the polymer matrix and is thus inaccessible for binding to its antigen. An alternative to entrapment is covalent attachment of the antibody to active groups on a pre-immobilized conducting polymer film. Recently, Darain *et al.* have reported an amperometric immunosensor based on the conducting polymer poly-terthiophene carboxylic acid (TTCA), deposited on a screen-printed carbon array (SPCA) for detecting the precursor protein vitellogenin (Vtg) [35]. Electropolymerization was achieved on the SPCA in a monomer solution of TTCA in 1:1 di(propylene glycol) methyl ether and tri(propylene glycol) methyl ether [36]. The potential was then cycled three times between 0.0 and 1.6 V (vs Ag|AgCl). The TTCA-coated SPCA was functionalized with N-(3-dimethylaminopropyl)-N′-ethylcarbodiimide (EDC) after immersing the array in a solution of EDC (10 mmol L^{-1}) for 4 h at room temperature. HRP and monoclonal anti-Vtg antibody were then coupled to the EDC-functionalized polymer. A competitive immunoassay was established by incubating the sensor in a solution containing Vtg and glucose oxidase (GOx)-labeled Vtg. When glucose was added, H_2O_2 was formed by GOx, and subsequently reduced via the immobilized HRP. The signal generation scheme is illustrated in Fig. 5.7. The presence of the TTCA layer ensures that only Vtg-GOx bound by the capture antibody was involved in the enzyme channeling, and that unbound Vtg-GOx in the bulk solution induced no significant electrocatalytic effect. This is often termed a separation-free immunosensor. The enzyme channeling has a catalytic effect on the amperometric signal, which was measured at −0.3 V (vs Ag|AgCl). A detection limit (based on four times the standard deviation of the blank) of 0.09 ng mL^{-1} of Vtg was achieved using this immunosensor.

5.4.4 Self-assembled monolayers

Self-assembled monolayers (SAMs) are another attractive method for immobilizing the capture antibody in immunoassay systems. By taking advantage of the spontaneous chemisorption of alkanethiols to such metals as gold or silver, highly ordered monolayers can be assembled. Gold electrodes are most frequently used because of the absence of stable oxide under ambient conditions. The first step in the formation of an alkanethiol SAM is the chemisorption of the sulfhydryl group of the alkanethiol to the gold surface through the formation of a gold-sulfur bond. One well-accepted model describing the formation of this bond suggests that it is an oxidative addition of the

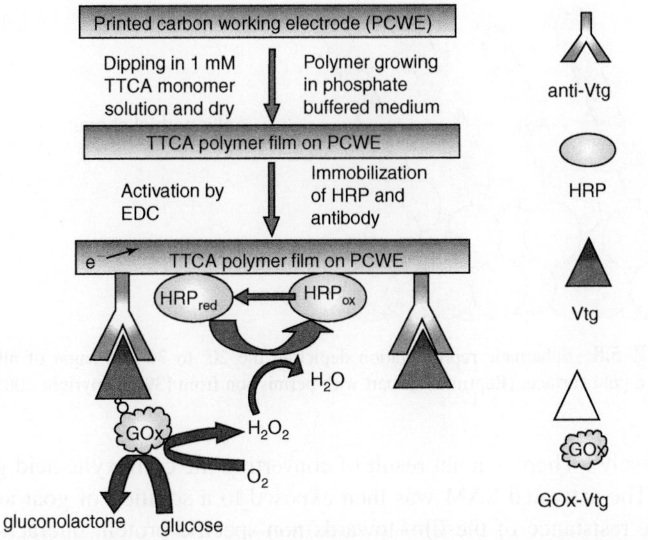

FIGURE 5.7 Schematic representation of electrode preparation and signal generation scheme for the immunoassay based on a film of TTCA. (Reprinted from [35] with permission from Elsevier.)

S-H bond to the gold surface, followed by the reductive elimination of hydrogen, as shown in Eq. (2) [37].

$$R - S - H + Au_n^0 \rightarrow R - S - Au^+Au_n^0 + \frac{1}{2}H_2 \quad (2)$$

The alkyl chains of the thiols are all in the *trans* conformation, and the second step in the formation of a highly ordered monolayer involves van der Waals interactions between the chains, which results in the chains having a final tilt angle of between 20° and 30° from the normal (Fig. 5.8) [38, 39]. It is the highly ordered nature of SAMs that is exploited to provide controlled orientation of antibodies in electrochemical immunoassay systems. By forming a SAM with an ω-carboxyfunctionalized alkanethiol, the exposed surface of the monolayer consists of free carboxylic acid groups that, upon activation, can be covalently bound to amine groups from the lysine residues of the capture antibody or other protein. SAMs with other terminal functionalities may also be used. For example, in Herrwerth *et al.*'s work, antibodies were covalently coupled to a SAM of carboxy-functionalized poly(ethylene glycol) alkanethiol [$HOOC—CH_2—(OCH_2—CH_2)_n—O—(CH_2)_{11}—SH$] [40]. This molecule has a dual function in which the carboxylic acid group is used for coupling to the antibody, and the poly(ethylene glycol) region provides resistance against non-specific binding. The SAM was formed by immersing thin films of polycrystalline gold in a 10 mL solution of the alkanethiol (0.5 mM) in absolute *N,N*-dimethylformamide for 18 h. The exposed carboxylic acid termini were activated by immersing the SAM-coated films in a solution containing EDC and N-hydroxysuccinimide (0.2 mol L^{-1} and 0.05 mol L^{-1}

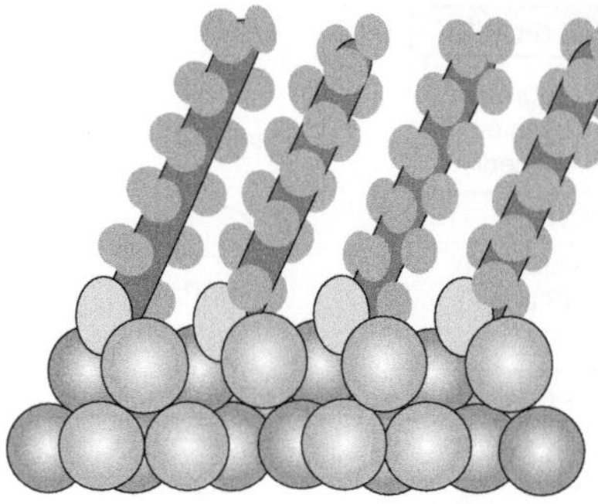

FIGURE 5.8 Schematic representation depicting the 20° to 30° tilt angle of alkanethiol chains in a SAM on a gold surface. (Reprinted in part with permission from [39]. Copyright 2005 American Chemical Society.)

respectively). There is a net result of converting the carboxylic acid groups into active esters. The activated SAM was then exposed to a solution of goat anti-rabbit IgG for 2h. The resistance of the film towards non-specific protein interactions and the ability of the immobilized capture antibody to partake in specific immunoreactions were investigated. The SAM was exposed to a solution of the protein fibrinogen. After thorough rinsing, the SAMs were characterized by Fourier transform-infrared reflection absorption spectroscopy (FT-IRRA). The lack of an amide I or amide II band in the spectra indicated that, despite the presence of terminal —COOH groups, the SAM was resistant to non-specific protein interactions. It was also demonstrated that the SAM was inert towards non-specific interactions after antibody immobilization. The immunoreactivity of the immobilized capture antibody was investigated by introducing rabbit anti-sheep IgG as an antigen. This resulted in significant growth of the amide bands in the FT-IRRA spectra, indicating the presence of a specific interaction.

In addition to covalently attaching antibodies to SAMs, they may also be adsorbed on charged SAMs via electrostatic interactions. Chen *et al.* have investigated the effects of varying surface and solution properties on the orientation of adsorbed antibodies [41]. In this work, the physisorption of anti-human chorionic gonadotrophin (hCG; which is a hormone used in pregnancy tests and can also act as a tumor marker) IgG to COOH- and NH_2-terminated SAMs was studied. Isoelectric focusing was used to determine the isoelectric points (IEPs) of IgG and its Fc and $F(ab')_2$ fragments to be 6.8, 8.3, and 6.0, respectively. Surface plasmon resonance was used to determine the surface coverage of adsorbed antibody to COOH- (negatively charged) and NH_2- (positively charged) terminated SAMs. A buffer with low ionic strength was used (2.1 mM PBS with 0.3 mM NaCl) in order to maximize electrostatic interaction between IgG and the SAMs. A much larger quantity of IgG was found to adsorb on the COOH-terminated SAMs. Maximum adsorption of IgG on COOH- and NH_2-terminated SAMs was expected to occur at pH values lower and higher than its IEP, respectively. The maximum adsorption on the COOH-terminated SAMs of approximately 5 mg m^{-2}, occurred at pH 6.25, whereas the maximum adsorption on the NH_2-terminated SAMs was approximately 2.3 mg m^{-2} and was also at pH 6.25. This particular result is inconsistent with what was expected and was attributed to the asymmetric structure of the IgG (Fc and $F(ab')_2$ regions). If a highly idealized packing of antibody is considered, the orientation of the adsorbed antibody can be estimated. In cases where antibodies are oriented "end on" (Fc region closer to the SAM) or "head on" ($F(ab')_2$ region closer to the SAM), the surface coverage is expected to range between 2.6 mg m^{-2} and 5.5 mg m^{-2}. On the other hand, when antibodies are adsorbed "side on" (Fc region and one Fab′ region bound to the SAM), a surface coverage of approximately 2.0 mg m^{-2} is estimated. Based on these values, IgG bound to the COOH-terminated SAMs is probably oriented head or end on. To further investigate the orientation of adsorbed antibodies, the ratio of hCG to immobilized anti-hCG IgG was investigated at various pHs. In these experiments, a higher ratio (0.48) was observed for IgG adsorbed to the positively charged NH_2 surface near the IEP of IgG. This suggests that there were more $F(ab')_2$ regions exposed and available for analyte binding. In summary, it was believed that IgG adsorbed to COOH-terminated SAMs has a head-on orientation, and on the NH_2-terminated SAMs the orientation is believed to vary between end on and side on. Therefore, for immunosensing systems where capture antibody adsorption to a SAM is the chosen method of immobilization, a positively charged NH_2-terminated SAM is the preferred surface. Additionally, adsorption should be carried out under low ionic strength conditions and at a pH similar to the IEP of the antibody.

Another way to control the orientation of a capture antibody on the surface of a SAM is to employ Protein A or Protein G, covalently bound to the SAM. As described in section 5.4.2, these proteins bind antibodies via their Fc regions and leave the antigen binding sites available for binding analyte. Recently, our laboratory has demonstrated the covalent coupling of Protein A to a thioctic acid SAM deposited on a gold disk electrode [42]. In this work, the thioctic acid SAM was activated to its *o*-acylurea intermediate with EDC. A solution of Protein A (containing 30 μg) was then applied to the electrode surface. The electrode was incubated overnight at 4ºC to facilitate the binding of Protein A via its lysine residues. The sensor surface was then ready for incubation with capture antibody and subsequent processing of an immunoassay.

5.4.5 Antibody fragments

An alternative to immobilizing whole capture antibody molecules in immunoassay systems is the immobilization of antibody fragments (Fab′). Fragmentation of an antibody is usually achieved enzymatically with proteolytic enzymes such as chymotrypsin, trypsin, and papain [43]. Following enzymatic digestion, the disulfide linkages holding the two chains of the resulting $F(ab')_2$ fragment together are typically reduced with reagents such as dithiothreitol or 2-mercaptoethalamine. This results in two Fab′ fragments, each with a terminal thiol group. The fragments thus have a high affinity for a gold surface, on which they can therefore self-assemble without the need of any additional reagent. The resulting layers have an ordered arrangement with the antibody-binding regions being oriented in such a way that they are more accessible to bind antigen. Recently, Zhang and Meyerhoff reported an immunoassay based upon Fab′ fragments immobilized on the surface of gold coated magnetic particles for detecting C-reactive protein (CRP) [44]. The immobilization of anti-CRP Fab′ fragments on the surface of the gold-plated particles was achieved by incubating the particles in a solution of the fragments overnight at 4ºC. The Fab′-coated beads were then incubated with different quantities of CRP. After thoroughly rinsing and resuspending the particles, excess HRP-labeled goat anti-human CRP was introduced. Following another washing and resuspension step, the two-site sandwich immunoassay was processed by introducing HRP substrate (3,3′,5,5′-tetramethylbenzidine). After terminating the enzymatic reaction with H_2SO_4, the absorbance of a portion of the supernatant was measured at 490 nm. A calibration plot was constructed by plotting absorbance against CRP concentration. A detection limit of 0.14 ng mL^{-1} was achieved using this immobilization strategy. The results were compared to an identical system, except that Fab′ was covalently linked to uncoated magnetic particles via a tosyl reaction with amine groups of the fragments. This assay yielded a detection limit of 1.9 ng mL^{-1}. The results reveal that the self-assembled layer of Fab′ fragments on the surface of gold-plated magnetic particles provides improved orientation for binding antigen, and is therefore an attractive mode of immobilization. Although this is not an electrochemical immunoassay system, this type of technology could be applied to a wide range of solid-phase immunoassays.

5.5 ELECTROCHEMICAL DETECTION TECHNIQUES

In the previous sections, we have discussed various methods for immobilizing capture antibody on the solid phase of a system. An immunocomplex is next to be constructed according to an adopted immunoassay format. A means of detection is then needed to quantitatively determine the amount of analyte present. Electrochemical techniques have often been used for detection in biosensor technology. This stems from a number of attributes of electrochemistry including the high sensitivity of electrochemical transducers, their compatibility with modern miniaturization/microfabrication technologies, minimal power requirements, economical cost, and independence of sample turbidity and color. In immunoassays and immunosensors, as most antibodies and antigens are intrinsically unable to act as redox partners, an appropriate label is often conjugated to a particular component of the immunocomplex to promote an electrochemical reaction. The electrochemical signal produced is then used to relate quantitatively to the amount of analyte present in a sample solution. Potentiometry, amperometry, voltammetry, and, more recently, electrochemical impedance spectroscopic measurements are among the electrochemical detection techniques often used in conjunction with immunoassay systems and immunosensors, leading to their respective categories according to the type of signal measured. The fundamental principles of each of the four techniques are presented below, followed by discussions based on some recent work that have specifically addressed problems encountered in these areas.

5.5.1 Potentiometric immunosensors

Potentiometric immunosensors rely upon a change in potential that occurs between an indicator and a reference electrode as a result of specific interaction between an antibody and its antigen [45]. There have hitherto been few reports of immunosensors relying on potentiometric detection. One of the main disadvantages of this type of detection is the relatively small change in potential that arises from the interaction between an antibody and its antigen. Moreover, interferences from the sample matrix may prevent this small signal from being successfully detected. Thus such sensors often have a compromised reliability and sensitivity. A recent example of a potentiometric immunosensor involves the detection of enzyme-labeled immunocomplexes formed at the surface of a polypyrrole coated screen-printed electrode [46]. In constructing this electrode, an electropolymerization strategy involving cyclic voltammetry in the presence of aqueous pyrrole and sodium dodecyl sulfate was carried out for at least four cycles. A constant potential was then applied to allow the polypyrrole film to settle into its final state. In addition to being stable at 37°C for at least 4 months, these polypyrrole coated electrodes showed improved sensitivity. Capture antibody was immobilized on the polypyrrole layer by either direct adsorption or by binding biotinylated antibody to streptavidin coated polypyrrole. The immunoelectrode was then incubated in a sample solution containing either hepatitis B surface antigen or the cardiac marker troponin I. A sandwich immunoassay was completed by introducing a signal antibody conjugated to HRP. Potentiometric measurements were conducted in 0.01 M PBS (pH 7.4). An active substrate, *o*-phenylenediamine dihydrochloride, was added to initiate an enzyme turnover, and the potential was recorded after 60 s. Note that a deliberate separation of the immunoreaction from the detection step was employed to achieve effective minimization of matrix interferences. The change in potential was found to be proportional to the extent of the antibody–analyte reaction, and therefore the concentration of analyte in the sample. However, the mechanism for the change in potential, induced by the conversion of *o*-phenylenediamine dihydrochloride to 2,3-diaminophenazine by HRP, is somewhat unclear and was referred to as a "charge-step procedure". It was proposed that a change in pH and ionic strength of the solution as a result of the enzymatic reaction altered the physical (porosity, density, thickness) and electrochemical (conductivity, charge) properties of the polypyrrole layer, resulting in the observed potential shifts. The sensitivities for hepatitis B surface antigen and troponin I were reported to be 50 fmol L^{-1} and 0.4 pmol L^{-1}, respectively.

5.5.2 Amperometric immunosensors

In amperometry, the current produced by the oxidation or reduction of an electroactive analyte species at an electrode surface is monitored under controlled potential conditions. The magnitude of the current is then related to the quantity of analyte present. However, as both antibody and antigen are not intrinsically electroactive, a suitable label must be introduced to the immunocomplex to promote an electrochemical reaction at the immunosensors. In this respect, enzyme labels including the oxidoreductase, HRP, and the hydrolytic enzyme AP are frequently used because of their ability to yield an electroactive product following the catalytic conversion of a substrate. The magnitude of the current arising from the redox reaction of the product can then be quantitatively related to the amount of analyte present. This signal-generation scheme is further illustrated by the example of the enzymatic reaction of AP on its substrate 4-aminophenyl phosphate (4-APP) to yield 4-aminophenol (4-AP) in Scheme 1. This is followed by the oxidation of 4-AP to 4-quinone imine (4-QI) to produce a Faradaic current that is proportional to the amount of analyte. Notably, a major advantage of enzyme labels is, as a result of the catalytic effect, the amplification of the signal that can be detected even when a minute quantity of enzyme is used.

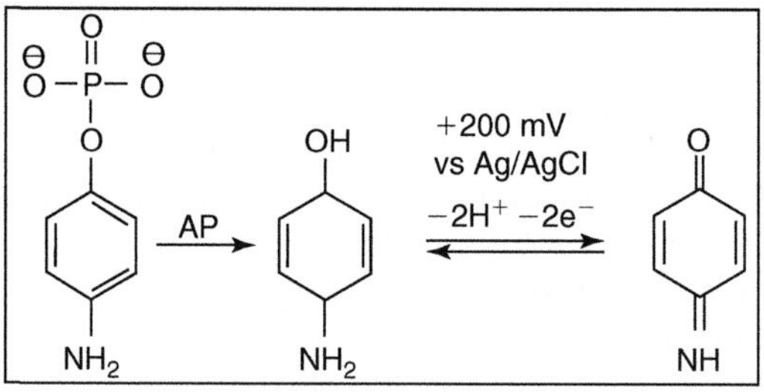

SCHEME 1

Due to the broad substrate specificity of AP, and the drive for higher efficiency, several studies have recently investigated the suitability of alternative substrates to the common 4-APP [47, 48]. For example, Pemberton *et al.* compared 4-APP and 1-naphthyl phosphate (1-NP) as AP substrates in an amperometric immunosensor for progesterone [47]. The signal generation scheme when 1-NP is used as a substrate is illustrated in Scheme 2.

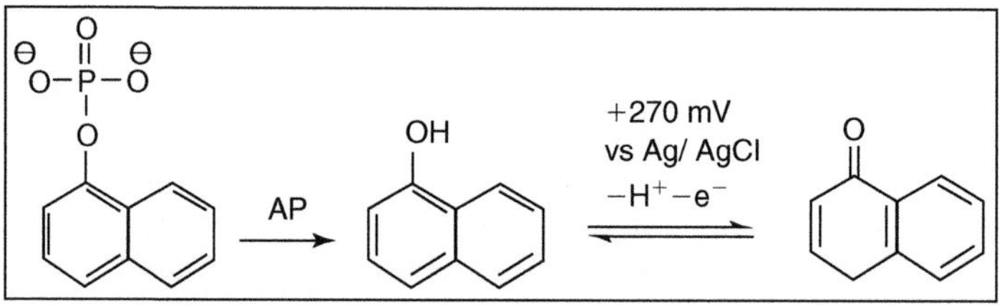

SCHEME 2

They found that the hydrolysis products of 4-AP and 1-naphthol produced well-defined anodic responses at low potentials at a bare SPCE. However, the presence of antibody immobilized on the electrode surface slowed the diffusion of 4-AP towards the electrode surface. In addition, 4-AP may interact with polyphenols on the electrode surface, thus reducing the electroactive working area of the electrode by fouling. In contrast, diffusion of 1-naphthol to the electrode surface was not hindered by immobilized antibody. This feature, along with its low cost, ease of availability, and high solubility, resulted in 1-NP being the preferred AP substrate in their work.

A similar study has also been conducted to determine the suitability of ascorbic acid 2-phosphate (AAP) as an alternative substrate to 4-AP for AP under identical conditions [48]. Although 4-APP and AAP were suitable substrates for amperometric immunosensors, 4-APP was superior owing to its sixfold faster enzymatic reaction and lower detection potential (approximately 200–400 mV). Notably, the lower detection potential for the hydrolysis product of 4-APP minimizes interferences from other species and hence improves the sensitivity of the immunosensor.

The hydrolysis products of common substrates for AP including 4-APP, 1-NP, and phenyl phosphate (PP) can cause passivation of the surface of the sensing electrode owing to poor diffusion characteristics, accumulation of electroinactive species, poor redox reversibility, and polymer formation on the electrode surface [49]. Electrode passivation will in turn result in inconsistencies in signal reproducibility and reliability, small amperometric responses, and possibly a complete quenching of the signal prior to a steady state current being achieved. A relatively new substrate involving hydroquinone diphosphate (HQDP) has been developed to minimize this fouling problem [49]. It is based upon the hydrolysis of HQDP to hydroquinone (HQ) by AP and is followed by the subsequent oxidation of HQ to benzoquinone (BQ) at the electrode surface (Scheme 3).

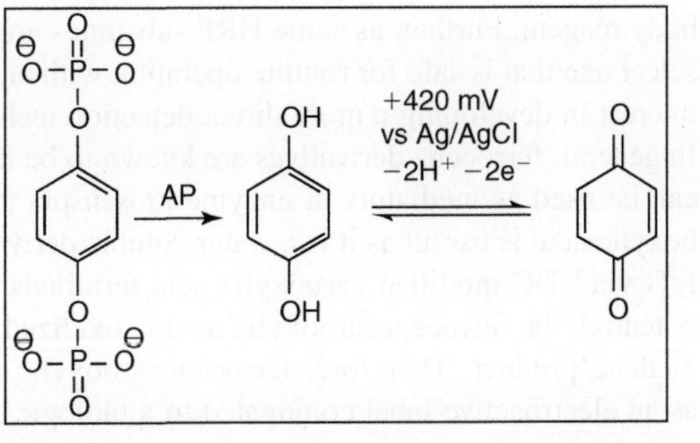

SCHEME 3

A comparison of the products of AP hydrolysis of HQDP (HQ), PP, and 1-NP using cyclic voltammetry revealed that HQ produced well-defined peaks, and that the oxidation of HQ is reversible. More importantly, no apparent passivation of the electrode surface was observed even at high millimolar concentrations after 50 scans. Following a series of investigations, this non-fouling nature of HQ was attributed to the non-accumulation of its oxidation products on the electrode surface and the good diffusional properties of HQ at the electrode–solution interface. Another positive feature of HQDP as a substrate for AP is a tenfold greater oxidation current response of HQ compared to those obtained in the presence of PP or 1-NP. Overall, HQDP provides a suitable and attractive alternative substrate system for AP in the development of amperometric immunosensors.

In some instances, the design of an amperometric immunosensor may be such that the enzyme is located some distance away from the electrode surface, or the presence of interfering substances in biological samples may require using an alternative electron transfer pathway. This usually involves a redox-active species with a small molecular weight (termed a mediator) used to shuttle electrons between the redox center of a peroxidase enzyme (commonly HRP or glucose oxidase) and the working electrode surface. Thus, a mediator must possess intrinsically fast electron transfer rates to facilitate an enhanced heterogeneous electron transfer for peroxidases. Furthermore, a mediator should exhibit reversible heterogeneous kinetics, a low overpotential for regeneration, and be stable under the required range of physical conditions [45]. Recently, a mediated electrochemical immunosensor using single walled carbon nanotube forests has been demonstrated for detecting biotin-HRP and unlabeled biotin, which, in a competitive immunoassay, compete for a limited number of antigen-binding sites on immobilized anti-biotin capture antibody [50]. The mediator of choice in this system is HQ, which is used to facilitate a coupled catalytic reduction of hydrogen peroxide by transferring electrons between HRP and the peroxide. Detection limits of 2.5 nM and 16 μM were achieved for biotin-HRP and unlabeled biotin, respectively. In some systems, leaching of the soluble mediators into the bulk solution may become a problem by reducing the sensitivity of the immunosensor.

A drawback of the enzymatic signal-generating scheme described above is that it is an indirect detection technique. Moreover, AP and other phosphatases present in biological samples such as serum and plasma could interfere in assays if not removed by washing steps following specific interaction of sample analyte with its specific antibody reagent. Further, as some HRP substrates are carcinogenic [51], it is important to select one that is safe for routine operation with an immunosensor. There has thus been interest in developing a more direct detection technique involving electroactive labels. In general, ferrocene derivatives are known to be fast, reversible redox compounds that can be used as mediators in enzyme biosensors [52–55]. Among them, ferrocenecarboxylic acid is useful as it is a water-soluble derivative and can be easily conjugated to IgG via EDC-modified carboxylic acid terminals [56]. Upon applying an appropriate potential, the ferrocenecarboxylic acid is oxidized rapidly without involving any intermediate product. Therefore, ferrocenecarboxylic acid is a superior candidate for use as an electroactive label conjugated to a biological component for developing a direct signal-generating scheme at an electrochemical immunosensor. In our laboratory, we have demonstrated ferrocenecarboxylic acid as an electroactive label, conjugated to the signal antibody of a sandwiched immunocomplex, for the quantitative determination of the tumor marker, human chorionic gonadotrophin with a detection limit of 2.2 IU L^{-1} [57]. Similarly, Okochi *et al.* have used antibodies conjugated with ferrocencarboaldehyde as an electrochemical probe for detecting IgG in an on-chip flow immunoassay [58]. Recently, Hromadová *et al.* demonstrated the organometallic compound (η^5-cyclopentadienyl)tricarbonylmanganese (or cymantrene), as a redox label bound to BSA [59]. Electrochemical detection was based on impedance measurements (see section 5.5.4) of a one-electron reduction of the organometallic label.

5.5.3 Voltammetric immunoassays

The development of electrochemical immunosensors generally involves the immobilization of an immunocomplex on a single electrode, followed by detection via the label on the immunocomplex at the same electrode. More recently, interdigitated array (IDA) microelectrodes have gained popularity as an alternative transducer in electrochemical immunoassays. In general, a simple design of an IDA consists of a pair of interdigitated microelectrode "fingers". When an IDA is used as a sensing electrode in a voltammetric experiment, the two interdigitated electrodes are usually held at different potentials to achieve "redox" cycling of the electroactive species to be detected. This is illustrated in Fig. 5.9 using the detection of 4-AP discussed in section 5.5.2 as an example. According to Scheme 1, an oxidizing potential is applied to one of the two IDA electrodes to promote a two-electron oxidation of 4-AP to 4-QI. Then, 4-QI diffuses to an adjacent electrode held at a more negative potential, where it is reduced to 4-AP, which can then undergo another oxidation at the next adjacent electrode. A major advantage of this redox cycling is that it improves the signal-to-noise ratio by enhancing the Faradaic current relative to the background current, resulting in lower detection limits and improved sensitivity. These features opened up many opportunities in which IDA electrodes were applied as electrochemical detectors in analytical chemistry and biosensor systems [60]. In a recent application, Thomas *et al.* used an IDA consisting of 25 pairs of platinum microelectrodes with 1.6 μm gaps and 2.4 μm widths as a detector in immunoassays for mouse IgG [61]. A fourfold amplification of signal was obtained compared to single-electrode detection, with redox cycling of 87% and a detection limit of 3.5 fmol mouse IgG.

5.5.4 Impedimetric immunoassays and immunosensors

Various strategies were described in section 5.4 for immobilizing antibody on an electrode surface, which is often the first step in building the desired immunocomplex on the surface of an electrochemical immunosensor. Upon the specific molecular recognition of the antigen by the immobilized antibody, there will be changes in the interfacial charge, capacitance, resistance, mass, and thickness at the immunosensor surface. There is thus an emerging interest in exploiting electrochemical techniques that follow such interfacial changes, which may yield quantitative determination of the targeted analyte. In this section, we shall focus on electrochemical impedance spectroscopy (EIS) as an effective method for probing the features of an electrode surface modified by immunocomplexes.

In an EIS experiment, a low amplitude (5 to 10 mV peak-to-peak) sine wave potential signal is superimposed on a fixed DC potential applied to an electrochemical system. Based on Ohm's law, the impedance can be computed from the applied sinusoidal potential and the measured sinusoidal current. As the sinusoidal potential and current will

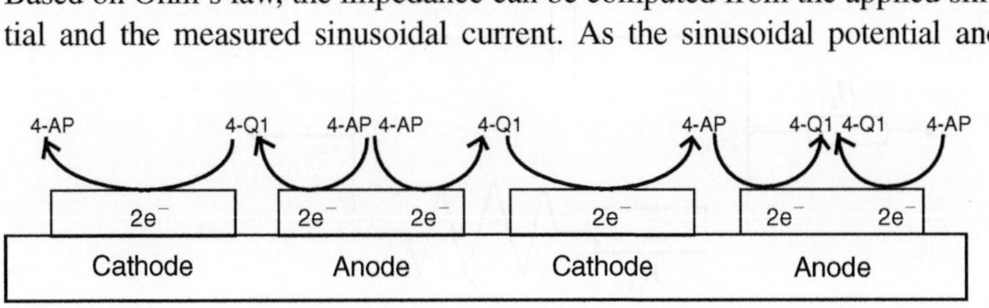

FIGURE 5.9 A schematic showing redox cycling between 4-AP and 4-QI (see Scheme 1) at adjacent "fingers" in an IDA electrode system.

differ in both magnitude and phase, the resultant impedance, Z, is usually represented by a vector with real (Z_{Re}) and imaginary (Z_{Im}) components as a function of frequency (ω):

$$Z(\omega) = Z_{Re} + \tilde{j}Z_{Im} \tag{3}$$

Very often, the electrode–solution interface can be represented by an equivalent circuit, as shown in Fig. 5.10, where R_S denotes the ohmic resistance of the electrolyte solution, C_{dl}, the double layer capacitance, R_{ct} the charge (or electron) transfer resistance that exists if a redox probe is present in the electrolyte solution, and Z_W the Warburg impedance arising from the diffusion of redox probe ions from the bulk electrolyte to the electrode interface. Note that both R_S and Z_W represent bulk properties and are not expected to be affected by an immunocomplex structure on an electrode surface. On the other hand, C_{dl} and R_{ct} depend on the dielectric and insulating properties of the electrode–electrolyte solution interface. For example, for an electrode surface immobilized with an immunocomplex, the double layer capacitance would consist of a constant capacitance of the bare electrode (C_{bare}) and a variable capacitance arising from the immunocomplex structure (C_{immun}), expressed as in Eq. (4).

$$\frac{1}{C_{dl}} = \frac{1}{C_{bare}} + \frac{1}{C_{immun}} \tag{4}$$

As the immunocomplex structure is generally electroinactive, its coverage on the electrode surface will decrease the double layer capacitance and retard the interfacial electron transfer kinetics of a redox probe present in the electrolyte solution. In this case, R_{ct} can be expressed as the sum of the electron transfer resistance of the bare electrode (R_{bare}) and that of the electrode immobilized with an immunocomplex (R_{immun}):

$$R_{ct} = R_{bare} + R_{immun} \tag{5}$$

There are several ways to present the Faradaic impedance data obtained at an electrode immobilized with an immunocomplex in the presence of a redox probe. For example, Z_{Im} is plotted vs Z_{Re} as a function of decreasing frequency to obtain a

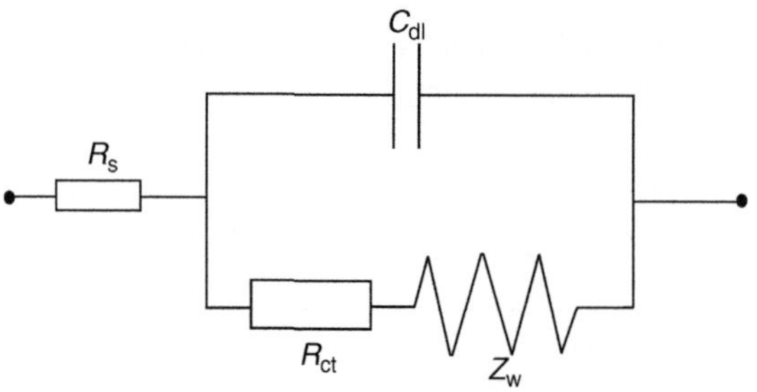

FIGURE 5.10 An equivalent circuit representing the interfacial features of an electrochemical immunosensor in the presence of a redox probe.

Nyquist plot, as shown in Fig. 5.11. A number of features can be readily identified in a Nyquist plot depending on the rate-determining processes at the immunoelectrode–solution interface. Trace (i) in the figure shows a semicircle with a centre located on the Z_{Re} axis at higher frequencies, followed by a straight line at lower frequencies. The semicircle feature is usually associated with the electron transfer-limited process, while the linear portion is related to the diffusion-limited process of the system. Also, the semicircle is offset on the Z_{Re} axis (as $\omega \rightarrow \infty$) by a value corresponding to the magnitude of R_S. For very rapid electron transfer processes, the Nyquist plot will only reveal the linear part of the impedance spectrum, as shown by Trace (ii). In contrast, if the electron transfer process is the rate-determining step, the impedance spectrum will show a large semicircle without a straight line, as illustrated by Trace (iii). Note that the diameter of the semicircle equals R_{ct} and extrapolation of the semicircle to lower frequencies will yield an intercept corresponding to $(R_S + R_{ct})$.

In recent years, several groups have relied on EIS to investigate antibody–antigen interactions on electrochemical immunosensors [62–65]. Lasseter *et al.* used EIS to investigate the intrinsic electrochemical response resulting from protein binding to a working electrode [62]. Gold, silicon or glassy carbon electrodes were coated with a monolayer of biotin that was used to bind avidin. EIS measurements revealed that avidin concentrations in the nanomolar range could be detected without an auxiliary redox species, leading to the development of label-free electrochemical immunosensing. Many of the reported impedimetric immunosensors use a conducting polymer (e.g. polypyrrole) to immobilize the capture antibody on the working electrode surface [65–67]. For example, Grant *et al.* reported a label-free impedimetric immunosensor incorporating a polypyrrole film to immobilize the capture antibody [65]. In this protocol, anti-BSA antibody was co-immobilized with the polypyrrole film on the surface of SPCE. These modified electrodes were then incubated in solutions of different BSA concentrations. Faradaic impedance was measured to construct Nyquist plots before and after binding BSA. In this case, Z_{Re} yielded a far smaller coefficient of variation than Z_{Im}. Based on Z_{Re} as a calibration signal, a linear response between 0 and 75 ppm of BSA was obtained. Miao and Guan

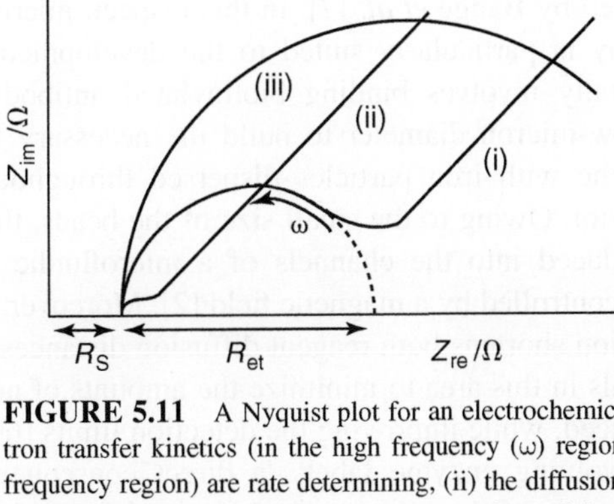

FIGURE 5.11 A Nyquist plot for an electrochemical immunosensor where (i) both the interfacial electron transfer kinetics (in the high frequency (ω) region) and the diffusion of the redox probe (in the low frequency region) are rate determining, (ii) the diffusion of the redox probe is rate determining, and (iii) the interfacial electron transfer kinetics are rate determining over the entire frequency range.

constructed an immunosensor by binding the antigen α-fetoprotein to a polyaniline–antibody backbone on an electrode [63]. Nyquist plots were obtained sequentially in the presence of $[Fe(CN)_6]^{3-/4-}$ at the bare electrode, followed by depositing polyaniline film, immobilizing antibody and binding antigen to the antibody. A semicircle feature was obtained when a bare electrode was used, indicating electron transfer kinetics at the electrode. The diameter of the semicircle increased after each of the remaining three steps in constructing the immunosensor. This was attributed to an increase in the charge transfer resistance for $[Fe(CN)_6]^{3-/4-}$ after each layer was applied sequentially to the electrode surface. Based on a calibration plot of R_{ct} vs α-fetoprotein concentration, a linear range from 200 to 800 ng mL^{-1} of α-fetoprotein was obtained. Similarly, Yang *et al.* examined the impedance data obtained in the presence of $[Fe(CN)_6]^{3-/4-}$ after immobilizing an immunocomplex involving *E. coli* at an IDA microelectrode [64]. The results showed that immobilizing antibodies and binding *E. coli* cells to the IDA microelectrode surface increased the electron transfer resistance, which was correlated with the concentration of *E. coli* from 4.36×10^5 to 4.36×10^8 cfu mL^{-1} with a detection limit of 10^6 cfu mL^{-1}.

In this section, we will make a brief mention of conductimetric immunosensors, which are essentially based on the same principles as impedimetric immunosensors, except that the change in conductance is measured as a result of antibody–antigen interaction rather than impedance. Therefore, a label-free conductimetric immunosensor has also been developed to detect the interaction between analyte rabbit IgG and immobilized anti-rabbit goat IgG [68]. The goat antibody was physically trapped in a layer of 3,4-ethylenedioxythiophene on gold coated polycarbonate membranes. Detection was based upon the change in conductance due to a change in the conformation of the conducting polymer before and after the interaction between the immobilized and analyte IgGs.

5.6 MICROFLUIDIC ELECTROCHEMICAL IMMUNOASSAY SYSTEMS

There have been vigorous efforts made to develop miniaturized flow systems in several different areas. Progress made in adapting microfluidic technology to immunoassays has recently been reviewed by Bange *et al.* [7]. In this respect, microbead-based electrochemical immunoassay is particularly suited to the development of microfluidic devices. Such work usually involves binding biotinylated antibody to streptavidin coated microbeads of low-micron diameter to build the necessary immunocomplex. The beads are polystyrene with iron particles dispersed throughout, giving rise to their paramagnetic behavior. Owing to the small size of the beads, they are extremely mobile and easily introduced into the channels of a microfluidic device and their movement can be easily controlled by a magnetic field [2]. Moreover, dispersion of the beads throughout a solution shortens both reagent diffusion distances and assay times. There are continuing goals in this area to minimize the amounts of necessary reagents used and the waste produced, while improving the detection limits [69]. In microbead-based immunoassays involving enzyme labels, a direct consequence of the small volumes used is that dilution of the enzyme product can be avoided, which aids in lowering detection limits. In most microbead-based immunoassays, after immobilizing an enzyme-conjugated immunocomplex on the beads, a sample of the bead solution is added to the enzyme substrate solution already making contact with an electrode (e.g. a rotating disk electrode, a microelectrode or an IDA) where detection is carried out. By separating the immunoreaction from the electrochemical detection steps, the working electrode surface is more accessible to the enzyme product as it diffuses to the bare electrode surface. In some cases, the separation of steps also reduces electrode fouling by biological species during analysis. For example, following the immobilization of a β-galactosidase-conjugated mouse IgG immunocomplex on microbeads, Thomas *et al.* obtained a detectable oxidation current when a 10 μL bead sample solution was added to 10 μL of the enzyme substrate solution (4-aminophenyl β-D-galactosidase) already placed on an IDA [70]. In this way, a detection limit of 26 ng mL^{-1} of mouse IgG was achieved. Similar applications of microbead-based immunoassays were more recently demonstrated for bacteriophage MS2 [71] and *E. coli* [72] with detection limits of 90 ng mL^{-1} and 20 cfu mL^{-1}, respectively. A disposable immunomagnetic electrochemical sensor was also developed for detecting polychlorinated biphenyls [73].

In addition to polystyrene beads, Zhang and Meyerhoff have also recently extended the type of magnetic beads by introducing a gold layer on polystyrene beads using an electroless gold plating procedure [44]. In this way, stable immunocomplexes can also be immobilized on gold coated magnetic beads via the thiol-gold pseudo covalent bond. In a different system, colloidal gold nanoparticles were used to anchor carcinoma antigen-125 in a cellulose acetate membrane on a glassy carbon electrode [74]. A competitive immunoassay was then established to detect carcinoma 125 down to approximately 1.8 U mL^{-1}.

5.7 CONCLUDING REMARKS

Although immunoassay techniques emerged over two decades ago, there are still vigorous research efforts and tremendous progress in the development of electrochemical immunoassays and immunosensors. An extraordinary feature of these immunosystems is their specificity. There are continuing studies examining various strategies that will aid in aligning antibodies on a solid phase in an optimal direction with minimal steric hindrance. Developments in this area will undoubtedly further enhance the degree of sensitivity achievable in analyses involving immunoassays and immunosensors. In conjunction with electrochemical detection, these systems will offer sensitive and selective analyses that are faster, simpler, and more economical. There is also continuing interest in developing and applying suitable labels for electrochemical immunoassays such that a more direct signal-generation scheme can be used. The application of electrochemical impedance spectroscopy has started to facilitate a label-free scheme, and this is definitely an attractive, simpler alternative to others involving a label. Nonetheless, much research effort is needed in this area to achieve the required sensitivity and dynamic range obtainable in amperometric detection. At the same time, there are also encouraging promises in the development of microfluidic electrochemical immunoassay systems, leading to further progress in the manufacture of miniaturized immunoassay devices. This will open up opportunities for developing hand-held tools for instant on-site pharmaceutical and clinical diagnosis, particularly in response to the gradual shift towards home-based diagnosis. Another challenge in this area is the desire to integrate immunosensors in an array format to perform simultaneous analysis of multiple analytes. In conclusion, new immunosensor technologies are anticipated in the near future in response to these exciting opportunities.

5.8 REFERENCES

1. A.J. Killard, B. Deasy, R. O'Kennedy, and M.R. Smyth, Antibodies: production, functions and applications in biosensors. *TrAC, Trends Anal. Chem.* **14**, 257–266 (1995).
2. N.J. Ronkainen-Matsuno, J.H. Thomas, H.B. Halsall, and W.R. Heineman, Electrochemical immunoassay moving into the fast lane. *TrAC, Trends Anal. Chem.* **21**, 213–225 (2002).
3. O.A. Sadik and J.M. van Emon, Applications of electrochemical immunosensors to environmental monitoring. *Biosens. Bioelectron.* **11**, i–xi (1996).
4. A.L. Ghindilis, P. Atanasov, M. Wilkins, and E. Wilkins, Immunosensors: electrochemical sensing and other engineering approaches. *Biosens. Bioelectron.* **13**, 113–131 (1998).
5. C.A. Wijayawardhana, H.B. Halsall, and W.R. Heineman, Electrochemical immunoassay. *Encyclopedia of Electrochemistry* **9**, 145, 147–174 (2002).
6. C.A. Wijayawardhana, H.B. Halsall, and W.R. Heineman, Milestones of electrochemical immunoassay at Cincinnati. *Electroanalytical Methods for Biological Materials* 329–365 (2002).
7. A. Bange, H.B. Halsall, and W.R. Heineman, Microfluidic immunosensor systems. *Biosens. Bioelectron.* **20**, 2488–2503 (2005).
8. I. Roit, J. Brostoff, and D. Male, *Immunology,* Mosby International Ltd, London (1998).
9. M. Diaz-Gonzalez, D. Hernandez-Santos, M.B. Gonzalez-Garcia, and A. Costa-Garcia, Development of an immunosensor for the determination of rabbit IgG using streptavidin modified screen-printed carbon electrodes. *Talanta* **65**, 565–573 (2005).
10. K. Matsumoto, A. Torimaru, S. Ishitobi, T. Sakai, H. Ishikawa, K. Toko, N. Miura, and T. Imato, Preparation and characterization of a polyclonal antibody from rabbit for detection of trinitrotoluene by a surface plasmon resonance biosensor. *Talanta* **68**, 305–311 (2005).
11. R.M. Pemberton, T.T. Mottram, and J.P. Hart, Development of a screen-printed carbon electrochemical immunosensor for picomolar concentrations of estradiol in human serum extracts. *J. Biochem. Biophys. Methods* **63**, 201–212 (2005).
12. Y.-M. Zhou, Z.-Y. Wu, G.-L. Shen, and R.-Q. Yu, An amperometric immunosensor based on Nafion-modified electrode for the determination of Schistosoma japonicum antibody. *Sensors and Actuators, B: Chemical* **89**, 292–298 (2003).
13. T.-S. Zhong and G. Liu, Silica sol-gel amperometric immunosensor for Schistosoma japonicum antibody assay. *Anal. Sci.* **20**, 537–541 (2004).
14. L. Ball, A. Jones, P. Boogaard, W. Will, and P. Aston, Development of a competitive immunoassay for the determination of N-(2-hydroxypropyl)valine adducts in human haemoglobin and its application in biological monitoring. *Biomarkers* **10**, 127–137 (2005).
15. Z.P. Aguilar, W.R.I.V. Vandaveer, and I. Fritsch, Self-contained microelectrochemical immunoassay for small volumes using mouse igg as a model system. *Anal. Chem.* **74**, 3321–3329 (2002).
16. S. Chemburu, E. Wilkins, and I. Abdel-Hamid, Detection of pathogenic bacteria in food samples using highly-dispersed carbon particles. *Biosens. Bioelectron.* **21**, 491–499 (2005).
17. K.L. Hoffman, G.H. Parsons, L.J. Allerdt, J.M. Brooks, and L.E. Miles, Elimination of "hook-effect" in two-site immunoradiometric assays by kinetic rate analysis. Clin. *Chem.* **30**, 1499–1501 (1984).
18. J.T. Wu and S.E. Christensen, Effect of different test designs of immunoassays on "hook effect" of CA 19-9 measurement. *J. Clin. Lab. Anal.* **5**, 228–232 (1991).

19. B.L. Haller, K.A. Fuller, W.S. Brown, J.W. Koenig, B.J. Eveland, and M.G. Scott, Two automated prolactin immunoassays evaluated with demonstration of a high-dose "hook effect" in one. *Clin. Chem.* **38**, 437–438 (1992).
20. S.A. Fernando and G.S. Wilson, Studies of the "hook" effect in the one-step sandwich immunoassay. *J. Immunol. Methods* **151**, 47–66 (1992).
21. C. Singh, G.S. Agarwal, G.P. Rai, L. Singh, and V.K. Rao, Specific detection of Salmonella typhi using renewable amperometric immunosensor. *Electroanalysis* **17**, 2062–2067 (2005).
22. R.E. Ionescu, C. Gondran, S. Cosnier, L.A. Gheber, and R.S. Marks, Comparison between the performances of amperometric immunosensors for cholera antitoxin based on three enzyme markers. *Talanta* **66**, 15–20 (2005).
23. S. Viswanathan, L.-C. Wu, M.-R. Huang, and J.-A.A. Ho, Electrochemical immunosensor for cholera toxin using liposomes and poly(3,4-ethylenedioxythiophene)-coated carbon nanotubes. *Anal. Chem.* **78**, 1115–1121 (2006).
24. E. Zacco, M.I. Pividori, and S. Alegret, Electrochemical biosensing based on universal affinity biocomposite platforms. *Biosens. Bioelectron.* **21**, 1291–1301 (2006).
25. G. Gitlin, E.A. Bayer, and M. Wilchek, Studies on the biotin-binding site of avidin. Lysine residues involved in the active site. *Biochem. J.* **242**, 923–926 (1987).
26. M.L. Jones and G.P. Kurzban, Noncooperativity of biotin binding to tetrameric streptavidin. *Biochemistry* **34**, 11750–11756 (1995).
27. M. Diaz-Gonzalez, M.B. Gonzalez-Garcia, and A. Costa-Garcia, Immunosensor for mycobacterium tuberculosis on screen-printed carbon electrodes. *Biosens. Bioelectron.* **20**, 2035–2043 (2005).
28. J.J. Langone, Protein A of Staphylococcus aureus and related immunoglobulin receptors produced by streptococci and pneumonococci. *Adv. Immunol.* **32**, 157–252 (1982).
29. L. Bjoerck and G. Kronvall, Purification and some properties of streptococcal protein G, a novel IgG-binding reagent. *J. Immunol.* **133**, 969–974 (1984).
30. B. Akerstrom, T. Brodin, K. Reis, and L. Bjorck, Protein G: a powerful tool for binding and detection of monoclonal and polyclonal antibodies. *J. Immunol.* **135**, 2589–2592 (1985).
31. E. Zacco, M.I. Pividori, X. Llopis, M. del Valle, and S. Alegret, Renewable Protein A modified graphite-epoxy composite for electrochemical immunosensing. *J. Immunol. Methods* **286**, 35–46 (2004).
32. J. Quinn, P. Patel, B. Fitzpatrick, B. Manning, P. Dillon, S. Daly, R. Okennedy, M. Alcocer, H. Lee, M. Morgan, and K. Lang, The use of regenerable, affinity ligand-based surfaces for immunosensor applications. *Biosens. Bioelectron.* **14**, 587–595 (1999).
33. J. Yakovleva, R. Davidsson, M. Bengtsson, T. Laurell, and J. Emneus, Microfluidic enzyme immunosensors with immobilised protein A and G using chemiluminescence detection. *Biosens. Bioelectron.* **19**, 21–34 (2003).
34. A. Ramanaviciene and A. Ramanavicius, Application of polypyrrole for the creation of immunosensors. *Crit. Rev. Anal. Chem.* **32**, 245–252 (2002).
35. F. Darain, D.S. Park, J.-S. Park, S.-C. Chang, and Y.-B. Shim, A separation-free amperometric immunosensor for vitellogenin based on screen-printed carbon arrays modified with a conductive polymer. *Biosens. Bioelectron.* **20**, 1780–1787 (2005).
36. F. Darain, S.-U. Park, and Y.-B. Shim, Disposable amperometric immunosensor system for rabbit IgG using a conducting polymer modified screen-printed electrode. *Biosens. Bioelectron.* **18**, 773–780 (2003).
37. A. Ulman, Formation and structure of self-assembled monolayers. *Chem. Rev. (Washington, D. C.)* **96**, 1533–1554 (1996).
38. J.J. Gooding, F. Mearns, W. Yang, and J. Liu, Self-assembled monolayers into the 21st century: Recent advances and applications. *Electroanalysis* **15**, 81–96 (2003).
39. J.C. Love, L.A. Estroff, J.K. Kriebel, R.G. Nuzzo, and G.M. Whitesides, Self-assembled monolayers of thiolates on metals as a form of nanotechnology. *Chem. Rev. (Washington, DC, United States)* **105**, 1103–1169 (2005).
40. S. Herrwerth, T. Rosendahl, C. Feng, J. Fick, W. Eck, M. Himmelhaus, R. Dahint, and M. Grunze, Covalent coupling of antibodies to self-assembled monolayers of carboxy-functionalized poly-(ethylene glycol): protein resistance and specific binding of biomolecules. *Langmuir* **19**, 1880–1887 (2003).

41. S. Chen, L. Liu, J. Zhou, and S. Jiang, Controlling antibody orientation on charged self-assembled monolayers. *Langmuir* **19**, 2859–2864 (2003).
42. M. Akram, M.C. Stuart, and D.K.Y. Wong, Direct application strategy to immobilize a thioctic acid self-assembled monolayer on a gold electrode. *Anal. Chim. Acta* **504**, 243–251 (2004).
43. L. Korecka, Z. Bilkova, M. Holeapek, J. Kralovsky, M. Benes, J. Lenfeld, N. Minc, R. Cecal, J.-L. Viovy, and M. Przybylski, Utilization of newly developed immobilized enzyme reactors for preparation and study of immunoglobulin G fragments. *Journal of Chromatography, B: Analytical Technologies in the Biomedical and Life Sciences* **808**, 15–24 (2004).
44. H. Zhang and M.E. Meyerhoff, Gold-coated magnetic particles for solid-phase immunoassays: enhancing immobilized antibody binding efficiency and analytical performance. *Anal. Chem.* **78**, 609–616 (2006).
45. M.-P. Marco and D. Barcelo, Environmental applications of analytical biosensors. *Meas. Sci. Technol.* **7**, 1547–1562 (1996).
46. D. Purvis, O. Leonardova, D. Farmakovsky, and V. Cherkasov, An ultrasensitive and stable potentiometric immunosensor. *Biosens. Bioelectron.* **18**, 1385–1390 (2003).
47. R.M. Pemberton, J.P. Hart, P. Stoddard, and J.A. Foulkes, A comparison of 1-naphthyl phosphate and 4 aminophenyl phosphate as enzyme substrates for use with a screen-printed amperometric immunosensor for progesterone in cows' milk. *Biosens. Bioelectron.* **14**, 495–503 (1999).
48. E.J. Moore, M. Pravda, M.P. Kreuzer, and G.G. Guilbault, Comparative study of 4-aminophenyl phosphate and ascorbic acid 2-phosphate, as substrates for alkaline phosphatase based amperometric immunosensor. *Anal. Lett.* **36**, 303–315 (2003).
49. M.S. Wilson and R.D. Rauh, Hydroquinone diphosphate: an alkaline phosphatase substrate that does not produce electrode fouling in electrochemical immunoassays. *Biosens. Bioelectron.* **20**, 276–283 (2004).
50. M. O'Connor, S.N. Kim, A.J. Killard, R.J. Forster, M.R. Smyth, F. Papadimitrakopoulos, and J.F. Rusling, Mediated amperometric immunosensing using single walled carbon nanotube forests. *Analyst (Cambridge, United Kingdom)* **129**, 1176–1180 (2004).
51. B. Law, *Immunoassay: A Practical Guide,* Taylor & Francis, London (1996).
52. K. di Gleria, H.A. Hill, C.J. McNeil, and M.J. Green, Homogeneous ferrocene-mediated amperometric immunoassay. *Anal. Chem.* **58**, 1203–1205 (1986).
53. M. Nakayama, T. Ihara, K. Nakano, and M. Maeda, DNA sensors using a ferrocene-oligonucleotide conjugate. *Talanta* **56**, 857–866 (2002).
54. R. Ojani, J.B. Raoof, and A. Alinezhad, Catalytic oxidation of sulfite by ferrocenemonocarboxylic acid at the glassy carbon electrode. Application to the catalytic determination of sulfite in real sample. *Electroanalysis* **14**, 1197–1203 (2002).
55. H.S. Mandal and H.-B. Kraatz, Ferrocene-histidine conjugates: N-ferrocenoyl-histidyl(imN-ferrocenoyl)methylester: synthesis and structure. *J. Organomet. Chem.* **674**, 32–37 (2003).
56. C. Duan and M.E. Meyerhoff, Immobilization of proteins on gold coated porous membranes via an activated self-assembled monolayer of thioctic acid. *Mikrochim. Acta* **117**, 195–206 (1995).
57. M. Akram, M.C. Stuart, and D.K.Y. Wong, Signal generation at an electrochemical immunosensor via the direct oxidation of an electroactive label. *Electroanalysis* **18**, 237–246 (2006).
58. M. Okochi, H. Ohta, T. Tanaka, and T. Matsunaga, Electrochemical probe for on-chip type flow immunoassay: immunoglobulin G labeled with ferrocenecarboaldehyde. *Biotechnol. Bioeng.* **90**, 14–19 (2005).
59. M. Hromadova, M. Salmain, N. Fischer-Durand, L. Pospisil, and G. Jaouen, Electrochemical microbead-based immunoassay using an (h5-cyclopentadienyl)tricarbonylmanganese redox marker bound to bovine serum albumin. *Langmuir* **22**, 506–511 (2006).
60. J. Min and A.J. Baeumner, Characterization and optimization of interdigitated ultramicroelectrode arrays as electrochemical biosensor transducers. *Electroanalysis* **16**, 724–729 (2004).
61. J.H. Thomas, S.K. Kim, P.J. Hesketh, H.B. Halsall, and W.R. Heineman, Microbead-based electrochemical immunoassay with interdigitated array electrodes. *Anal. Biochem.* **328**, 113–122 (2004).
62. T.L. Lasseter, W. Cai, and R.J. Hamers, Frequency-dependent electrical detection of protein binding events. *Analyst (Cambridge, United Kingdom)* **129**, 3–8 (2004).

63. Y.-Q. Miao and J.-G. Guan, Probing of antibody–antigen reactions at electropolymerized polyaniline immunosensors using impedance spectroscopy. *Anal. Lett.* **37**, 1053–1062 (2004).
64. L. Yang, Y. Li, and G.F. Erf, Interdigitated array microelectrode-based electrochemical impedance immunosensor for detection of Escherichia coli O157:H7. *Anal. Chem.* **76**, 1107–1113 (2004).
65. S. Grant, F. Davis, K.A. Law, A.C. Barton, S.D. Collyer, S.P.J. Higson, and T.D. Gibson, Label-free and reversible immunosensor based upon an AC impedance interrogation protocol. *Anal. Chim. Acta* **537**, 163–168 (2005).
66. P. Cooreman, R. Thoelen, J. Manca, M. vande Ven, V. Vermeeren, L. Michiels, M. Ameloot, and P. Wagner, Impedimetric immunosensors based on the conjugated polymer PPV. *Biosens. Bioelectron.* **20**, 2151–2156 (2005).
67. O. Ouerghi, A. Touhami, N. Jaffrezic-Renault, C. Martelet, H. Ben Ouada, and S. Cosnier, Impedimetric immunosensor using avidin-biotin for antibody immobilization. *Bioelectrochemistry* **56**, 131–133 (2002).
68. M. Kanungo, D.N. Srivastava, A. Kumar, and A.Q. Contractor, Conductimetric immunosensor based on poly(3,4-ethylenedioxythiophene). *Chem. Comm.* **7**, 680–681 (2002).
69. A. Bange, D.K.Y. Wong, C.J. Seliskar, H.B. Halsall, and W.R. Heineman, Microscale immunosensors for biological agents. *Proceedings of SPIE – The International Society for Optical Engineering* **5718**, 142–150 (2005).
70. J.H. Thomas, N.J. Ronkainen-Matsuno, S. Farrell, H. Brian Halsall, and W.R. Heineman, Microdrop analysis of a bead-based immunoassay. *Microchem. J.* **74**, 267–276 (2003).
71. J.H. Thomas, S.K. Kim, P.J. Hesketh, H.B. Halsall, and W.R. Heineman, Bead-based electrochemical immunoassay for bacteriophage MS2. *Anal. Chem.* **76**, 2700–2707 (2004).
72. I.H. Boyaci, Z.P. Aguilar, M. Hossain, H.B. Halsall, C.J. Seliskar, and W.R. Heineman, Amperometric determination of live Escherichia coli using antibody-coated paramagnetic beads. *Anal. Bioanal. Chem.* **382**, 1234–1241 (2005).
73. S. Centi, S. Laschi, M. Franek, and M. Mascini, A disposable immunomagnetic electrochemical sensor based on functionalized magnetic beads and carbon-based screen-printed electrodes (SPCEs) for the detection of polychlorinated biphenyls (PCBs). *Anal. Chim. Acta* **538**, 205–212 (2005).
74. L. Wu, J. Chen, D. Du, and H. Ju, Electrochemical immunoassay for CA125 based on cellulose acetate stabilized antigen/colloidal gold nanoparticles membrane. *Electrochim. Acta* **51**, 1208–1214 (2006).

CHAPTER 6

Superoxide electrochemical sensors and biosensors: principles, development and applications

Lanqun Mao, Yang Tian, and Takeo Ohsaka

6.1 CHEMISTRY AND BIOCHEMISTRY OF SUPEROXIDE

Basically, the complete reduction of one molecule of dioxygen to water essentially requires four electrons and the intermediates in this reduction process mainly include superoxide radical ($O_2^{\bullet -}$), hydrogen peroxide (H_2O_2), and hydroxyl radical ($^{\bullet}OH$) [1, 2], of which H_2O_2 is the most stable and the only species that could accumulate appreciably in neutral aqueous media. It has long been recognized as one of the products of the biological reduction of dioxygen and as a possible explanation for the toxicity of dioxygen [3, 4]. In contrast, the radical intermediates of the reduction of dioxygen including $O_2^{\bullet -}$ and $^{\bullet}OH$ have fleeting lifetimes under ordinary conditions and thus have only recently received a serious consideration [5]. It is now accepted that the superoxide radical is an important agent of the toxicity of dioxygen and superoxide dismutases (SODs) constitute the primary defense against this radical [6–9].

Univalently reduced oxygen is called the hydroperoxyl radical ($HO_2^{\bullet}$) in its protonated form and the $O_2^{\bullet -}$ radical in its ionized form. The $HO_2^{\bullet}$ is a weak acid with a p*Ka* of 4.8 [10]. The $O_2^{\bullet -}$ radical can be produced either by the univalent reduction of dioxygen or by the univalent oxidation of H_2O_2. For example, the $O_2^{\bullet -}$ can be produced from the electrochemical reduction of dioxygen in aprotic solvents [11–15] or in alkaline aqueous media [16, 17] or from the chemical reduction of dioxygen by hydrated electrons or by hydrogen atoms generated during the photolysis [18–21], radiolysis [10, 22–24], or ultrasonication of water [25, 26]. It could also be produced from the reduction of dioxygen by carbanions [27, 28], reduced dyes or flavins [29–33], catecholamines [34], ferredoxins [35–37], or hemoproteins [38, 39]; and the oxidation of H_2O_2 by ceric ions [40]. The $O_2^{\bullet -}$ radical has been detected by a number of physical methods including conductimetry [41], optical spectroscopy [42, 43], electron-spin-resonance spectroscopy (ESR) [44–49], and mass spectrometry [50].

In biological systems, $O_2^{\bullet -}$ is generated as a reduced intermediate of molecular dioxygen in significant quantities and is a primary species of the so-called reactive oxygen species (ROS). As one part of the host defense systems and as cell/cell signaling molecules, $O_2^{\bullet -}$ performs an essential function. Under normal physiological conditions, $O_2^{\bullet -}$ undergoes a disproportionation by non-catalytic or enzymatic reactions, leading to a rather low and undetectable endogenous physiological concentration. An increase in the activity of $O_2^{\bullet -}$ has been found to occur in response to traumatic brain injury ischemia reperfusion and hypoxia [51] and $O_2^{\bullet -}$ may be involved in the etiology of aging, cancer, and progressive neurodegenerative diseases, such as Parkinson's disease [52–56].

6.2 O_2^- BIOASSAY: AN OVERVIEW

Due to its great roles in biological process, quantitative information on the $O_2^{\bullet -}$ level in a variety of *in-vitro* and *in-vivo* models has become very essential in understanding pathology and physiology of diseases relevant to ROS and free-radical biochemistry. However, the short lifetime, low concentration, and high reactivity of $O_2^{\bullet -}$ substantially make it relatively difficult to measure the concentration of $O_2^{\bullet -}$. Actually, $O_2^{\bullet -}$ is not so reactive itself, but it can be disproportionated rapidly in the presence of SODs or react with other biomolecules to produce H_2O_2. H_2O_2 can be easily turned to hydroxyl radical, which is the most potent radical, by the Fenton reaction in the presence of metal ions such as Fe^{2+} and Cu^{2+} [57–59].

Due to the rapidity of the spontaneous dismutation reactions, the steady-state concentrations of $O_2^{\bullet -}$ achieved by chemical or by enzymatic reactions are usually quite low. The physical methods for detecting $O_2^{\bullet -}$, although direct and unequivocal, are restricted to measurements of steady-state concentrations and are thus often found to lack of sensitivity. For distince, due to the reason mentioned above, when the EPR method was employed for studying the $O_2^{\bullet -}$ production by xanthine oxidase, it was necessary to use a high concentration of the reactants and to work at elevated pH so as to suppress the dismutation reaction and thereby to obtain detectable levels of $O_2^{\bullet-}$. Chemical methods for the detection of $O_2^{\bullet-}$ are integrative and offer the advantages of sensitivity and simplicity. In those methods, $O_2^{\bullet-}$ can be trapped with an indicating scavenger. The reaction between $O_2^{\bullet-}$ and the trapping agents can be followed by a suitable optical, manometric, or polarographic method to constitute analytical protocols for the $O_2^{\bullet-}$ measurements. The scavenger can be used at concentrations that compete effectively with the dismutation reactions such that the produced $O_2^{\bullet-}$ will be trapped and thus detected. Thus far, some chemical methods have been used for the detection of $O_2^{\bullet-}$. For example, the $O_2^{\bullet-}$ radical can reduce ferricytochrome *c* [60] (reaction shown below), tetranitromethane, and nitroblue tetrazolium [61] and these reductions can be followed spectrophotometrically in terms of the accumulations of ferrocytochrome *c*, the nitroformate anion, and blue formazan, respectively, to establish spectrophotometric methods for the $O_2^{\bullet-}$ determination. There are, of course, other agents that can be involved in these reductions and thus interfere with the $O_2^{\bullet-}$ determination. In this case, the net response for $O_2^{\bullet-}$ can be differentiated by using SOD since SOD can specifically catalyze the dismutation of $O_2^{\bullet-}$.

$$\text{Cytochrome } c\ (\text{Fe(III)}) + O_2^{\bullet-} \rightarrow \text{Cytochrome } c\ (\text{Fe(II)}) + O_2$$

The $O_2^{\bullet-}$ radical can act as an oxidant as well as a reductant and chemical estimates of its production can also be based on its ability to oxidize epinephrine to adrenochrome [62]. These chemical methods have the additional advantage of not requiring highly specialized equipments. Also based on its redox property, the $O_2^{\bullet-}$ radical can be determined by chemiluminescence methods through the measurement of the intensity of the fluorescence radiation emitted after chemical oxidation of $O_2^{\bullet-}$ by, e.g., lucigenin [63–67]. These methods, however, are limited by the poor selectivity and lack of capability for *in-vivo* performance.

Considerable interest has been devoted to the application of electrochemical method for $O_2^{\bullet-}$ determination because of its high selectivity, sensitivity, and capability for *in-vivo* use [68–83]. In this chapter, we will mainly focus our attention on electrochemical methodologies for $O_2^{\bullet-}$ determination, highlighting recent attempts on this aspect by using SODs. As will be described, a combination of the promoted direct electron transfer of the SODs with the biomolecular recognition by virtue of specific and significant enzyme–substrate reactivity of the SODs toward $O_2^{\bullet-}$ essentially results in a sensitive measurement of $O_2^{\bullet-}$ without a virtual interference from physiological levels of H_2O_2, ascorbic acid (AA), uric acid (UA), and metabolites of neurotransmitters. Furthermore, these strategies could be further accomplished with carbon fiber microelectrodes, which can be readily employed for *in-vivo* determination of $O_2^{\bullet-}$ in biological systems.

6.3 $O_2^{\bullet-}$ ELECTROCHEMISTRY AND $O_2^{\bullet-}$ ELECTROCHEMICAL SENSORS

$O_2^{\bullet-}$ is not stable in aqueous media, especially in acidic solutions. This poor stability has essentially made it difficult to study $O_2^{\bullet-}$ electrochemistry in aqueous media. By carefully designing the electrode–solution interface of a hanging mercury drop electrode with hydrophobic surfactants, such as quinoline and isoquinoline, Ohsaka *et al.* have successfully achieved the redox reaction of the $O_2^{\bullet -}/O_2$ redox couple and thus could determine the biomolecular reactivity of $O_2^{\bullet -}$ in aqueous media [84, 85]. The $O_2^{\bullet -}/O_2$ couple exhibits a reversible electrochemical behavior with a formal potential of −135 mV vs Ag/AgCl in 0.10 M NaOH and 0.5 M KCl solution. Such an electrochemical property of $O_2^{\bullet -}$ has been used for the development of electrochemical sensors for $O_2^{\bullet -}$ determination. For example, on the basis of the direct oxidation of $O_2^{\bullet -}$ at carbon fiber microelectrode, Tanaka *et al.* have successfully determined periodic fluctuations in $O_2^{\bullet -}$ production by a single phagocytic cell [68]. In that case, the carbon fiber was sealed into a glass capillary by heating the glass to melt it around the fiber. Also, by virtue of the direct oxidation of $O_2^{\bullet -}$ on carbon fiber microelectrodes at +0.12 V vs SCE (saturated calomel electrode), Privat and coworkers simultaneously determined $O_2^{\bullet -}$ and nitric oxide [86]. This method has been demonstrated to be useful for real-time monitoring of extracellular $O_2^{\bullet -}$ production in stimulated human vascular cells. Although the determination of $O_2^{\bullet -}$ based on the direct oxidation of $O_2^{\bullet -}$ has been proved to be mechanistically simple and to possess a low detection limit and quick response, its application has been limited by its poor selectivity. Such a problem could be expected to be solved by a combination of suitable electrochemical methods with specific biomolecular recognition of enzymes toward substrate, such as SODs toward $O_2^{\bullet -}$, as will be illustrated in the following sections.

6.4 ELECTROCHEMICAL SENSORS FOR $O_2^{\bullet -}$

6.4.1 Biosensors with enzymes other than SODs

In addition to a family of SODs, several other kinds of enzymes and proteins, including tyrosinase [87], galactose oxidase [87], hemin, and cytochrome *c* (Cyt. *c*), have been employed to construct enzyme-based biosensors for the $O_2^{\bullet -}$ determination. Here, we will use Cyt. *c* as an example to illustrate the analytical mechanism of such a kind of $O_2^{\bullet -}$ biosensors. For constructing a Cyt. *c*-based biosensor, Cyt. *c* is normally immobilized on the electrode surface and acts as an electron transfer mediator between the electrode and $O_2^{\bullet -}$. The $O_2^{\bullet -}$ radical reduces the immobilized Cyt. *c* (Fe(III)) to Cyt. *c* (Fe(II)) and the Cyt. *c* (Fe(II)) is reoxidized on the electrode at a potential of 0.15–0.25 V (vs Ag/AgCl). In this regard, the electron transfer between Cyt. *c* and the electrode becomes essential. However, such an electron transfer could not be readily obtained at electrodes frequently used in electrochemistry, such as glassy carbon, gold, and platinum. The electron transfer could be achieved at the platinized activated carbon electrode (PACE) and such a property has been further exploited for $O_2^{\bullet -}$ determination [88–90]. The sensitivity of the Cyt. *c*-modified PACE electrode toward $O_2^{\bullet -}$ was evaluated to be 34.0 ($\mu A\ cm^{-2}$)/(μM^{-1}), which was greater than that obtained at a planar Au electrode. The increase in sensitivity was ascribed to the larger effective surface area of the PACE compared with that of the planar Au electrode because the PACE is an extremely porous material capable of binding a large amount of Cyt. *c*.

The electron transfer property of Cyt. *c* can also be obtained at Au electrodes modified with self-assembled monolayers of, for example, N-acetyl cysteine [91], or 4,4-dithiobipyridine [92] or 3,3-dithiobis (sulfosuccinimidylpropionate) [93–95]. These modifiers were employed as promoters for facilitating the electron transfer of Cyt. *c*. From a practical application point of view, a free orientation of the Cyt. *c* without denaturation is generally required for achieving an effective electron transfer between Cyt. *c* and the electrode and for the subsequent biosensing of $O_2^{\bullet -}$. In this case, self-assembled monolayers of alkanethiols formed onto Au electrodes are remarkable because they can not only facilitate direct electron transfer of Cyt. *c* but also prevent the electrode from fouling by the potential interferents in the solution. As a consequence, the self-assembled monolayer (SAM) of $HS(CH_2)_{10}COOH$ confined on the Au electrode has been used for constructing a Cyt. *c*-based amperometric $O_2^{\bullet -}$ biosensor [72, 76]. In addition, mixed SAMs, e.g. those consisting of short alkanethiol SAM, such as 3-mercaptopropinic acid and 3-mercaptopropanol, and long alkanethiol SAM, such as 11-mercaptoundecanoic acid and 11-mercaptoundecanol [96], have been also used for facilitating the electron transfer of Cyt. *c* and this direct electron transfer property has been further developed for $O_2^{\bullet -}$ biosensing. In a different way, Campanella *et al.* have developed a Cyt. *c*-based $O_2^{\bullet -}$ biosensor by using hemin as an electron transfer mediator between Cyt. *c* and a carbon paste electrode [87]. The oxidation current of hemin (Fe(II)) constituted the analytical current for continuous $O_2^{\bullet -}$ determination at +0.8 V vs SCE. The response time and the detection limit of the $O_2^{\bullet -}$ biosensor was 2 min and 0.2 mM, respectively. The lifetime of the biosensor was 3 days. More recently, a multilayer Cyt. *c*-modified electrode was used for biosensing of $O_2^{\bullet -}$, of which the electrode assembled with six layers of Cyt. *c* exhibits the highest sensitivity toward $O_2^{\bullet -}$ [97].

Although the Cyt. *c*-based electrochemical biosensors have been demonstrated to be useful for the determination of $O_2^{\bullet -}$ in biological samples, it is known that Cyt. *c* is not an enzyme specific for $O_2^{\bullet -}$. For example, Cyt. *c* also shows an inherent catalytic activity like the peroxidase to reduce oxidants including H_2O_2 and $ONOO^-$. This non-specific catalytic activity of Cyt. *c* somewhat limits the application of Cyt. *c*-based electrochemical biosensors for selective determination of $O_2^{\bullet -}$ in biological systems even though the peroxidase activity of Cyt. *c* has been reported to be controlled by electrode design [75]. In this aspect, the utilization of SODs would be a good alternative because SODs are the enzymes for catalyzing the dismutation of $O_2^{\bullet -}$ into O_2 and H_2O_2 with a strong activity and great specificity. As such, the SOD-based electrochemical biosensors have been recently studied and developed for selective and sensitive determination of $O_2^{\bullet -}$.

6.4.2 Brief introduction to SODs

SOD comprises a family of metalloproteins primarily classified into four groups: copper, zinc-containing SOD (Cu, Zn-SOD), manganese-containing SOD (Mn-SOD), iron-containing SOD (Fe-SOD) and nickel-containing SOD (Ni-SOD). In the following studies, we will only focus on the uses of the former three kinds of SODs to construct SOD-based $O_2^{\bullet -}$ biosensors since the last one, Ni-SOD, is not commercially available.

The SODs are ubiquitous metallo-enzymes in oxygen-tolerant organisms and protect the organism against the toxic effects of $O_2^{\bullet -}$ by efficiently catalyzing its dismutation into O_2 and H_2O_2 via a cyclic oxidation-reduction electron transfer mechanism as shown in Scheme 1, with Cu, Zn-SOD with an example. Two steps have been proposed in the dismutation reaction, in which the oxidized form of the metal central ions (i.e. Cu (II), Fe(III), Mn(III) for Cu, Zn-SOD, Fe-SOD and Mn-SOD, respectively) were first reduced to the reduced form (i.e. Cu (I), Fe(II), Mn(II) for Cu, Zn-SOD, Fe-SOD and Mn-SOD, respectively) with the formation of O_2 followed by a subsequent equivalent oxidation of the reduced form of the metal ions into their oxidized form by $O_2^{\bullet -}$ with the release of H_2O_2 (Scheme 6.1) [98].

Valentine *et al.* demonstrated that Cu^{2+} ions are in the same site in native Cu, Zn-SOD protein and CuESOD (E = empty, i.e. with removal of Zn ion from the native Cu, Zn-SOD) and the activity of the CuESOD is at least 80 ± 5% that of the native Cu, Zn-SOD [99]. However, the EZnSOD (E = empty, i.e. with removal of Cu ion from the native Cu, Zn-SOD) does not show any activity for catalytic dismutation of $O_2^{\bullet -}$. This suggests Cu^{2+} present in the Cu, Zn-SOD is essential for $O_2^{\bullet -}$ dismutation even though Zn^{2+} is also required for dismutase activity [100–102].

The structures of the SODs (i.e. Cu, Zn-SOD, Fe-SOD, and Mn-SOD) in the SOD family are known to be relatively different. For instance, the Cu, Zn-SOD, which is the first SOD characterized and analyzed by X-ray methods [103], is a homodimer with a molecular weight of 31 200 Da and containing one Cu (II) and Zn (II) per monomer subunit. Its fundamental structural motif is a β-barrel. In the Cu, Zn-SOD, the metals are bound by sequences connecting the barrel strands and are on opposite sides of the dimer with the Cu atoms separated by 33.8 Å. The dimer is an elongated ellipsoid about 33 Å wide, 67 Å long and 36 Å deep [103].

The second family of SODs, which utilizes either Fe or Mn to catalyze the dismutation of $O_2^{\bullet -}$ (i.e. Fe-SOD and Mn-SOD), constitutes a close-knit group of proteins in which the sequences are highly conserved [104, 105]. Fe-SOD and Mn-SOD have been isolated as either dimers or tetramers [67]. Crystallographic analyses of both

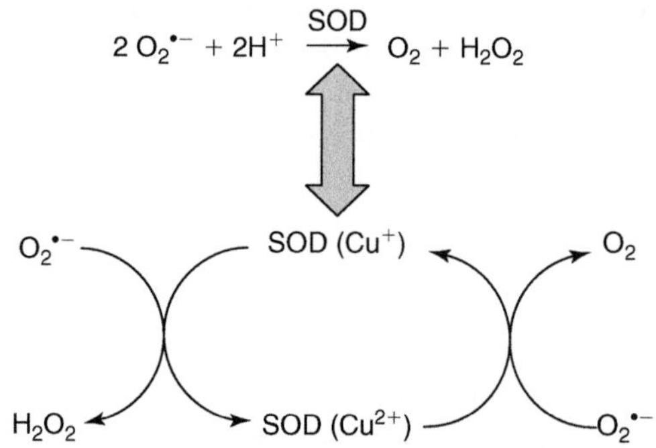

SCHEME 1 Schematic illustration of the biological process of $O_2^{\bullet -}$ dismutation into O_2 and H_2O_2 catalyzed by Cu, Zn-SOD via a cyclic oxidation-reduction electron transfer mechanism. (Reprinted from [98], with permission from Elsevier.)

enzymes indicate that they are structurally similar, but unrelated to the Cu, Zn-SOD proteins [106–114]. The monomers fold into two helix-rich domains with Mn or Fe bound by two residues from each domain. Dimer contacts occur at an interface bridging two metal sites that are separated by about 18 Å. In both Fe-SOD and Mn-SOD, the central metal is coordinated to three histidine nitrogens and one aspartate oxygen, with several conserved aromatic residues enveloping the active site. While the accurate size of the SODs is difficult to determine, previous works have suggested that Fe-SOD and Mn-SOD are a little larger than Cu, Zn-SOD [115, 116].

6.4.3 Electrochemistry of SODs

It is known that the direct electron transfer between redox enzymes and proteins and an electrode is relatively difficult to obtain because of the existence of a thick insulating protein shell around the active sites of the enzymes and proteins [117–119]. Similarly, it has been a long-standing challenge to realize the direct electron transfer properties of the SODs. Efforts in this field have been motivated by the facts that the information on direct electron transfer is very useful in understanding the intrinsic thermodynamic and kinetic properties of the SODs and, more importantly, in practical development of the SOD-based third-generation biosensors for $O_2^{\bullet-}$. Iyer observed a direct and irreversible oxidation of Cu, Zn-SOD at a bare Au electrode in phosphate buffer solution (pH 4.0) and suggested that a conformational change occurs at the active sites via its adsorption on the electrode surface, which thus facilitated the direct electron transfer [120]. Borsari and Azab and their coworkers [121, 122] and Wu *et al.* [123, 124] have observed the reversible redox response of bovine and human Cu, Zn-SOD at an Au electrode in the presence of so-called promoters for direct electron transfer. Ohsaka *et al.* found that the electron transfer between a glassy carbon electrode and polyethylene oxide-modified Cu, Zn-SOD could be successfully accomplished by using methyl viologen as a redox mediator [125].

Recently, much effort has been made on the facilitation of direct electron transfer of the SODs by self-assembled monolayers (SAMs) confined onto Au electrodes. For instance, Ohsaka *et al.* have formed various kinds of SAMs of alkanethiols onto an Au electrode and studied the electron transfer properties of the SODs [98]. Here, we will use the SAM of cysteine as an example to demonstrate the electron transfer of the SODs promoted by the SAMs of alkanethiols. Figure 6.1 depicts cyclic voltammograms (CVs) obtained at a cysteine-modified Au electrode (curves a and b) in 25 mM phosphate buffer containing 0.56 mM Cu, Zn-SOD (the concentration used represents that of the Cu^{2+} or Zn^{2+} site of Cu, Zn-SOD). For comparison, the CV obtained at a bare Au electrode (curve c) under the same conditions was also given. As shown, the cysteine-modified electrode exhibits one pair of well-defined voltammetric peaks in the SOD-containing phosphate buffer (curve a). These redox peaks were not obtained at the bare Au electrode (curve c). This observation suggests that the direct electron transfer between Cu, Zn-SOD and Au electrode does not occur actually at the bare electrode, but it can be significantly promoted at Au electrode modified with the SAM of cysteine.

As described above, Cu, Zn-SOD contains one Cu (II) and Zn (II) per monomer subunit, of which both Cu (II) and Zn (II) are redox active. Wang *et al.* have observed two

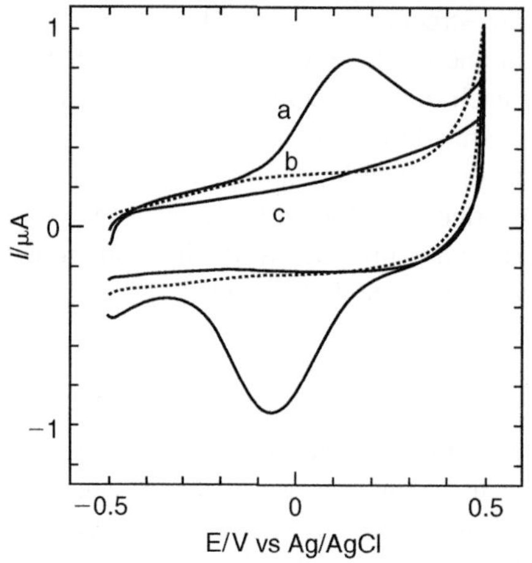

FIGURE 6.1 CVs obtained at cysteine-modified (curves a and b) and bare Au (curve c) electrodes in 25 mM phosphate buffer in the presence (curves a and c) and absence (curve b) of 0.56 mM Cu, Zn-SOD. Potential scan rate, 100 mV s^{-1}. (Reprinted from [98], with permission from Elsevier.)

redox waves for the Cu, Zn-SOD at a mercury electrode and further ascribed to the voltammetric peaks at −0.642 V and −0.98 V (vs SCE) to the redox reactions of Cu (II) and Zn (II), respectively [126]. The electron transfer of the Cu, Zn-SOD promoted by the SAM of cysteine on an Au electrode shown in Fig. 6.1 has been attributed to the electrochemical redox response of Cu (II), the active site of the Cu, Zn-SOD for the dismutation of $O_2^{\bullet -}$. This was conducted by comparing the electrochemical responses between Cu-free derivative EZnSOD (E = empty), Zn-free derivative CuESOD, and the reconstituted Cu, Zn-SOD from EZnSOD and Cu^{2+}, as shown in Fig. 6.2. The Cu-free derivative SOD (i.e. EZnSOD) and the Zn-free derivative SOD (i.e. CuESOD) were prepared by removing Cu (II) and Zn (II) from the native Cu, Zn-SOD, respectively, with a method described previously [127]. The reconstituted SOD was prepared by adding Cu^{2+} into the solution of EZnSOD and incubating the mixture at room temperature for 1 hr. As shown in Fig. 6.2 (Panel A), the absorption spectrum of the reconstituted SOD is in good agreement with that of native Cu, Zn-SOD with λ_{max} = 680 nm [128], but obviously different from that of Cu^{2+} aqua ion (inset in Fig. 6.2) and that of Cu-free SOD (EZnSOD) reported by Cocco [127]. This confirms that the reconstituted SOD can be prepared by the procedure mentioned above (i.e. mixing Cu^{2+} into the solution of Cu-free SOD and incubating the mixture at room temperature for 1 h). As can be evident from Fig. 6.2 (Panel B), no voltammetric response of EZnSOD (curve c) was observed at the cysteine-modified electrode in a potential range from −0.5 to 0.5 V. On the other hand, similar to that of native Cu, Zn-SOD, one pair of well-defined voltammetric peaks appeared in the presence of CuESOD (curve b). This implies that the cysteine-promoted direct electron transfer of Cu, Zn-SOD depicted in Fig. 6.1 can be attributed to the electrochemical redox reaction of the Cu moiety (not of the Zn moiety) in the potential range from −0.5 to 0.5 V. This can be further confirmed by the same formal potentials of the reconstituted SOD as that of the native Cu, Zn-SOD as shown in Fig. 6.3 [98].

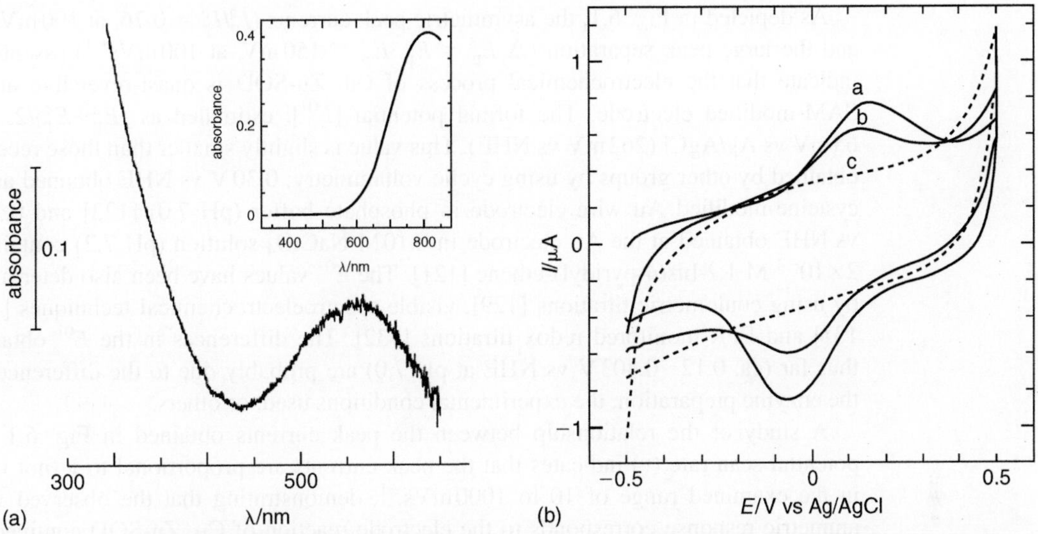

FIGURE 6.2 Panel A, UV-visible spectrum of the reconstituted SOD. The inset shows the spectrum of $CuSO_4$ solution (0.56 mM). Panel B, CVs obtained at a cysteine-modified Au electrode in 25 mM phosphate buffer containing 0.56 mM native Cu, Zn-SOD (curve a), 0.56 mM CuESOD (curve b), and 0.56 mM EZnSOD (curve c) at 100 mV s^{-1}. (Reprinted from [98], with permission from Elsevier.)

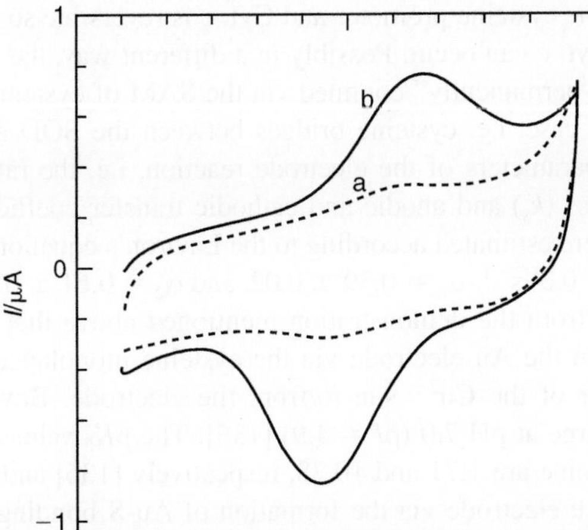

FIGURE 6.3 CVs obtained at a cysteine-modified Au electrode in 25 mM phosphate buffer containing 0.56 mM reconstituted SOD (curve a) or 0.56 mM native Cu, Zn-SOD (curve b). Potential scan rate, 100 mV s^{-1}. (Reprinted from [98], with permission from Elsevier.)

As depicted in Fig. 6.1, the asymmetric peak currents (I_p^a/I_p^c = 0.76, at 100 mVs^{-1}) and the large peak separation ($\Delta E_p = E_p^a - E_p^c$ = 150 mV, at 100 mVs^{-1}) essentially indicate that the electrochemical process of Cu, Zn-SOD is quasi-reversible at the SAM-modified electrode. The formal potential [$E^{0\prime}$], estimated as $(E_p^a + E_p^c)/2$, was 65 mV vs Ag/AgCl (262 mV vs NHE). This value is slightly smaller than those recently obtained by other groups by using cyclic voltammetry; 0.30 V vs NHE obtained at the cysteine-modified Au wire electrode in phosphate buffer (pH 7.0) [123] and 0.32 V vs NHE obtained at the Au electrode in 0.10 M $NaClO_4$ solution (pH 7.2) containing 2×10^{-4} M 1,2-bis(4-pyridyl)-ethene [121]. The $E^{0\prime}$ values have been also determined by using coulometric titrations [129], visible spectroelectrochemical techniques [130, 131] and EPR-monitored redox titrations [132]. The differences in the $E^{0\prime}$ obtained thus far (ca. 0.12~0.403 V vs NHE at pH 7.0) are probably due to the differences in the enzyme preparation, the experimental conditions used, or others.

A study of the relationship between the peak currents obtained in Fig. 6.1 and potential scan rate (v) indicates that the peak currents are proportional to v (not $v^{1/2}$) in the examined range of 10 to 1000 mVs^{-1}, demonstrating that the observed voltammetric response corresponds to the electrode reaction of Cu, Zn-SOD confined on the SAM of cysteine. This is very different from the redox reaction of Cyt. *c* in solution phase at the same cysteine-modified Au electrode, in which the peak current increased linearly with $v^{1/2}$ and thus the redox process is expected to be a diffusion-controlled electrode reaction of solution-phase species [133]. In the latter case, the cysteine promoter can be considered to be able to "transiently" orient Cyt. *c* onto the electrode surface without a deactivation in such a way that the prosthetic group is disposed towards the electrode surface and, as a result, the direct electron transfer takes place. In addition, the binding between the cysteine promoter and Cyt. *c* is reversible so that the exchange with solution-phase Cyt. *c* can occur. Possibly in a different way, the SOD enzyme can be regarded as being "permanently" confined via the SAM of cysteine to the electrode surface in the present case, i.e. cysteine bridges between the SOD and the electrode. The relevant kinetic parameters of the electrode reaction, i.e. the rate constant of the electrochemical process (k_s) and anodic and cathodic transfer coefficients (α_a and α_c) of the Cu, Zn-SOD were estimated according to the Laviron's equation [134] and calculated to be $k_s = (1.2 \pm 0.2)$ s^{-1}, $\alpha_a = 0.39 \pm 0.02$, and $\alpha_c = 0.61 \pm 0.02$ [98].

It can be deduced from the demonstration mentioned above that the Cu, Zn-SOD is properly oriented on the Au electrode via the cysteine monolayer so as to allow a rapid electron transfer of the Cu^{2+} site to/from the electrode. Bovine Cu, Zn-SOD has a net negative charge at pH 7.0 (pI = 4.9) [135]. The pK_a values of —COO^- and —NH_2 groups of cysteine are 1.71 and 10.78, respectively [136] and thus the cysteine immobilized on the Au electrode via the formation of Au-S bonding can be expected to behave as a zwitter ion at pH 7.0. Therefore, the orientation of Cu, Zn-SOD on the Au electrode through the SAM of cysteine and the resulting facilitated electron transfer may be considered to be not only simply due to an electrostatic interaction between the SOD molecule and these functional groups, but also due to a unique interaction on a molecular level. For example, the —NH_2 and —COO^- groups of cysteine are considered to cooperatively function to bind the SOD; the hydrogen bonding between the —NH_2 group and —COO^- group of Thr 135 occurs and the —COO^- group favorably interacts with the positively charged —NH_3^+ of amino acid residues (e.g. Lys 134 and Arg 141) surrounding the Cu^{2+} site, as schematically illustrated in Fig. 6.4. The surface coverage of cysteine confined on the Au electrode surface was estimated to be 6.7×10^{-10} mol cm^{-2} and the amount of the Cu, Zn-SOD confined on the SAM of cysteine was estimated to be 1.3×10^{-11} mol cm^{-2}. Thus, the ratio of the Cu, Zn-SOD to cysteine on the electrode surface is ca. 1:50 [98].

These demonstrations essentially suggest that the facilitated electron transfer of the SOD primarily relies on the interactions between the SOD and the promoters used, which is significantly dependent on the structure of the promoters. As a subsequent work, Tian *et al.* have systematically studied a large variety of thiols and disulfides with different structures with respect to their ability for facilitating the electron transfer between the Cu, Zn-SOD and the electrode, aiming at figuring out a facile, but effective, route to electron transfer promoters for the SOD and providing a common knowledge at a molecular level on the structure-associated essences of the promoted electron transfer of the SODs with the SAMs of alkanethiols [137].

It is widely accepted that an ideal promoter used for facilitating electron transfer of the enzymes and proteins should provide a suitable interface between the enzymes and proteins and electrode surface to somewhat or completely eliminate the denaturation of the biomacromolecules on the electrode surface and, more importantly, to facilitate (promote) the electron transfer between the biomacromolecules and the electrode. Moreover, the promoter molecules should bear functional groups such that they can be stably anchored onto the electrode surface. Generally, the typical structure of the promoters can be presented as X~Y, where X and Y are functional groups, and "~" represents the central part linking X and Y. In most cases, X represents —SH and —S-S groups which can form the SAMs of thiols or disulfides on Au electrodes via the formation of an Au-S bond. Y is the group that can interact with the biomacromolecules and thus facilitate the electron transfer between the proteins and the electrode. Various molecules with different structures of terminal groups and central moieties linking X and Y have been studied with cyclic voltammetry with respect to their efficiency for promoting the electron transfer of Cu, Zn-SOD at the Au electrode. The voltammetric results revealed that the electron transfer between the Cu, Zn-SOD and the Au electrode is largely dependent on the terminal groups of the SAMs examined. Typically, the SAMs terminated with —COOH groups, as displayed in Table 6.1, were found to be able to efficiently promote the direct electron transfer of Cu, Zn-SOD, while those terminated with groups of —NH_2, —CH_3, —SO_3^- and —OH, as listed in Table 6.2, could not.

Table 6.3 displays the solution pK_a values of the promoters capable of facilitating the electron transfer of Cu, Zn-SOD listed in Table 6.1. The —COOH-terminated SAMs are mostly negatively charged in phosphate buffer (pH 7.0). Although the bovine erythrocyte Cu, Zn-SOD has a net negative charge at pH 7.0 (pI = 4.9), an electrostatic interaction is still expected to occur between the SAMs and the positively charged amino acid moieties (typically —NH_3^+). Besides, the hydrogen bonding between —COOH groups and the amino acid residues is believed to comprehensively

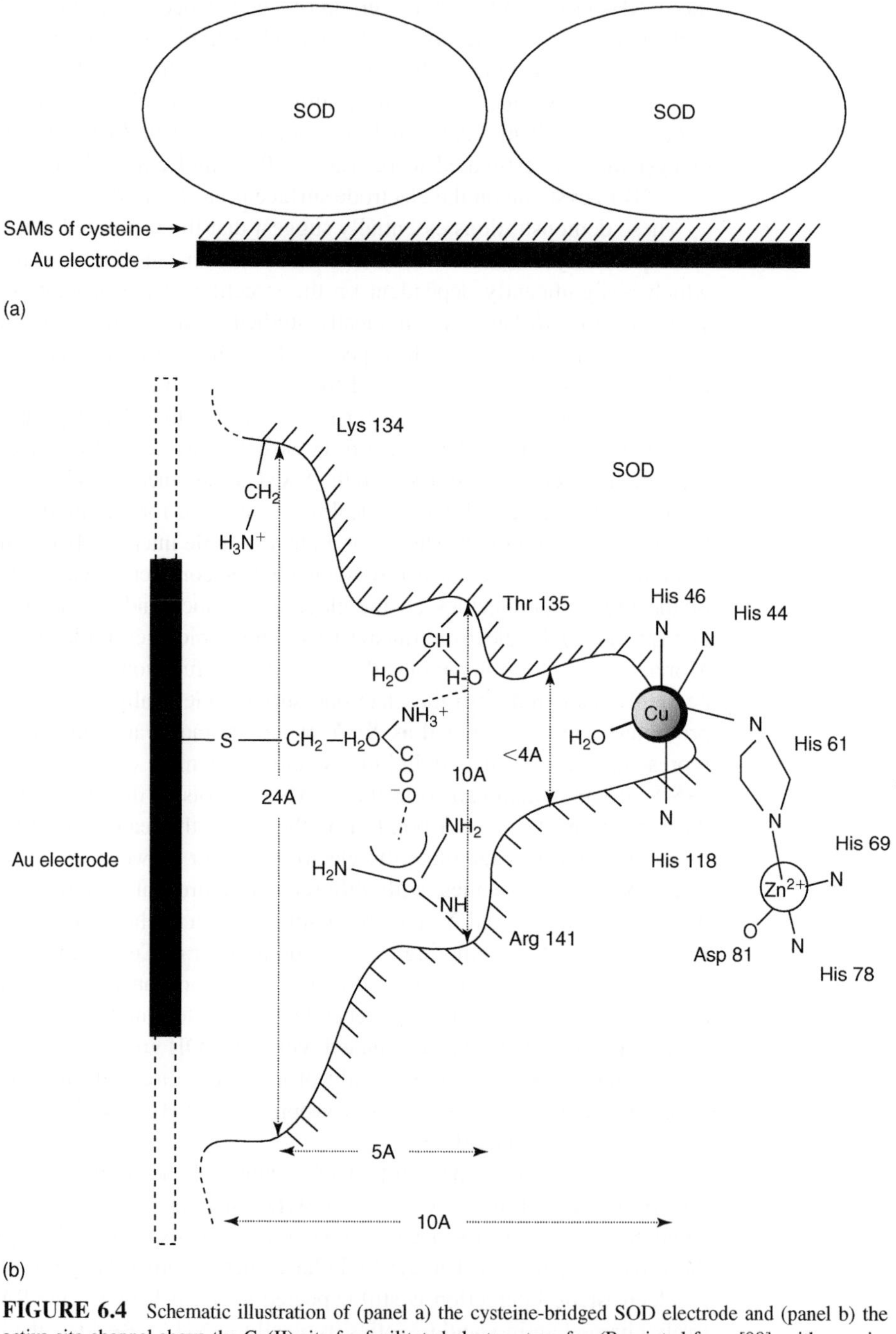

FIGURE 6.4 Schematic illustration of (panel a) the cysteine-bridged SOD electrode and (panel b) the active-site channel above the Cu(II) site for facilitated electron transfer. (Reprinted from [98], with permission from Elsevier.)

TABLE 6.1
Electrochemical parameters of Cu, Zn-SOD promoted by COOH-terminated SAMs

Molecules	$\Gamma_{modifier}$ (10^{-10} mol cm^{-2})	Γ_{SOD} (10^{-11}mol cm^{-2})	$E^{0\prime}$ (mV vs Ag/AgCl)	α_c	k_s (s^{-1})
Cysteine	6.71 ± 0.21	1.31 ± 0.21	65 ± 3	0.61 ± 0.02	1.2 ± 0.1
D,L-Homo-cysteine	4.09 ± 0.78	0.51 ± 0.09	135 ± 3	0.38 ± 0.02	0.4 ± 0.1
Mercapto-succinic acid	5.65 ± 0.60	1.23 ± 0.35	205 ± 3	0.44 ± 0.02	0.8 ± 0.1
N-Acetyle-L-cysteine	4.42 ± 0.38	1.02 ± 0.14	197 ± 3	0.46 ± 0.02	0.9 ± 0.1
Thiolactic acid	9.04 ± 0.38	0.53 ± 0.05	195 ± 3	0.74 ± 0.02	<0.1
Mercapto-acetic acid	13.2 ± 1.2	1.45 ± 0.36	202 ± 2	0.59 ± 0.02	1.3 ± 0.1
3-Mercapto-propionic acid	8.11 ± 0.32	1.26 ± 0.42	175 ± 3	0.61 ± 0.02	1.1 ± 0.1
Dimercapto-succinic acid	6.63 ± 0.56	0.55 ± 0.08	201 ± 2	0.69 ± 0.02	<0.1
L-Cystine	5.41 ± 0.24	1.42 ± 0.32	40 ± 3	0.52 ± 0.02	2.1 ± 0.1

$E^{0\prime}$ was estimated as $(E_p^a+E_p^c)/2$ in the cyclic voltammograms obtained with the SAM-modified Au electrodes in phosphate buffer at 100 mVs^{-1}. E_p^a and E_p^c are anodic and cathodic peak potentials of the Cu, Zn-SOD, respectively. (Reprinted from Y. Tian, T. Ariga, N. Takashima, T. Okajima, L. Mao, and T. Ohsaka, *Electrochem. Commun.* **6** (2004) 609–614, with permission from Elsevier.)

TABLE 6.2
SAMs unable to facilitate the electron transfer of Cu, Zn-SOD

Molecules	$\Gamma_{modifier}$ (10^{-10} mol cm^{-2})	Molecules	$\Gamma_{modifier}$ (10^{-10} mol cm^{-2})
2-Amino ethanthiol	8.97 ± 0.52	4-Aminothiophenol	4.12 ± 0.41
Cystamine	5.25 ± 0.53	4-Hydroxythiophenol	3.81 ± 0.36
1-Ethanethiol	8.54 ± 0.65	1-Butanethiol	9.86 ± 0.96
1-Decanethiol	11.2 ± 1.8	MESA	12.4 ± 2.8
4-Mercaptopyridine	3.68 ± 0.23	2-Mercaptopyridine	0.62 ± 0.06
Diphenyl disufide	3.42 ± 0.25	4,4′ -Dithiodipyridine	1.23 ± 0.14

MESA: 2-Mercaptoethanesulfonic acid (sodium salt).

function to bond the Cu, Zn-SOD on the COOH-terminated SAMs. These interactions could be relatively favorable for the orientation of the Cu, Zn-SOD on the SAM-modified Au electrodes and further for facilitating the observed electron transfer. While more evidence is still needed to understand the interactions between the SAMs and the Cu, Zn-SOD, the above demonstrations may primarily conclude that the interactions between —COOH groups of the SAMs and the Cu, Zn-SOD are probably the determinative interactions, which are mainly responsible for the observed promoted electron transfer between the Cu, Zn-SOD and the Au electrode.

TABLE 6.3
p*Ka* values of the compounds capable of facilitating electron transfer of Cu, Zn-SOD

Compounds	p*Ka*	Compounds	p*Ka*
Mercaptoacetic acid	3.73 ± 0.20	3-Mercaptopropionic acid	4.34 ± 0.20
Cysteine	2.07 ± 0.10, 11.05 ± 0.16	Dimercaptosuccinic acid	2.74 ± 0.25
L-Cystine	1.70 ± 0.10, 8.72 ± 0.16	Mercaptosuccinic acid	3.49 ± 0.23
Thiolactic acid	3.74 ± 0.20	N-Acetyle-L-cysteine	3.25 ± 0.10
D,L-Homocysteine	2.21 ± 0.10, 9.33 ± 0.16		

While the interactions between the —COOH groups of the SAMs (as listed in Table 6.1) and the Cu, Zn-SOD are critical to the direct electron transfer of the SOD as addressed above, the constituents of the central moieties linking X and Y are found to be also of great importance for the electron transfer because of their remarkable influences on the kinetics of electron transfer and the interaction of the SAM with the SOD. Generally speaking, the alkylene chains linking X and Y act as a potential energy barrier for the electron transfer between the SOD and the Au electrode, and consequently the electron transfer rate decreases with increasing the chain length. The increase in the length of the alkylene chain linking X and Y, ($—(CH_2)—_n$), was found to obviously tune the reversibility of electrode reactions of the SOD from reversible ($n = 1$) to quasi-reversible ($n = 2, 3$), and they are actually suppressed when the chain length is longer than $—(CH_2)—_3$, as listed in Table 6.4, although the interaction between the —COOH groups of the SAMs and the SOD is expected to occur in these cases [137].

Besides the dependency on the length of the alkylene chain linking X and Y, the electron transfer of the Cu, Zn-SOD was also found to be greatly dependent on the constitution (aliphatic or aromatic) of such a section. As shown in Table 6.4, the SAMs with an aromatic group linking X and Y could not promote the electron transfer of the Cu, Zn-SOD although some of them could facilitate the electron transfer of Cyt. *c*. Perhaps this is due to the fact that the aromatic —COOH-terminated SAMs are not favorable for interactions between the SAMs and the Cu, Zn-SOD, probably because of the spatial barrier of these aromatic SAMs.

The uses of SAMs of alkanethiols for facilitating electron transfer of the Cu, Zn-SOD can be extended for other kinds of the SODs, such as Fe-SOD and Mn-SOD in the SOD family [138]. Figure 6.5 shows typical CVs obtained at 3-mercaptopropionic acid (MPA)-modified Au electrodes in 5 mM phosphate buffer (pH 7.0) containing Fe-SOD (1), Mn-SOD (2) or Cu, Zn-SOD (3). The concentrations of the SODs used here represent those of the Cu^{2+} site of Cu, Zn-SOD, Fe^{3+} site of Fe-SOD, or Mn^{3+} site of Mn-SOD, respectively.

A pair of redox peaks, which could be ascribed to the electron transfer of the SODs, was observed for the three kinds of SODs at the MPA-modified Au electrode, while it was not obtained at a bare Au electrode (not shown in Fig. 6.5). This result demonstrates that the electron transfer between the SODs and the Au electrode can be well promoted by the

TABLE 6.4
—COOH-terminated SAMs unable to facilitate electron transfer of Cu, Zn-SOD

Molecules	Molecular formula	$\Gamma_{modifier}$ (10^{-10} mol cm^{-2})
6-Mercaptohexanoic acid	$HS—(CH_2)_5—COOH$	12.2 ± 3.2
8-Mercaptooctanoic acid	$HS—(CH_2)_7—COOH$	12.0 ± 2.8
11-Mercaptoundecanoic acid	$HS—(CH_2)_{10}—COOH$	11.6 ± 2.2
16-Mercaptohexadecanoic acid	$HS—(CH_2)_{15}—COOH$	11.9 ± 2.5
4, 4′-Dithiodibutyric acid	$S—CH_2—CH_2—CH_2—COOH$ / $S—CH_2—CH_2—CH_2—COOH$ (S–S bonded)	5.07 ± 0.39
D,L-Penicillamine	$HS—C(CH_3)_2—CH(NH_2)COOH$	5.23 ± 0.28
s-Metyl-L-cysteine	$H_3C—S—CH_2—CH(NH_2)COOH$	2.16 ± 0.34
L-Methionine	$H_3C—S—CH_2—CH(NH_2)COOH$	1.32 ± 0.32
4-Mercaptobenzoic acid	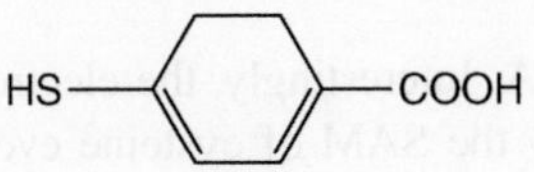	4.71 ± 0.61
4-Mercaptohydrocinnamic acid	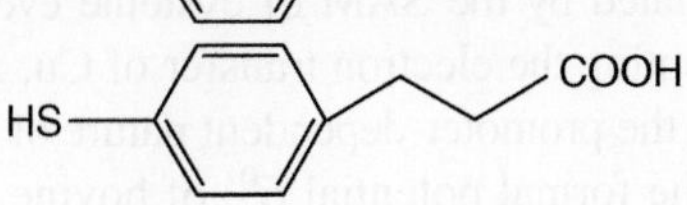	4.25 ± 0.73
Thioctic acid	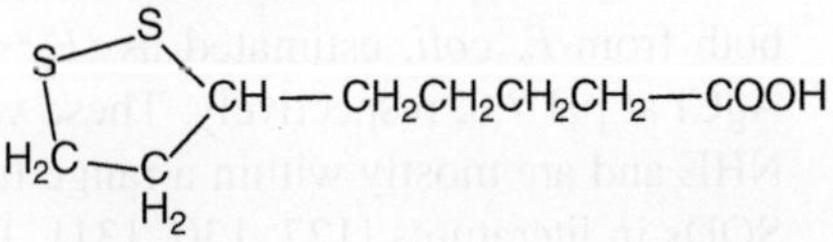	8.63 ± 0.72
5,5′-Dithiobis-(2-nitrobenzoic acid)	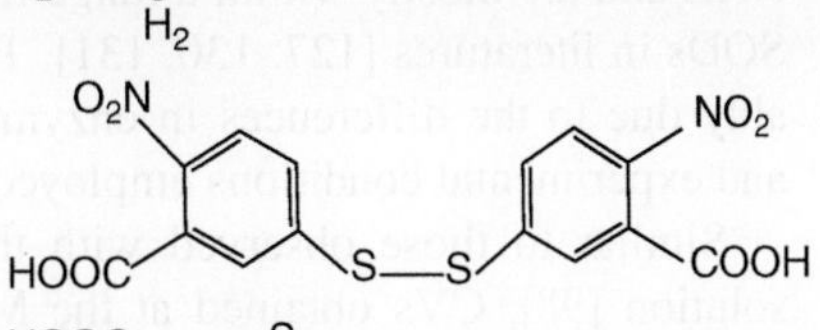	0.89 ± 0.12
L-Thioproline	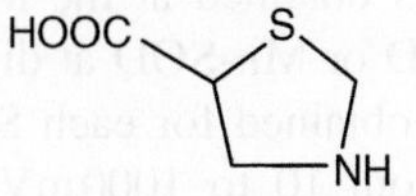	4.22 ± 0.11
Rhodanine-3-acetic acid	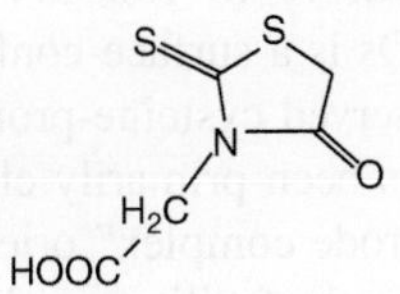	4.03 ± 0.16
2-Thiophenecarboxylic acid	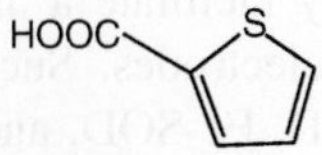	0.51 ± 0.08
3-Thiophenecarboxylic acid	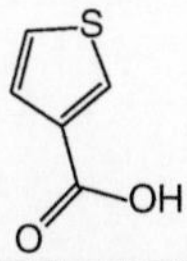	0.42 ± 0.08

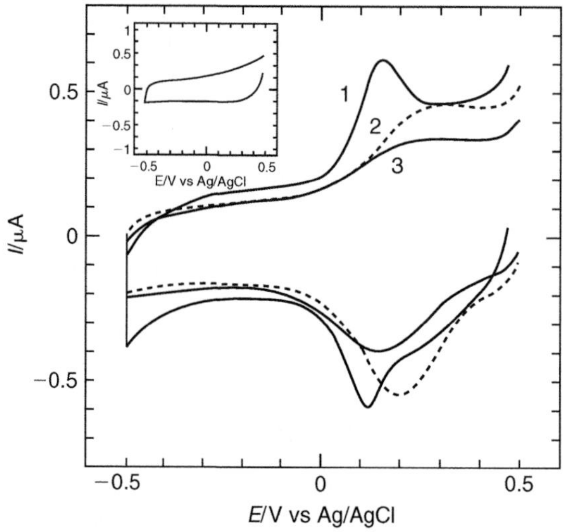

FIGURE 6.5 CVs at MPA-modified Au electrode in 5 mM phosphate buffer (pH 7.0) containing (1) Fe-SOD (0.32 mM), (2) Mn-SOD (0.40 mM), and (3) Cu/Zn-SOD (0.20 mM). Inset: CV of the MPA-modified Au electrode in pure phosphate buffer. Potential scan rate, 100 mV s^{-1}. (Reprinted from [138], with permission from the American Chemical Society.)

SAM of MPA. Interestingly, the electron transfer of Fe-SOD and Mn-SOD could not be facilitated by the SAM of cysteine even though the SAM of cysteine could be used for promoting the electron transfer of Cu, Zn-SOD [98], as described above. This again suggests the promoter-dependent nature of the electron transfer properties of the SODs.

The formal potential $E^{0\prime}$ of bovine erythrocyte Cu, Zn-SOD, Fe-SOD and Mn-SOD both from *E. coli*, estimated as $(E_p^a+E_p^c)/2$, were 0.21 V, 0.14 V, and 0.23 V vs Ag/AgCl at pH 7.0, respectively. These values correspond to 0.41 V, 0.33 V, and 0.42 V vs NHE and are mostly within a range from ca. 0.04 to 0.403 V vs NHE reported for the SODs in literatures [127, 130, 131]. The diversity in these documented values is probably due to the differences in enzyme preparation, electron transfer promoters used, and experimental conditions employed.

Similar to those observed with the cysteine-modified electrode in Cu, Zn-SOD solution [98], CVs obtained at the MPA-modified Au electrode in phosphate buffer containing Fe-SOD or Mn-SOD at different potential scan rates (v) clearly show that the peak currents obtained for each SOD are linear with v (not $v^{1/2}$) over the potential scan range from 10 to 1000 mVs^{-1}. This observation reveals that the electron transfer of the SODs is a surface-confined process and not a diffusion-controlled one. The previously observed cysteine-promoted surface-confined electron transfer process of Cu, Zn-SOD has been primarily elucidated based on the formation of "a cysteine-bridged SOD-electrode complex" oriented at an electrode–solution interface, which is expected to sufficiently facilitate a direct electron transfer between the metal active site in SOD and Au electrodes. Such a model appears to be also suitable for the SODs (i.e. Cu, Zn-SOD, Fe-SOD, and Mn-SOD) with MPA promoter. The so-called

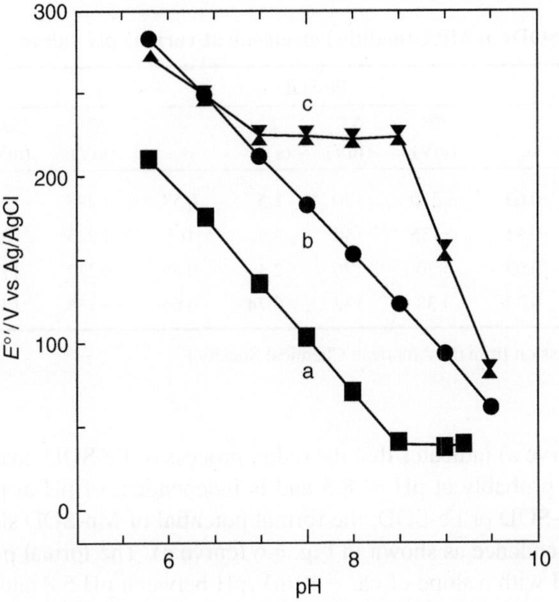

FIGURE 6.6 Plots of the formal potentials of (a) Fe-SOD, (b) Cu, Zn-SOD, and (c) Mn-SOD at the MPA-modified Au electrode vs solution pH. (Reprinted from [138], with permission from the American Chemical Society.)

"MPA-bridged SOD-electrode complex" could be formed via a variety of interactions between MPA and the SODs, such as electrostatic, hydrophobic, and/or hydrogen bonding interactions, which is believed to be responsible for the observed direct electron transfer properties of the SODs. Besides, such interactions substantially enable the SODs to be stably confined at the MPA-modified Au electrode, which can be further evident from the re-observation of the redox responses of SODs in a pure electrolyte solution containing no SOD with the MPA-modified electrode previously used in SOD solutions.

The formal potentials ($E^{0\prime}$) of the three kinds of SODs were found to be dependent on solution pH as displayed in Fig. 6.6. As shown, the formal potential of bovine erythrocyte Cu, Zn-SOD decreases linearly with increasing solution pH with a slope of ca. $-60\,\mathrm{mV/pH}$ from pH 5.8 to pH 9.5 (curve b), indicating one proton and one electron are included in the electrode reaction of Cu, Zn-SOD, which is similar to previously proposed enzymatic catalytic mechanistic scheme of the Cu, Zn-SOD [139–144]. In contrast, the pH dependency of Fe-SOD from *E. coli* was complicated (curve a); the formal potential changes linearly with solution pH in a range from pH 5.8 to 8.5 with a slope of ca. $-60\,\mathrm{mV/pH}$, and becomes pH-independent at above pH > 8.5. Previous studies have observed that the Fe (III) form of the protein ionizes with an apparent pK_a of 9.0 ± 0.3 and such ionization effect has been interpreted in terms of hydrolysis of a bound water molecule with pK_a of ca. 8.5 [145]. The $E^{0\prime}$–pH profile of

TABLE 6.5
Electrochemical Parameters of SODs at MPA-modified electrode at various pH values

	Cu, Zn-SOD				Fe-SOD				Mn-SOD			
pH	$E^{0\prime}$ (mV)	ΔE_p (mV)	k_s (s^{-1})	α_c	$E^{0\prime}$ (mV)	ΔE_p (mV)	k_s (s^{-1})	α_c	$E^{0\prime}$ (mV)	ΔE_p (mV)	k_s (s^{-1})	α_c
5.8	+282	121	0.98	0.63	+210	120	1.5	0.59	+275	104	1.2	0.61
7.0	+212	100	1.1	0.61	+135	40	3.9	0.5	+225	85	1.9	0.58
8.0	+153	115	0.94	0.63	+70	76	2.4	0.55	+222	112	1.6	0.59
9.0	+93	146	0.46	0.74	+38	132	0.74	0.65	+155	133	0.35	0.76

(Reprinted from [138], with permission from the American Chemical Society.)

the Fe-SOD (curve a) indicates that the redox process of Fe-SOD involves one electron and one proton probably at pH < 8.5 and is independent of pH at pH > 8.5. Unlike those of Cu, Zn-SOD or Fe-SOD, the formal potential of Mn-SOD showed more complicated pH dependence as shown in Fig. 6.6 (curve c). The formal potential decreases linearly with pH with a slope of ca. -40 mV/pH between pH 5.8 and pH 7.0, retains a constant between 7.0 and 8.5, and then decreases sharply between pH 8.5 and 9.5 (the slope is ca. -140 mV/pH). This $E^{0\prime}$–pH profile likely suggests that the Mn-SOD has two pK_as; one around 7.0 and the other about 8.5. This almost coincides with earlier results obtained with optical titrations, in which two pK_as of 6.7 $\pm$ 0.1 and 8.5 $\pm$ 0.3 were suggested for Mn-SOD [146, 147].

The rate constant of electron transfer (k_s) and anodic and cathodic electron transfer coefficients (α_a and α_c) of the SODs at various pH values were estimated with Laviron's equation and summarized in Table 6.5. Interestingly, the fastest electron transfer of the SODs was essentially achieved in a neutral solution, probably in agreement with the biological conditions for the inherent catalytic mechanisms of the SODs for $O_2^{\bullet -}$dismutation, although the electrode processes of the SODs follow a different mechanism.

6.4.4 SOD-based electrochemical biosensors for O_2^{-}

Generally, the enzyme-based biosensors can be divided into three categories, i.e. first-, second-, and third-generation biosensors. The first-generation enzyme-based biosensors are preliminarily based on the measurement of compounds involved in the enzymatic reactions, such as hydrogen peroxide and dissolved oxygen. These kinds of biosensors unfortunately have their essential limits in sensitivity, selectivity, response time, and so on. The development of electrochemical studies on electron transfer mediation provides the possibility to construct the second-generation biosensors, of which the electron transfer between the enzymes and electrode is shuttled by artificial electron transfer mediators. The mediators can be retained at the electrode surface by a discrete membrane or mixed with the enzyme in a carbon paste or entrapped in a film. The third-generation biosensors are

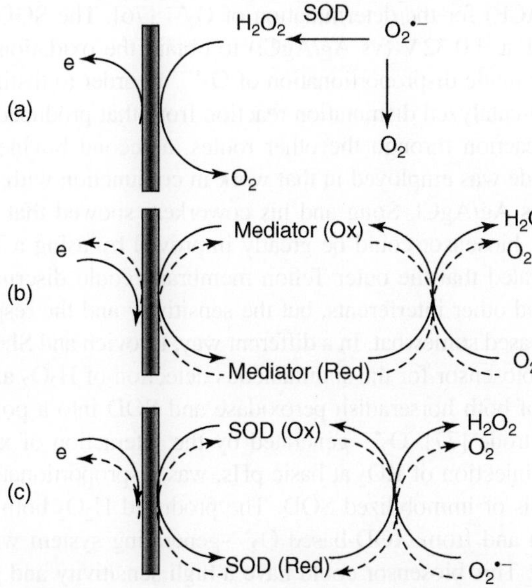

SCHEME 2 Schematic illustration of analytical mechanism of (a) first-, (b) second-, and (c) third-generation $O_2^{\bullet-}$ biosensors. Note that the reactions shown in (b) and (c) are bi-directional since SODs are enzymes specifically catalyzing the $O_2^{\bullet-}$ dismutation, i.e. oxidation into O_2 and reduction into H_2O_2.

fabricated based on the direct electron transfer of enzymes, that is, by creating an electrode surface that can interact with the active center of the redox enzymes.

Similarly, the SOD-based enzymatic biosensors for the $O_2^{\bullet-}$ determination can also fall into the three categories mentioned above, i.e. the first-, second-, and third-generation biosensors with the analytical mechanism as schematically illustrated in Scheme 2. Most of the SOD-based first-generation biosensors reported thus far have been based on the detection of H_2O_2 produced from the SOD dismutation of $O_2^{\bullet-}$ as shown in Scheme 2(a). Mesaros and coworkers have designed an amperometic $O_2^{\bullet-}$ biosensor at the platinum electrode with electropolymerized pyrrole film containing SOD [79, 80]. The $O_2^{\bullet-}$ penetrates from the bulk solution into the polypyrrole film, where it is dismutated into H_2O_2 and O_2 under the catalysis of SOD. The generated H_2O_2 is then oxidized at platinum electrode at a potential of +0.7 V (vs SCE). The biosensor was demonstrated to have a low detection limit (15 nM), good temperature stability, and short response time (<5 s). However, the high potential required for the oxidation of H_2O_2 may cause the simultaneous oxidation of some coexisting electroactive species in biological samples, e.g. ascorbate and uric acid. Moreover, H_2O_2 is a metabolite of the dismutation of $O_2^{\bullet-}$ and a product from the enzymatic reaction of endogenous oxidases, such as monoamine oxidase and L-amino acid oxidase. Thus, the first-generation biosensors based on the detection of H_2O_2 seem to be limited in distinguishing H_2O_2 produced by the SOD-catalyzed dismutation of $O_2^{\bullet-}$ from that endogenously produced in the biological systems. McNeil and his coworkers have fabricated the SOD-coated platinized activated carbon electrode (PACE) for the determination of $O_2^{\bullet-}$ [76]. The SOD-based $O_2^{\bullet-}$ biosensor was polarized at +0.32 V (vs Ag/AgCl) to obtain the oxidation current of H_2O_2 produced by the enzymatic disproportionation of $O_2^{\bullet-}$. In order to distinguish H_2O_2 produced from the SOD-catalyzed dismutation reaction from that produced from the natural disproportionation reaction through the other routes, a second bovine serum albumin-coated PACE electrode was employed in that work in conjunction with the bipotentiostat poised at +0.32 V vs Ag/AgCl. Song and his coworkers showed that the selectivity of the SOD-based $O_2^{\bullet-}$ biosensor could be greatly improved by using a Teflon membrane [82]. They demonstrated that the outer Teflon membrane could discriminate $O_2^{\bullet-}$ from endogenous H_2O_2 and other interferents, but the sensitivity and the response time of the biosensor were decreased somewhat. In a different way, Lvovich and Sheeline have developed a two-channel biosensor for the simultaneous detection of H_2O_2 and $O_2^{\bullet-}$ based on the immobilization of both horseradish peroxidase and SOD into a polypyrrole layer at a glassy carbon electrode [77]. $O_2^{\bullet-}$ generated by the interaction of xanthine and xanthine oxidase, or by injection of KO_2 at basic pHs, was disproportionated into H_2O_2 and O_2 under the catalysis of immobilized SOD. The produced H_2O_2 both from the disproportionation reaction and from XOD-based $O_2^{\bullet-}$-generating system was reduced under the catalysis of HRP. This biosensor could have a high sensitivity and good stability for ten days.

The second-generation $O_2^{\bullet-}$ biosensors are mainly based on the electron transfer of SOD shuttled by surface-confined or solution-phase mediators, as shown in Scheme 2(b). In 1995, Ohsaka *et al.* found that methyl viologen could efficiently shuttle the electron transfer between SOD and the glassy carbon electrode and proposed that such a protocol could be useful for developing $O_2^{\bullet-}$ biosensors [125]. Recently, Endo *et al.* reported an $O_2^{\bullet-}$ biosensor based on mediated electrochemistry of SOD [148]. In that case, ferrocene-carboxaldehyde was used as the mediator for the redox process of SOD. The as-developed $O_2^{\bullet-}$ biosensor showed a high sensitivity, reproducibility, and durability. A good linearity was obtained in the range of 0 ~ 100 μM. In the flow cell system, tissue-derived $O_2^{\bullet-}$ was measured.

Besides the mediator-based second-generation SOD biosensors, much attention is being devoted to the development of SODs-based third-generation $O_2^{\bullet-}$ biosensors on the basis of the direct electron transfer properties of the SODs as shown in Scheme 2(c). This is because the third-generation biosensors are advantageous over the other two types of biosensors in, for example, the simple procedure required for the biosensor design and for mechanistic understanding. Di *et al.* developed a third-generation $O_2^{\bullet-}$ biosensor by immobilizing SOD into sol-gel thin film confined on an Au electrode [149]. The uniform porous structure of the silica-PVA sol-gel matrix resulted in a very low mass transport barrier and a rapid and direct electron transfer of SOD. Based on biomolecular recognition for specific reactivity of SOD toward $O_2^{\bullet-}$, the SOD-immobilized electrode enabled a sensitive and selective detection of $O_2^{\bullet-}$ with a low potential of −0.15 vs SCE.

On the basis of the direct electron transfer properties of SODs at the SAM-modified Au electrodes as described in the previous section, Ohsaka *et al.* have developed SOD-based third-generation $O_2^{\bullet-}$ biosensors by immobilizing SODs including Cu, Zn-SOD, Fe-SOD, and Mn-SOD onto SAM-modified Au electrodes [138]. The concept of the biosensing technique for the determination of $O_2^{\bullet -}$ can be illustrated in Scheme 3 together with the mechanism of $O_2^{\bullet -}$ dismutation catalyzed by the SODs in biological systems [150], by using Cu, Zn-SOD as an example. In biological systems, SODs efficiently catalyzes the dismutation of $O_2^{\bullet -}$ into O_2 and H_2O_2 via a redox cycle of the metal complex moiety [M = Cu, Fe or Mn] in the SODs. During the dismutation reactions, two $O_2^{\bullet -}$ ions are stoichiometrically converted to one O_2 molecule and one H_2O_2 molecule with consumption of two H^+ ions. That is, one $O_2^{\bullet -}$ reduces the SOD [$M_{\text{(oxidized form)}}$] to produce O_2 and the SOD [$M_{\text{(reduced form)}}$], while another $O_2^{\bullet -}$ oxidizes the SOD [$M_{\text{(reduced form)}}$] to produce H_2O_2 and the SOD [$M_{\text{(oxidized form)}}$]. The biological homogeneous catalytic reactions can be split into two independent heterogeneous electrode reactions, in which the SODs are immobilized on the electrode surface, as shown in Scheme 3. In the cathodic process, the redox reaction between $O_2^{\bullet -}$ and SOD [$M_{\text{(reduced form)}}$] takes place to produce H_2O_2 and SOD [$M_{\text{(oxidized form)}}$]. The generated SOD [$M_{\text{(oxidized form)}}$] can be reduced at the electrode. On the other hand, in the anodic process, $O_2^{\bullet -}$ reduces SOD [$M_{\text{(oxidized form)}}$] to produce SOD [$M_{\text{(reduced form)}}$], which can be reoxidized at the electrode. Thus, by measuring the oxidation or reduction current at the SOD-modified electrode in the presence of $O_2^{\bullet -}$, one may detect $O_2^{\bullet -}$. This kind of third-generation SOD-based $O_2^{\bullet -}$ biosensor is essentially based on the direct electron transfer properties and the catalytic activities of the SODs.

The thing to be noted here is that the E^0 values of the O_2/ $O_2^{\bullet -}$ and $O_2^{\bullet -}$ H_2O_2 redox couples are −0.35 and 0.68 V vs Ag/AgCl at pH 7.4 and thus the SODs, for example, Cu, Zn-SOD (Cu (I/II)) with $E^{0\prime}$ = 65 mV can mediate both the oxidation of $O_2^{\bullet -}$ to O_2 and the reduction of $O_2^{\bullet -}$ to H_2O_2. Such a bi-directional electromediation (electrocatalysis) by the SOD/SAM electrode is essentially based on the inherent specificity of the SOD enzyme which catalyzes the dismutation of $O_2^{\bullet -}$ to O_2 and H_2O_2 via a redox cycle of their metal complex moiety (Scheme 3).

Figure 6.7 illustrates the voltammetric response of the third-generation SOD-based $O_2^{\bullet -}$ biosensors with Cu, Zn-SOD confined onto cystein-modified Au electrode as an example. The presence of $O_2^{\bullet -}$ in solution essentially increases both the cathodic and anodic peak currents of the SOD compared with its absence [150]. Such a redox response was not observed at the bare Au or cysteine-modified Au electrodes in the presence of $O_2^{\bullet -}$. The observed increase in the anodic and cathodic current response of the Cu, Zn-SOD/cysteine-modified Au electrode in the presence of $O_2^{\bullet -}$ can be considered to result from the oxidation and reduction of $O_2^{\bullet -}$, respectively, which are effectively mediated by the SOD confined on the electrode as shown in Scheme 3. Such a bi-directional electromediation (electrocatalysis) by the SOD/cysteine-modified Au electrode is essentially based on the inherent specificity of SOD for the dismutation of $O_2^{\bullet -}$, i.e. SOD catalyzes both the reduction of $O_2^{\bullet -}$ to H_2O_2 and the oxidation to O_2 via a redox cycle of its Cu (II/I) complex moiety as well as the direct electron transfer of SOD realized at the cysteine-modified Au electrode. Thus, this coupling between the electrode and enzyme reactions of SOD could facilitate the development of the third-generation biosensor for $O_2^{\bullet -}$.

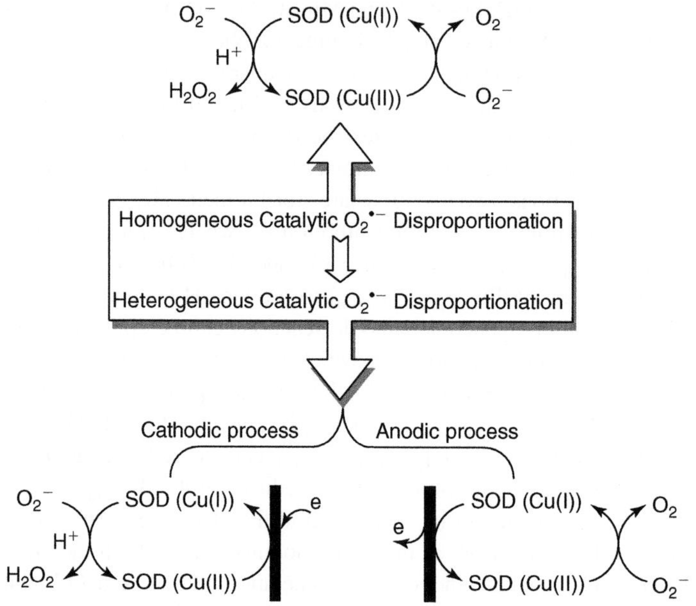

SCHEME 3 Schematic diagram illustrating the concept of the third-generation $O_2^{\bullet-}$ biosensors based on the biocatalytic activity inherent in SODs toward the dismutation of $O_2^{\bullet-}$. (Reprinted from [150], with permission from the Royal Society of Chemistry.)

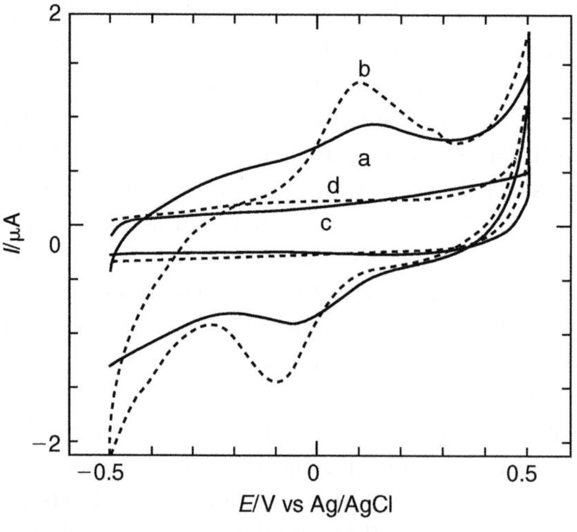

FIGURE 6.7 CVs obtained at (a, b) Cu,Zn-SOD/cysteine-modified, (c) bare, and (d) cysteine-modified Au electrodes in 25 mM PBS (pH 7.4) saturated by N_2 (a, c, d) or O_2 (b). Solutions (b) and (c) contain 0.002 U ml^{-1} XOD and 25 mM xanthine. Potential scan rate, 100 mV s^{-1}. (Reprinted from [150], with permission from the Royal Society of Chemistry.)

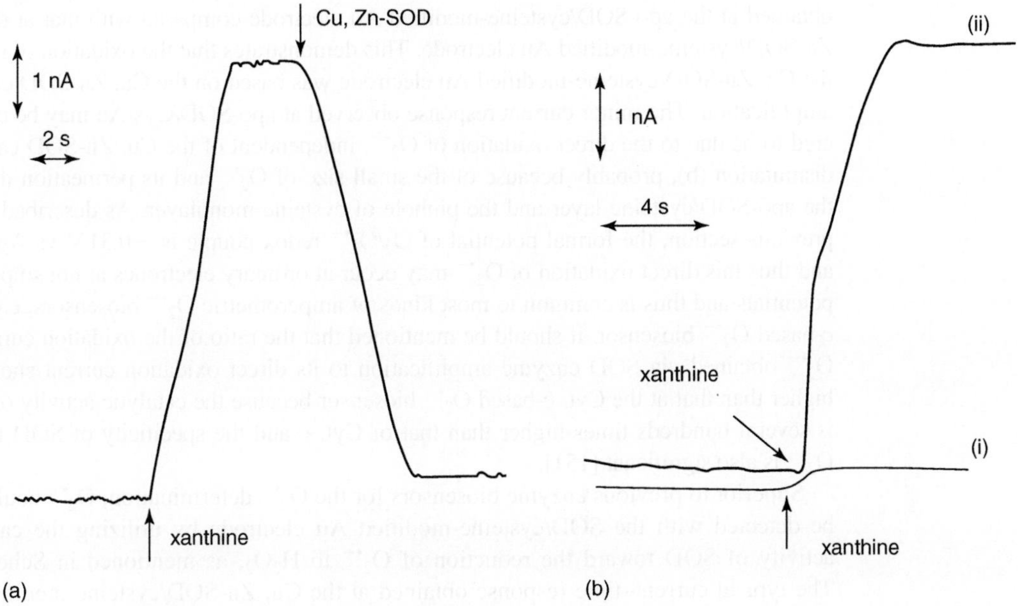

FIGURE 6.8 (a) Typical current–time response obtained at the Cu, Zn-SOD/cysteine-modified Au electrode in phosphate buffer (O_2-saturated) solution containing 0.002 unit of XOD upon the addition of 50 nM xanthine and the subsequent addition of 6 μM Cu, Zn-SOD. (b) Current–time responses of (i) the apo-SOD/cysteine-modified Au electrode and (ii) the Cu, Zn-SOD/cysteine-modified Au electrode toward $O_2^{\bullet-}$ in phosphate buffer containing 0.002 unit of XOD upon the addition of 40 nM xanthine. The electrode was polarized at +300 mV, and the solution was gently stirred with a magnetic stirrer at 200 rpm. (Reprinted from [151], with permission from the American Chemical Society.)

Figure 6.8a shows the typical current–time response of the Cu, Zn-SOD/cysteine-modified Au electrode at +300 mV on addition of xanthine and Cu, Zn-SOD to the O_2-saturated phosphate buffer containing 0.002 U XOD. The introduction of 50 nM xanthine into the solution produced a rapid and obvious increase in the anodic current. To verify that the observed anodic current is attributed to the oxidation of $O_2^{\bullet-}$ rather than that of the co-products of the XOD/xanthine-based $O_2^{\bullet-}$ generating reaction, i.e. uric acid and H_2O_2, 6 μM Cu, Zn-SOD was added into the solution since Cu, Zn-SOD, a selective scanvenger of $O_2^{\bullet-}$, can specifically dismutate $O_2^{\bullet-}$. This addition caused the anodic current to decrease by >95% within 6 s (Fig. 6.8a). The addition of more Cu, Zn-SOD to the solution resulted in the anodic current decreasing to almost the background current, strongly indicating that only $O_2^{\bullet-}$ generated by the XOD-xanthine system is oxidized at the Cu, Zn-SOD/cysteine-modified Au electrode to give an amperometric response. In order to examine the direct oxidation of $O_2^{\bullet-}$ at the Cu, Zn-SOD/cysteine-modified Au electrode without the above-mentioned redox mediation via Cu, Zn-SOD, the response of the apo-SOD/cysteine-modified Au electrode toward $O_2^{\bullet-}$ was also measured. The apo-SOD (Cu-free derivative EZnSOD, E = empty) was prepared according to a method described by Cocco and Calabrese [127]. As shown in Fig. 6.8b, much less response was obtained at the apo-SOD/cysteine-modified Au electrode compared with that at the Cu, Zn-SOD/cysteine-modified Au electrode. This demonstrates that the oxidation of $O_2^{\bullet -}$ at the Cu, Zn-SOD/cysteine-modified Au electrode was based on the Cu, Zn-SOD enzyme amplification. The minor current response observed at apo-SOD/Cys/Au may be considered to be due to the direct oxidation of $O_2^{\bullet -}$, independent of the Cu, Zn-SOD catalytic dismutation (b), probably because of the small size of $O_2^{\bullet -}$ and its permeation through the apo-SOD/cysteine layer and the pinhole of cysteine monolayer. As described in the previous section, the formal potential of $O_2/O_2^{\bullet -}$ redox couple is -0.31 V vs Ag/AgCl and thus this direct oxidation of $O_2^{\bullet -}$ may occur at ordinary electrodes at not so positive potentials and thus is common to most kinds of amperometric $O_2^{\bullet -}$ biosensors, e.g. Cyt. *c*-based $O_2^{\bullet -}$ biosensor. It should be mentioned that the ratio of the oxidation current of $O_2^{\bullet -}$ obtained via SOD enzyme amplification to its direct oxidation current should be higher than that at the Cyt. *c*-based $O_2^{\bullet -}$ biosensor because the catalytic activity of SOD is several hundreds times higher than that of Cyt. *c* and the specificity of SOD toward $O_2^{\bullet -}$ is also significant [151].

Superior to previous enzyme biosensors for the $O_2^{\bullet -}$ determination, $O_2^{\bullet -}$ could also be detected with the SOD/cysteine-modified Au electrode by utilizing the catalytic activity of SOD toward the reduction of $O_2^{\bullet -}$ to H_2O_2, as mentioned in Scheme 3. The typical current–time response obtained at the Cu, Zn-SOD/cysteine-modified Au electrode at -200 mV is shown in Fig. 6.9. The addition of xanthine to the solution resulted in an obvious increase in the cathodic current, but such a cathodic current was not observed at the bare or cysteine-modified Au electrodes. As can be seen from Fig. 6.9, the introduction of catalase, which has a high catalytic activity toward the dismutation of H_2O_2, to the solution did not result in any change in the current response, suggesting that the observed cathodic current is not due to the reduction of H_2O_2 coproduced in the XOD/xanthine-based $O_2^{\bullet -}$ generating reaction. On the contrary, the

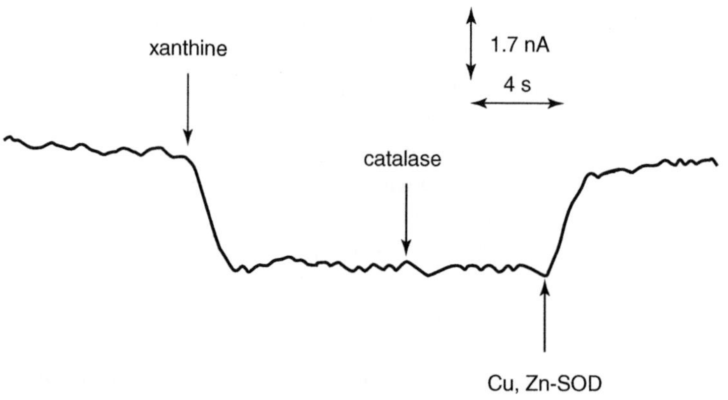

FIGURE 6.9 Typical current–time responses obtained at the Cu, Zn-SOD/cysteine-modified Au electrode in phosphate buffer (O_2-saturated) containing 0.002 unit of XOD upon the addition of 30 nM xanthine and the subsequent addition of 590 units of catalase and 6 μM Cu, Zn-SOD. The electrode was polarized at -200 mV, and other conditions are the same as those in Fig. 6.8.

addition of Cu, Zn-SOD caused the cathodic current to decrease to the background level. Thus, the observed cathodic response is reasonably ascribable to the reduction of $O_2^{\bullet -}$ mediated by the Cu, Zn-SOD confined on the electrode [151].

The steady-state amperometric responses of the Cu, Zn-SOD/cysteine-modified Au electrode to successive addition of $O_2^{\bullet -}$ in solution examined at +300 mV and −200 mV were found to be proportional to $O_2^{\bullet -}$ concentration in a range of ca. 13 to 130 nM min^{-1}. The sensitivity of the SOD/cysteine-modified Au electrode was 24 and 22 nA cm^{-2}/(nM min^{-1}) at +300 and −200 mV, respectively. The detection limit was evaluated based on a signal-to-noise ratio of 3 to 1 and calculated to be 5 nM at +300 mV and 6 nM at −200 mV, respectively. The response time of the biosensor was measured as the time to reach 95% of the maximum change in response to a step injection of xanthine and found to be less than 6 s. For the stability test, the anodic and cathodic responses for $O_2^{\bullet -}$ generated by the XOD-xanthine system were recorded four times each day and the current responses were reported to be constant for at least one week [151].

The protocol demonstrated for the development of Cu, Zn-SOD-based third-generation $O_2^{\bullet -}$ biosensors is also sutiable for other kinds of SODs, such as Fe-SOD and Mn-SOD in the SOD family [138]. In those cases, MPA was used as promoter for the SODs instead of cysteine because cysteine was unable to promote electron transfer of Fe-SOD and Mn-SOD. Figure 6.10 compares the CVs obtained at the Cu, Zn-SOD- (a), Fe-SOD- (b), and Mn-SOD- (c) based electrodes in the absence and presence of $O_2^{\bullet -}$. As shown, the presence of $O_2^{\bullet -}$ in solution obviously increases both anodic and cathodic peak currents of the SODs confined on the electrodes, suggesting the good bifunctional catalytic activity for the reduction and oxidation of $O_2^{\bullet -}$ at the SODs, which is similar to the results obtained with the Cu,Zn-SOD/cysteine-modified Au electrode. It should be mentioned that the same response was observed neither at the MPA-modified Au electrode nor at the bare Au electrode under the same conditions. Such a bidirectional electromediation of the SOD-based biosensors is essentially based on the inherent specificity

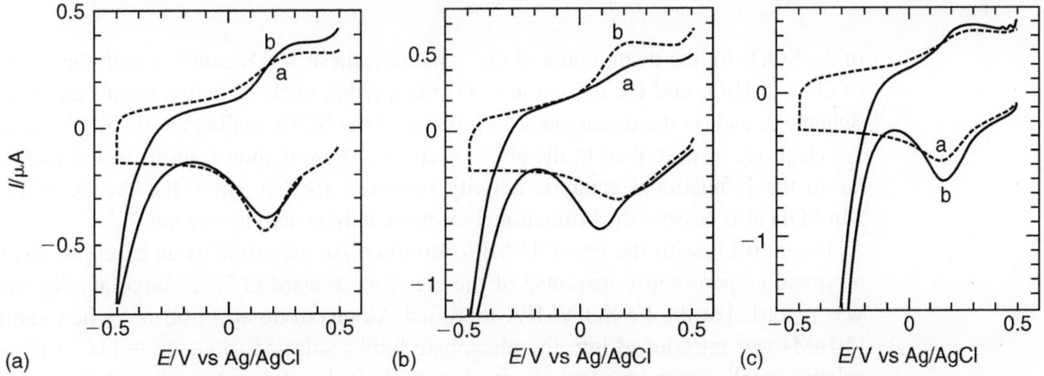

FIGURE 6.10 CVs obtained at (a) Cu, Zn-SOD/MPA-modified, (b) Fe-SOD/MPA-modified, and (c) Mn-SOD/MPA-modified Au electrodes in 25 mM phosphate buffer (pH 7.5) in the absence (dotted lines) and presence (solid lines) of 1.8 μM min^{-1} $O_2^{\bullet -}$. Potential scan rate: 100 mV s^{-1}. (Reprinted from [138], with permission from the American Chemical Society.)

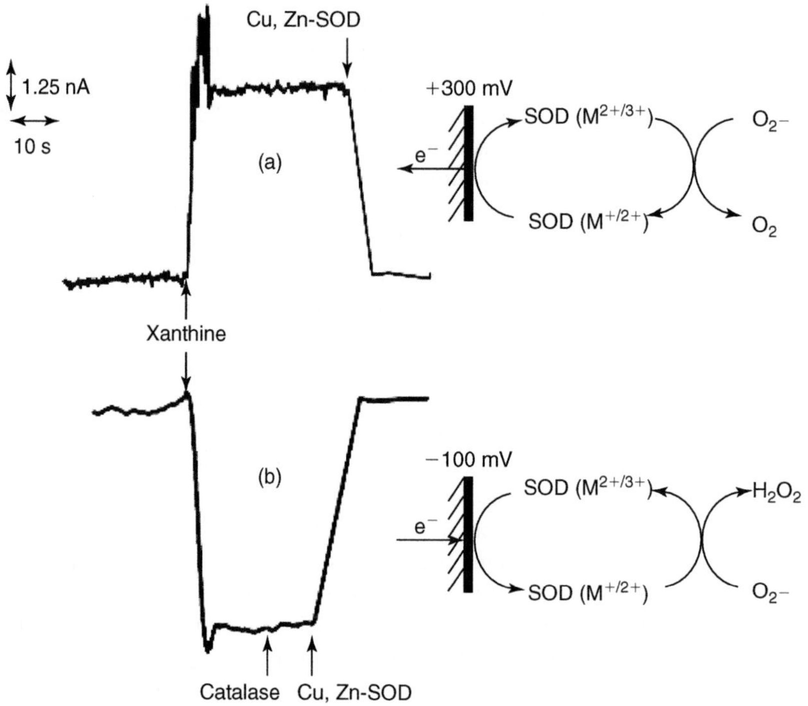

FIGURE 6.11 Typical current–time responses of Fe-SOD/MPA-modified Au electrode toward $O_2^{\bullet-}$ in 25 mM phosphate buffer (O_2-saturated, pH 7.5) containing 0.002 unit of XOD upon the addition of 50 nM xanthine at +300 (a) and −100 mV (b). The arrows represent the addition of 10 μM of Cu, Zn-SOD (a) and 580 units of catalase and 10 μM of Cu, Zn-SOD to the solution (b). The solution was stirred with a magnetic stirrer at 200 rpm. Inset: mechanism for the amperometric response of SODs/MPA-modified Au electrodes to $O_2^{\bullet-}$ based on enzymatic catalytic oxidation (a) and reduction (b) of $O_2^{\bullet-}$ (M: metal ions of SODs). (Reprinted from [138], with permission from the American Chemical Society.)

of the SODs for the dismutation of $O_2^{\bullet-}$; namely, these SODs catalyze both the reduction of $O_2^{\bullet-}$ to H_2O_2 and the oxidation to O_2 via a redox cycle of active metals as shown in Scheme 3, and on the direct electron transfer of the SODs realized at the MPA-modified Au electrode as described in the above sections. These demonstrations reveal that, similar to the bifunctional catalytic activity observed for Cu, Zn-SOD, the Fe-SOD and Mn-SOD also possess the bifunctional electro-catalytic activity toward $O_2^{\bullet-}$.

Figure 6.11, with the Fe-SOD/MPA-modified Au electrode as an example, displays a typical amperometric response of the electrode toward $O_2^{\bullet-}$. A large anodic current was recorded at the Fe-SOD/MPA-modified Au electrode at +300 mV when xanthine (50 nM) was introduced into the phosphate buffer solution to generate $O_2^{\bullet-}$ (a), while relative small responses were obtained at the bare Au electrode and the MPA-modified Au electrode for the same concentration of $O_2^{\bullet-}$ (not shown), indicating the enzymatic amplification nature of $O_2^{\bullet-}$ oxidation at the Fe-SOD/MPA-modified Au electrode. The assignment of the observed large anodic current to the oxidation of $O_2^{\bullet-}$, rather than

TABLE 6.6
Analytical properties of three kinds of $O_2^{\bullet-}$ biosensors

	Cu, Zn-SOD		Fe-SOD		Mn-SOD	
Applied potential (mV)	300	−100	300	−10	300	−100
Surface coverage ($\times 10^{-1}$ mol cm^{-2})	1.1	1.1	1.9	1.9	1.2	1.2
Sensitivity (nA cm^{-2}/nM min^{-1})	19	25	25	31	17	30
Detection limit (nA/nM min^{-1})	0.38	0.49	0.49	0.63	0.34	0.61
Linear range (nM min^{-1})	13–130	13–130	13–130	13–130	13–130	13–130

(Reprinted from [138], with permission from the American Chemical Society.)

those of the species co-produced with $O_2^{\bullet-}$ in the xanthine-XOD generating system, e.g. uric acid and H_2O_2, was evident by adding Cu, Zn-SOD, a selective scavenger of $O_2^{\bullet-}$, into the solution containing $O_2^{\bullet-}$. As expected, the presence of Cu, Zn-SOD in solution greatly decreases the anodic current by >96% (a).

On the other hand, an obvious cathodic current was clearly recorded with the addition of xanthine into PBS containing XOD when the Fe-SOD/MPA/Au electrode was polarized at −100 mV (b). The introduction of catalase, an enzyme specifically catalyzing the dismutation of H_2O_2, resulted in no change in the current response, precluding the originality of the recorded current response from H_2O_2 co-produced in the xanthine-XOD system. In contrast, the addition of Cu, Zn-SOD yielded a large decrease in the cathodic current almost to the background level. These observations may allow us to ascribe cathodic response to the reduction of $O_2^{\bullet-}$ at the Fe-SOD/MPA-modified Au electrode [138].

The sensitivity of the SOD-based biosensors for $O_2^{\bullet-}$ determination was found to be dependent on the operation potential and the surface coverage of each kind of SOD. The analytical properties of three kinds of $O_2^{\bullet-}$ biosensors under optimized conditions are summarized in Table 6.6.

The main purpose of the development of $O_2^{\bullet-}$ biosensors lies in their applications for monitoring $O_2^{\bullet-}$ in biological systems and thus the developed biosensors should be studied with respect to their relevance for *in-vivo* biological measurements. It is known that there are a variety of interferents coexisting in biological samples, suggesting that the biosensors used for the practical measurements should have significant specificity against potential interferents. In biological systems, H_2O_2 is a metabolite in the degradation of $O_2^{\bullet-}$ and a product of enzyme reactions of endogenous oxidases, such as monoamine oxidase and L-amino acid oxidase. In addition, H_2O_2 is one of the main byproducts in the XOD-xanthine system to generate $O_2^{\bullet-}$. Therefore, the specificity of the SOD-based biosensors against H_2O_2 is of great importance in their applications for the determination of $O_2^{\bullet-}$ in biological systems as well as in their calibration with the XOD-xanthine $O_2^{\bullet-}$-generating system. Thus, the interference from H_2O_2 was first examined. In addition, the current responses of the SOD-based biosensors against other potential interferents, such as the principal metabolites (DOPAC, HVA, and 5-HIAA)

TABLE 6.7
Current responses of the biosensor toward $O_2^{\bullet-}$ and potential interferents

	Current/nA	
Interferents (concentration)	300 mV	−200 mV
O_2- (13 nM min^{-1})[a]	6.2	−5.4
H_2O_2 (20 μM)[b]	0	−0.09
5-HIAA (10 μM)[c]	0	−0.004
HVA (10 μM)	0	−0.002
DOPAC (10 μM)	0.03	0
UA (50 μM)	1.2	0
AA (500 μM)	2.3	0

[a] Anodic (positive values) and cathodic (negative values) currents observed on the addition of 50 nM xanthine into 25 mM phosphate buffer (pH 7.4, O_2-saturated) containing 0.002 unit of XOD. The rate of $O_2^{\bullet-}$ generation is 13 nM min^{-1}. The solution was stirred with a magnetic stirrer at 200 rpm.
[b] Anodic (positive values) and cathodic (negative values) currents observed on the addition of 20 μM H_2O_2, 10 μ M 5-HIAA, 10 μ M HVA, 10 μ M DOPAC, and 50 μM UA or 500 μM AA into 25 mM phosphate buffer (pH 7.4). The solution was stirred with a magnetic stirrer at 200 rpm.
[c] 5-HIAA, 5-hydroxyindole-3-acetic acid; DOPAC, 3,4-dihydroxyphenylacetic acid; HVA, homovanillic acid; AA, ascorbic acid; and UA, uric acid.
(Reprinted from [151], with permission from the American Chemical Society.)

of some neurotransmitters, AA and UA, were also investigated. Table 6.7, with the Cu, Zn-SOD/cystine-modified Au electrode as an example, summarizes the steady-state current responses of the biosensor toward these interferents in comparison with those measured toward $O_2^{\bullet-}$ at +300 and −200 mV, in which the concentrations of the interferents approximate their extracellular fluid levels [152, 153]. As seen from Table 6.7, no current response for H_2O_2 was observed at the Cu, Zn-SOD/cysteine-modified Au electrode at +300 mV and the response was also negligible at −200 mV [151]. This point is remarkable in comparison with the Cyt. *c*-based $O_2^{\bullet-}$ biosensor, that is, it shows a good response for H_2O_2 because of the peroxidase activity of Cyt. *c*. In other words, interference from H_2O_2 cannot be avoided at the Cyt. *c*-based $O_2^{\bullet-}$ biosensor. AA is another major interferent perplexing most electrochemical techniques for biological measurements because of its high concentration and low oxidation potential. At +300 mV, the interference level of 37% was obtained for AA, indicating that further improvement will be needed to enhance the specificity of the biosensor for $O_2^{\bullet-}$ determination in the presence of AA of a high concentration. On the contrary, no response of AA was obtained at −200 mV. In addition, UA is found to be the interferent for the measurement of the oxidation current of $O_2^{\bullet-}$ at +300 mV (the degree of its interference is 19%), but the interference of 5-HIAA, HVA, and DOPAC is negligible at both +300 and −200 mV. The other two kinds of SOD-based biosensors, i.e. Fe-SOD/MPA/Au and Mn-SOD/MPA/Au electrodes, also have a good selectivity against the interference mentioned above. For example, the interferences from H_2O_2,

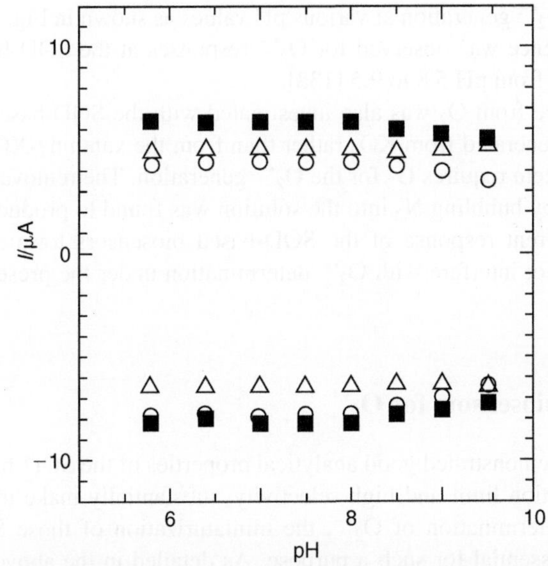

FIGURE 6.12 Plots of amperometric responses of Cu, Zn-SOD/MPA-modified (△), Mn-SOD/MPA-modified (○), and Fe-SOD/MPA-modified (■) Au electrodes toward 13 nM min^{-1} $O_2^{\bullet-}$ in 25 mM phosphate buffer at various pH values from pH 5.8 to 9.5. The solution was stirred with a magnetic stirrer at 200 rpm. (Reprinted from [138], with permission from the American Chemical Society.)

UA, AA, and DOPAC, with the concentrations approximating their extracellular fluid levels were investigated at +300 and −100 mV at the SOD-based biosensors; at +300 mV, the interferences from AA and UA were considerable, for instance 15% and 23% current responses were obtained for 500 μM AA relative to 13 nM $O_2^{\bullet-}$ with Fe-SOD/MPA-modified and Mn-SOD/MPA-modified Au electrodes, respectively. In addition, 10% current response was obtained for 50 μM UA relative to 13 nM $O_2^{\bullet-}$ at both electrodes. Fortunately, such interferences were well suppressed when the electrodes were polarized at −100 mV. Besides, the interferences of H_2O_2, 5-HIAA, HVA, and DOPAC were negligible at both +300 and −100 mV at both electrodes [138].

On the other hand, *in-vivo* formation of physiologically inappropriate levels of free radicals occurs in response to low blood flow, low oxygen levels, and low pH [154, 155]. The probable interference from pH and O_2 was consequently investigated over the biologically relevant range. Figure 6.12 shows the steady-state amperometric responses for $O_2^{\bullet-}$ at the SOD-based biosensors at various pH values. It should be noted here that the rate of $O_2^{\bullet-}$ generation in the xanthine-XOD system depends on solution pH because of the pH dependence of the enzymatic activity of xanthine oxidase. Therefore, the rate of $O_2^{\bullet-}$ generation under various pH values was determined by recording the reduction of ferricytochrome *c* spectrophotometrically and using the extinction coefficient (21.1 mM^{-1} cm^{-1}) of ferrocytochrome *c* at 550 nm to guarantee the same rate of $O_2^{\bullet-}$ generation at various pH values as shown in Fig. 6.12. As shown, slight pH dependence was observed for $O_2^{\bullet-}$ responses at the SOD-based biosensors within a pH range from pH 5.8 to 9.5 [138].

The interference from O_2 was also investigated with the SOD-based biosensors, in which $O_2^{\bullet-}$ was generated from KO_2 rather than from the xanthine-XOD system since the enzymatic system requires O_2 for the $O_2^{\bullet-}$ generation. The removal of O_2 from the phosphate buffer by bubbling N_2 into the solution was found to produce no observable change in the current response of the SOD-based biosensors toward KO_2, suggesting that O_2 does not interfere with $O_2^{\bullet-}$ determination under the present experimental conditions.

6.4.5 SOD-based micro-sized biosensors for $O_2^{\bullet-}$

Even though the demonstrated good analytical properties of the SOD-based biosensors, such as low detection limit and high selectivity, substantially make them very potential for *in-vivo* determination of $O_2^{\bullet-}$, the miniaturization of those SOD-based biosensors remains essential for such a purpose. As detailed in the above sections, other groups have successfully fabricated the third-generation biosensors for $O_2^{\bullet-}$ based on the direct electron transfer of SODs on SAM-modified Au electrodes. However, this concept can not be readily realized on carbon fiber microelectrodes because the promoters could hardly be stably anchored at the carbon-based electrode directly. From a practical point of view, the utilization of carbon fiber microelectrodes (CFMEs) is one of the most powerful analytical protocols, because CFMEs possess a number of unique features. Their small dimensions result in an increased mass transport of an electroactive species to the electrode surface, a low double-layer capacitance and ohmic loss. Moreover, carbon fibers have better mechanical characteristics than noble-metal electrodes of a micrometer size. In addition to the capability for insertion in a single cell and for implantation into biological tissues with a relatively good biocompatibility of the material itself [156, 157], they are rather easily prepared and handled.

To fulfill both the requirement of CFME for the practical applications and the necessity of Au substrate to assemble so-called promoters to construct the third-generation biosensor, Tian *et al.* have combined the electrochemical deposition of Au nanoparticles (Au-NPs) onto carbon fiber microelectrodes with the self-assembly of a monolayer on these Au-NPs to facilitate the direct electron transfer of SOD at the carbon fiber microelectrode. The strategy enabled a third-generation amperometric $O_2^{\bullet-}$ biosensor to be readily fabricated on the carbon fiber microelectrode. This CFME-based biosensor is envisaged to have great potential for the detection of $O_2^{\bullet-}$ in biological systems [158].

As shown in Scheme 4, all electrodes were fabricated from a single carbon fiber (10 μm in diameter) connected to the copper wire using a silver paste. After cutting the tip of the prepulled glass to obtain a smooth blunt opening of 20 $\sim$ 30 μm diameter, each carbon fiber was inserted into a pulled glass capillary. Diluted epoxy solution was used to make the seal between the carbon fiber and glass by capillary action. After the epoxy had cured, the extruded tip portion of carbon fiber was cut to the desired length

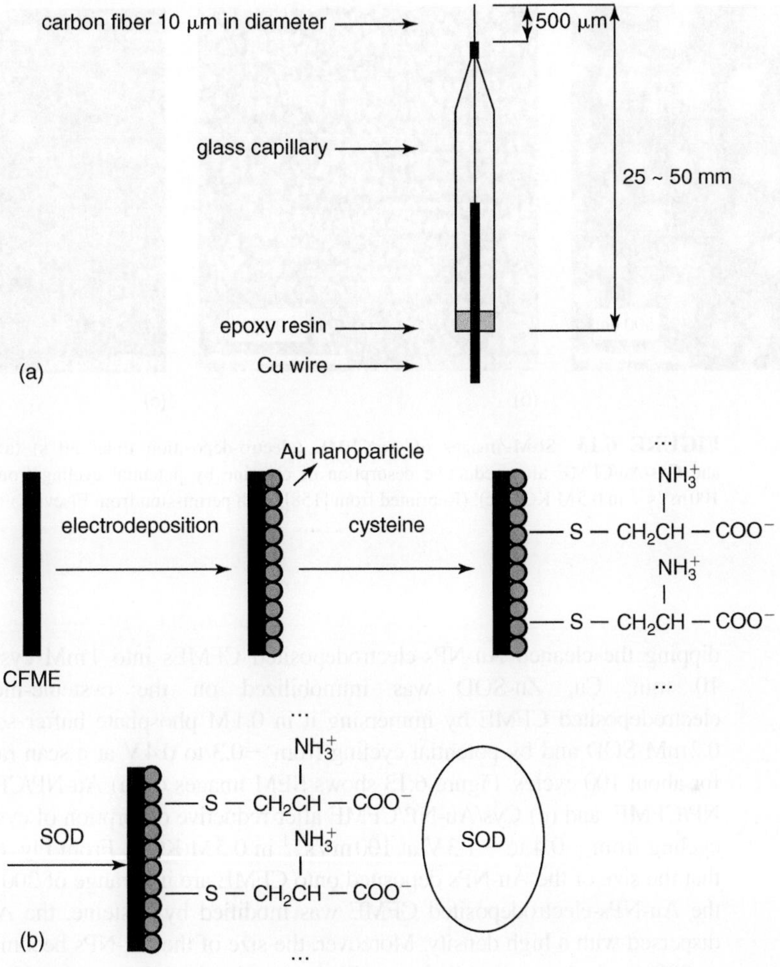

SCHEME 4 (a) Carbon fiber microelectrode, and (b) the process of electrode modification. (Reprinted from [158], with permission from Elsevier.)

(about 500 μm) with a scalpel under a microscope. Prior to electrochemical activation, the fabricated carbon fiber microelectrodes were ultrasonically washed with acetone, 3.0 M HNO_3, 1.0 M NaOH, and distilled water sequentially. The electrode pretreatment was performed in a 0.50 M sulfuric acid solution by applying a constant potential at the electrodes, first, +2.0 V for 30 s, then −1.0 V for 10 s, and then the electrodes were treated by using cyclic voltammetry in the potential range from 0.0 to +1.0 V at a scan rate of 100 mV/s until a stable cyclic voltammogram was obtained [159, 160].

The Au-NPs were electrodeposited on the carbon fiber microelectrodes from 0.5 M H_2SO_4 solution containing 1.0 mM Na[$AuCl_4$] by applying a potential step from 1.1 V to 0 V for 30 s. Cysteine-modified Au-NPs-electrodeposited CFMEs were prepared by

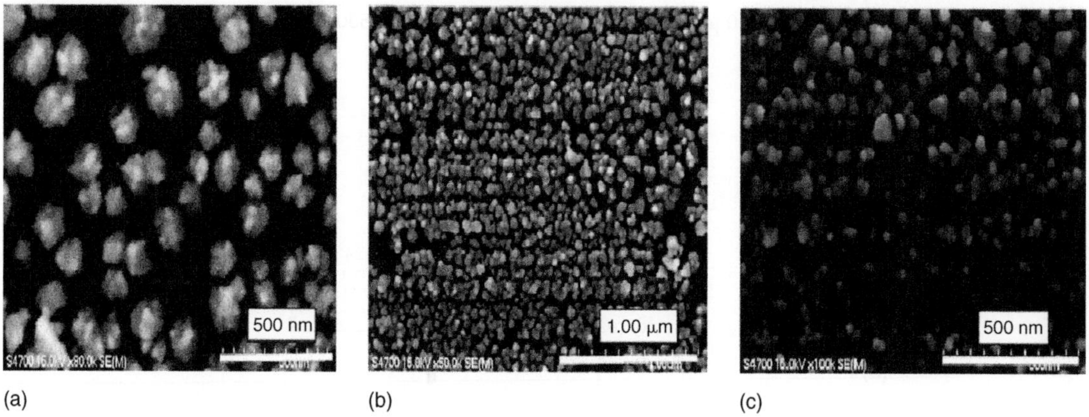

FIGURE 6.13 SEM images of Au/CFME (electro-deposition time: 30 s) (a), Cys/Au/CFME (b) and Cys/Au/CFME after reductive desorption of cysteine by potential cycling from −0.4 to −1.3 V at 100 mVs^{-1} in 0.5 M KOH (c). (Reprinted from [158], with permission from Elsevier.)

dipping the cleaned Au-NPs-electrodeposited CFMEs into 1 mM cysteine solution for 10 min. Cu, Zn-SOD was immobilized on the cysteine-modified Au-NPs-electrodeposited CFME by immersing it in 0.1 M phosphate buffer solution containing 0.2 mM SOD and by potential cycling from −0.3 to 0.4 V at a scan rate of 100 mV s^{-1} for about 100 cycles. Figure 6.13 shows SEM images of (a) Au-NP/CFME, (b) Cys/Au-NP/CFME, and (c) Cys/Au-NP/CFME after reductive desorption of cysteine by potential cycling from −0.4 to −1.3 V at 100 mVs^{-1} in 0.5 M KOH. From Fig. 6.13a, one can see that the size of the Au-NPs deposited onto CFME are in a range of 200 ~ 500 nm. When the Au-NPs-electrodeposited CFME was modified by cysteine, the Au-NPs were well dispersed with a high density. Moreover, the size of the Au-NPs becomes smaller (i.e. 20 ~ 200 nm as shown in (b)). It was likely that the Au-NPs were "self-assembled" by the formation of the SAM of cysteine. This dispersion is very stable and there have been no obvious changes even after the cysteine was desorped in 0.5 M KOH, as shown in (c).

Figure 6.14 (left panel) displays amperometric responses of the Cu, Zn-SOD/Cys/Au-NP/CFME to successive addition of $O_2^{\bullet -}$ at (a) +250 mV and (b) −150 mV. The right panel in Fig. 6.14 shows the calibration curve of the micro-sized $O_2^{\bullet -}$ biosensor. Well-defined steady-state current responses were obtained at both +250 mV and −150 mV, and the currents increased stepwise with successive additions of xanthine (i.e. $O_2^{\bullet -}$). The steady-state currents at +250 and −150 mV were proportional to $O_2^{\bullet -}$ concentration in the examined range of ca. 13 to 104 nM min^{-1}. The sensitivity of the Cu, Zn-SOD/Cys/Au-NP/CFME was found to be 30 and 25 nA cm^{-2} nM^{-1} min^{-1} at +250 and −150 mV, respectively. The response time of the micro-sized biosensor was measured as the time to reach 95% of the maximum change in response to a step injection of xanthine and found to be less than 5 s.

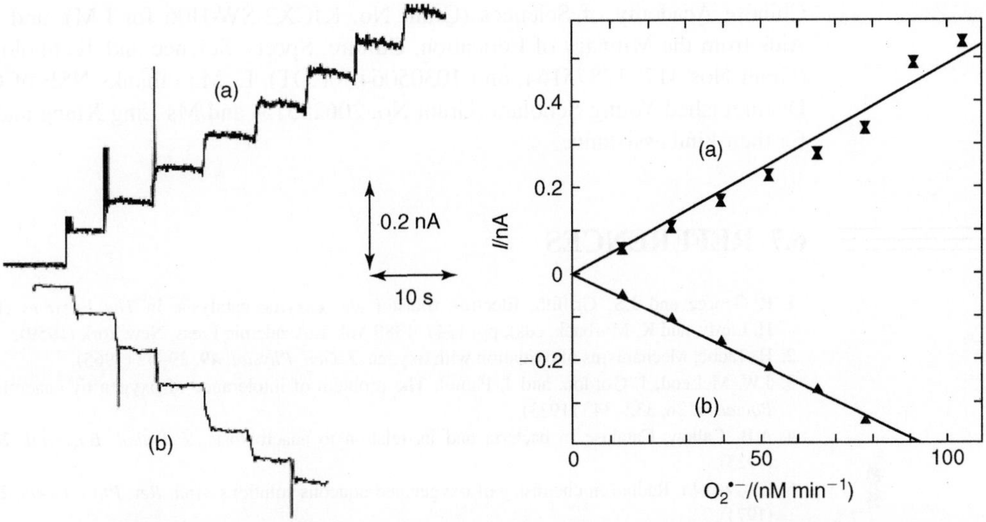

FIGURE 6.14 (Left panel) Amperometric responses of the Cu, Zn-SOD/Cys/Au/CFME to successive xanthine injection at applied potentials of (a) +250 mV and (b) −150 mV in O_2-saturated 0.10 M phosphate buffer (pH 7.4) containing 0.002 U XOD. (Right panel) Calibration curves for $O_2^{\bullet -}$ at (a) +250 mV, and (b) −150 mV. (Reprinted from [158], with permission from Elsevier.)

6.5 CONCLUDING REMARKS AND OTHER DIRECTIONS

Electrochemical biosensors, especially those based on superoxide dismutases, are very competent for real-time monitoring of $O_2^{\bullet -}$ production and consumption in biological systems and would pave a facile, but direct, approach to physiological and pathological processes related with ROS and to free radical chemistry. SODs, enzymes specific for $O_2^{\bullet -}$ dismutation, are considered to be the best choice among all the enzymes employed for construction of $O_2^{\bullet -}$ biosensors because of their great catalytic activity and high specificity and realizable direct electron transfer properties. These properties substantially endow the as-prepared SOD-based third-generation $O_2^{\bullet -}$ biosensors with excellent analytical properties, such as high sensitivity and selectivity, rapid response time, and good linearity with nanomolar detection limit. Moreover, the SOD-based third-generation $O_2^{\bullet -}$ biosensors could be miniaturized to meet the requirements of *in-vivo* measurements. The excellent properties of the electrochemical $O_2^{\bullet -}$ biosensors substantially offer them great potential for real-time monitoring of $O_2^{\bullet -}$ in biological tissues, such as brain tissues, which will be the challenge for the future.

6.6 ACKNOWLEDGMENTS

We gratefully acknowledge the financial support from National Natural Science Foundation of China (Grant Nos 20375043, 20435030, and 20575071 for LM),

Chinese Academy of Sciences (Grant No. KJCX2-SW-H06 for LM), and Grant-in-Aids from the Ministry of Education, Culture, Sports, Science and Technology, Japan (Grant Nos 417, 12875164, and 10305064 for OT). L. Mao thanks NSF of China for Distinguished Young Scholars (Grant No. 20625515) and Ms Ling Xiang and Ping Yu for their kind assistance.

6.7 REFERENCES

1. P. George and J.S. Griffith, Electron transfer and enzyme catalysis, in *The Enzymes* (P.D. Boyer, H. Lardy, and K. Myrback, eds), pp. 1347–1389 Vol. 1, Academic Press, New York (1959).
2. H. Taube, Mechanisms of oxidation with oxygen. *J. Gen. Physiol.* **49**, 29–52 (1965).
3. J.W. McLeod, J. Gordon, and J. Pathol, The problem of intolerance of oxygen by anaerobic bacteria. *Bacteriol.* **26**, 332–343 (1923).
4. A.B. Callow, Catalase in bacteria and its relation to anaerobiosis. *J. Pathol. Bacteriol.* **26**, 320–325 (1923).
5. G. Czapski, Radiation chemistry of oxygenated aqueous solutions. *Ann. Rev. Phys. Chem.* **22**, 171–208 (1971).
6. J.M. McCord, B.B. Keele Jr, and I. Fridovich, Enzyme-based theory of obligate anaerobiosis: physiological function of superoxide dismutase. *Proc. Natl. Acad. Sci. U.S.A.* **68**, 1024–1027 (1971).
7. I. Fridovich, Superoxide radical and superoxide dismutase. *Acc. Chem. Res.* 5, 321–326 (1972).
8. E.M. Gregory and I. Fridovich, Induction of superoxide dismutase by molecular oxygen. *J. Bacteriol.* **114**, 543–548 (1973).
9. E.M. Gregory and I. Fridovich, Oxygen toxicity and the superoxide dismutase. *J. Bacteriol.* **114**, 1193–1197 (1973).
10. D. Behar, G. Czapski, J. Rabani, L.M. Dorfman, and H.A. Schwarz, Acid dissociation constant and decay kinetics of the perhydroxyl radical. *J. Phys. Chem.* **74**, 3209–3213 (1970).
11. M.E. Poever and B.S. White, Electrolytic reduction of oxygen in aprotic solvents: the superoxide ion. *Electrochim. Acta.* **11**, 1061–1067 (1966).
12. A.D. McElroy and J.S. Hashman, Synthesis of tetramethylammonium superoxide. *Inorg. Chem.* **3**, 1798–1799 (1964).
13. D.L. Maricle and W.G. Hodgson, Reduction of oxygen to superoxide anion in aprotic solvents. *Anal. Chem.* **37**, 1562–1565 (1965).
14. D.T. Sawyer and J.L. Roberts, Electrochemistry of oxygen and superoxide ion in dimethyl sulfoxide at platinum, gold, and mercury electrodes. *J. Electroanal. Chem.* **12**, 90–101 (1966).
15. T. Odajima and I. Yamazaki, Myeloneperoxidase of the leukocyte of normal blood. 3. The reaction of ferric myeloperoxidase with superoxide anion. *Biochim. Biophys. Acta.* **284**, 355–359 (1972).
16. J. Chevalet, F. Roulle, L. Gierst, and J.P. Lambert, Electrogeneration and some properties of the superoxide ion in aqueous solutions. *J. Electroanal. Chem. Interfacial Electrochem.* **390**, 201–216 (1972).
17. H.J. Forman and I. Fridovich, Electrolytic univalent reduction of oxygen in aqueous solution demonstrated with superoxide dismutase. *Science.* **175**, 339 (1972).
18. J.H. Baxendale, The flash photolysis of water and aqueous solutions. *Radiat. Res.* **17**, 312–326 (1962).
19. G.E. Adams, J.W. Boag, and B.D. Michael, The flash photolysis of water and aqueous solutions. *Proc. Roy. Soc. (London), A289*, 321–326 (1965).
20. E. Hayon and J. McGarvey, Flash photolysis in the vacuum ultraviolet region of sulfate, carbonate, and hydroxyl ions in aqueous solutions. *J. Phys. Chem.* **71**, 1472–1477 (1967).
21. J.M. McCord and I. Fridovich, Production of O_2^- in photolyzed water demonstrated through the use of superoxide dismutase. *Photochem. Photobiol.* **17**, 115–121 (1973).
22. G.H. Czapski and B.H.J. Bielski, The formation and decay of H_2O_3 and HO_2 in electronirradiated aqueous solutions. *J. Phys. Chem.* **67**, 2180–2184 (1963).

23. J. Rabani, W.A. Mulac, and M.S. Matheson, The pulse radiolysis of aqueous tetranitromethane. I. Rate constants and the extinction coefficient [absorptivity] of aq-. II. Oxygenated solutions. *J. Phys. Chem.* **69**, 53–70 (1965).
24. B.H.J. Bielski and A.O. Allen, Radiation chemistry of aqueous tetranitromethane solutions in the presence of air. *J. Phys. Chem.* **71**, 4544–4549 (1967).
25. M. Anbar and I. Pecht, On the sonochemical formation of hydrogen peroxide in water. *J. Phys. Chem.* **68**, 352–355 (1964).
26. B. Lippitt, J.M. McCord, and I. Fridovich, The sonochemical reduction of cytochrome c and its inhibition by superoxide dismutase. *J. Biol. Chem.* **247**, 4688–4690 (1972).
27. G.A. Russell, Fundamental processes of autoxidation. *J. Chem. Ed.* **36**, 111–118 (1959).
28. A. le Berre and P. Goasguen, Autoxidation of metal ketyls and related compounds. *Compt. Rend.* **254**, 1306–1308 (1962).
29. V. Massey, S. Strickland, S.G. Mayhew, L.G. Howell, P.C. Engel, R.G. Matthews, M. Schuman, and P.A. Sullivan, Production of superoxide anion radicals in the reaction of reduced flavins and flavoproteins with molecular oxygen. *Biochem. Biophys. Res. Commun.* **36**, 891–897 (1969).
30. D. Ballou, G. Palmer, and V. Massey, Direct demonstration of superoxide anion production during the oxidation of reduced flavin and of its catalytic decomposition by erythrocuprein. *Biochem. Biophys. Res. Commun.* **36**, 898–904 (1969).
31. J.M. McCord and I. Fridovich, Utility of superoxide dismutase in studying free radical reactions. II. Mechanism of the mediation of cytochrome c reduction by a variety of electron carriers. *J. Biol. Chem.* **245**, 1374–1377 (1970).
32. H.P. Misra and I. Fridovich, Role of superoxide anion in the autoxidation of epinephrine and a simple assay for superoxide dismutase. *J. Biol. Chem.* **247**, 3170–3175 (1972).
33. M. Nishikimi, N.A. Rao, and K. Yagi, Occurrence of superoxide anion in the reaction of reduced phenazine methosulfate and molecular oxygen. *Biochem. Biophys. Res. Commun.* **46**, 849–854 (1972).
34. H.P. Misra and I. Fridovich, Role of superoxide anion in the autoxidation of epinephrine and a simple assay for superoxide dismutase. *J. Biol. Chem.* **247**, 3170–3175 (1972).
35. W.H. Orme-Johnson and H. Beinert, Formation of the superoxide anion radical during the reaction of reduced iron–sulfur proteins with oxygen. *Biochem. Biophys. Res. Commun.* **36**, 905–911 (1969).
36. R. Nilsson, F.M. Pick, and R.C. Bray, Electron paramagnetic resonance studies on reduction of oxygen to superoxide by some biochemical systems. *Biochim. Biophys. Acta.* **192**, 145–148 (1969).
37. H.P. Misra and I. Fridovich, Generation of superoxide radical during the autoxidation of ferredoxins. *J. Biol. Chem.* **246**, 6886–6890 (1971).
38. H.P. Misra and I. Fridovich, Generation of superoxide radical during the autoxidation of hemoglobin. *J. Biol. Chem.* **247**, 6960–6962 (1972).
39. R. Wever, B. Oudega, and B.F. van Gelder, Generation of superoxide radicals during the autoxidation of mammalian oxyhemoglobin. *Biochim. Biophys. Acta.* **302**, 475–478 (1973).
40. E. Satio and B.H.J. Bielski, Electron paramagnetic resonance spectrum of the HO_2 radical in aqueous solution. *J. Am. Chem. Soc.* **83**, 4467–4468 (1961).
41. S. Ander, Studie on UV and x-irradiated water. *Strahlentherapie* **132**, 135–142 (1967).
42. G. Gzapski and L.M. Dorfman, Pulse radiolysis studies. V. Transient spectra and rate constants in oxygenated aqueous solutions. *J. Phys. Chem.* **68**, 1169–1177 (1964).
43. L. Andrews, Infrared spectra and bonding in the sodium superoxide and sodium peroxide molecules. *J. Phys. Chem.* **73**, 3922–3928 (1969).
44. J. Kroh, B.C. Green, and J.W.J. Spinks, Electron paramagnetic resonance (EPR) studies on the production of free radicals in hydrogen peroxide at liquid nitrogen temperature. *J. Am. Chem. Soc.* **83**, 2201–2202 (1961).
45. J. Kroh, B.C. Green, and J.W.J. Spinks, Electron paramagnetic resonance studies on free radicals produced by Tβ-particles in frozen H_2O and D_2O media at liquid nitrogen temperature. *Can. J. Chem.* **40**, 413–425 (1962).
46. B.H.J. Bielski and E. Satio, Activation energy for the disproportionation of the HO_2 radical in acid solutions. *J. Phys. Chem.* **66**, 2266–2268 (1962).

47. P.F. Knowles, J.F. Gibson, F.M. Pick, and R.C. Bray, Electron-spin-resonance evidence for enzymic reduction of oxygen to a free radical, *the superoxide ion. Biochem. J.* **111**, 53–58 (1969).
48. R. Nilsson, F.M. Pick, R.C. Bray, and M. Fielden, ESR evidence for O_2^- as a longlived transient in irradiated oxygenated alkaline aqueous solutions. *Acta Chem. Scand.* **23**, 2554–2556 (1969).
49. R.C. Bray, F.M. Pick, and D. Samuel, Oxygen-17 hyperfine splitting in the electron paramagnetic resonance spectrum of enzymically generated superoxide. *Eur. J. Biochem.* **15**, 352–355 (1970).
50. S.N. Foner and R.L. Hudson, Mass spectrometry of inorganic free radicals. *Adv. Chem. Series.* **36**, 34–49 (1962).
51. H.A. Kontos and E.P. Wei, Superoxide production in experimental brain injury. *J. Neurosurgery.* **64**, 803–807 (1986).
52. M.Y. Globus, O. Alonso, W.A. Dietrich, R. Busto, and M.D. Ginsberg, Glutamate release and free radical production following brain injury: effects of posttraumatic hypothermia. *J. Neurochem.* **65**, 1704–1711 (1995).
53. E.D. Hall and J.M. Braughler, Central nervous system trauma and stroke. II. Physiological and pharmacological evidence for involvement of oxygen radicals and lipid peroxidation. *Free Radical Biol. Med.* **6**, 303–313 (1989).
54. A.Vanella, C. di Giacomo, V. Sorrenti, A. Russo, C. Castorina, A. Campisi, M. Renis, and J.R. Perez-Polo, Free radical scavenger depletion in post-ischemic reperfusion brain damage. *Neurochem. Res.* **18**, 1337–1340 (1993).
55. R.A. Floyd, Role of oxygen free radicals in carcinogenesis and brain ischemia. *FASEB J.* **4**, 2587–2597 (1990).
56. B.N. Ames, M.K. Shigenaga Hagen, and M. Tory, Oxidants, antioxidants, and the degenerative diseases of aging. *Proc. Natl. Acad. Sci. U.S.A.* **90**, 7915–7922 (1993).
57. B. Halliwell, Reactive oxygen species and the central nervous system. *J. Neurochem.* **59**, 1609–1623 (1992).
58. G. Benzi and A. Moretti, Age- and peroxidative stress-related modifications of the cerebral enzymatic activities linked to mitochondria and the glutathione system. *Free Radical Biol. Med.* **19**, 77–101 (1995).
59. S.J. Stohs and D. Bagchi, Oxidative mechanisms in the toxicity of metal ions. *Free Radical Biol. Med.* **18**, 321–336 (1995).
60. E.J. Land and A.J. Swallow, One-electron reactions in biochemical systems as studied by pulse radiolysis. V. Cytochrome c. *Arch. Biochem. Biophys.* **145**, 365–372 (1971).
61. C. Beauchamp and I. Fridovich, Superoxide dismutase. Improved assays and an assay applicable to acrylamide gels. *Anal. Biochem.* **44**, 276–287 (1971).
62. J.M. McCord and I. Fridovich, Superoxide dismutase. Enzymic function for erythrocuprein (hemocuprein). *J. Biol. Chem.* **244**, 6049–6055 (1969).
63. J.L. Zweier, J.H. Flaherty, and M.L.Weisfeldt, Direct measurement of free radical generation following reperfusion of ischemic myocardium. *Proc. Natl. Acad. Sci. U.S.A.* **84**, 1404–1407 (1987).
64. Y. Ohara, T.E. Peterson, and D.G. Harrison, Hypercholesterolemia increases endothelial superoxide anion production. *J. Clin. Invest.* **91**, 2546–2551 (1993).
65. J. Vasquez-Vivar, N. Hogg, K.A. Pritchard, and B. Kalyanaraman, Superoxide anion formation from lucigenin: an electron spin resonance spin-trapping study. *FEBS Lett.* **403**, 127–130 (1997).
66. H. Zhang, J. Joseph, J. Vasquez-Vivar, H. Karoui, C. Nsanzumuhire, P. Martasek, P. Tordo, and B. Kalyanaraman, Detection of superoxide anion using an isotopically labeled nitrone spin trap: potential biological applications. *FEBS Lett.* **473**, 58–62 (2000).
67. T. Ohyashiki, M. Nunomura, and T. Katoh, Detection of superoxide anion radical in phospholipid liposomal membrane by fluorescence quenching method using 1,3-diphenylisobenzofuran. *Biochim. Biophys. Acta.* **1421**, 131–139 (1999).
68. K. Tanaka, F. Kobayashi, Y. Isogai, and T. Iizuka, Electrochemical determination of superoxide anions generated from a single neutrophil. *Bioelectrochem. Bioenerg.* **26**, 413–421 (1991).
69. C. Privat, S. Trevin, F. Bedioui, and J. Devynck, Direct electrochemical characterization of superoxide anion production and its reactivity toward nitric oxide in solution. *J. Electroanal. Chem.* **436**, 261–265 (1997).
70. F. Lisdat, B. Ge, E. Ehrentreich-Forster, R. Reszka, and F.W. Scheller, Superoxide dismutase activity measurement using cytochrome c-modified electrode. *Anal. Chem.* **71**, 1359–1365 (1999).

71. K. Tammeveski, T. Tenno, A.A. Mashirin, E.W. Hillhouse, P. Manning, and C.J. McNeil, Superoxide electrode based on covalently immobilized cytochrome c: modeling studies. *Free Radical Biol. Med.* **25**, 973–978 (1998).
72. J. Chen, U. Wollenberger, F. Lisdat, B. Ge, and F.W. Scheller, Superoxide sensor based on hemin modified electrode. *Sens. Actuators B.* **70**, 115–120 (2000).
73. J. Xue, X. Xian, X. Ying, J. Chen, L. Wang, and L. Jin, Fabrication of an ultramicrosensor for measurement of extracellular myocardial superoxide. *Anal. Chim. Acta.* **405**, 77–85 (2000).
74. W. Scheller, W. Jin, E. Ehrentreich-Forster, B. Ge, F. Lisdat, R. Buttemeier, U. Wollenberger, and F.W. Scheller, Cytochrome c-based superoxide sensor for in vivo application. *Electroanalysis.* **11**, 703–706 (1999).
75. K.V. Gobi and F. Mizutani, Efficient mediatorless superoxide sensors using cytochrome c-modified electrodes. Surface nano-organization for selectivity and controlled peroxidase activity. *J. Electroanal. Chem.* **484**, 172–181 (2000).
76. C.J. McNeil, D. Athey, and W.O. Ho, Direct electron transfer bioelectronic interfaces: application to clinical analysis. *Biosens. Bioelectron.* **10**, 75–83 (1995).
77. V. Lvovich and A. Scheeline, Amperometric sensors for simultaneous superoxide and hydrogen peroxide detection. *Anal. Chem.* **69**, 454–462 (1997).
78. L. Campanella, L. Persi, and M. Tomassetti, A new tool for superoxide and nitric oxide radicals determination using suitable enzymatic sensors. *Sens. Actuators B.* **68**, 351–359 (2000).
79. S. Mesaros, Z. Vankova, A. Mesarosova, P. Tomcik, and S. Grunfeld, Electrochemical determination of superoxide and nitric oxide generated from biological samples. *Bioelectrochem. Bioenerg.* **46**, 33–37 (1998).
80. S. Mesaros, Z. Vankova, S. Grunfeld, A. Mesarosova, and T. Malinski, Preparation and optimization of superoxide microbiosensor. *Anal. Chim. Acta.* **358**, 27–33 (1998).
81. L. Campanella, G. Favero, L. Persi, and M. Tomassetti, New biosensor for superoxide radical used to evidence molecules of biomedical and pharmaceutical interest having radical scavenging properties. *J. Pharm. Biomed. Anal.* **23**, 69–76 (2000).
82. M.I. Song, F.F. Bier, and F.W. Scheller, A method to detect superoxide radicals using Teflon membrane and superoxide dismutase. *Bioelectrochem. Bioenerg.* **38**, 419–422 (1995).
83. S. Descroix and F. Bedioui, Evaluation of the selectivity of overoxidized polypyrrole/superoxide dismutase based microsensor for the electrochemical measurement of superoxide anion in solution. *Electroanalysis.* **13**, 524–528 (2001).
84. F. Matsumoto, K. Tokuda, and T. Ohsaka, Electrogeneration of superoxide ion at mercury electrodes with a hydrophobic adsorption film in aqueous media. *Electroanalysis.* **8**, 648–653 (1996).
85. T. Ohsaka, F. Matsumoto, and K. Tokuda, An electrochemical approach to dismutation of superoxide ion using a biological model system with a hydrophobic/hydrophilic interface, in *Frontiers of Reactive Oxygen Species in Biological and Medicine* (K. Asaka and T. Yoshikawa, eds), pp. 91–93. Elsevier Science B.V.: Oxford (1994).
86. C. Privat, O. Stepien, M. David-Dufilho, A. Brunet, F. Bedioui, P. Marche, J. Devynck, and M.-A. Devynck, Superoxide release from interleukin-1 β-stimulated human vascular cells: in situ electrochemical measurement. *Free Radicals Bio. Med.* **27**, 554–559 (1999).
87. L. Campanella, G. Favero, and M. Tomassetti, A modified amperometric electrode for the determination of free radicals. *Sens. Actuators B.* **44**, 559–565 (1997).
88. C.J. McNeil, K.R. Greenough, P.A. Weeks, and C.H. Self, Electrochemical sensors for direct reagentless measurement of superoxide production by human neutrophils. *Free Rad. Res. Comm.* **17**, 399–406 (1992).
89. C.M. Tolias, J.C. McNeil, J. Kazlauskate, and E.W. Hillhouse, Superoxide generation from constitutive nitric oxide synthase in astrocytes in vitro regulates extracellular nitric oxide availability. *Free Radical Biol. Med.* **26**, 99–106 (1999).
90. R.H. Fabian, D.S. deWitt, and T.A. Kent, The 21-aminosteroid U-74389G reduces cerebral superoxide anion concentration following fluid percussion injury of the brain. *J. Neurotroma.* **15**, 433–440 (1998).
91. J.C. Cooper, G. Thompson, and C.J. McNeil, Direct electron transfer between immobilized cytochrome c and gold electrodes. *Mol. Cryst. Liq. Cryst.* **235**, 127–132(1993).
92. C.J. McNeil, K.A. Smith, P. Bellavite, and J. Bannister, Application of the electrochemistry of cytochrome c to the measurement of superoxide radical production. *Free Rad. Res. Comm.* **7**, 89–96 (1989).

93. H.K. Datte, H. Rathod, P. Manning, Y. Turnbull, and C.J. McNeil, Parathyroid hormone induces superoxide anion burst in the osteoclast: evidence for the direct instantaneous activation of the osteoclast by the hormone. *J. Endocrinology.* **149**, 269–275 (1996).
94. P. Manning, C.J. McNeil, J.M. Cooper, and E.W. Hillhouse, Direct, real-time sensing of free radical production by activated human glioblastoma cells. *Free Radical Bio. Med.* **24**, 1304–1309 (1998).
95. K. Tammeveski, T. Tenno, A.A. Mashirin, E.W. Hillhouse, P. Manning, and C.J. McNeil, Superoxide electrode based on covalently immobilized cytochrome c: modeling studies. *Free Radical Bio. Med.* **25**, 973–978 (1998).
96. B. Ge and F. Lisdat, Superoxide sensor based on cytochrome c immobilized on mixed-thiol SAM with a new calibration method. *Anal. Chim. Acta.* **454**, 53–64 (2002).
97. M.K. Beissenhirtz, F.W. Scheller, and F. Lisdat, A superoxide sensor based on a multilayer cytochrome c electrode. *Anal. Chem.* **76**, 4665–4671 (2004).
98. Y. Tian, M. Shioda, S. Kasahara, T. Okajima, L. Mao, T. Hisabori, and T. Ohsaka, A facilitated electron transfer of copper-zinc superoxide dismutase (SOD) based on a cysteine-bridged SOD electrode. *Biochim. Biophys. Acta.* **1569**, 151–158 (2002).
99. J.S. Valentine, M.W. Pantoliano, P.J. Mcdonnell, A.R. Burger, and S.J. Lippard, pH-dependent migration of copper(II) to the vacant zinc-binding site of zinc-free bovine erythrocyte superoxide dismutase. *Proc. Natl. Acad. Sci. U.S.A.* **76**, 4245–4249 (1979).
100. J.A. Fee, R. Natter, and G.S.T. Baker, Reconstitution of bovine erythrocyte superoxide dismutase. II. Observations on the nature of catalyzed superoxide. *Biochim. Biophy. Acta.* **295**, 96–106 (1973).
101. J.V. Bannister, W.H. Bannister, and E.J. Wood, Bovine erythrocyte cupro-zinc protein. 1. Isolation and general characterization. *Eur. J. Biochem.* **18**, 178–186 (1971).
102. J.A. Fee and R.G. Briggs, Reconstitution of bovine erythrocyte superoxide dismutase. V. Preparation and properties of derivatives in which both zinc and copper sites contain copper. *Biochim. Biophy. Acta.* **400**, 439–450 (1975).
103. J.A. Tainer, E.D. Getzoff, K.M. Beem, and D.C. Richardson, Determination and analysis of the 2.ANG. structure of copper, zinc superoxide dismutase. *J. Mol. Biol.* **160**, 181–217 (1982).
104. V.W.F. Chan, M.J. Bjerrum, and C.L. Borders, Jr, Evidence that chemical modification of a positively charged residue at position 189 causes the loss of catalytic activity of iron-containing and manganese-containing superoxide dismutases. *Arch. Biochem. Biophys.* **279**, 195–201 (1990).
105. M.E. Schinia, L. Maffey, D. Barra, F. Bossa, K. Puget, and A.M. Michelson, The primary structure of iron superoxide dismutase from Escherichia coli. *FEBS Lett.* **221**, 87–90 (1987).
106. D. Ringe, G.A. Petsko, F. Yamakura, K. Suzuki, and D. Ohmori, Structure of iron superoxide dismutase from Pseudomonas ovalis at 2.9 .ANG. resolution. *Proc. Natl. Acad. Sci. U.S.A.* **80**, 3879–3883 (1983).
107. W.C. Stallings, T.B. Powers, K.A. Pattridge, J.A. Fee, and M.L. Ludwig, Iron superoxide dismutase from Escherichia coli at 3.1 .ANG. resolution: a structure unlike that of copper/zinc protein at both monomer and dimer levels. *Proc. Natl. Acad. Sci. U.S.A.* **80**, 3884–3888 (1983).
108. W.C. Stallings, K.A. Pattridge, R.K. Strong, and M.L. Ludwig, The structure of manganese superoxide dismutase from Thermus thermophilus HB8 at 2.4 .ANG. resolution. *J. Biol. Chem.* **260**, 16424–16432 (1985).
109. W.C. Stallings, A.L. Metzger, K.A. Pattridge, J.A. Fee, and M.L. Ludwig, Structure-function relationships in iron and manganese superoxide dismutases. *Free Rad. Res. Commun.* **12–13**, 259–268 (1991).
110. W.C. Stallings, C. Bull, J.A. Fee, M.S. Lah, and M.L. Ludwig, Iron and manganese superoxide dismutases: catalytic inferences from the structures. *Current Communications in Cell & Molecular Biology.* **5**, 193–211 (1992).
111. A. Carlioz, M.L. Ludwig, W.C. Stallings, J.A. Fee, H.M. Steinman, and D. Touati, Iron superoxide dismutase. Nucleotide sequence of the gene from Escherichia coli K12 and correlations with crystal structures. *J. Biol. Chem.* **263**, 1555–1562 (1988).
112. B.L. Stoddard, P.L. Howell, D. Ringe, and G.A. Petsko, The 2.1 .ANG. resolution structure of iron superoxide dismutase from Pseudomonas ovalis. *Biochemisry.* **29**, 8885–8893 (1990).
113. B.L. Stoddard, D. Ringe, and G.A. Petsko, The structure of iron superoxide dismutase from Pseudomonas ovalis complexed with the inhibitor azide. *Protein Eng.* **4**, 113–119 (1990).

114. M.L. Ludwig, A.L. Metzger, K.A. Pattridge, and W.C. Stallings, The structure of iron superoxide dismutase from Pseudomonas ovalis complexed with the inhibitor azide. *J. Mol. Biol.* **219**, 335–358 (1991).

115. M.S. Lah, M.M. Dixon, K.A. Pattridge, W.C. Stallings, J.A. Fee, and M.L. Ludwig, Structure-function in Escherichia coli iron superoxide dismutase: comparisons with the manganese enzyme from Thermus thermophilus. *Biochemistry.* **34**, 1646–1660 (1995).

116. K.M. Beem, J.S. Richardson, and D.C. Richardson, Manganese, superoxide dismutases from Escherichia coli and from yeast mitochondria: preliminary x-ray crystallographic studies. *J. Mol. Biol.* **105**, 327–332 (1976).

117. R.J. Carrico and H.F. Deutsch, Presence of zinc in human cytocuprein and some properties of the apoprotein. *J. Biol. Chem.* **245**, 723–727 (1970).

118. H. Porter and S. Ainsworth, The isolation of the copper-containing protein cerebrocuprein I from normal human brain. *J. Neurochem.* **5**, 91–98 (1959).

119. J.R. Kimmel, H. Markowitz, and D.M. Brown, Some chemical and physical properties of erythrocuprein. *J. Biol. Chem.* **234**, 46–50 (1959).

120. R.N. Iyer and W.E. Schmidt, Observations on the direct electrochemistry of bovine copper-zinc superoxide dismutase. *Bioelectrochem. Bioenerg.* **27**, 393–404 (1992).

121. M. Borsari and H.A. Azab, Voltammetric behavior of bovine erythrocyte superoxide dismutase. *Bioelectrochem. Bioenerg.* **27**, 229–233 (1992).

122. H.A. Azab, L. Banci, M. Borsari, C. Luchinat, M. Sola, and M.S. Viezzoli, Redox chemistry of superoxide dismutase. Cyclic voltammetry of wild-type enzymes and mutants on functionally relevant residues. *Inorg, Chem.* **31**, 4649–4655 (1992).

123. X. Wu, X. Meng, Z. Wang, and Z. Zhang, Study on the direct electron transfer process of superoxide dismutase. *Bioelectrochem. Bioenerg.* **48**, 227–231 (1999).

124. X. Wu, X. Meng, Z. Wang, and Z. Zhang, Study on the ET process of SOD at cysteine modified gold electrode. *Chem. Lett.* 1271–1272 (1999).

125. T. Ohsaka, Y. Shintani, F. Matsumoto, T. Okajima, and K. Tokuda, Mediated electron transfer of polyethylene oxide-modified superoxide dismutase by methyl viologen. *Bioelectrochem. Bioenerg.* **37**, 73–76 (1995).

126. Z. Wang, W. Qian, Q. Luo, and M. Shen, Abnormal electrochemical behavior of copper-zinc superoxide dismutase on mercury electrodes. *J. Electroanal. Chem.* **482**, 87–91 (2000).

127. D. Cocco, L. Calabrese, A. Rigo, F. Marmocchi, and G. Rotitlio, Preparation of selectively metal-free and metal-substituted derivatives by reaction of Cu–Zn superoxide dismutase with diethyldithiocarbamate. *Biochem. J.* **199**, 675–680 (1981).

128. M.W. Pantoliano, P.J. McDonnell, and J.S. Valentine, Reversible loss of metal ions from the zinc binding site of copper-zinc superoxide dismutase. The low pH transition. *J. Amer. Chem. Soc.* **101**, 6454–6456 (1979).

129. C.A. Widrig, C. Chung, and M.D. Porter, The electrochemical desorption of n-alkanethiol monolayers from polycrystalline gold and silver electrodes. *J. Electroanal. Chem.* **310**, 335–359 (1991).

130. J.A. Fee and P.E. DiCorleto, Oxidation-reduction properties of bovine erythrocyte superoxide dismutase. *Biochemistry.* **12**, 4893–4899 (1973).

131. C.S. StClair, H.B. Gray, and J.S. Valentine, Spectroelectrochemistry of copper-zinc superoxide dismutase. *Inorg. Chem.* **31**, 925–927 (1992).

132. M.F.J.M. Verhagen, E.T.M. Meussen, and W.R. Hagen, On the reduction potentials of Fe and Cu-Zn containing superoxide dismutases. *Biochem. Biophys. Acta.* **1244**, 99–103 (1995).

133. K.D. Gleria, H.A.O. Hill, V.J. Lowe, and D.J. Page, Direct electrochemistry of horse-heart cytochrome c at amino acid-modified gold electrodes. *J. Electroanal. Chem.* **213**, 333–338 (1986).

134. E. Laviron, General expression of the linear potential sweep voltammogram in the case of diffusionless electrochemical systems. *J. Electroanal. Chem.* **101**, 19–28 (1979).

135. M.L. Salin and W.W. Wilson, Porcine superoxide dismutase. Isolation and characterization of a relatively basic cuprozinc enzyme. *Molec. Cell Biochem.* **36**, 157–161 (1981).

136. H. Dugas and C. Penney, *Bioorganic Chemistry*, Springer-Verlag, New York, Heidelberg, Berlin (1981), p. 23.

137. Y. Tian, T. Ariga, N. Takashima, T. Okajima, L. Mao, and T. Ohsaka, Self-assembled monolayers suitable for electron-transfer promotion of copper, zinc-superoxide dismutase. *Electrochem. Commun.* **6**, 609–614 (2004).
138. Y. Tian, L. Mao, T. Okajima, and T. Ohsaka, Electrochemistry and electrocatalytic activities of superoxide dismutases at gold electrodes modified with a self-assembled monolayer. *Anal. Chem.* **76**, 4162–4168 (2004).
139. M.J. Stansell and H.F. Deutsch, Preparation of crystalline erythrocuprein and catalase from human erythrocytes. *J. Biol. Chem.* **240**, 4299–4305 (1965).
140. H. Markowitz, G.E. Cartwright, and M.M. Wintrobe, Copper metabolism. XXVII. Isolation and properties of an erythrocyte cuproprotein (erythrocuprein). *J. Biol. Chem.* **234**, 40–45 (1959).
141. B.G. Malmstrom and T. Vanngard, Electron spin resonance of copper proteins and some model complexes. *J. Mol. Biol.* **2**, 118–124 (1960).
142. E.K. Hodgson and I. Fridovich, The interaction of bovine erythrocyte superoxide dismutase with hydrogen peroxide: inactivation of the enzyme. *Biochemistry*. **14**, 5294–5299 (1975).
143. J.A. Fee and J.S. Valentine, Chemical and physical properties of superoxide, in *Superoxide and Superoxide Dismutases* (A.M. Michelson, J.M. McCord, and I. Fridovich, eds), pp. 19–60, Academic Press, New York (1977).
144. M.E. McAdam, E.M. Fielden, F. Lavelle, L. Calabrese, D. Cocco, and G. Rotilio, The involvement of the bridging imidazolate in the catalytic mechanism of action of bovine superoxide dismutase. *Biochem. J.* **167**, 271–274 (1977).
145. J.A. Fee, The copper/zinc superoxide dismutase. *Ions Biol. Syst.* **13**, 259–298 (1981).
146. F. Yamakura, K. Kobayashi, H. Ue, and M. Konno, The pH-dependent changes of the enzymic activity and spectroscopic properties of iron-substituted manganese superoxide dismutase. A study on the metal-specific activity of Mn-containing superoxide dismutase. *Eur. J. Biochem.* **227**, 700–706 (1995).
147. C.K. Vance and A.F. Miller, Spectroscopic comparisons of the pH dependencies of Fe-substituted (Mn)superoxide dismutase and Fe-superoxide dismutase. *Biochemistry*. **37**, 5518–5527 (1998).
148. K. Endo, T. Miyasaka, S. Mochizuki, N. Himi, H. Asahara, K. Tsujioka, and K. Sakai, Development of a superoxide sensor by immobilization of superoxide dismutase. *Sens. Actuators B*. **83**, 30–34 (2002).
149. J. Di, S. Bi, and M. Zhang, Third-generation superoxide anion sensor based on superoxide dismutase directly immobilized by sol-gel thin film on gold electrode. *Biosen. Bioelectron.* **19**, 1479–1486 (2004).
150. T. Ohsaka, Y. Tian, M. Shioda, S. Kasahara, and T. Okajima, A superoxide dismutase-modified electrode that detects superoxide ion. *Chem. Commun.* 990–991 (2002).
151. Y. Tian, L. Mao, T. Okajima, and T. Ohsaka, Superoxide dismutase-based third-generation biosensor for superoxide anion. *Anal. Chem.* **74**, 2428–2434 (2002).
152. M. Miele and M.J. Fillenz, In vivo determination of extracellular brain ascorbate. *Neurosci. Methods.* **70**, 15–19 (1996).
153. T. Zetterstrom, L. Vernet, U. Ungerstedt, T.B. Jonzon, and B.B. Fredholm, Purine levels in the intact rat brain. Studies with an implanted perfused hollow fibre. *Neurosci. Lett.* **29**, 111–115 (1982).
154. E. Zilkha, T.P. Obrenovitch, A. Koshy, H. Kusakabe, and H.P. Bennetto, Extracellular glutamate: online monitoring using microdialysis coupled to enzyme-amperometric analysis. *J. Neurosci. Methods.* **60**, 1–9 (1995).
155. S.A.M. Marzouk, S. Ufer, R.P. Buck, T.A. Johnson, L.A. Dunlap, and W.E. Cascio, Electrodeposited iridium oxide pH electrode for measurement of extracellular myocardial acidosis during acute ischemia. *Anal. Chem.* **70**, 5054–5061 (1998).
156. F. Gonon, M.F. Suaud-Chagny, and M. Buda, Fast in vivo monitoring of electrically evoked dopamine release by differential pulse amperometry with untreated carbon fibre electrodes, in *Proceedings of Satellite Symposium on Neuroscience and Technology* (A. Dittmar and J.C. Froment, eds), p. 215. Lyon (1992).
157. R. Cespuglio, H. Faradji, Z. Hahn, and M. Jouvet, Voltammetric detection of brain 5-hydroxyindolamines by means of electrochemically treated carbon fiber electrodes: chronic recordings for up to one month with movable cerebral electrodes in the sleeping or waking rat, in *Measurements of Neurotransmitters Release in Vivo* (C.A. Marsden, ed.), Wiley, Chichester (1984).

158. Y. Tian, L. Mao, T. Okajima, and T. Ohsaka, A carbon fiber microelectrode-based third-generation biosensor for superoxide anion. *Biosens. Bioelectron.* **21**, 557–564 (2005).
159. L. Mao, J. Jin, L. Song, K. Yamamoto, and L. Jin, Electrochemical microsensor for in vivo measurements of oxygen based on Nafion and methylviologen modified carbon fiber microelectrode. *Electroanalysis.* **11**, 499–504 (1999).
160. L. Mao, F. Xu, Q. Xu, and L. Jin, Miniaturized amperometric biosensor based on xanthine oxidase for monitoring hypoxanthine in cell culture media. *Anal. Biochem.* **292**, 94–101 (2001).

CHAPTER 7

Detection of charged macromolecules by means of field-effect devices (FEDs): possibilities and limitations

Michael J. Schöning and Arshak Poghossian

7.1 INTRODUCTORY PART AND STATUS REPORT

The field of (bio-)chemical sensor research represents one of the most interesting and exciting multi-disciplinary topics with a broad range of applications, like environmental monitoring, biomedicine, biotechnology, food and drug industry, process technology, security, antibioterrorism, etc. Nowadays, semiconductor field-effect devices (FEDs) such as ISFET (ion-sensitive field-effect transistor), capacitive EIS (electrolyte-insulator-semiconductor) structures, and LAPS (light-addressable potentiometric sensor) represent one of the key structural elements of a new generation of electronic chips for chemical and/or biological sensing with a direct electronic readout. This type of sensor has been developed using different sensor configurations, sensitive materials, and fabrication technologies; the transducer principle of using an electric field to

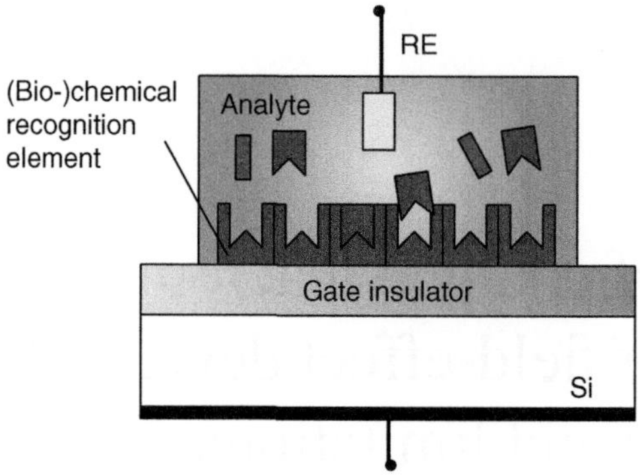

FIGURE 7.1 Schematic of an EIS hetero-structure. For the (bio-)chemical sensing, suitable chemical or biological recognition elements need to be functionally coupled to the sensitive surface of the FED. Chemical and/or biological recognition elements are, e.g., ionophores, enzymes, immunospecies, DNA, living cells, microorganisms and receptors. RE: reference electrode.

create regions of excess charge in a semiconductor substrate is common to all of them. FEDs are sensors with an external modulation possibility of the threshold voltage (V_{th}) or flat-band voltage (V_{fb}) by means of the interface potential analyte (test sample)–sensor chip. This finally varies the electric field inside the insulator of the FED yielding a modulation of the space-charge region in the semiconductor at the silicon/insulator interface (e.g. change of the capacitance of an EIS structure or conductance of the inversion channel of an ISFET). The basic common structure of all FEDs for (bio-)chemical sensing is the electrolyte-insulator-semiconductor system that is schematically shown in Fig. 7.1.

Originally, FEDs are derived from either MIS (metal-insulator-semiconductor) capacitors or IGFETs (insulated-gate field-effect transistors), where an analyte and a reference electrode have replaced the gate electrode. FEDs are basically surface-charge measuring devices (they detect the charge in a capacitive way) and therefore they are very sensitive for any kind of electrical interaction at or nearby the gate insulator/electrolyte interface. Therefore, nearly each (bio-)chemical reaction leading to chemical or electrical changes at this interface can be measured by means of an ISFET, capacitive EIS sensor or LAPS. For this purpose, suitable chemical or biological recognition elements need to be functionally coupled to the sensitive surface of the respective FED. Changes in the chemical composition will induce changes in the surface charge of the gate insulator and in the potential drop at the electrolyte/insulator interface, consequently modulating the current in the ISFET's channel, the capacitance of the EIS sensor or the photocurrent of the LAPS. These devices have been shown to be versatile tools for detecting pH, ion concentrations, enzymatic reactions, cellular metabolisms and action potentials of living cells, etc. More recently, the possibility of application of FEDs for the detection of charged macromolecules, like DNA, proteins, and polyelectrolytes, has been shown. For more detailed information concerning the operation principle and different types of ISFET, EIS and LAPS, see, e.g., [1–8].

The detection of molecular interactions at the solid–liquid interface is of great interest for a wide variety of applications, ranging from biomedical implants and drug-carrier systems up to biosensors, DNA arrays, and protein-chip technology. Moreover, a deep understanding of the adsorption and binding of charged macromolecules onto charged surfaces is important not only for sensor applications, but also for the fundamental understanding of many key physiological processes. More than 400 diseases can be detected by molecular analysis of nucleic acids, and a growing demand of DNA diagnosis in genetics, medicine, and drug discovery can be prediced [9]. Typically, DNA-detection principles are based on a DNA-hybridization event, where an unknown single-stranded DNA (ssDNA) is identified by its complementary DNA (cDNA) molecule. As a result of the hybridization event, a double-stranded DNA (dsDNA) helix structure with the two complementary strands is formed: Due to the unique complementary nature of the bases pairs' binding reaction (adenine-thymine (A-T), cytosine-guanine (C-G)), the hybridization event is highly efficient and specific, even in the presence of a mixture of additional non-complementary nucleic acids. The hybridization reaction occurs best if all the bases along both the probe and complementary target DNA molecules are fully matching. Therefore, genosensors and DNA chips should have a sensitivity high enough to detect even a single mismatch (single nucleotide polymorphism) with high reliability.

In the techniques actually employed for DNA-hybridization detection, the readout of the DNA-hybridization event requires the labeling of DNA molecules (either the analyte DNA or the immobilized ssDNA) with various markers (radiochemical, enzymatic, fluorescent, redox, etc.). In spite of their high sensitivity, selectivity and low detection limits, all these techniques, however, suffer from being time-consuming, expensive, and complex to implement (e.g. [10]). A label-free detection is thus highly desirable. The direct electrical detection of intrinsic charges of biomolecules with bio-functionalized semiconductor devices would circumvent the obstacles of labeling. Therefore, recently different research groups have devoted considerable effort to realize a label-free electronic detection of charged biomolecules, such as DNA, proteins, and peptides, by their intrinsic molecular charge using the field-effect platform [11–47]. Moreover, a possibility of potentiometric detection of single nucleotide polymorthisms by means of a genetic field-effect transistor has been experimentally demonstrated [29, 30]. Due to the inherent miniaturization and compatibility with advanced micro- and nano-fabrication technologies, these devices offer a new challenge of DNA chips with direct electronic readout for a label-free, fast, simple, and inexpensive real-time analysis of nucleic acid samples.

Results from a DNA-hybridization detection achieved with FEDs differ in:

- the set-up of the FED (capacitive EIS and MIS structure, depletion-/enhancement-mode FET, Au or Pt floating-gate FET, extending-gate FET, FET devices with or without reference electrode, poly-Si and hydrogenated amorphous Si (a-Si:H) thin film transistor),
- the different gate-insulator materials (SiO_2, silanized SiO_2, SiO_2-Si_3N_4, SiO_2-Ta_2O_5, SiO_2-poly-L-lysine) with different thicknesses (from 2 nm to $\sim$ 100 nm),

- various ssDNA-immobilization methods (adsorption, covalent attachment, biotin–avidin complexation, linker molecules),
- various densities of the immobilized ssDNA from 3.6×10^5 to 5×10^{13} molecules/cm^2,
- hybridization-buffer solutions with different electrolyte concentrations (10 μM to 1 M), and
- divergent sensor signals reaching from several mV up to ~2 V with hybridization times between several seconds up to several hours.

Most of the experiments for detecting charged macromolecules with FEDs, reported in literature, have been realized using a transistor structure [11–36]. Recent successful experiments on the detection of charged biomolecules as well as polyelectrolytes with other types of FEDs, namely semiconductor thin film resistors [39–41], capacitive MIS [42] and EIS structures [43–50], have demonstrated the potential of these structures – more simple in layout, easy, and cost effective in fabrication – for studying the molecular interactions at the solid–liquid interface. A summary of results for the DNA detection with different types of FEDs is given in Table 7.1.

The large diversity of sensor configurations and experimental results as well as the absence of a detailed theory explaining their working principles makes their comparison, however, quite difficult. Controversial effects such as higher signals for sensors with less density of immobilized ssDNA (0.87 V with 3.8×10^8 molecules/cm^2) compared to sensors with more densely packed ssDNA (3 mV with 5×10^{13} molecules/cm^2), and higher sensor signals that are observed when floating-gate transistors or FEDs without a reference electrode are used, are only representing two examples in this context.

From the experiments presented in the literature, it is obvious that the adsorption and binding of charged macromolecules onto the gate surface changes the flat-band voltage of FEDs, thus generating a sensor signal. However, which mechanism is responsible for such a shift of the flat-band voltage is still under discussion [39, 51, 52]. In most cases, the experimentally observed sensor response is interpreted as a direct electrostatic detection of charged macromolecules by their intrinsic molecular charge, fully ignoring the ion concentration and charge redistribution within the intermolecular spaces (or in the molecular layer) as well as the possible interaction of small ions with the surface of the underlying gate (insulator) material. An alternative mechanism based on the detection of the DNA hybridization-induced redistribution of the counter-ion concentration within the intermolecular spaces or in the molecular layer has been recently proposed [51]. Thus, there are some open questions regarding the functional principle of FEDs functionalized with charged macromolecules, and the source of the experimentally observed signal generation.

In this work, a critical evaluation of the possibilities and limitations of a direct electrical detection of charged macromolecules using a field-effect platform will be given, mainly focusing on capacitive EIS devices. With these devices it is possible to study both the geometric capacitance and charge effects induced by the adsorption or binding of charged macromolecules. Theoretical calculations of the physical model for the bio-functionalized EIS sensor and experimental results for the detection of the DNA

TABLE 7.1
Summary of field-effect-based DNA sensors reported in literature

Sensor type	Probe ssDNA; density	Electrolyte	Target cDNA	Sensor signal	Hybridization time	RE	Ref.
FET; EIS p-Si-SiO_2 (10 nm, silanization with APTES)	oligo(dT_{20}); poly(dT) (1000 bases); brominated oligonucleotides	50 mM NaCl; 10 mM Tris-HCl, pH 7.1	oligo(dA_{18}); poly(dA) (1000 bases); calf thymus DNA	$\Delta V_{fb} \sim 100$ mV; $\Delta V_{th} \sim 120$ mV at 2 μg/ml target DNA	4 h	Ag/AgCl	[11]
n-channel FET; Si-SiO_2-Si_3N_4; amino-silanized	17-mer ssDNA; 1.7×10^8 ssDNA/cm^2; glutaraldehyde as cross-linking agent	25 mM Na_2HPO_4 pH 6.86	17-mer cDNA	$\Delta V_{th} \sim 11$ mV	15 h	Ag/AgCl	[27]
EIS; Si-SiO_2 (30 nm)-Si_3N_4 (50 nm); amino-silanized	17-mer ssDNA; 3.7×10^8 ssDNA/cm^2; glutaraldehyde as cross-linking agent	25 mM Na_2HPO_4 pH 6.86	17-mer cDNA	$\Delta V_{fb} \sim 20$ mV	12 h	Ag/AgCl	[45]
n-Si-SiO_2 (2 nm)-poly-L-lysine	12-mer ssDNA; adsorption 5×10^{13} ssDNA/cm^2	23 mM	12-mer cDNA	$\Delta V_{fb} \sim 3$ mV;	10–15 min	Ag/AgCl	[43]
Au-gate n-channel depletion FET; p-Si-SiO_2 (63 nm)-Si_3N_4 (30 nm)-Au	thiolated 15- or 25-mer ssDNA	1 M NaCl pH 7.4	15-mer cDNA 25-mer cDNA	$\Delta V_{th} \sim 10$ mV $\Delta V_{th} \sim 3$ mV	1000 s	Ag wire	[13]
Ta_2O_5-gate FET; silanized	amino-modified 21-mer ssDNA; 3.6×10^5 ssDNA/cm^2	0.5 M $MgCl_2$	21-mer cDNA	$\Delta V_{th} \sim 10$ mV	overnight	Ag/AgCl	[24]
n-channel FET; p-Si-SiO_2 (50 nm)	20-mer ssDNA; adsorption	phosphate buffer, pH 7	detection of probe ssDNA	$\Delta V_{th} \sim 1.9$ V	3.5 h	without RE	[18]
p-channel FET; n-Si-SiO_2 (8–12 nm); silanization with APTES	adsorption $\sim 4 \times 10^{11}$ ssDNA/cm^2	1 mM NaCl	PolyA, 20 or 45 bases	$\Delta V_{th} \sim 5$ mV	≈ 1 h	Ag/AgCl liquid-junction	[14]

(continued)

TABLE 7.1
(continued)

Sensor type	Probe ssDNA; density	Electrolyte	Target cDNA	Sensor signal	Hybridization time	RE	Ref.
p-channel Au-gate (floating) FET; n-Si-SiO_2 (50 nm)	12-mer thiol-modified ssDNA, spacer; 3.5 × 10^8 molecules/cm^2		12-mer cDNA	$\Delta V_{th} \sim 0.87$ V	several minutes	without RE	[19]
p-channel Au-gate (floating) FET; n-Si-SiO_2 (50 nm)	15-mer thiol-modified ssDNA, spacer; 2.4 × 10^8 molecules/cm^2	phosphate buffer saline, pH 7.4	15-mer cDNA	$\Delta V_{th} \sim 0.23$ V		Pt	[22]
Au-gate poly-Si thin film transistor; Au-gate MIS capacitance	18-mer ssDNA, ~ $10^{12} - 10^{13}$ ssDNA/cm^2	5 mM phosphate buffer; pH 7.2	18-mer cDNA	$\Delta V_{th} \sim 355$ mV $\Delta V_{fb} \sim 140$ mV	1 h 30 min	Ag/AgCl	[36]
p-channel FET; n-Si-SiO_2 (10 nm)-poly-L-lysine	20-mer dsDNA; 2 × 10^{13} molecules/cm^2	0.01 mM KCl 1 mM >10 mM	detection of dsDNA	80 mV 40 mV <10 mV		Ag/AgCl wire	[16]

ΔV_{fb}: flat-band-voltage change; ΔV_{th}: threshold-voltage change; RE: reference electrode; APTES: 3-aminopropyltriethoxysilane.

immobilization and hybridization as well as layer-by-layer adsorbed polyelectrolyte multilayers are presented.

7.2 CAPACITANCE–VOLTAGE CHARACTERISTICS OF A BARE AND FUNCTIONALIZED EIS STRUCTURE

A DNA-FED is obtained by immobilizing well-defined sequences of an ssDNA as a biological recognition layer on top of the field-effect transducer, which should convert the specific recognition process between the two complementary DNA strands into a measurable signal (in this case, changes in the capacitance of the functionalized EIS structure). Figure 7.2 depicts the schematic structure and measuring set-up of a functionalized capacitive EIS sensor for the DNA immobilization and hybridization detection. For operation, a DC (direct current) polarization voltage (V_G) is applied via the reference electrode to set the working point of the EIS sensor, and a small AC (alternating current) voltage ($V_{\sim}$ = 10–50 mV) is applied to the system in order to measure the capacitance of the sensor.

The complete AC equivalent circuit of a bare EIS hetero-system is complex and combines components, like the bulk resistance and space-charge capacitance of the semiconductor, the capacitance of the gate insulator, the double-layer capacitance at the electrolyte–insulator interface, the resistance of the bulk-electrolyte solution and the impedance of the reference electrode, all related to the semiconductor, gate insulator, different interfaces, electrolyte, and reference electrode, respectively (see, e.g., [53, 54]). However, for usual values of insulator thickness (~30–100 nm), ionic strength of the electrolyte solution ($>10^{-4}$ M) and measurement frequencies (below ~1 kHz), the equivalent circuit of an EIS structure (see Fig. 7.3a) can be simplified as a series connection of the insulator capacitance, $C_i = \varepsilon_i/d$ (ε_i and d are permittivity and thickness of the insulator, respectively) and the space-charge capacitance of the semiconductor, $C_{sc}(\varphi)$, which is among other things a function of the voltage applied to the system and the electrolyte–gate insulator interfacial potential (the electrochemical double-layer capacitance is assumed to be much greater than C_i and C_{sc} (φ) and therefore can be neglected) [54].

Thus, the expression for the total capacitance of the bare EIS structure, $C(\varphi)$, is similar to the equation for an MIS capacitance, but with a modulation possibility of the space-charge capacitance by means of the electrolyte solution–insulator interface potential (φ) (the capacitances $C(\varphi)$, C_i, and C_{sc} (φ) are usually defined per cm^2 surface area):

$$\mathrm{C}(\varphi) = \frac{C_i C_{sc}(\varphi)}{C_i + C_{sc}(\varphi)} \qquad (1)$$

The typical shape of a capacitance–voltage (C–V) curve for a p-type EIS structure is given in Fig. 7.4. As can be seen from Fig. 7.4, dependent on the magnitude and polarity of the applied gate voltage, V_G, three regions in the C–V curve can be distinguished: accumulation, depletion and inversion (an n-type EIS structure shows an identical

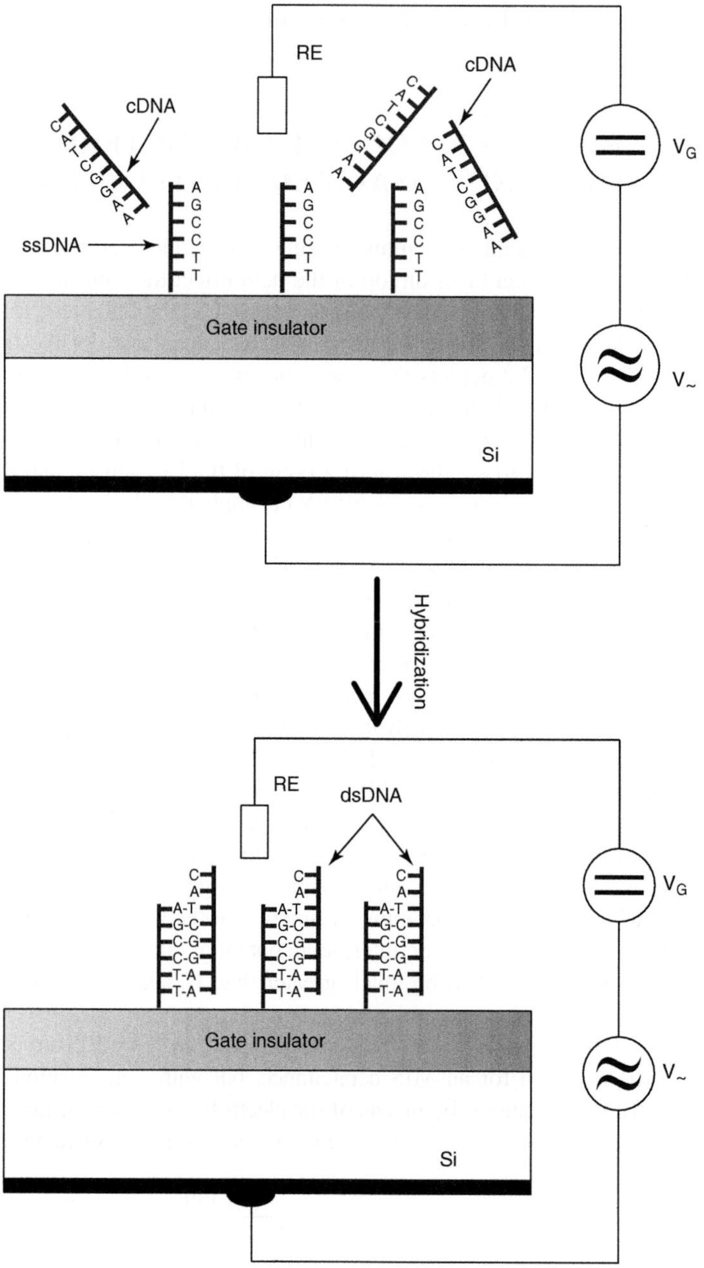

FIGURE 7.2 Schematic of a DNA sensor based on a capacitive EIS structure. For operation, a DC (direct current) polarization voltage (V_G) is applied via the reference electrode (RE) to set the working point of the EIS sensor, and a small AC (alternating current) voltage ($V_\sim$) is applied to the system in order to measure the capacitance of the sensor. ssDNA – single-stranded DNA, cDNA – complementary DNA, dsDNA – double-stranded DNA.

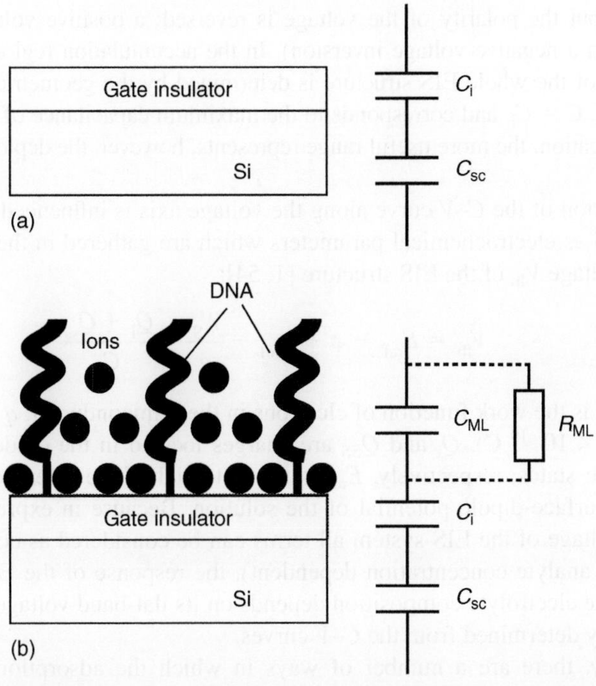

FIGURE 7.3 Simplified equivalent circuit of an original (unmodified) EIS structure (a) and EIS biosensor functionalized with charged macromolecules (b). C_i, C_{sc} and C_{ML} are capacitances of the gate insulator, the space-charge region in the semiconductor, and the molecular layer, respectively; R_{ML} is the resistance of the molecular layer.

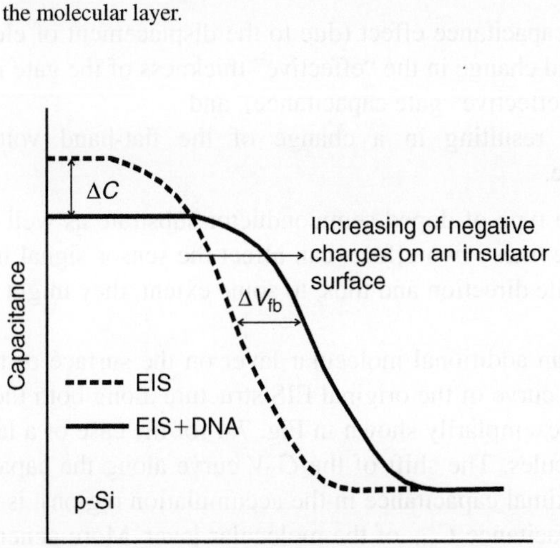

FIGURE 7.4 Capacitance–voltage (C–V) curve for a bare (unmodified) EIS sensor and EIS sensor with a molecular layer (here, DNA). The presence of the additional molecular layer shifts the C–V curve of the original EIS structure along both the capacitance (ΔC) and voltage axis (ΔV_{fb}).

C–V curve but the polarity of the voltage is reversed; a positive voltage causes accumulation and a negative voltage inversion). In the accumulation region ($C_i \ll C_{sc}$), the capacitance of the whole EIS structure is determined by the geometrical capacitance of the insulator, $C = C_i$, and corresponds to the maximum capacitance of the system. For a sensor application, the more useful range represents, however, the depletion region of the C–V curve.

The position of the C–V curve along the voltage axis is influenced by several solid-state as well as electrochemical parameters which are gathered in the equation for the flat-band voltage V_{fb} of the EIS structure [1, 54]:

$$V_{fb} = E_{ref} - \varphi + \chi_{sol} - \frac{W_S}{q} - \frac{Q_i + Q_{ss}}{C_i} \tag{2}$$

Here, W_S is the work function of electrons in the semiconductor, q is the elementary charge (1.6×10^{-19} C), Q_i and Q_{SS} are charges located in the oxide and the surface and interface states, respectively, E_{ref} is the potential of the reference electrode, and χ_{sol} is the surface-dipole potential of the solution. Because in expression (2) for the flat-band voltage of the EIS system all terms can be considered as constant except for φ (which is analyte concentration dependent), the response of the EIS structure with respect to the electrolyte composition depends on its flat-band voltage shift, which can be accurately determined from the C–V curves.

Generally, there are a number of ways in which the adsorption and binding of charged macromolecules (in particular, DNA immobilization and hybridization) can affect the electrochemical properties of the analyte–FED interface. In the case of field-effect devices, two basic effects are usually considered:

- a geometrical capacitance effect (due to the displacement of electrolyte by macromolecules and change in the "effective" thickness of the gate insulator and thus change in the "effective" gate capacitance), and
- charge effects resulting in a change of the flat-band voltage of the EIS hetero-structure.

Dependent on the type of doped semiconductor substrate as well as on the sign of the molecular charge, these two effects can affect the sensor signal in the same direction, or in the opposite direction and thus, to some extent, they might even compensate each other.

The presence of an additional molecular layer on the surface of the FED can lead to a shift of the C–V curve of the original EIS structure along both the capacitance and voltage axis, as it is exemplarily shown in Fig. 7.4 for the case of a layer of negatively charged macromolecules. The shift of the C–V curve along the capacitance axis, ΔC (decrease of the maximal capacitance in the accumulation region), is usually due to an additional series capacitance C_{ML} of the molecular layer. More generally, it is a series impedance, because usually monolayers of charged macromolecules do not represent a perfectly homogeneous and tightly packed, electrically blocked layer, but a rather much less dense layer with interstitial spaces permeable to ions, and therefore their resistivity is much lower than that of the underlying gate insulator. For a particular case of DNA brushes (polyelectrolyte brushes consist of charged macromolecules densely end-grafted to the surface, see Fig. 7.3b), the capacitance of the brush (i.e. the diffuse layer capacitance inside the DNA layer at the underlying layer–brush interface) depends among other things on the DNA coverage, the interfacial potential as well as on the ionic strength inside the brush, which is much higher than that of the bulk solution [55].

As can be seen in Fig. 7.4, the *C–V* curve of the functionalized EIS structure is also shifted along the voltage axis with respect to the *C–V* curve of the original (unmodified) EIS structure. The direction of the shift of the *C–V* curves along the voltage axis in Fig. 7.4 corresponds to an additional negative charging of the gate-insulator surface. This indicates that the molecular layer may also induce an interfacial potential change (change in flat-band voltage ΔV_{fb}) at the electrolyte side and/or gate-insulator side of the molecular layer, in series to the applied gate voltage V_G.

Such a simultaneous shift of the *C–V* curve along both the capacitance and voltage axis makes the *C–V* measurements more interesting and informative than static DC measurements with the transistor structure. If these two effects are independent, the capacitance change and flat-band voltage change induced by adsorption or binding of charged macromolecules can be obtained from one and the same measurement. However, it should be noted that in the case of the presence of charges in the gate insulator and surface and interface states (see Eq. (2) for the flat-band voltage), changes in the flat-band voltage can also be caused due to the series capacitance, thus coinciding with the effect of modulation of the flat-band voltage induced by the molecular charge.

In the following sections, we will discuss the origin of possible mechanisms of flat-band voltage changes induced by charged macromolecules, mainly focusing on a direct electrostatic detection of charged macromolecules by their intrinsic molecular charge, and the mechanism that utilizes the DNA hybridization-induced charge redistribution within the intermolecular spaces.

7.3 DIRECT ELECTROSTATIC DNA DETECTION BY ITS INTRINSIC MOLECULAR CHARGE

Since FEDs are surface-charge measuring devices detecting the charge in a capacitive way, they are principally able to measure the charge of adsorbed macromolecules such as DNA or the charge change due to a hybridization event. The electric field in the gate insulator depends, among other parameters, on the net surface charge at the electrolyte–insulator interface. Any charge changes at the insulator surface will result in an equal change in the charge density of opposite sign in the semiconductor space-charge region. Since DNA molecules are polyanions with negative charges at their phosphate backbone, it can be expected that during the event of hybridization of ssDNA molecules with their complementary strands (cDNA), the charge associated with the target molecule effectively changes the charge applied to the gate and thus modulates the flat-band voltage and capacitance of the EIS sensor as well as the threshold voltage and the drain current of the FED.

However, all reported results on DNA-FEDs that are based on the mechanism of a direct electrostatic detection by their intrinsic molecular charge have some principal limitations due to the so-called counter-ion screening effect, which is also well known from immuno-modified FETs proposed by Schenck already in 1978 (see, e.g., [56]). It has been intensively debated whether it would be possible to detect an antibody–antigen affinity-binding reaction with an FET, or not [3, 54]. As a result of these discussions, it was generally accepted that the screening of protein charges by small inorganic counter-ions present in the electrolyte solution will result in macroscopically nearly uncharged layers and prevent successful measurements of immunospecies with FETs. Under ideal conditions (i.e. a truly capacitive interface at which the immunological binding sites can be immobilized, a nearly complete antibody coverage, highly charged antigens, and a very low ionic strength), the theoretically expected signal should be on the order of 10 mV or less [3].

The charge distribution in the immediate vicinity of the interface will play a critical role in transferring the hybridization-induced signal to the FED. Only effects of charge-density changes that occur directly at the surface of the FED or within the order of the Debye length λ_D from the surface can be detected as a measurable biosensor signal (see also Eq. (3)):

$$\lambda_{\mathrm{D}} = \sqrt{\frac{\varepsilon_{\mathrm{el}}\varepsilon_0 kT}{2z^2q^2I}} \tag{3}$$

Here, ε_0 is the permittivity of vacuum, ε_{el} the dielectric constant of the electrolyte, z is the valency of the ions in the electrolyte, and I represents the ionic strength, which for a 1:1 salt, can be replaced by the electrolyte concentration n_0.

Figure 7.5 clarifies this effect for a 10-bases DNA molecule attached normally to the surface of the FED. Note, with increasing ionic strength of the electrolyte, the fraction of DNA charge, which remains in the double layer and thus will be mirrored in the space-charge region of the FED, is decreased. For example, under physiological conditions with $\lambda_D \sim 0.8$ nm, most of the DNA charge will be at a distance greater than the Debye length from the surface, which makes its detection more difficult or even impossible. If ssDNA molecules are, furthermore, immobilized using additional linker molecules or spacers extending from the insulating layer of the FED, the DNA hybridization-induced charge changes will be still smaller. On the other hand, if the DNA molecules lie more or less flat on the surface, a higher hybridization signal can be expected.

The screening of the charge associated with the probe- or target-DNA molecules by small counter-ions (cations) present in the solution is a major obstacle also in detecting the DNA hybridization. According to Manning's counter-ion condensation theory [57, 58], monovalent cations reduce the DNA charge by 76% and divalent cations by 88%: the more diffuse ionic layer will compensate the remaining charge. Consequently, the counter-ion condensation effect will mask or reduce the expected hybridization signal and prevent successful measurements, especially in high ionic strength solutions. Higher salt concentrations of the electrolyte solution should result in a bigger screening of the molecular charges and thus in a smaller sensor signal. In the most unfortunate

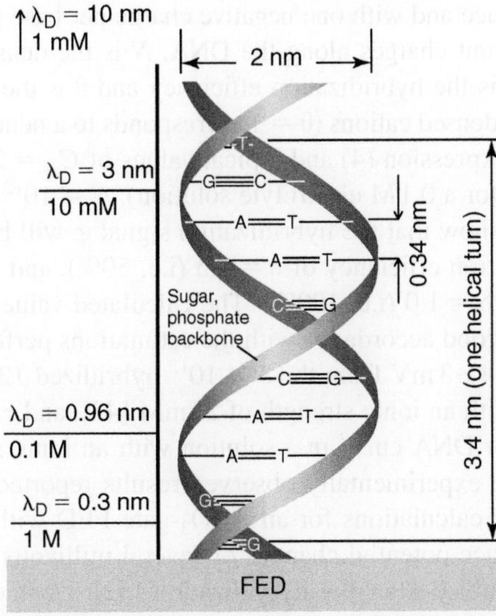

FIGURE 7.5 FED with 10-bases DNA molecule oriented normal to the sensor surface and Debye length λ_D in the electrolyte with different ionic strengths (schematically). Increasing ionic strength of the electrolyte solution decreases the fraction of DNA charge that can be detected by the FED.

case, the hybridized pair (cDNA-ssDNA) produces a net-reduced or even zero charge, although the charged target molecule has bound to the immobilized probe molecule. As a consequence, the underlying field-effect transducer is not able to deliver a measurable sensor signal.

In order to estimate the sensor signal that is induced upon the DNA-hybridization process, in a first approximation, the hybridization of the probe molecules with their complementary target molecules can be modeled as a transfer of a certain quantity of charge from the test sample to the surface of the gate insulator. Since the requirement of electroneutrality must be realized in the system, an equal quantity of the opposite charge must either enter in the inversion layer of the FET or enter the double layer from the solution. The estimations for typical values of double-layer and gate-insulator capacitances ($C_{dl} = 20\,\mu F\ cm^{-2}$ and $C_i = 0.35\,\mu F\ cm^{-2}$ for a 10 nm thick SiO_2 layer, respectively) imply, that only about 1.7% of the hybridization-induced charge will be mirrored in, e.g., an FET [51]. The remaining charge will be compensated by ions in the solution. The potential changes induced by hybridization (i.e. the expected sensor signal) can be defined as [51]:

$$\varphi = \frac{Q_h}{C_i + C_{dl}} = \frac{mN\delta(1-\theta)}{C_i + C_{dl}} \tag{4}$$

Here, Q_h is the charge change induced upon the hybridization, $m = q\,\lambda_D/b$ is the fraction of DNA charge in the double layer (for simplicity, DNA molecules are considered arranged normal to the surface and with one negative charge per base pair), b is the distance between the nearest unit charges along the DNA, N is the density of the immobilized probe molecules, δ is the hybridization efficiency and θ is the fraction of DNA charge compensated by condensed cations ($\theta = 1$ corresponds to a neutral molecule).

The calculations using expression (4) and typical values of $C_{dl} = 20\,\mu F\,cm^{-2}$, $C_i = 0.35\,\mu F\,cm^{-2}$, $\lambda_D = 1\,nm$ (for a 0.1 M electrolyte solution), $N = 10^{12}$ molecules cm^{-2}, $b = 0.34\,nm$ and $\theta = 0.76$ show that the hybridization signal φ will be on the order of about 3 mV for a hybridization efficiency of $\delta = 0.5$ (i.e. 50%), and about 6 mV for a hybridization efficiency of $\delta = 1.0$ (i.e. 100%). The calculated values of the expected hybridization signal are in good accordance with the estimations performed in [14, 43] using the Graham equation (~3 mV from the 3×10^{12} hybridized 12-mer DNA cm^{-2} in an electrolyte solution with an ionic strength of 23 mM [43], and ~0.8 mV from the 4×10^{11} hybridized 20-mer DNA cm^{-2} in a solution with an ionic strength of 1 mM [14]) as well as with some experimentally observed results reported in the literature (see Table 7.1). The model calculations for an Al_2O_3-gate FED with charged macromolecules also predict surface potential changes of several millivolts when the charge density of molecules is doubled after the hybridization [52]. On the other hand, the same model predicts larger signals of several tens of mV for FEDs with an uncharged gate-insulator surface [52].

To reduce the counter-ion screening effect and thus to enhance the sensitivity of the sensor, FEDs must be operated in:

- very low ionic-strength solutions, and
- have to use a high densitiy of the immobilized probe molecules ($N > 10^{12}$ molecules/cm^{-2}).

For these cases, however, a reduced probability of hybridization and therefore an extended hybridization time and reduced sensor signals can be expected. Thus, the theoretical basis of the sometimes experimentally observed "large" sensor signals (see Table 7.1) still remains unclear.

A further task for the correct functioning of FEDs for DNA detection by its intrinsic charge is given by the fact that the surface interaction should only occur between the immobilized ssDNA and its complementary cDNA. There should be no interference by any background interaction of small ions with the underlying gate surface of the FED. Ideally, to "insulate" the underlying gate insulator from the solution, the immobilized ssDNA molecules should form a perfectly homogeneous and tightly packed monolayer without any pores or interstitial spaces. This demand, however, is controversy to reported values for the theoretical maximum surface coverage of 25–30% for a random-sequential adsorption of rod-like molecules [59]. Many experimental results support this fact, i.e. the molecular layers are much less dense allowing an interstitial penetration of small inorganic ions and water molecules to the underlying layer.

In summary, a practical realization of FEDs for the pure electrostatic detection of charged macromolecules by their intrinsic molecular charge, especially in high ionic-strength solutions such as physiological conditions, seems to be problematic. All the above discussed "disturbing" factors, together with a possible undesired adsorption or unspecific binding of DNA molecules at the reference electrode used in the experimental set-up, can reduce or even mask the expected biosensor signal. Moreover, sensor drift as well as leakage currents can also yield a falsified sensor signal, or at least interfere with it. It is always favorable to use differential measuring set-ups [15, 28, 43] to at least exclude some of these disturbing and interfering effects. Nonetheless, much more theoretical modeling and experimental research has to be done in order to understand and correctly interpret the DNA-detection experiments by means of FEDs.

7.4 NEW METHOD FOR LABEL-FREE ELECTRICAL DNA DETECTION

As a new mechanism for a direct label-free DNA detection using an FED, the ion-concentration redistribution in the intermolecular spaces of the immobilized ssDNA (due to the DNA-hybridization event) and the alteration of the ion sensitivity has been proposed recently [51]. In this approach, the top surface of the ion-sensitive FED is modified with immobilized ssDNA probe molecules arranged normal to the surface with a center-to-center average interprobe distance a_s. The remaining surface of the ion-sensitive layer between the immobilized molecules is in contact with the electrolyte solution. In such a DNA brush-like model (see also Fig. 7.3b), the mobile ions pass freely between the DNA layer and the external electrolyte. The probe ssDNA molecules should be arranged on the surface with enough interstitial space to allow a rapid hybridization and to provide a high hybridization efficiency. A preferable average center-to-center separation distance could be in the range from $\approx$2.5 nm to $\approx$10 nm which corresponds to a probe density from about 2×10^{13} to 1.3×10^{12} molecules cm^{-2}, typically reported in the literature (see, e.g., [60–63]).

Since ssDNA and dsDNA molecules are negatively charged via their phosphate groups, such negatively charged molecules will attract positively charged counter-ions (including protons) from the solution and repel the co-ions. As a result, the DNA charge is effectively compensated by the surrounding small counter-ions. This may result in a local ion-concentration redistribution within the intermolecular spaces (increasing the cation concentration and decreasing the anion concentration) which can substantially differ from the concentration in the bulk electrolyte, n_0. After hybridization, because the charge of the dsDNA is nearly doubled, a new distribution of the electrostatic potential and of the ions within the intermolecular spaces will be reached. The hybridization-induced ion-concentration (including proton concentration) redistribution in the intermolecular spaces (or in the DNA layer) can be detected by the underlying ion-sensitive FED. Thus, in contrast to the above-discussed FEDs for the DNA-hybridization detection by the intrinsic molecular charge, where the screening of the molecular charge by small counter-ions is considered as a major obstacle, here the counter-ion condensation effect is used to detect the DNA-hybridization event.

For sensor applications, a more interesting parameter is the degree of change in the average ion concentration in the intermolecular spaces upon the hybridization event. The model for the theoretical calculations of the average concentration of cations and

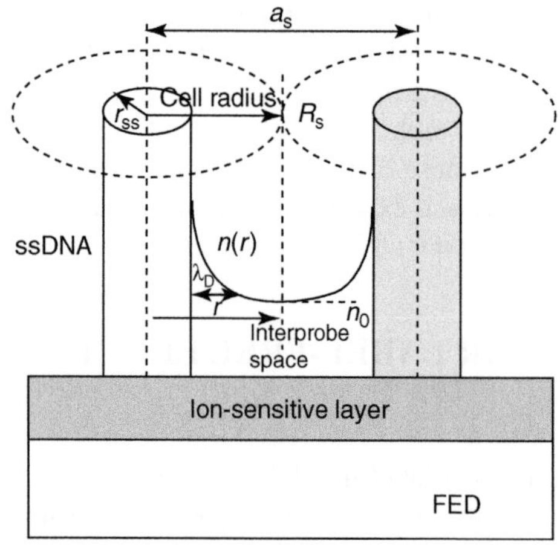

FIGURE 7.6 Cell model for the theoretical calculations of the DNA hybridization-induced ion-concentration redistribution as well as the average concentration of cations and anions in the intermolecular spaces. Both the ssDNA and dsDNA have been modeled as negatively charged cylinders with a radius $r_{ss} = 0.5\,\text{nm}$ and $r_{ds} = 1\,\text{nm}$, respectively, which form a hexagonal lattice with a cell radius of R_s; the DNA molecules are arranged normal to surface of the FED with a center-to-center average separation distance of $a_s \approx 2R_s$; $n(r)$ is the ion concentration as a function of the coordinate r from the DNA axis and n_0 is the bulk-ion concentration.

anions in the intermolecular spaces is sketched in Fig. 7.6. The ssDNA and dsDNA have been modeled as uniformly negatively charged, infinitely long cylinders with a radius $r_{ss} = 0.5\,\text{nm}$ and $r_{ds} = 1\,\text{nm}$, respectively, which form a hexagonal lattice with a cell radius of R_s. The effect of DNA hybridization is modeled by doubling the linear charge density of the cylinder and by increasing the cylinder radius from 0.5 nm to 1 nm. For simplicity, the model does not take into account the charge state of the sensor surface.

For the above-described model, the average concentration of cations and anions in the intermolecular spaces can be calculated using equations derived in [51]:

$$\langle n_{\pm}(R_s)\rangle = n_0\left(\pm\zeta + \sqrt{\zeta^2 + 1}\right) \tag{5}$$

with

$$\zeta = \frac{1-\theta}{2\pi b n_0 (R_s^{\,2} - a^2)} \tag{6}$$

Here, b is the distance between the nearest unit charges along the cylinder ($b = 0.34\,\text{nm}$ for the ssDNA and $b = 0.17\,\text{nm}$ for the dsDNA), $(+)$ and $(-)$ are related to cations and anions, respectively, and $a = r_{ss}$ for the ssDNA and $a = r_{ds}$ for the dsDNA. The expressions (5) and (6) have been obtained using the equations for the electrostatic potential derived in [64, 65], where a linearization of the Poisson–Boltzmann equation near the Donnan potential in the hexagonal DNA cell was implemented.

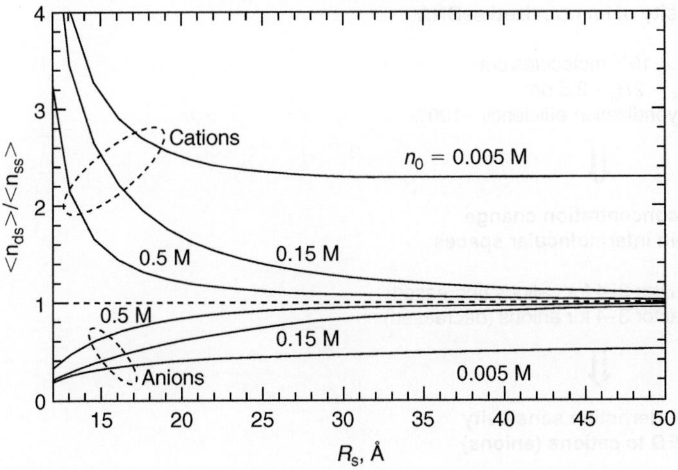

FIGURE 7.7 Ratio $<n_{ds}>/<n_{ss}>$ of average concentration of cations (upper part of diagram) and anions (lower part of diagram) within the intermolecular spaces before hybridization $<n_{ss}>$ and after hybridization $<n_{ds}>$ in 1:1 salt solutions with bulk ion concentration of n_0 = 0.005, 0.15 and 0.5 M, respectively. R_s: DNA cell radius.

Figure 7.7 shows the calculated ratio $<n_{ds}>/<n_{ss}>$ of the average concentration for cations and anions within the intermolecular spaces before hybridization $<n_{ss}>$ and after hybridization $<n_{ds}>$. The ratio $<n_{ds}>/<n_{ss}>$ is plotted as a function of the DNA cell radius R_s and for 1:1 salt solutions of different bulk-ion concentrations of n_0 = 0.005, 0.15 (physiological solution) and 0.5 M, respectively. The fraction of both the ssDNA and dsDNA charge compensated by the condensed cations was taken as θ = 0.8. Two effects can be recognized from Fig. 7.7:

- the local concentration of cations and anions within the intermolecular spaces differs strongly from that of the bulk solution (a similar effect has been experimentally observed for DNA brushes in [55]), and
- the average ion concentration within the intermolecular spaces after the hybridization differs distinctly from that of before the hybridization; the effect is dependent on the ionic strength of the bulk electrolyte solution and is stronger in a low ionic-strength electrolyte.

In contrast to the DNA detection by its intrinsic molecular charge, even in high ionic-strength solutions (0.5 M), where the hybridization efficiency is high and the hybridization event can be faster, a detectable sensor signal can be achieved. The estimations performed under the assumptions presented in Fig. 7.8 predict signal values of about 28–35 mV.

The achieved biosensor signal will be influenced by the following factors:

- an increase of R_S implies a decreased density of immobilized ssDNA yielding a smaller $<n_{ds}>/<n_{ss}>$ ratio (more strongly in high ionic-strength solutions),

Density of immobilized ssDNA

- 2×10^{13} molecules/cm^2
- $a_s = 2R_s \sim 2.5$ nm
- Hybridization efficiency ~100%

⇓

Ion-concentration change within intermolecular spaces

- Factor 3–4 for cations (increased)
- Factor 3–4 for anions (decreased)

⇓

Ideal Nernstian sensitivity of FED to cations (anions)

⇓

Upon hybridization induced sensor signal

- 28–35 mV

FIGURE 7.8 Theoretically expected values of an FED signal due to the DNA hybridization-induced ion-concentration redistribution within the intermolecular spaces.

- high hybridization signals at low density of immobilized ssDNA require solutions with lower ionic strength, and
- a strong increase in $\langle n_{ds} \rangle / \langle n_{ss} \rangle$ at a small separation distance (high density of the immobilized ssDNA), on the other hand, decreases the sensor area available for the ion interaction.

Generally, dependent on the sensor design and working conditions, the optimum separation distance or optimum density of the immobilized ssDNA should be found in order to achieve a maximal hybridization signal. A twice increase in the hybridization signal can be obtained when combining a cation- and anion-sensitive FED in a differential measuring set-up.

In summary, it should be noted that in the model described above, the ssDNA is presented as a rod-like molecule oriented normal to the surface. In a more realistic picture, the ssDNA is a flexible coil-like molecule. If those coil-like molecules lie preferentially flat on the ion-sensitive gate of the FED, they can partially cover the surface active sites for ion-binding or ion-exchange processes as well as prevent or hinder potential-determining ions to reach the transducer surface (see Fig. 7.9a). In contrast, after the hybridization event, a rigid rod-like dsDNA is formed. Now, the surface of the FED is opened for ion interaction (Fig. 7.9b), resulting in an hybridization-induced alteration of the ion sensitivity of the FED and generation of an additional sensor signal.

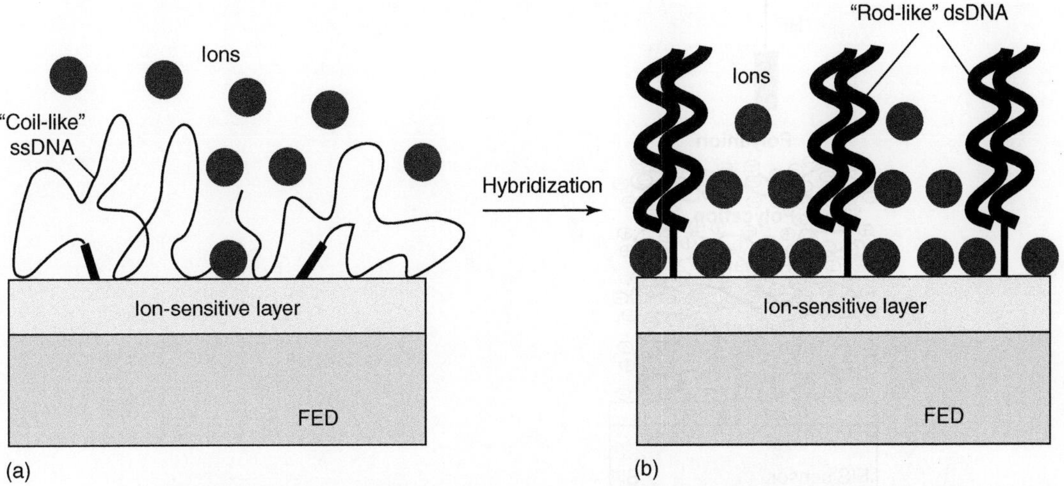

FIGURE 7.9 Model for the DNA detection based on an alteration of the ion sensitivity of an FED induced upon the DNA hybridization. If flexible ssDNA lie preferentially flat on the ion-sensitive gate of the FED, they can partially cover the surface active sites for ion-binding or ion-exchange processes as well as prevent or hinder potential-determining ions to reach the transducer surface (a). In contrast, after hybridization a rigid rod-like dsDNA is formed. Now, the surface of the FED is opened for ion interaction (b), resulting in the hybridization-induced signal generation.

7.5 MEASUREMENT RESULTS UTILIZING POLYELECTROLYTE LAYERS AND SYNTHETIC DNA

In this chapter, exemplary experiments with capacitive EIS sensors that have been utilized for DNA immobilization and hybridization detection as well as for monitoring the layer-by-layer adsorption of polyelectrolytes anionic poly(sodium 4-styrene sulfonate): PSS and cationic poly(allylamine hydrochloride: PAH) are presented. Polyelectrolyte (PE) multilayers represent a very useful model system for studying effects induced in FEDs by the adsorption of charged macromolecules [41, 46–50]. They are linear macromolecule chains bearing a large number of charged or chargeable groups when dissolved in a suitable polar solvent, e.g. in water. The formation of PE multilayers is based on the consecutive adsorption of polyions with alternating charge. Each adsorption step of a charged PE layer leads to a charge inversion of the surface, the so-called charge overcompensation effect [66–68]. The subsequent deposition, finally, results in a PE multilayer stabilized by strong electrostatic forces. Figure 7.10 sketches the FED build-up with a PE multilayer arrangement.

The preparation of the FEDs, the experimental set-up and measuring conditions for the detection of DNA immobilization and hybridization as well as for the monitoring of the layer-by-layer adsorption of the polyelectrolyte multilayers are described in detail elsewhere [46–50]. The attachment of these charged macromolecules to the FED surfaces has been systematically characterized by means of capacitance–voltage,

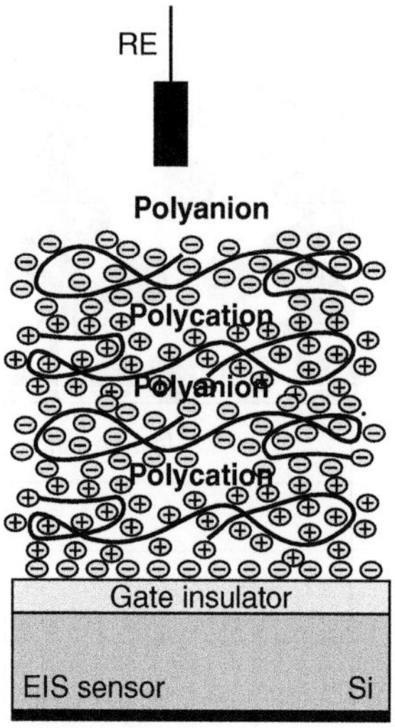

FIGURE 7.10 Schematic sensor build-up of an FED with layer-by-layer adsorbed PE multilayers of polycationic and polyanionic composition.

constant–capacitance, impedance–spectroscopy and atomic–force microscopy methods. Both the effect of the number and polarity of the polyelectrolyte layers and the influence of the DNA-hybridization event on the shift of the capacitance–voltage curves along the capacitance and voltage axis have been investigated.

Figure 7.11a shows a typical set of *C–V* curves for a p-Si-SiO_2-Ta_2O_5 sensor – as prepared and after the adsorption of each PE layer (PAH and PSS, respectively). As it has been discussed in section 7.2, the *C–V* measurements allow to observe both the flat-band voltage change (ΔV_{fb}) and the changes in the maximum capacitance (ΔC) in the accumulation region of the *C–V* curve, induced by charged macromolecules. The shifts along the voltage axis can be clearly recognized from the zoomed graph in the depletion region (~60% of the C_{max}) (see Fig. 7.11a, right); their direction depends on the sign of the PE layer's charge.

Alternating potential shifts of about 30–90 mV have been observed after the adsorption of each polycation and polyanion layer. In the case of adsorption of the positively charged PAH, the potential shifts into the direction as for an additional positive charge of the Ta_2O_5 surface of the FED, whereas the direction of the potential change after adsorption of the negatively charged PSS corresponds to the case of a negative Ta_2O_5 surface charge.

At the same time, Fig. 7.11a also shows some small changes in the maximum (or geometrical) capacitance in the accumulation range of the *C–V* curve (decrease of about up to ~2.5%). When assuming the adsorbed PE layers as homogeneous dielectric films of defined thickness and dielectric constant, the geometrical capacitance effect due to the additional eight PE layers should be somewhat larger as achieved in this experiment. This can be attributed to the not completely dense structure of the

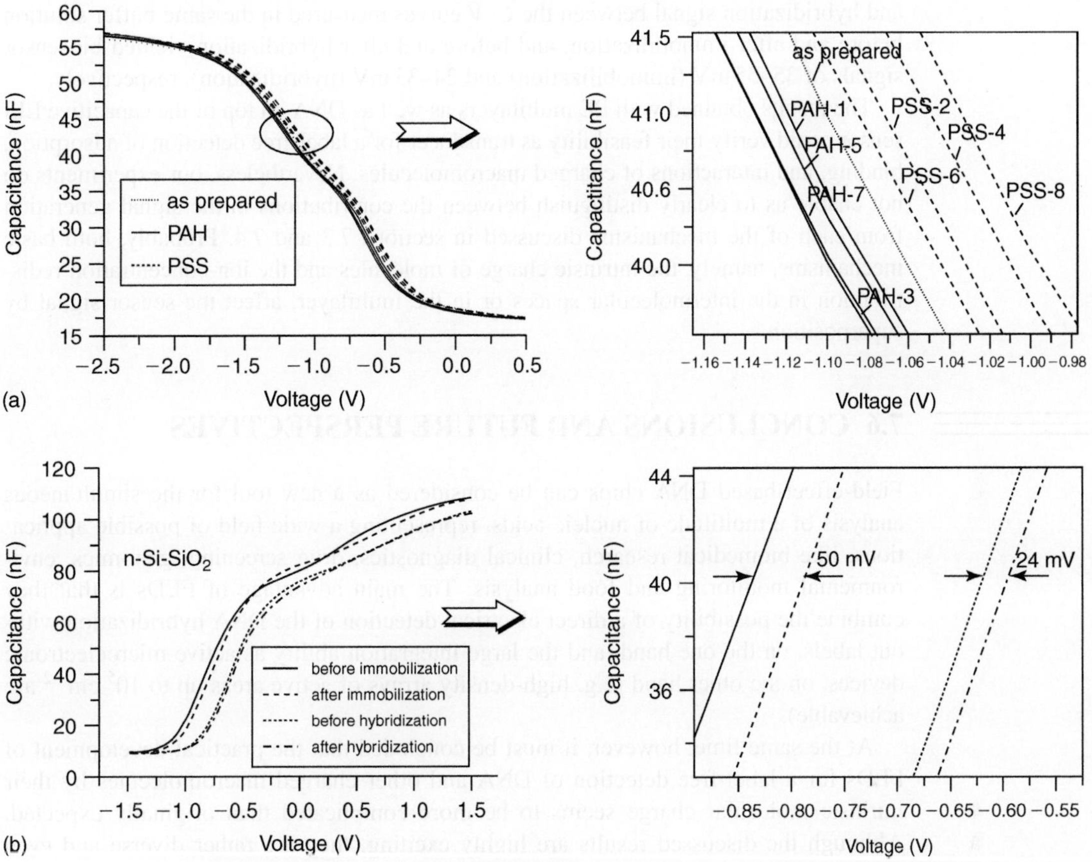

FIGURE 7.11 (a) *C–V* curves of a capacitive p-Si-SiO_2 (30 nm)-Ta_2O_5 (60 nm) FED – as prepared, and after adsorption of PAH and PSS polyelectrolyte layers (left), and resulting potential shifts in the depletion region (right); measurement frequency: 1 kHz; 0.1 M NaCl, pH 6 solution. (b) *C–V* curves of a capacitive n-Si-SiO_2 (30 nm) FED before and after ssDNA immobilization in buffer pH 4.6 and before and after cDNA hybridization in buffer pH 7.24 (left), and resulting potential shifts in the depletion region (right); measurement frequency: 1 kHz.

adsorbed PE multilayer containing a relatively large amount of water molecules and small ions.

The label-free DNA immobilization and hybridization detection for an n-Si-SiO_2 FED is presented in Fig. 7.11b. The *C–V* curves of the FED have been recorded before and after the ssDNA immobilization as well as before and after cDNA hybridization. The observed large shifts in the *C–V* curve along both the capacitance and voltage axis can be attributed to the different values of pH and electrolyte conductivity of immobilization (pH 4.6, ionic strength 0.5 M) and hybridization (pH 7.24, ionic strength 0.02 M) buffers. When considering these effects, the remaining shifts of the immobilization and hybridization signal between the *C–V* curves measured in the same buffer solution before and after immobilization, and before and after hybridization yielded biosensor signals of 35–55 mV (immobilization) and 24–33 mV (hybridization), respectively.

The results obtained with PE multilayers as well as DNA on top of the capacitive EIS sensor could verify their feasibility as transducer for a label-free detection of adsorption, binding, and interactions of charged macromolecules. Nevertheless, our experiments do not enable us to clearly distinguish between the contributions in the signal generation from each of the mechanisms discussed in sections 7.3 and 7.4. Probably, both basic mechanisms, namely, the intrinsic charge of molecules and the ion-concentration redistribution in the intermolecular spaces or in the multilayer, affect the sensor signal by superposition.

7.6 CONCLUSIONS AND FUTURE PERSPECTIVES

Field-effect-based DNA chips can be considered as a new tool for the simultaneous analysis of a multitude of nucleic acids, representing a wide field of possible applications, like biomedical research, clinical diagnostics, drug screening, genomics, environmental monitoring and food analysis. The main advantage of FEDs is that they combine the possibility of a direct electrical detection of the DNA hybridization without labels, on the one hand, and the large integration ability as active microelectronic devices, on the other hand (e.g. high-density arrays of active areas up to 10^5 cm^{-2} are achievable).

At the same time, however, it must be concluded that the practical development of FEDs for a label-free detection of DNA and other charged macromolecules by their intrinsic molecular charge seems to be more complicated than originally expected. Although the discussed results are highly exciting, they are rather diverse and even sometimes inconsistent. Factors influencing the DNA immobilization and hybridization detection by FEDs are:

- the gate material of the FED (metal, insulator, polymer) and its surface charge or surface modification,
- the immobilization method of the ssDNA (passive or electrochemical adsorption, covalent binding, binding via cross-linker),
- the immobilization, hybridization as well as measuring conditions (pH, ionic strength, temperature), and
- the packing density and length of the immobilized ss-DNA, the length of the spacer or linker molecule, etc.

Therefore, a deep understanding of the adsorption and interaction of charged macromolecules such as DNA onto (charged) surfaces of FEDs is of great interest not only for biosensor applications but also for the fundamental characterization of many key physiological processes. Further experiments are required to figure out the achieved results with DNA-based FEDs and to develop accompanying correct theoretical models. In this context, utilizing new devices and transducer strategies might be a promising way to guarantee reproducible and reliable biosensor signals. One approach taking this into consideration has been recently introduced by the authors: in contrast to "conventional" field-effect-based DNA sensors, the new approach takes advantage of the DNA hybridization-induced redistribution of the ion concentration within the intermolecular spaces as a detection mechanism. Both theoretical calculations as well as preliminary experiments with synthetic ssDNA and polyelectrolyte multilayers could demonstrate a substantial sensor signal on the order of several tens of millivolts. Moreover, the new biosensor concept is capable of working in low ionic-strength as well as high ionic-strength solutions.

To solve and at the same time to deeply understand the quite complicated and multidisciplinary task of field-effect-based DNA biosensors, many disciplines and research fields, scientists from bio- and electrochemistry, biophysics, device engineering, and analytics should work hand in hand.

7.7 ACKNOWLEDGMENTS

The authors thank A. Cherstvy for theoretical calculations, S. Ingebrandt for valuable discussions, and M. Abouzar for technical support. Part of the work was supported by the Ministerium für Innovation, Wissenschaft, Forschung und Technologie des Landes NRW (Germany).

7.8 REFERENCES

1. A. Poghossian and M.J. Schöning, Silicon-based chemical and biological field-effect sensors, in *Encyclopedia of Sensors* (C.A. Grimes, E.C. Dickey, and M.V. Pishko, eds), Vol. 9, pp. 463–533. American Scientific Publisher, Stevenson Ranch, 2006.
2. M.J. Schöning and A. Poghossian, Recent advances in biologically sensitive field-effect transistors. *Analyst* **127**, 1137–1151 (2002).
3. G.F. Blackburn, Chemically sensitive field-effect transistors, in *Biosensors: Fundamentals and Applications* (A.P.F. Turner, I. Karube, and G.S. Wilson, eds), pp. 481–530. Oxford University Press, Oxford (1987).
4. P. Bergveld and A. Sibbald, *Analytical and Biomedical Applications of Ion-Selective Field-Effect Transistors*. Elsevier, Amsterdam (1988).
5. M. Grattarola and G. Massobrio, *Bioelectronics Handbook: MOSFETs, Biosensors and Neurons.* McGraw-Hill, New York (1998).
6. P. Bergveld, Thirty years of ISFETOLOGY: what happened in the past 30 years and what may happen in the next 30 years. *Sens. Actuators B* **88**, 1–20 (2003).
7. M.J. Schöning, "Playing around" with field-effect sensors on the basis of EIS structures, LAPS and ISFETs. *Sensors* **5**, 126–138 (2005).
8. A. Poghossian, T. Yoshinobu, A. Simonis, H. Ecken, H. Lüth and M.J. Schöning, Penicillin detection by means of field-effect based sensors: EnFET, capacitive EIS sensor or LAPS? *Sens. Actuators B* **78**, 237–242 (2001).
9. C.H. Mastrangelo, DNA analysis systems on a chip, in *Solid-State Chemical and Biochemical Sensors* (P. Vincenzini and L. Dori, eds), pp. 465–476. Techna, Faenza (1999).
10. P. de-los-Santos-Alvarez, M. Jesus Lobo-Castanon, A.J. Miranda-Ordieres, and P. Tunon-Blanco, Current strategies for electrochemical detection of DNA with solid electrodes. *Anal. Bioanal. Chem.* **378**, 104–118 (2004).

11. E. Souteyrand, J.P. Cloarec, J.R. Martin, C. Wilson, I. Lawrence, S. Mikkelson, and M.F. Lawrence, Direct detection of the hybridization of synthetic homo-oligomer DNA sequences by field effect. *J. Phys. Chem. B* **101,** 2980–2985 (1997).
12. G. Libo, H. Jinghong, Z. Hong, and C. Xiang, DNA field-effect transistor. *Proc. SPIE* **4414**, 47–49 (2001).
13. F.K. Perkins, L.M. Tender, S.J. Fertig, and M.C. Peckerar, Sensing macromolecules with microelectronics. *Proc. SPIE* **4608**, 251–265 (2002).
14. F. Uslu, S. Ingebrandt, D. Mayer, S. Böcker-Meffert, M. Odenthal, and A. Offenhäusser, Labelfree fully electronic nucleic acid detection system based on a field-effect transistor device. *Biosens. Bioelectron.* **19**, 1723–1731 (2004).
15. Y. Han, A. Offenhäuser, and S. Ingebrandt, Detection of DNA hybridisation by a field-effect based sensor with covalently attached catcher molecules. *Surf. Interface Anal.* **38**, 176–181 (2006).
16. F. Pouthas, C. Gentil, D. Cote, and U. Bockelmann, DNA detection on transistor arrays following mutation-specific enzymatic amplification. *Appl. Phys. Lett.* **84**, 1594–1596 (2004).
17. F. Pouthas, C. Gentil, D. Cote, G. Zeck, B. Straub, and U. Bockelmann, Spatially resolved electronic detection of biopolymers. *Phys. Rev. E* **70**, 031906-1-8 (2004).
18. M.W. Dashiell, A.T. Kalambur, R. Leeson, K.J. Roe, J.F. Rabolt, and J. Kolodzey, The electrical effects of DNA as the gate electrode of MOS transistors in *Proc. IEEE Lester Eastman Conference*, pp. 259–264. Delaware (2002).
19. D.S. Kim, Y.T. Jeong, H.K. Lyu, H.J. Park, H.S. Kim, J.K. Shin, P. Choi, J.H. Lee, G. Lim, and M. Ishida, Fabrication and characteristics of a field effect transistor-type charge sensor for detecting deoxyribonucleic acid sequence. *Jpn. J. Appl. Phys.* **42**, 4111–4115 (2003).
20. J.K. Shin, D.S. Kim, H.J. Park, and G. Lim, Detection of DNA and protein molecules using an FET-type biosensor with gold as a gate metal. *Electroanalysis* **16**, 1912–1918 (2004).
21. D.S. Kim, Y.T. Jeong, H.K. Lyu, H.J. Park, J.K. Shin, P. Choi, J.H. Lee, and G. Lim, An FET-type charge sensor for highly sensitive detection of DNA sequence. *Biosens. Bioelectron.* **20**, 69–74 (2004).
22. D.S. Kim, H.J. Park, H.M. Jung, J.K. Shin, Y.T. Jeong, P. Choi, J.H. Lee, and G. Lim, Field-effect transistor-based biomolecular sensor employing a Pt reference electrode for the detection of deoxyribonucleic acid sequence. *Jpn. J. Appl. Phys.* **43**, 3855–3859 (2004).
23. K.Y. Park, M.S. Kim, and S.Y. Choi, Fabrication and characteristics of MOSFET protein chip for detection of ribosomal protein. *Biosens. Bioelectron.* **20**, 2111–2115 (2005).
24. T. Ohtake, C. Hamai, T. Uno, H. Tabata, and T. Kawai, Immobilisation of probe DNA on Ta_2O_5 thin film and detection of hybridised helix DNA using ISFET. *Jpn. J. Appl. Phys.* **43**, L1137–L1139 (2004).
25. T. Uno, T. Ohtake, H. Tabata, and T. Kawai, Direct deoxyribonucleic acid detection using ion-sensitive field-effect transistors. *Jpn. J. Appl. Phys.* **43**, L1584–L1587 (2004).
26. T. Sakata, M. Kamahori, and Y. Miyahara, Immobilisation of oligonucleotide probes on Si_3N_4 surface and its application to genetic field effect transistor. *Mat. Sci. Eng. C* **24**, 827–832 (2004).
27. T. Sakata, M. Kamahori, and Y. Miyahara, DNA analysis chip based on field-effect transistors. *Jpn. J. Appl. Phys.* **44**, 2854–2859 (2005).
28. T. Sakata, S. Matsumoto, Y. Nakajima, and Y. Miyahara, Potential behaviour of biochemically modified gold electrode for extended-gate field-effect transistor. *Jpn. J. Appl. Phys.* **44**, 2860–2863 (2005).
29. T. Sakata and Y. Miyahara, Potentiometric detection of single nucleotide polymorphism by using a genetic field-effect transistor. *ChemPhysChem.* **6**, 703–710 (2005).
30. T. Sakata and Y. Miyahara, DNA sequencing based on intrinsic molecular charges. *Angew. Chem. Int. Ed.* **45**, 2225–2228 (2006).
31. Y. Ishige, M. Shimoda, and M. Kamahori, Immobilization of DNA probes onto gold surface and its application to fully electric detection of DNA hybridization using field-effect transistor sensor. *Jpn. J. Appl. Phys.* **45**, 3776–3783 (2006).
32. G.. Xuan, J. Kolodzey, V. Kapoor, and G.. Gonye, Characteristics of field-effect devices with gate oxide modification by DNA. *Appl. Phys. Lett.* **87**, 103903-1-3 (2005).
33. D. Goncalves, D.M.F. Prazeres, V. Chu, and J.P. Conde, Label-free electronic detection of biomolecules using a-Si:H field-effect devices. *J. Non-Crystalline Solids* **352**, 2007–2010 (2006).

34. M. Barbaro, A. Bonfiglio, L. Raffo, A. Alessandrini, P. Facci, and I. Barak, A SMOS fully integrated sensor for electronic detection of DNA hybridization. *IEEE Electron Device Lett.* **27**, 595–597 (2006).
35. M. Barbaro, A. Bonfiglio, and L. Raffo, A charge-modulated FET for detection of biomolecular processes: conception, modelling, and simulation. *IEEE Trans. Electron Devices* **53**, 158–166 (2006).
36. P. Estrela, A.G. Stewart, F. Yan, and P. Migliorato, Field effect detection of biomolecular interactions. *Electrochim. Acta* **50**, 4995–5000 (2005).
37. F. Wei, B. Sun, Y. Guo, and X.S. Zhao, Monitoring DNA hybridization on alkyl modified silicon surface through capacitance measurement. *Biosens. Bioelectron* **18**, 1157–1163 (2003).
38. H. Berney, J. West, E. Haefele, J. Alderman, W. Lane, and J.K. Collins, A DNA diagnostic biosensor: development, characterisation and performance. *Sens. Actuators B* **68**, 100–108 (2000).
39. S.Q. Lud, M.G.. Nikolaides, I. Haase, M. Fischer, and A.R. Bausch, Field effect of screened charges: electrical detection of peptides and proteins by a thin-film resistor. *ChemPhysChem.* **7**, 379–384 (2006).
40. M.G. Nikolaides, S. Rauschenbach, S. Luber, K. Buchholz, M. Tornow, G. Abstreiter, and A.R. Bausch, Silicon-on-insulator based thin-film resistor for chemical and biological sensor applications. *ChemPhysChem.* **4**, 1104–1106 (2003).
41. P.A. Neff, A. Naji, C. Ecker, B. Nickel, R. Klitzing, and A.R. Bausch, Electrical detection of self-assembled polyelectrolyte multilayers by a thin film resistor. *Macromolecules* **39**, 463–466 (2006).
42. P. Estrela, P. Migliorato, H. Takiguchi, H. Fukushima, and S. Nebashi, Electrical detection of biomolecular interactions with metal-insulator-semiconductor diodes. *Biosens. Bioelectron.* **20**, 1580–1586 (2005).
43. J. Fritz, E.B. Cooper, S. Gaudet, P.K. Sorger, and S.R. Manalis, Electronic detection of DNA by its intrinsic molecular charge. *PNAS* **99**, 14 142–14 146 (2002).
44. J.P. Cloarec, N. Deligianis, J.R. Martin, I. Lawrence, E. Souteyrand, C. Polychronakos, and M.F. Lawrence, Immobilisation of homooligonucleotide probe layers onto Si/SiO_2 substrates: characterisation by electrochemical impedance measurements and radiolabelling. *Biosens. Bioelectron.* **17**, 405–412 (2002).
45. T. Sakata and Y. Miyahara, Detection of DNA recognition events using multi-well field effect devices. *Biosens. Bioelectron.* **21**, 827–832 (2005).
46. M.J. Schöning, M.H. Abouzar, Y. Han, S. Ingebrandt, A. Offenhäusser, and A. Poghossian, Markierungsfreie DNA-Detektion mit Silizium-Feldeffektsensoren-Messeffekte oder Artefakte? in *Sensoren und Messsysteme*, pp. 443–446. Berlin, VDE Verlag (2006).
47. A. Poghossian, M.H. Abouzar, F. Amberger, D. Mayer, Y. Han, S. Ingebrandt, A. Offenhäuser, and M.J. Schöning, Field-effect sensors with charged macromolecules: characterisation by capacitance–voltage, constant capacitance, impedance spectroscopy and atomic-force microscopy methods. *Biosens. Bioelectron.* **22**, 2100–2107 (2007).
48. T. Kassab, Y. Han, A. Poghossian, S. Ingebrandt, A. Offenhäuser, and M.J. Schöning, Detection of layer-by-layer adsorbed polyelectrolytes by means of field-effect-based capacitive EIS structures. *Biomedizinische Technik* **49**, 1034–1035 (2004).
49. M.J. Schöning, M.H. Abouzar, S. Ingebrandt, J. Platen, A. Offenhäuser, and A. Poghossian, Towards label-free detection of charged macromolecules using field-effect-based structures: scaling down from capacitive EIS sensor over ISFET to nano-scale devices, in *Mater. Res. Soc. Symp. Proc.* **915**, 0915-R05-04 (2006).
50. A. Poghossian, M.H. Abouzar, M. Sakkari, T. Kassab, Y. Han, S. Ingebrandt, A. Offenhäuser, and M.J. Schöning, Field-effect sensors for monitoring the layer-by-layer adsorption of charged macromolecules. *Sens. Actuators B* **118**, 163–170 (2006).
51. A. Poghossian, A. Cherstvy, S. Ingebrandt, A. Offenhäuser, and M.J. Schöning, Possibilities and limitations of label-free detection of DANN hybridisation with field-effect-based devices. *Sens. Actuators B* **111**, 470–480 (2005).
52. D. Landheer, G. Aers, W.R. Mckinnon, M.J. Deen, and J.C. Ranuarez, Model for the field effect from layers of biological macromolecules on the gates of metal-oxide-semiconductor transistors. *J. Appl. Phys.* **98**, 044701-1-15 (2005).
53. L.J. Bousse, N.F. de Rooij, and P. Bergveld, Operation of chemically sensitive field-effect sensors as a function of the insulator–electrolyte interface. *IEEE Trans. Electron Dev.* **30**, 1263–1270 (1983).

54. P. Bergveld, A critical evaluation of direct electrical protein detection methods. *Biosens. Bioelectron.* **6**, 55–72 (1991).
55. G. Shen, N. Tercero, M.A. Gaspar, B. Varughese, K. Shepard, and R. Levicky, Charging behaviour of single-stranded DNA polyelectrolyte brushes. *J. Am. Chem. Soc.* **128**, 8427–8433 (2006).
56. J.F. Schenck, Technical difficulties remaining to the application of ISFET devices, in *Theory, Design and Biomedical Applications of Solid State Chemical Sensors* (P.W. Cheung, ed.), pp. 165–173. CRC Press, Boca Raton (1978).
57. G.S. Manning, The molecular theory of polyelectrolyte solution with applications to the properties of polynucleotides. *Quart. Rev. Biophys.* **II**, 179–246 (1978).
58. G.S. Manning, Counterion condensation on a helical charge lattice. *Macromolecules* **34**, 4650–4655 (2001).
59. K.A. Melzak, C.S. Sherwood, R.F.B. Turner, and C.A. Haynes, Driving forces for DNA adsorption to silica in perchlorate solutions. *J. Coll. Interf. Sci.* **181**, 635–644 (1996).
60. M.C. Pirrung, How to make a DNA chip. *Angew. Chem. Int. Ed.* **41**, 1276–1289 (2002).
61. V. Chan, S.E. McKenzie, S. Surrey, P. Fortina, and D.J. Graves, Effect of hydrophobicity and electrostatics on adsorption of DNA oligonucleotides at liquid/solid interfaces. *J. Coll. Interf. Sci.* **203**, 197–207 (1998).
62. Z. Lin, T. Strother, W. Cai, X. Cao, L.M. Smith, and R.J. Hamers, DNA attachment and hybridization at the silicon (100) surface. *Langmuir* **18**, 788–796 (2002).
63. Y. Belosludtsev, B. Iverson, S. Lemeshko, R. Eggers, R. Wiese, S. Lee, T. Powdrill, and M. Hogan, DNA microarrays based on noncovalent oligonucleotide attachment and hybridization in two dimensions. *Anal. Biochem.* **292**, 250–256 (2001).
64. A.G.. Cherstvy, A.A. Kornyshev, and S. Leikin, Temperature dependent DNA condensation triggered by rearrangement of adsorbed cations. *J. Phys. Chem. B* **106**, 13 362–13 369 (2002).
65. A.G. Cherstvy and R.G. Winkler, Complexation of semiflexible chains with oppositely charged cylinder. *J. Chem. Phys.* **120**, 9394–9400 (2004).
66. G. Decher, M. Eckle, J. Schmitt, and B. Struth, Layer-by-layer assembled multicomposite films. *Curr. Opin. Coll. Interface Sci.* **3**, 32–39 (1998).
67. M. Schönhoff, Self-assembled polyelectrolyte multilayers. *Curr. Opin. Coll. Interface Sci.* **8**, 86–95 (2003).
68. P.T. Hammond, Recent explorations in electrostatic multilayer thin film assembly. *Curr. Opin. Coll. Interface Sci.* **4**, 430–442 (2000).

CHAPTER 8

Electrochemical sensors for the determination of hydrogen sulfide production in biological samples

David W. Kraus, PhD, Jeannette E. Doeller, PhD, and Xueji Zhang, PhD

8.1 INTRODUCTION

Hydrogen sulfide (H_2S) is most commonly known as a toxic gas with the repulsive odor of rotten eggs. Although H_2S can be toxic through its inhibitory interactions with hemoglobin and cytochrome *c* oxidase [1, 2], many prokaryotic and eukaryotic organisms thrive in sulfidic habitats and have been shown to possess putative H_2S oxidases and alternative oxidases that operate in concert with mitochondria to gain energy from H_2S oxidation as well as to prevent inhibition of aerobic metabolism [3, 4]. In addition, micromolar tissue H_2S levels have been observed in animals from H_2S-free environments [5, 6] as well as in human brain tissue and blood [7–9], suggesting that H_2S is a constituent of the cellular milieu.

H_2S can be produced via the metabolism of sulfhydryl-bearing amino acids, specifically by several enzymes found in the methionine-homocysteine-cysteine pathway such as cystathionine β synthase (CBS) and cystathionine γ lyase (CGL) (Fig. 8.1) [6, 10, 11]. The sequence of CBS has been identified in genomes from bacteria to humans [12–14], and a gene similar to the sulfide:quinone oxidoreductase gene has been identified in the genome of flies, worms, mice, rats, and humans [15], indicating that cellular H_2S and its regulation may be widespread and essential.

The apparent paradox of a toxic molecule in living aerobic tissues has precedent in the presence of other small rapidly diffusing molecules that are both critical in cell signaling and also potentially toxic, namely nitric oxide (NO) and carbon monoxide (CO). In the past two decades, the demonstration that NO is a critically important biological molecule has greatly expanded our understanding of cell signaling. H_2S shares many properties with NO, and the potential for H_2S to participate in cell signaling is clear. However, its broader biological role is not well understood partially as a result of the lack of sensitive and specific methods for the real time measurement of this molecule in complex biological milieu. In this review, we describe a novel polarographic H_2S

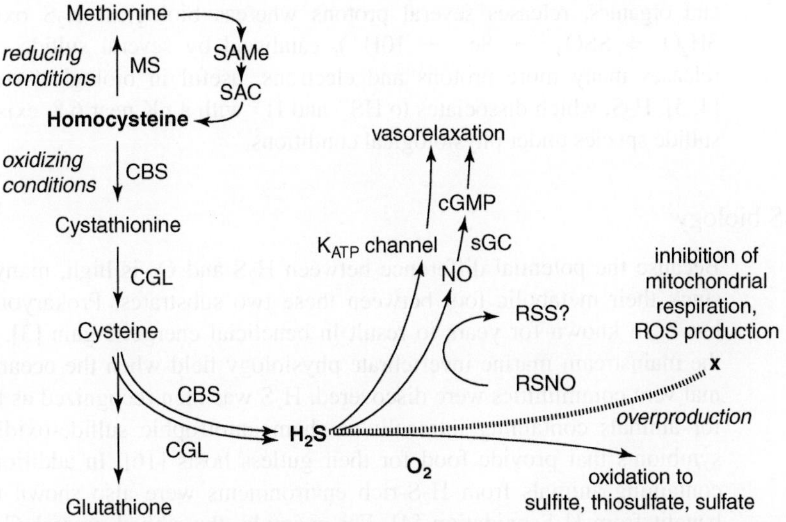

FIGURE 8.1 H_2S metabolic pathways. Under oxidative stress, homocysteine is metabolized toward H_2S production by the enzymes cystathionine β synthase (CBS) and cystathionine γ lyase (CGL). H_2S affects vascular smooth muscle cells directly via K_{ATP} channels as well as indirectly by reacting with S-nitrosated thiols (RSNO) to release NO and produce persulfides (RSS?), with NO causing vasodilation by soluble guanylate cyclase (sGC) activation of the cyclic guanidine monophosphate (cGMP) pathway. Under reducing conditions, homocysteine is metabolized to methionine by methionine synthase (MS) and can be subsequently converted back to homocysteine through intermediates S-adenosyl methionine (SAMe) and S-adenosyl cysteine (SAC). An overproduction of H_2S not adequately removed by oxidation pathways may result in mitochondrial respiration inhibition and reactive oxygen species (ROS) generation (after [41]).

sensor (PHSS) for real-time physiological measurement of H_2S *in vivo* and *in vitro* and discuss its usefulness to better understand the role of H_2S in cell signaling.

8.1.1 Significance of H_2S in the life sciences

8.1.1.1 H_2S chemistry

H_2S was present in the reducing atmosphere of early earth, along with CO_2, NH_3, CH_4, CO, and H_2. Early organisms made energy using limited oxidative reactions, and in some reactions H_2S served as a source of reducing equivalents, for example in anoxygenic photosynthesis ($2H_2S + CO_2 \Rightarrow 2S + CH_2O + H_2O$). The low energy wavelengths of light (870 nm and 840 nm) necessary for this reaction carried out by purple/green bacteria indicate the ease of H_2S oxidation.

Environmental H_2S is formed mainly by geothermal and biological activity. Solubility is high, 2 mM total sulfide kPa^{-1}, and the potential difference between H_2S and O_2 is high, −270 mV to +800 mV, making H_2S oxidation reactions energetically favorable. Abiotic H_2S oxidation ($H_2S + 2O_2 \Rightarrow SO_4^{-2} + 2H^+$), catalyzed by metals and organics, releases several protons whereas biological H_2S oxidation ($2H_2S + 3H_2O \Rightarrow SSO_4^{-2} + 8e^- + 10H^+$), catalyzed by several sulfide oxidase enzymes, releases many more protons and electrons, useful in biological oxidation reactions [4, 5]. H_2S, which dissociates to HS^- and H^+ with a pK near 6.8, exists as two reactive sulfide species under physiological conditions.

8.1.1.2 H_2S biology

Because the potential difference between H_2S and O_2 is high, many organisms have stuck their metabolic foot between these two substrates. Prokaryotic H_2S oxidation has been known for years to result in beneficial energetic gain [3], and H_2S entered the mainstream marine invertebrate physiology field when the ocean floor hydrothermal vent communities were discovered. H_2S was then recognized as the energy source for animals containing intracellular chemoautotrophic sulfide-oxidizing prokaryotic symbionts that provide food for their gutless hosts [16]. In addition, non-symbiont-containing animals from H_2S-rich environments were also shown to gain energetic benefit from H_2S oxidation [4]. For example, the ribbed mussel *Geukensia demissa* from H_2S-rich sediment was shown to use H_2S for mitochondrial oxidative phosphorylation in a process called metazoan chemolithoheterotrophy [5].

When H_2S was found to be a common constituent of mammalian tissues and cells, researchers began to show that H_2S could operate in manners similar to NO [17, 18]. For example, H_2S modulates the function of heme proteins such as cytochrome *c* oxidase, hemoglobin and myoglobin similar to NO, and H_2S interacts with thiol groups as a reductant [2]. Physiological effects of H_2S include smooth muscle relaxation and K_{ATP} channel conductance [19–22], long-term neuronal potentiation via changes in NMDA receptor [23, 24], and regulation of enzyme activity [2, 25–27] and metabolic state [28]. Pathological effects of decreased H_2S are implicated in cardiovascular disease, Alzheimer's disease, and diabetes [29–33], and excess H_2S production is implicated in trisomy 21 [34]. Accordingly, cellular H_2S concentration is most likely tightly controlled, highlighting the important regulation of H_2S production and consumption pathways by cellular redox status, O_2, and other factors [35, 36].

Although it is clear that H_2S plays an important physiological role in mammalian systems, it must be realized that many investigations have been carried out at O_2 and H_2S concentrations that do not represent physiological levels. *In-situ* H_2S levels have not been monitored partly because continuous physiological H_2S measurements have not been possible until recently. Without the benefit of real time H_2S measurements, the effects of H_2S under physiological conditions remain largely undefined.

8.1.2 H_2S measurement in biological samples

8.1.2.1 Stability of sulfur

Like NO, the stability of H_2S and HS^- under physiological conditions is influenced by a number of inorganic and organic components that catalyze oxidation reactions, and this reactivity limits the usefulness of most standard analytical methods for real-time H_2S measurements. Several analytical methods are sufficiently sensitive to detect the micromolar to submicromolar H_2S levels found in biological samples, but they often require multiple chemical steps and/or chromatographic procedures, and rely on the conversion of all sulfide species to H_2S or S^{2-} under acidic or alkaline conditions, respectively [37]. Sulfur can exist in several different oxidation states, and in animal tissues is more commonly found as reduced divalent or fully oxidized hexavalent states. Sulfur not liberated by acid or reducing agents such as dithiothreitol is considered stable. Reduced organic thiols and sulfane sulfur atoms, existing as divalent anions covalently bound to other sulfur atoms, are more easily liberated as inorganic hydrosulfide anion or hydrogen sulfide by enzymatic cleavage or mild reducing conditions [37]. Determination of free H_2S in biological samples may be influenced by these sulfur sources especially if measurement conditions are not similar to the physiological conditions of the sample. Many of the standard methods used to measure free H_2S rely on sample manipulations to generate products for detection by spectrophotometry, fluorescence, ion-specific electrodes or titration. For the most part it is unclear if these methods also measure H_2S that existed as a persulfide or other labile species.

8.1.2.2 Methods for H_2S measurements

Measurements of H_2S production rate, reported for a variety of homogenized tissues [6, 20, 23, 31, 38], are often performed by placing the homogenized tissue in the outer well of a flask that is flushed with N_2 and sealed to limit spontaneous H_2S oxidation. A separate center well contains H_2S-trapping agents such as alkaline zinc acetate. After a specified time, enzymatic H_2S production is halted by the addition of trichloroacetic acid to precipitate protein and convert any remaining S^{2-} and HS^- to H_2S but perhaps also resulting in the uncontrolled liberation of acid-labile H_2S. The amount of H_2S trapped as ZnS at a single time point is determined with an assay that produces a proportional amount of methylene blue dye, and the H_2S production rate is calculated assuming zero H_2S at time zero. Alternative methods used to determine H_2S production rate or tissue H_2S levels include gas chromatography following acid or alkaline extraction, and the silver/silver-sulfide (Ag_2S) electrode to measure the concentration of S^{2-} in tissue samples placed in a pH 14 sulfide anti-oxidant buffer [20, 31]. The bare Ag_2S electrode in combination with a reference electrode was developed to detect S^{2-} in solution at alkaline pH, and recent work suggests that it may also report HS^- down to submicromolar concentrations [39]. However, the bare Ag_2S coating requires daily reconditioning to remove interfering deposits that accumulate from constituents present in biologically relevant solutions [39].

With both the acidic and alkaline single time point assays, it remains unclear if acid labile H_2S or desulfuration of proteins [40], respectively, creates artifactual overestimates of free H_2S levels, as described earlier. Tissue H_2S concentrations determined by thiol derivatization using bromobimane near neutral pH followed by HPLC also indicate micromolar levels, but it again remains unclear if the use of reagents with strong affinity for H_2S cause a significant equilibrium shift to release H_2S that was previously bound as persulfides.

A recent comprehensive review of the methods used to determine tissue H_2S levels details the multiple sources of measured H_2S as acid labile, bound or free [37]. Mammalian tissue H_2S concentrations tabulated in the review cover a large range from not detectable to $>100\,\mu M$, indicating that method, tissue, and metabolic state all have a strong influence on the actual measurement. Although the presence of tissue H_2S is generally accepted, how much is freely dissolved is an important unanswered question. Furthermore, kinetic interactions of H_2S with any other cellular signals in real time are not captured with single time point determinations. It is clear that available H_2S measurement techniques are not applicable to dynamic biological experiments. To address the problems of H_2S stability and kinetics, we have developed a novel polarographic H_2S sensor (PHSS) that allows real-time determination of H_2S levels in biological samples under simulated physiological conditions of pH, temperature, and O_2 tension [36, 41].

8.2 ADVANTAGES OF ELECTROCHEMICAL SENSORS FOR H_2S DETERMINATION

8.2.1 Electrochemistry

The development of sensitive and selective polarographic NO sensors has greatly advanced investigations by allowing the continuous real-time measurement of the highly labile NO under physiological conditions. Ideally, an H_2S sensor appropriate for biological samples would report H_2S continuously, not be contaminated by numerous biological salts and compounds, and be sufficiently sensitive to detect H_2S at physiological levels. Our polarographic H_2S sensor (PHSS) has these characteristics. The PHSS anode, cathode, and electrolyte are protected from solution constituents by an H_2S-permeable polymer membrane so that only free H_2S is able to diffuse across the membrane and interact electrochemically with the appropriately polarized anode. The PHSS electrolyte consists of 0.05 M $K_3[Fe(CN)_6]$ in alkaline 0.5 M carbonate buffer pH 10 (Fig. 8.2). The electrochemical reaction is initiated as H_2S diffuses from the sample solution through the membrane and dissociates to HS^-. HS^- then reduces ferricyanide to ferrocyanide, which subsequently donates electrons to the anode, polarized at 100 to 200 mV, creating a current proportional to sample H_2S concentration. The PHSS background current resulting from electrolytic conduction of current from cathode to anode occurs as ferricyanide is reduced at the cathode and oxidized at the anode. Because the cathode potential, 0 mV, is not as negative as the equilibrium potential of HS^-, -270 mV, the ferricyanide reduction rate and hence background current is less than signal current resulting from sample H_2S. The background current, which is subtracted from signal current, dictates the PHSS lower limit of detection.

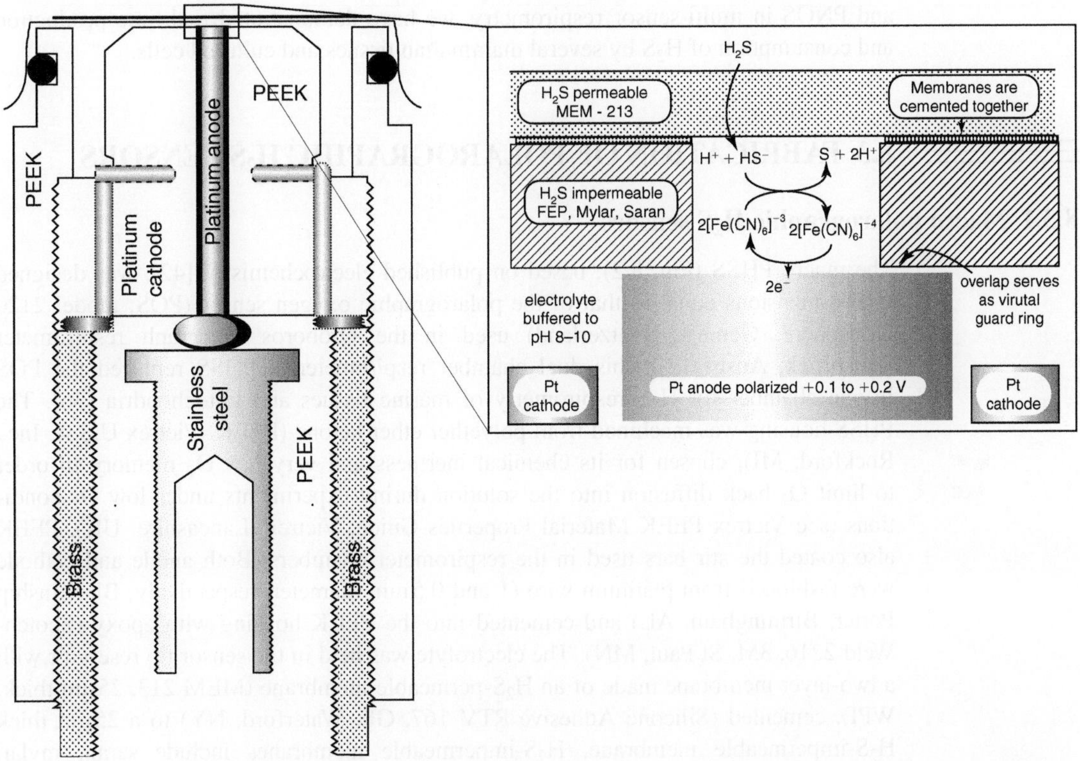

FIGURE 8.2 Diagram of the macro polarographic H_2S sensor (PHSS). A side view and expanded tip view of the macro PHSS, 25 mm long and 12 mm diameter, illustrates the component parts. The platinum anode and cathode are cemented into the polyether-ether ketone (PEEK) housing with epoxy. The membranes are cemented together with a silicone cement. The 0.5 mm hole in the H_2S-impermeable membrane is concentric with the anode and provides a small reservoir of electrolyte between the anode and the H_2S-permeable membrane. Lateral H_2S diffusion from this region at the tip into the bulk electrolyte is limited and greatly shortens response time as the impermeable membrane serves as a virtual guard ring. Electrochemical reactions within the electrolyte are described in the text (after [36]; reproduced with permission of the Company of Biologists).

8.2.2 Multi-sensor respirometry

The PHSS operates over a concentration range and with a response time ideal for kinetic studies of H_2S metabolism in broken cell systems, cell suspensions, intact tissues, and whole organisms. The PHSS selectivity enables it to be combined with other real-time polarographic sensors for O_2 (POS) and NO (PNOS) to determine the interaction of these factors with H_2S in biological systems. Because the existence of free H_2S may be in rapid equilibrium with thiol and sulfane sources and sinks through enzymatic and spontaneous reactions, only the PHSS can follow concentration changes in a physiological time frame and under physiological conditions. As a result, complex kinetic responses of tissues, cells, and mitochondria are observed as O_2 and NO levels are adjusted within physiological limits. Using the PHSS in combination with POS and PNOS in multi-sensor respirometry, we have demonstrated real-time production and consumption of H_2S by several mammalian tissues and cultured cells.

8.3 FABRICATION OF POLAROGRAPHIC H_2S SENSORS

8.3.1 Macro polarographic H_2S sensors

The macro PHSS (Fig. 8.2), based on published electrochemistry [42], was designed with dimensions equal to that of the polarographic oxygen sensor (POS; model 2120 Orbisphere, Geneva, Switzerland) used in the Oroboros Oxygraph respirometer (Innsbruck, Austria). In this dual chamber respirometer, a PHSS replaced the POS in one chamber for H_2S respirometry of marine tissues and mitochondria [36]. The PHSS housing was machined from polyether ether ketone (PEEK, Victrex U.S.A. Inc., Rockford, MI), chosen for its chemical inertness and very low O_2 memory in order to limit O_2 back diffusion into the solution during experiments under low O_2 conditions (see Victrex PEEK Material Properties Guide, Victrex, Lancashire, UK). PEEK also coated the stir bars used in the respirometer chambers. Both anode and cathode were fashioned from platinum wire (1 and 0.5 mm diameter, respectively; Blankinship Porter, Birmingham, AL) and cemented into the PEEK housing with epoxy (Scotch-Weld 2216, 3M, St Paul, MN). The electrolyte was held in the sensor tip reservoir with a two-layer membrane made of an H_2S-permeable membrane (MEM 213, 25 μm thick, WPI), cemented (Silicone Adhesive RTV 167, GE, Waterford, NY) to a 25 μm thick H_2S-impermeable membrane. H_2S-impermeable membranes include saran, mylar, perfuoroalkyl-tetrafluoroethylene copolymer (PFA), and fluorinated ethylene/propylene (FEP). In our experience, only silicone polymer membranes allow rapid diffusion of H_2S, whereas polyethylene and polypropylene permit only minor H_2S diffusion (see Table 1 in [36]). The 0.5 mm diameter hole in the impermeable membrane, concentrically located above the 1 mm platinum anode, was covered by the permeable membrane. In this configuration, the impermeable membrane ring served as a virtual guard ring, greatly accelerating the sensor response time by preventing the diffusing H_2S from equilibrating the entire electrolyte volume. The bilayered membrane was held onto the PHSS tip between an O-ring and an adaptor ring. H_2S-dependent changes in anode current were converted to proportional voltage with a modified POS meter, and the output voltage was recorded digitally (Virtual Bench, National Instruments, Austin, TX).

8.3.2 Miniature polarographic H_2S sensors

The miniature PHSS design was based on the original design of the macro PHSS [41]. The 2 mm diameter PHSS utilized a cylindrical polished-tip platinum anode that was electrically insulated along the side from a platinum cathode that was coiled around the core (Fig. 8.3). The core was housed in a 2 mm outside diameter stainless steel cylindrical sleeve that served as an electrical shield and to which a 25 μm thick H_2S-permeable membrane was fixed to cover the tip. The reservoir between the sleeve and core

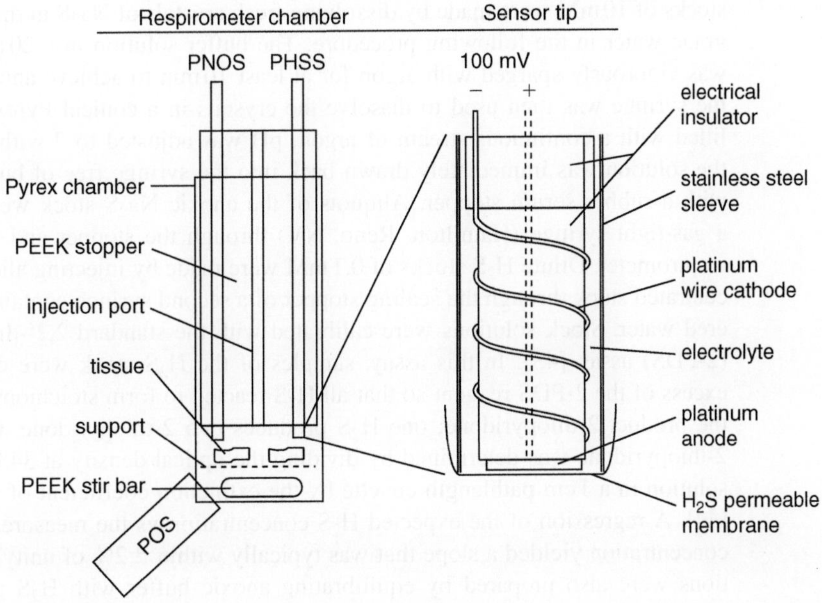

FIGURE 8.3 Diagram of the miniature PHSS and respirometer system. The respirometer chamber (left) is equipped with three sensors, the PHSS and PNOS in the chamber stopper and the POS diagonal to the chamber bottom, and has a minimum volume of 2 ml. Tissue is held on a support just above the stir bar for adequate perfusion. The miniature PHSS (right) is approximately 5 mm long and 2 mm outside diameter. The anode is polished at the tip, the cathode is coiled around the insulated anode, the H_2S-permeable membrane is cemented to the stainless steel sleeve, and the 30 μL electrolyte volume covers both anode and cathode (after [41]).

was filled with the ferricyanide electrolyte. The end of the sensor contained within the sleeve, approximately 50 mm long, was fitted into a port of a PEEK stopper designed for the oxygraph respirometer chamber. The stopper contained two sensor ports and an injection port. The PHSS polarizing voltage was set at 150 mV with a multichannel analyzer (Apollo 4000, WPI, Sarasota, FL) that also recorded sensor signals. A newer miniature PHSS model is currently available from World Precision Instruments (model ISO-H2S-2, Sarasota, FL).

8.4 CALIBRATION OF POLAROGRAPHIC H_2S SENSORS

8.4.1 H_2S stock solutions

The PHSS was calibrated with step-wise additions of H_2S stock while positioned in the respirometer chamber containing 3 ml stirred (500 rpm) 20 mM Tris-buffered zero-grade argon (<0.0003 kPa O_2) equilibrated analytical grade purified water (Solution 2000, Jasper, GA), at the pH and temperature of the subsequent experiment. H_2S

stocks of 10 mM were made by dissolving fresh crystals of Na_2S in the same analytical grade water in the following procedure. The buffer solution in a 20 ml Pyrex syringe was vigorously sparged with argon for at least 10 min to achieve anoxia. Liquid from the syringe was then used to dissolve the crystals in a conical Pyrex centrifuge tube filled with a continuous stream of argon, pH was adjusted to 7 with dilute HCl, and the solution was immediately drawn back into the syringe free of bubbles and sealed with a rubber serum stopper. Aliquots of the anoxic Na_2S stock were obtained with a gas-tight syringe (Hamilton, Reno, NV) through the stopper and injected into the respirometer. Dilute H_2S stocks of 0.1 mM were made by injecting aliquots of the concentrated stock through the sealing stopper of a second syringe containing anoxic buffered water. Stock solutions were calibrated with the standard 2,2′-dipyridyl disulfide (2-PDS) assay [43]. In this assay, samples of the H_2S stock were dissolved into an excess of the 2-PDS reagent so that all H_2S reacted to form stoichiometric amounts of the product 2-thiopyridone; one H_2S produces two 2-thiopyridone. Concentration of 2-thiopyridone was determined by dividing the optical density at 343 nm of the assay solution in a 1 cm pathlength cuvette by the extinction coefficient of 8.08 $mM^{-1} cm^{-1}$ [44]. A regression of the expected H_2S concentration vs the measured 2-thiopyridone concentration yielded a slope that was typically within $\pm 2\%$ of unity. H_2S stock solutions were also prepared by equilibrating anoxic buffer with H_2S gas (Mattheson), yielding concentrations near 140 mM H_2S. These stocks were used for very small volume injections with negligible effects on pH.

8.4.2 Chemical sources of H_2S

In studies in which NaHS is used as the sulfide source, effective concentrations are typically higher than with other sulfide sources [20]. Stock solutions prepared with yellow NaHS flakes are pale yellow in color compared to the clear solutions prepared with Na_2S or equilibrated with H_2S gas, suggesting the presence of elemental sulfur in NaHS solutions. The PHSS consistently produced a twofold higher signal in response to Na_2S injections compared to injections of the same concentration of NaHS [41], indicating that, although the yellow NaHS solution produced a linear signal, approximately one half of the sulfur is present in a form not able to diffuse through the H_2S-permeable membrane to chemically react with the ferricyanide electrolyte. Spectral features of the anoxic stock solutions, specifically a strong 380 nm absorbance band exhibited by the NaHS solution compared to little 380 nm absorption by the Na_2S solution, suggested the presence of scattering components such as polysulfides or elemental sulfur in NaHS solutions.

8.5 CHARACTERIZATION OF POLAROGRAPHIC H_2S SENSORS

PHSS performance characteristics are described below and compiled in Table 8.1. PHSS performance and calibration were evaluated in the temperature-controlled

TABLE 8.1
Performance characteristics of the miniature PHSS[a]

Sulfide species detected	Hydrogen sulfide (H_2S), dissolved or gaseous
Sensitivity	1 nA/μM H_2S near neutral pH
Detection limits[b]	10 nM (0.3 ppb) to 200 μM (6000 ppb) linear
Stability	<5 pA/h over 10 hours with continuous flow of sulfide sample
Accuracy	±2% of concentration determined by 2-PDS[c]
Precision	±2% with repeated 10 μM steps within linear range
Response time to >90% new sulfide level	10 s (stirring speed 500 rpm)
Sulfide consumption	<1 pmol s^{-1} at 10 μM, based on current
Operational pH range[d]	pH 8.5 and below
Interference by other compounds	None with: S_2O_3, SO_2, SO_4, cysteine, glutathione, ascorbate, O_2, NO, NO_2, NO_3, H_2O_2 HCN gas will cause minor increase in signal (pH sensitive)

[a] PHSS configured in respirometer with 2 ml phosphate buffered saline at 37ºC
[b] samples in pH 7
[c] standard colorimetric method [43]
[d] pH-dependent signal results from H_2S/HS^- pK of approximately 6.8 under the conditions tested

respirometer chamber containing 3 ml stirred (500 rpm) phosphate buffered saline (PBS; 10 mM sodium phosphate, 150 mM NaCl) at pH 7.3 and 37ºC, with 50 μM diethylenetriaminepentaacetic acid (DTPA) added to chelate metal contaminants that could otherwise catalyze H_2S oxidation [45].

Polarizing voltages for the H_2S microsensor and macrosensor have been given as +85 to 150 mV [41] and +100 to 200 mV [36], respectively. The miniature PHSS was tested for the signal strength of a 20 μM sulfide injection in anoxic PBS, pH 7.3 and 37ºC, as a function of polarizing voltage over a range of +30 to 300 mV. The representative polarogram shown in Fig. 8.4 exhibited a signal plateau from +100 to 200 mV, within which the PHSS signal current was relatively insensitive to changes in polarizing voltage as the ferricyanide to ferrocyanide ratio was maintained near unity. Polarizing potentials below or above this plateau region destabilized this equilibrium towards ferrocyanide or ferricyanide, respectively, so that even small changes in polarizing voltage further shifted the ratio, leading to marked increases in signal fluctuations. A polarizing voltage of 100 to 200 mV was insufficient for O_2 reduction, thus preventing the PHSS from responding to changes in solution O_2 content.

8.5.1 Selectivity

Calibration of the miniature PHSS performed in the respirometer chamber in combination with a POS and PNOS (model ISO-NOP, WPI, Sarasota, FL) is shown in Fig. 8.5. The selectivity of these sensors is clearly demonstrated, as the PHSS and

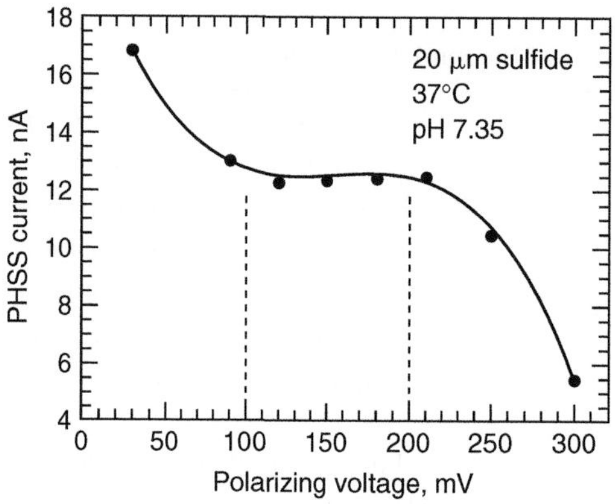

FIGURE 8.4 PHSS polarogram. A representative graph of PHSS current obtained for separate 20 μM sulfide samples as a function of polarizing voltage under anoxic conditions. Signal was stable between 100 and 200 mV as the ferricyanide to ferrocyanide ratio remained near unity (after [41]).

PNOS responded linearly to increasing concentrations of solution H_2S and NO, respectively, while the POS, unresponsive to either H_2S or NO, reported a stable O_2 concentration near 2 μM. In addition, the PHSS signal did not exhibit interference from O_2, NO (Fig. 8.5), millimolar concentrations of hydrogen peroxide, sulfite, sulfate, ascorbate, other sulfhydryl-bearing compounds such as L-cysteine, glutathione, homocysteine, or thiosulfate, or organosulfur compounds such as diallyl disulfide, diallyl trisulfide, methyl sulfide, or propyl disulfide (not shown). The addition of 1 mM KCN caused a small (relative to 10 μM H_2S) positive offset of the baseline signal (not shown), probably the result of HCN diffusion across the membrane increasing the electrolyte conductance, which can be subtracted. The addition of *Lucina pectinata* metHb I, which binds specifically to H_2S [46], resulted in an abrupt stoichiometric drop of the PHSS signal (Fig. 8.5). The rate of H_2S consumption by the PHSS, calculated from signal current at a 10 μM H_2S concentration using Faraday's constant, was in the subpicomole s^- range and was thus negligible compared to the biological consumption rate.

8.5.2 Sensitivity

Because the pK for the H_2S/HS couple is near 6.8, the PHSS amperometric signal for H_2S is highly dependent on solution pH. Under physiological conditions near neutral pH, small changes in pH can alter the PHSS signal. A pH titration of the PHSS signal shown in Fig. 8.6 indicates that the sensor signal increased from 24 to 440 nA for an 80 μM H_2S injection over the pH range of 8 to 6, demonstrating that H_2S becomes the predominant sulfide species at lowered solution pH. The pK for this protonation is approximately 6.8 under the conditions tested. The PHSS sensitivity at pH 6, where H_2S is approximately 90% of the total sulfide present, is 5.5 nA/μM H_2S. Although the PHSS signal is less at pH above the pK, the sensor is responsive and can be calibrated

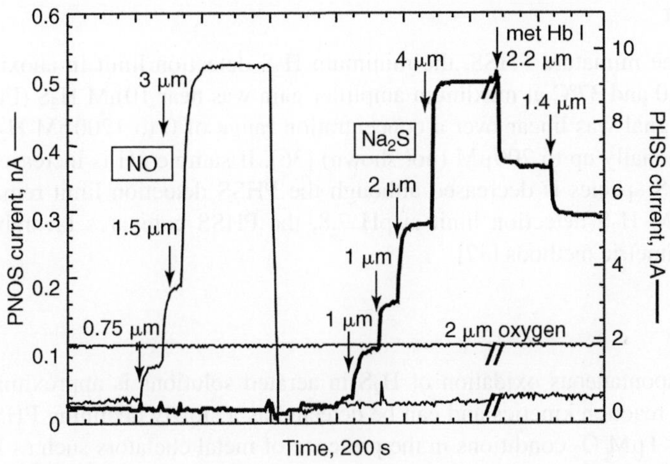

FIGURE 8.5 Multiple-sensor respirometry. Representative calibration traces of PNOS (thin line, left ordinate) and PHSS (thick line, right ordinate) operating simultaneously in PBS, pH 7.3 at 37°C, with 50μM DTPA in a closed chamber respirometer. After NO additions were made, the chamber solution was replaced with fresh buffer, to which Na_2S stock solutions were then injected in a stepwise manner. The stable POS signal shown at 2μM O_2 demonstrates that the POS does not respond to NO or H_2S. Injections of anoxic buffered NO and H_2S stocks are shown with concentrations at arrows, as are additions of *Lucina pectinata* ferric hemoglobin I (metHb I), which stoichiometrically binds to H_2S (after [41]).

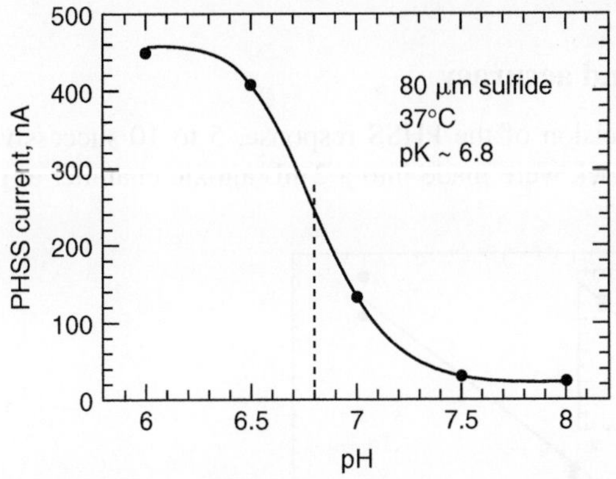

FIGURE 8.6 pH titration of PHSS signal. A representative graph of pH titration in PBS at 37°C demonstrates that the sulfide species detected is H_2S and that the pK for H_2S/HS^- is approximately 6.8 under the conditions tested (after [41]).

at pH 8. However, it is important to carefully control solution pH and to perform PHSS calibrations at the same pH as the experimental solution.

Solution H_2S can also be lowered by spontaneous oxidation catalyzed by a number of compounds in the presence of O_2. Accordingly, calibration solutions should contain DTPA and should be sparged with nitrogen or argon to remove O_2.

8.5.3 Detection limit

For the miniature PHSS, the minimum H_2S detection limit in anoxic 10 mM PBS at pH 7.0 and 37°C at maximum amplifier gain was near 10 nM H_2S (Fig. 8.7 inset), and the signal was linear over a concentration range of 0 to 1200 nM H_2S (Fig. 8.7), and additionally up to 200 μM (not shown) [36]. If sample pH is increased, the proportion of H_2S species is decreased although the PHSS detection limit remains constant. At 100 nM H_2S detection limit at pH 7.8, the PHSS compares favorably with standard colorimetric methods [37].

8.5.4 Stability

The spontaneous oxidation of H_2S in aerated solutions is approximated with second order reaction kinetics and can be detected as a slowly declining PHSS signal even in low <1 μM O_2 conditions in the presence of metal chelators such as DTPA. Therefore to determine PHSS signal drift, a flow-through system was designed to limit spontaneous H_2S oxidation. To do this, a syringe pump (model 22, Harvard Apparatus, Holliston, MA) delivered a concentrated H_2S stock from a Pyrex syringe at a constant rate to a 3 ml constant volume flow-through chamber. The H_2S stock solution was 7.5 mM Na_2S in 500 mM Tris-buffered anoxic analytical grade water, pH 7.0. Delivery rates were adjusted between 0.1 and 0.2 ml/h to create a steady-state H_2S concentration between 20 and 100 μM, determined by 2-PDS [43]. PHSS signal drift was measured at <5 pA/h for up to 10 hours, corresponding to <5 nM/h.

8.5.5 Reproducibility, precision and accuracy

To determine the precision of the PHSS response, 5 to 10 successive 2 μL injections of a 10 mM anoxic stock were made into a 2 mL anoxic chamber at pH 7.3 and 37°C,

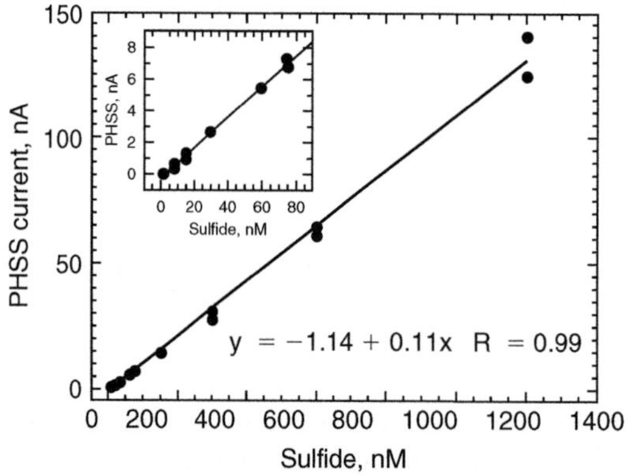

FIGURE 8.7 PHSS calibration. A representative PHSS calibration regression shows linearity of the signal current from 0 to 1200 nM H_2S. Inset: An expanded segment of the calibration trace shows the PHSS detection limit near 10 nM H_2S with a sensor that exhibited low background current at high amplifier gain (after [41]).

yielding 10 μM H_2S with each step. These stepwise injections produced a stepwise increase in current that differed by less than ±2% for each injection. Response accuracy was determined by periodically sampling the chamber solution and determining the H_2S concentration with 2-PDS [43]. Calibration of the PHSS was within ±1% of the 2-PDS assay.

8.5.6 Linearity and dynamic response range

From the lower detection limit near 10 nM to approximately 200 μM H_2S, the PHSS signal is best fit by a linear regression (see Fig. 8.7), with coefficient of regression typically >0.98. However, the amperometric signal loses linearity at higher H_2S concentrations. This behavior has also been observed in other micro polarographic sulfide sensors [42]. It is likely that the small electrolyte volume at the sensor tip becomes saturated at higher H_2S concentrations as the rate of H_2S diffusion becomes more rapid than the rate at which ferrocyanide is reoxidized at the platinum anode.

8.5.7 Response time

In response to step H_2S changes by addition or removal, the PHSS typically takes less than 10 s to reach 90% of the new signal (see Fig. 8.5), in agreement with previous experiments performed with the macro PHSS at 20ºC [36]. With a freshly prepared PHSS, we often observe that the first H_2S injection to the chamber results in a somewhat slower response time, approximately 30 s, to reach 90% of the new signal. This slower response is not observed with subsequent injections, suggesting that for a newly prepared PHSS, initial electrolyte conditioning occurs upon first exposure to H_2S.

8.5.8 Reliability (maintenance-free working time)

The time frame of PHSS use prior to requiring new electrolyte and membrane replacement depends on solution type and H_2S concentrations to which the PHSS has been exposed. Generally, using the PHSS on a daily basis with physiological buffer solutions and H_2S concentrations less than about 50 μM provides about 2 to 3 weeks of service, after which the membrane and electrolyte require replacement to refurbish the PHSS. A loss in H_2S sensitivity and an elevated background current compared to earlier calibrations are symptoms of needed refurbishment. Performance decline can be hastened if higher H_2S concentrations are routinely used since elemental sulfur will more rapidly accumulate in the electrolyte. Likewise if experimental solutions promote rapid bacterial growth coating chamber surfaces as well as the PHSS membrane with a biofilm, the refurbishment time frame will be shortened.

8.5.9 Biocompatibility

Because the PHSS electrochemical components are separated from sample solution by the H_2S-permeable polymer membrane which is essentially inert with respect to biological components, we observe no indications of altered biological processes such as oxygen consumption rates, signaling events or enzymatic activities, or loss of viability of isolated organelles, cells or tissues. However, if biological samples contain lipid suspensions, the polymer membrane may accumulate a fine lipid film layer that may lengthen response time.

8.6 APPLICATIONS OF POLAROGRAPHIC H_2S SENSORS IN BIOLOGICAL SAMPLES

8.6.1 Measurement of H_2S production

8.6.1.1 Tissue homogenates

H_2S production in rat aorta homogenate was demonstrated by adding the substrate L-cysteine (L-cys) and the cofactor pyridoxal phosphate (PLP) for the enzymes cystathionine β•synthase (CBS) and cystathionine γ lyase (CGL) to dilute homogenate supernatant in the respirometer chamber at $<5\,\mu M$ O_2, and recording the rise in PHSS signal as a function of time [5] (Fig. 8.8). In all experiments, the PHSS was calibrated prior to and after the experiment by adding either an internal Na_2S stock standard to check calibration or a known concentration of *Lucina pectinata* metHb I [46] to scavenge H_2S. H_2S production rate was determined at the initial steepest slope of each trace. In all experiments, H_2S was not produced in the absence of L-cysteine and PLP, nor was it produced by L-cysteine and PLP in solution alone or by heat-denatured supernatants or heat-killed cells following addition of substrate (not shown). The CGL inhibitor β-cyano-L-alanine (BCA) at 10 mM was effective in inhibiting aorta H_2S production [19, 23].

8.6.1.2 Cultured and isolated cells

Cultured rat vascular smooth muscle cells (VSMCs), grown and prepared for respirometry as described in Doeller *et al.*, 2005 [41], were injected into the respirometer chamber to a concentration of between 10^5 and 10^6 cells ml^{-1}. Cell viability remained at >90% throughout experiments. Near $4\,\mu M$ O_2, H_2S production was stimulated by the addition of L-cysteine and PLP (Fig. 8.8). The initial H_2S production rate was approximately 20% of the rat aorta homogenate rate. H_2S production rate decreased after the initial rise in H_2S concentration, perhaps the result of product feedback inhibition. The addition of the CGL inhibitor BCA showed an effect similar to aorta homogenate.

Isolated primary rat hepatocytes in the respirometer chamber at 10^6 cells ml^{-1} exhibited robust H_2S production, achieving concentrations of $>20\,\mu M$ in 30 min as the O_2 concentration decreased and in the presence of L-cysteine and PLP (Kraus *et al.*, preliminary data, not shown). Inhibitors of both CBS and CGL (amino oxyacetic acid and propargylglycine, respectively) inhibited H_2S production, indicating the presence of both enzymes in hepatocytes.

8.6.1.3 Intact tissues and organs

Intact rat aorta, combined fresh weight average of 24 mg, in the respirometer chamber at approximately 5 μM O_2 exhibited H_2S production when supplied with PLP and L-cysteine or homocysteine (Fig. 8.8). Aorta H_2S production typically exhibited a pronounced plateau that could be sustained for >30 min, suggesting the establishment of a dynamic steady state between H_2S production and consumption at approximately 1 to 2 μM H_2S. The addition of 1 μM *Lucina pectinata* metHb I to the respirometer chamber caused the PHSS signal to abruptly drop, again indicating signal specificity to H_2S (Fig. 8.8).

8.6.2 Measurement of H_2S consumption

8.6.2.1 Isolated mussel gill mitochondria

H_2S consumption was exhibited by mitochondria isolated from the gills of the marine ribbed mussel *Geukensia demissa*, an inhabitant of sulfide-rich intertidal sediment, in an

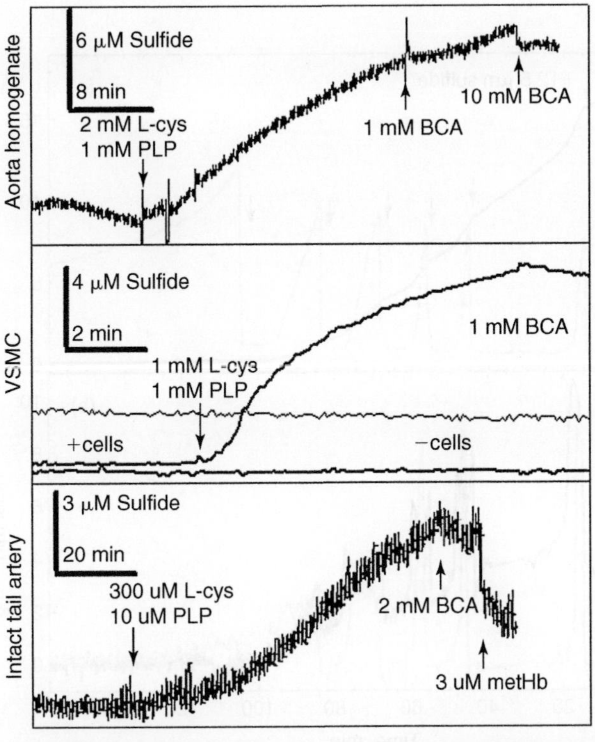

FIGURE 8.8 H_2S production in vascular tissues. H_2S production by aorta homogenate (upper panel), cultured rat vascular smooth muscle cells (VSMCs; middle panel), and intact rat aorta occurs after the addition of substrate L-cysteine (L-cys) and cofactor pyridoxal L-phosphate (PLP) for the enzyme CGL located in vascular tissue. H_2S production is inhibited after the CGL. β cyano-L-alanine (BCA) is added. Ferric *Lucina pectinata* hemoglobin I (metHb) is added to confirm H_2S production. The quantity of metHb-sulfide produced, determined spectrophotometrically, matched the levels of H_2S detected by the PHSS (after [41]).

O_2-dependent manner but also in the absence of O_2 (Fig. 8.9), leading to the hypothesis that while this process may prevent H_2S toxicity of cytochrome oxidase, it may also provide electrons and protons for oxidative phosphorylation, thus perhaps sparing carbohydrates for other use [5]. Under air-equilibrated conditions, mitochondrial H_2S consumption was accompanied by simultaneous increased O_2 consumption, leading some investigators to use H_2S-stimulated O_2 consumption as an indirect and largely qualitative measure of H_2S consumption. Using the PHSS to measure H_2S consumption directly in combination with O_2 consumption provided a dynamic metabolic stoichiometry of H_2S to O_2 consumption over a range of O_2 levels as well as demonstrated that H_2S can be consumed in the absence of O_2. This detailed assessment of mitochondrial H_2S metabolism provided evidence of specific enzymatic steps and insight into mammalian H_2S metabolism.

8.6.2.2 Cultured cells, intact tissues and organs

To determine cellular H_2S consumption rates, bolus additions of known H_2S concentrations were added to the respirometer chamber with or without rat aorta smooth muscle

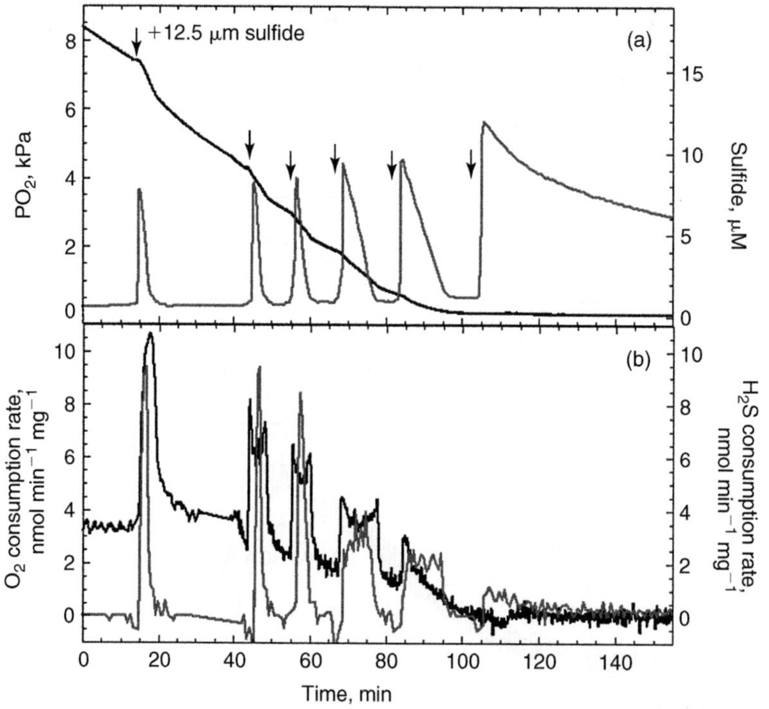

FIGURE 8.9 Mitochondrial O_2 and H_2S consumption from non-limiting O_2 to anoxic conditions. (a) Isolated mitochondria were exposed to repeated bouts of 12.5μM H_2S until anoxia was achieved. (b) At higher O_2 levels, both O_2 and H_2S consumption events are coincident, but as the O_2 levels decline the events become uncoupled and O_2 consumption is limited first. The multiphasic kinetics of O_2 consumption may result from transient inhibition of cytochrome *c* oxidase by H_2S. Under anoxia, H_2S consumption continues at a low level (after [36]; reproduced with permission of the Company of Biologists).

cells (RASMCs) or with heat-killed cells (Fig. 8.10). Cellular H_2S consumption rates were calculated from the slope of the PHSS signal between H_2S injections, minus the background rate without cells, and divided by mg protein ml^{-1} over the 5–200 μM H_2S range (Fig. 8.10). Aliquots of heat-killed cells were used to determine the non-enzyme-catalyzed H_2S oxidation rate associated with cellular constituents. Cellular H_2S consumption rate may rely in part on H_2S-sensitive mitochondrial respiration and accordingly may exhibit saturation kinetics as a function of H_2S concentration. In contrast, spontaneous H_2S oxidation, which can be catalyzed by several inorganic and organic trace contaminants in solution [45], exhibits second order kinetics as a function of both H_2S and O_2 concentrations, and will become more rapid as O_2 concentration increases [36]. Accordingly, to limit spontaneous H_2S oxidation while maintaining mitochondrial operation, these experiments were performed at 5 to 15 μM O_2. Consumption rates were dependent on H_2S concentration but exhibited saturation kinetics. At low H_2S concentrations, cellular consumption rates were approximately tenfold greater than spontaneous oxidation rates in the absence of cells or in the presence of heat-killed cells (Fig. 8.10). However, at higher H_2S concentrations, the spontaneous H_2S oxidation rate approximated the measured rate with cells, indicating that H_2S consumption was inhibited at higher H_2S levels.

H_2S is both produced and consumed, and consumption pathways may be coupled to mitochondrial O_2 consumption. However, H_2S is also toxic to mitochondria, indicating that the steady-state cellular level is under tight control. To determine the H_2S

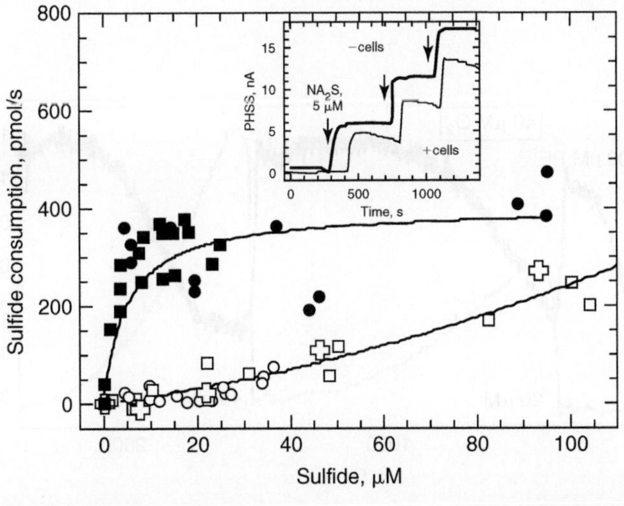

FIGURE 8.10 H_2S consumption in rat aorta smooth muscle cells (RASMCs). Accumulated data from several experiments showing RASMC H_2S consumption rates (filled circles and squares) as a function of H_2S concentration, compared to H_2S oxidation rates in solution without cells (open circles and squares). Heat-inactivated RASMC H_2S consumption rates (open plus symbols) were equivalent to background rates without cells. Inset: Representative PHSS traces showing stepwise additions of Na_2S stock, at arrows, in the presence (thin line) and absence (thick line) of RASMCs (after [41]).

concentration range over which isolated intact aorta consume H_2S without exhibiting inhibition of either O_2 or H_2S consumption, aorta segments stimulated with phenylephrine to about 50–75% of maximum vasoconstriction to contract were held in the respirometer at 40 to 50 μM O_2, above the limiting O_2 concentration (called critical O_2), and H_2S was added to titrate both rates (Kraus *et al.*, preliminary data, not shown). H_2S stimulated both O_2 and H_2S consumption until rates became inhibited at higher H_2S levels, as was observed with RASMCs. The H_2S concentration that stimulated maximal consumption rates, approximately 20 μM, is also the concentration that elicits maximal vasorelaxation of the precontracted aorta (see Fig. 8.11), demonstrating that H_2S-mediated vasorelaxation is not the result of inhibited mitochondrial respiration.

8.6.3 Simultaneous measurement of H_2S level and vessel tension

Demonstration of the rapid relaxation and contraction response of an intact blood vessel to simulated changes in vascular H_2S levels was made possible by placing the PHSS in the organ bath of a vessel bioassay system (Radnoti, Monrovia, CA). At physiological O_2 concentrations, rat aortic vessel tension responded to H_2S in a concentration-dependent manner, and the magnitude of these responses was highly O_2 dependent (Fig. 8.11). Moreover, with the use of chemical inhibitors of specific signaling pathways in the organ bath, it was also possible to demonstrate that H_2S-mediated changes in vascular tone resulted from multiple signaling pathways operating at different rates. Thus, the ability to follow and dissect the complex vascular effects of H_2S is only possible with a real-time PHSS.

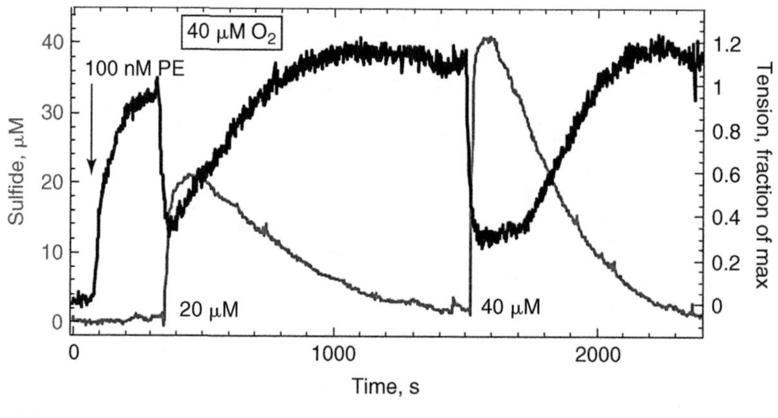

FIGURE 8.11 H_2S-mediated vasorelaxation. Rat aorta segments suspended in an organ bath containing the miniature PHSS and equilibrated with 40 μM O_2 are stimulated to constrict with 100 nM phenylephrine (PE). Subsequent addition of H_2S causes an immediate relaxation event that gradually recovers as the H_2S is oxidized or removed by the gas perfusion stream. Repeated additions of H_2S at physiologically relevant concentrations demonstrate a predictable kinetic response.

8.6.4 Measurement of steady-state H_2S levels in blood and tissue

A major challenge to understanding the multiple roles that H_2S plays in normal cellular processes has been to accurately define physiological vs pathological H_2S levels and the kinetics of H_2S adjustments. Since any single-point H_2S measurement is only a snapshot of a dynamic steady state, the optimal method for this would be to use the PHSS as an indwelling catheter. This approach is currently in development. An alternative method also in development places the PHSS in a flow-through micro cell to sample μL volumes of blood in <1 min sampling time. Integrating the H_2S peak would provide an accurate measure of the total free H_2S in the sample. Although rodent model blood H_2S levels are often reported in the 40 to 100 μM range as determined with standard colorimetric assays which may reflect the inclusion of other sulfide sources, we find blood H_2S levels much lower, typically less than 10 μM at pH 7.3 and 37ºC (Kraus *et al.*, preliminary data).

8.7 CONCLUDING REMARKS AND FUTURE DIRECTIONS

The PHSS method of real-time H_2S measurement allows for investigating the potentially complex H_2S kinetic responses of organs, tissues, cells, and mitochondria as levels of O_2 and NO as well as metabolic state are adjusted within physiological limits. Kinetic changes in H_2S concentration continuously reported by the PHSS, which are not seen with other H_2S measurement techniques, suggest potentially complex interactions of H_2S production and consumption mechanisms. H_2S may likely exist as a cellular pool of free and labile persulfides able to rapidly respond to redox challenges with production and consumption pathways that operate to maintain the pool. This possible scenario reinforces the need for the PHSS as a valuable tool to provide a continual report of H_2S throughout the course of an experimental treatment or to accurately determine H_2S levels *in situ*.

The PHSS design is continually subject to change based on measurement needs. For example, miniaturization will allow the PHSS to function as an indwelling catheter for continuous *in-situ* measurement of H_2S in blood and tissues. At the other end of the spectrum, for continuous measurement of H_2S at hydrothermal vent communities during ocean floor investigations, the PHSS will need modifications to withstand pressure. As our understanding of H_2S biology and physiology expands, the demand for different geometries and applications of the PHSS will increase.

8.8 ACKNOWLEDGMENTS

This research is supported in part by the American Heart Association grant-in-aid 0455296B and NIGMS grant 08-RGM073049A to DWK and WPI R&D research fund for XZ.

8.9 REFERENCES

1. P. Nicholls, *Biochem. Soc. Trans.* **3**, 316 (1975).
2. National Research Council. Committee on Medical and Biological Effects of Environmental Pollutants: Subcommittee on Hydrogen Sulfide: *Hydrogen Sulfide*. University Park Press, Baltimore, MD (1979).
3. D.P. Kelly, J.K. Shergill, W.-P. Lu, and A.P. Wood, Oxidative metabolism of inorganic sulfur compounds by bacteria. *Antonie van Leeuwenhoek* **71**, 95–107 (1997).
4. M.K. Grieshaber and S. Volkel, Animal adaptations for tolerance and exploitation of poisonous sulfide. *Ann. Rev. Physiol.* **60**, 33–53 (1998).
5. J.E. Doeller, M.K. Grieshaber, and D.W. Kraus, Chemolithoheterotrophy in a metazoan tissue: thiosulfate production matches ATP demand in ciliated mussel gills. *J. Exp. Biol.* **204**, 3755–3764 (2001).
6. D. Julian, J.L. Statile, S.E. Wohlgemuth, and A.J. Arp, Enzymatic hydrogen sulfide production in marine invertebrate tissues. *Comp. Biochem. Physiol. A Mol. Integr. Physiol.* **133**, 105–115 (2002).
7. L.R. Goodwin, D. Francom, F.P. Dieken, J.D. Taylor, M.W. Warenycia, R.J. Reiffenstein, and G. Dowling, Determination of sulfide in brain tissue by gas dialysis/ion chromatography: postmortem studies and two case reports. *J. Anal. Toxicol.* **13**, 105–109 (1989).
8. M.W. Warenycia, L.R. Goodwin, C.G. Benishin, R.J. Reiffenstein, D.M. Francom, J.D. Taylor, and F.P. Dieken, Acute hydrogen sulfide poisoning: demonstration of selective uptake of sulfide by the brainstem by measurement of brain sulfide levels. *Biochem. Pharmacol.* **38**, 973–981 (1989).
9. Y. Ogasawara, K. Ishii, T. Togawa, and S. Tanabe, Determination of sulfur in serum by gas dialysis/high performance liquid chromatography. *Anal. Biochem.* **215**, 73–78 (1993).
10. C. Dello Russo, G. Tringali, E. Ragazzoni, N. Maggiano, E. Menini, M. Vairano, P. Preziosi, and P. Navarra, Evidence that hydrogen sulphide can modulate hypothalamo-pituitary-adrenal axis function: in vitro and in vivo studies in the rat. *J. Neuroendocrinol.* **12**, 225–233 (2000).
11. P. Kamoun, Endogenous production of hydrogen sulfide in mammals. *Amino Acids* **26**, 243–254 (2004).
12. M. Swaroop, K. Bradley, T. Ohura, T. Tahara, M.D. Roper, L.E. Rosenberg, and J.P. Kraus, Rat cystathione beta-synthase: gene organization and alternative splicing. *J. Biol. Chem.* **267**, 11455–11461 (1992).
13. F.W. Alexander, E. Sandmeier, P.K. Mehta, and P. Christen, Evolutionary relationships among pyridoxal-5′-phosphate-dependent enzymes: regio-specific alpha, beta, and gamma families. *Eur. J. Biochem.* **219**, 953–960 (1994).
14. M. Meier, M. Janosik, V. Kery, J.P. Kraus, and P. Burkhard, Structure of human cystathione beta-synthase: a unique pyridozal 5′-phosphate-dependent heme protein. *EMBO J.* **20**, 3910–3916 (2001).
15. J.G. Vande Weghe and D.W. Ow, A fission yeast gene for mitochondrial sulfide oxidation. *J. Biol. Chem.* **274**, 13250–13257 (1999).
16. F.J. Stewart, I.L.G. Newton, and C.M. Cavanaugh, Chemosynthetic endosymbioses: adaptations to oxic-anoxic interfaces. *TRENDS Microbiol.* **13**, 439–448 (2005).
17. C.W. Leffler, H. Parfenova, J.H. Jaggar, and R. Wang, Carbon monoxide and hydrogen sulfide: gaseous messengers in cerebrovascular circulation. *J. Appl. Physiol.* **100**, 1065–1076 (2006).
18. P.K. Moore, M. Bhatia, and S. Moochhala, Hydrogen sulfide: from the smell of the past to the mediator of the future? *TRENDS Pharma. Sci.* **24**, 609–611 (2003).
19. R. Hosoki, N. Matsuki, and H. Kimura, The possible role of hydrogen sulfide as an endogenous smooth muscle relaxant in synergy with nitric oxide. *Biochem. Biophys. Res. Commun.* **237**, 527–531 (1997).
20. W. Zhao, J. Zhang, Y. Lu, and R. Wang, The vasorelaxant effect of H_2S as a novel endogenous gaseous K(ATP) channel opener. *EMBO J.* **20**, 6008–6016 (2001).
21. W. Zhao and R. Wang, H_2S-induced vasorelaxation and underlying cellular and molecular mechanisms. *Am. J. Physiol. Heart Circ. Physiol.* **283**, H474–H480 (2002).
22. H. Kimura, Y. Nagai, K. Umemure, and Y. Kimura, Physiological roles of hydrogen sulfide: synaptic modulation, neuroprotection, and smooth muscle relaxation. *Antioxidants Redox Signaling* **7**, 795–803 (2005).
23. K. Abe and H. Kimura, The possible role of hydrogen sulfide as an endogenous neuromodulator. *J. Neurosci.* **16**, 1066–1071 (1996).
24. H. Kimura, Hydrogen sulfide induces cyclic AMP and modulates the NMDA receptor. *Biochem. Biophys. Res. Commun.* **267**, 129–133 (2000).

25. A.A. Khan, M.M. Schuler, and R.W. Coppock, Inhibitory effects of various sulfur compounds on the activity of bovine erythrocyte enzymes. *J. Toxicol. Environ. Health* **22**, 481–490 (1987).
26. A.A. Khan, M.M. Schuler, M.G. Prior, S. Yong, R.W. Coppock, L.Z. Florence, and L.E. Lillie, Effects of hydrogen sulfide exposure on lung mitochondrial respiratory chain enzymes in rats. *Toxicol. Appl. Pharmacol.* **103**, 482–490 (1990).
27. R.A. Nicholson, S.H. Roth, A. Zhang, J. Zheng, J. Brookes, B. Skrajny, and R. Bennington, Inhibition of respiratory and bioenergetic mechanisms by hydrogen sulfide in mammalian brain. *J. Toxicol. Environ. Health* 54, 491–507 (1998).
28. E. Blackstone, M. Morrison, and M.B. Roth, H_2S induces a suspended animation-like state in mice. *Science* **308**, 518 (2005).
29. M. Nordstrom and T. Kjellstrom, Age dependency of cystathione beta-synthase activity in human fibroblasts in homocysteinemia and atherosclerotic vascular disease. *Atherosclerosis* **94**, 213–221 (1992).
30. S. Yla-Herttuala, J. Luoma, T. Nikkari, and T. Kivimaki, Down's syndrome and atherosclerosis. *Atherosclerosis* **76**, 269–272 (1989).
31. K. Eto, T. Asada, K. Arima, T. Makifuchi, and H. Kimura, Brain hydrogen sulfide is severely decreased in Alzheimer's disease. *Biochem. Biophys. Res. Commun.* **293**, 1485–1488 (2002).
32. M. Yusuf, B.T. Kwong Huat, A. Hsu, M. Whiteman, M. Bhatia, and P.K. Moore, Streptozotocin-induced diabetes in the rat is associated with enhanced tissue hydrogen sulfide biosynthesis. *Biochem. Biophys. Res. Comm.* **333**, 1146–1152 (2005).
33. J.C. Koster, M.A. Permutt, and C.G. Nichols, Diabetes and insulin secretion: the ATP-sensitive K^+ channel (K ATP) connection. *Diabetes* **54**, 3065–3072 (2006).
34. P. Kamoun, M.C. Belardinelli, A. Chabli, K. Lallouchi, and B. Chadefaux-Vekemans, Endogenous hydrogen sulfide overproduction in Down syndrome. *Am. J. Med. Genet. A* **116**, 310–311 (2003).
35. S. Taoka, S. Ohja, X. Shan, W. D. Kruger, and R. Banerjee, Evidence for heme-mediated redox regulation of human cystathionine beta-synthase activity. *J. Biol. Chem.* **273**, 25179–25184 (1998).
36. D.W. Kraus and J.E. Doeller, Sulfide consumption by mussel gill mitochondria is not strictly tied to oxygen reduction: measurements using a novel polarographic sulfide sensor. *J. Exp. Biol.* **207**, 3667–3679 (2004).
37. T. Ubuka, Assay methods and biological roles of labile sulfur in animal tissues. *B. Analyt. Technol. Biomed. Life Sci.* **781**, 227–249 (2002).
38. K. Eto and H. Kimura, The production of hydrogen sulfide is regulated by testosterone and *S*-adenosyl-L-methionine in mouse brain. *J. Neurochem.* **83**, 80–86 (2002).
39. D.G. Searcy and M.A. Peterson, Hydrogen sulfide consumption measured at low steady state concentrations using a sulfidostat. *Anal. Biochem.* **324**, 269–275 (2004).
40. S.U. Khan, G.F. Morris, and M. Hidiroglou, Rapid estimation of sulfide in rumen and blood with a sulfide-specific ion electrode. *Microchem. J.* **25**, 388–395 (1980).
41. J.E. Doeller, T.S. Isbell, G. Benavides, J. Koenitzer, H. Patel, R.P. Patel, J.R. Lancaster Jr, V.M. Darley-Usmar, and D.W. Kraus, Polarographic measurement of hydrogen sulfide production and consumption by mammalian tissues. *Anal. Biochem.* **341**, 40–51 (2005).
42. P. Jeroschewski, C. Steuckart, and M. Kuhl, An amperometric microsensor for the determination of H_2S in aquatic environments. *Anal. Chem.* **68**, 4351–4357 (1996).
43. A. Svenson, A rapid and sensitive spectrophotometric method for determination of hydrogen sulfide with 2,2′-dipyridyl disulfide. *Anal. Biochem.* **107**, 51–55 (1980).
44. P.E. Jensen, J.D. Reid, and C.N. Hunter, Modification of cysteine residues in the ChlI and ChlH subunits of magnesium chelatase results in enzyme inactivation. *Biochem. J.* **352**, 435–441 (2000).
45. K.Y. Chen and J.C. Morris, Oxidation of sulfide by O_2: catalysis and inhibition. *J. Sanit. Eng. Div. Amer. Soc. Civil Eng.* **98**, 215–227 (1972).
46. D.W. Kraus and J.B. Wittenberg, Hemoglobins of the *Lucina pectinata* bacteria symbiosis: I. Molecular properties, kinetics, and equilibria of reactions with ligands. *J. Biol. Chem.* **265**, 16 043–16 053 (1990).

CHAPTER 9

Aspects of recent development of immunosensors

Hua Wang, Guoli Shen, and Ruqin Yu

9.1 INTRODUCTION

9.1.1 General working principle of immunosensors

Immunosensors are affinity ligand-based biosensing devices that involve the coupling of immunochemical reactions to appropriate transducers. In recent decades, immunosensors have received rapid development and wide applications with various detection formats [1–2]. The general working principle of the immunosensors is based on the fact that the specific immunochemical recognition of antibodies (antigens) immobilized on a transducer to antigens (antibodies) in the sample media can produce analytical signals dynamically varying with the concentrations of analytes of interest. Here, the highly specific reaction between the variable regions of an antibody and the epitopes of an antigen involves different types of bonding, basically hydrophobic and electrostatic interactions, van der Waals force, and hydrogen bonding. The antigen–antibody reaction is reversible and, owing to the relative weakness of the forces holding the antibody and antigen together, the complex formed would dissociate in dependence upon the reaction environment (e.g. pH and ion strength). The strength of the binding of an antibody to an antigen could be characterized by its affinity constant (*K*), which is of the order between 5×10^4 and 1×10^{12} L mol^{-1}. The high affinity and specificity of this antigen–antibody binding reaction defines the unique immunosensor characteristics.

The general immunosensor design consists of three individual parts in close contact: a biological recognition element, a physicochemical transducer, and an electronic part. Antibodies or antibody derivatives (antigens or haptens) usually serve as the biological recognition elements, which are either integrated within or intimately associated with a physicochemical transducer. This recognition reaction defines the high selectivity and sensitivity of the transducer device. The electronic part is used to amplify and digitalize the physicochemical output signal from the transducer devices such as electrochemical (potentiometric, conductometric, capacitative, impedance, amperometric), optical (fluorescence, luminescence, refractive index), and microgravimetric devices. Gizeli and Lowe [3] suggested that an ideal immunosensor design should possess the following specifications: the ability to detect and quantify the antigens (antibodies), the capacity to transform the binding event without externally added reagents, the ability to repeat the measurement on the same device, and the capacity to detect the specific binding of the antigens (antibodies) in real samples. All of these specifications have been the main issues to pursue in developing immunosensors applied in various fields.

9.1.2 Main performance characteristics of immunosensors in clinical analysis

As an important branch of immunoassay techniques, immunosensors possess all essential performance characteristics of immunoassays. They show high selectivity, sensitivity, reversibility and efficient reagent usage. At the same time, the immunosensors are generally simple to operate, and easy to realize automation, digitization, and miniaturization. They may bypass some inherent problems of traditional analytical methods. Therefore, immunosensors have been the subject of expanding interest in the immunochemical studies with enormous potential in clinical diagnosis [1–2, 4], environmental analysis [5–6], and biological process monitoring [7]. As for the medical diagnosis of some diseases, herein considerable efforts have been devoted to the development of precise, rapid, sensitive, and selective immunosensors by measurement of the markers or pathogenic microorganisms responsible for the diseases, such as proteins, enzymes, viruses, bacteria, and hormones [1, 8–9]. Chagas' disease, an American trypanosomiasis caused by the hemoflagellate *Trypanosoma cruzi*, is an example. An amperometric immunosensor has been recently proposed to probe the presence of antibodies against *T. cruzi* in blood donors, and to follow the antibody decay during treatment of chagasic patients with the available drugs [10]. Yuan *et al.* reported a novel potentiometric immunosensor for detection of hepatitis B surface antigen by immobilizing hepatitis B surface antibody on a platinum electrode [11]. A piezoelectric immunosensor was developed for the on-line detection of severe acute respiratory syndrome (SARS)-associated coronavirus (SARS-CoV) in sputum in the gas phase. Compared to other SARS detection techniques, this method can rapidly test SARS-CoV at low cost [12]. Moreover, the determination of some tumor markers plays an important role in diagnosing, screening, and determining the prognosis of a cancer disease. Such tumor markers to be detected are often found in abnormally high amounts in the blood, urine, or tissue of patients with certain types of cancers. The examples include carcinoembryonic antigen (CEA), carbohydrate antigen 19-9 (CA19-9), carcinoma antigen 125 (CA125), alpha-fetoprotein (AFP), prostate specific antigen (PSA), CA15-3 and human chorionic gonadotropin (HCG) [13–15]. Wilson proposed an electrochemical immunosensor for the simultaneous detection of two tumor markers of CEA and AFP [15]. An increasing number of immunosensors have been utilized to analyze a series of biochemical targets for diagnosing infectious diseases, although there are still problems concerning the assay of analytes in real sample matrixes [1].

9.2 IMMOBILIZATION OF IMMUNOACTIVE ELEMENTS

Since immunosensors usually measure the signals resulting from the specific immunoreactions between the analytes and the antibodies or antigens immobilized, it is clear that the immobilization procedures of the antibodies (antigens) on the surfaces of base transducers should play an important role in the construction of immunosensors. Numerous immobilization procedures have been employed for diverse immunosensors, such as electrostatic adsorption, entrapment, cross-linking, and covalent bonding procedures. They may be appropriately divided into two kinds of non-covalent interaction-based and covalent interaction-based immobilization procedures.

9.2.1 Non-covalent interaction-based immobilization procedures

This type of immobilization of immunoactive entities is based on the non-covalent interactions between the antibody or antigen molecules and the transducer substrates, and usually refers to hydrophobic interaction, electrostatic interaction, van der Waals force, and hydrogen bonding. One notices that besides pure physical adsorption, some weak chemical interactions are also involved here. The non-covalent interactions may vary from the different substrates of transducers. For a non-polarity sensing substrate, the antibody or antigen molecules can be adsorbed through the hydrophobic interaction and van der Waals force. Wenmeyer *et al.* attached anti-digoxin antibodies at the surfaces of polystyrene microtubes by direct adsorption interaction, achieving the determination of digoxin with a detection limit of 50 pg mL^{-1} [16]. While for the charged substrates, the non-covalent interactions are mainly associated with the electrostatic interactions. The most typical layer-by-layer technique of self-assembly has attracted considerable attention in biomolecular immobilizations [17–21]. Caruso and coworkers assembled polyallylamine hydrochloride/polystyrene sulfonate layers on the self-assembled monolayer of mercaptopropionic acid, providing a charged polyelectrolyte layer on the transducer surface [19]. The biomolecules of avidin and anti-immunoglobulin (IgG) antibodies were then well immobilized through electrostatic interaction. A novel biosensing interfacial design strategy has been developed for immobilizing the antibodies onto the positively charged surfaces of plasma-polymerized film (PPF) via electrostatic interaction through a polyelectrolyte-mediated layer [20]. The immunosensors so prepared exhibited excellent response sensitivity due to the low disturbance of the electrostatic adsorption immobilization to the activity of antibody. The PPF surfaces can be regenerated repetitively by changing the pH of the buffer solutions to remove the polyelectrolyte-mediated layer.

Moreover, antibodies or antigens may be physically entrapped into the films of organic high polymers or inorganic materials (e.g. sol-gel, graphite powder) with stereo meshy structures. Of these entrapment immobilizations, the sol-gel-based immobilizations have recently attracted much attention due to their ability to encapsulate biomolecules at low temperature, as well as the physical tenability, optical transparency, mechanical rigidity, and low chemical reactivity [22–23]. Most applications of the sol-gel-based immobilizations have been primarily directed to the optical immunosensors [14, 22, 24] and the electrochemical immunosensors [23, 25–28]. Martínez-Fàbregas *et al.* proposed a polishable entrapment immobilization based on rigid biocomposite materials consisting of graphite powder, rabbit IgG, and methacrylate (or epoxy resins) [28]. The surface of the immunosensor can be regenerated by simply polishing to obtain a fresh layer of immunocomposite ready for next immunoassay. The aforementioned physical interaction-based immobilization procedures are demonstrated to be operated simply and rapidly. However, their immobilization stability might be influenced by the bulk metal surfaces and environmental factors such as temperature, pH, and ion strength of solution, resulting in a loss of bioactivity or denaturation of the proteins. Moreover, the gradual elution of proteins physically adsorbed may occur during the analytical performances, which may in turn bring about some problems associated with loss of detection sensitivity and low reproducibility of the sensors.

In recent years, nanomaterials (e.g. noble metals, magnetic oxides, and carbon nanoparticles or nanotubes) with unique physical and chemical properties have been successfully applied to modify immunosensing interfaces to achieve greatly improved immobilization of antibodies or antigens [29–31]. Some pioneering works have shown that the assembly of the gold nanoparticle layer on an electrode would lead to substantially increased electrode surface areas available for direct adsorption of biological entities, thus offering the possibility of the great enhancement of analytical sensitivity [11, 32–36]. For example, a new immobilization procedure of antibodies for capacitive immunosensor has been recently proposed using thiol compound and gold nanoparticles [36]. It was here demonstrated that the proposed immobilization procedure could retain the high biological activity of immobilized entities and provide favorable sensing performances. Moreover, magnetic nanoparticles as special carriers for immobilizing biomolecules have also been the current hot subject of a series of investigations for the construction of different immunosensors [37–39]. The easy localization of magnetic beads was used to generate a sensing layer at the surface of a piezoelectric sensor, where the magnetic beads bearing antibodies were immobilized with the help of a permanent magnet at the surface of the crystal [37]. More recently, an amperometric immunosensor has been developed by employing a kind of core–shell magnetic nanoparticle of ($CdFe_2O_4$–SiO_2) to immobilize antibody onto the electrode surface with a magnet field [38]. Additionally, magnetic beads may be applied to label or attach antibodies (antigens) for the magneto-detection of the immune complex based on the perturbation of a magnetic field, which could be quantified using a suitable electronic device [39]. Compared with the conventional immobilization methods, these magnetism-driven immobilization procedures may have some merits such as simple manipulation, easy biomolecule modification, low cost, and repeatable regeneration.

9.2.2 Covalent interaction-based immobilization procedures

The covalent interaction procedures, typically the cross-linking methods, are the most popular immobilization manipulation for fabricating various immunosensors. Due to the lack of an amount of active covalently binding sites at some transducer substrates (e.g. metals, semiconductor, or optical fibers), the precoatings of the base transducers with thin films are generally necessary for covalently binding the antibodies or antigens by using the functional reagents such as glutaraldehyde, carbodiimide succinimide ester, maleinimide, and periodate. Many traditional coating materials, such as polyethyleneimine [40–41], (γ-aminopropyl) trimethoxysilane [42–43], and copolymer of hydroxyethyl- and methyl-methacrylate [44], are often used as the mediate layers for immunoactive molecule immobilization. In recent decades, however, some new coating or functionalized film techniques (materials) have been introduced into this field.

Self-assembled monolayers (SAMs) offer promising functionalized films for the immobilization of antibodies or antigens [45–46]. Since sulfur donor atoms strongly coordinate on noble metal substrates (e.g. Au, Ag, and Pt), various sulfur-containing molecules such as disulfides (R-SS-R), sulfides (R-S-R), and thiols can form various functionalized SAMs of highly organized and compact construction. The applications of the SAM technique in the immobilization of biomolecules have been widely documented [47–49]. Knoll and coworkers presented a versatile biotin-functionalized SAM, on which the biotinylated antibodies can be readily immobilized through an avidin mediator [48]. Mixed SAMs composed of long-chain thiols with carboxylic and hydroxyl groups are also used to attain a specific and stable affinity interface of immunosensors [50–51]. Langmuir-Blodgett (LB) films are other useful alternatives to traditional mediate layers [52–54]. LB films, which are usually prepared by transferring a monolayer on a solid substrate, have great potential in helping to control the orientation and surface density of the antibodies. Hirata *et al.* [52] successfully prepared the lipid-tagged antibody/phospholipid monolayers with high immobilization properties using the LB technique. Vikholm *et al.* demonstrated the incorporation of lipid-tagged single-chain antibodies into lipid monolayers obtaining desirable retention of antibody activities [53–54]. Moreover, recent years witness a newly emerged ultra-thin polymer film, plasma-polymerized film (PPF), which is reported with successful applications in various immunosensor designs [19, 55–57]. PPFs, which are generally prepared by using glow discharge or plasma of organic vapors, are extremely thin, homogeneous, mechanically and chemically stable, with strong adhesion to the substrates. Karube's group first reported the application of PPF to QCM immunosensors [55]. They verified that the resultant sensors were more reproducible from batch to batch, and might have lower noise and higher sensitivity than sensors using some conventional organic coatings (e.g. polyethylenimine). This kind of functionalized film may offer promising alternatives in interfacial design of immunosensors of various transducers.

In recent years, various nanomaterials are found to be skillfully applied in combination with the covalent interaction-based immobilization procedures for immunosensors. Carbon nanotubes (CNTs), for example, have been recognized as the quintessential nano-sized materials since their discovery in 1991 [58]. These nanotubes are now chemically functionalized for the immobilization of biological entities for different biosensors, i.e. electrochemical devices [59–61]. Pantarotto *et al.* successfully bound a model pentapeptide and a virus epitope of foot-and-mouth disease onto single-walled CNTs [61]. They found that the CNTs-loaded peptide might retain the structural integrity to be well recognized by monoclonal or polyclonal antibodies, indicating the potential applications for diagnostic purposes and vaccine delivery. A silica nanoparticles-based immobilization strategy was also proposed by Wang *et al.* for direct immunosensing determination of *Toxoplasma gondii*-specific IgG [62]. Herein, the preparation strategy could allow for antigens covalently bound with higher loading amount and better retained immunoactivity compared to the commonly applied cross-linking methods.

The aforementioned covalent interaction-based procedures may usually allow for the immunoactive proteins immobilized with high stability and repeatability, and the robust covalent bonds may favor the low noise of detection. Nevertheless, problems associated with these covalent bond immobilizations are the decrease of binding capacity of antibodies (antigens) in the immobilization process. Such a phenomenon may be presumably contributed to the partial loss of the immunoactive sites and the random orientation of antibody molecules bound on the transducer surfaces. In addition, cross-linking can produce a three-dimensional multilayer matrix that creates diffusion barriers and transport limitations, resulting in long immunoreaction time and low sensitivity [63]. It is established that the oriented immobilization of antibodies has low influence on their immunological activities to a certain degree [64–68], which antigen binding capacity was demonstrated with a factor of 2–8 higher than that of antibodies randomly immobilized [68]. Therefore, special interest has been given to the development of the orientation-controlled immobilization techniques for antibodies, i.e. mostly through proteins A or G to specifically bind the antibody Fc fragment, or by directly binding the chemical groups at antibody Fc region [64, 69–71].

Lee *et al.* utilized the self-assembled layer of thiol group-modified protein A for the oriented immobilization of antibodies [64]. An increased binding capacity was further observed. As another illustrative instance, a protein A-based orientation-controlled immobilization strategy for antibodies was proposed for the fabrication of a QCM immunosensor using nanometer-sized gold particles and amine-terminated PPF [65]. Moreover, in recent years, there has emerged another oriented immobilization methodology for antibodies through their native thiol (-SH) groups, which were liberated after the splitting of the intact IgG into two antibody fragments [72–73]. Karyakin *et al.* reported a site-oriented immobilization strategy of antibodies on the gold electrode surfaces by use of native sulfide groups of IgG fragments obtained by reduction of intact IgG [72]. They found that antibodies immobilized by this procedure showed an antigen binding capacity 20–30 times higher than that of non-specifically adsorbed intact ones traditionally used.

In general, an ideal immobilization should have the following characteristics: (i) a sufficient loading amount of active antigens or antibodies at the transducer surface; (ii) the immobilized antigens or antibodies staying stable during the measurement process; (iii) the immobilization process having no influence to the sensing behavior of the transducer; and (v) the ability of sensor regeneration. An effective dissociation of the antigen and regeneration of antibody, i.e. by using Gly-HCl buffer (pH 2.3), for cost effectiveness is of practical interest in real immunosensor applications [74].

9.3 MAJOR TYPES OF IMMUNOSENSORS

There are mainly three types of transducers used in immunosensors: electrochemical, optical, and microgravimetric transducers. The immunosensors may operate either as direct immunosensors or as indirect ones. For direct immunosensors, the transducers directly detect the physical or chemical effects resulting from the immunocomplex formation at the interfaces, with no additional labels used. The direct immunosensors detect the analytes in real time. For indirect immunosensors, one or multiple labeled bio-reagents are commonly used during the detection processes, and the transducers should detect the signals from the labels. These indirect detections used to need several washing and separation steps and are sometimes called immunoassays. Compared with the direct immunosensors, the indirect immunosensors may have higher sensitivity and better ability to defend interference from non-specific adsorption.

9.3.1 Electrochemical immunosensors

The majority of known immunosensor devices belong to the group of electrochemical immunosensors. Electrochemical immunosensors may possess several advantages, for example high sensitivity, low cost, and portable design. The principle of their operation is based on the electrochemical detection of the labeled immunoagents or markers such as enzymes, metal ions, or other electroactive compounds, thus providing an opportunity to analyze complex multicomponent mixtures for diagnosing diseases or monitoring the status of patients [75]. The kinds of detection transducer for electrochemical immunosensors can be mainly subdivided into potentiometric, conductometric, capacitive, impeditive, and amperometric (metal and graphite electrodes) devices.

Potentiometric transducers now belong to the most mature transducers with numerous commercial products. For potentiometric transducers, a local equilibrium is established at the transducer interface at near-zero current flow, where the change in electrode or membrane potential is logarithmically proportional to the specific ion activity. The relationship of logarithmical proportionality constitutes the fundamental principle of all potentiometric transducers such as the ion-selective electrodes (ISE). The groups of biosensors are characterized as simple in preparation, robust in operation, and moderately selective in analytical performance [76–81]. Janata first proposed a potentiometric transducer for immunosensing and named it "immuno-electrode" [78], using the immuno-electrode to detect Concanavalin A through covalent attachment to the surface of a PVC membrane deposited on a platinum electrode. The incorporation of ISEs, pH electrodes or gas-sensing electrodes into potentiometric immunosensors to improve their assay sensitivity has been extensively investigated by Rechnitz and coworkers, i.e. for immunochemical measurements of digoxin and human IgG [79–80]. D'Orazlo *et al.* reported the indirect measurements of immunoagents using ion-selective electrodes [81]. A potentiometric immunosensor based on a molecularly imprinted polymer was prepared as a detecting element in micro total analysis systems with the intent of providing easy clinical analysis [82]. Moreover, the ion-selective field-effect transistor (ISFET) as a semiconductor device is generally constructed by substituting an ion-sensing membrane for the metal gate of a field-effect transistor (FET) [83]. The ISFET is able to respond to the surface potential change resulting from the specific immunochemical reaction between the immobilized antibodies and the free antigens. The pH-sensitive ISFETs, as the most widely used sensor of this type, are fabricated with a large range of possible insulators (i.e. SiO_2, Si_3N_4, and Al_2O_3) and enzyme labels (i.e. urease, peroxidase, and glucose oxidase) [84–85]. Nevertheless, only a few examples of ISFET-based immunosensors could be found in the literature [86–88]. For example, Zayats *et al.* report the impedance measurements on an ISFET device that can be used to detect antigen–antibody interactions on the gate surface [88]. In the meantime, they performed complementary surface plasmon resonance (SPR; see above) experiments to illustrate that the ISFET impedance measurements and the SPR reveal comparable sensitivities.

Conductometric transducers, as the oldest electrochemical devices, seem not to enjoy wide applications due to their poor selectivity. For example, Yagiuda *et al.* proposed a conductometric immunosensor for the determination of methamphetamine (MA) in urine [89]. The decrease in the conductivity between a pair of platinum electrodes might result from the direct attachment of MA onto the anti-MA antibodies immobilized on the electrode surface. The system was claimed to be a useful detection technique of MA in comparison with a gas chromatography–mass spectrometry method.

Capacitance and impedance transducers with high sensitivity are widely employed for various immunosensing assays [90–102]. The capacitance sensors are essentially based on the principle that the electrolyte capacitance of an electrode depends on the thickness and dielectric behavior of the dielectric layer on the electrode surface and the solid/solution interface. Dijksma *et al.* designed an immunosensor for the direct detection of interferon-γ at the attomolar level by using the AC impedance approach [90]. The immobilization processes of antibodies (antigens) play an important role in these immunosensors, and the sensitivity of a capacitive immunosensor increases with the decreasing thickness of the insulating layer. Shen and coworkers fabricated a heterostructure of Au/*o*-aminobenzenethiol layer/covalent-coupling antibody/electrode for the direct detection of the antibody–antigen interaction by capacitance measurements [91]. A capacitive immunoassay based on antibody-embedded ultra-thin alumina sol-gel films (~20 to 40 nm) was reported and used for direct determination of antigens with a detection limit as low as ~1 ng mL^{-1} [92]. Fernandez-Sanchez *et al.* reported a successful integration of the lateral flow immunoassay format and impedance detection for prostate-specific antigen of tumor marker, where the electrochemical transducer was coated with a pH-sensitive polymer layer [93].

Although capacitance and impedance immunosensors can directly be utilized to investigate the antibody–antigen interaction without the need of other reagents and a separation step, their analytical sensitivity is limited in clinical applications [14]. In order to amplify the capacitance or impedance response to immunoreaction for the sensitive detection of various clinical markers, different labels have been used including enzymes, fluorophores, and metal chelates [103–104]. Ruan *et al.* developed an immunosensor based on enzyme-stimulated precipitation for the detection of *Escherichia coli* O157:H7 using an electrochemical impedance spectroscopy [103]. Another illustrative example was the sensitized immunosensor proposed by Chen *et al.* [104]. In their study, a receptor protein was directly adsorbed on a porous nanostructure gold film to perform a sandwich immunoreaction with the precipitation of insoluble product on the electrode. The impedance signals so amplified showed good linearity with the content of IgG in the range 0.011–11 ng mL^{-1} with a detection limit of 0.009 ng mL^{-1}. A new strategy of signal amplification was also introduced for highly sensitive impedance measurements using biotin-labeled protein–streptavidin network complex [105].

Amperometric immunosensors, as the most popular immunosensing formats, are based on the measurement of the currents resulting from the electrochemical oxidation or reduction of electroactive species at a certain constant voltage. This kind of immunosensor usually uses a complex three-electrode measuring system consisting of a working electrode (e.g. gold, glassy carbon, or carbon paste), a reference electrode (e.g. Ag/AgCl), and a conducting auxiliary electrode (e.g. platinum). Since most antibodies and antigens are not electrochemically active, there are only a few applications available for direct amperometric sensing. Therefore, most amperometric immunosensors are indirect ones which can detect mainly the redox currents associated with electroactive or catalytic labels [25–26, 28, 106–116]. Aizawa *et al.* first developed an amperometric immunosensor for the determination of human chorionic gonadotropin using an amperometric oxygen electrode [106]. Among the labels used, enzymes are the most popular ones in different types of immunoassays, such as horseradish peroxidase (HRP) or glucose oxidase. An immunosensor was designed for determining isopentenyl adenosine based on the electro-polymerization of polypyrrole and poly(m-phenylenediamine) entrapped with HRP on the glassy carbon electrode [108]. A design strategy of reagentless immunosensor was reported for the detection of carcinoma antigen-125 antibodies by direct HRP-labeled electrochemistry [109]. Due to the high sensitivity inherent in these transducers by enzymatic catalysis, amperometric immunosensors can obtain a much higher sensitivity than the classical ELISA. For the immunosensors used in clinical applications, their surfaces should be capable of renewal. Yu *et al.* developed a renewable amperometric immunosensor for the determination of *Schistosoma japonium* (Sj) antibody by using the paraffin graphite–Sj antigen biocomposite paste electrodes which might be regenerated by polishing the surface [25]. Ionescu and his collaborators have developed two similar amperometric immunosensors for cholera antitoxin immunoglobulins, where the cholera toxin biorecognition entities were bound to a biotinylated polypyrrole film or pyrrole–biotin and lactitobionamide electropolymerized copolymer [117–118]. Moreover, nano-sized particles or sol-gel matrixes have also been increasingly employed in the design of amperometric immunosensors with enhanced analytical performance [23, 119–120]. For example, an electrochemical immunosensor has been developed for probing complement III (C_3) by use of nanogold particle monolayer as the sensing interface [119]. With the coupling of sol-gel and screen-printing technologies, a sensitive thick film immunosensor was fabricated by dispersion of rabbit immunoglobulin G, graphite powder, and a binder in the sol-gel solution [23]. A new HRP-labeled amperometric immunosensor for determination of chorionic gonadotrophin in human serum was constructed by immobilizing HCG within titania sol-gel on a glassy carbon electrode [120]. Anodic stripping voltammetry as an electrochemical assay technique has been well adopted for sensitive measurements of heavy metals such as copper and silver, which may also offer an attractive way of sensitive immunosensor development [121–122]. An immunosensor was designed by coupling immunoassay with the square wave anodic stripping voltammetry technique involving copper ion-labeled antigen in the competitive immunoreaction [121]. This immunosensor might allow rapid, accurate, and inexpensive detection of gibberellin acid with a concentration as low as $1\,\mu g\ mL^{-1}$. Chu *et al.* designed a silver-enhanced colloidal gold metalloimmunoassay for the determination of *Schistosoma japonicum* antibody (SjAb) in rabbit serum [122]. In their study, after the immunoreaction of SjAb target with immobilized Sj antigens, colloidal gold-labeled secondary antibody was introduced to favor the silver enhancement process. An acidic solution was further used to dissolute silver metal atoms, followed by the sensitive determination of dissolved silver ions using anodic stripping voltammetry. In addition, many immunoreaction signal-amplified methods or processes have also been adopted for the development of sensitive amperometric immunosensors. Willner's group reported an amplified immunosensing scheme of chronopotentiometry and Faradaic impedance spectroscopy by way of a bio-catalyzed precipitation of the insoluble product onto the gold electrode [123]. They also designed a variation of this scheme with signal amplification by employing liposomes labeled with biotin and HRP as a probe to amplify the sensing of antigen–antibody interactions [124]. In this case, the electrode with the antigen–antibody complex was exposed to the biotinylated anti-IgG antibody, and further the biotin-labeled HRP-liposomes through an avidin bridge to achieve the biocatalyzed precipitation of an insoluble product on the conductive support.

9.3.2 Optical immunosensors

Since almost all optical phenomena at sensing surfaces (e.g. adsorption, fluorescence, luminescence, scatter or refractive index, etc.) can be used for biochemical sensing designs, optical immunosensors are considered as one of the most promising alternatives to the traditional immunoassays in clinic diagnosis and environmental analysis. In recent years, there has been an increased trend in the use of optical transduction techniques in immunosensor technologies due to the advantages of applying visible radiation, non-destructive operation mode, and the rapid signal generation and reading [1, 125–126]. The optical immunosensors may be divided into two types of approaches: direct optical immunosensors and indirect immunosensors depending upon the use of labeled signaling molecules.

Surface plasmon resonance (SPR) as a direct and reliable optical transducer is commonly based on the evanescent wave, in which a thin gold layer is generally deposited on a prism serving as an optically rarer medium [127–128]. Not requiring additional labels and separation steps, the direct SPR immunosensors have been proven to be powerful analytical tools for rapid real-time monitoring the immunological targets. Schofield and Dimmock developed a SPR system in combination with the flow system for detection of influenza virus by use of carboxylated dextran polymer matrix to couple monoclonal antibody of HC10 [129]. In order to validate the feasibility of SPR immunosensor as a tool for diagnosing type I diabetes, Choi *et al.* modified mixed SAMs onto the optical substrate achieving the immuno-response detection for monoclonal antibodies of anti-glutamic acid decarboxylase [130]. Moreover, the fatty acid-binding protein assay has an application potential in clinical analysis for diagnosis of myocardial infarction. A direct optical immunosensor based on SPR was developed for detecting the human heart-type fatty acid binding protein with a detection limit of 200 ng mL^{-1} [131]. Highly sensitive SPR-based immunosensors using self-assembled protein G have also been successfully applied for the detection of microbes such as *Salmonella typhimurium* and *Legionella pneumophila* [132–133]. More importantly, several instrument systems using SPR technology have been commercially available, such as the BIAcore™ system from Pharmacia Biosensor, the Iasys™ system from Affinity Sensors, and so on. Nevertheless, at present, there are still some unsolved problems for these SPR devices, such as non-specific adsorption and poor analytical sensitivity to analytes of low molecular weight.

Fluorescence immunosensors, as the total internal reflection fluorescence devices, continue to prove themselves as another promising type of sensitive and selective optical immunoassay technique, in which labels are sometimes used [134]. When the fluorescence-labeled antibodies or antigens are attached to the transducer surface and enter the evanescent field, the incident light will excite fluorescent molecules producing a fluorescent evanescent wave signal to be detected. The optic-fiber immunosensor system by fluorescence enhancement or quenching is separation-free, reagentless and applicable to the determination of various proteins by antigen–antibody reactions [134–138]. Maragos *et al.* described the development of a fluorescence polarization-based competition immunoassay for fumonisins in maize using fumonisin-specific monoclonal antibodies [135]. A fluorescence-based immunosensor array for simultaneous determination of multiple clinical analytes was developed by Rowe *et al.* [137]. In their study, the patterned array of recognition elements was immobilized onto the planar waveguide to "capture" the analytes from the samples to be quantified by means of fluorescent detector molecules. Moreover, in recent years, quantum dots as the most suitable fluorescence labels have received increasing applications for developing fluorescence immunosensors due to their high fluorescence quantum yield and sensitivity to environmental changes upon binding proteins. Aoyagi *et al.* proposed a reagentless, regenerable, and portable optic immunosensor for the ultra-sensitive detection of a model sample of IgG based on changes in fluorescent intensity of fluorescent quantum dot-labeled protein A [138]. An antibody for leukemia cell recognition was attached to the luminophore-doped nanoparticle through silica chemistry, yielding an optical microscopy imaging technique for the identification of leukemia cells [139]. Experimental results in this report showed that the new technique using the antibody-coated luminophore nanoparticles could allow leukemia cells to be easily and clearly identified with high efficiency.

Chemiluminescence sensors have also been extensively applied in routine clinical analysis as well as biomedical research due to the advantages of no radioactive wastes, simple instrumentation, low detection limit, and wide dynamic range [14, 140–144]. A chemiluminescent immunosensor for carbohydrate antigen 19-9 (CA19-9) was described by Lin *et al.*, with CA19-9 immobilized on the cross-linked chitosan membrane [141]. The decrease of the immunosensor chemiluminescent signal was proportional to the CA19-9 concentration in the range 2.0–25 U mL^{-1}, with the detection limit of 1.0 U mL^{-1}. Pandian *et al.* developed an automated chemiluminometric immunoassay for the measurement of HCG [142]. It was demonstrated that the immunoassay might facilitate exploration of HCG utility for Down syndrome screening, early pregnancy detection, and differentiation of invasive from non-invasive trophoblastic disease. An optical microbiosensor has been newly designed for the diagnosis of hepatitis C virus (HCV) by using a novel photo-immobilization methodology based on a photo-activable electro-generated polymer film [143]. Herein, the immunosensor using optical fiber photochemically modified was tested for the determination of anti-E_2 protein antibodies through chemiluminescence reaction. Another published study presented the use of electrogenerated luminol chemiluminescence in a homogeneous immunosensor, where digoxin was labeled with luminol through a luminol-BSA-digoxin conjugate [144]. The prepared chemiluminescence immunosensor in a competitive format was shown allowing for the detection of free digoxin with the concentration as low as 0.3 $\mu g\ L^{-1}$.

9.3.3 Microgravimetric immunosensors

Microgravimetric immunosensors may incorporate high sensitivity of piezoelectric response and high specificity of antibody–antigen immunoreaction. The detection principle of these devices is generally based on adsorbate recognition where the selective binding may cause the changes in mass loading and interfacial properties (i.e. viscoelasticity and surface roughness), which can be recognized by a corresponding shift in the oscillation frequency [145–148]. Outstanding features of these sensors include low cost, simple usage, high sensitivity, and real-time output. Microgravimetric immunosensors have two kinds of sensing formats, gas phase and solution phase sensing. The sensitivity to the mass change in air on the transducer surface is about 1 Hz ng^{-1}

for a bulk acoustic wave device with 9 MHz of fundamental frequency, which can be described by the Sauerbrey equation [145]. The microgravimetric transducer is thus mainly known as the quartz crystal microbalance (QCM). Microgravimetric immunosensors in solution phase sensing were used for the quantification of a number of biological targets [42, 65, 146–153]. Nüsslein's group reported a QCM assay for bacteria using a cell-selective polymer film, with desirably low detection limit and no need for prior sample treatment [149]. Wang and coworkers initially developed an integrated QCM immunosensor array composed of four kinds of leukemic lineage-associated probes to explore the differentiated leukocyte antigens for immunophenotyping of acute leukemia [150]. In their study, the probes (crystals) of the array were immobilized separately with Fab fragments of leukemic lineage-associated monoclonal antibodies (markers). The developed immunosensor array was demonstrated to be able to rapidly identify normal cells from leukemic blasts and define the leukemic blasts within certain phenotypic groups (lineages). Recently, a QCM immunosensor using protein A for antibody immobilization has been described for the detection of *Salmonella typhimurium* in chicken meat sample by simultaneous measurements of the resonant frequency and motional resistance [152]. Based on the modification of mixed SAMs on gold electrodes for covalently binding antigens, another piezoelectric immunosensor has been recently developed to detect antisperm antibody [153]. The analytical results for evaluating several clinical specimens by the developed method were found to be in satisfactory agreement with those given by the classical ELISA.

Despite many salient successes, the use of QCM-based immunosensors for trace biological target detection is still challenged by its relatively low intrinsic sensitivity. Kim *et al.* incorporated the immunomagnetic separation with the QCM-based impedance technique achieving a new immunoassay for quantifying *Salmonella typhimurium* with very high sensitivity of cell detection [154]. Herein, antibodies immobilized on magnetic particles were delivered into the sample medium to capture the targets. The resultant immunocomplex was further magnetically collected onto the piezoelectric crystal to be quantified with impedance spectroscopy. Through the enzyme-catalyzed formation of a precipitate on the QCM surface, a mass-amplified microgravimetric immunosensor was proposed in combination with a sandwich enzyme-linked immunoassay [155]. Su and his coworkers successfully used QCM for detection of dengue virus [156]. The authors immobilized two monoclonal antibodies on the crystal that act specifically against the dengue virus envelope protein and non-structural protein. The sensitivity reported for the fabricated piezoelectric immuno-chip was 100-fold greater than the conventional sandwich ELISA method. A highly sensitive microgravimetric biosensor has been developed incorporating noble metal particle-amplified sandwiched immunoassay and silver enhancement reaction [157]. Upon the formation of the sandwiched immunocomplex, the sensor surfaces were coated with gold nanoparticles serving as the nucleation sites to catalyze silver ion reduction. The silver metal deposition would result in a large change in frequency responses, achieving approximately two orders of magnitude improvement in human IgG quantification.

Moreover, there is another important type of microgravimetric immunosensor which is based on the immunological agglutination events. The agglutination immunoreaction of antibody-bearing suspensoids such as polymers, microbeads, and naoparticles may induce a corresponding change in the solution parameters (i.e. density and viscosity) and the interfacial properties of the crystal monitored by the QCM device [158–161]. In contrast to the common conventional piezoelectric assays, the QCM sensing format offers a unique advantage in that the immobilization of antibodies or antigens on the crystal is not necessary. The kind of QCM-sensing methods are widely recognized to be simple, sensitive, and feasible for detecting relevant targets responsible for many clinical diseases [158–163]. Kurosawa *et al.* first developed an agglutination-based piezoelectric immunoassay using antibody-bearing latex, termed as the latex piezoelectric immunoassay (LPEIA), for detecting C-reactive protein [158]. Recently, it has been demonstrated that the LPEIA could be greatly improved by using gold nanoparticles as replacements for latex particles, resulting in a novel agglutination-based piezoelectric immunoassay for directly detecting anti-*T. gondii* immunoglobulins in infected rabbit sera and bloods [159].

9.3.4 Other kinds of immunosensors

In recent years, considerable efforts have been devoted to the development of cantilever-based immunosensors with unique enantio-selective antibodies [164–165]. These devices are mainly used for quality and process control, and diagnostic biosensing for medical analysis. They may have fast responses and high sensitivity and are suitable for mass production. Lee *et al.* fabricated a piezoelectric nanomechanical cantilever by a novel electrical measurement. They found that this technique might allow for the label-free detection of a prostate-specific antigen (PSA) with a detection sensitivity as low as 10 pg mL^{-1} [164]. A microfabricated cantilever was utilized to perform the direct (label-free) stereo-selective detection of trace amounts of an important class of chiral analytes, the r-amino acids, based on immunomechanical responses involving nanoscale bending of the cantilever. The major advantages of the microcantilever sensors over more traditional scale transducers such as the QCM reside in the superior sensitivity to minute quantities of analytes and the ability to micro-fabricate compact arrays of cantilevers to facilitate simultaneous and high throughput measurements [165].

Moreover, mass-sensitive magnetoelastic immunosensors are exploited to design extraordinarily versatile and useful sensor platforms [166]. Magnetoelastic sensors are well established and benefit from mass sensitivity compared to that of a surface acoustic wave (SAW) sensor. However, they may cost much less and are much smaller in size than SAW devices. Ruan *et al.* proposed a mass-sensitive magnetoelastic immunosensor based on the immobilization of affinity-purified antibodies on the surface of a micrometer-scale magnetoelastic cantilever achieving the highly sensitive detection of *Escherichia coli* O157:H7 [167]. In addition, imaging ellipsometry (IE) has also been developed as a new kind of immunosensor, i.e. for the detection of pathogens of *Yersinia enterocolitica* [168]. As another example, a label-free multi-sensing immunosensor based on the combination of IE and the protein chip was reported to be able to detect multiple analytes simultaneously, and even to monitor multiple biological interaction processes *in situ* and in real-time conditions [169].

9.4 CONCLUSION AND FUTURE TRENDS

Immunosensors incorporate the specific immunochemical reaction with the modern transducers including electrochemical (potentiometric, conductometric, capacitative, impedance, amperometric), optical (fluorescence, luminescence, refractive index), and microgravimetric transducers, etc. [1]. These immunosensor devices with dramatic improvements in the sensitivity and selectivity possess the abilities to investigate the reaction dynamics of antibody–antigen binding and the potential to revolutionize conventional immunoassay techniques. With the rapid development of immunological reagents and detection equipments, immunosensors have allowed an increasing range of analytes to be identified and quantified. In particular, simple-to-use, inexpensive and reliable immunosensing systems have been developed to bring immunoassay technology to much more diverse areas, such as outpatient monitoring, large screening programs, and remote environmental surveillance [9]. However, there are still some unsolved problems associated with the immobilization of immunoactive entities, nonspecific adsorption from sample backgrounds (e.g. blood, serum, plasma, urine, and saliva) and practical applications of various transducer devices.

The current development of new immunosensors should aim at solving the problems of clinical analysis in medicine and of chemical analysis in the food industry and biotechnology. The development trends of immunosensors are likely to be primarily driven by the requirements of analytical practice on the improvement in sensitivity, selectivity, rapidity, and especially efficiency of assays (i.e. immunosensing array or microfluidic system). Immunosensors with lowered detection limits and increased sensitivities have been developed in various fields, particularly in clinical analysis. For example, the sandwich immunoassay using enzyme-functionalized liposomes as the catalytic label is proposed to obtain the substantially improved assay sensitivity, as validated in the immunoassay of cholera toxin [170]. Meanwhile, as the latest paradigm of development topic, nanomaterials with unique chemical and physical properties should continue to be exploited to offer important possibilities for new immunosensor designs [29]. A noticeable development trend is also observed in the development of immunosensors combining with other techniques such as flow injection analysis (FIA) or capillary electrophoretic (CE) analysis, to complement and improve the present immunoassay methods [171–172]. Moreover, the miniaturization and automation of immunosensing devices should be another important intention of development to facilitate the significantly shortened analysis time and simplified analytical procedure (i.e. one-step analysis). Of note, protein and antibody array technologies are envisaged to have potential for biomedical and diagnostic applications in recent years [173–177]. Belov *et al.* have proposed a novel immunophenotyping method for leukemias using a cluster of differentiation antibody microarray [174]. A microarray of enzyme-linked immunosorbent assay has been developed for autoimmune diagnosis of systematic rheumatic disease, where the high titers of antinuclear antibodies against various nuclear proteins and nucleoprotein complexes might be detected with high throughput [177]. At the same time, the screen-printing techniques may also appear to be the most promising technology for immunosensor array to be commercialized on a large scale and widely applied in clinical diagnosis. Moreover, there have been increasing reports focusing on the development of microfluidic immunosensor systems for proteomics and drug discovery in recent years [178]. Microfluidic system integrating multiple processes in a single device generally seeks to improve analytical performance by reducing the reagent consumption and the analysis time, and increasing reliability and sensitivity through automation. The micro total analysis systems (μTAS) are already under development and should represent the future of high throughput immuno-tests [179]. In addition, with the development of protein engineering technology and molecular biology techniques, more flexible antibodies suitable for immunosensing applications may be expected. For example, the recombinant or fusion approach is powerful in the production of antibodies and antibody derivates. Use of various new generations of antibodies should lead to the enhancement of activity and stability of the immobilized bio-species and even the improvement of the regeneration and sensitivity of the immunosensors. As an inspiringly illustrative instance, aptamers are beginning to emerge as a class of synthetic oligonucleotides or molecules that rival antibodies in both therapeutic and diagnostic applications [180–182]. Baldrich and coworkers first demonstrated the exploitation of an aptamer in an extremely rapid and highly sensitive displacement assay, the displacement enzyme-linked aptamer assay, using enzyme-labeled target as a suboptimal displaceable molecule [182].

To sum up, immunosensors are now becoming one of the most widely used analytical techniques, embracing a vast repertoire of analytes that are detected by a diverse range of transducer devices. The enormous potential of immunosensors in clinical diagnosis, environmental analysis, and biological process monitoring has been widely accepted and increasing efforts have been devoted to these fields. In particular, with the continual development of transducer technology, laser technology, nano-sized material technology, and antibody engineering technology, immunosensors based on the application of these technologies should be inevitably powerful tools in increasingly wide analytical areas [9].

9.5 REFERENCES

1. R.I. Stefan, J.F. van Staden, and H.Y. Aboul-Enein, Immunosensors in clinical analysis. *Fresenius J. Anal. Chem.* **366**, 659–668 (2000).
2. P.B. Luppa, L.J. Sokoll, and D.W. Chan, Immunosensor principles and applications to clinical chemistry. *Clin. Chim. Acta* **314**, 1–26 (2001).
3. E. Gizeli and C.R Lowe, Immunosensors. *Current Opin. Biotech.* **7**, 66–71 (1996).
4. P. D'Orazio, Biosensors in clinical chemistry. *Clin. Chim. Acta* **334**, 41–59 (2003).
5. E. Mallat, D. Barceló, C. Barzen, G. Gauglitz, and R. Abuknesha, Immunosensors for pesticide determination in natural waters. *Trac-Trends Anal. Chem.* **20**, 124–132 (2001).
6. M.P. Marco, S. Gee, and B.D. Hammock, Immunochemical techniques for environmental analysis: I. Immunosensors. *Trends Anal. Chem.* **14**, 341–350, (1995).
7. A. Sadana and T. Vo-Dinh, Single- and dual-fractal analysis of hybridization binding kinetics: biosensor application. *Biotechnol. Prog.* **14**, 782–790 (1998).
8. A.F.P. Turner, I. Karube, and G.S. Wilson, in *Biosensors: Fundamentals and Applications,* pp. 5–8. Oxford University Press (1987).

9. C.L. Morgan, D.J. Newman, and C.P. Price, Immunosensors: technology and opportunities in laboratory medicine. *Clin. Chem.* **42**, 193–209 (1996).
10. A.A.P. Ferreira, W. Colli, P.I. Costac, and H. Yamanaka, Immunosensor for the diagnosis of Chagas' disease. *Biosens. Bioelectron.* **21**, 175–181 (2005).
11. R. Yuan, D. Tang, Y.Q. Cai, X. Zhong, Y. Liu, and J. Dai, Ultrasensitive potentiometric immunosensor based on SA and OCA techniques for immobilization of HBsAb with colloidal Au and polyvinyl butyral as matrixes. *Langmuir* **20**, 7240–7245 (2004).
12. B. Zuo, S. Li, Z. Guo, J. Zhang, and C. Chen, Piezoelectric immunosensor for SARS-associated coronavirus in sputum. *Anal. Chem.* **76**, 3536–3540 (2004).
13. D. Faraggi and A. Kramar, Some methodological issues associated with tumour marker development: biostatistical aspects. *Urol. Oncol.* **5**, 211–213 (2000).
14. J.H. Lin and H.X. Ju, Electrochemical and chemiluminescent immunosensors for tumor markers. *Biosens. Bioelectron.* **20**, 1461–1470 (2005).
15. S.M. Wilson, Electrochemical immunosensors for the simultaneous detection of two tumor markers. *Anal. Chem.* **77**, 1496–1502 (2005).
16. K.R. Wenmeyer, H.B. Halsall, W.R. Heineman, C.P. Volle, and I.W. Chen, Competitive heterogeneous enzyme immunoassay for digoxin with electrochemical detection. *Anal. Chem.* **58**, 135–139 (1986).
17. R. Tilton, E. Blomberg, and P. Claesson, Effect of anionic surfactant on interactions between lysozyme layers adsorbed on mica. *Langmuir* **9**, 2102–2108 (1993).
18. E. Blomberg, P. Claesson, J. Froberg, and R. Tilton, Interaction between adsorbed layers of lysozyme studied with the surface force technique. *Langmuir* **10**, 2325–2334 (1994).
19. F. Caruso, E. Rodda, D.F. Furlong, K. Niikura, and Y. Okahata, Quartz crystal microbalance study of DNA immobilization and hybridization for nucleic acid sensor development. *Anal. Chem.* **69**, 2043–2049 (1997).
20. Z.Y. Wu, Y.H. Yan, G.L. Shen, and R.Q. Yu, A novel approach of antibody immobilization based on n-butyl amine plasma-polymerized films for immunosensors. *Anal. Chim. Acta* **412**, 29–35 (2000).
21. T. Deng, H. Wang, J.S. Li, G.L. Shen, and R.Q. Yu, A novel biosensing interfacial design based on the assembled multilayers of the oppositely charged polyelectrolytes. *Anal. Chim. Acta* **532**, 137–144 (2005).
22. R. Wang, U. Narang, P.N. Prasad, and F.V. Bright, Affinity of antifluorescein antibodies encapsulated within a transparent sol-gel glass. *Anal. Chem.* **65**, 2671–2675 (1993).
23. J. Wang, P. Pamidi, and K.R. Rogers, Sol-gel-derived thick-film amperometric immunosensors. *Anal. Chem.* **70**, 1171–1175 (1998).
24. F.C. Gong, Z.J. Zhou, G.L. Shen, and R.Q. Yu, Schistosoma japonicum antibody assay by immunosensing with fluorescence detection using 3,3′,5,5′-tetramethylbenzidine as substrate. *Talanta* **58**, 611–618 (2002).
25. G.D. Liu, Z.Y. Wu, S.P. Wang, G.L. Shen, and R.Q. Yu, Renewable amperometric immunosensor for *Schistosoma japonium* antibody assay. *Anal. Chem.* **73**, 3219–3226 (2001).
26. Y.M. Zhou, S.Q. Hu, G.L. Shen, and R.Q. Yu, An amperometric immunosensor based on an electrochemically pretreated carbon–paraffin electrode for complement III (C_3) assay. *Biosens. Bioelectron.* **18**, 473–481 (2003).
27. J. Li, L.T. Xiao, G.M. Zeng, G.H. Huang, G.L. Shen, and R.Q. Yu, A renewable amperometric immunosensor for phytohormone β-indole acetic acid assay. *Anal. Chim. Acta* **494**, 177–185 (2003).
28. M. Santandreu, F. Céspedes, S. Alegret, and E. Martínez-Fàbregas, Amperometric immunosensors based on rigid conducting immunocomposites. *Anal. Chem.* **69**, 2080–2085 (1997).
29. C. R. Martin and D. T. Mitchell, Nanomaterials in analytical chemistry. *Anal. Chem.* **70**, 322A–327A (1998).
30. N.T.K. Thanh and Z. Rosenzweig, Development of an aggregation-based immunoassay for anti-protein A using gold nanoparticles. *Anal. Chem.* **74**, 1624–1628 (2002).
31. J. Gascón, E. Martinez, and D. Barceló, Determination of atrazine and alachlor in natural waters by a rapid-magnetic particle-based ELISA influence of common cross-reactants: deethylatrazine, deisopropylatrazine, simazine and metolachlor. *Anal. Chim. Acta* **311**, 357–364 (1995).

32. A.L. Crumbliss, S.C. Perine, J. Stonehuerner, K.R. Tubergen, Junguo Zhao, R.W. Henkens, and J.P. O'Daly, Colloidal gold as a biocompatible immobilization matrix suitable for the fabrication of the enzyme electrode by electrodeposition. *Biotechnol. Bioeng.* **40**, 483–490 (1992).
33. K.R. Brown, A.P. Fox, and M.J. Natan, Morphology-dependent electrochemistry of cytochrome *c* at Au colloid-modified SnO2 electrodes. *J. Am. Chem. Soc.* **118**, 1154–1157 (1996).
34. H. Wang, C.C. Wang, C.X. Lei, Z.Y. Wu, G.L. Shen, and R.Q. Yu, A novel biosensing interfacial design produced by assembling nano-Au particles on amine-terminated plasma-polymerized films. *Anal. Bioanal. Chem.* **377**, 632–638 (2003).
35. M.J. Wang, L.Y. Wang, G. Wang, X.H. Ji, Y.B. Bai, T.J. Li, S.Y. Gong, and J.H. Li, Application of impedance spectroscopy for monitoring colloid Au-enhanced antibody immobilization and antibody–antigen reactions. *Biosens. Bioelectron.* **19**, 575–582 (2004).
36. J.S. Li, Z.S. Wu, H. Wang, G.L. Shen, and R.Q. Yu, A reusable capacitive immunosensor with a novel immobilization procedure based on 1,6-hexanedithiol and nano-Au self-assembled layers. *Sens. Actuators B* **110**, 327–334 (2005).
37. J.S. Li, X.X. He, Z.Y. Wu, K.M. Wang, G.L. Shen, and R.Q. Yu, Piezoelectric immunosensor based on magnetic nanoparticles with simple immobilization procedures. *Anal. Chim. Acta* **481**, 191–198 (2003).
38. Z.M. Liu, H.F. Yang, Y.F. Li, Y.L. Liu, G.L. Shen, and R.Q. Yu, Core–shell magnetic nanoparticles applied for immobilization of antibody on carbon paste electrode and amperometric immunosensing. *Sens, Actuators B* **113**, 956–962 (2006).
39. J. Richardson, P. Hawkins, and R. Luxton, The use of coated paramagnetic particles as a physical label in a magneto-immunoassay. *Biosens. Bioelectron.* **16**, 989–993 (2001).
40. B. König and M. Grätzel, Development of a piezoelectric immunosensor for the detection of human erythrocytes. *Anal. Chim. Acta* **276**, 329–333 (1993).
41. B. König and M. Grätzel, Detection of viruses and bacteria with piezoimmunosensors. *Anal. Lett.* **26**, 1567–1585 (1993).
42. H. Muramatsu, K. Kajiwara, E. Tamiya, and I. Karube, Piezoelectric immunosensor for the detection of candida albicans microbes. *Anal. Chim. Acta* **188**, 257–261 (1986).
43. H. Muramatsu, E. Tamiya, and I. Karube, Piezoelectric crystal biosensor system for detection of *Escherichia coli. Anal. Lett.* **22**, 2155–2166 (1989).
44. X. Chu, Z.H. Lin, G.L. Shen, and R.Q. Yu, Piezoelectric immunosensor for the detection of immunoglobulin M. *Analyst* **120**, 2829–2832 (1995).
45. A. Ulman, Formation and structure of self-assembled monolayers. *Chem. Rev.* **96**, 1533–1554 (1996).
46. T. Wink, S.J. van Zuilen, A. Bult, and W.P. van Bemekom, Self-assembled monolayers for biosensors: a tutorial review. *Analyst* **122**, 43R–50R (1997).
47. Z. Y. Wu, S. P. Wang, G. L. Shen and R. Q. Yu, Quartz-crystal microbalance immunosensor for *Schistsoma-Japonicum*-infected rabbit serum. *Anal. Sci.* **19**, 437–440 (2003).
48. L. Häussling, H. Ringsdorf, F.J. Schimitt, and W. Knoll, Biotin-functionalized self-assembled monolayers on gold-surface-plasmon optical studies of specific recognition reactions. *Langmuir* **7**, 1837–184 (1991).
49. C.C. Wang, H. Wang, Z.Y. Wu, G.L. Shen, and R.Q. Yu, A piezoelectric immunoassay based on self-assembled monolayers of cystamine and polystyrene sulfonate for determination of *Schistosoma japonicum* antibodies. *Anal. Bioanal. Chem.* **373**, 803–809 (2002).
50. V.H. Perez-Luna, M.J. O'Brien, K.A. Opperman, P.D. Hampton, G.P. Lopez, L.A. Klumb, and P.S. Stayton, Molecular recognition between genetically engineered streptavidin and surface-bound biotin. *J. Am. Chem. Soc.* **121**, 6469–6478 (1999).
51. F. Frederix, K. Bonroy, W. Laureyn, G. Reekmans, A. Campitelli, W. Dehaen, and G. Maes, Enhanced performance of an affinity biosensor interface based on mixed self-assembled monolayers of thiols on gold. *Langmuir* **19**, 4351–4357 (2003).
52. Y. Hirata, M.-L. Laukkanen, K. Keinänen, H. Shigematsu, M. Aizawa, and F. Mizutani, Microscopic characterization of Langmuir–Blodgett films incorporating biosynthetically lipid-tagged antibody. *Sens. Actuators B* **76**, 181–186 (2001).
53. I. Vikholm and J. Peltonen, Layer formation of a lipid-tagged single-chain antibody and the interaction with antigen. *Thin Solid Films* **284/285**, 924–926 (1996).

54. I. Vikholm, E. Györvary, and J. Peltonen, Incorporation of lipid-tagged single-chain antibodies into lipid monolayers and the interaction with antigen. *Langmuir* **12**, 3276–3281 (1996).
55. K. Nakanishi, H. Muguruma, K. Ikebukuro, and I. Karube, A novel method of immobilizing antibodies on a quartz crystal microbalance using plasma-polymerized films for immunosensors. *Anal. Chem.* **68**, 1695–1700 (1996).
56. R. Nakamura, H. Muguruma, K. Ikebukuro, S. Sasaki, R. Nagata, I. Karube, and H. Pedersen, A plasma-polymerized film for surface plasmon resonance immunosensing. *Anal. Chem.* **69**, 4649–4652 (1997).
57. H. Wang, D. Li, Z.Y. Wu, G.L. Shen, and R.Q. Yu, A reusable piezo-immunosensor with amplified sensitivity for ceruloplasmin based on plasma-polymerized film. *Talanta* **62**, 201–206 (2004).
58. S. Iijima, Helical microtubules of graphitic carbon. *Nature* **354**, 56–58 (1991).
59. J. Wang and M. Musameh, Carbon nanotube/teflon composite electrochemical sensors and biosensors. *Anal. Chem.* **75**, 2075–2079 (2003).
60. S. Hrapovic, Y. Liu, K.B. Male, and J.H.T. Luong, Electrochemical biosensing platforms using platinum nanoparticles and carbon nanotubes. *Anal. Chem.* **76**, 1083–1088 (2004).
61. D. Pantarotto, C.D. Partidos, R. Graff, J. Hoebeke, J.P. Briand, M. Prato, and A. Bianco, Synthesis, structural characterization, and immunological properties of carbon nanotubes functionalized with peptides. *J. Am. Chem. Soc.* **125**, 6160–6164 (2003).
62. H. Wang, J.S. Li, Y.J. Ding, C.X. Lei, G.L. Shen, and R.Q. Yu, Novel immunoassay for Toxoplasma gondii-specific immunoglobulin G using a silica nanoparticle-based biomolecular immobilization method. *Anal. Chim. Acta* **501**, 37–43 (2004).
63. Y.S. Fung and Y.Y. Wong, Self-assembled monolayers as the coating in a quartz piezoelectric crystal immunosensor to detect salmonella in aqueous solution. *Anal. Chem.* **73**, 5302–5309 (2001).
64. W. Lee, B.K. Oh, Y.M. Bae, S.H. Paek, W.H. Lee, and J.W. Choi, Fabrication of self-assembled protein A monolayer and its application as an immunosensor. *Biosens. Bioelectron.* **19**, 185–192 (2003).
65. H. Wang, Y. Liu, Y. Yang, T. Deng, G.L. Shen, and R.Q. Yu, A protein A-based orientation-controlled immobilization strategy for antibodies using nanometer-sized gold particles and plasma-polymerized film. *Anal. Biochem.* **324**, 219–226 (2004).
66. J. Kaur, K.V. Singh, A.H. Schmid, G.C. Varshney, C.R. Suri, and M. Raje, Atomic force spectroscopy-based study of antibody pesticide interactions for characterization of immunosensor surface. *Biosens. Bioelectron.* **20**, 284–293 (2004).
67. I. Vikholm, W.M. Albers, H. Välimäki, and H. Helle, In situ quartz crystal microbalance monitoring of Fab′-fragment binding to linker lipids in a phosphatidylcholine monolayer matrix: application to immunosensors. *Thin Solid Films* **327**, 643–646 (1998).
68. B. Lu, M.R. Smyth, and R. O'Kennedy, Oriented immobilization of antibodies and its applications in immunoassays and immunosensors. *Analyst* **121**, 29R–32R (1996).
69. D.A. Palmer, M.T. French, and J.N. Miller, Use of protein A as an immunological reagent and its application using flow injection: a review. *Analyst* **119**, 2769–2776 (1994).
70. K. Saha, F. Bender, and E. Gizeli, Comparative study of IgG binding to proteins G and A: nonequilibrium kinetic and binding constant determination with the acoustic waveguide device. *Anal. Chem.* **75**, 835–842 (2003).
71. D.J.O. Shannessy and W.L. Hoffman, Site-directed immobilization of glycoproteins on hydrazide-containing solid supports. *Biotechnol. Appl. Biochem.* **9**, 488–496 (1987).
72. A.A. Karyakin, G.V. Presnova, M.Y. Rubtsova, and A.M. Egorov, Oriented immobilization of antibodies onto the gold surfaces via their native thiol groups. *Anal. Chem.* **72**, 3805–3811 (2000).
73. H. Wang, J. Wu, J.S. Li, Y.J. Ding, G.L. Shen, and R.Q. Yu, Nanogold particle-enhanced oriented adsorption of antibody fragments for immunosensing platforms. *Biosens. Bioelectron.* **20**, 2210–2217 (2005).
74. V.B. Kandimalla, N.S. Neeta, N.G. Karanth, M.S. Thakur, K.R. Roshini, B.E.A. Rani, A. Pasha, and N.G.K. Karanth, Regeneration of ethyl parathion antibodies for repeated use in immunosensor: a study on dissociation of antigens from antibodies. *Biosens. Bioelectron.* **20**, 903–906 (2004).
75. E.P. Medyantseva, E.V. Khaldeeva, and G.K. Budnikov, Immunosensors in biology and medicine: analytical capabilities, problems, and prospects. *J. Anal. Chem.* **56**, 886–900 (2001).
76. R.Q. Yu, Z.R. Zhang, and G.L. Shen, Potentiometric sensors: aspects of the recent development. *Sens. Actuators B* **65**, 150–153 (2000).

77. D.L. Simpson and R.K. Kobos, Potentiometric microbiological assay of gentamicin, streptomycin, and neomycin with a carbon dioxide gas-sensing electrode. *Anal. Chem.* **55**, 1974–1977 (1983).
78. J. Janata, Immunoelectrode. *J. Am. Chem. Soc.* **97**, 2914–2916 (1975).
79. M.Y. Keating and G.A. Rechnitz, Potentiometric enzyme immunoassay for digoxin using polystyrene beads. *Anal. Lett.* **18**, 1–10 (1985).
80. T. Fonong and G.A. Rechnitz, Homogeneous potentiometric enzyme immunoassay for human immunoglobulin G. *Anal. Chem.* **56**, 2586–2590 (1984).
81. P. D'Orazlo and G.A. Rechnitz, Ion electrode measurements of complement and antibody levels using marker-loaded sheep red blood cell ghosts. *Anal. Chem.* **49**, 2083–2086 (1977).
82. T. Kitade, K. Kitamura, T. Konishi, S. Takegami, T. Okuno, M. Ishikawa, M. Wakabayashi, K. Nishikawa, and Y. Muramatsu, Potentiometric immunosensor using artificial antibody based on molecularly imprinted polymers. *Anal. Chem.* **76**, 6802–6807 (2004).
83. S. Koch, P. Woias, L.K. Meixner, S. Drost, and H. Wolf, Protein detection with a novel ISFET-based zeta potential analyzer. *Biosens. Bioelectron.* **14**, 413–421 (1999).
84. Y.Q. Miao, J.G. Guan, and J.R. Chen, Ion sensitive field effect transducer-based biosensors. *Biotech. Adv.* **21**, 527–534 (2003).
85. A.P. Soldatkin, V. Volotovsky, A.V. Elskaya, N. Jaffrezic-Renault, and C. Martelet, Improvement of urease based biosensor characteristics using additional layers of charged polymers. *Anal. Chim. Acta* **403**, 25–29 (2000).
86. Z.E. Selvanayagam, P. Neuzil, P. Gopalakrishnakone, U. Sridhar, M. Singh, and L.C. Ho, An ISFET-based immunosensor for the detection of β-bungarotoxin. *Biosens. Bioelectron.* **17**, 821–826 (2002).
87. Y.V. Plekhanova, A.N. Reshetilov, E.V. Yazynina, A.V. Zherdev, and B.B. Dzantiev, A new assay format for electrochemical immunosensors: polyelectrolyte-based separation on membrane carriers combined with detection of peroxidase activity by pH-sensitive field-effect transistor. *Biosens. Bioelectron.* **19**, 109–114 (2003).
88. M. Zayats, O.A. Raitman, V.I. Chegel, A.B. Kharitonov, and I. Willner, Probing antigen–antibody binding processes by impedance measurements on ion-sensitive field-effect transistor devices and complementary surface plasmon resonance analyses: development of cholera toxin sensors. *Anal. Chem.* **74**, 4763–4773 (2002).
89. K. Yagiuda, A. Hemmi, S. Ito, Y. Asano, Y. Fushinuki, C. Chen, and I. Karube, Development of a conductivity-based immunosensor for sensitive detection of methamphetamine (stimulant drug) in human urine. *Biosens. Bioelectron.* **11**, 703–707 (1996).
90. M. Dijksma, B. Kamp, J.C. Hoogvliet, and W.P. van Bennekom, Development of an electrochemical immunosensor for direct detection of interferon-γ at the attomolar level. *Anal. Chem.* **73**, 901–907 (2001).
91. S.Q. Hu, Z.X. Cao, G.L. Shen, and R.Q. Yu, Capacitive immunosensor for transferrin based on an o-aminobenzenthiol oligomer layer. *Anal. Chim. Acta* **458**, 297–304 (2002).
92. D. Jiang, J. Tang, B. Liu, P. Yang, and J. Kong, Ultrathin alumina sol-gel-derived films: allowing direct detection of the liver fibrosis markers by capacitance measurement. *Anal. Chem.* **75**, 4578–4584 (2003).
93. C. Fernandez-Sanchez, C.J. McNeil, K. Rawso, and O. Nilsson, Disposable noncompetitive immunosensor for free and total prostate-specific antigen based on capacitance measurement. *Anal. Chem.* **76**, 5649–5656 (2004).
94. P. Bataillard, F. Gardies, N.J. Renault, C. Martelet, B. Colin, and B. Mandrand, Direct detection of immunospecies by capacitance measurements. *Anal. Chem.* **60**, 2374–2379 (1988).
95. M. Vladimir, M. Mirsky, O. Riepl, and O. Wolfbeis, Capacitive monitoring of protein immobilization and antigen–antibody reactions on monomolecular alkylthiol films on gold electrodes. *Biosens. Bioelectron.* **12**, 977–989 (1997).
96. C.J. Mcneil, D. Athey, M. Ball, W.O. Ho, S. Krause, R.D. Armstrong, J.D. Wright, and K. Rawson, Electrochemical sensors based on impedance measurement of enzyme-catalyzed polymer dissolution: theory and applications. *Anal. Chem.* **67**, 3928–3935 (1995).
97. A. Gebbert, M. Alvarez-lcaza, W. Stocklein, and R.D. Schmid, Real-time monitoring of immunochemical interactions with a tantalum capacitance flow-through cell. *Anal. Chem.* **64**, 997–1003 (1992).

98. E. Souteyrand, J.R. Martin, and C. Martelet, Direct detection of biomolecules by electrochemical impedance measurements. *Sens. Actuators B* **20**, 63–69 (1994).
99. A. Battaud, H. Perrot, V. Billard, C. Martelet, and J. Therasse, Study of immunoglobulin G thin layers obtained by the Langmuir-Blodgett method: application to immunosensors. *Biosens. Bioelectron.* **8**, 39–48 (1993).
100. A. Gebbert, M. Alvarez-Icaza, H. Peters, V. Jager, U. Bilitewski, and R.D. Schmid, On-line monitoring of monoclonal antibody production with regenerable flow-injection immuno systems. *J. Biotechnol.* **32**, 213–220 (1994).
101. B. Prasad and R. Lai, A capacitative immunosensor measurement system with a lock-in amplifier and potentiostatic control by software. *Meas. Sci. Technol.* **10**, 1097–1104 (1999).
102. S.Q. Hu, J. Xie, Q. Xu, K. Rong, G.L. Shen, and R.Q. Yu, A label-free electrochemical immunosensor based on gold nanoparticles for detection of paraoxon. *Talanta* **61**, 769–777 (2003).
103. C.M. Ruan, L. Yang, and Y. Li, Immunobiosensor chips for detection of *Escherichia coli*O157:H7 using electrochemical impedance spectroscopy. *Anal. Chem.* **74**, 4814–4820 (2002).
104. Z.P. Chen, J.H. Jiang, G.L. Shen, and R.Q. Yu, Impedance immunosensor based on receptor protein adsorbed directly on porous gold film. *Anal. Chim. Acta* **553**, 190–195 (2005).
105. R.J. Pei, Z.L. Cheng, E.K. Wang, and X.R. Yang, Amplification of antigen–antibody interactions based on biotin labeled protein–streptavidin network complex using impedance spectroscopy. *Biosens. Bioelectron.* **16**, 355–361 (2001).
106. M. Aizawa, A. Morioka, S. Suzuki, and Y. Nagamura, Enzyme immunosenser: III. Amperometric determination of human cherienic gonadotropin by membrane-bound antibody. *Anal. Biochem.* **94**, 22–28 (1979).
107. R. Blonder, E. Katz, Y. Cohen, N. Itzhak, A. Riklin, and I. Willner, Application of redox enzymes for probing the antigen–antibody association at monolayer interfaces: development of amperometric immunosensor electrodes. *Anal. Chem.* **68**, 3151–3157 (1996).
108. J. Li, L.T. Xiao, G.M. Zeng, G.H. Huang, G.L. Shen, and R.Q. Yu, Amperometric immunosensor based on polypyrrole/poly(m-phenylenediamine) multilayer on glassy carbon electrode for cytokinin N^6-(D^2-isopentenyl) adenosine assay. *Anal. Biochem.* **321**, 89–95 (2003).
109. Z. Dai, F. Yan, J. Chen, and H.X. Ju, Reagentless amperometric immunosensors based on direct electrochemistry of horseradish peroxidase for determination of carcinoma antigen-125. *Anal. Chem.* **75**, 5429–5434 (2003).
110. C.H. Liu, K.T. Liao, and H.J. Huang, Amperometric immunosensors based on protein A coupled polyaniline-perfluorosulfonated ionomer composite electrodes. *Anal. Chem.* **72**, 2925–2929 (2000).
111. I. Willner, R. Blonder, and A. Dagan, Application of photoisomerizable antigenic monolayer electrodes as reversible amperometric immunosensors. *J. Am. Chem. Soc.* **116**, 9365–9366 (1994).
112. H.S. Jung, J.M. Kim, J.W. Park, H.Y. Lee, and T. Kawai, Amperometric immunosensor for direct detection based upon functional lipid vesicles immobilized on nanowell array electrode. *Langmuir* **21**, 6025–6029 (2005).
113. I. Willner and B. Willner, Electronic transduction of photostimulated binding interactions at photoisomerizable monolayer electrodes: novel approaches for optobioelectronic systems and reversible immunosensor devices. *Biotechnol. Prog.* **15**, 991–1002 (1999).
114. A.F. Chetcuti, D.K.Y. Wong, and M.C. Stuart, An indirect perfluorosulfonated ionomer-coated electrochemical immunosensor for the detection of the protein human chorionic gonadotrophin. *Anal. Chem.* **71**, 4088–4094 (1999).
115. Y.M. Zhou, Z.Y. Wu, G.L. Shen, and R.Q. Yu, An amperometric immunosensor based on Nafion-modified electrode for the determination of Schistosoma japonicum antibody. *Sens. Actuators B* **89**, 292–298 (2003).
116. R. Blonder, S. Levi, G. Tao, I. Ben-Dov, and I. Willner, Development of amperometric and microgravimetric immunosensors and reversible immunosensors using antigen and photoisomerizable antigen monolayer electrodes. *J. Am. Chem. Soc.* **119**, 10467–10478 (1997).
117. R.E. Ionescu, C. Gondran, S. Cosnierc, L.A. Gheber, and R.S. Marks, Comparison between the performances of amperometric immunosensors for cholera antitoxin based on three enzyme markers. *Talanta* **66**, 15–20 (2005).

118. R.E. Ionescu, C. Gondran, L.A. Gheber, S. Cosnier, and R.S. Marks, Construction of amperometric immunosensors based on the electrogeneration of a permeable biotinylated polypyrrole film. *Anal. Chem.* **76**, 6808–6813 (2004).
119. C.X. Lei, Y. Yang, H. Wang, G.L. Shen, and R.Q. Yu, Amperometric immunosensor for probing complement III (C_3) based on immobilizing C_3 antibody to a nano-Au monolayer supported by sol-gel-derived carbon ceramic electrode. *Anal. Chim. Acta* **513**, 379–384 (2004).
120. J. Chen, F. Yan, Z. Dai, and H.X. Ju, Reagentless amperometric immunosensor for human chorionic gonadotrophin based on direct electrochemistry of horseradish peroxidase. *Biosens. Bioelectron.* **21**, 330–336 (2005).
121. J. Li, L.T. Xiao, G.M. Zeng, G.H. Huang, G.L. Shen, and R.Q. Yu, Immunosensor for rapid detection of gibberellin acid in the rice grain. *J. Agric. Food Chem.* **53**, 1348–1353 (2005).
122. X. Chu, Z.F. Xiang, X. Fu, S.P. Wang, G.L. Shen, and R.Q. Yu, Silver-enhanced colloidal gold metalloimmunoassay for *Schistosoma japonicum* antibody detection. *J. Immunol. Methods* **301**, 77–88 (2005).
123. E. Katz, L. Alfonta, and I. Willner, Chronopotentiometry and Faradaic impedance spectroscopy as methods for signal transduction in immunosensors. *Sens. Actuators B* **76**, 134–141 (2001).
124. L. Alfonta, A.K. Singh, and I. Willner, Liposomes labeled with biotin and horseradish peroxidase: a probe for the enhanced amplification of antigen–antibody or oligonucleotide–DNA sensing processes by the precipitation of an insoluble product on electrodes. *Anal. Chem.* **73**, 91–102 (2001).
125. O.S. Wolfbies, Fiber-optic chemical sensors and biosensors. *Anal. Chem.* **72**, 81–90 (2000).
126. O.S. Wolfbies, Fiber-optic chemical sensors and biosensors. *Anal. Chem.* **74**, 2663–2678 (2002).
127. J. Homola, S.S. Yee, and G. Gauglitz, Surface plasmon resonance sensors: a review. *Sens. Actuators B* **54**, 3–15 (1999).
128. W.M. Mullett, E.P.C. Lai, and J.M. Yeung, Surface plasmon resonance-based immunoassays. *Methods* **22**, 77–91 (2000).
129. D.J. Schofield and N.J. Dimmock, Determination of affinities of a panel of IgGs and Fabs for whole enveloped (influenza A) virions using surface plasmon resonance. *J. Virol. Methods* **62**, 33–42 (1996).
130. S.H. Choi, J.W. Lee, and S.J. Sim, Enhanced performance of a surface plasmon resonance immunosensor for detecting Ab–GAD antibody based on the modified self-assembled monolayers. *Biosens. Bioelectron.* **21**, 378–383 (2005).
131. U. Kunz, A. Katerkamp, R. Renneberg, F. Spener, and K. Cammann, Sensing fatty acid binding protein with planar and fiber-optical surface plasmon resonance spectroscopy devices. *Sens. Actuators B* **32**, 149–155 (1996).
132. B.K. Oh, Y.K. Kim, K.W. Park, W.H. Lee, and J.W. Choi, Surface plasmon resonance immunosensor for the detection of *Salmonella typhimurium*. *Biosens. Bioelectron.* **19**, 1497–1054 (2004).
133. B.K. Oh, Y.K. Kim, W. Lee, Y.M. Bae, W.H. Lee, and J.W. Choi, Immunosensor for detection of *Legionella pneumophila* using surface plasmon resonance. *Biosens. Bioelectron.* **18**, 605–611 (2003).
134. A.I. Hemmila, in *Applications of Fluorescence in Immunoassay*, pp. 146–148. John Wiley & Sons Press (1991).
135. C.M. Maragos, M.E. Jolley, R.D. Plattner, and M.S. Nasir, Fluorescence polarization as a means for determination of fumonisins in maize. *J. Agric. Food Chem.* **49**, 596–602 (2001).
136. S. Aoyagi, R. Imai, K. Sakai, and M. Kudo, Reagentless and regenerable immunosensor for monitoring of immunoglobulin G based on non-separation immunoassay. *Biosens. Bioelectron.* **18**, 791–795 (2003).
137. C.A. Rowe, S.B. Scruggs, M.J. Feldstein, J.P. Golden, and F.S. Ligler, An array immunosensor for simultaneous detection of clinical analytes. *Anal. Chem.* **71**, 433–439 (1999).
138. S. Aoyagi and M. Kudo, Development of fluorescence change-based, reagent-less optic immunosensor. *Biosens. Bioelectron.* **20**, 1680–1684 (2005).
139. S. Santra, P. Zhang, K.M. Wang, R. Tapec, and W.H. Tan, Conjugation of biomolecules with luminophore-doped silica nanoparticles for photostable biomarkers. *Anal. Chem.* **73**, 4988–4993 (2001).
140. L.J. Kricka, Clinical applications of chemiluminescence. *Anal. Chim. Acta* **500**, 279–286 (2003).
141. J.H. Lin, F. Yan, X.Y. Hu, and H.X. Ju, Chemiluminescent immunosensor for CA19-9 based on antigen immobilization on a cross-linked chitosan membrane. *J. Immunol. Methods* **291**, 165–174 (2004).

142. R. Pandian, J.L. Lu, and J. Ossolinska-Plewnia, Fully automated chemiluminometric assay for hyperglycosylated human chorionic gonadotropin (invasive trophoblast antigen). *Clin. Chem.* **49**, 808–810 (2003).
143. T. Konry, A. Novoa, Y. Shemer-Avni, N. Hanuka, S. Cosnier, A. Lepellec, and R.S. Marks, Optical fiber immunosensor based on a poly(pyrrole-benzophenone) film for the detection of antibodies to viral antigen. *Anal. Chem.* **77**, 1771–1779 (2005).
144. H. Qi and C. Zhang, Homogeneous electrogenerated chemiluminescence immunoassay for the determination of digoxin. *Anal. Chim. Acta* **501**, 31–35 (2004).
145. G. Sauerbrey, Use of a quartz vibrator for weighing thin layers on a microbalance. *Z. Phys.* **155**, 206–222 (1959).
146. C.K. O'Sullivan, R. Vaughan, and G.G. Guibault, Piezoelectric immunosensors – theory and applications. *Anal. Lett.* **32**, 2353–2377 (1999).
147. R.L. Bunde, E.J. Jarvi, and J.J. Rosentreter, Piezoelectric quartz crystal biosensors. *Talanta* **46**, 1223–1236 (1998).
148. S.J. Martin, V.E. Granstaff, and G.C. Frye, Characterization of a quartz crystal microbalance with simultaneous mass and liquid loading. *Anal. Chem.* **63**, 2272–2281 (1991).
149. K. Das, J. Penelle, V.M. Rotello, and K. Nüsslein, Specific recognition of bacteria by surface-templated polymer films. *Langmuir* **19**, 6226–6229 (2003).
150. H. Wang, H. Zeng, Z. Liu, Y. Yang, T. Deng, G.L. Shen, and R.Q. Yu, Immunophenotyping of acute leukemia using an integrated piezoelectric immunosensor array. *Anal. Chem.* **76**, 2203–2209 (2004).
151. S. Tombelli, M. Mascini, and A.P.F. Turner, Improved procedures for immobilisation of oligonucleotides on gold-coated piezoelectric quartz crystals. *Biosens. Bioelectron.* **17**, 929–936 (2002).
152. X.L. Su and Y.B. Li, Micromechanical cantilever array sensors for selective fungal immobilization and fast growth detection. *Biosens. Bioelectron.* **21**, 840–856 (2005).
153. G.Y. Shen, H. Wang, S.Z. Tan, J.S. Li, G.L. Shen, and R.Q. Yu, Detection of antisperm antibody in human serum using a piezoelectric immunosensor based on mixed self-assembled monolayers. *Anal. Chim. Acta* **540**, 279–284 (2005).
154. G.H. Kim, A.G. Rand, and S.V. Letcher, Impedance characterization of a piezoelectric immunosensor. Part II: *Salmonella typhimurium* detection using magnetic enhancement. *Biosens. Bioelectron.* **18**, 91–99 (2003).
155. R.C. Ebersole and M.D. Ward, Amplified mass immunosorbent assay with a quartz crystal microbalance. *J. Am. Chem. Soc.* **110**, 8623–8628 (1988).
156. C.C. Su, T.Z. Wu, L.K. Chen, H.H. Yang, and D.F. Tai, Development of immunochips for the detection of dengue viral antigens. *Anal. Chim. Acta* **479**, 117–123 (2003).
157. X.D. Su, S.F.Y. Li, and J. O'Shea, Au nanoparticle- and silver-enhancement reaction-amplified microgravimetric biosensor. *Chem. Commun.* **8**, 755–756 (2001).
158. S. Kurosawa, E. Tawara, N. Kamo, F. Ohta, and T. Hosokawa, Latex piezoelectric immunoassay detection of agglutination of antibody-bearing latex using a piezoelectric quartz crystal. *Chem. Pharm. Bull.* **38**, 1117–1120 (1990).
159. H. Wang, C. Lei, J. Li, Z. Wu, G.L. Shen, and R.Q. Yu, A piezoelectric immunoagglutination assay for Toxoplasma gondii antibodies using gold nanoparticles. *Biosen. Bioelectron.* **19**, 701–709 (2004).
160. M. Muratsugu, S. Kurosawa, and N. Kamo, Detection of antistreptolysin O antibody: application of an initial rate method of latex piezoelectric immunoassay. *Anal. Chem.* **64**, 2483–2487 (1992).
161. H.O. Ghourchian and N. Kamo, Improvement of latex piezoelectric immunoassay: detection of rheumatoid factor. *Talanta* **41**, 401–406 (1994).
162. H.O. Ghourchian and N. Kamo, Latex piezoelectric immunoassay: effect of interfacial properties. *Anal. Chim. Acta* **300**, 99–105 (1995).
163. X. Chu, G.L. Shen, F.Y. Xie, and R.Q. Yu, Polymer agglutination-based piezoelectric immunoassay for the determination of human serum albumin. *Anal. Lett.* **30**, 1783–1796 (1997).
164. J.H. Lee, K.S. Hwang, J. Park, K.H. Yoon, D.S. Yoon, and T.S. Kim, Immunoassay of prostate-specific antigen (PSA) using resonant frequency shift of piezoelectric nanomechanical microcantilever. *Biosens. Bioelectron.* **20**, 2157–2162 (2005).

165. P. Dutta, C.A. Tipple, N.V. Lavrik, P.G. Datskos, H. Hofstetter, O. Hofstetter, and M.J. Sepaniak, Enantioselective sensors based on antibody-mediated nanomechanics. *Anal. Chem.* **75**, 2342–2348 (2003).
166. C.A. Grimes, C.S. Mungle, K. Zeng, M.K. Jain, W.R. Dreschel, M. Paulose, and K.G. Ong, Invited paper: wireless magnetoelastic resonance sensors: a critical review. *Sensors* **2**, 289–308 (2002).
167. C.M. Ruan, K.F. Zeng, O.K. Varghese, and C.A. Grimes, Magnetoelastic immunosensors: amplified mass immunosorbent assay for detection of *Escherichia coli* O157:H7. *Anal. Chem.* **75**, 6494–6498 (2003).
168. Y.M. Bae, B.K. Oh, W. Lee, W.H. Lee, and J.W. Choi, Immunosensor for detection of *Yersinia enterocolitica* based on imaging ellipsometry. *Anal. Chem.* **76**, 1799–1803 (2004).
169. Z.H. Wang and G. Jin, A label-free multisensing immunosensor based on imaging ellipsometry. *Anal. Chem.* **75**, 6119–6123 (2003).
170. L. Alfonta, I. Willner, D.J. Throckmorton, and A.K. Singh, Electrochemical and quartz crystal microbalance detection of the cholera toxin employing horseradish peroxidase and GM1-functionalized liposomes. *Anal. Chem.* **73**, 5287–5295 (2001).
171. J.G. Guan, Y.Q. Miao, and J.R. Chen, Prussian blue modified amperometric FIA biosensor: one-step immunoassay for α-fetoprotein. *Biosens. Bioelectron.* **19**, 789–794 (2004).
172. Z.H. He, N. Gao, and W.R. Jin, Capillary electrophoretic enzyme immunoassay with electrochemical detection using a noncompetitive format. *J. Chromatogr. B* **784**, 343–350 (2003).
173. D.J. Cahill, Protein and antibody arrays and their medical applications. *J. Immunol. Methods* **250**, 81–91 (2001).
174. L. Belov, O. de la Vega, C.G. dos Remedios, S.P. Mulligan, and R.I. Christopherson, Immunophenotyping of leukemias using a cluster of differentiation antibody microarray. *Cancer Res.* **61**, 4483–4489 (2001).
175. L.J. Yang, Y.B. Li, and G.F. Erf, Interdigitated array microelectrode-based electrochemical impedance immunosensor for detection of *Escherichia coli* O157:H7. *Anal. Chem.* **76**, 1107–1113 (2004).
176. R.M.T. de Wildt, C.R. Mundy, B.D. Gorick, and I.M. Tomlinson, Antibody arrays for high-throughput screening of antibody–antigen interactions. *Nature Biotech.* **18**, 989–994 (2000).
177. T.A. Joos, M. Schrenk, P. Hopfl, K. Kroger, U. Chowdhury, D. Stoll, D. Schorner, M. Durr, K. Herick, S. Rupp, K. Sohn, and H. Hammerle, A microarray enzyme-linked immunosorbent assay for autoimmune diagnostics. *Electrophoresis* **21**, 2641–2650 (2000).
178. A. Bange, H.B. Halsall, and W.R. Heineman, Microfluidic immunosensor systems, *Biosens. Bioelectron.* **20**, 2488–2503 (2005).
179. S.J. Lee and S.Y. Lee, Micro total analysis system (μ-TAS) in biotechnology. *Appl. Microbiol. Biotechnol.* **64**, 289–299 (2004).
180. S. Tombelli, M. Minunni, and M. Mascini, Analytical applications of aptamers. *Biosens. Bioelectron.* **20**, 2424–2434 (2005).
181. S.D. Jayasena, Aptamers: an emerging class of molecules that rival antibodies in diagnostics. *Clin. Chem.* **45**, 1628–1650 (1999).
182. E. Baldrich, J.L. Acero, G. Reekmans, W. Laureyn, and C.K. O'Sullivan, Displacement enzyme linked aptamer assay. *Anal. Chem.* **77**, 4774–4784 (2005).

CHAPTER 10

Microelectrodes for *in-vivo* determination of pH

David Daomin Zhou

10.1 INTRODUCTION

pH measurement techniques have been widely used in chemical, biological, biomedical, and clinical laboratories. Reliable and accurate analytical methods are essential for successful pH monitoring *in vivo*. The method should require only a small volume of sample, and be rapid, simple, and reliable. Potentiometric measurement using microelectrodes is the technique of choice for most *in-vivo* applications of pH determination.

10.1.1 Significance of pH measurement *in vivo*

pH is a critical measurement for our human health. The monitoring of pH levels can serve as a marker for diagnosing diseases, as a guide for optimizing medical treatments, and as an indicator for monitoring biochemical and biological processes. When the body's ability to regulate its pH is compromised, the consequences are serious. In cancer research, an elevated intracellular pH is an important trigger for cell proliferation [1, 2]. Intracellular pH in tumor cells is normal or slightly alkaline; in contrast, extracellular pH, which varies with the type of tumor, is usually acidic. Intracellular pH has been shown to be alkaline in cancer cells [3]. Rich *et al.* [4] noted that the intracellular pH was approximately 0.4 units higher in various leukemia cell lines and leukemia cells than in peripheral blood mononuclear cells from healthy donors. The reduction of intracellular pH resulted in inhibition of overexpression of cyclooxygenase-2 enzyme in human colon cancer cells [5]. Measurement of extracellular pH can provide guidance not only for cancer diagnosis but also for choice of therapy. Most chemotherapeutic agents are weak bases or weak acids, so the measurement of pH gradient in cancer cells can help to determine their therapeutic efficacy [3].

Furthermore, pH determination has been used in other clinical research, both alone and in combination with other measurements. This research includes studies into the relationship between extracellular and intracellular pH in an ischemic heart [6, 7], the pH of airway lining fluid in respiratory disease [8], the study of pH as a marker for pyloric stenosis [9], malnutrition in alkalotic peritoneal dialysis patients [10], pH modulation of heterosexual HIV transmission [11, 12], and wound prevention and treatment [13]. In addition, pH changes due to blood acidosis have been used to trigger and pace the ventricular rate of an implanted cardiac pacemaker [14]. Research using pH measurements have also shown that ischemia from reduced cerebral blood flow can lead to brain tissue acidosis, which can cause brain damage [15]. Finally, pH measurements in the eye have revealed that hyperglycemia induce an acidification of 0.12 pH in the retina, which may contribute to the development of diabetic retinal disease [16].

The importance of monitoring pH changes during neural stimulation has been demonstrated in studies for cochlear [17], retina [18, 19], spinal dorsal horns [20], and dopamine cell bodies in the brain [21]. A shift in pH due to electrical stimulation will change the electrode's corrosion potential and cause electrode materials to dissolve. More importantly, pH changes will also affect cell function and cause tissue damage, by altering the structure and activity of proteins, ionic conductance of the neural membrane, and neuronal excitability. Measurements of pH changes are important for studying the tissue/electrode interaction. This aids in the development of neural prostheses and stimulation protocols to ensure minimal pH changes and provide safe electrical stimulations.

10.1.2 Techniques of measurement of pH *in vivo*

A variety of techniques have been used for the measurement of pH. These techniques range from simple pH dye indicator, to conventional pH electrodes, to very sophisticated pH systems employing spectrometers such as nuclear magnetic resonance (NMR) spectroscope [22]. Based on the nature of the physical detection used in transducers, pH measurement systems can be mainly classified as optical, gravimetric, and electrochemical. Optical detection systems are based on light-sensitive elements. Gravimetric transducers are based on sensitive detection of mass changes following concentration changes. Electrochemical detection using pH sensing electrodes is based on detection of changes in potential, current or impedance.

The optical signal detection of pH can be conducted by spectrophotometric, spectrofluorimetric, or other related techniques [23]. Spectrometric pH sensors consist of functional organic or inorganic indicator dye incorporated in porous glass beads (e.g. sol-gel glasses). These sensors are interfaced with a spectrometer, allowing colorimetric changes with pH to be measured. The indicator dyes are incorporated into the sensor device by non-covalent entrapment within the sol-gel matrix [24] or by covalent binding [25]. However, many of these devices suffer from drawbacks such as dye leaching and slow response times [26].

Various optical detection methods have been used to measure pH *in vivo*. Fluorescence ratio imaging microscopy using an inverted microscope was used to determine intracellular pH in tumor cells [5]. NMR spectroscopy was used to continuously monitor temperature-induced pH changes in fish to study the role of intracellular pH in the maintenance of protein function [27]. Additionally, NMR spectroscopy was used to map *in-vivo* extracellular pH in rat brain gliomas [3]. Electron spin resonance (ESR), which is operated at a lower resonance, has been adapted for *in-vivo* pH measurements because it provides a sufficient RF penetration for deep body organs [28]. The non-destructive determination of tissue pH using near-infrared diffuse reflectance spectroscopy (NIRS) has been employed for pH measurements in the muscle during ischemia [29], in human atherosclerotic plaques during myocardial infarction [30], and in the intestines during hemorrhagic shock [31].

The electrochemical detection of pH can be carried out by voltammetry (amperometry) or potentiometry. Voltammetry is the measurement of the current potential relationship in an electrochemical cell. In voltammetry, the potential is applied to the electrochemical cell to force electrochemical reactions at the electrode–electrolyte interface. In potentiometry, the potential is measured between a pH electrode and a reference electrode of an electrochemical cell in response to the activity of an electrolyte in a solution under the condition of zero current. Since no current passes through the cell while the potential is measured, potentiometry is an equilibrium method.

Although a few amperometric pH sensors are reported [32], most pH electrodes are potentiometric sensors. Among various potentiometric pH sensors, conventional glass pH electrodes are widely used and the pH value measured using a glass electrode is often considered as a gold standard in the development and calibration of other novel pH sensors *in vivo* and *in vitro* [33]. Other pH electrodes, such as metal/metal oxide and ISFETs have received more and more attention in recent years due to their robustness, fast response, all-solid format and capability for miniaturization. Potentiometric microelectrodes for pH measurements will be the focus of this chapter.

10.1.3 Advantages of microelectrodes for the determination of pH

In general, optical-based pH measurement techniques require relatively expensive and cumbersome instruments, and their sophisticated method cannot be easily carried out for routine assay. Interfering contact and reactions of the dye molecules, particularly considering *in-vivo* measurements, cannot be excluded [34]. Some other possible factors, such as a weaker signal at shorter response times, complications in microfabrication, and difficulties in attaching the chemical or biological agents to the small fiber tip, are potential limitations for the application of these optical sensors to *in-vivo* measurements in micro environments [35].

Microelectrode pH sensors offer advantages over other techniques, specifically in biology and medicine, as they allow sensitive detection in spatially and temporally resolved measurements in small sample volumes. Furthermore, those that are potentiometric particularly have the advantage over other sensor types, being simpler in both sensor design and measurement device. Generally speaking, microelectrode pH sensors offer higher sensitivity and selectivity for fine variation of pH measurement *in vivo* than optical pH sensors. They are very reliable, provide a wide dynamic pH response, and are convenient to use.

10.2 CHARACTERIZATION OF pH MICROELECTRODES

10.2.1 pH and pH measurements

pH is the measurement of the acidity or alkalinity of a solution. Concepts of acids and bases in analytical chemistry are commonly based on the theories developed by

Arrhenius, Bronsted, and Lewis. Arrhenius was the first person to give a definition of an acid and a base in 1887. According to Arrhenius, acids were substances that contained hydrogen and yielded hydrogen ions in aqueous solutions; bases contained the OH group and yielded hydroxide ions in aqueous solutions. Bronsted-Lowry in 1923 introduced the familiar concept of the conjugate acid–base pair and developed the more generalized concept that an acid is a hydrogen ion or proton donor, and that a base is a hydroxide ion or proton acceptor. An even broader theory of the acid–base theory was introduced in 1923 by Lewis, who extended the concept to defining an acid as an electron acceptor and a base as an electron donor. In the body where all the fluids are basically water solutions, the concepts of Arrhenius and Bronsted are preferred.

In 1909, Danish biochemist Soren Sorenson first introduced the pH concept. pH is defined by the negative logarithm of the hydrogen ion activity:

$$pH = -\log a_H \quad (1)$$

where a_H is the hydrogen ion activity. For most purposes, the difference between the hydrogen ion activity and the hydrogen ion concentration for dilute aqueous solutions can be ignored. Therefore, it is more convenient to define pH as

$$pH = -\log[H^+] \quad (2)$$

where $[H^+]$ is the hydrogen ion concentration in mol/L.

A schematic diagram of a typical pH electrode system is shown in Fig. 10.1. The cell potential, i.e. the electromotive force, is measured between a pH electrode and a reference electrode in a test solution. The pH electrode responds to the activity or concentration of hydrogen ions in the solution. The reference electrode has a very stable half-cell potential. The most commonly used reference electrodes for potentiometry are the silver/silver chloride electrodes (Ag/AgCl) and the saturated calomel electrodes (SCE).

This cell voltage measurement is made with a pH/mV meter under equilibrium conditions, i.e. at open circuit potential with zero current. The pH/mV meter is actually a high input impedance voltmeter. The first commercial pH meter (Beckman G) was developed by Beckman in the 1930s. The meter was built with a vacuum tube and housed in a walnut box. It was designed for testing citrus juice [36]. Taking advantage of the fast progress made in IC circuit technology, today's advanced digital pH meters are much smaller in size, have more functions, and provide reliable and accurate pH measurements.

The properties of a pH electrode are characterized by parameters like linear response slope, response time, sensitivity, selectivity, reproducibility/accuracy, stability and biocompatibility. Most of these properties are related to each other, and an optimization process of sensor properties often leads to a compromised result. For the development of pH sensors for *in-vivo* measurements or implantable applications, both reproducibility and biocompatibility are crucial. Recommendations about using ion-selective electrodes for blood electrolyte analysis have been made by the International Federation of Clinical Chemistry and Laboratory Medicine (IFCC) [37]. IUPAC working party on pH has published IUPAC's recommendations on the definition, standards, and procedures

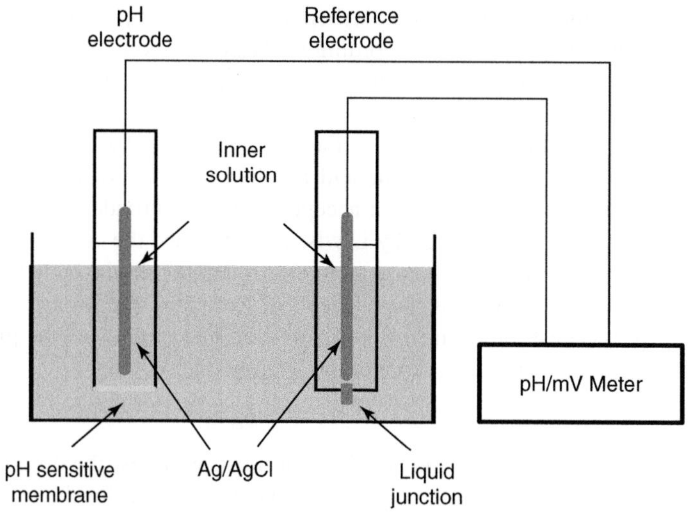

FIGURE 10.1 A schematic diagram for a typical electrode system for potentiometric pH measurements. A potential is established on the pH sensitive membrane–solution interface of a pH electrode that responds to the activity or concentration of hydrogen ions in the solution. The reference electrode has a very stable half-cell potential. The cell potential, which is proportional to the pH in the test solution, is measured using a high input impedance voltmeter between the pH electrode and the reference electrode.

of pH measurement [38]. A pH electrode is selective towards hydrogen ions and these documents can serve as a guideline for the characterization of pH sensing properties.

10.2.2 Calibration curve and linear response slope of pH microelectrodes

The principle of pH electrode sensing mechanisms which are based on glass or polymer membranes is well investigated and understood. Common to all potentiometric ion selective sensors, a pH sensitive membrane is the key component for a sensing mechanism. When the pH sensitive membrane separates the internal standard solution with a constant pH from the test solution, the potential difference E across the membrane is determined by the Nernst equation:

$$E = \text{constant} + (RT/F) \ln [\text{H}^+] \tag{3}$$

where R is the gas constant, T is the absolute temperature (K), F is the Faraday constant, and $[\text{H}^+]$ is the hydrogen ion concentration in mol/L. Replaced with $\text{pH} = -\log[\text{H}^+]$, the equation has a form as follows:

$$E = \text{constant} - (2.303RT/F)\,\text{pH} = \text{constant} - (\text{slope})\,\text{pH} \tag{4}$$

where $2.303RT/F$ is the slope of the line plot of E vs pH (also known as the slope factor), which is the basis of the pH electrode calibration curve. Strictly speaking, the activity of hydrogen ions should be used in the Nernst equation. However, in dilute solutions, activity of H^+ almost equals its molar concentration. In *in-vivo* and clinical applications, the molar concentration is used rather than the activity of H^+.

Using a series of calibration solutions, the response curve or calibration curve of a pH electrode can be experimentally determined by plotting the cell voltage vs the pH of the calibration solution. The linear range of the calibration curve is applied to determine the pH in any unknown solution. The slope of the calibration curve within the linear range is used to determine the response slope or electrode sensitivity in mV/pH. This response slope is an important diagnostic characteristic of the electrode; generally, the slope gets lower as the electrode gets old or contaminated [39].

For a pH sensor with a Nerstian response, the slope, calculated from $2.303RT/F$, is 59.16 mV/pH at 298 K. A useful slope range can be regarded as 50–70 mV/pH. Super-Nerstian slopes are mostly reported from electrodes based on metal/metal oxide sensing materials. It has been reported that some electrodes present two linear response ranges. For example, IrOx microelectrodes prepared by Wipf *et al.* [40] have shown two linear slopes in the range of pH 2–6 and 6–12.

It is clear from the Nernst equation that the temperature of the solution affects the response slope ($2.303RT/F$) of the calibration curve. The electrode voltage changes linearly in relationship to changes in temperature at a given pH; therefore, the pH of any solution is a function of its temperature. For example, the electrode response slope increases from 59.2 mV/pH at 25ºC to 61.5 mV/pH at a body temperature of 37ºC. For modern pH sensing systems, a temperature probe is normally combined with the pH electrode. The pH meter with an automatic temperature compensation (ATC) function automatically corrects the pH value based on the temperature of the solution detected with the temperature probe.

10.2.3 Sensitivity

The sensitivity of a pH electrode is determined by the linear response slope of the pH electrode as defined by the Nernst equation. Typically, the electrode calibration curve exhibits a linear response range between a pH of 2 and 9. At very high and very low pH, there are deviations from linearity. The high detection end (high $[H^+]$, low pH) of most pH sensors is limited by the so-called *acid error*. The electrode reads higher than the actual pH in very acidic solutions. The mechanism of such an error is not well understood. The lower detection limit (low $[H^+]$, high pH) is often governed by the selectivity of the sensor. At high pH, alkaline interfering ions such Na^+ or K^+ are about 8 to 9 orders of magnitude higher than H^+ in the solution. Electrodes respond slightly to Na^+ or K^+, giving a lower reading than the actual pH. This deviation from the actual pH is often referred to as *alkaline error*.

10.2.4 Response time

Response time is defined as the time at which the pH concentration in a solution is changed on contact with a pH sensor and a reference electrode has reached 95% (or 90%)

of the final value. The response time is reported as $t_{95\%}$ or $t_{90\%}$ in seconds or minutes. In many investigations, the response time of the overall measuring system is determined, which influences the response time of the pH sensor. Generally speaking, metal/metal oxide-based pH sensors present a faster response time than that of glass membrane electrodes. However, when covered with a polymer membrane to eliminate or minimize interference, the response time of oxide-based electrodes was adversely affected [41].

10.2.5 Reproducibility/accuracy

Accuracy is a measure of how close the result is to the true value; while reproducibility or precision is a measure of how close a series of measurements on the same sample are to each other. The accuracy and reproducibility of pH measurements can be highly variable and are dependent on several factors: electrode stability (drift and hysteresis), response slope/calibration curve, and accuracy of the pH meters. While some of these factors are determined by the properties of electrodes, some measures can be taken to improve measurement accuracy and reproducibility.

The concentration is proportional to the measured voltage and so any error in voltage measurement will cause an error in the solution concentration. The measured voltage is the cell voltage including different potentials generated at all-solid–solid, solid–liquid and liquid–liquid interfaces of both sensing and reference electrodes. The potential of the electrochemical cell, E_{cell}, is mainly given by

$$E_{cell} = E_{pH} - E_{ref} + E_j + E_{xy} \tag{5}$$

where E_{pH} is the half-cell potential of the pH electrode, E_{ref} is the half-cell potential of the reference electrode, E_j is the liquid-junction potential, and E_{xy} is the interfering ions induced potential which affects the electrode's selectivity. Any variations in these potentials will cause changes in the overall cell voltages. Frequent recalibration can minimize potential drift, while using protection membranes can reduce the effect from interfering ions. Liquid-junction potentials develop at the interface between two electrolytes because of the differences in the migration rates of charged species across the interface [42]. By using reference filling solutions with nearly equitransferrent electrolytes such as KCl, in which both ions have similar mobility when diffusing through the liquid junction, E_j can be minimized. E_{pH} is more strongly dependent on temperature in most cases than E_{ref} and E_j are. Calibrations and measurements should therefore be carried out under temperature controlled conditions [38].

The error from pH measurement systems depends on the slope of the calibration line. For a pH sensor with a slope of 59 mV/pH, an error of 1 mV in measuring the electrode potential will cause a 0.017 pH change in the concentration. The lower the slope, the higher the errors are on the sample measurements [39].

Several factors need to be considered to reduce the pH measurement error. First, an electrode with a high response slope should be used. Second, it is important to use a meter that is capable of measuring the millivolts or microvolts accurately and precisely. With modern meter technology, this is not normally a limiting factor. The potential measuring devices regularly used generally have an accuracy of 0.1 mV or higher (0.01 mV).

Third, the electrode should be calibrated in the pH range close to the intended applications. Although the calibration curve may show a straight line over several decades of concentration with an average slope, it is unlikely that this will be exactly the same across the whole range. For example, when multi-points (five points) calibrations are made, there may be a variation of several millivolts between the individual slopes calculated from two adjacent points at different concentration ranges or between the individual slopes and the overall slope. Therefore, for the most accurate results, it is recommended that the electrode slope be determined by using two-point calibration (also called bracketing) using two standard buffer solutions. These two buffer solutions are often selected that closely cover the expected range of the samples. This is especially important for some *in-vivo* measurements, where the pH variation is within a very narrow range.

In some *in-vivo* applications, one-point calibration is used. Although a one-point calibration is insufficient to determine both the slope and one-point pH value, by assuming that the slope of prior calibration is unchanged, the electrode's performance *in vivo* can be assessed during operation. For example, in measuring esophageal acid exposure, since pH 4 is accepted as a cut-off value, one-point calibration was carried out during *in-vivo* ambulatory pH monitoring in which patients swallowed a juice with a predetermined pH close to 4 [33].

The desired results for a practical measurement vary from application to application. In many pH measurements, accuracy within 0.1 pH unit is more than sufficient while 0.01 pH is required in some other more accurate tests. Reaching an accuracy of less than 0.01 pH will be a challenge for most measurement systems.

10.2.6 Selectivity

Selectivity is one of the most important characteristics of an electrode, as it often determines whether a reliable measurement in the sample is possible or not. In practice, most pH sensitive membranes will also respond slightly to some interfering ions. As can be seen in Eq. (5), the potential of such a membrane is governed mainly by the activity of the hydrogen ion and also by the concentration of other interfering ions. To improve selectivity, advanced membrane compositions or protection membranes with size-exclusion or ion-exchange properties are often utilized.

It is reported that the greatest interference that affects selectivity of an oxide-based pH electrode is from redox couples. For practical applications, some oxide-based sensors are covered with a thin layer of a size-exclusive protection membrane such as Nafion or polyphenol [43]. However, the improvement in selectivity is at the expense of the response slopes and, sometimes, the response times.

10.2.7 Stability and reliability

Poor operational stability due to drift has largely limited the long-term or implantable application of pH sensors. Some oxide-based electrodes present very high initial potential drift but their stability is improved after soaking for a certain amount of time ranging from hours to days [44]. Shelf-life also affects the storage stability of some sensors, such as glass electrodes. When all or most of the above properties are optimized, a reliable pH sensor system can be achieved for *in-vivo* measurements.

10.2.8 Biocompatibility

Biomaterials are inert substances that are used in contact with living tissue, resulting in an interface between living and non-living substances [45, 46]. Biocompatibility of this interface is achieved by using such biomaterials for encapsulation in the construction of sensor devices.

The interaction in an interface of device/tissue is limited by two factors. There is the corrosive environment, such as biological fluid, which contains salts and proteins among other cellular structures in which the sensor device must survive [47, 48]. Second, there is the encapsulation material which may induce a toxic reaction due to poor biocompatibility and hemocompatibility [49, 50]. It is crucial to use a biomaterial that can overcome both limiting factors to maintain the lifetime of the sensor device and protect the body [51, 52].

10.3 FABRICATION OF MICROELECTRODES FOR pH DETERMINATION

Various pH microelectrodes and numerous fabrication techniques have been developed. Some selected examples of microelectrodes based on the formats of pH sensitive materials are discussed in this section.

10.3.1 Glass-based pH microelectrodes

Glass electrodes are the most commonly used potentiometric sensors for pH measurements [53]. The pH response of glass electrodes was first observed by Max Cremer in 1906. While studying liquid–solid interactions, Cremer discovered that the solid–liquid interface, which was separated by blowing a thin bubble of glass, created an electric potential that could be measured. Later, Fritz Haber and Zygmunt Klemensiewicz applied this potential to hydrogen ion activity in 1909. They used the glass bulb to measure hydrogen ion activity and named it the glass electrode [54, 55].

A schematic diagram of a combination glass electrode is shown in Fig. 10.2. The glass electrode consists of a glass tube with a thin glass bulb at the tip. Inside is a known solution of potassium chloride (KCl), buffered at a pH of 7.0. A silver wire with an Ag/AgCl electrode tip makes contact with the inner solution. An outer jacket contains a second Ag/AgCl in an electrolyte with known Cl^- concentration, such as saturated KCl. This Ag/AgCl serves as the reference electrode during pH measurements. A combined thermocouple is used to measure and compensate for temperature effects.

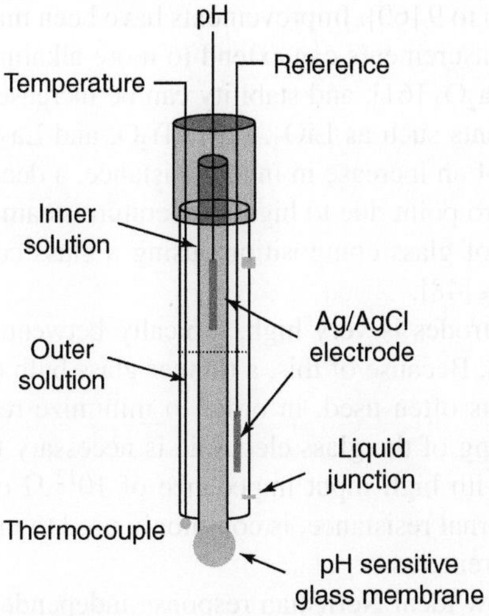

FIGURE 10.2 A schematic diagram of a combination glass pH electrode. A thin glass bulb with an inner Ag/AgCl electrode responds to pH changes in the test solution. A second Ag/AgCl in an outer jacket with a liquid junction serves the reference electrode for potentiometric measurement. An attached temperature probe is used to compensate for temperature effects.

When the glass membrane is immersed in an aqueous solution, a thin hydrated layer (~10 nm to 10 μm) is formed on the glass–solution interface, depending on the composition of the glass and soak environments [56]. Monovalent ions such as Na^+ from the glass exchange with the H^+ ions of the solution in a diffusion process. At equilibrium, there is a build-up of charge, therefore a potential is established across the membrane that is proportional to the hydrogen ion concentration in the test solution. According to the Nernst equation as described in Eqs (3) and (4), the glass electrode potential is related to the pH as follows:

$$E = k - 59\ (\mathrm{mV/pH})\ \mathrm{pH} \qquad (6)$$

where k is a constant and the slope 59 (mV/pH) is calculated at 25ºC.

Generally accepted theories about the origin of the pH response of glasses are based on the assumption that hydrogen ions do not cross the glass membrane. Studies reported by Abe and Maeda [57] suggest that some mobile hydrogen ions exist in glass and are responsible for pH response. The diffusion coefficient of hydrogen ions in pH sensitive glass membranes is estimated based on a stimulated model and experimental data to be 1×10^{-8} cm²/s [58]. This value might be overestimated as it is three orders of magnitude higher than that of non-plasticized membranes (10^{-11} to 10^{-12} cm²/s) but in the same order of magnitude as the pH ion carrier in ionophore doped plasticized polymer membranes [59].

The pH sensing glass consists of mainly SiO_2 with some non-silicon components. The most widely known example of pH sensitive glass is Corning 015, which contains

SiO_2, Na_2O, and CaO for pH up to 9 [60]. Improvements have been made in the composition of the glass such that measurements can extend to more alkaline ranges with the aid of LiO_2, Cs_2O, BaO, and La_2O_3 [61], and stability can be increased and impedance can be lowered using components such as LiO_2, BaO, Ti_2O, and La_2O_3 [62]. Damage of pH sensitive glass in terms of an increase in inner resistance, a decrease in electrode sensitivity, and a shift of the zero point due to high temperature steam sterilization, was overcome by the optimization of glass compositions using a glass consisting of SiO_2, Li_2O, La_2O_3, Nd_2O_3, and Nb_2O_5 [48].

The resistance of glass electrodes is very high, typically between 10 and 100 MΩ, with some as high as 4 GΩ [60]. Because of this, a thinner glass bulb (50–200 μm) with sufficient mechanical strength is often used, in order to minimize resistance as much as possible. In addition, shielding of the glass electrode is necessary to reduce electromagnetic noise. A pH meter with high input impedance of 10^{12} Ω or more, which is much higher than electrode internal resistance, is commonly used to minimize electrode polarization for accurate measurements.

The glass pH electrodes show ideal Nernstian response independent of redox interferences and have a long lifetime; however, they do not measure correct values when hydrogen ion concentration is either high or low. This is known as acid error or alkaline error [63]. While acid error starts approximately at pH = 2 for most pH glasses, the alkaline error starts at pH = 8 to 11 depending on the sources and composition of the glass membrane. At these pH values the electrode is sensitive to sodium and potassium ions [54]. However, some researchers attributed the deviation to two non-linear electrode response slopes, one for H^+ and one for OH^- [64]. The temperature effect for glass electrodes is reduced or compensated by using a stand-alone or built-in thermistor, or a thermocouple as a temperature probe and an automatic temperature correction (ATC) function, found in most modern pH meters.

10.3.2 Polymer membrane-based pH microelectrodes

Although glass pH electrodes are, in general, simple to use and available at a reasonable cost, they are limited by the potential problems of glass breakage [65] and miniaturization difficulties [60, 66]. One of the alternative approaches to preparation of non-glass pH sensors is to use polymer-based pH sensitive membranes to replace solid glass membranes.

Unlike the glass membrane, which contains fixed binding sites, polymer membranes contain incorporated mobile ion exchangers or ionophores (ion carriers) in the membrane matrix. The ionophore is usually present in small amounts (approximately 1% or 10^{-2} M), which is relatively low when compared to the glass electrode. Such ion carriers are able to complex with ions reversibly and transfer them through a polymer membrane by carrier translocation [67]. A variety of ion carriers (ion exchanger, neutral or charged carrier) have been used in the preparation of pH sensitive microelectrodes showing Nerstian potentiometric pH response [63].

The dynamic pH response range of a membrane-based pH electrode can be tailored by the incorporation of different functional groups in the ion carriers. Yuan *et al.* [68]

reported a PVC membrane electrode with a wide pH linear range of 1.7 to 13.2 based on 4,4′-bis[(N,N-dialyl-amino)methyl]azobenzene. The long alkyl chains in the tertiary amino group of the ionophore are sensitive in the basic range; while the azo group is sensitive in the acidic range. In addition to pH ionophore/polymer membranes, conductive polymers have been used in the fabrication of pH sensors [69].

Among all the polymers used in preparing ion-selective membranes, poly(vinylchloride) (PVC) is the most widely used matrix due to its simplicity of membrane preparation [32, 70]. In order to ensure the mobility of the trapped ionophore, a large amount of plasticizer (approximately 66%) is used to modify the PVC membrane matrix (approximately 33%). Such a membrane is quite similar to the liquid phase, because diffusion coefficients for dissolved low molecular weight ionophores are high, on the order of 10^{-7}–10^{-8}cm^2/s [59].

Polymer-based pH sensors are not suitable for continuous *in-vivo* measurements due to the poor biocompatibility of plasticizers used in the polymer membrane. To minimize such a problem, surface treatment or using a reduced amount of plasticizers has been proposed [71]. In order to improve stability and adhesion, polyurethane has been used as an alternative to PVC membranes in the construction of pH sensing membranes [72, 73].

Some pH sensitive ionophores have been incorporated into liquid membranes in the construction of capillary glass micro pH sensors. Apart from ETH 1907 tridodecylamine [74], and ETH 2418 [4′-(dipropylamino)-2-azobenzene-carboxylic acid octadecylester] [75], the Fluka 95291 hydrogen ion sensitive resin cocktail A, a well-known commercial pH sensitive ionophore, has been used in many studies [76].

10.3.3 Silicon-based pH microelectrodes

The fabrication techniques for IC circuits and MEMs have been adapted in the development of silicon (Si) wafer-based pH sensors. The ion sensitive field-effect transistor (ISFET) for pH measurements is developed on the basis of the metal oxide semiconductor field-effect transistor (MOSFET). The first pH sensitive ISFET reported in 1970 by Bergveld was developed for electrophysiological measurements on neural responses. He used a FET without a metal gate and exposed the bare gate oxide layer to a saline solution [56]. While ISFET finds continuing applications in electrophysiological measurements, such as bacterial activity detection [77] and determination of the pH gradient across a cell membrane [78], it has also become increasingly useful for in-line process monitoring applications, especially for food, pharmaceutical, cosmetic, and water purification industries [52, 65].

A schematic diagram of an ISFET is shown in Fig. 10.3. In the ISFET, the MOSFET's metallic gate is replaced by an insulator layer that is pH sensitive. The ISFET uses a different pH sensing mechanism than that of conventional glass electrodes [66]. When the gate is exposed to an electrolyte solution, there is a pH dependent charge at the double-layer gate insulator–solution interface. Under a constant gate voltage, the induced electric field from the pH dependent surface charge electrostatically influences the current ($i_{s\text{–}d}$) flow between the source and drain. In this way, the current change between the source and

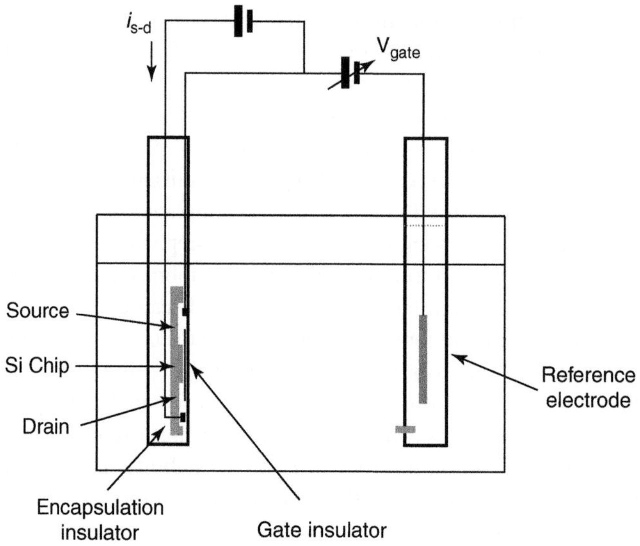

FIGURE 10.3 A schematic diagram of an ISFET pH measurement system. A potential responding to the hydrogen ions concentration in the solution is established on the pH sensitive gate–solution interface of the ISFET. It electrostatically influences the current (i_{s-d}) flow between the source and drain, therefore, the current i_{s-d} change is directly related to the pH change in the solution.

drain is directly related to the pH change in solution. A reference electrode with a stable voltage, independent of pH changes, is required to complete the voltage circuit between the gate and source [79] and to control the gate voltage. This voltage (V_{gate}) circuit has extremely high impedance, while the current circuit via source and drain has a low impedance of a few ohms. For such low impedance, a pre-amplifier, which is required for glass electrodes, may not be necessary, and a long cable can be used for in-line pH measurements.

The heart of an ISFET is the double-layer gate [80]. The ideal gate material should provide a Nernstian pH response, a low alkali metal ion sensitivity, a small hysteresis, a low drift, and a high corrosion resistance [81]. When a native SiO_2 gate surface from the MOSFET process is used, the pH sensitivity is only about 30–40 mV/pH depending on the electrolyte concentration, and the gate is sensitive to both sodium and hydrogen ion concentrations in the solution. Early ISFETs use a silicon nitride (Si_3N_4) coated gate, which has good insulating properties and provides a pH sensitive interface with a Nerstian response of 40–60 mV/pH [56]. However, it suffers from problems of light sensitivity, and both short- and long-term drift [52]. To overcome such problems, research efforts have been made to develop other pH sensitive gate materials, such as Al_2O_3, and Ta_2O_5 [82].

Another example of an Si-based pH sensing device is an electrolyte-insulator-semiconductor (EIS) structure. In a typical EIS sensor, a voltage applied to the semiconductor back gate (*V*gate) attracts ions in the electrolyte to the oxide surface. The

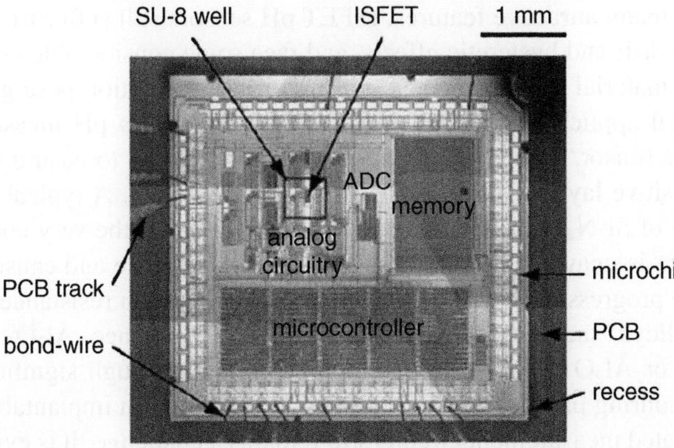

FIGURE 10.4 Microphotograph of the pH ISFET with an on-chip pH meter. The whole device was coated with SU-8 photoresist using standard UV photolithography, leaving a 500 μm square well as the ISFET opening. (Reproduced from [85], with permission from Elsevier.)

arrangement of ions in response to a specific *V*gate is determined by the size and charge of the ions. By measuring the gate capacitance (*C*gate) vs *V*gate, ion species and concentration can be determined [65, 81, 82]. EIS sensors have a response behavior comparable to ISFETs due to their common detection principle and common transducer material. In contrast to ISFETs, the sensitive area of an EIS sensor, in general, is larger than the gate region of an ISFET, and therefore is less suitable for miniaturization [83].

As a supplement for widely used glass electrodes, ISFET sensors provide some unique features. One of them is the capability of dry storage, which helps to avoid the problem of shelf-life and the hydration requirement time for glass electrodes. Fast response is another feature of an ISFET pH sensor. In a comparison study of commercial pH sensors reported by Smit *et al.* [84], the ISFET sensor showed about a ten times faster response than that of glass electrodes, and a response similar to that of an iridium oxide-based electrode [100, 104]. This fast response time of ISFET sensors was attributed to its sensing mechanism, which is based on the electrostatic interaction of H^+ ions with surface charge at the gate surface [66].

A third feature for an ISFET sensor is its capability of miniaturization. Fabrication of ISFET is mainly based on various CMOS (complementary metal-oxide-semiconductor) technologies developed for IC manufacturing with some modifications for applying gate materials [79]. This provides the possibility of miniaturization and opens an avenue for integration of pH ISFET with other sensors as well as measurement circuits. Recently, Hammond and Cumming [85] reported an ISFET that has an integrated digital pH meter (Fig. 10.4). The device was fabricated using a commercial CMOS process with a built-in on-chip drift compensation function. However, the device has a low pH sensitivity of 48 mV/pH. This may be a result of the sensor's unmodified gate, which is limited by the commercial CMOS process used.

Despite many attractive features, ISFET pH sensors still suffer from problems, such as inherent drift and hysteretic effects, and face some considerable challenges. Among them, gate material stability and Si sensor chip encapsulation pose great obstacles for their clinical applications, especially in long-term *in-vivo* pH measurement or as an implantable sensor. Since a thin gate material is required to ensure high capacitance, the pH sensitive layer of ISFET sensors is extremely thin. A typical ISFET has about only 50 nm of Si_3N_4. Dissolution of such gate material in the very corrosive biological environment is unavoidable. This limits the sensor lifetime and causes long-term drift [48]. Some progress has been made in improving corrosion resistance of gate materials by using alloys, such as Al_2O_3—Ta_2O_5 for acidic resistance, Al_2O_3—ZrO_2 for alkali resistance, or Al_2O_3—Ta_2O_5—ZrO_2 for both [81]. Although significant progress has been made during past decades, encapsulation of long-term implantable ISEFT devices with integrated measurement circuits still remains a challenge. It is evident that the protection of the ISFET chip with a standard passivation layer provided by CMOS technology is not enough. Various additional insulation materials such as epoxy, silicone, and polyimide have been reported for the ISFET encapsulation [52]. In fact, encapsulation has been considered a major cost factor in the development of commercial medical devices including chemical sensors [86].

10.3.4 Metal/metal oxide-based pH microelectrodes

A considerable amount of study has been focused on the development, fabrication, and characterization of metal/metal oxide pH electrodes. Typically, the pH electrodes employing metal/metal oxide as sensing materials are all-solid-state, and have several advantages in comparison to conventional glass electrodes. Unlike the glass pH electrodes, they require a high input impedance pH meter; the metal oxide-based pH sensor has low electrode impedance. In contrast to the sluggish response of glass pH electrodes, the solid-state metal oxide pH sensor presents a faster pH response. The method to prepare a metal/metal oxide-based pH electrode is compatible with thin film and MEMS manufacturing technologies and provides capability of mass production and miniaturization.

A variety of metal/metal oxide materials show ideal or near-ideal Nernstian responses; these materials have been explored for use as pH sensing layers [87, 88]. Some examples are IrOx [41, 89], RuO_2 [90, 91], nanoporous PtO_2 [43], RuO_2—TiO_2 [92], TiO_2—PVC [93], PaO_2 [94], Sb/Sb_2O_3 [95], WO_3 [96], PbO_2 [97], Co_3O_4 [44], and SnO_2 [98, 99]. Among these many pH sensitive oxides, iridium oxide (IrOx) shows high conductivity, good chemical stability, and most importantly superior biocompatibility, making it the most popular material for the fabrication of all-solid-state pH microelectrodes. In addition to IrOx, antimony has received renewed interest recently. Antimony has been used as a pH sensing material in commercially available pH catheters for esophageal pH monitoring, as will be discussed later.

Fabrication methods and conditions that determine the structure and composition of iridium oxide affect pH response characteristics of resultant pH sensors. A comparison of some typical iridium oxide-based pH electrodes with respect to the fabrication method and pH sensing characteristics has been made by Madou's group [100]. Marzouk [101] described a modified method to prepare the iridium oxide layer on etched titanium substrate to enhance the stability of the pH sensitive layer. Micro-tip pH electrodes based on iridium oxide on a tapered glass micropipette with a tip of 3–10 μm in diameter [102] and on a carbon fiber [40] have been reported. The iridium oxide coated glass micropipette electrode has a short response time ranging from 1 to 17 s (t_{90}) in the pH range of 4–10 with an accuracy of 0.05 pH. The iridium oxide coated carbon fiber electrode has a fast response time of only 50 ms, and an equilibrium value is reached within 30 s. In a recent study on a planar thin film iridium oxide-based pH sensor, Ges *et al.* [41] reported aging effects of IrOx plating solution on the sensor's sensitivity. Iridium oxide films deposited from fresh plating solution showed maximum sensitivity, while those from a one-month-old solution showed decreases of 10–15% in sensitivity.

Yao *et al.* [100] reported a pH electrode based on lithium carbonate melt-oxidized iridium oxide film with the composition of $Li_xIrO_y \cdot nH_2O$. The electrode based on this oxide film exhibits promising pH sensing performance and high chemical stability, with an ideal Nernstian response 58.9 mV/pH over the pH range of 1 to 13. The electrode also shows a fast potential response with a 90% response time less than 0.2 s, and a low open-circuit potential drift 0.1 mV/day measured in pH 6.6 solution. The reproducibility in terms of the Nernst slopes and the apparent standard electrode potentials has been improved among electrodes within the same batch. However, the biocompatibility due to inclusion of lithium salt was not assessed.

The iridium oxide-based solid-state micro-pH electrodes have been used in many clinical research studies, such as measuring extracellular pH in ischemic hearts [6, 103], in biofilms [104] and in esophagus tubes [105]. Grant *et al.* [106] evaluated an electrochemical pH sensor along with a fiber optic sensor to monitor brain tissue *in vitro* as well as *in vivo*. The electrochemical pH sensor was based on sputtered iridium oxide thin films. The electrochemical sensor was fabricated on a flexible kapton substrate with sputter coating of 200 A titanium for adhesion, followed by 2000–4000 A of iridium, and then finally 2000–4000 A iridium oxide. The Ag/AgCl reference electrode was also fabricated on kapton substrate with a sputtered silver layer, and then the silver chloride top layer was formed electrochemically. The kapton substrates were sectioned into strips with 1 mm in width and 8 cm in length. The iridium oxide and silver/silver chloride electrodes were adhered back-to-back using an epoxy. Tests *in vitro* and *in vivo* indicated that the sensors showed a linear response of 57.9 ± 0.3 (before) and 57.8 ± 1.5 mV/pH (after 3 h *in-vivo* test) between pH 6.8 to 8.0. The t_{90} response time of the electrochemical sensor was less than 5 s. The potential drift was less than 0.4 mV/h in PBS; however, an increased drift was observed in *in-vivo* tests. Protein absorption may account for this increased drift in *in-vivo* tests.

Although the redox reaction mechanisms of iridium oxide are still not clear, most researchers believe that the proton exchange associated with oxidation states of metal oxides is one of the possible pH sensing mechanisms [41, 87, 100, 105]. During electrochemical reactions, oxidation state changes in the hydrated iridium oxide layer are accompanied by the injection/ejection of H^+. The possible reactions involved in the electrode processes are:

Hydration of iridium oxide

$$Ir_2O_3 + 3H_2O => 2Ir(OH)_3 \quad (7)$$

Charge transfer and injection/ejection of hydrogen ion

$$2Ir(OH)_3 \Leftrightarrow 2IrO_2 \cdot H_2O + 2H^+ + 2e^- \quad (8)$$

Depending on the fabrication techniques and deposition parameters, the pH sensitive slope of IrOx electrodes varies from near-Nernstian (about 59 mV/pH) to super-Nernstian (about 70 mV/pH or higher). Since the compounds in the oxide layers are possibly mixed in stoichiometry and oxidation states, most reported iridium oxide reactions use x, y in the chemical formulas, such as $Ir_2O_3 \cdot xH_2O$ and $IrOx(OH)y$. Such mixed oxidation states in IrOx compounds may induce more H^+ ion transfer per electron, which has been attributed to causing super-Nerstian pH responses [41].

Considering the H^+ dependent redox reaction between two oxidation states of the iridium oxide as the basis of the pH sensing mechanism, the electrode potential changes to the hydrogen ion concentration are expressed by Nernstian equation:

$$E = E_o + 2.303RT/F\,(\log[H^+]) = E^o - 61.54\ \text{pH} \quad (9)$$

where E_o is the standard electrode potential with the value of 729 mV vs Ag/AgCl reference electrode. The potential/pH slope is expected to be ~61.5 mV/pH at 37ºC.

10.3.5 Ag/AgCl reference microelectrodes

Reference electrodes provide a standard for the electrochemical measurements. For potentiometric sensors, an accurate and stable reference electrode that acts as a half-cell in the measurement circuit is critical to providing a stable reference potential and for measuring the change in potential difference across the pH sensitive membrane as the pH concentration changes. This is especially important in clinical applications such as pH measurements in the blood, heart, and brain, where the relevant physiological pH range is restricted to a very small range, usually less than one unit.

The three main requirements for a satisfactory reference electrode, given by Ives and Janz [107], are reversibility (non-polarizability), reproducibility, and stability. Hydrogen electrodes have been chosen as the primary reference electrode due to their excellent reproducibility [54]. The electrode is represented schematically as

$$\text{Pt/Pt black}|H_2\ (1\ \text{atm}),\ 1\,\text{M}\ H_2SO_4 \quad (10)$$

which consists of a platinum foil with electroplated Pt black to catalyze the hydrogen electron transfer reaction. The standard potential of the hydrogen electrode is conventionally set to zero at all temperatures, thus establishing a hydrogen scale of standard potential. However, this electrode is impractical in routine usages, thus secondary reference electrodes, such as calomel, sulphate, and silver/silver chloride (Ag/AgCl) are used [80].

As a secondary reference electrode, the Ag/AgCl electrode is the most common due to its simplicity, stability, and capability of miniaturization. A conventional Ag/AgCl reference electrode is a silver wire that is coated with a thin layer of silver chloride either by electroplating or by dipping the wire in molten silver chloride. The electrode reaction is as follows

$$\text{Ag (solid)} + \text{Cl}^- = \text{AgCl (solid)} + \text{e}^- \tag{11}$$

The potential developed is determined by the chloride concentration of the inner solution, as defined by the Nernst equation. As can been seen from the above reaction, the potential of the electrode remains constant as long as the chloride concentration remains constant. Potassium chloride is widely used for the inner solution because it does not generally interfere with pH measurements, and the mobility of the potassium and chloride ions is nearly equal. Thus, it minimizes liquid-junction potentials. The saturated potassium chloride is mainly used, but lower concentrations such as 1 M potassium chloride can also be used. When the electrode is placed in a saturated potassium chloride solution, it develops a potential of 199 mV vs the standard hydrogen electrode.

Silver chloride is slightly soluble (about 6×10^{-3} mol/L at 25°C) in strong potassium chloride solutions [54]. It is a common practice to saturate the potassium chloride with AgCl salt to minimize stripping of the silver chloride off the silver wire [107]. When used in thin film format as a microreference electrode, such AgCl dissolution will certainly limit its lifetime. The usable lifetime of thin film all-solid-state Ag/AgCl electrodes was reported to be in only minutes to hours, as opposed to the lifetime in months or years for conventional electrodes [108]. Dissolution of AgCl in the body, when used *in vivo* or as an implantable sensor, was responsible for the toxic reactions with surrounding tissue [109]. Polymer coatings such as Nafion and polyurethane have been used to prolong the lifetime of implanted reference electrodes [110].

Research work that focuses on the development of all-solid-state or microreference electrodes to match a miniaturized pH measurement system, such as lab-on-a-chip, has received increased attention in the past decades. Several approaches have been pursued, such as scaling down a macroscopic reference electrode to form a miniature conventional reference electrode [111], using electrolyte loaded polymers or hydrogel to replace inner electrolyte solution [112, 113], and using solid contact to eliminate inner reference solution [108, 114, 115]. The first two approaches produced functional reference electrodes, but their lifetimes were limited by the volume of the inner solution and the leach rate of the electrolyte from polymer. The outflow of the inner solution through micro salt bridges and the variable leach rate from the gel layer were attributed to unstable liquid-junction potential in practical applications [80]. The last approach produced all-solid-state reference electrodes. However, because no constant chloride concentration was maintained at the Ag/AgCl–solution interface, its potential changed with the composition of the solution.

Covering the Ag/AgCl reference electrode surface with hydrophobic polymer, such as a solvent-processible polyurethane (PU), was reported to provide a stable potential

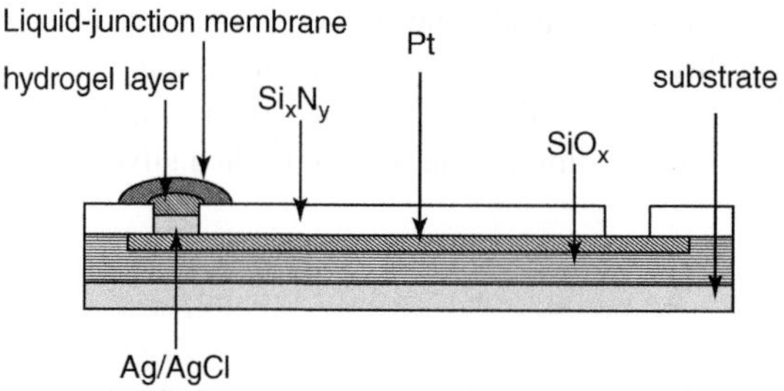

FIGURE 10.5 A schematic diagram of an all-solid-state polymer membrane-based Ag/AgCl reference electrode formed on silicon substrate. Platinum trace lines and silver layer were deposited on 4 inch silicon wafers with a pre-coated 1.2 μm of thermal oxide using a lift-off process. PECVD Si3N4 was used as a top insulation layer and contacts were opened using a reactive ion etch. The exposed silver electrodes were chlorinated with 0.1 M FeCl3 for 5 min. A drop of hydrogel (~3 μL) was dispensed and dried on the Ag/AgCl electrode to form an internal electrolyte layer, and 3–5 μL of the CA/PU membrane cocktail was deposited to cover the hydrogel layer. (Reproduced from [113], with permission from Elsevier.)

comparable to a solid-state inner reference electrode for about 2 weeks [112]. By introducing hydrophilic cellulose acetate (CA) to the PU to form a CA/PU membrane, the reference electrode was reported to reach a 5-month operational time [74]. Ha *et al.* [113] have demonstrated a mass produced solid-state miniaturized Ag/AgCl reference electrode using the CA/PU membrane (30/70 wt.%) as a polymeric junction, and an electrolyte loaded hydrogel layer as internal solution. The electrodes were microfabricated on ceramic or silicon chips. A schematic diagram of one such electrode on Si wafer is shown in Fig. 10.5. The hydrogel layer contains 3 M KCl with a 6 wt.% water soluble polymer PVP. The microporous membrane provided micro-channels for the hydration and diffusion of the inner electrolyte. The electrode provided a stable reference potential for about 25 mins for Si chip-based electrodes, while this stability was held for about 90 mins for a ceramic chip-based electrode after a hydration time of 100–200 s. After these stable times, the electrode potential drifted rapidly at a rate of +40 to +90 mV/h. The authors attributed the shorter lifetime for Si chip-based electrodes to their thinner membranes and smaller internal electrolyte volumes, as compared to those based on ceramic chips. Another possible cause for such a short lifetime, although not discussed in their paper [113], might be due to partial membrane adhesion failure. Poorer adhesion of membrane on smooth Si chips compared to adhesion on rough ceramic chips may contribute to the shorter lifetime of Si chip-based electrodes.

Such electrodes should be sufficient as a reference electrode for short-term usage or as a disposable electrode. However, the requirement of a pre-hydration time may limit its applications for fast measurements, such as POCT (the point-of-care testing), due to its slow response time. In fact, the lack of long-term stable microreference electrodes will continue to hamper the development of integrated pH sensing systems.

10.4 ADVANCED MICROELECTRODE SYSTEMS FOR pH DETERMINATION

Advances in both microfabrication technology and electronics have accelerated the development of advanced microelectrode systems for pH determination. The microfabrication technology, taking advantage of the thin film and thick film techniques developed originally for the semiconductor industry, are increasingly applied for making all-solid-state sensors and complete analytical microsystems, either in a lab-on-a-chip format for a micro-total analysis systems (μTAS) or in a multisensing array format.

10.4.1 All-solid-state pH microelectrodes

The miniaturized pH sensors are of great importance in biomedical and clinical applications. By taking advantage of all-solid-state configurations, the sensor size can be greatly reduced and the sensors can be mass produced with improved reproducibility. An all-solid-state pH sensor uses direct electrical contact between the pH sensitive membrane and the inner metal contact. Among pH sensitive membranes used in the fabrication of all-solid-state pH sensors, metal and metal oxides show clear advantages over other membranes, such as glass membranes [60] and polymer membranes [112], due to their well-defined metal/metal oxide interface and compatibility with microfabrication techniques. As an all-solid-state pH sensor, ISFET has an insulator/metal interface.

Poor adhesion of membrane to metal is the leading cause of failure in solid-state potentiometric sensors [116]. For glass membranes, the mismatch of thermal coefficients of expansion between thin glass membrane and metal (mostly Pt) has been attributed to premature failure due to hairline crack formations in the glass layer [60]. For polymer-based membranes, water vapor penetration was reported to compromise the membrane–metal interface, therefore affecting the sensor's performance.

Yoon *et al.* [112] reported an all-solid-state sensor for blood analysis. The sensor consists of a set of ion-selective membranes for the measurement of H^+, K^+, Na^+, Ca^{2+}, and Cl^-. The metal electrodes were patterned on a ceramic substrate and covered with a layer of solvent-processible polyurethane (PU) membrane. However, the pH measurement was reported to suffer severe unstable drift due to the permeation of water vapor and carbon dioxide through the membrane to the membrane–electrode interface. For conducting polymer-modified electrodes, the adhesion of conducting polymer to the membrane has been improved by introducing an adhesion layer. For example, polypyrrole (PPy) to membrane adhesion is improved by using an adhesion layer, such as Nafion [60] or a composite of PPy and Nafion [117].

Another problem that is common for all membrane-based solid-state sensors is the ill-defined membrane–metal interface. A large exchange current density is required to produce a reversible interface for a stable potentiometric sensor response. One approach to improving this interface is to use conducting polymers. Conducting polymers are electroactive π-conjugated polymers with mixed ionic and electronic conductivity. They are able to transduce an ionic signal into an electrical signal. Such polymers provide an interface between pH selective membranes and metallic transducers, which replaces the internal electrolyte of a conventional pH sensor. Research efforts have been made to utilize conducting polymers in a number of designs of all-solid-state miniaturized sensors [118, 119, 120]. Electroactive π-conjugated polymers, like polyaniline (PANI) and polypyrrole (PPy), are most commonly used in sensor fabrications.

10.4.2 pH Microelectrode for a lab-on-a-chip

Multisensing devices such as lab-on-a-chip sensing systems are becoming an important area of sensor research and development because of the advantages they hold over conventional single analyte sensors [121]. The technology now exists to enable fabrication of miniature microfluidic systems capable of switching, regulating, mixing, and separating samples. Integrating sample preparation and separation with sensing elements in a single miniaturized device has led to the concept of μTAS.

A good example of such planar multisensor systems arranged in a lab-on-a-chip format was described by Vonau *et al.* [91] and shown in Fig. 10.6. The system contains planar sensors on ceramic substrates which determine pH, dissolved oxygen, and electrical conductivity in biological liquids. The sensors were combined in a microfluidic system to characterize metabolic processes in biological cells. A four-electrode impedance sensor operated at 300 Hz was used to evaluate metabolism of growth and cell adhesion. A thick film RuO_2 electrode was used for pH determination. A layer of 20 μm ruthenium paste was screen-printed onto the 5 μm thick Pt seed layer, and then sintered at about 900°C to form the pH sensitive layer. RuO_2 electrodes show significantly lower internal resistances of 1–5 MΩ at 25°C and at 1 MHz compared to thick film glass electrodes with values greater than 800 MΩ. The pH sensitivity determined from calibration curves was about 52 mV/pH between pH 4 and 9.2. The integrated oxygen sensor consists of two Pt electrodes and an Ag/AgCl reference electrode in a three-electrode measuring system with the Pt cathode less than 100 μm.

The integrated planar silver chloride electrode uses a thin layer of 150 μm polymer that consists of a heat curing epoxy resin poly-hydroxy-ethylmethacrylate (PHEMA) to immobilize the KCl electrolyte. The potential drift of the reference electrode reduced to 59 μV/h after a conditioning phase of several hours. However, this reference electrode was only used for PO_2 measurement, while an external reference electrode was used for pH measurement.

Challenges remain in the development of lab-on-a-chip sensing systems. The overall lifetime of a sensor chip is always determined by the sensor with the shortest lifetime, which in most cases is the depletion of reference electrolytes. Measures to minimize cross-talking among sensors, especially when biosensors are integrated in the system, also should be implemented [122]. The development of compatible deposition methods of various polymeric membranes on the same chip is another key step in the realization of multisensing devices.

By incorporating on-chip electronics or using external analyzers with advanced control and signal processing functions, the lab-on-a-chip or μTAS can act as "smart sensors",

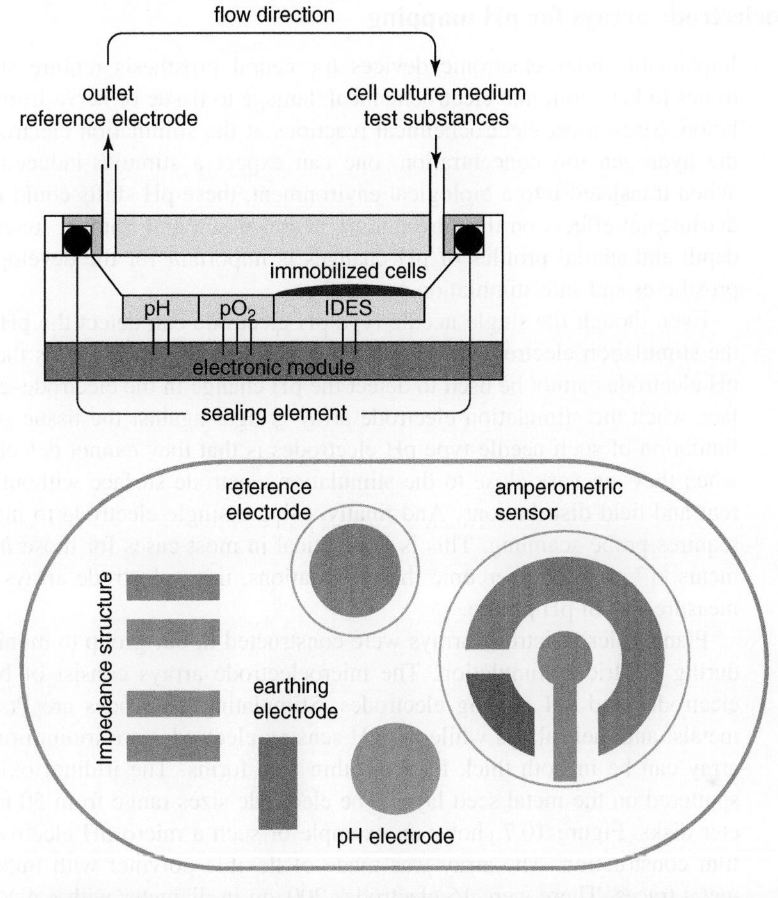

FIGURE 10.6 Schematic drawings of a lab-on-a-chip system. (Top) Multisensor chip consists of pH, pO2, and conductivity (impedance) electrodes incorporated in a microfluidic cell. (Bottom) The layout of electrodes on the chip. (Reproduced from [91], with permission from Elsevier.)

offering enhanced measurement capabilities for many clinical applications [123]. Such smart sensors can even be used in closed-loop controlled medical devices [14, 124].

The commercially available handheld i-STAT system (Abbott Laboratories) is a typical example of such μTAS or lab-on-chip devices (www.i-stat.com). The self-contained test cartridge consists of microfabricated thin film multisensors with a pH sensor on silicon platform, on-chip reference electrodes, washing and calibration solutions, and fluid transportation system. Together with their handheld electrochemical analyzer and appropriate cartridges, the system is capable of performing a comprehensive panel of critical tests for blood chemistries, coagulation, hematology, glucose, or cardiac markers. The chemical sensors and biosensors in the cartridge provide good diagnostic characteristics such as high sensitivity and reproducibility, short analysis time, and a small sample size requirement [125].

10.4.3 Microelectrode arrays for pH mapping

Implantable microelectronic devices for neural prosthesis require stimulation electrodes to have minimal electrochemical damage to tissue or nerve from chronic stimulation. Since most electrochemical reactions at the stimulation electrode surface alter the hydrogen ion concentration, one can expect a stimulus-induced pH shift [17]. When translated into a biological environment, these pH shifts could potentially have detrimental effects on the surrounding neural tissue and implant function. Measuring depth and spatial profiles of pH changes is important for the development of neural prostheses and safe stimulation protocols.

Even though the single needle type pH electrode can detect the pH change around the stimulation electrode, it has limitations. One such limitation is that a needle type pH electrode cannot be used to detect the pH change in the electrode–electrolyte interface when the stimulation electrode array is tight against the tissue surface. Another limitation of such needle type pH electrodes is that they cannot detect the pH change when they are very close to the stimulation electrode surface without disturbing current and field distributions. And finally, using a single electrode to map a pH change requires probe scanning. This is impractical in most cases for those *in-vivo* measurements [17, 18]. To overcome these limitations, microelectrode arrays can be used to measure spatial pH profile.

Planar microelectrode arrays were constructed in our group to monitor pH changes during electrical stimulation. The microelectrode arrays consist of both stimulating electrodes and pH sensing electrodes. Stimulating electrodes are Pt, or other noble metals, and their alloys while the pH sensing electrodes are iridium oxide based. The array can be in both thick film and thin film forms. The iridium oxide is reactively sputtered on the metal seed layer. The electrode sizes range from 50 to 350 μm diameter disks. Figure 10.7 shows an example of such a micro pH electrode array in thin film construction. The array was made of flexible polymer with imbedded thin film metal traces. There were 16 electrodes 200 μm in diameter with a 4×4 arrangement on the array. The needle type pH electrode can be seen on the left corner of the array as a control. The pH response of the iridium oxide electrodes was calibrated in three buffer solutions with a nearly ideal Nernstian response 56.9 ± 1.1 mV/pH in the tested pH range of 4 to 10.

The pH changes due to electric stimulation were recorded successfully in saline by the planar micro pH electrode array [19]. A two-dimensional distribution of pH change can be established by using such combined microelectrode arrays. Figure 10.8 shows a typical 2D pH change profile after stimulation for only one minute. The electrode site in dark color is the Pt stimulating electrode. All other 15 electrodes are IrOx pH sensing electrodes. The charge density of stimulation pulses applied on the Pt electrode was 0.14 mC/cm^2. The pH changes on electrodes presented a clear 2D distribution. The four electrodes closest to the Pt stimulation electrode had the most pH changes (2.9 ± 0.3 pH), while the four corner ones had slightly fewer pH changes (2.0 ± 0.4 pH). The electrodes far away from the stimulating site showed relatively small pH changes (0.2 ± 0.15 pH). These studies have revealed that the pH changes around the electrode–electrolyte interface

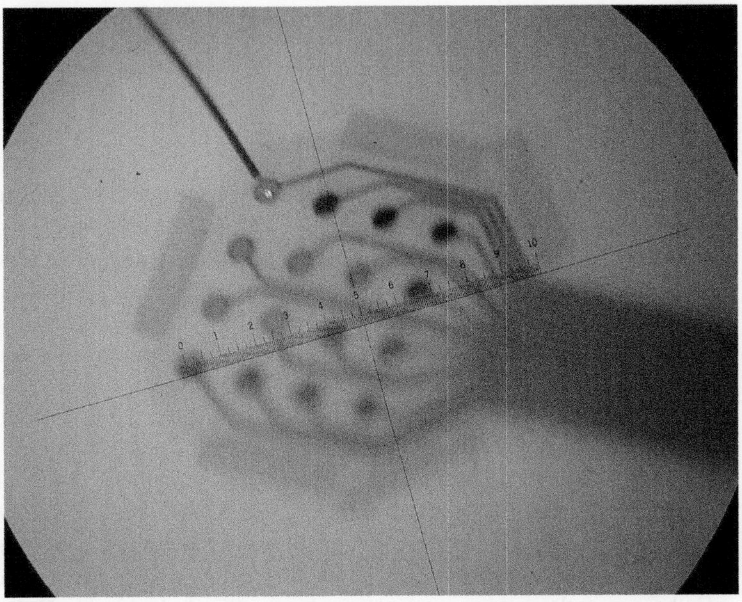

FIGURE 10.7 A thin film planar pH microelectrode array with 16 sputtered IrO_x electrodes in a 4 × 4 arrangement. The needle type pH electrode as a control can be seen on the upper left corner of the array. (Reproduced from [19], with permission from the Electrochemical Society, Inc.)

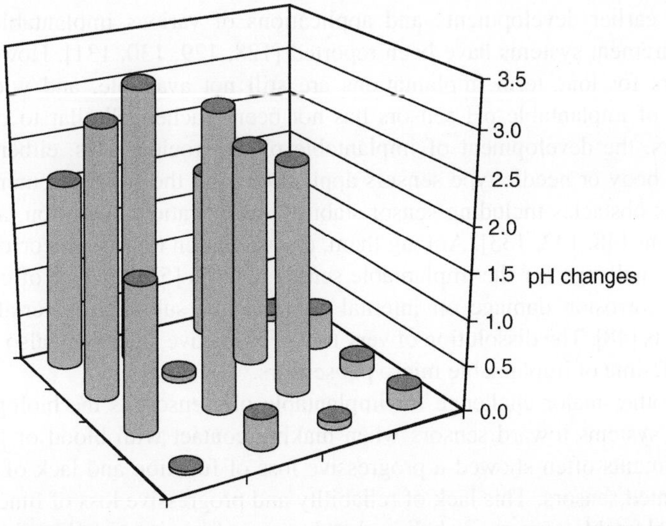

FIGURE 10.8 Two-dimensional pH distribution measured by an iridium oxide-based planar micro-pH electrode array after stimulation for only one minute. The electrode site in dark color is the Pt stimulating electrode. All other 15 electrodes are iridium oxide pH sensing electrodes. (Reproduced from [19], with permission from the Electrochemical Society, Inc.)

were found to have a typical depth and spatial profile and increased with the stimulation frequency, intensity, and duration [18, 19].

10.4.4 Microelectrodes for continuous recording of pH *in vivo*

Microelectrodes capable of continuously measuring pH have proven to be useful in various applications including medical diagnoses and in regulating operative and post-operative management of patients. Arterial blood gas analyses are essential to monitor gas exchange in critically ill patients and during anesthesia for major surgery. Usually, intermittent sampling of arterial blood samples is made and the samples are analyzed in a central laboratory or by a point-of-care blood gas analyzer. Such intermittently performed measurements may miss some adverse events and delay therapeutic responses during surgery [126].

Although sensor breakage, drift, and reference electrode failure have limited the widespread use of continuous pH monitoring, some successful studies have been reported. Papeschi *et al.* [105] described an IrOx electrode for *in-vivo* esophageal and gastric pH measurement performed on 15 healthy volunteers. The electrode was small and flexible, so it was well accepted by patients for 12–24 hours' continuous monitoring pH changes in the upper gastrointestinal tract. Similarly, blood pH, PO_2, and PCO_2 are also continuously monitored using microelectrodes during cardiac surgery [103, 127].

10.4.5 Implantable pH microelectrodes

Some earlier developments and applications of various implantable pH sensors or measurement systems have been reported [128, 129, 130, 131]. However, reliable pH sensors for long-term implantations are still not available, and widespread clinical usage of implantable pH sensors has not been reached. Similar to other implantable sensors, the development of implantable pH microelectrodes, either fully implanted in the body or needle type sensors applied through the skin (percutaneous), has faced serious obstacles including sensor stability deterioration, corrosion, and adverse body reactions [48, 132, 133]. Among them, encapsulation to prevent corrosion represents a major challenge for the implantable sensor devices [51]. Failure of encapsulation can cause corrosion damage on internal components, substrate materials, and electrical contacts [48]. The dissolution of very thin pH sensitive layers will also limit the stability and lifetime of implantable micro pH sensors.

Another major challenge for implantable pH sensors is the biological response of living systems toward sensors when making contact with blood or tissue [134]. The experiments often showed a progressive loss of function and lack of reliability of the implanted sensors. This lack of reliability and progressive loss of function, common to all implantable sensors, is believed to be caused by tissue or blood reactions such as inflammatory response, fibrosis, and loss of vasculature and thrombus formation [135, 136]. The delivery of anti-inflammatory drugs to the sensor site can minimize tissue reactions and extend the lifetime of the device [50, 137].

10.4.6 Wireless pH measurement systems

Wireless pH measurement systems have been used in clinical studies. Unlike a stand-alone pH electrode, a power source, control electronics, and a signal transmitter have to be incorporated into a wireless system. Watanabe *et al.* [138] have reported a wireless pH sensor to record salivary pH continuously. The sensor system transmits pH data via a telemetry system for about 19 hours with a 3 V lithium battery (190 mAh). The error of transmitted pH data was less than 0.15 pH in the range of pH 5.0 to 9.0.

In another study by Okij *et al.* [139], a wireless pH sensing system developed for *in-vivo* pH measurement in the digestive system consists of a pH measurement chip, a wireless communication chip, and a power management chip. The radio frequency (RF) for signal transmission was 2.45 GHz and the signal attenuation was found to be 60–90 dB through the human body equivalent. An ISFET was used as the pH sensor and a capacitor type battery as the energy source. The power consumption was reported to have a low 50 nJ with the help of a differential amplification and a pulse operation design.

Liao *et al.* [62] reported a different telemetric system that was designed to measure pH potentiometrically or ethanol amperometrically. In this system, the pH was measured from an all-solid-state pH sensitive glass membrane electrode with a slope of 45.19 mV/pH using a homemade potentiostat. The very thin planar glass membrane (only 0.7 μm) was sputtered on top of Pt seed layer on a ceramic substrate. The measured pH data was wirelessly transmitted through a PDA-based telemeter to a remote receiver. A fast A/D converter showed benefits in transferring pH signal with less damping for peak pH response.

Rao *et al.* [140] described a study using a commercial pH sensitive radio-telemetry capsule (RTC) to evaluate small bowel and colonic transit time in athletes with gastrointestinal symptoms. The RTC (type 7006 Remote Control Systems, London, UK) consists of a glass electrode with an integral reference cap and battery. RF transmissions from the capsule are detected by a solid-state receiver worn on the belt of the patient. The recorder samples the pH from the capsule at 6 second intervals for a period of 24 hours. They used pH changes as an indication of the pH capsule's movement. A sharp rise in pH from around pH 2 to pH 6 indicates that the capsule has moved into the duodenum from the stomach. Then the pH progressively rises to a plateau around pH 8, which indicates that the capsule has moved into the terminal ileum. Another commercially available wireless pH sensor (Bravo) from Medtronic Inc. has been used to measure esophageal pH for a period of 48 hours and it will be discussed later.

10.5 *IN-VIVO* APPLICATIONS OF pH MICROELECTRODES

10.5.1 pH in the body

The normal pH range for most body chemistry is small and close to neutral. Extracellular fluid, which is about 20% of body water (typically 14 liters), has a normal pH of 7.4 [141]. This value is close to that of blood and is slightly alkaline. On the other hand, the intracellular pH, which varies from one part to another, is about 7.0 on average [142]. Certain body organs require specific low pH, e.g. ~pH 2 in the stomach, or high pH, e.g. ~pH 8 in the small intestine environments, for normal function. Abnormalities in pH are assumed to be of primary physiological significance [143].

The two main organs regulating the pH disturbance (acid–base balance) are the lungs and kidneys. The body is able to respond to changes in the pH either immediately through respiratory control of O_2 and CO_2 gases, or by slowly adapting to the pH changes through the kidneys through regulation of the dissolved CO_2 or bicarbonate in the plasma [143].

Moreover, several buffer systems exist in the body, such as proteins, phosphates, and bicarbonates. Proteins are the most important buffers in the body. Protein molecules contain multiple acidic and basic groups that make protein solution a buffer that covers a wide pH range. Phosphate buffers ($HPO_4^{2-}/H_2PO_4^-$) are mainly intracellular. The pK of this system is 6.8 so that it is moderately efficient at a physiological pH of 7.4. The concentration of phosphate is low in the extracellular fluid but the phosphate buffer system is an important urinary buffer. Bicarbonate ($H_2CO_3/HCO^-{}_3$) is also involved in pH control but it is not an important buffer system because normal blood pH 7.4 is too far from its pK 6.1 [144].

Acid–base balance involves chemical and physiological processes responsible for the maintenance of the pH of body fluids at levels that allow optimal function of the whole individual. The ability for the body to regulate pH is critically important in maintaining the operation of many cellular enzymes and the function of vital organs, such as the brain and the heart [143].

The importance of *in-vivo* pH measurement has been demonstrated in many clinical studies. Some examples of these studies, which involved the applications of pH measurements in several body organs, are discussed in this section.

10.5.2 Measurement of pH in blood

Our human health depends on a balanced and buffered blood pH. For instance, human blood is slightly basic with a pH between 7.3 and 7.5. If the blood pH drops below 7.3, acidosis occurs. If the blood pH rises above 7.5, alkalosis occurs. Adverse events including death will occur if a significant change in the blood pH goes below 6.8 or above 7.8. pH levels in blood are frequently determined in clinical laboratories [145].

pH, pO_2, pCO_2, and bicarbonate [$HCO^-{}_3$], known as blood gases, are four important parameters in blood chemistries. Clinical problems of pH are all related to the pH of the plasma of whole blood. pH in extracellular fluid is always close to that of blood. pH inside cells differs from that of blood but it is not recognized as being an important clinical problem apart from blood pH changes [144]. Changes in plasma pH reflect pH changes in other compartments, such as extracellular fluid and intracellular fluid.

The *in-vivo* pH of red blood cell hemolysates was measured in 20 normal subjects and it was determined that the mean arterial red cell was a pH of 7.19 (7.15 to 7.22). The mean difference in pH between plasma and cells was determined to be 0.21, ranging from 0.15 to 0.23 [146].

With the advancements in the fabrication of miniaturized sensors and analyzers, pH measurement as part of blood gas analysis using the point-of-care testing (POCT) [147, 148] or the continuous monitoring devices [126, 149] has been carried out in laboratory testing and patient care. The i-STAT hand-held clinical analyzer (Abbott Laboratories), as described earlier, uses a cartridge with multiple thin film-based electrodes as sensors or biosensors for the detection of many parameters, e.g. pH, pCO_2, pO_2, base excess, bicarbonate, sodium, potassium, ionized calcium, hemoglobin, hematocrit, and glucose [125].

For pH sensors used in *in-vivo* applications, especially those in continuous pH monitor or implantable applications, hemocompatibility is a key area of importance [150]. The interaction of plasma proteins with sensor surface will affect sensor functions. Thrombus formation on the device surface due to accelerated coagulation, promoted by protein adsorption, provided platelet adhesion and activation. In addition, variation in the blood flow rate due to vasoconstriction (constriction of a blood vessel) and sensor attachment to vessel walls, known as "wall effect", can cause significant errors during blood pH monitoring [50, 126].

In practice, some anticoagulation agents such as heparin or antiplatelet agents, e.g. nitric oxide (NO) are delivered to sensor sites in order to reduce the risk of thrombus formation. Nitric oxide (NO), which is a potent inhibitor of platelet adhesion and activation as well as a promoter of wound healing in tissue, has been incorporated in various polymer metrics including PVC (poly(vinyl-chloride)), PDMS (poly-dimethyl-siloxane) and PU (poly-urethanes). Those NO release polymers have been tested in animals as outer protection coatings and have shown promising effects for the analytical response characteristics of the sensor devices [137].

10.5.3 Measurement of pH in the brain

Brain tissue acidosis resulting from ischemia can cause brain damage when cerebral blood flow reduction reaches a critical value. Continuous pH monitoring is critical for the treatment of patients with stroke or severe head injury [15]. A group of researchers have measured pH changes in the rabbit brain during cerebral ischemia induced by bilateral common carotid artery occlusion [132]. They found that there were significant differences in pH over time between the control and occlusion group. The brain pH in the control rabbits was between 6.69 and 6.78 throughout the 60-minute observation period, whereas the brain pH in the group of rabbits with carotid occlusion was found to decline steadily from 6.82 to 6.46 with increasing occlusion time. In another study, Doppenberg *et al.* [15] examined correlations between brain pH, pO_2, pCO_2, and cerebral blood flow (CBF) in a group of 25 patients with severe traumatic brain injuries. The pH measurements were made using an optical pH sensor (Paratrend 7, Biomedical Sensors, Malvern, PA) based on absorbance of phenol red in bicarbonate solution. In a similar study, the extracellular pH in a rat brain glioma was mapped *in vivo* by a probe with pH dependent ^{1}H resonances detectable by ^{1}H NMR spectroscopy [3].

Terminal activity causes an increase in local cerebral blood flow that can be quantified by measuring the accompanying increase in tissue oxygen. Alkaline pH changes can also follow neuronal activation [151]. Studies have been carried out by Venton *et al.* [21] to relate these changes in extracellular oxygen with pH in brain. pH and oxygen levels were monitored simultaneously after electrical stimulation of the dopamine cell bodies. A biphasic increase in both oxygen and pH (alkaline shift) was observed after terminal activation from the immediate increase in dopamine. Their results indicated that the alkaline pH shifts induced by the increased terminal activity in local cerebral blood flow was due to the rapid removal of carbon dioxide, a component of the extracellular brain buffering system [21].

Various pH electrodes have been used for the brain pH studies. Classically, double-barreled, pH selective glass capillary microelectrodes were used in brain pH measurements. The microelectrodes were fabricated from thin-walled capillary glass pipettes (A & M system), which bound together with shrink wrap tubing, heated until soft, twisted 180°, and then drawn on a vertical pipette puller to a point. Reference barrels were filled with 0.15 M or 1 M NaCl. The pH sensitive barrels were backfilled with 150 mM phosphate buffered saline at pH 7.4. The tips were then broken to 3–5 μm, and the pH sensing barrel was silanized to be hydrophobic by repeated suction and ejection of 4% trimethyl-chlorosilane in xylene in order to allow the formation of a high resistance seal between the glass and the hydrophobic pH selective membrane. A column of pH sensitive ionophore (hydrogen ionophore cocktail A, 95291 from Fluka) was incorporated into the tip by suction. Once ionophore was present, the open end was sealed with dental wax [21, 152, 153].

A commercially available needle type glass membrane pH electrode with a battery powered, portable measurement system (Khuri pH monitor from Vascular Technology Inc., Chelmsford, MA), which is designed for pH measurement in the heart (as will be discussed later), was adapted to monitor the pH change in rabbit brain [132]. The sensor tip of the pH/temperature electrode is 4 mm in length, 1 mm in diameter, and consists of a hollow glass needle filled with inner electrolyte containing ethylene glycol and surrounding an Ag/AgCl internal electrode. A separate single reference electrode was placed in the subcutaneous tissue. The pH values were sampled at 2–3 second intervals and corrected to a 37°C temperature using a built-in temperature sensor.

10.5.4 Measurement of pH in the heart

Ischemia occurs when the blood supply to the heart muscle is temporarily or permanently reduced. The events which may cause ischemia include occlusion of a coronary artery, cardiac arrest, heart failure, a variety of arrhythmias, cardiopulmonary bypass, and aortic clamping during various cardiac operations. Such ischemia can possibly lead to infarction of the heart muscle and impairing of the heart [127].

Myocardial tissue pH measurement has been used in the studies of different approaches to myocardial protection in various cardiac operations. A needle type glass membrane miniature electrode has been used in studies for pH-guided myocardial management [127]. As described in the previous section, this electrode was also adapted to measure brain pH [132]. The electrode has a right-angled glass electrode tip that is 1 mm in diameter and 10 mm in depth. When the electrode was fully inserted perpendicularly into myocardial tissue, a 10 mm depth was reached. The electrode that was previously developed in conjunction with Vascular Technology Inc. (VTI) was discontinued in 2001. A new system has been under development for the myocardial tissue pH measurement in conjunction with Terumo Cardiovascular Systems (http://www.terumo-cvs.com).

In a study published by Skrobik and Filep [154], two such electrodes were used. One electrode was surgically implanted in the right ventricular myocardium (epicardial) and the other electrode was implanted in the jugular vein to the level of the endocardium of the right ventricle in pigs, to monitor endomyocardial and epicardial tissue pH, respectively. Khuri *et al.* [155] have reported a study using these glass electrodes in a large number of patients to monitor the intramyocardial pH to examine the relationship between intraoperative regional myocardial acidosis and long-term survival of patients undergoing cardiac surgery with cardiopulmonary bypass and aortic clamping. They reported the findings based on data from 496 patients who were followed up for an average of 10 years postoperatively. They concluded that regional myocardial acidosis, as measured with tissue pH electrodes during cardiac surgery, was predictive of long-term survival of these patients, and reducing acidosis during operations improved long-term patient survival rate.

Simultaneous and continuous measurements of extracellular pH, potassium K^+, and lactate in an ischemic heart were carried out to study lactic acid production, intracellular acidification, and cellular K^+ loss and their quantitative relationships [6, 7]. The pH sensor was fabricated on a flexible kapton substrate and the pH sensitive iridium oxide layer was electrodeposited on a planar platinum electrode. Antimony-based pH electrodes have also been used for the measurement of myocardial pH in addition to their application in esophageal acid reflux detection.

Rosenfeldt *et al.* [156] evaluated a miniature antimony electrode for the measurement of myocardial pH. The electrode was in a coated wire format with about 0.7 mm in diameter and a layer of polycrystalline antimony was coated on a copper wire. The antimony electrodes had low impedance at sub MΩ range, which is much lower than a glass electrode. The electrode response was calibrated in a 100 mM phosphate buffer of pH 6.2–7.8 and the linear response slopes were 45 to 52 mV/pH at 25°C. The electrode properties were compared to those of a conventional glass electrode *in vitro* and an optical pH sensor *in vivo* (Paratrend 7, Biomedical Sensors Ltd, UK). They found that antimony electrodes were more sensitive to temperature changes than glass electrodes. In fact, the temperature effect was observed to be about three times higher in the antimony electrode than that in the glass electrode, within the temperature range of 10°C to 40°C. Their results indicated that such temperature effects were fairly reproducible on all the antimony electrodes tested and suggested the possibility of thermal compensation to overcome such limitation.

The antimony electrode used by Rosenfeldt *et al.* [156] did respond proportionally to the myocardial pH changes in dogs produced by infusion of sodium bicarbonate or inhalation of carbon dioxide. However, pH measured by the antimony electrodes was consistently about 0.26 units higher than that which was measured with the Paratrend optical sensors. The authors stated that this difference could be explained by the influence of oxygen. It is clear that the oxygen interference with the antimony electrode would result in the inaccurate measurement of pH in cardiac surgery, since aortic clamping and cardioplegia produce dramatic changes in myocardial oxygen levels.

The corrosion of antimony electrodes was also measured using ICP-MS (inductively coupled plasma mass spectrometry) for dissolved antimony *in vivo* [156]. After the electrodes were inserted in the plasma, the antimony concentration showed a linear rise with time at a rate approximately of 94 μg/L/h ($r^2 = 0.997$). Although the projected antimony concentration is lower than the safe limit, accumulation of dissolved antimony and localized toxic effects in tissue may prevent the antimony electrode from long-term implantable applications.

10.5.5 Measurement of pH in the esophagus

The esophagus is the muscular tube that carries food and liquid from the throat to the stomach. Gastroesophageal reflux disease (GERD) is a common chronic acid-related disorder caused by reflux of gastric contents into the esophagus. This reflux is responsible for the discomfort of heartburn, a common symptom of GERD. This condition affects an estimated 35–40% of Western populations, with approximately 7% experiencing daily symptoms [157, 158]. An ambulatory 24 hour continuous pH monitoring *in vivo* is the method of choice for diagnosis of GERD and determination of the effectiveness of treatments [159, 160].

The pH measurement device consists of a small, thin polymer catheter (about 1.5–2 mm in diameter) with a pH sensor probe at the end of the tubing. The thin tubing is gently inserted through one nostril, down to the end of the esophagus as the patient swallows. The catheter is connected to a waist worn recorder. The patient is sent home with the catheter and recorder in place. Over 12 to 24 hours, the pH in the lower esophagus is recorded and the reflux or other symptoms the patient experiences are recorded by pushing buttons on the recorder. A decrease in esophageal pH below 4.0 indicates that acid gastric contents reflux in the esophagus. The percentage of time spent below pH 4 over total pH monitoring time is defined as the reflux index (RI) and a cut-off RI value for "normal" vs "abnormal" is set at 5% [159].

In some studies, a probe with two pH sensors in the proximal and distal esophagus has been used to monitor pH simultaneously [161]. Such a catheter probe for esophageal pH can also be used for monitoring the hypopharynx for laryngopharyngeal reflux (LPR). In this case, one pH sensor is positioned in the hypopharynx and the other in the esophagus [162]. To further accommodate variable esophageal lengths in different patients, a triple-pH sensor catheter probe has recently been developed [163].

In addition to glass electrodes [159], the pH sensors incorporated in the catheter for esophageal pH monitoring were reported to be mostly antimony electrodes such as the Slimline from Medtronic Inc. [95, 158, 164]. In a recent study, Pandolfino *et al.* [33] compared the accuracy of the Slimline antimony pH monitoring system to that of a conventional glass electrode catheter pH system during ambulatory conditions in 18 patients. They reported that the antimony electrodes had acceptable performance compared with the glass electrodes. Their results suggested that the electrochemical limitations of antimony catheters could be overcome by careful calibration and thermal compensation.

Wise *et al.* [165] reviewed the potential drift of antimony pH electrodes in 100 consecutive 24 hour ambulatory esophageal pH studies. Their findings suggested that drift in pH during 24 hour pH studies using antimony electrodes was common, but large degrees of drift are less frequent. Drifts of at least 0.1 pH occurred in 88% of studies while drifts of 0.4 pH or higher occurred in only 9% of studies.

Although they are effective, catheter-based testing systems may be limited by patient discomfort and interference with normal diet and activities. The probe can cause a great deal of discomfort and sometimes patients do not tolerate it. Wireless pH measurement devices have been used in studies to reduce these limitations and to provide prolonged, continuous recordings of esophageal pH for up to 48 hours [166, 167].

The wireless pH capsule (Medtronic Inc.) is oblong in shape and contains an antimony pH electrode, a reference electrode at its distal tip, a battery, and a RF transmitter. The whole device is encapsulated in epoxy. The capsule is introduced into the esophagus on a catheter through the nose or mouth and is attached to the lining of the esophagus with a clip. The probe monitors the pH in the esophagus and transmits the information via RF telemetry at a rate of 6 per second (0.17 Hz) to a pager-sized receiver that is worn by the patient on a belt. Prior to implantation, the capsule is calibrated with its receiver in pH buffer solutions of pH 1.07 and pH 7.01 [168].

Kahrilas and Pandolfino [160] have reviewed certain clinical performances, such as safety, tolerability, and diagnostic accuracy of the Bravo wireless pH system in ambulatory esophageal pH measurements. Des Varannes *et al.* [168] analyzed the correlation between wireless pH systems and catheter-based systems. They reported that strong correlations between esophageal acid exposures recorded with the two devices were established during simultaneous recordings of esophageal acid exposure. They observed that the wireless pH system reported a lower pH and significantly underrecorded acid exposure (especially for brief reflux events 12 s or less) compared to the catheter-based pH device. The discrepancy in lower pH readings was attributable to the pH sensor calibration. The discrepancy in recorded acid exposure was explained by the different response characteristics of the two electrodes. The sampling frequency of the wireless Bravo electrode was lower (0.17 vs 0.25 Hz), and the response time was longer than that of the catheter electrode. In addition, poor reproducibility of the catheter system due to the movement of the pH sensor tip may also contribute to such detection differences [160, 169].

Studies to determine tolerance of such implanted devices have been carried out [167, 170]. Generally speaking, the devices are well tolerated by the patients, although some mild symptoms are reported. The symptoms range from foreign body sensation to chest discomfort or pain. It is reported that such symptoms are more obvious in women and younger patients [171]. Although more studies are still needed to examine the correlation between reflux events and symptoms, the Bravo pH capsule represents a significant advancement in the development of wireless, implantable, and solid-state pH measurement systems.

10.5.6 Measurement of pH under skin

The pH of intact skin ranges from about 4.8 to 6.0, while interstitial fluid exhibits a pH that is near neutral. The low pH on skin is attributed mainly to the presence of the so-called "acid mantle", a natural skin barrier to the external environment [172]. Wagner *et al.* [173] measured both *in-vivo* and *in-vitro* pH profiles across human stratum corneum (SC) using the tape stripping technique and a flat surface pH electrode (InLab 426 from Mettler Toledo). They found a steep pH increase from pH ~6 to 8 in the first 100 μm after the removal of the SC.

Transdermal drug delivery based on iontophoresis delivers drugs across the skin and into the body through the application of an electric current or voltage. Iontophoretic transportation of drugs in ionized (charged) or molecular (neutral) form through the skin may occur through diffusion driven by a concentration gradient, migration driven by an electric field, and convection driven by electroosmotic solvent flow [174]. The pH in the drug carrier, as well as the pH across the skin, are both important for the effective penetration of the drugs.

The effect of pH on skin permeability, accumulation, and penetration was studied for some important drugs such as insulin for glucose control [175], acyclovir for HIV-related conditions [176] and 5-fluorouracil for cancer treatment [177].

Safety is an important factor when determining the quality of any iontophoresis electrode. During transdermal iontophoretic delivery using metal electrodes, an applied DC current will induce pH changes on the electrode/skin interface [178]. pH measurement is used to eliminate the possibility of unsafe pH changes (chemical burns). It has been reported that the pH shift caused by platinum electrodes has a significant influence on the permeation and stability of insulin [175].

Another application of pH measurement in skin is for the development of non-invasive glucose biosensor. Reverse iontophoresis has been used to extract glucose from interstitial fluid across skin for non-invasive glucose measurement in diabetes management. The amount of glucose extracted through electroosmotic flow at the glucose biosensor has been demonstrated to correlate with blood glucose in diabetic patients [179]. Tamada and Comyns [174] have investigated the impact of the buffer type, pH, ionic strength, and buffer concentration on the magnitude of electroosmotic extraction through the skin in human subjects. They demonstrated that electroosmotic extraction was significantly enhanced with an increase in pH or a decrease in the ionic strength of a carrier solution.

10.5.7 Measurement of pH in the eye

pH is strictly regulated in the retina to maintain normal functions. Neuronal activity and energy metabolism activities can result in significant pH shifts. Severe pH fluctuations or changes in retinas can have devastating consequences [180, 181].

pH sensitive microelectrodes have been used to obtain spatial pH profiles in retina and to examine the effects of acute hypoxemia and hyperglycemia on retinal pH, to

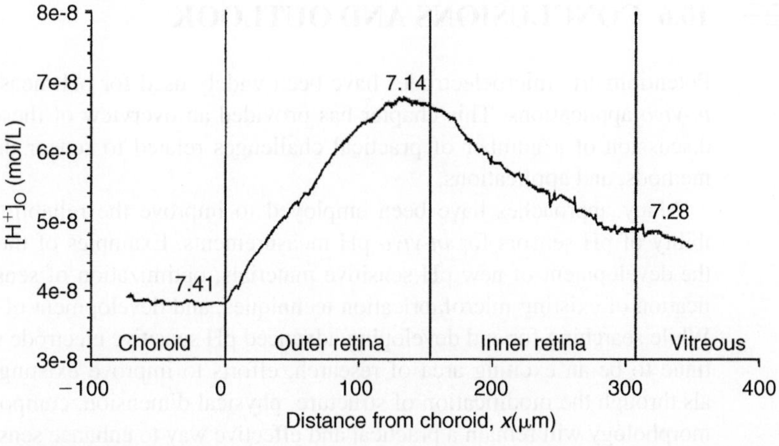

FIGURE 10.9 A typical intraretinal proton concentration spatial distribution measured by a double-barreled pH microelectrode (tip size of ~3 μm) in cat retina. The cell voltage was continuously recorded at 20 readings/s and successive ten readings were averaged for plots. The electrode has an average sensitivity of 48.6 ± 5.9 mV/pH from calibration in three Ringer's solutions of known pH (6.0, 7.0, and 8.0 pH) at 37ºC. (Reproduced from [76], with permission from Cambridge University Press.)

understand hyperglycemia-induced changes in the normal intact cat retina [76, 153]. Hyperglycemia induced an acute panretinal acidification of 0.12 pH, which may contribute to the development of diabetic retinal disease [16]. A similar acidification of 0.1 pH in the inner retinal layer was also detected during retinal artery occlusion [181].

A biological circadian clock with a period of 24 h synchronizes biological activities and behavior in the day/night cycle. Circadian clocks in the vertebrate retina optimize retinal function by driving rhythms in gene expression, photoreceptor outer segment membrane turnover, and visual sensitivity [182]. Circadian changes in retinal extracelluar pH have been demonstrated in retinas *in vivo* as well as *in vitro* [180]. It has been suggested that circadian clock regulation of retinal pH reflects circadian regulation of retinal energy metabolism [76, 183].

The electrodes used in the above studies were double-barreled glass pH sensitive microelectrodes, and the spatial retinal pH profile was recorded by withdrawing the microelectrode tip at a rate of 1 μm/s or 100 μm/step across the retina *in vivo* or *in vitro*. In a typical retina pH profile (Fig. 10.9), measured in cat retina by the microelectrode, started from the choroids (Ph = 7.41, at distance 0 μm). The pH steadily decreased to a minimum value (a maximum $[H^+]$ concentration) in the proximal portion of the outer nuclear layer (pH = 7.14 at ~140 μm), then increased to ~7.28 (at ~310 μm) at the vitreous retinal border. The peak $[H^+]$ concentration in this layer indicated that a net production of proton occurred across the avascular outer retina [76].

10.6 CONCLUSIONS AND OUTLOOK

Potentiometric microelectrodes have been widely used for pH measurements in many *in-vivo* applications. This chapter has provided an overview of these electrodes and a discussion of a number of practical challenges related to sensor designs, fabrication methods, and applications.

Many approaches have been employed to improve the reliability and biocompatibility of pH sensors for *in-vivo* pH measurements. Examples of these approaches are the development of new pH sensitive materials, optimization of sensor designs, modification of existing microfabrication techniques, and development of new technologies. While searching for and developing advanced pH sensitive electrode materials will continue to be an exciting area of research, efforts to improve existing electrode materials through the modification of structure, physical dimension, composition, and surface morphology will remain a practical and effective way to enhance sensor performance.

Microfabrication of sensing devices using thin film and thick film technology has led to miniaturized sensors and certain integration of various sensors and microfluidic devices to form so-called μTAS or lab-on-a-chip systems. Such microfabricated sensors and microsystems will play an important role in the future of pH measurements. In addition to adapting standard microfabrication technologies to fit the needs of chemical sensors, new techniques are also under development, e.g. precision liquid handling and dispensing technologies, and electropolymerization of conducting polymer membranes on electrode surfaces.

Microelectrodes with all-solid-state designs will receive more and more attention. However, ill-defined pH sensitive membrane–metal interfaces will limit conventional membrane based all-solid-state sensors in applications requiring high accuracy and reproducibility. Lack of reliable solid-state reference electrodes will continue to restrain the development of all-solid-state μTAS or lab-on-a-chip systems. It has been envisioned that with the advancement in digital instrumentation technology, hand-held analyzers together with such μTAS systems will provide fast and reliable point-of-care testing (POCT) for the determination of various clinical parameters including pH.

Some progress in implantable pH sensors has been made; however, reliability and biocompatibility remain as main challenges for the development of long-term implantable pH sensors. Sensor technology is also progressing towards using nano-scaled sensor devices in analytical applications. Such progress in nanotechnologies and device packaging technologies show high potential for the development of implantable sensors.

Metal/metal oxides are the materials of choice for construction of all-solid-state pH microelectrodes. A further understanding of pH sensing mechanisms for metal/metal oxide electrodes will have a significant impact on sensor development. This will help in understanding which factors control Nernstian responses and how to reduce interference of the potentiometric detection of pH by redox reactions at the metal–metal oxide interface. While glass pH electrodes will remain as a gold standard for many applications, all-solid-state pH sensors, especially those that are metal/metal oxide-based microelectrodes, will continue to make potentiometric *in-vivo* pH determination an attractive analytical method in the future.

10.7 ACKNOWLEDGMENTS

The author is grateful to Second Sight Medical Products, Inc. for the support and to Amy Hines for her contributions to this work.

10.8 REFERENCES

1. S. Harguindey, J.L. Pedraz, R. Garcia Canero, J. Perez de Diego, and E.J. Cragoe, Hydrogen ion-dependent oncogenesis and parallel new avenues to cancer prevention and treatment using a H^+-mediated unifying approach: pH-related and pH-unrelated mechanisms. *Crit. Rev. Oncog.* **6**, 1–33 (1995).
2. S. Grinstein, D. Rotin, and M.J. Mason, Na^+/H^+ exchange and growth factor-induced cytosolic pH changes. Role in cellular proliferation. *Biochim. Biophys. Acta* **988**, 73–97 (1989).
3. M. Garcia-Martin, G. Herigault, C. Remy, R. Farion, P. Ballesteros, J. Coles, S. Cerdan, and A. Ziegler, Mapping extracellular pH in rat brain gliomas in vivo by 1h magnetic resonance spectroscopic imaging: comparison with maps of metabolites. *Cancer Research* **61**, 6524–6531 (2001).
4. I.N. Rich, D. Worthington-White, O.A. Garden, and P. Musk, Apoptosis of leukemic cells accompanies reduction in intracellular pH after targeted inhibition of the Na^+/H^+ exchanger. *Blood* **95**, 1427–1434 (2000).
5. D. Pirkebner, M. Fuetschi, W. Wittmann, H. Weiss, T. Haller, H. Schramer, R. Margereiter, and A. Amberger, Reduction of intracellular pH inhibits constitutive expression of cyclooxygenase-2 in human colon cancer cells. *J. Cell. Physiol.* **198**, 295–301 (2004).
6. S. Marzouk, S. Ufer, R. Buck, T. Johnson, L. Dunlap, and W. Cascio, Electrodeposited iridium oxide pH electrode for measurement of extracellular myocardial acidosis during acute ischemia. *Anal. Chem.* **70**, 5054–5061 (1998).
7. A.M. Sayed, S. Marzouk, R. Buck, L. Dunlap, T. Johnson, and W. Cascio, Measurement of extracellular pH, K^+, and lactate in ischemic heart. *Anal. Biochem.* **308**, 52–60 (2002).
8. J.C. Ojoo, S.A. Mulrennan, J.A. Kastelik, A.H. Morice, and A.E. Redington, Exhaled breath condensate pH and exhaled nitric oxide in allergic asthma and in cystic fibrosis. *Thorax* **60**, 22–26 (2005).
9. E.A. Oakley and P.L. Barnett, Is acid base determination an accurate predictor of pyloric stenosis? *J. Paediatr. Child Health* **36**, 587–589 (2000).
10. S.C. Kung, S.A. Morse, E. Bloom, and R.M. Raja, Acid–base balance and nutrition in peritoneal dialysis. *Adv. Perit. Dial.* **17**, 235–237 (2001).
11. C. Tevi-benissan, L. Belec, M. Levy, V. Schneider-Fauveau, A.I. Mohamed, M. Hallouin, M. Matta, and G. Gresenguet, In vivo semen-associated pH neutralization of cervicovaginal secretions. *Clin. Diagn. Lab. Immunol.* **4**, 367–374 (1997).
12. J. Ongradi, L. Ceccherini-Nelli, M. Pistello, S. Specter, and M. Bendinelli, Acid sensitivity of cell-free and cell-associated HIV-1: clinical implications. *AIDS Res. Hum. Retroviruses* **6**, 1433–1436 (1990).
13. M. Romanelli, G. Gaggio, A. Piaggesi, M. Coluccia, and F. Rizzello, Technological advances in wound bed measurements. *Wounds* **14**, 58–66 (2002).
14. L. Cammilli, L. Alcidi, G. Papeschi, V. Wiechmann, L. Padeletti, and G. Grassi, Preliminary experience with the pH-triggered pacemaker. *Pacing Clin. Electrophysiol.* **1**, 448–457 (1978).
15. E.M.R. Doppenberg, A. Zauner, R. Bullock, J.D. Ward, P.P. Fatouros, and H.F. Young, Correlations between brain tissue oxygen tension, carbon dioxide tension, pH, and cerebral blood flow – a better way of monitoring the severely injured brain? *Surg. Neurol.* **49**, 650–654 (1998).
16. E. Budzynski, N.D. Wangsa-Wirawan, L. Padnick-Silver, D.L. Hatchell, and R.A. Linsenmeier, Intraretinal pH in diabetic cats. *Cur. Eye Res.* **30**, 229–240 (2005).
17. C.Q. Huang, P.M. Carter, and R.K. Shepherd, Stimulus induced pH changes in cochlear implants: an in vitro and in vivo study. *Ann. Biomed. Eng.* **29**, 791–802 (2001).
18. A.P. Chu, K. Morris, R.J. Greenberg, and D.M. Zhou, Stimulus induced pH changes in retinal implants, in *Proceedings of 26th Annual International Conference of the IEEE Engineering in Medicine and Biology Society*, pp. 4160–4163. San Francisco (2004).

19. D. Zhou, A. Chu, A. Agazaryan, A. Istomin, and R. Greenberg, Towards an implantable micro pH electrode array for visual prostheses, in *Nanoscale Devices, Materials, and Biological Systems: Fundamentals and Applications* (M. Cahay, ed.), pp. 563–576. Electrochemical Society (2004).
20. A. Chvatal, P. Jendelova, N. Krizi, and E. Sykova, Stimulation-evoked changes in extracellular pH, calcium and potassium activity in the frog spinal cord. *Physiologia Bohemoslovaca.* **37**, 203–212 (1988).
21. B.J. Venton, D.J. Michael, and R.M. Wightman, Correlation of local changes in extracellular oxygen and pH that accompany dopaminergic terminal activity in the rat caudate–putamen. *J. Neurochem.* **84**, 373–381 (2003).
22. Y. Miao, J. Chen, and K. Fang, New technology for the detection of pH. *J. Biochem. Biophys. Methods* **63**, 1–9 (2005).
23. K.J. Albert, C.L. Schauer, and D.R. Walt, Optical sensor arrays for medical diagnostics, in *Biomedical Diagnostic Science and Technology* (W. Law, N. Akmal, and A Usmani, eds), pp. 121–138. Marcel Dekker, NY (2002).
24. T.M. Butler, B. MacCraith, and C. McDonagh, Leaching in sol-gel derived silica films for optical sensing. *J. Non-Cryst. Solids* **224**, 249–258 (1998).
25. A. Lobnik, I. Oehme, I. Murkovic, and O. Wolfbeis, pH optical sensors based on sol-gels: chemical doping versus covalent immobilization. *Anal. Chim. Acta* **367**, 159–165 (1998).
26. L.S. Wong, W.S. Brocklesby, and M. Bradleya, Fibre optic pH sensors employing tethered non-fluorescent indicators on macroporous glass. *Sens. Actuators B.* **107**, 957–962 (2005).
27. C. Bock, F.J. Sartoris, R.M. Wittig, and H.O. Portner, Temperature-dependent pH regulation in stenothermal antarctic and eurythermal temperate eelpout (Zoarcidae): an in-vivo NMR study. *Polar Biol.* **24**, 869–874 (2001).
28. M. Foster, I. Grigorev, D. Lurie, V. Khramtsov, S. McCallum, I. Panagiotelis, J. Hutchison, A. Koptioug, and I. Nicholson, In vivo detection of a pH-sensitive nitroxide in the rat stomach by low-field ESR-based techniques. *Magn. Reson. Med.* **49**, 558–567 (2003).
29. S. Zhang, B. Soller, and R. Micheels, Partial leastsquares modeling of near infrared reflectance data for noninvasive in-vivo determination of deep tissue pH. *Appl. Spectroscopy* **52**, 400–406 (1998).
30. T. Khan, B. Soller, M. Naghavi, and W. Casscells, Tissue pH determination for the detection of metabolically active, inflamed vulnerable plaques using near-infrared spectroscopy: an in-vitro feasibility study. *Cardiol.* **103**, 10–16 (2005).
31. J. Puyana, B. Soller, S. Zhang, and S. Heard, Continuous measurement of gut pH with near infrared spectroscopy during hemorrhagic shock. *J. Trauma* **46**, 9–15 (1999).
32. G. Horvai and E. Pungor, Amperometric determination of hydrogen and hydroxyl ion concentrations in unbuffered solutions in the pH range 5–9. *Anal. Chim. Acta* **243**, 55–59 (1991).
33. J.E. Pandolfino, S. Ghosh, Q. Zhang, M. Heath, T. Bombeck, and P.J. Kahrilas, Slimline vs. glass pH electrodes: what degree of accuracy should we expect? *Alimentary Pharmacol. & Therapeutics* **23**, 331–340 (2006).
34. G.T. Hanson, T.B. McAnaney, E.S. Park, M.E. Rendell, D.K. Yarbrough, S. Chu, L. Xi, S. Boxer, M. Montrose, and S. Remington, Green fluorescent protein variants as ratiometric dual emission pH sensors. 1. Structural characterization and preliminary application. *Biochem.* **41**, 15 477–15 488 (2002).
35. X.J. Zhang, B. Ogorevc, and J. Wang, Solid-state pH nanoelectrode based on polyaniline thin film electrodeposited onto ion-beam etched carbon fiber. *Anal. Chim. Acta* **452**, 1–10 (2002).
36. M.S. Frant, History of the early commercialization of ion-selective electrodes. *Analyst* **199**, 2293–2301 (1994).
37. R.W. Burnett, A.K. Covington, N.F. Andersen, W.R. Kulpmann, A. Lewenstam, A.H.J. Maas, O. Muller-Plathe, A.L.VanKessel, and W.G. Zijlstra, Use of ion-selective electrodes for blood-electrolyte analysis. Recommendations for nomenclature, definitions and conventions. *Clin. Chem. Lab. Med.* **38**, 363–370 (2000).
38. R.P. Buck, S. Rondinini, A.K. Covington, F.G. Baucke, C.M. Brett, M.F. Camoes, M.J. Milton, T. Mussini, R. Naumann, K.W. Pratt, P. Spitzer, and G.S. Wilson, The measurement of pH – definition, standards and procedures (IUPAC Recommendations 2002). *Pure Appl. Chem.* **74**, 2169–2200 (2002).

39. C. Rundle, A Beginner's Guide to Ion-Selective Electrode Measurements, http://www.nico2000.net/Book/Guide1.html. (2005).
40. D. Wipf, F. Ge, T. Spaine, and J. Baur, Microscopic measurement of pH with iridium oxide microelectrodes. *Anal. Chem.* **72**, 4921–4927 (2000).
41. I.A. Ges, B.L. Ivanov, D.K. Schaffer, F.A. Lima, A.A. Werdich, and F.J. Baudenbacher, Thin-film IrOx pH microelectrode for microfluidic-based microsystems. *Biosens. Bioelectron.* **21**, 248–256 (2005).
42. C. Wijayawardhana and W. Heineman, Electrochemcial biosesnors, *in Biomedical Diagnostic Science and Technology* (W. Law, N. Akmal, and A. Usmani, eds), pp. 1–27. Marcel Dekker, NY (2002).
43. S. Park, H. Boo, Y. Kim, J. Han, H.C. Kim, and T.D. Chung, pH-sensitive solid-state electrode based on electrodeposited nanoporous platinum. *Anal. Chem.* **77**, 7695–7701 (2005).
44. Q. Li, G. Luo, and Y. Shu, Response of nanosized cobalt oxide electrodes as pH sensors. *Anal. Chim. Acta* **409**, 137–142 (2000).
45. T. Yuen, W. Agnew, L. Bullara, and D. McCreery, Biocompatibility of electrodes and materials in the central nervous system, in *Neural Prostheses: Fundamental Studies* (W. Agnew and D. McCreery, eds), pp. 171–321. Prentice Hall, Englewood Cliffs, NJ (1990).
46. M. Schlosser and M. Ziegler, Biocompatibility of active implantable devices, in *Biosensors in the Body: Continuous In Vivo Monitoring* (D.M. Fraser, ed.), pp. 139–170. John Wiley, NY (1997).
47. B. Ziaie, J.A. von Arx, M.R. Dokmeci, and K. Najafi, A hermetic glass-silicon micropackage with high-density on-chip feedthroughs for sensors and actuators. *J. Microelectromechan. Sys.* **5**, 66–177 (1996).
48. U. Guth, W. Oelbner, and W. Vonau, Investigation of corrosion phenomena on chemical microsensors. *Electrochim. Acta* **47**, 201–210 (2001).
49. Y. Kanda, R. Aoshinma, and A. Takada, Blood compatibility of components and materials in silicon integrated circuits. *Electron. Lett.* **17**, 558–559 (1981).
50. M.C. Frost, M.M. Batchelor, Y. Lee, H. Zhang, Y. Kang, B. Oha, G.S. Wilson, R. Gifford, S.M. Rudich, and M.E. Meyerhoff, Preparation and characterization of implantable sensors with nitric oxide release coatings. *Microchem. J.* **74**, 277–288 (2003).
51. M.F. Nichols, The challenges for hermetic encapsulation of implanted devices – a review. *Biomed. Eng.* **22**, 39–67 (1994).
52. W. Oelbner, J. Zosel, U. Guth, T. Pechstein, W. Babel, J.G. Connery, C. Demuth, M.G. Gansey, and J.B. Verburg, Encapsulation of ISFET sensor chips. *Sens. Actuators B.* **105**, 104–117 (2005).
53. R.G. Bates, *Determination of pH*, Wiley, New York (1973).
54. A.K. Covington, *Ion-selective Electrode Methodology*, CRC Press, Boca Raton, Florida (1979).
55. W. Gopel, T.A. Jones, M. Kleitz, J. Lundstrom, and T. Seiyama, *Chemical and Biochemical Sensors*, Chap 2, VCH, Weinham (1991).
56. E.A.H. Hall, *Biosensors*, Open University Press, Buckingham (1990).
57. Y. Abe and M. Maeda, Origin of pH-glass electrode potentials and development of pNa-responsive glasses. *J. Electrochem. Soc.* **147**, 787–791 (2000).
58. G.T. Yu, The diffusivity of mobile hydrogen ion in pH-glass membrane. *Chem. Phys. Lett.* **384**, 124–127 (2004).
59. L.Y. Heng, K. Toth, and E.A.H. Hall, Ion-transport and diffusion coefficients of non-plasticised methacrylic-acrylic ion-selective membranes. *Talanta* **63**, 73–87 (2004).
60. W. Vonau, J. Gabel, and H. Jahn, Potentiometric all solid-state pH glass sensors. *Electrochim. Acta* **50**, 4981–4987 (2005).
61. J. Ylen, Measuring, modeling and controlling the pH value and the dynamic chemical state, Dissertation for the degree of Doctor of Science in Technology, Helsinki University of Technology (2001).
62. W.Y. Liao, Y.G. Lee, C.Y. Huang, H.Y. Lin, Y.C. Weng, and T.C. Chou, Telemetric electrochemical sensor. *Biosens. Bioelectron.* **20**, 482–490 (2004).
63. R.Q. Yu, Z.R. Zhang, and G.L. Shen, Potentiometric sensors: aspects of the recent development. *Sens. Actuators B.* **65**, 150–153 (2000).
64. K.L. Cheng and D.M. Zhu, On calibration of pH meters. *Sensors* **5**, 209–219 (2005).
65. M.J. Schoning, D. Brinkmann, D. Rolka, C. Demuth, and A. Poghossian, CIP (cleaning-in-place) suitable "non-glass" pH sensor based on a Ta_2O_5-gate EIS structure. *Sens. Actuators B.* **111–112**, 423–429 (2005).

66. P. Bergveld, ISFETs for physiological measurements, in *Implantable Sensors for Closed-Loop Prosthetic Systems* (W.H. Ko, ed.), Futura Publishing Co., New York (1985).
67. Y.M. Mi, C. Green, and E. Bakker, Polymeric membrane pH electrodes based on electrically charged ionophores. *Anal. Chem.* **70**, 5252–5258 (1998).
68. R. Yuan, Y.Q. Chai, and R.Q. Yu, Poly(vinyl chloride) matrix membrane pH electrode based on 4,4′-bis[(N,N-diakylamino)-methyl]azobenzene with a wide linear pH response range. *Analyst* **117**, 1891–1893 (1991).
69. K. Hamdani and K.L. Cheng, Polyaniline pH electrodes. *Microchem. J.* **61**, 198–217 (1999).
70. R.B. Brown, Solid-state liquid chemical sensors, in *Proc. Chemistry Forum 4th Int. Symp*, pp. 120–126 (1997).
71. C. Espadas-Torre and M.E. Meyerhoff, Thrombogenic properties of untreated and poly(ethylene oxide)-modified polymeric matrices useful for preparing intraarterial ion-selective electrodes. *Anal. Chem.* **67**, 3108–3114 (1995).
72. S. Ryu, J. Shin, G. Cha, R. Hower, and R. Brown, Polymer membrane matrices for fabricating potentiometric ion sensors, in *Technical Digest 5th Int. Mtg. on Chemical Sensors*, *vol. 2*, pp. 961–964. Rome, Italy, July 11–14 (1994).
73. K. Joung, H.J. Yoon, H. Nam, and K. Paeng, Development of pH sensor based on aromatic polyurethane matrix. *Microchem. J.* **68**, 115–120 (2001).
74. H. Nam, G. Cha, T. Strong, J. Ha, J. Sim, R. Hower, S. Martin, and R. Brown, Micropotentometric sensors. *Proceedings of the IEEE* **91**, 870–880 (2003).
75. X.J. Zhang, A. Fakler, and U.E. Spichiger, Design of pH microelectrodes based on ETHT 2418 and measurement of pH profile in instant noodles. *Anal. Chim. Acta* **445**, 57–65 (2001).
76. L. Padnick-Silver and R.A. Linsenmeier, Quantification of in vivo anaerobic metabolism in the normal cat retina through intraretinal pH measurements. *Visual Neurosci.* **19**, 793–806 (2002).
77. M.L. Pourciel-Gouzy, W. Sant, I. Humenyuk, L. Malaquin, X. Dollat, and P. Temple-Boyer, Development of pH-ISFET sensors for the detection of bacterial activity. *Sens. Actuators B.* **103**, 247–251 (2004).
78. M. Lehmann, W. Baumann, M. Brischwein, R. Ehret, M. Kraus, A. Schwinde, M. Bitzenhofer, I. Freund, and B. Wolf, Non-invasive measurement of cell membrane associated proton gradients by ion-sensitive field effect transistor arrays for microphysiological and bioelectronical applications. *Biosens. Bioelectron.* **15**, 117–124 (2000).
79. S. Martinoia, G. Massobrio, and L. Lorenzelli, Modeling ISFET microsensor and ISFET-based microsystems: a review. *Sens. Actuators B.* **105**, 14–27 (2005).
80. J. Janata, *Principles of Chemical Sensors*, pp. 81–107. Plenum Press, New York (1989).
81. S. Katayama, N. Akao, N. Hara, and K. Sugimotob, Al_2O_3 —Ta_2O_5 —ZrO_2 thin films having high corrosion resistance to strong acid and alkali solutions. *J. Electrochem. Soc.* **152**, B286–B290 (2005).
82. P. Bergveld, Thirty years of ISFETOLOGY, what happened in the past 30 years and what may happen in the next 30 years. *Sens. Actuators B.* **88**, 1–20 (2003).
83. D. Rolka, A. Poghossian, and M.J. Schoning, Integration of a capacitive EIS sensor into a FIA system for pH and penicillin determination. *Sensors* **4**, 84–94 (2004).
84. A. Smit, M. Pollard, P. Cleaton-Jones, and A. Preston, A comparison of three electrodes for the measurement of pH in small volumes. *Caries Res.* **31**, 55–59 (1997).
85. P.A. Hammond and D.R.S. Cumming, Performance and system-on-chip integration of an unmodified CMOS ISFET. *Sens. Actuators B.* **111–112**, 254–258 (2005).
86. T. Velten, H.H. Ruf, D. Barrow, N. Aspragathos, P. Lazarou, E. Jung, C.K. Malek, M. Richter, and J. Kruckow, Packaging of bio-MEMS: strategies, technologies, and applications. *IEEE Trans. Adv. Packaging* **28**, 533–546 (2005).
87. A. Fog and R.P. Buck, Electronic semiconducting oxides as pH sensors. *Sens. Actuators* **5**, 137–146 (1984).
88. K.G. Kreider, M.J. Tarlov, and J.P. Cline, Sputtered thin-film pH electrodes of platinum, palladium, ruthenium, and iridium oxides. *Sens. Actuators B.* **28**, 167–172 (1995).
89. M.L. Hitchman and S. Ramanathan, Evaluation of iridium oxide electrodes formed by potential cycling as pH probes. *Analyst* **113**, 35–39 (1988).

90. R. Koncki and M. Mascini, Screen-printed ruthenium dioxide electrodes for pH measurements. *Anal. Chim. Acta* **351**, 143–149 (1997).
91. W. Vonau, U. Enseleit, F. Gerlach, and S. Herrmann, Conceptions, materials, processing of miniaturized electrochemical sensors with planar membranes. *Electrochim. Acta* **49**, 3745–3750 (2004).
92. L.A. Pocrifka, C. Goncalves, P. Grossi, P.C. Colpa, and E.C. Pereira, Development of RuO_2—TiO_2 (70—30) mol% for pH measurements. *Sens. Actuators B*. **113**, 1012–1016 (2006).
93. A. Gac, J.K. Atkinson, Z. Zhang, C.J. Sexton, S.M. Lewis, C.P. Please, and R. Sion, Investigation of the fabrication parameters of thick film titanium oxide-PVC pH electrodes using experimental designs. *Microelectronics Int*. **21**, 44–53 (2004).
94. C.C. Liu, B.C. Bocchicchio, P.A. Overmeyer, and M.R Neuman, A palladium oxide miniature pH electrode. *Science* **207**, 188–189 (1980).
95. R.D. Jones, M.R. Neuman, G. Sanders, and F.T. Cross, Miniature antimony pH electrodes for measuring gastro esophageal reflux. *Ann. Thoracic Surg*. **33**, 491–495 (1982).
96. K. Yamamoto, G.Y. Shi, T.S. Zhou, F. Xu, M. Zhu, M. Liu, T. Kato, J.Y. Jin, and L.T. Jin, Solid-state pH ultramicrosensor based on a tungstic oxide film fabricated on a tungsten nanoelectrode and its application to the study of endothelial cells. *Anal. Chim. Acta* **480**, 109–117 (2003).
97. A. Eftekhari, pH sensor based on deposited film of lead oxide on aluminum substrate electrode. *Sens. Actuators B*. **88**, 234–238 (2003).
98. C.W. Pan, J.C. Chou, T.P. Sun and S.K. Hsiung, Development of the real-time pH sensing system for array sensors. *Sens. Actuators B*. **108**, 870–876 (2005).
99. C.N. Tsai, J.C. Chou, T.P. Sun, and S.K. Huiung, Study on the sensing characteristics and hysteresis effect of the tin oxide pH electrode. *Sens. Actuators B*. **108**, 877–882 (2005).
100. S. Yao, M. Wang, and M. Madou, A pH electrode based on melt-oxidized iridium oxide. *J. Electrochem. Soc*. **148**, H29–H36 (2001).
101. S.A. Marzouk, Improved electrodeposited iridium oxide pH sensor fabricated on etched titanium substrates. *Anal. Chem*. **75**, 1258–1266 (2003).
102. A. Bezbaruah and T. Zhang, Fabrication of anodically electrodeposited iridium oxide film pH microelectrodes for microenvironmental studies. *Anal. Chem*. **74**, 5726–5733 (2002).
103. S. Marzouk, R. Buck, L. Dunlap, T. Johnson, and W. Cascio, Measurement of extracellular pH, K(+), and lactate in ischemic heart. *Anal. Biochem*. **308**, 52–60 (2002).
104. P. VanHoudt, Z. Lewandowski, and B. Little, Iridium oxide pH microelectrode. *Biotechnol. Bioeng*. **40**, 601–608 (1992).
105. G. Papeschi, S. Merigliano, G. Zaninotto, M. Baessato, E. Ancona, and M. Larini, The iridium/iridium oxide electrode for in vivo measurement of oesophageal and gastric pH. *J. Med. Eng. Technol*. **8**, 221–223 (1984).
106. S.A. Grant, K. Bettencourt, P. Krulevitch, J. Hamilton, and R. Glass, Development of fiber optic and electrochemical pH sensors to monitor brain tissue. *Crit. Rev. Biomed. Eng*. **28**, 159–163 (2000).
107. D.J.G. Ives and G.J. Janz, *Reference Electrodes*, pp. 1–230. Academic Press, New York (1961).
108. H. Suzuki, A. Hiratsuka, S. Sasaki, and I. Karu, Problems associated with the thin-film Ag:AgCl reference electrode and a novel structure with improved durability. *Sens. Actuators B*. **46**, 104–113 (1998).
109. T.G. Yuen, W.F. Agnew, and L.A. Bullara, Tissue response to potential neuroprosthetic materials implanted subdurally. *Biomaterials* 8, 138–141 (1987).
110. F. Moussy and D.J. Harrison, Prevention of the rapid degradation of subcutaneously implanted Ag/AgCl reference electrodes using polymer coatings. *Anal. Chem*. **66**, 674–679 (1994).
111. S. Kojima, M. Loughran, and H. Suzuki, Microanalysis system for pO_2, pCO_2, and pH constructed with stacked modules. *IEEE Sens*. J. **5**, 1120–1126 (2005).
112. H.J. Yoon, J.H. Shin, S.D. Lee, H. Nam, G.S. Cha, T.D. Strong, and R.B. Brown, Solid-state ion sensors with a liquid junction-free polymer membrane-based reference electrode for blood analysis. *Sens. Actuators B*. **64**, 8–14 (2000).
113. J. Ha, S.M. Martin, Y. Jeon, I.J. Yoon, R.B. Brown, H. Nam, and G.S. Cha, A polymeric junction membrane for solid-state reference electrodes. *Anal. Chim. Acta* **549**, 59–66 (2005).

114. J. Gabel, W. Vonau, P. Shuk, and U. Guth, New reference electrodes based on tungsten-substituted molybdenum bronzes. *Solid State Ionics* **169**, 75–80 (2004).
115. G. Valdes-Ramirez, G.A. Alvarez-Romero, C.A. Galan-Vidal, P.R. Hernandez-Rodrguez, and M.T. Ramrez-Silva, Composites: a novel alternative to construct solid state Ag/AgCl reference electrodes. *Sens. Actuators B.* **110**, 264–270 (2005).
116. S. Martin, J. Ha, J. Kim, T. Strong, G. Cha, and R. Brown, ISE arrays with improved dynamic response and lifetime, in *Tech. Digest of the Solid-State Sensor, Actuator, and Microsystems Workshop*, pp. 396–399. Hilton Head, SC (2004).
117. H. Kaden, H. Jahn, and M. Berthold, Study of the glass/polypyrrole interface in an all-solid-state pH sensor. *Solid State Ionics* **169**, 129–133 (2004).
118. M. Trojanowicz, Application of conducting polymers in chemical analysis. *Microchim. Acta* **143**, 75–91 (2003).
119. J.E. Zachara, R. Toczylowska, R. Pokrop, M. Zagorska, A. Dybko, and W. Wroblewski, Miniaturised all-solid-state potentiometric ion sensors based on PVC-membranes containing conducting polymers. *Sens. Actuators B.* **101**, 207–212 (2004).
120. L. He and C.S. Toh, Recent advances in analytical chemistry – a material approach. *Anal. Chim. Acta* **556**, 1–15 (2006).
121. A.J. Tudos, G.A.J. Besselink, and R.B.M. Schasfoort, Trends in miniaturized total analysis systems for point-of-care testing in clinical chemistry. *Lab on a Chip* **1**, 83–95 (2001).
122. I. Moser, G. Jobst, and G.A. Urban, Biosensor array for simultaneous measurement of glucose, lactate, glutamate, and glutamine. *Biosens. Bioelectron.* **17**, 297–302 (2002).
123. D. Diamond, Overview, in *Principles of Chemical and Biological Sensors Chemical Analysis: A Series of Monographs on Analytical Chemistry and Its Applications* (D. Diamond, ed.), pp. 1–18. John Wiley & Sons, Inc., New York (1998).
124. D. Zhou and R. Greenberg, Microsensors and microbiosensors for retinal implants. *Frontiers in Biosci.* **10**, 166–179 (2005).
125. D. Bingham, J. Kendall, and M. Clancy, The portable laboratory: an evaluation of the accuracy and reproducibility of i-STAT. *Ann. Clin. Biochem.* **36**, 66–71 (1999).
126. M. Ganter and A. Zollinger, Continuous intravascular blood gas monitoring: development, current techniques, and clinical use of a commercial device. *Bri. J. Anaesthesia* **91**, 397–407 (2003).
127. K.R. Khabbaz, F. Zankoul, and K.G. Warner, Intraoperative metabolic monitoring of the heart: II. Online measurement of myocardial tissue pH. *Ann. Thorac. Surg.* **72**, S2227–S2234 (2001).
128. C.C. Liu, B.K. Ahn, E.G. Brown, and M.R. Neuman, Engineering development & evaluation of implantable pO_2, pH and pCO_2 sensors, in *Proceedings of the First Pacific Chemical Eng. Conference*, p. 144. Kyoto, Japan (1972).
129. M.A. Afromowitz and S.S. Yee, Fabrication of pH-sensitive implantable electrode by thick-film hybrid technology. *J. Bioeng.* **1**, 55–60 (1977).
130. M. Telting-Diaz, M.E. Collison, and M.E. Meyerhoff, Simplified dual lumen catheter design for simultaneous potentiometric monitoring of carbon dioxide and pH. *Anal. Chem.* **66**, 576–583 (1994).
131. J.F. Martinez-Tica, R. Berbarie, P. Davenport, and M.H. Zornow, Monitoring brain pO_2, pCO_2, and pH during graded levels of hypoxemia in rabbits. *J. Neurosurg. Anesthesiol.* **11**, 260–263 (1999).
132. A. Jabre, Y. Bao, and E.L. Spatz, Brain pH monitoring during ischemia. *Surg. Neurol.* **54**, 55–58 (2000).
133. M.C. Frost and M.E. Meyerhoff, Implantable chemical sensors for real-time clinical monitoring: progress and challenges. *Curr. Opin. Chem. Biol.* **6**, 633–641 (2002).
134. M.C. Frost, M.M. Reynolds, and M.E. Meyerhoff, Polymers incorporating nitric oxide releasing/generating substances for improved biocompatibility of blood-contacting medical devices. *Biomaterials* **26**, 1685–1693 (2005).
135. D.A. Gough and J.C. Armour, Development of the implantable glucose sensor. What are the prospects and why is it taking so long? *Diabetes* **44**, 1005–1009 (1995).
136. A. Heller, Implanted electrochemical glucose sensors for the management of diabetes. *Annu. Rev. Biomed. Eng.* **1**, 153–175 (1999).

137. M.M. Reynolds, M.C. Frost, and M.E. Meyerhoff, Nitric oxide-releasing hydrophobic polymers; preparation, characterization, and potential biomedical applications. *Free Radical Biol. Med.* **37**, 926–936 (2004).
138. T. Watanabe, K. Kobayashi, T. Suzuki, M. Oizumi, and G. Clark, A preliminary report on continuous recording of salivary pH using telemetry in an edentulous patient. *Int. J. Prosthodont.* **12**, 313–317 (1999).
139. A. Okij, T. Yamada, H. Nakase, H. Uesugi, K. Tsubou, K. Masu, and Y. Horiike, In-vivo measurement of pH in digestive system through wireless communication, 7th Internatonal Conference on Miniaturized Chemical and Biochemical Analysts Systems, October 5–9, Squaw Valley, California (2003).
140. K.A. Rao, E. Yazaki, D. F. Evans and R. Carbon, Objective evaluation of small bowel and colonic transit time using pH telemetry in athletes with gastrointestinal symptoms. *J. Sports Med.* **38**, 482-487 (2004).
141. A.W. Grogono, Acid-Base Physiology, http://www.acid-base.com/ physiology.-php (2005).
142. R.B. Reeves and H. Rahn, Patterns in vertebrate acid–base regulation, evolution of respiratory processes: a comparative approach (S.C. Wood, C. Lenfant, eds), pp. 225–252. New York, Marcel Dekker (1979).
143. H.E. Adrogue and H.J. Adrogue, Acid-base physiology. *Respir. Care* **46**, 328–341 (2001).
144. M.J. Bookallil, pH of the blood: acid-base balance, Anaesthetics Royal Prince Alfred Hospital, the University of Sydney, http://www.usyd.edu.au/-su/anaes/-lectures/acidbase_mjb/description.html (2005).
145. P.H. Breen, Arterial blood gas and pH analysis. Clinical approach and interpretation. *Clinics of N. Am.* **4**, 885–906 (2001).
146. M.K. Purcell, G.M. Still, T. Rodman, and H.P. Close, Determination of the pH of hemolyzed packed red cells from arterial blood. *Clin. Chem.* **7**, 536–541 (1961).
147. G. Billman, A.B. Hughes, G.G. Dudell, E. Waldman, L.M. Adcock, D.M. Hall, E.N. Orsini, A.J. Koska, L.J. van Marter, N.N. Finer, J.C. Kulhavy, R.D. Feld, and J.A. Widness, Clinical performance of an in-line, ex vivo point-of-care monitor: a multicenter study. *Clin. Chem.* **48**, 2030–2043 (2002).
148. M. Weiss, A. Dullenkopf, and U. Moehrlen, Evaluation of an improved blood-conserving POCT sampling system. *Clin. Biochem.* **37**, 977–984 (2004).
149. M.A. Pakulla, D. Obal, and S.A. Loer, Continuous intra-arterial blood gas monitoring in rats. *Lab. Animals* **38**, 133–137 (2004).
150. Y. Benmakroha, S. Zhang, and P. Rolfe, Hemocompatibility of invasive sensors. *Med. Eng. Compu.* **33**, 811–821 (1995).
151. J.B. Zimmerman and R.M. Wightman, Simultaneous electrochemical measurements of oxygen and dopamine in vivo. *Anal. Chem.* **63**, 24–28 (1991).
152. C.K. Tong and M. Chesler, Activity-evoked extracellular pH shifts in slices of rat dorsal lateral geniculate nucleus. *Brain Res.* **815**, 373–381 (1999).
153. L. Padnick-Silver and R.A. Linsenmeier, Effect of hypoxemia and hyperglycemia on pH in the intact cat retina. *Archives of Ophthalmol.* **123**, 1684–1690 (2005).
154. Y.K. Skrobik and J.G. Filep, EPI and endomyocardial pH allows the detection of acute right ventricular ischemia in pigs: a new evaluation method. *Can. J. Anaesth.* **49**, 84–89 (2002).
155. S.F. Khuri, N.A. Healey, M. Hossain, V. Birjiniuk, M.D. Crittenden, J. Miguel, P. R. Treanor, S.F. Najjar, and D.J. Kumbhani, Intraoperative regional myocardial acidosis and reduction in long-term survival after cardiac surgery. *J. Thorac. Cardiovasc. Surg.* **129**, 372–381 (2005).
156. F.L. Rosenfeldt, R. Ou, J.A. Smith, D.E. Mulcahy, J.T. Bannigan, and M.R. Haskard, Evaluation of a miniature antimony electrode for measurement of myocardial pH. *J. Med. Eng. Technol.* **23**, 119–126 (1999).
157. M. Pettit, Treatment of gastroesophageal reflux disease. *Pharm World Sci.* **27**, 432–435 (2005).
158. S. Charbel, F. Khandwala, and M.F. Vaezi, The role of esophageal pH monitoring in symptomatic patients on PPI therapy. *Am. J. Gastroenterol.* **100**, 283–289 (2005).
159. Y. Vandenplas, H. Badriul, M. Verghote, B. Hauser, and L. Kaufman, Oesophageal pH monitoring and reflux oesophagitis in irritable infants. *Eur J. Pediatr.* **163**, 300–304 (2004).

160. P.J. Kahrilas and J. Pandolfino, Oesophageal pH monitoring – technologies, interpretation and correlation with clinical outcomes. *Alimentary Pharmacol. Therapeutics* **22**, 2–9 (2005).
161. F. Radaelli, E. Strocchi, S. Passaretti, E. Strada, R. Frego, M. Dinelli, D. Fossati, F. Barzaghi, E. Limido, A. Bortoli, D. Casa, G. Missale, L. Snider, R. Noris, G. Viviani, and G. Minoli, Is esophageal pH monitoring used appropriately in an open-access system? A prospective multicenter study. *Am. J. Gastroenterol.* **99**, 2115–2120 (2004).
162. G.N. Postma, P.C. Belafsky, J.E. Aviv, and J.A. Koufman, Laryngopharyngeal reflux testing. *Ear Nose Throat J.* **81**, 14–18 (2002).
163. S. Harrell, B. Evans, S. Goudy, W. Winstead, E. Lentsch, J. Koopman, and J.M. Wo, Design and implementation of an ambulatory pH monitoring protocol in patients with suspected laryngopharyngeal reflux. *Laryngoscope* **115**, 89–92. (2005)
164. A. Arana, B. Bagucka, B. Hauser, B. Hegar, D. Urbain, L. Kaufman, and Y. Vandenplas, pH monitoring in the distal and proximal esophagus in symptomatic infants. *J. Pediatric Gastroenterol. Nutrition* **32**, 259–264 (2000).
165. J.L. Wise, P.K. Kammer, and J.A Murray, Post-test calibration of single-use, antimony, 24-hour ambulatory esophageal pH probes is necessary. *Dig. Dis. Sci.* **49**, 688–692 (2004).
166. M.D. Crowell, A.G. Decker, V.A. Schettler, M.M. Moirano, H.J. Kim, and V.K. Sharma, Continuous 48-hr esophageal pH-metry in obese patients with gastroesophageal reflux symptoms compared to non-obese patients. *Am. J. Gastroenterol.* **100**, S9–S23 (2005).
167. S.K. Ahlawat, D.J. Novak, D.C. Williams, K.A. Maher, F. Barton, and S.B. Benjamin, Day-to-day variability in acid reflux patterns using the BRAVO pH monitoring system. *J. Clin. Gastroenterol.* **40**, 20–24 (2006).
168. S.B. des Varannes, F. Mion, P. Ducrotte, F. Zerbib, P. Denis, T. Ponchon, R. Thibault, and J.P. Galmiche, Simultaneous recordings of oesophageal acid exposure with conventional pH monitoring and a wireless system (Bravo). *Gut* **54**, 1682–1686 (2005).
169. J.E. Pandolfino, Q. Zhang, M.A Schreiner, S. Ghosh, M.P. Roth, and P.J. Kahrilas, Acid reflux event detection using the Bravo wireless versus the Slimline catheter pH systems: why are the numbers so different? *Gut* **54**, 1672–1681 (2005).
170. J.M. Remes-Troche, J. Ibarra-Palomino, R.I. Carmona-Sanchez, and M.A. Valdovinos, Performance, tolerability, and symptoms related to prolonged pH monitoring using the Bravo system in Mexico. *Am. J. Gastroenterol.* **100**, 2382–2386 (2005).
171. J.A. Hochman and T. Favaloro-Sabatier, Tolerance and reliability of wireless pH monitoring in children. *Nutrition* **41**, 411–415 (2005).
172. S. Dikstein and A. Zlotogorsky, Skin surface hydrogen ion concentration (pH), in Cutaneous *Investigation in Health and Disease – Non-Invasive Methods and Instrumentation* (J.L. Leveque, ed.), pp. 59–62. Marcel Dekker, New York (1989).
173. H. Wagner, K. Kostka, C. Lehr, and U.F. Schaefer, pH profiles in human skin: influence of two in vitro test systems for drug delivery testing. *Eur. J. Pharmaceutics Biopharmaceutics* **55**, 57–65 (2003).
174. J.A. Tamada and K. Comyns, Effect of formulation factors on electroosmotic glucose transport through human skin in vivo. *J. Pharmaceutical Sci.* **94**, 1839–1849 (2005).
175. O. Pillai, N. Kumar, C.S. Dey, S. Borkute, S. Nagalingam, and R. Panchagnula, Transdermal iontophoresis of insulin. Part 1: A study on the issues associated with the use of platinum electrodes on rat skin. *J. Pharmacy Pharmacol.* **55**, 1505–1513 (2003).
176. C. Padula, F. Sartori, F. Marra, and P. Santi, The influence of iontophoresis on acyclovir transport and accumulation in rabbit ear skin. *Pharmaceutical Res.* **22**, 1519–1524 (2005).
177. V. Merino, A. Lopez, Y. Kalia, and R.H. Guy, Electrorepulsion versus electroosmosis: effect of pH on the iontophoretic flux of 5-fluorouracil. *Pharmaceutical Res.* **16**, 758–761 (1999).
178. L.A. Geddes and R.A. Roeder, Direct-current injury: electrochemical aspects. *J. Clin. Monit.* **18**, 157–161 (2004).
179. S. Garg, R. Potts, N. Ackerman, S. Fermi, J. Tamada, and H. Chase, Correlation of fingerstick blood glucose measurements with GlucoWatch biographer glucose results in young subjects with type 1 diabetes. *Diabetes Care* **22**, 1708–1714 (1999).

180. A.V. Dmitriev and S.C. Mangel, Circadian clock regulation of pH in the rabbit retina. J. Neurosci. 21, 2897–2902 (2001).
181. G. Birol, E. Budzynski, N.D. Wangsa-Wirawan, and R.A. Linsenmeier, Retinal arterial occlusion leads to acidosis in the cat. *Exp. Eye Res.* **80**, 527–533 (2005).
182. P.M. Iuvone, G. Tosini, N. Pozdeyev, R. Haque, D.C. Klein, and S.S. Chaurasia, Circadian clocks, clock networks, arylalkylarnine N-acetyltransferase, and melatonin in the retina. *Prog. Retinal Eye Res.* **24**, 433–456 (2005).
183. A.V. Dmitriev and S.C. Mangel, Retinal pH reflects retinal energy metabolism in the day and night. *J. Neurophysiol.* **91**, 2404–2412 (2004).

CHAPTER 11

Biochips – fundamentals and applications

Chang Ming Li, Hua Dong, Qin Zhou, and Kai H. Goh

11.1 INTRODUCTION

The biochip, a bio-microarray device, has been extensively studied and developed to enable large-scale genomic, proteomic and functional genomic analyses. A biochip comprises mainly three types: DNA microarray, protein microarray, and microfluidic chip. The promise of microarrays, a miniaturized device, lies in the spatially addressable grid of specific binding sites, implying that hundreds or thousands of unique binding events can be analyzed simultaneously. The microfluidic chip is used to process analyte sample such as transportation, separation, and purification. With the integration of microarray and microfluidic systems, a micro total analysis system, sometimes called a lab-on-chip system, is produced. Advances of nanotechnology continuously reduce the size of the biochip. This in turn reduces the manufacturing cost and increases the high throughput capability. Due to the benefits of low expense, high throughput and miniaturization, this technology has great potential to be a crucial and powerful tool for clinical research, diagnostics, drug development, toxicology studies, and patient selection for clinical trials.

It was the work of Gilbert and Sanger that turned the biochip concepts into a reality through their DNA sequencing approach which is still widely used today. As biochips evolve throughout the years, their size decreases. This decrease is largely due to the combination of DNA sequencing chemistry with electric current and micropore agarose gels. Due to the small size of the hybridization array and the small amount of target present, it is a challenge to acquire the signals from a DNA array. The signals must first be amplified before they can be detected by the imaging devices. Polymerase Chain Reaction, invented by Kary Mullis in 1983, provides a means to amplify signals, thus solving the problems of small size hybridization array and small amount of target present. Three years later, Leroy Hood brought the biochip technology to another new height by using fluorescence-based DNA sequencing to facilitate the automation of reading DNA sequence. In the early 1990s, companies like Affymetrix, Motorola, and Hyseq enabled mass production of DNA microarray chips by using photolithographic synthesis of oligonucleotide probe arrays or robotic microprinting nanodespensing machine and developed miniaturized and automated versions of DNA sequencing and analysis through microfluidic systems.

Research in proteomics is the next logical step after genomics in the understanding of life processes at the molecular level. Clark [1], in his article, mentioned that historically, one can point back to over 20 years ago, when scientists first thought about mapping the entire set of human proteins. But this effort was perhaps ahead of its time, given the lack of suitable technologies. Although automation is being applied to what has traditionally been the workhorse of protein analysis – 2D-gel electrophoresis – many limitations, remain; such as the speed, time-consumed, sensitivity, and reproducibility of this decade-old method. Protein microarrays may be used to examine many protein–protein, protein–ligand, or enzyme–substrate interactions on a single biochip. This has the possibility of supplementing, or possibly replacing, the current 2D-gel technology with the protein chips. The protein array can offer potential applications in diagnostics, drug discovery, healthcare, environmental protection, and homeland security. In the last five years, protein array chips have been extensively investigated and developed. Some products are already available on the market today. The protein biochip provides one of the most efficient methods, due to its high throughput and simple operation. Expression profiling – measuring the location, timing, causative factors, and level of protein expression – is everything to proteomics researchers and is the main application of the protein chip. Proteomics is the analysis of protein mixtures from tissues or within a cell, both specific to a disease/condition and are controlled. The objective of the protein chip is to rapidly identify new or previously known proteins associated with the condition, and to understand how their level of expression and interactions with other proteins are important, thus forming a basis for new tools in diagnosis and drug discovery.

The biochip has great potential in biomedical, pharmaceutical, environmental, and biodefense applications. The DNA biochip is often used in sequencing, gene expression profiling, and single polymorphism nucleotide (SNP) discovery. The protein biochip that explores proteomics encompasses knowledge of the structure, function, and expression of all proteins in the biochemical or biological contexts of all organisms. Biochips allow scientists and researchers to observe thousands of biochemical changes that are occurring in the biochips in parallel, presenting a holistic view of what is actually taking place during a disease process. Personalized medication to cater for individual genetic makeup is then possible.

Although the manufacture and use of DNA arrays have been commercialized with automatic operation, there are some important challenges for researchers and developers of the biochips.

Standardization in the biochip industry is an important issue and is one of the key challenges faced. In the case of genetic diagnostic applications, clinical decisions are based on the interpretation of gene chips readout, which are dependent on the manufacturer of the biochips. Hence, interfacing between assays and the ancillary instruments is essential for the integration of data into existing equipments. Zipkin recorded that the formation of the Genetic Analysis Technology Consortium (GATC) aims to address the concern for standardization. The decision whether or not to join this consortium solely lies in companies' marketing strategies. If it is a company's interest to have a highly customized application in a niche area, there is a high possibility that the company will not spend the time and effort to standardize its products. On the other hand, if companies desire to have a broad use for their products, there is a high tendency that such companies are willing to invest time, effort, and money on standardization.

Many novel platforms to read and decipher biochips output have been developed. As summarized by Jing [2], these platforms have the potential to give higher and higher throughputs as well as to give an accurate sequence analysis when integrated with detection and analysis software. An example of such a novel platform is the "optical mapping" of DNA, which preserves the biochemical accessibility of the DNA molecules through elongating and attaching them onto derivatized glass slides. At the moment, this system, which maps DNA optically, is not widely used and it may take some time before this system emerges as a system of choice.

For protein chips, more challenging hurdles exist. Proteins are not easy to attach to surfaces, at least not if the hope is to offer a consistent density and orientation of binding sites for ligands. Some proteins are easily denatured at solid–liquid and air–liquid interfaces, rendering protein arrays much more unstable than DNA arrays. To obtain good information content in a protein probe array, it requires that the epitopes bind to the specific targets at nanomolar concentrations. This will result in a more efficient way of detecting the biomolecules. A US patent has been filed by Li *et al.* [3]. This invention provides an efficient method, leading to a high throughput electrical or electrochemical detection of biomolecules.

Without highly specific binding to proteins on the chip, thorough washing of the array may remove specific binders as well as non-specific binders. Protein bindings cannot offer a universal detection scheme like hybridization used in DNA arrays, which makes it difficult to discover enough probes for proteomics.

11.2 DNA ARRAYS

DNA chips – often referred to as DNA microarrays – vastly increase the number of genes that can be studied in a single experiment. They enable thousands of genes-specific sequences to be immobilized on a wafer, nylon or glass array substrate which are then queried by labeled (radioactive or fluorescent labels) copies of biological samples. High density DNA arrays allow the hybridizations between the probes and the targets to be done in parallel. The high throughput enhances the analytical power of the array for complicated genome analyses; thus reducing time, money, and effort. In recent years, advances in the DNA biochips have overcome the problems of low hybridization efficiency, poor sequence discrimination, long process time, and tedious procedures that are inherent in the previous technologies. With the current technology, it is now possible to mass produce miniaturized arrays to sample small volumes with the ability to perform multisequence detection simultaneously.

Demand for these tools is being driven by the need to identify genetic polymorphisms that may be associated with disease, as well as gene functions in human, animal, and microorganism genomes. The DNA microarray technology can be used to identify pathogens, infectious species, and drug resistant mutants at molecular level. This facilitates a better understanding of the disease, so that a more appropriate diagnosis and detection method can take place. Single nucleotide polymorphism (SNP) study is one such application of DNA arrays. With high density DNA arrays, it is now possible to screen for SNPs throughout the genome as this will permit the parallel sequencing of human DNA at thousands of distinct polymorphisms. Li *et al.* [4] came out with methods and compositions for determining single nucleotide polymorphisms (SNPs) in P450 genes. This invention provides a unique collection of P450 SNP probes on one assay, primer sequences for specific amplification of each of the seven P450 genes, and amplicon control probes to evaluate whether the intended P450 gene targets were amplified successfully.

DNA arrays are also useful in the studies of transcriptional dysregulation for many diseases so as to discover how different genes express themselves under certain pathological and pharmacological conditions.

In the near future, there will be increasing demands for DNA arrays in the pharmaceutical, academic, and diagnostic research and development areas. This surge in demand is due largely to the decrease in time, effort, and expanses required in the molecular-biological R&D sectors. In pharmaceutical R&D, DNA arrays shorten the screening time of potential drug targets and enable high speed arrays for compound synthesis and new drugs testing. DNA arrays enhance the identification and sequencing of new genes and the identification of new biological mechanisms for controlling diseases. At such, new drugs are developed and marketed at a higher frequency, leading to a reduction in the development cost.

Scientists at academic research institutions aim to develop novel methods for gene expression and protein function analysis. DNA arrays help scientists to achieve their goals by providing the tools to simultaneously measure all genes in the human genome so as to discover the signaling mechanisms that result in diseases. Scientists using DNA arrays are able to generate a massive volume of data, thus accelerating the rate of discovery.

The need for DNA arrays in diagnostics is largely due to its cost saving applications, leading to an overall reduction in healthcare expenditure. These cost saving applications include prescription of drugs at greater accuracy and identification of patients at risk for certain diseases so that early treatments and monitoring the progression of viral diseases are possible.

11.2.1 Types of DNA arrays

In a DNA array, gene-specific probes are created and immobilized on a chip (silicon wafer, nylon or glass array substrate). Biological samples are labeled with fluorescent dyes or radioactivity. These labeled samples are then incubated with the probes to allow hybridizations to take place in a high fidelity manner. After incubation, non-hybridized samples are washed away and spot fluorescent or radioactivity signals resulting from hybridization can be detected.

Many formats of DNA arrays are currently available in today's market. These formats include microarrays, oligonucleotide arrays, macroarrays, and microelectronic arrays. Choice of usage of any one of these formats would be very much dependent on the users' research applications and budget. The different array formats can be differentiated by the

TABLE 11.1
A summary of the different DNA hybridization array formats

	Probe generations method	Array size	Labeling and detection method	Hybridization method	Commercial suppliers
Microarrays [5]	Robotic printing or piezoelectric inkjet printing of PCR products	2.5 cm by 7.5 cm slide with approximately 10000 genes	Fluorescent tag labeling prior to hybridization; fluorophore added after hybridization and washing	Passive	Agilent Technologies, Genometrix, Operon Technologies, Stratagene
Oligonucleotide arrays [6]	*In-situ* on the surface of the matrix	1 cm by 1 cm slide with approximately 40000 genes; Affymetrix's GeneChip can contain up to 400000 different oligonucleotides and is the densest array	Fluorescent tag labeling; fluorophore detector is added after hybridization	Passive	Affymetrix
Macroarrays [7]	Probes are spotted onto nylon, plastic or nitrocellulose solid matrix	8 cm by 12 cm with approximately 200 to 5000 genes	Radioactivity tag labeling; phosphorimager detector	Passive	Clontech Laboratories, Research Genetics
Microelectronics arrays [8]	Probes are drawn by electric current to chip surface	Number of genes is dependent on the number of electrodes that can be made onto the surface of the array	Fluorescent tag labeling and fluorescent detection	Active	Nanogen

probe generation methods, the size of the arrays, labeling and detection methods, and the hybridization methods. A summary of the different DNA hybridization array formats is shown in Table 11.1. Understanding the different categories of the array formats is essential as this will enable the researchers to choose the correct array to meet their research criteria.

11.2.2 Fabrication of DNA arrays

DNA arrays are fabricated by immobilizing the complementary DNA (cDNA) onto a solid substrate such as silicon, nylon or glass. This can be achieved by robotic printing of polymerase chain reaction (PCR) products (also known as direct-deposition approach), photolithographical synthesis of complementary oligonucleotides or piezoelectric inkjet printing of PCR products (also known as indirect-deposition approach).

Three factors have to be considered during the probe immobilization: first, it is very crucial for the immobilization chemistry to be stable during the subsequent assay steps; second, the probes have to remain functional after the attachment; and last, to prevent base pairing restrain, biomolecules have to be immobilized with an appropriate orientation and configuration. Signals are obtained from the arrays caused by the

TABLE 11.2
A summary of the different array generation approaches

	Spatial resolution	Cost	Probe length	Ease of use
Robotic microprinting	Poorest	Most cost effective	Not restricted	Requires cloning and PCR steps
Photolithography	Highest	Highest as expensive equipments and particular expertise are required	Limited to 25-mers or less	Photolithography method is protected by patent and currently only Affymetrix has the rights to use this method
Inkjet printing	In between robotic printing and photolithography	In between robotic printing and photolithography	5–75-mers	Equipments need strict maintenance and experiment must be performed in a clean and uncontaminated environment

different level of hybridizations; due to this, the effects of cross-contaminations and immobilization errors must be kept to a minimum. Therefore, a good resolution of the DNA immobilization techniques is necessary for arraying and miniaturization.

A quick summary of each approach is listed in Table 11.2 and each of these methods of creating the DNA arrays will be discussed in greater details in the subsequent chapters.

11.2.2.1 Fabrication by robotic microprinting (direct-deposition approach)

Most robotic microprinting systems make use of the XYZ coordinate motion control to enable the printhead to move in all three directions. The printhead consists of a series of pins as shown in Fig. 11.1. These pins are dipped into a tray of probe solution consisting of pre-synthesized oligonucleotides or PCR products. The probe solution at the tip of the pins is then transferred to a predetermined position on a solid substrate such as nylon or glass slides. The amount transferred is typically in the range of nanoliter or picoliter and is dependent on the geometry of the printhead. Cheung *et al.* [9] have reported that with a DNA concentration of 2 μg/μl, the spot size from the microprinting is approximately 100–150 μm in diameter. These spots must be placed very close together (less than 200 μm apart) without overlapping each other. This is a challenge as machines with high accuracy and fine resolution must be used to deliver these spots onto the substrate.

The nylon or glass substrate on which the probe solution is deposited must be pretreated with coupling agents such as lysine or acrylamide functional groups. Coupling agents bind the probe cDNA to the substrate. After hybridization, the array is washed in heated isopropanol and water to remove any unhybridized sample DNA. It is essential

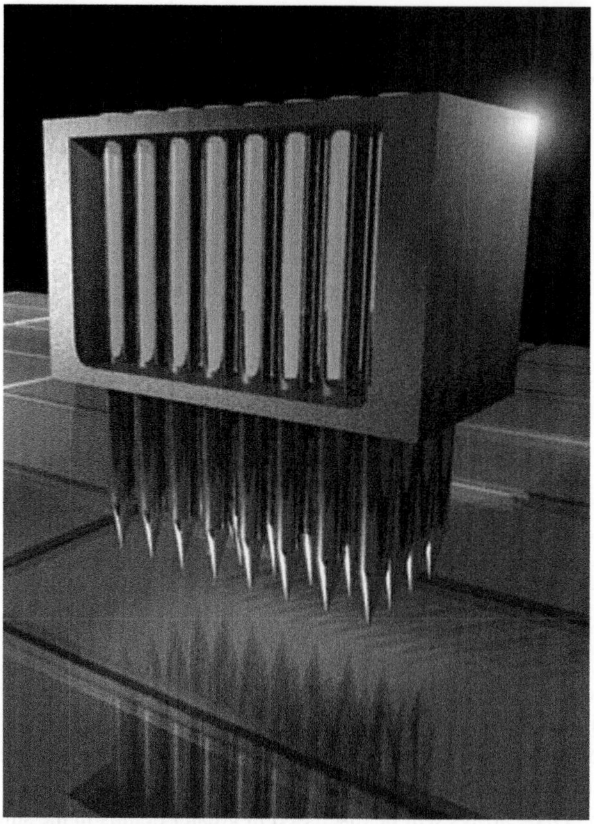

FIGURE 11.1 Printhead with a series of pins. Source: http://cmgm.stanford.edu/pbrown/

that the coupling agent on the substrate has high affinity for DNA in order to prevent it from being washed away during the hybridization process.

11.2.2.2 Fabrication by photolithography

Photolithography is a technique for manufacturing high density oligonucleotide probe arrays which involve the parallel synthesis of a large number of cDNA sequence. Each set of probes has a distinct "address" on the surface of the substrate and in a single hybridization process, these probes are capable of acquiring massive genetic information from biological samples. The spatial resolution of the photolithographic process determines the maximum density of the array. This will eventually determine the amount of sequence information that can be encoded on a single array. To acquire a large quantity of high spatial resolution genetic information from high density oligonucleotide arrays, expensive lithography masks are needed. This increases the overall cost of the equipments. Also, this technique is currently protected by patent and only Affymetrix has the rights to use it.

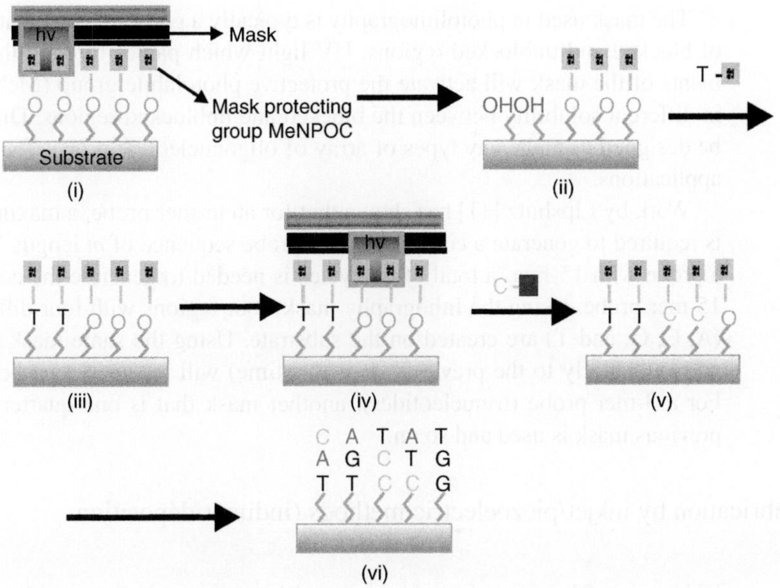

FIGURE 11.2 Photolithographic synthesis of oligonucleotide probe arrays.
(i) Glass substrate coated with silane coupling agent is illuminated through a lithography mask with many segments of blocked and unblocked regions;
(ii) Areas under the unblocked segment of the mask are activated, resulting in the addition of nucleoside phosphoramidite monomers;
(iii) The phosphoramidite group reacts with the hydroxyl group on the substrate in the presence of the silica coupling agent;
(iv) A new mask is used and the surface is illuminated once again;
(v) A new nucleoside phosphoramidite monomers is coupled. Procedure is repeated as many times as need to achieve the required length of the nucleotide chain (usually 25-mer or less).
Source: www.affymetrix.com

In DNA array fabrication by photolithography, a fused silica substrate is coated with a silane coupling agent with a hydroxyl functional group in order to serve as the initial synthesis site. In a report by Pease *et al.* [10], this substrate reacts with the DMT-hexa-ethyloxy-O-cyanoethyl phosphoramidite linker. This linker is protected with a protective photolabile group (α-methyl-2-nitropiperonyl oxycarbonyl) (MeNPOC) which is activated when regions of the surface are exposed to ultraviolet or deep ultraviolet light, resulting in the addition of nucleoside phosphoramidite monomers. The phosphoramidite group reacts with the hydroxyl group on the substrate in the presence of the silica coupling agent. Under the UV radiation, the photolabile MeNPOC group produces the $-$OH group. The next MeNPOC protected nucleotide is added and coupled to the free $-$OH group of the grafted molecule. The MeNPOC group protecting the 5′ end of the added nucleotide is removed by UV radiation and this procedure is repeated as many times as required to achieve the required length of the nucleotide chain (usually 25-mer or less). The photolithographic synthesis of oligonucleotide probe arrays is shown in Fig. 11.2.

The mask used in photolithography is typically a glass substrate with many segments of blocked and unblocked regions. UV light which passes through the unblocked segments of the mask will activate the protective photolabile group (MeNPOC) and result in different solubility between the blocked and unblocked regions. Different masks can be designed to make any types of array of oligonucleotide probes for a wide variety of applications.

Work by Lipshutz [11] has shown that for an m-mer probe, a maximum of $4m$ cycles is required to generate a complete set of probe sequence of m length. That is to say, for a probe with 15-mer, a total of 60 cycles is needed to generate the complete array of a 15-mer probe. Using the lithography mask, four regions with four different DNA bases (A, C, G, and T) are created on the substrate. Using the same mask again (but placed perpendicularly to the previous steps this time) will result in a probe length of 2-mer. For a 3-mer probe (trinucleotides), another mask that is one quarter the width of the previous mask is used and so on.

11.2.2.3 Fabrication by inkjet/piezoelectric methods (indirect-deposition approach)

DNA array fabrication by inkjet method has a very similar working principle to that of an inkjet printer which sprays ink onto paper that is to be printed. Inkjet printing is a non-contact printing technology where oligonucleotide probes are sprayed onto the array surface without touching it; thus reducing the risk of cross-contaminations. This is different from the robotic microprinting and photolithography methods which apply the probes directly onto the substrate.

In his work, Wallace [12] formed inkjet printheads from the rectangular blocks made of piezoelectric material. A diamond saw is used to create fluid channel grooves and channel actuator structures. These grooves are approximately 1 μm apart, 360 μm deep and 170 μm wide. Next, a cover plate is attached to the top of the grooves to form an enclosed rectangle channel for the working fluids. A polymer orifice plate (see Fig. 11.3) with many 40 μm diameter orifices is attached to the other end of the grooves.

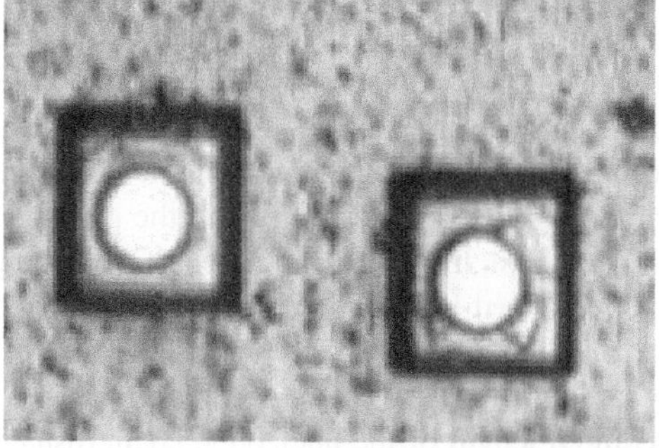

FIGURE 11.3 Orifice plate with 40 μm diameter orifices.

The number of orifices is dependent on the number of grooves. The back of the grooves is then sealed with an acoustic energy absorbing polymer. Figure 11.4 shows the complete printhead fabrication process.

When the grooves are formed by sawing the channels in the piezoelectric material rectangular block, the walls are formed into actuators as well. The voltage applied through the walls causes an electric current to flow through. The electric current is transformed into a mechanical moving force to drive a small amount of droplets from the fluid channel out of the orifice. The mechanical motion is shown in Fig. 11.5. The amount of current will determine the size of the droplets. This droplet is then dispensed from the printhead. Table 11.3 shows some of the data obtained from the inkjet printing method.

Any forms of contamination at the orifice will lead to the misdirection of the droplets; hence, the inkjet printing method requires that the orifice be free of contaminations. Damages at the orifice must be repaired in order to prevent the altering of the size of the

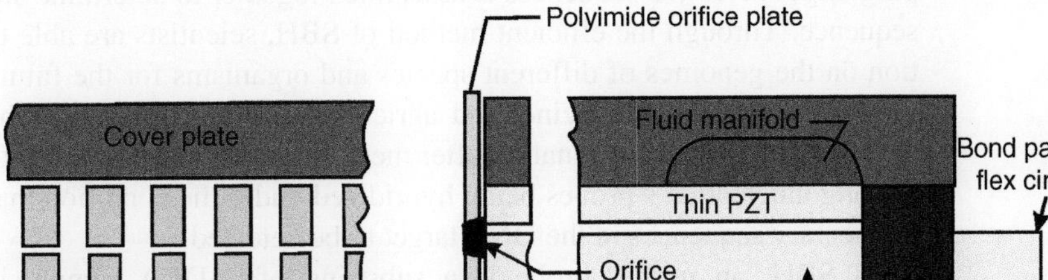

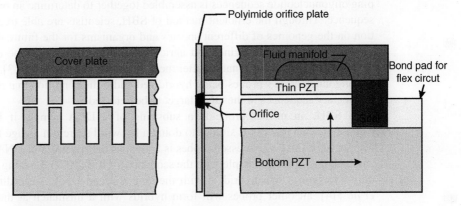

FIGURE 11.4 Printhead with cover, orifice plate and channel sealing.

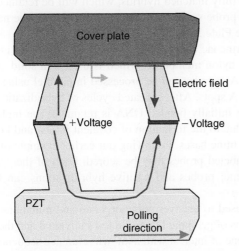

FIGURE 11.5 Mechanical motion of the actuator.

TABLE 11.3
Data from inkjet printing method

Dispense volume	Spot sizes	Spot densities	Delivery speed
50 pL	125–175 μm	500–2500 spots/cm^2	100–500 spots/s

droplets and also cause crossover contamination between probes. Due to air bubbles in inkjet printing, repeatability and reliability of the system may also be greatly reduced.

11.2.3 Sequencing by hybridization

Sequencing by hybridization (SBH) is an approach whereby a collection of overlapping oligonucleotide sequences is assembled together to determine an organism's DNA sequence. Through the efficient method of SBH, scientists are able to gather information on the genomes of different species and organisms for the future development of biological sciences, medicine, and agriculture. SBH is based on the complementary DNA strands' ability to renature after melting as reported by Doty [13]. This results in the oligonucleotide's probes being hybridized under the condition that allows the complementary sequences in the DNA target to be detected.

In SBH, an m-mer probe is a substring of a DNA sample if it is positively expressed. This process is similar to doing a keyword search in a page filled with text. The set of positively expressed probes is known as the spectrum of DNA sample. For a single-strand of DNA sample with the sequence 5′GGTCTCG 3′, using a 4-mer probe, only five probes will hybridize with the single strand DNA. According to Drmanac *et al.* [14], all other probes will form hybrids with a mismatch at the end base and will be denatured during selective washing. The five probes that are of good match at the end base will result in fully matched hybrids, which will be retained and detected. Each positively expressed probe serves as a platform to decipher the next base as can be seen from Fig. 11.6 (see Plate 2 for color version).

In another report by Drmanac *et al.* [15], Format 1 SBH, a large number of DNA samples are attached onto nylon membrane, which is used as a solid support to form the DNA array. These DNA samples can be processed in parallel using a high density array containing many DNA spots. After repeated cycles of hybridization and washing, approximately 75% of the initially bonded DNA is retained. The high proportion of DNA sample retention is due to the formation of covalent interstrand bonds caused by the UV crosslinking of thymine bases. Following our earlier example of 4-mer probes, a full set of 256 (or 4^4) labeled probes may be scored. Each of these DNA arrays is exposed to the labeled 4-mer probes and positive hybridizations can be observed at various DNA spots.

Format 1 SBH can be used to uncover polymorphism and mutations in a particular gene. The sample amplicons of genomic DNA of a test individual and the amplicons for the control DNA for the gene of interest with a known reference sequence are both prepared by polymerase chain reaction (PCR). A subset of probes that is complementary

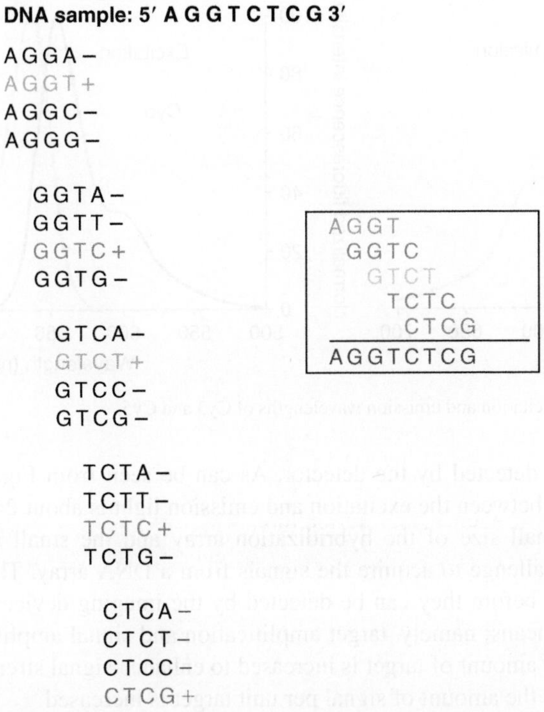

FIGURE 11.6 Sequence assembly using overlapping oligonucleotide probes (see Plate 2 for color version).

to the reference sequence is retrieved from a stock of all possible probes of a given length. Hybridization between the probes and the control and patient samples then takes place. In the event where a mutation is present, there is a mismatch between the probe and the DNA sequence; hence, the probes do not bind to the test sample. The overlapping of probes results in a low percentage of positively expressed probes at the mutation sites, indicating that the sequence of the test sample differs from the control DNA.

11.2.4 Labeling

Labeling must be done on the target materials so that during hybridization, the degree of hybridization can be detected by the imaging devices. Labeling is commonly done by polymerase reaction, in which a fluorescent or a radioactive label is used. A labeled cDNA is created by reverse transcription, allowing for the creation of a label copy of the target to be hybridized onto the array.

Two of the common labeling dyes used are Cyanine3 (Cy3) and Cyanine5 (Cy5). These fluorescence dye molecules emit light when stimulated by a laser. Since there is a wavelength difference between the excited light and the emitted light, the fluorescence detector is able to filter out the excitation wavelength. At such conditions, the emitted light scattered from the slide is separated from the excitation light and it will

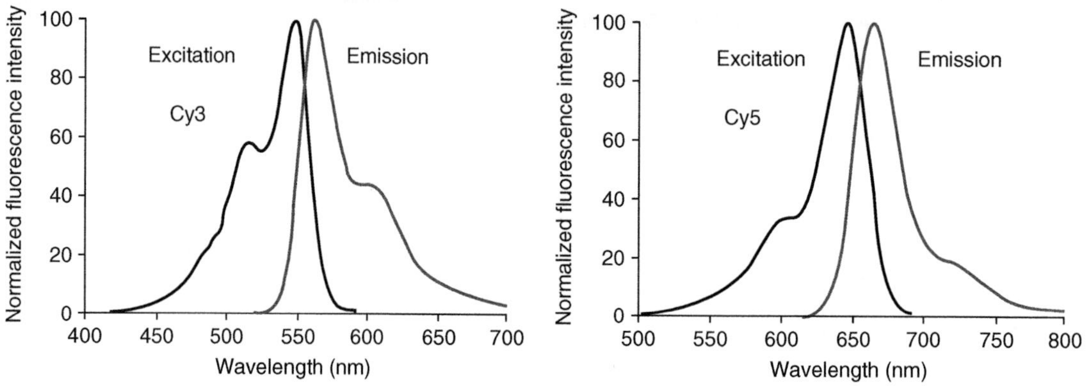

FIGURE 11.7 Excitation and emission wavelengths of Cy3 and Cy5.

be the only light detected by the detector. As can be seen from Fig. 11.7, the wavelength difference between the excitation and emission light is about 24 nm.

Due to the small size of the hybridization array and the small amount of target present, it is a challenge to acquire the signals from a DNA array. These signals must first be amplified before they can be detected by the imaging devices. Signals can be boosted by two means; namely, target amplification and signal amplification. In target amplification, the amount of target is increased to enhance signal strength while in signal amplification; the amount of signal per unit target is increased.

11.2.4.1 Target amplification

Low gene expression or small sample size results in signal strength that is too weak to be detected. Target amplification assists in boosting the signal strength by using techniques like polymerase chain reaction and rolling circle amplification technology to increase the amount of target DNA. This leads to an increase in signal.

Polymerase chain reaction (PCR)

Polymerase chain reaction is a method invented by Kary Mullis [16] for creating multiple copies of DNA through repeated cycles of denaturing, annealing, and synthesizing driven by DNA polymerase. Specific nucleotide sequences are amplified without the use of living organisms. PCR is a quick, easy, and efficient technique for amplifying any particular segments of the DNA. Initially, DNA polymerase was taken from bacterium *E. coli*. The doubled stranded DNA is heated at 96°C in order to separate it into two single strands. At this temperature, the DNA polymerase is destroyed and new enzymes have to be added at the start of every new cycle. Replenishing the destroyed *E. coli* requires much time and effort, thus rendering PCR through DNA polymerase taken from bacterium *E. coli* inefficient. Saiki *et al.* [17] experimented and report that bacterium *T. aquaticus* (Taq) is subsequently selected to amplify nucleotide sequences as the Taq polymerase thrives well at a temperature over 110°C. Taq polymerase does not breakdown when it is heated with the double-stranded DNA during denaturing. PCR takes place in a thermal cycler (as shown in Fig. 11.8) which is able to change

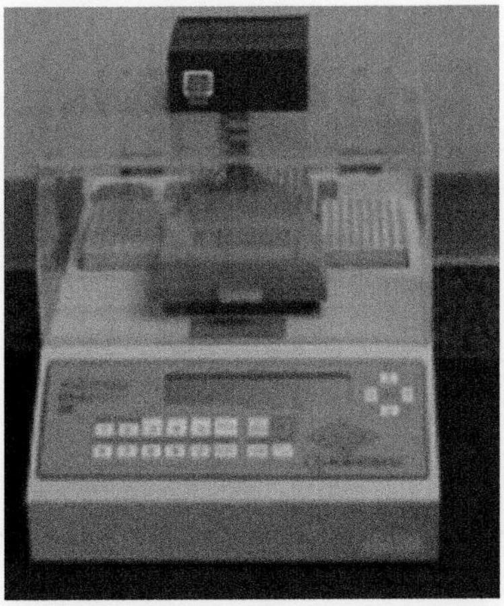

FIGURE 11.8 Thermal cycler for PCR. Source: http://www.intlmiss.com/

the temperature inside the reaction tubes to the temperature required for the different parts of the reaction to take place. For polymerase chain reaction to be possible, five components, namely, a DNA template, two primers, DNA polymerase, nucleotides, and a buffer, are needed. The region of DNA to be amplified must be identified and this region will be contained in a DNA template. Primers are short, artificial DNA strands that match exactly to the region of the DNA that is to be amplified. For a start, two primers are needed because the double strand DNA is separated into two single strands. The DNA polymerase will generate copies of any fragments of DNA that are to be amplified in a suitable environment provided by the buffer. Nucleotides are needed to build the new DNA.

The PCR process consists of about 20–30 cycles. Each cycle consists of three steps, namely, denaturing, annealing, and synthesizing. In denaturing, the double-stranded DNA is heated at 95ºC. At this temperature, the hydrogen bonds between the double helix are broken, resulting in the complete separation of the DNA template and the primers. After separation, the temperature is lowered. The primers bind to their complementary bases on the single strand DNA. Temperature at this step is approximately 55ºC. Finally, in the synthesizing step, the DNA polymerase starts reading a template strand from the beginning of the primer and matches it with a complementary nucleotide. Each cycle will take about 1–3 min; the time taken for the third step will depend on the length of the DNA to be amplified. Repeating the three steps will result in more DNA as every cycle will double the amount of DNA from the previous cycle. A schematic of the PCR and its timeline is shown in Fig. 11.9 and Fig. 11.10 (see Plate 3 for color version), respectively.

Rolling cycle amplification (RCA)

Rolling circle amplification (RCA) is an alternative method to polymerase chain reaction; it is also a generic amplification technique that can be used in antibody assays. Using a replication process similar to that used by viruses, RCA allows the recognition,

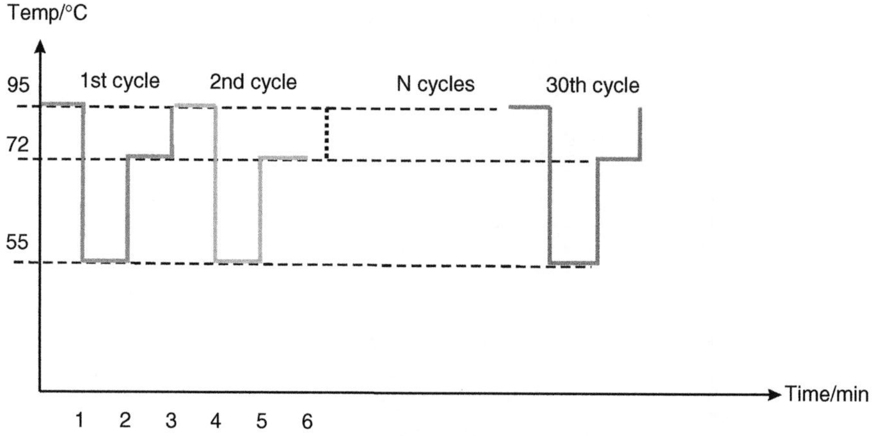

FIGURE 11.9 Timeline for PCR.

amplification, and detection of signals from the array. One of the inventions by Li *et al.* [18] is a novel method of amplifying and detecting DNA, making use of variations of rolling circle amplification to several detection platforms. In their report, Fire *et al.* [19] stated that in RCA, single-strand DNA minicircles are used as templates to perform the rolling nucleic acid synthesis by certain DNA polymerase at a constant temperature. Later, in 1998, Lizardi *et al.* found that the amplification of RCA is rapid, technically simple, and in the presence of a large excess of wild-type DNA, it allows the detection of mutations. There are two priming approaches for RCA in DNA diagnostics: the single-primed and double-primed approaches. In the single-primed approach, DNA polymerase extends a circle-hybridized primer by continuously synthesizing the DNA minicircle to replicate its sequence. A report by Liu *et al.* [20] mentioned that the RCA products usually consist of about 10^2–10^3 repeats of sequence that are complementary to the sequence of the DNA minicircles. As suggested in the name, double-primed RCA operates with two different pairs of primers. One primer is complementary to the DNA minicircle. The second primer is targeted to the repeated single-strand DNA sequences of the original RCA product. Through the series event of multiple hybridization, extension of primers, and strand displacement, a double-stranded DNA segment is formed. A study by Thomas *et al.* [21] found that double-primed RCA gives a greater degree of amplification as compared to it single-primed RCA counterpart; in a time frame of one hour, the double-primed RCA is able to yield 10^9 or more copies greater degree of amplification as compared to it single-primed RCA counterpart; in a time frame of one hour, the double-primed RCA is able to yield 10^9 or more copies circular sequence.

Since rolling circle amplification takes place at a constant temperature, there is no need for the target amplification process to take place in a thermal cycler, which is required to regulate the temperature for different parts of the reaction. The type of DNA polymerase to be used in RCA is not limited to thermostable enzymes, like the PCR-based diagnostics. On the other hand, the RCA method requires the environment to be free of contaminations as the RCA arrays are highly sensitive. Wiltshire [22]

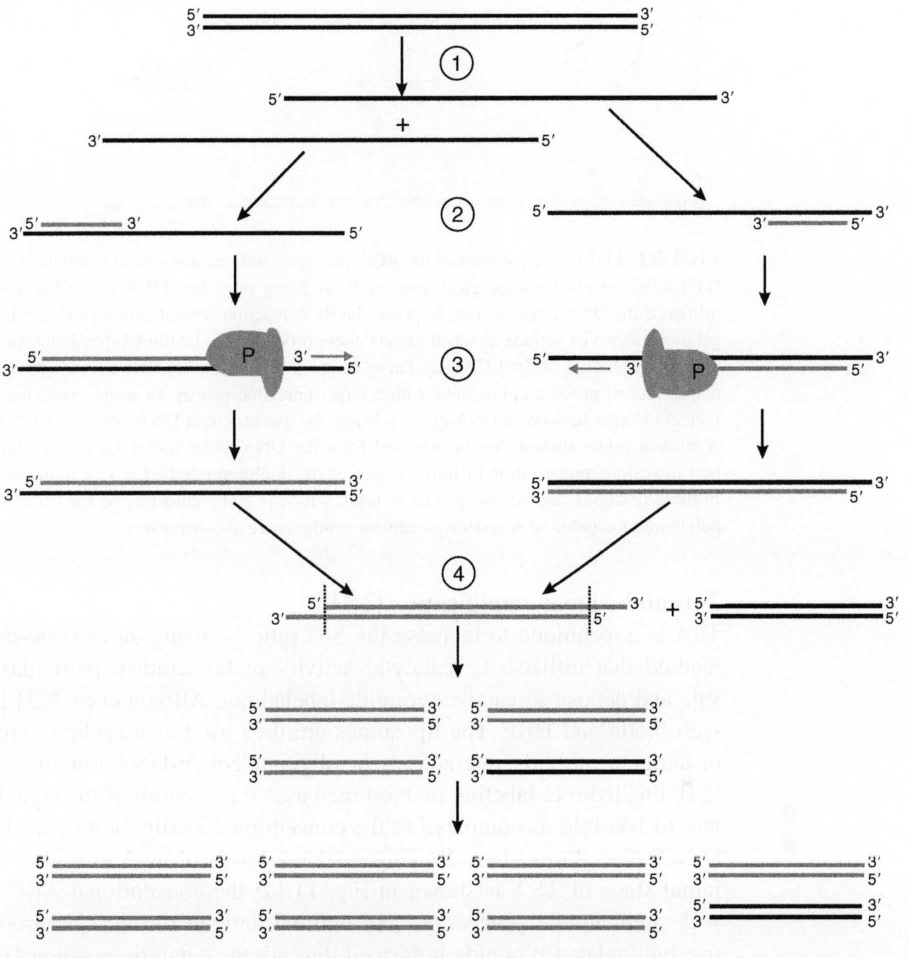

FIGURE 11.10 Schematic drawing of the PCR cycle. (1) Denaturing at 95ºC. (2) Annealing at 55ºC. (3) Synthesizing at 72ºC (P = Polymerase). (4) The first cycle is complete. The two resulting DNA strands make up the template DNA for the next cycle, thus doubling the amount of DNA duplicated for each new cycle. Source: http://en.wikipedia.org/wiki/Polymerasechainreaction (see Plate 3 for color version).

came up with a new technique of combining RCA with immunochemistry, whereby an oligonucleotide primer is attached to an antibody covalently. After bonding with the antigen, the antigen can be detected due to amplification resulting in a long chain with repeated sequences that remains attached to the antibody. All these are done in the presence of circular DNA and nucleotides as shown in Fig. 11.11.

11.2.4.2 Signal amplification

Signals obtained by methods like tyramide signal amplification can enhance signal intensity, reduce background fluorescence, and then increase sensitivity.

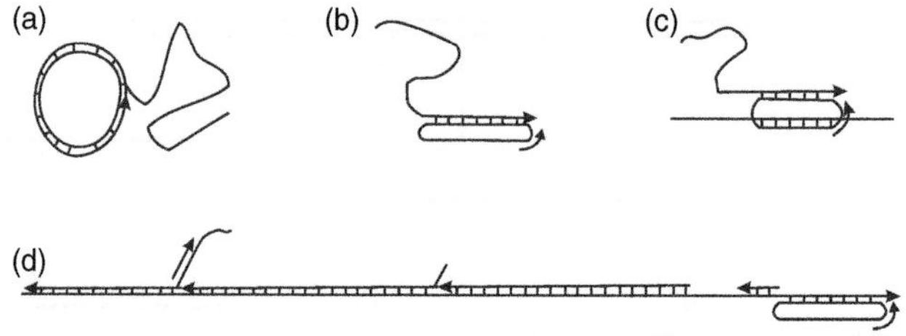

FIGURE 11.11 Schematics of the RCA processes with an arrowhead symbolizing the DNA polymerase. (a) Traditional phenomenological view of RCA going on a free DNA minicircle with the use of a single primer. If the DNA target is used to prime the RCA reaction, amplification products are fixedly linked to target molecules. For surface-attached targets, these products will be immobilized on solid-phase. (b) More realistic representation of the RCA-generating complex that resembles here a bow with a string. (c) RCA-based diagnostics of probe amplification with a target-unrelated primer. In some cases, the pseudo- or true topological linkages between a DNA minicircle and the marker/target DNA site may affect the rolling replication. A circular probe should then be released from the DNA target following the hybridization. (d) Simplified (not in scale) representation of initial stages of the double-primed RCA (for more detailed schematics see). In these reactions, the second primer is used, which is complementary to the original RCA product; DNA polymerases capable of strand-displacement synthesis are also necessary.

Tyramide signal amplification (TSA)

TSA is a technique to increase the S/N ratio by using an enzyme-mediated detection method that utilizes the catalytic activity of horseradish peroxidase (HRP) to activate and deposit a reactive tyramide-labeled tag. Alfonta *et al.* [23] labeled liposomes with biotin and HRP. The liposomes are then used as a probe to amplify the sensing of antigen–antibody interactions or oligonucleotide-DNA binding. According to Am [24], this indirect labeling method increases the strength of the signal amplification by ten- to 100-fold as compared to the conventional avidin–biotinylated enzyme complex (ABC) procedures, thus allowing very low levels of nucleic acid to be discerned. The initial stage of TSA is shown in Fig. 11.12; the conventional ABC method, together with pretreatment procedures and optimal antigen retrieval, is used. Next, the reactive biotinylated tyramide is formed through the catalytic reaction HRP and hydrogen peroxide catalyses in Fig. 11.13. Covalent coupling of the produced, highly reactive, short-lived tyramide radicals to the electron-rich moiety of protein tyrosine residues in the vicinity of the HRP reaction site results in minimal diffusion-related loss of signal localization. The fluorescent signal can be immediately detected, resulting in both excellent spatial resolution and high signal intensity.

11.2.5 Detection and data analysis

11.2.5.1 Detection technologies

Fluorescent detection technology applicable to biochips is evolving rapidly, resulting in detection instruments that are more powerful, user-friendly, and less expensive. Most systems employ photomultiplier tube (PMT) technology in conjunction with multiple colors, lasers, and a variety of filters. It is essentially a fluorescent microscope that

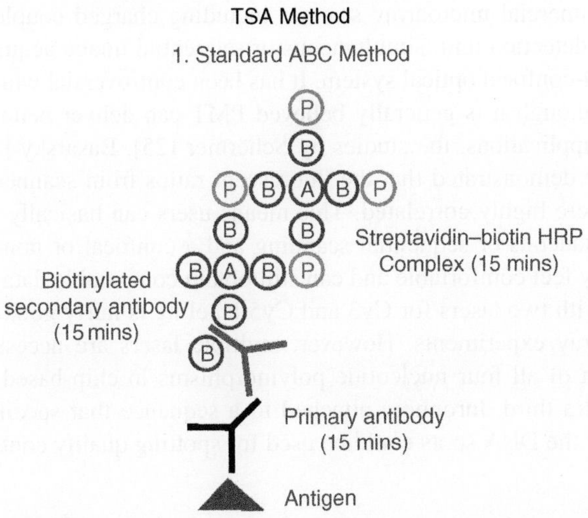

FIGURE 11.12 TSA method using the conventional ABC method, together with pretreatment procedures and optimal antigen retrieval. Source: www.hmds.org.uk/ histology.html

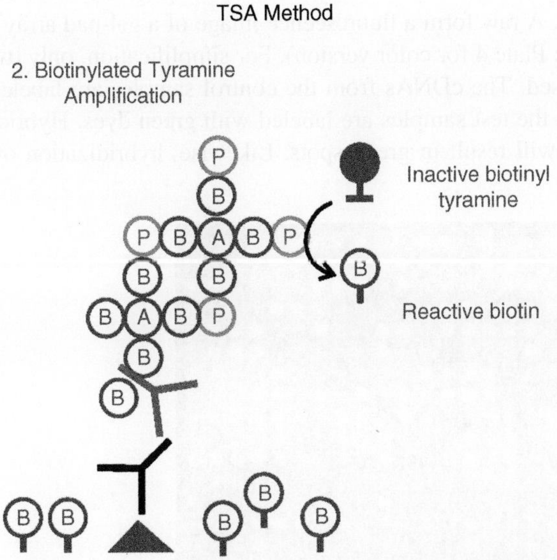

FIGURE 11.13 Formation of reactive biotinylated tyramide through the catalytic reaction HRP and hydrogen peroxide catalyses. Source: www.hmds.org.uk/histology.html

is specialized for acquiring microarray fluorescent images for arrays detection. The three most widely used scanner brands from the ABRF survey 2001 are Packard/GSI Lumonics (34%), Axon (24%), and Affymetrix/GMS (13%). There are many factors that determine the performance of a microarray scanner, like sensitivity, resolution, and dynamic range. In general, there are three major considerations in technologies to select a commercial microarray scanner including charged coupled device (CCD) or PMT as the detection unit, simultaneous or sequential image acquisition mode, and confocal or non-confocal optical system. It has been controversial which is the superior technology. Although it is generally believed PMT can deliver better data than CCD in microarray applications, the studies by Schermer [25], Basarsky [26], and Ramdas *et al.* [27] have demonstrated that the fluorescent ratios from scanners using different technologies were highly correlated. This means users can basically choose the scanner with simultaneous or sequential scanning and a confocal or non-confocal system with which they feel comfortable and can still obtain comparable data with others.

A scanner with two lasers for Cy3 and Cy5 labeling is fairly good enough for most of the microarray experiments. However, multiple lasers are necessary for simultaneous detection of all four nucleotide polymorphisms in chip-based SNPs detection. Besides, an extra third flurophore attached to a sequence that specifically binds to a linker region of the DNA spots could be used for spotting quality control.

11.2.5.2 Data analysis

In the microarray technology, the abundance of cDNA is measured indirectly by measuring the intensity of the fluorescent dye in order to generate more data for interpretation and analysis. A raw form a fluorescence image of a gel-pad array biochip is shown in Fig. 11.14 (see Plate 4 for color version). For simplification, only two color dyes (red and green) are used. The cDNAs from the control samples are labeled with a red dye, while those from the test samples are labeled with green dyes. Hybridization of a gene with the control will result in green spots. Likewise, hybridization of a gene with the

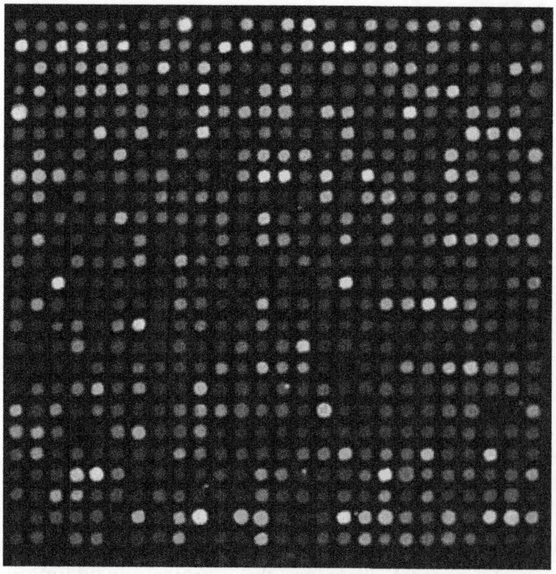

FIGURE 11.14 Fluorescence image of a gel-pad array biochip (see Plate 4 for color version).

test sample will result in red spots. Yellow spots occur when the gene hybridizes with both control and test samples.

Figure 11.15 (see Plate 5 for color version) shows ideal microarray spots. However, such ideal round spots are in fact impossible to achieve. As suggested by Yang *et al.* [28], the imperfect spot size and shape occurs during the hybridization and printing process and is also due to the conditions of the slide surface, making it more tedious to extract the raw data. Some of the imperfect microarray spots appear in the form of a "comet tail", as shown in Fig. 11.16 (see Plate 6 for color version), which is due to

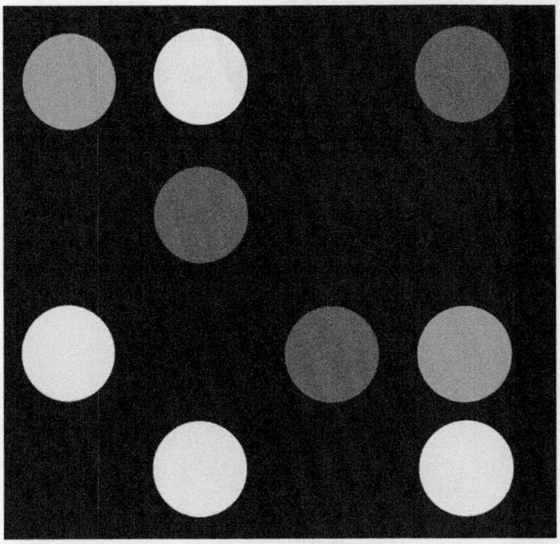

FIGURE 11.15 Ideal microarray spots (see Plate 5 for color version).

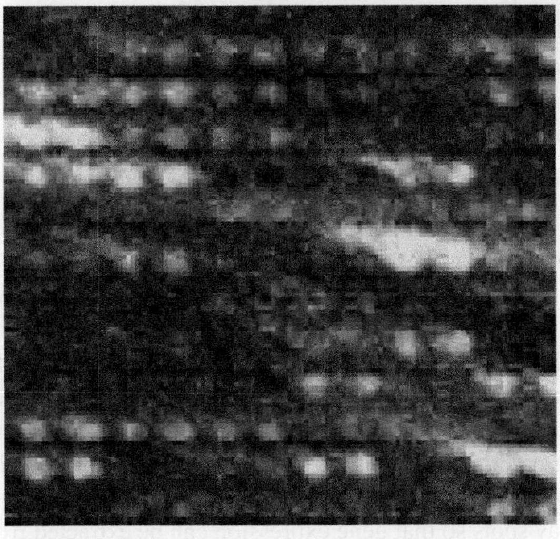

FIGURE 11.16 Comet tails (see Plate 6 for color version).

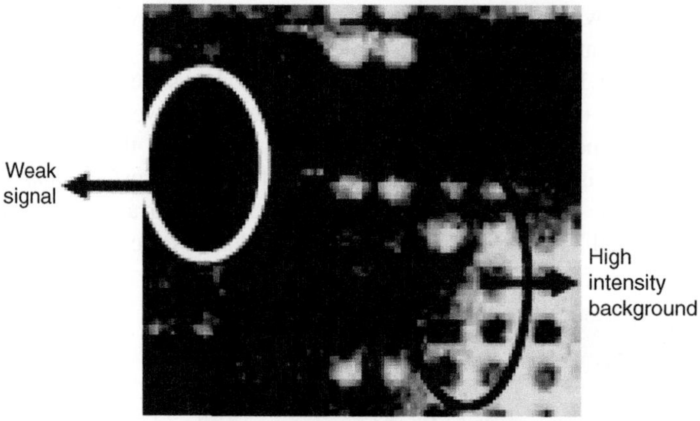

FIGURE 11.17 High intensity background and weak signal (see Plate 7 for color version).

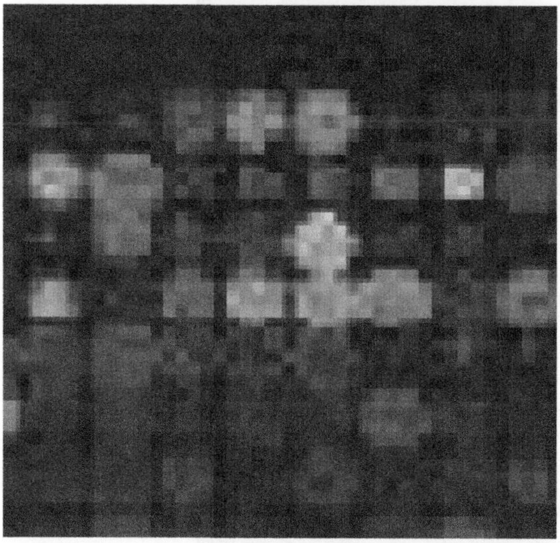

FIGURE 11.18 Spot overlap (see Plate 8 for color version).

insufficient rapid immersion of the nylon slides into the succinic anhydride blocking solution. Insufficient blocking and precipitation of the labeled probe results in high intensity background illumination. Hybridization that produces weak signals does not allow any image detector to read its spot location, size, and shape. High background intensity and weak signal spot array is shown in Fig. 11.17 (see Plate 7 for color version). Overlapping of spots is shown in Fig. 11.18 (see Plate 8 for color version). A possible cause for spots overlap is excess rehydration during post-processing. It is therefore necessary that sophisticated software programs are utilized to solve the problems of non-ideal microarray spots so that gene expression can be extracted from the raw data.

TABLE 11.4
Software for different segmentation methods

Fixed circle	ScanAlyze, GenePix, QuantArray
Adaptive circle	GenePix, Dapple
Adaptive shape	Spot, region growing, and watershed
Histogram	ImaGene, QuantArraym DeArray, and adaptive thresholding

The procedure of processing the images of the microarray consists of addressing, segmentation, and information extraction.

1. Addressing

Since the microarray spot images have many different sizes and shapes and some of the spots may not be located at the central position, addressing is essential to find the centre of the spots. In order to ensure accuracy of the measurements, an automatic spot detector is used to calculate the spacing between rows and columns of spots, the overall position of the array in the image and the spot size so that coordinates can be assigned to each of the spots in the array.

2. Segmentation

Segmentation is the process for classifying the image pixels as either foreground or background. The fluorescence intensities are calculated for each spot as a measure of transcript abundance. A bright spot occurs due to the sufficient hybridization of the probe material. The detector is able to detect the fluorescence of the bound labeled probe material and record the fluorescence intensities. The regions occupied by the bright spot are the foreground. In regions where no hybridization occurs, the intensity is equal to or less than the background values. Proposed methods for segmentation include fixed circle, adaptive circle, adaptive shape, and histogram segmentation. The various software used to analyze the different segmentation methods are shown in Table 11.4.

In fixed circle segmentation method, a circle of constant diameter is drawn on all the spots in the microarray images (Fig. 11.19, see Plate 9 for color version). This method is easy to implement; however, the fixed circle segmentation is inefficient as spots of same size and shape rarely occur.

In reality, spots are of different shapes and size. In adaptive circle segmentation, different circle diameters are assigned to individual spot diameters in the microarray as can be seen in Fig. 11.20 (see Plate 10 for color version). By doing this, the circle would fit the spot perfectly, regardless of the size of the spot. However, the tradeoff is that much time is required to estimate and compute each of the thousands of spots in the microarray.

Dapple is the image analysis software used for adaptive circle segmentation. It performs the analysis by estimating the circle diameter for each spot. By taking the second derivative of the pixel intensity vs coordinate graph, the edges of the spots can be

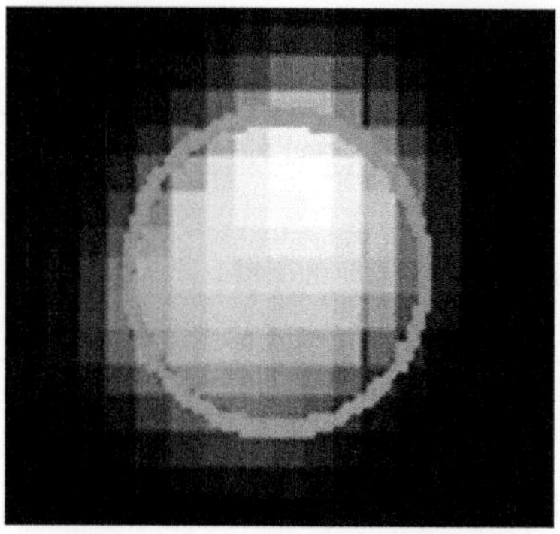

FIGURE 11.19 Fixed circle segmentation (see Plate 9 for color version).

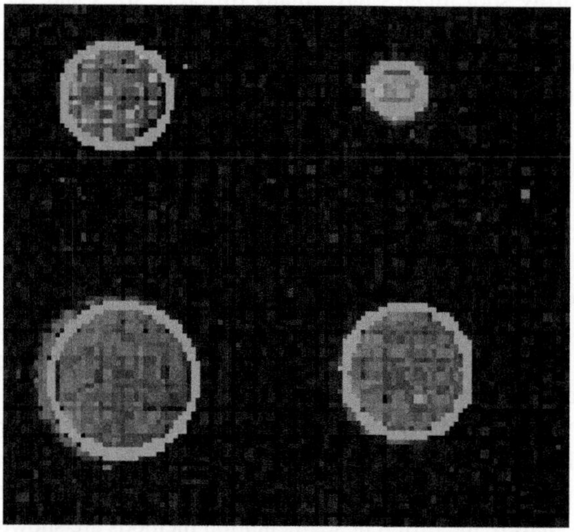

FIGURE 11.20 Spots with different sizes (see Plate 10 for color version).

obtained (Fig. 11.21). This method will be a little tricky if the shape of the spot is non-circular as shown in Fig. 11.22. For an irregularly shaped spot, a circular spot mask is impractical as it provides a poor fit to the spot. In the adaptive shape segmentation method, no assumption of the shape of the spot is made. A small number of pixels, known as seeds, are selected from within the spot (Fig. 11.22). At each step of the algorithm, the selected region is grown outward by adding a pixel to the existing seed according to the difference between the seed pixel's value and the running mean of the values in an adjoining region. In histogram segmentation, a target mask larger than any spot is used. Within the target mask, the foreground and background intensities are determined. For pixel values greater than a predetermined threshold, the pixel is classified

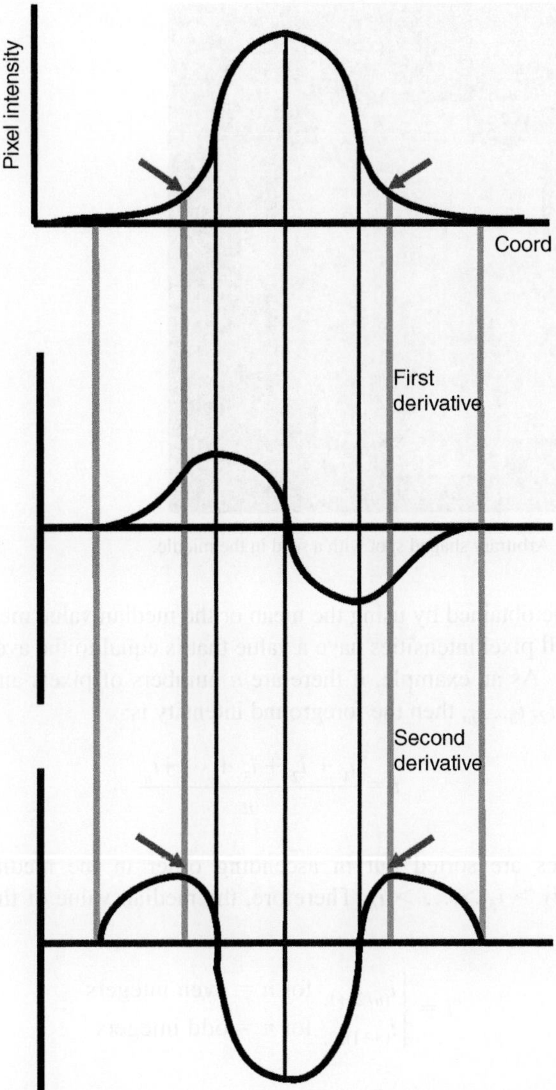

FIGURE 11.21 Spot finding using adaptive circle segmentation method.

as a foreground. On the other hand, if the pixel value is smaller than the predetermined value, the pixel is classified as the background.

3. Information extraction [29]

After determining the shape, size, and address of each spot in the microarray, the next step is to calculate the foreground and the background intensities. The foreground

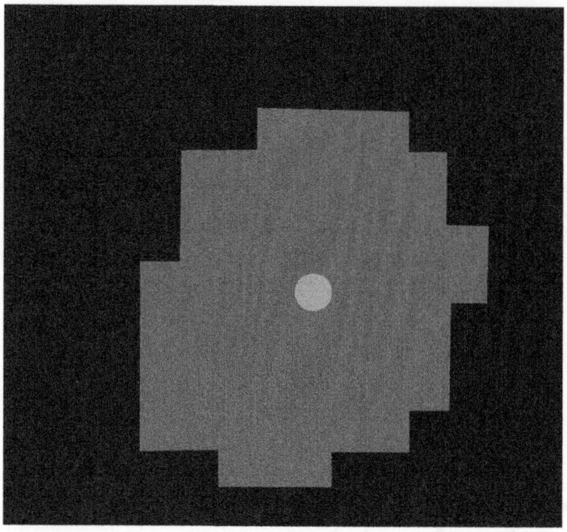

FIGURE 11.22 Arbitrary shaped spot with a seed in the middle.

intensities can be obtained by using the mean or the median value method. In the mean value method, all pixel intensities have a value that is equal to the average of the entire pixel intensities. As an example, if there are n numbers of pixels and the intensity of each pixel is $i_1, i_2, i_3 \ldots i_n$, then the foreground intensity is:

$$i = \frac{i_1 + i_2 + i_3 + \cdots + i_n}{n}$$

The pixel values are sorted out in ascending order in the median value method whereby: $i_1 > i_2 > i_3 > \ldots > i_n$. Therefore, the median value of the intensity of the foreground is:

$$i = \begin{cases} i_{(n/2)+1}, & \text{for } n = \text{even integers} \\ i_{(n+1)/2}, & \text{for } n = \text{odd integers} \end{cases}$$

If there is no hybridization, the background will consist of regions that are not occupied by any DNA, and hence, no spot detected. Pixels that are near the edge of a spot are taken as background pixels. Taking an average of these values will result in local background intensities. Local background, morphological opening, and constant background are methods used to calculate the background intensities. The local background method focuses on a small area around the spot mask and determines the median value of the intensities in this area. The background estimate is less sensitive to the performance of the segmentation procedure as the pixels right next to the spots are not considered. Some of the common approaches to local background methods are shown in Fig. 11.23.

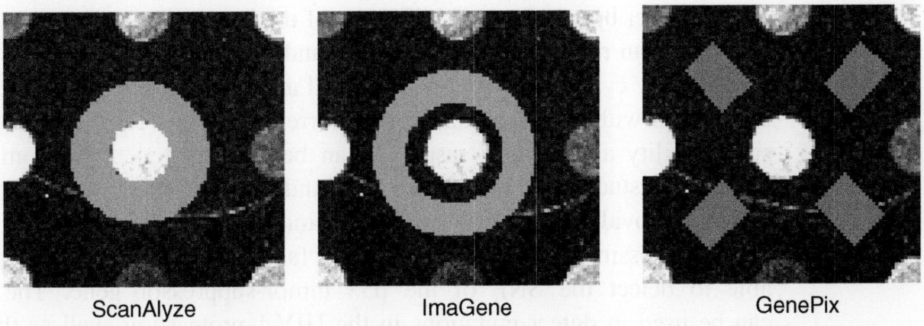

FIGURE 11.23 Common approaches to local background methods.

11.2.6 Applications

DNA arrays are useful for a variety of genomic applications, including single nucleotide polymorphism (SNP) analysis, gene expression studies, disease classification, function prediction, pathway identification, new drug development, clinic diagnostics, and toxicology studies. Perhaps the greatest driver of the DNA array market is the demand for speed and high throughput. The ability to look at thousands of genes in a single experiment will increase the speed of drug-target identification and validation, as well as decrease the time required for optimizing lead drugs.

Gene expression profiling is the leading application in the biomedical research category, due to biochips' potential to simultaneously monitor genes expression in diseased cells and cells that are healthy. Gene expression profiling has enabled the measurements of thousands of genes in a single RNA sample. RNAs are labeled with fluorescent dyes and hybridized on the biochip. Hybridization between the sample RNAs and the reference RNAs under different environmental conditions allows scientists to compare gene expression under different conditions such that groups of genes that play a role in disease processes can be identified.

In the drug discovery process, researchers may be interested in the gene expressions under the conditions of the presence or absence of drugs. Tens of thousands of mRNAs in a large number of samples are monitored. The differences in gene expression also provide a drug marker with unique targets that are present only in diseased cells. Using DNA biochips, labor and reagent cost can be reduced by a great magnitude. In the case of cancer, oncogenes and proto-oncogenes are activated. On activation, proto-oncogenes become tumor-inducing agents – an oncogene, leading to the possibility of cancer. Oncogenes and proto-oncogenes do not occur in healthy cells; therefore, targeting these genes may lead to new therapeutic approaches.

Some of the gene expression profiling analysis biochips on the market include Affymetrix's standardized GeneChip for a variety of human and yeast genes and HyX Gene Discovery Modules for genes from tissues of the cardiovascular and from tissues that are germ infected.

Single nucleotide polymorphism (SNP) is a variation of the genetic sequence (point mutation) when a single nucleotide in the DNA is altered. It has a 0.1% chance of occurring per bp within the 3 billion bp of the human DNA. Point mutations can affect how a human reacts to bacteria, drugs, and chemicals and may also result in genetic diseases like cystic fibrosis and sickle-cell anemia. Biochips allow the analysis of multiple SNPs within individuals so that correlations between SNP patterns and disease susceptibility and/or drug response can be investigated. Drug companies are using SNP-based studies on gene candidates and will be able to achieve delivery systems that can provide accurate, inexpensive, and rapid automated genotyping of as many SNPs as possible. From the Affymetrix family of GeneChips, the p53 GeneChip is able to detect the SNP of the p53 tumor-suppressor gene. The HIV GeneChip can be used to detect mutations in the HIV-1 protease as well as the virus's reverse transcriptase genes.

Apparently, biomedical applications are the main demands of the DNA biochip. The selected applications are shown in Table 11.5.

TABLE 11.5
Selected DNA arrays applications. Source: 3rd Millennium

Applications	Major goals	Comments
Disease classification	Identify disease subgroups based on their unique gene expression profiles	Such classification can shed light on the biological basis for different clinical outcomes. It may be useful for creating sensitive diagnostic tests, thereby permitting more effective, targeted treatment. This approach has been used with several kinds of cancers.
Function prediction	Predict the function of unknown genes based on the similarity of their gene expression profiles to those of known genes	These experiments require a large database of "known" expression profiles that can be used for comparisons. This approach will become powerful as such databases continue to grow over the next several years.
Pathway identification	Find biomolecular pathways that are affected by disease and disease treatment	This task is crucial for finding targets for potential drugs or for more clearly understanding how existing drugs work.
Gene network identification	Identify prevalent expression patterns (gene clusters) and then identify DNA sequence patterns	This approach has been used to find regulatory elements in yeast. It will soon be used to discover coregulated genes in other organisms (including humans).
Toxicology	Test drug-treated tissue samples for toxicological effects	These studies help drug companies eliminate from their pipelines – at an early stage of development – drugs that are likely to have poor side-effect profiles. This approach will make drug development more efficient and cost effective.
Transcript discovery	Find transcripts (genes) in genomic sequence	In these experiments, arrays are made from genomic DNA sequences, rather than from complementary DNAs (cDNAs). Such studies can be used to confirm computational gene predictions and characterize alternative splicing and the boundaries of exons (DNA sequences destined to become part of the mature messenger RNA [mRNA]).

11.3 PROTEIN CHIPS

11.3.1 Protein array and proteome

Protein arrays (or chips), in comparison to DNA chips, represent a great technical advance, allowing thousands of proteins to be studied in a single experiment for the understanding of biological systems or system biology, in which protein plays a central role as most biological functions in organisms are mainly executed by protein. In addition, proteins are extremely important in human physiology and in medicine because most diseases occur at the level including those which are considered genetic or heritable. Therefore, when we wish to identify specific protein functions, various diseases in diagnostics or discover new drugs in medicine, we should eventually focus on the proteome.

Proteins are created, or translated on the basis of the information contained in genes, namely, the DNA sequence of genes. While gene expression data provide important information regarding the underlying proteins, they cannot necessarily predict the myriad changes that occur to a protein subsequent to translation. This inability to accurately characterize proteins based on gene expression data is a result of protein control mechanisms that occur after the protein is synthesized from the gene. Consequently, there is no direct way at present to correlate the genomic sequence and protein function although protein components are encoded by gene (DNA or sometimes RNA).

Compared with genome, proteome is much more complicated and thus more difficult to explore. First, proteins are composed of 20 amino acids while genes contain only five nucleic acids. Moreover, proteins possess complex 3D structure (primary, secondary, tertiary, and quaternary) and their functionality is often dependent on the state of proteins, such as post-translational modifications, partnership with other proteins, protein subcellular localization, and reversible covalent modifications. As a result, there might be as many as 1 million different human proteins compared with only 30 000 or so genes. Second, proteins require more delicate handling than DNA, because they can easily unfold and get denatured when coming in contact with the improper surface or environment. Third, strand complementarity makes identification of DNA a simple task, whereas proteins must be detected using mass spectrometric analysis in conjunction with sophisticated software or using molecules (such as antibodies) that specifically recognize their molecular structure. Fourth, protein detection should be highly sensitive for the lack of an effective amplifying technology such as PCR. All these make proteome far from the level of precision available to users and purveyors of genomic technology.

Proteome may help biologists to study basic cell functions and molecular organizations, another big field in microbiology for various research areas. Proteomics is also applicable to plant research for many different purposes such as breeding plants against higher bacterial, heat, cold, drought, and other resistances, increasing the yield of crops, and many more. In such fields, proteomes are usually combined with genomes.

Traditional methodologies such as 2D-gel electrophoresis and mass spectrometry have been considerably improved to resolve thousands of proteins in a single experiment. However, these approaches are both time consuming and unsuitable for the analysis of low abundant proteins. As a result, they cannot meet the requirements of automation and high throughput analysis. Protein biochips or protein microarrays are an emerging alternative technology to increase the overall throughput at reduced cost. By providing an addressable array of spots, with analytes detected by methods such as mass spectrometry or fluorescence, protein chips eliminate much of the irreproducibility and complexity of 2D-gel analysis.

11.3.2 Fabrication of protein chips

11.3.2.1 Types of protein chips

Currently, protein microarrays can be divided into various types depending on the strategies to be chosen. For example, according to the array structure and shape, protein microarrays include 3D-surface structure [30, 31], nanowell [32], and plain chips [33–36]. Meanwhile, considering the field of application, protein microarrays can be classified into five categories: antibody array, antigen or reserve array, functional array, capture array, and solute array. Table 11.6 shows the differences among them.

The principle of antibody arrays, and their converse counterparts, antigen or reverse arrays, is to use high affinity ligands to detect the presence of specific proteins and biomarkers in a complex mixture. Knezevic *et al.* [38] selected 368 antibodies specific to cancer-related proteins. Exposure of the printed arrays to six categories of histologic lysates identified 11 proteins that showed consistent change in expression or state of phosphorylation. Sreekumar *et al.* [39] used an array of 146 distinct antibodies to monitor alterations of protein levels in colon carcinoma cells. Paweletz *et al.* [40] used this technique to reproducibly quantify the status of signal proteins in 3 nL spots containing whole protein repertoires of lysed cell proteins, and peptides and nucleic

TABLE 11.6
Types of protein microarrays [37]

Array type	Description
Antibody array	Polyclonal or monoclonal antibodies are arrayed and used to detect and quantify specific proteins in a biological sample. An antibody array is effectively a parallel series of miniature immunoassays
Antigen or reverse array	The converse of an antibody array, this chip has immobilized antigens that are used to detect and quantify antibodies in a biological sample
Functional array	Purified proteins are arrayed on the surface and used to detect and characterize protein – protein, protein – DNA or protein – small molecule interactions
Capture array	Non-protein molecules that interact with proteins are immobilized on the surface. These may be broad capture agents based on surface chemistries such as the Ciphergen Protein Chip, or may be highly specific such as molecular imprinted polymers or oligonucleotide aptamers
Solute array	The potential next generation of arrays is to have nanowells containing coded microspheres or barcoded nanoparticles in solution

acids were incubated with patient serum to detect and characterize autoantigens in the serum. Bacarese-Hamilton and coworkers used antigen arrays to detect serum antibodies against allergens and infectious antigens [41,42].

Functional arrays are concerned with elucidating novel protein interactions, and are thus more analogous to the yeast two-hybrid system [43] and co-immunoprecipitation studies [44, 45]. Compared with antigen – antibody array, functional arrays show great potential in the mapping of interacting proteins on a system-wide or genome-wide scale. Major advantages of this *in-vitro* technique are that we can control the conditions of the experiment, modify the state of the proteins under investigation, and study the interaction of proteins with non-proteinaceous molecules. Several pioneer investigators have conducted a number of studies in this field. Ge [46] arrayed 48 purified human proteins onto a nitrocellulose membrane and probed with different proteins, nucleic acids, and small organic compounds. Results show that a double-stranded DNA probe bound more tightly to a phosphorylated form of protein PC4 than to the unmodified form. Zhu *et al.* [35] used functional arrays to identify 150 putative phospholipid-binding proteins. MacBeath and Schreiber [34] proved that the activities of purified proteins can be retained when spotted onto chemically derivatized glass slides which simplifies the mass production of functional microarrays. Indeed, functional arrays are not limited to whole proteins. Espejo *et al.* [47] used immobilized glutathione S-transferase-fused protein interaction domains to fish out proteins from total cell lysates.

Capture array involves the immobilization of non-protein molecules onto the surface which can interact with proteins in the solute phase. Generally, capture molecules may be broad capture agents based on chromatography type surface chemistries such as ion exchange, hydrophobic and metal affinity functionality, or they may be highly specific such as molecular imprinted polymers or oligonucleotide aptamers.

11.3.2.2 Surface functionalization for protein arrays

No matter what kind of biochip is chosen for use, the first and sometimes the most important step to construct a biochip is to effectively immobilize proteins or biomolecules on a solid surface. The ideal properties of protein biochip surface are:

- High immobilization density
- No or little influence on the activity of immobilized biomolecules
- Effective inhibition of non-specific adsorption
- Strong suppression of matrix effect of complex biological solution on the detection of the one component of interest
- Orientation of immobilized protein with binding site towards the solution.

At present, a wide range of solid substrates are available for protein immobilization. According to the protein attachment strategies, namely, adsorption, affinity binding, and covalent binding, all these substrates can be separated into three main parts. Surfaces like ploy(vinylidene fluoride) (PVDF), poly(dimethylsiloxane) (PDMS), nitrocellulose, polystyrene, and poly-1-lysine coated glass can adsorb proteins by electrostatic or hydrophobic forces. A potential drawback of such substrates is the difficulty in preventing non-specific binding. To solve this problem, various blocking agents have been employed to saturate the free binding sites on the surface and thus reduce non-specific binding. For example, the use of high salt conditions [48], high detergent concentrations [49], cocoating of surfaces with anionic proteins [50], milk proteins [51] or denatured proteins [52] can effectively passivate the surface while maintaining specific binding. However, despite rigorous blocking, it seems that the level of non-specific binding still depends on the intrinsic protein-binding capability of a surface, reflecting the fact that it is impossible to completely block a surface. Another drawback in using an adsorption method is that some proteins may not bind depending on the nature of the protein. A weakly bound protein may be washed away during the washing step especially if stringent washes such as those with high salt or high detergent concentrations are used. All the above indicate that adsorption can only provide low protein loading and therefore is unsuitable for high throughput applications.

Compared with the adsorption method, affinity binding can offer a strong and highly selective attachment through the use of specific biological interactions, for instance between biotin and avidin or between protein A and IgG. Note that both the surface and the protein to be attached should be derivatized with the components of the interaction. Moreover, if the derivatized surface is resistant to non-specific protein adsorption, a lower background can be achieved while keeping specific binding. It has been reported that a polyethylene glycol (PEG) coated surface which is functionalized with biotin and streptavidin can adsorb spotted antibodies containing biotin easily but non-specific background proteins can be repelled by PEG. A main drawback of attachment by affinity binding is that proteins to be attached have to be modified before spotting, adding steps to protein preparation and sometimes changing the protein structure.

Covalent binding is the most promising method for microarray technology for the reason that it provides the strongest attachment of proteins to surfaces [53]. For example, PDMS surface with (3-aminopropyl)-triethoxysilane (APTES) treatment offered efficient protein immobilization and high sensitivity through succinic acid anhydride (SAA) and 1-ethyl-3-(3-dimethylaminopropyl) carbodiimide hydrochloride (EDC), or glutaraldehyde (GA) for covalent binding [54]. In such a mood, proteins are attached to the surface through the reaction between the amino or carboxyl groups on proteins and immobilized functional groups on the substrate such as aldehydes or succinimides. Methods to form aldehyde groups on surfaces include oxidation of the carbohydrate groups of adsorbed polyacrylamide or agarose and cross-linking glutaraldehyde to immobilize polyacrylamide or aminosilanes. Similarly, succinimide groups were formed using a cross-linker attached to aminosilanes or a bovine serum albumin (BSA) coating. However, the requirement of accessible amino groups on the spotted proteins may, in some cases, limit this approach. Besides, other covalent attachment techniques that have been popular in immunosensors and chromatography can also be used for protein microarrays. IgG molecules can be covalently bound by oxidizing the carbohydrate group on the Fc region, creating an aldehyde and reacting the aldehyde with hydrazide-activated surface. Attention should be paid in this treatment process because oxidation of the antibody may occasionally damage the antigen-binding site.

Most materials developed for covalent binding of proteins cannot be directly used unless their surfaces are functionalized. Normally, inorganic oxides, particularly silica, porous glass, and oxidized metals, are derivatized with organosilanes [55]. The most popular examples are aminopropyltrimethoxysilane (APTES) treated glasses, silica, and quartz. A typical procedure is to treat the surface with 10% APTES in chloroform, toluene or in aqueous silane at PH 3–4. The alkoxy groups in APTES are hydrolyzed by the surface-free water and then connected to the surface silanol groups (Si—OH) via Si—O—Si bonds. Post-thermal curing is sometimes necessary to cross-link the residual free silanol groups and thus form a uniform saline film. If the matrix of silane is absolutely anhydrous, a monolayer can be obtained. Otherwise, a multiple layer would exist on the surface (Fig. 11.24).

Gold-based substrates are also widely investigated for covalent binding of probes. Instead of using polycrystalline gold surfaces, thin gold films are now preferable since they can be easily deposited on solid supports like glass and silicon wafers by evaporation or sputtering in an ultra-high vacuum. A thin layer of Cr or Ti is generally required to deposit prior to gold coating. After careful cleaning, a self-assembly monolayer (SAM), namely, gold-thiol monolayer can be prepared by exposing gold surface to the vapor or solution of thiols. The affinity of thiols to gold is so strong that they can replace most of other adsorbates on the gold surface in virtually any solvent, which make this preparation very simple. Depending on the end group located on the monolayer surface, 1-ethyl-3-(3-dimethylaminopropyl)carbodiimide (EDC) or glutaraldehyde (GA) can be served as cross-linker to couple protein to the gold-thiol surface. Because of the excellent electronic properties of gold, these surfaces can be applied in more advanced and label-free detection technologies such as surface plasmon resonance (SPR) and mass spectrometry.

Except inorganic substrates, functionalized polymer membranes are another choice to immobilize proteins. For instance, commercial SPR instrument producer Biacore adopts epoxy-terminated thiol to couple carboxy-methyl-modified dextran polymers, which contains extensive reactive sites for activation and covalent attachment of proteins of interest. Another new approach is to use the copolymerization of methacrylate, styrene or vinyl alcohol monomers on glass surface followed by subsequent functionalization using photografting. The optical properties of the surface can be tailored to the reflectance of light or scattering effects. This technology has been used to make highly reflective microarray surfaces at applied wavelength for potential applications in highly sensitive luminescent-based assay systems.

Although functionalized surface technology has achieved much success in the past few decades, there are still challenges in building and using functionalized protein microarrays. First, the notorious liability of proteins raises concerns about their stability and integrity on the functionalized surface. Second, it is time consuming and costly to produce proteins of good purity and yield, and not many proteins can be purified at all. Third, functionalized surface lacks the ability to prevent non-specific adsorption. Finally, the methods used to attach proteins to the substrate surface may affect the behavior of the proteins. Improvements in these aspects are crucial to enhance sensitivity and to avoid false positives.

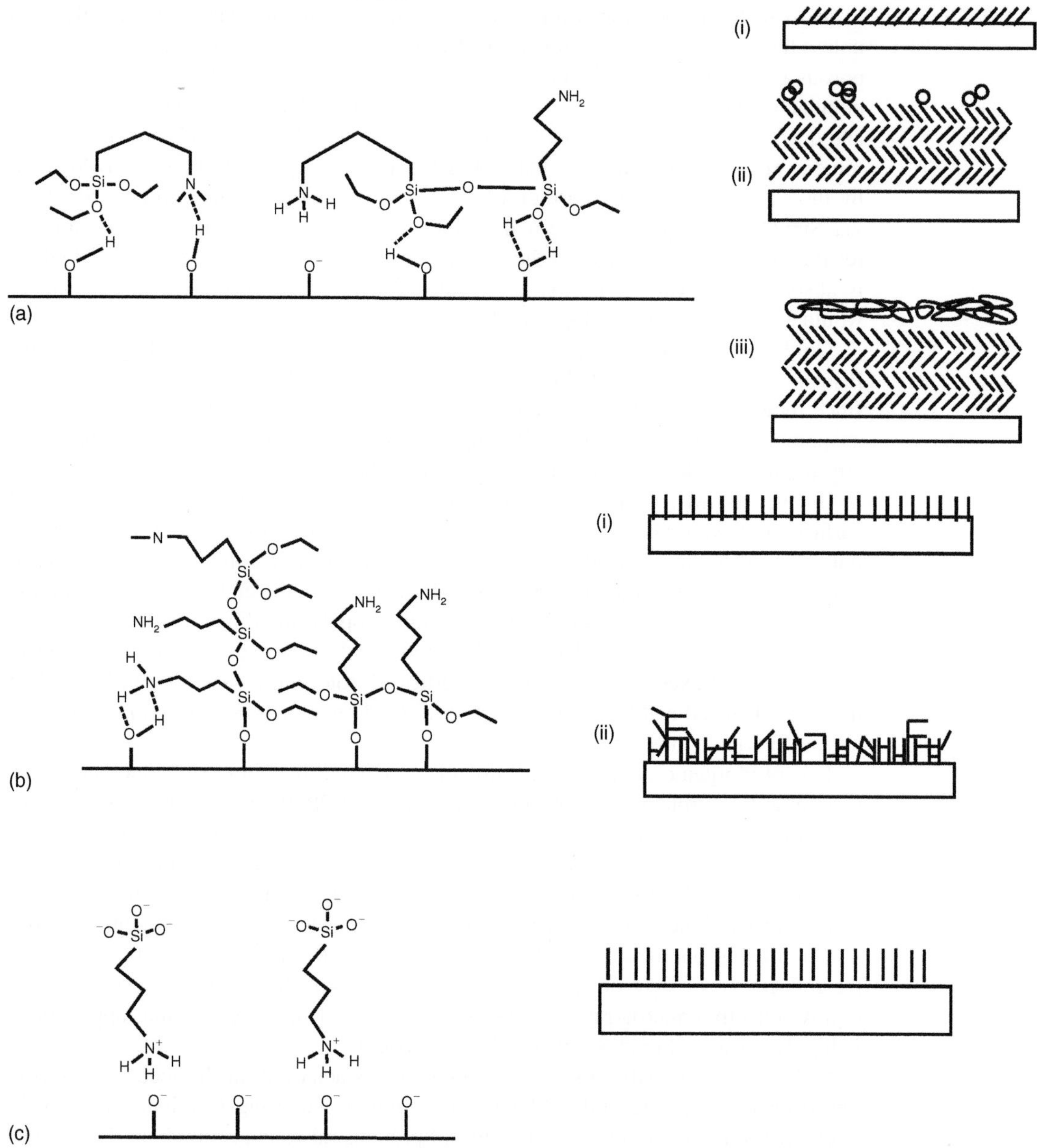

FIGURE 11.24 Schematic representation of the structure of APTES after various deposition conditions. Microscopic orientation is indicated on the left side of the figure. A schematic, not-to-scale diagram of macroscopic orientation is on the right side, showing the underlying solid surface, APTES monolayers (each about 0.7 nm thick), and APTES aggregates (10s to 100s of nanometers thick). (a) Deposition from toluene without curing: (i) reaction time of minutes at room temperature, (ii) reaction time of hours at room temperature, and (iii) reaction time of hours under reflux. (b) Refluxed deposition from toluene-RT with curing: (i) reaction time of minutes and (ii) reaction time of hours. (c) Deposition from water without curing. Adapted from [56].

11.3.2.3 Fabrication of gel pad, ELISA, and SELDI protein biochips

Different fabrication technologies have been reported in the literature [34, 57–64]. The most straightforward way to construct the protein biochip is to follow the approach which has already been successfully applied in DNA biochip, namely, gel pad. Actually, gel-pad technology provides three-dimensional (3D) structure for protein immobilization which includes larger surface area and greater probe density. As we know, the functional properties of a protein are highly dependent on its conformation and structure. Therefore, it is possible that delicate proteins may not be able to survive the harsh conditions of 2D surfaces. Since gel pad closely mimics the biological environments of various proteins, the conformation and activity of immobilized protein are preserved in the homogeneous surrounding water. In the past years, researchers in this field have mainly focused on polyacrylamide gels by virtue of its unique properties such as low fluorescence background, low non-specific absorption, high immobilization density and high porosity for easy access of macromolecules into the gel pad [65]. Normally, polyacrylamide gels deposited on certain substrate (for example, microscope slides) are synthesized by exposing acrylamide solution to UV radiation and then washing to remove non-polymerized monomer. After that, polyacylamide gels are further activated with glutaraldehyde or partially substituting the amide groups with hydrazide groups. Hydrazine-activated microchips demonstrated higher capacity than glutaraldhyde-activated microchips and gave signals 1.8-fold higher after immobilization of the same antibodies.

Protein array chips can be fabricated by using well-established protein detection methods such as enzyme-linked immunosorbent assay (ELISA). Since the signals can be dramatically amplified by enzyme, this method can detect very low abundance of antigen or antibody. In general, ELISA contains direct and secondary detection. The direct detection directly labels the target proteins for measurements while secondary detection recognizes the analyte proteins through a secondary labeled antibody binding to the captured target proteins in a sandwich mode after an initial incubation. Due to its highly inherent specificity and sensitivity, secondary detection is preferable. In sandwich ELISA mode (Fig. 11.25), the secondary antibodies are labeled either with a fluorophore or with an enzyme that produces a detectable luminescent product. The sandwich ELISA allows an assay with little or no purification. The primary drawback, however, is that well-characterized antibodies generated against a single antigen are necessary to confer a high degree of specificity. If an antibody binds a common antigen across multiple protein species, a "composite" signal will be generated and additional assays are necessary to decipher the binding event.

Nevertheless, methodologies that combine genomic arraying techniques with ELISA-based methodologies are now gaining considerable attention [66]. This approach strives to deposit nanoliter quantities of bait reagents, such as antibodies, recombinant peptides or small drug libraries on an addressable, high density (potentially thousands of features) microarray in a process analogous to genomic microarrays, which is now possible due to advances in recombinant protein technologies that allow the rapid production and purification of affinity-based reagents such as antibodies, ScFv, aptamers, and peptide libraries. Although many groups have demonstrated the utilities of protein microarrays,

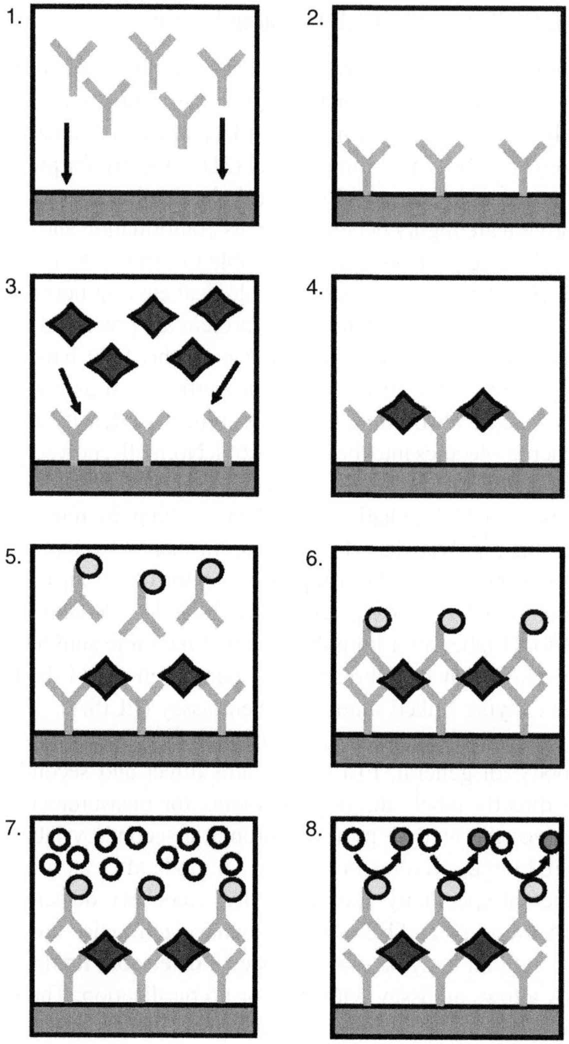

FIGURE 11.25 Schematic illustration of sandwich ELISA mode. Adapted from http://pl.wikipedia.org/wiki/ELISA

there still exist considerable obstacles associated with commercialization. Attachment of the bait protein to the array surface can be accomplished through either non-specific adsorption or through direct covalent binding. Critical limitations include protein denaturation upon interaction with the surface, disruption of the interaction domain through poor orientation or steric hindrance caused by physical closeness to the support itself (particularly important with small bait molecules), or a lack of sufficient density to allow for subsequent detection. Additionally, the range of biologically significant affinities (mM to pM) that must be considered when optimizing target binding conditions to

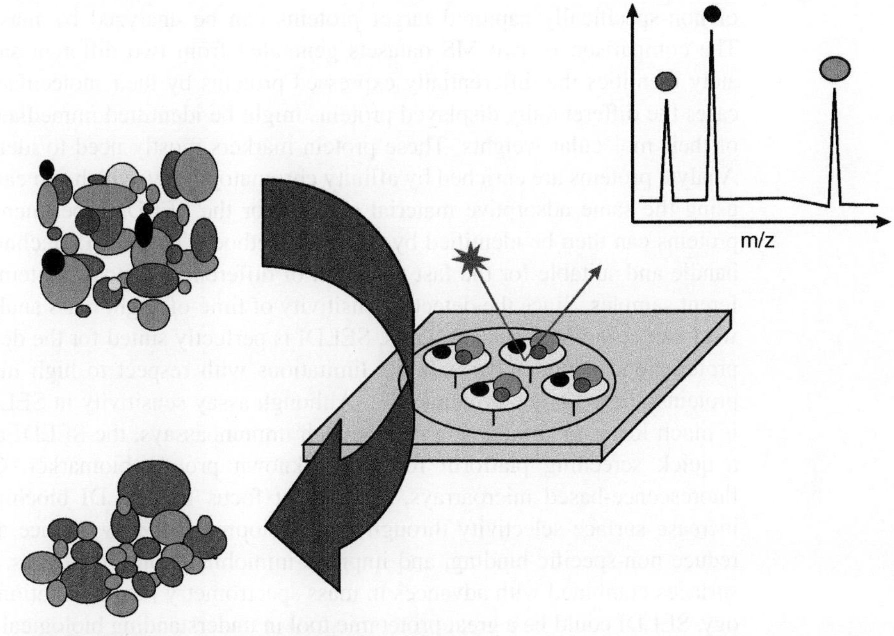

FIGURE 11.26 Schematic illustration of SELDI.

retain all relevant low affinity interactants (to prevent false negatives) while eliminating non-specific binding (to prevent false positives) is critical when working with potentially hundreds of proteins simultaneously. Moreover, although macroarrays (96-well) with polystyrene as solid phase and multi-channel spectrophotometers are commercially available for ELISA, when it comes to microarray, the situation is more difficult. Efforts should be focused on how to avoid cross-interference at different test spots caused by diffusion of the indicator.

Undoubtedly, ELISA is a powerful tool to detect known proteins or enzyme activity. However, this approach will never suffice as a method of discovering new interactants. Protein biochips which integrate a broad range of surface affinities with direct detection methodologies will ultimately prevail as true discovery tools. Currently SELDI (surface enhanced laser desorption ionization)-based biochips (Fig. 11.26) are the only commercial products that accommodate this strategy. SELDI incorporate a solid-phase extraction adsorbent on an electrically conductive support that can be interrogated directly by time-of-flight mass spectrometry (TOF-MS). The extraction adsorbents are functionalized with a number of different chemistries to yield complementary selection characteristics across a wide physicochemical range. Biochips can be provided as preactivated surfaces for the covalent binding of biomolecules, or chromatographic surfaces for the selective capture of proteins via charge, hydrophobicity or metal–chelate interactions. No matter what kind of surface is utilized, the methodology for capturing specific classes of proteins from a crude biological sample is similar. After incubation, unbound or weakly bound proteins can be washed away whereas the whole variety of non-specifically captured target proteins can be analyzed by mass spectrometry. The comparison of two MS datasets generated from two different samples immediately identifies the differentially expressed proteins by their molecular mass. In some cases the differentially displayed proteins might be identified immediately on the basis of their molecular weights. These protein markers mostly need to identify separately. Analyte proteins are enriched by affinity chromatography, which can easily be achieved using the same adsorptive material as used for the SELDI experiment. The enriched proteins can then be identified by standard methods. The SELDI technology is easy to handle and suitable for the fast detection of differences in total protein content of different samples. Since the detector sensitivity of time-of-flight mass analyzers decreases with increasing molecular weights, SELDI is perfectly suited for the detection of small proteins and peptides but exhibits limitations with respect to high molecular-weight proteins or membrane proteins [67]. Although assay sensitivity in SELDI experiments is much lower in comparison to sandwich immunoassays, the SELDI approach is still a quick screening platform for any unknown protein biomarker. Compared with fluorescence-based microarrays, the current focus for SELDI biochips continues to increase surface selectivity through the development of new surface affinity legends, reduce non-specific binding, and improve immobilization efficiencies. With improved surfaces combined with advances in mass spectrometry instrumentation and methodology, SELDI could be a great proteomic tool in understanding biological functions.

11.3.3 Protein chip applications

The protein chip technology has become a powerful tool for large-scale and high throughput biology. Further advances in this field will lead to convergence of the physical, chemical, and biological sciences with crucial impact on biotechnology and improving quality of life. Adoption of a protein chip concept to understand the global expression profile and detection of proteins in different physiological states will accelerate better understanding of biology and disease mechanisms that can lead to more effective therapeutic strategies. The protein chip is mainly used in biomedical and pharmaceutical applications. The protein chip accelerates pharmaceutical research for new drug protein targets identification in transformed cell lines or diseased tissues. The validation of the detected targets, *in-vitro* and *in-vivo* toxicology studies, and checks for side effects can be performed with this approach. Clinical researchers can use the chip to compare normal vs disease samples, disease vs treated samples, find molecular markers in body fluids for diagnosis, monitor diseases and their treatments, and determine and characterize post-translational modifications. In clinical chemistry it would be interesting to subtype individuals to predict response to therapy. The great potential of the protein chip in basic research, diagnostics, and drug discovery has been demonstrated.

11.3.3.1 Basic research

The use of biochips in basic biological research has provided information that was heretofore inaccessible. Protein chips, especially functional microarrays, are used to study basic biological properties like examining protein interactions with other ligands such as proteins, peptides, lipids or other molecules. In addition, functional microarrays are also used to determine enzyme activity and substrate specificity. These biochips are typically produced by printing the proteins or other molecules of interest on the array surface using various methods to maintain their integrity and activity, which allows hundreds to thousands of target proteins to be simultaneously screened for function [68]. For example, Espejo *et al.* [47] proved that protein interaction domains would retain function and specificity when interacting with their corresponding ligands by arraying several kinds of proteins onto nitrocellulose coated microarrays. Hall *et al.* [69] probed yeast protein microarrays with dye-labeled genomic single and double-stranded DNA probes and identified more than 200 proteins that bound DNA. Fang *et al.* [70] demonstrated specific binding of small molecule ligands to G-protein-coupled receptors, indicating that receptors in model membranes retain membrane-like properties on the slide. Except for the reported results, the potential applications of such microarrays are still large. An array of a particular class of enzymes could be screened with a candidate inhibitor to examine binding selectivity. Protein interaction networks, including the assembly of multiprotein complexes, could shed light on biochemical pathways. Eventually, it may even be of great possibility to use these high density microarrays as a MELDI source for mass spectrometry, allowing the possibility to probe complex samples for binding partners to many proteins simultaneously.

In addition, molecular profiles of cell types provided by tissue microarrays and the layered expression scanning technique are relevant in understanding biological processes. Many other types of experiments performed on microchips from assays to mimicking cellular migration in tissues and neuronal behavior *in vivo* give new insight into processes relevant in cellular biology and medicine. From running the DNA arrays to profiling proteins encoded by differentially expressed cDNA clones requires a high throughput approach to parallel protein expression analysis. This implies expressing a large number of cDNA clones simultaneously, having the appropriate vector system to do this, and arraying the resulting proteins rapidly [71].

11.3.3.2 Clinical diagnostics

The most common use of protein microarrays is in immunoassays. In particular, antibody-based immunoassays are the main stream of diagnostic assays due to their specificity. The assay usually runs in a multiplexed mode where the antibodies or other capture agents are immobilized and then exposed to a biological sample. There are four immunoassay formats: direct binding, sandwich (ELISA), competitive, and displacement. Direct-binding and sandwich assays are the most common. There are some reports on the use of competitive assays and displacement assays, which are usually associated with high surface area/volume systems [72–76].

Apart from immunoassays, enzyme assays can also be used to detect certain substrates in a clinical diagnostic setting. The benefits of performing enzymatic assays on microchips are the analytical power and minimal reagent use in microfluidic systems combined with the selectivity and amplification factors that come with biocatalysis.

Therefore, for clinical testing of substrates, on-chip enzymatic assays have advantages over conventional systems in terms of speed and performance, not to mention sample volume since reagents and the enzymes themselves can be costly. At nanoliter volumes, these chips reduce their enzyme consumption by about four orders of magnitude over standard assays. Enzymes can convert 10^2–10^5 substrate molecules to product within 1 s and thus have been used in FIA and CE. By changing corresponding enzymes, various substrates can be used to produce either optical or electrochemical signals. An enzymatic assay procedurally involves a derivatization reaction where a non-detectable species is transformed into a detectable one and hence has steps including mixing, reactions, and separations. Fluid control was via electrokinetic pumping with voltages applied at the enzyme, sample, and buffer reservoirs. The application of this type of on-chip assay to complex samples such as biological fluids would benefit from additional sample processing functions on chip such as cleanup and preconcentration [77, 78].

The discovery of biomarkers, for both prognostics and diagnostics, is a very active research field. Biomarkers can be used as indicators of disease pathophysiology, such as blood pressure, cholesterol levels, and viral load in HIV. Moreover, biomarkers can also be used as a substitute for a clinical endpoint. Since screening single biomarkers cannot properly diagnose the disease and their clinical value is questionable, investigators use high throughput platforms such as protein array and other approaches to identify large numbers of candidate biomarkers. The reason for using high throughput technologies is that they provide a large number of correlative data on protein expression in relation to disease. Such data are then analyzed for their association to the disease, which provide major impetus for the molecular profiling approaches to find patterns or profiles for a clinical test based on high dimensional protein expression panels. These protein array techniques can help not only to detect potential novel biomarkers, but also to generate a greater understanding of the signaling pathways associated with the printed proteins. This understanding can extend the use of a potential biomarker from being merely a mechanism to actually defining a characteristic of a disease or a potential drug target [79, 80].

Another application of protein microarray in clinical diagnostics is disease monitoring. Organ and disease-specific protein arrays can be used in expression profiling to identify disease-related proteins. These arrays can also quantify specific sample proteins in terms of abundance, location, and modification as the disease progresses. By studying protein regulation and expression, clinicians and researchers can predict predisposition to disease. Once the disease is manifested, gathered data can be used for monitoring disease progression, determining response to treatment, and providing overall prognosis. It is also possible to screen for molecular markers and diagnostic and/or therapeutic targets in patient-matched tissue during disease progression [81].

11.3.3.3 Drug discovery

Protein microchips have a significant impact on the development of safer drugs through the comprehensive profiling of drugs or lead compounds for effects. Zhu *et al.* [35] used a microarray of an entire eukaryotic proteome to screen different biochemical activities. First, protein chips can screen potential lead compounds at high throughput and can be used for toxicological testing as well. A multiparametric cell monitoring system may detect side effects more easily [82]. Arrays are well suited for this application when there is only a limited supply of cells or amount of drugs available for testing [83], especially in the high throughput screening of lead compounds. Although most protein arrays are based on soluble proteins, progress has been made in arraying of functional membrane proteins, which represent the majority of drug targets. By direct screening of entire proteomes for protein – drug interactions, protein arrays can aid in determining the selectivity and specificity of drug leads in further downstream testing. Second, protein microarray, especially tissue microarray, can be used to predict the probable toxicity of lead compounds by evaluating the distribution of drug targets in normal tissue. Finally, protein biochips may be utilized in monitoring drug metabolism and toxicity especially during preclinical pharmacokinetic studies, to supplement *in-vitro* and *in-vivo* model systems. That means biochips can serve as tools to optimally design clinical trials and exclude individuals with potentially deleterious drug-metabolizing enzyme (DME) profiles.

11.4 ELECTRONIC AND ELECTROCHEMICAL MICROARRAY BIOCHIPS

Electronic biochips based on the detection of alterations in the electrical properties of an electrode arising from DNA hybridization and protein bindings have been extensively studied and developed. The electronic detection-based microarrays provide many advantages inherent in comparison to radioactive or fluorescent labeling techniques to discern molecular interactions [3, 84–89]. This process offers a detection technique that is safe, inexpensive, and sensitive, not burdened with complex and onerous regulatory requirements, and can be a miniaturized, small sample-required and portable device. Electronic biomolecular or cellular sensing transducers, which convert the biomolecular interaction events into analytical signals, can be amperometric [90], potentiometric [91], and impedimetric devices [23]. These sensors are mainly based on detecting electrochemically active labels such as organic dyes, metal complexes, enzymes or metal nanoparticles [92]. Labelless electronic DNA and protein biosensors are even more attractive for fabrication of electronic biochip arrays. Research and development of the label-free techniques has been conducted on direct recognition of target DNA as a dopant within conductive polymer films [93] or by measurements of impedance spectroscopy [94–96]. Among these label-free techniques, the direct impedance detection of DNA hybridization provides prominent advantages due to its simplicity and low cost.

11.4.1 Theoretical consideration

Most electronic biosensors involve the measurements in solutions, and thus the detection schemes greatly rely on electrochemical methods, which can be used to monitor electronic signals produced from electrochemical reactions or changes of electronic properties caused by biomolecular interactions or specific cell functions on the electrodes. The electronic biochip, namely electronic microarray biosensors, is composed of patterned multiple ultramicroelectrodes (UMEs). The electrode materials can be platinum, gold, carbon, conducting polymers, metal oxides, carbides, nitrides, etc. In order to increase the detection of electronic signals, the arrayed ultramicroelectrodes are often porous for high specific surface area by specific fabrication processes. Electrochemical processes occurring on arrayed UMEs are different from that on conventional electrodes, due to the small size and porous properties. Theoretical considerations of the electrochemical behaviors of the electronic microarrays can greatly provide data analysis methods and optimized design.

The UMEs used in bioarrays can be divided into three types; disk, ring, and strip electrodes. The theory of the disk, ring, and strip UMEs has been extensively studied [97–100]. Due to the edge effect, the profile of the mass diffusion to the ultramicroelectrode surface is three dimensional, and can significantly enhance the mass transportation in comparison to the conventional large electrode with one-dimensional mass transportation. The steady-state measurement at a planar UME can be expressed as

$$i_L = 4nFDC_0r \tag{1}$$

where n is the number of electron transfers involved in the electrochemical reaction, F is the Faraday constant, D is the diffusion coefficient of the reactant species in solution, and C_0 and r are the bulk concentration of the reactant species and the radius of the UME, respectively. For band UMEs, the steady-state currents are expressed as the following:

$$i_L = \frac{2\pi nFADC_o}{w} \ln \frac{64Dt}{w^2} \tag{2}$$

where w is the width of the band electrode and A is the surface area of the band UME, which equals to $w \cdot l$ (l: the length of the band). The theory of transient measurements on different UMEs is described in [97, 98].

The porous UME can be formed by coating a porous layer such as conductive polymers, or by etching and growing porous materials such as nano carbontubes on a solid UME surface. A porous UME can significantly improve the sensitivity and detection limit, due to its high specific surface area. Li and Cha reported that a porous UME remarkably increased the amount of loading of enzyme and electron mediator of a biosensor for great sensitivity improvement [101]. The theory and applications of the porous UME have been investigated by Li and Cha [102–106]. Theoretical analysis and experimental results of porous UME demonstrated that the porous UME could be considered as a combination of two different types of electrochemical device. The outer surface (end-surface) of the porous UME behaves as a planar microdisk electrode of the same diameter, while the porous matrix and the electrolyte in the matrix pores behave as a thin-layer device with a high reaction surface area. Thus, the steady-state current of the thin-layer device is zero at any fixed potential, and the steady-state rate of diffusion at the powder microelectrode is also controlled by a mass transfer process in the solution outside the end-surface of the porous UME; hence the same Eq. (1) is applicable to both planar and porous UME. However, the transient response I of the porous UME can be calculated as the sum of the responses I_D, and I_T, of the microdisk electrode and the thin-layer device. Thus, the measured current using the porous UME can be described by

$$I = I_D + I_T = K_D v^{1/2} + K_T v \; \Delta E_{P(D)} > \Delta E_P > \Delta E_{P(T)} \qquad (3)$$

where

$$I_D = (nF)^{\frac{3}{2}} \left(\frac{\pi D v}{RT} \right)^{\frac{1}{2}} A c^o \left[\frac{nF}{RT}(E_i - E) \right] = K_D v^{\frac{1}{2}}$$

$$\Delta E_{P(D)} = 59/n \text{ mV} \qquad (4)$$

$$I_T = \frac{n^2 F^2 v V c^o}{RT} \frac{\exp[(nF/RT)(E - E^{o\prime})]}{\{1 + \exp[(nF/RT(E - E^{o\prime})]\}^2} = K_T v$$

$$\Delta E_{P(T)} = 0 \qquad (5)$$

where $E^{0\prime}$ is the formal potential of the redox couple, E_i is the initial scanning potential, $\Delta E_{P(D)}$ and $\Delta E_{P(T)}$ are the differences between the anodic and cathodic peak potentials for the disk and the thin-layer device, respectively, $\chi(E)$ is the normalized dimensionless current function for a sweep experiment with a reversible system, K_D and K_T are constants independent of the scan rate, A is the surface area of the disk electrode, v is the scan rate, and V is the volume of electrolyte within the porous matrix of the porous UME. It is possible to achieve a significantly enhanced electrode surface and homogeneous polarization simultaneously at a porous UME if the porous layer is under 50 μm. Improved reversibility, a better developed limiting diffusion current and easier analysis of experimental data should render the porous UME applicable in many areas of electroanalytical chemistry and electrode kinetics studies in terms of Eqs (1)–(5).

The effects of the array pattern on UME current output of different packing densities have been studied by using steady-state and chronoamperometric responses [107]. When the interelectrode distance $d >> 2r$, the array was considered loosely packed and Eqs (1)–(5) can be applied. When $d \leqslant 2r$, the array was considered densely packed and more closely resembled the behavior from an UME of a similar surface area to a large electrode. The implication is that only when d is large enough, the advantages of using an array pattern can be achieved. The convection-dependent contribution to UME array is also affected by the interelectrode distance. It was suggested that the theoretical current response is in agreement with the experimental data when $d = 10r$ [107, 108].

11.4.2 Fabrication technologies

11.4.2.1 Overview

Since the late 1970s, electroanalytical array sensors have been constructed by methods similar to the well-established semiconductor fabrication processes, which use a sequence of process steps to fabricate silicon wafer-based microarray devices [109]. However, the fabrication of the electronic biochip comes only partly from semiconductor fabrication technology and some specific fabrication process for micromachining, assembled together in a given order to produce a physical structure in the range of micrometers to millimeters, has to be developed to meet detection requirements of the biomolecular interaction. Printed circuit board (PCB) and screen-printing technology have been exploited to fabricate low cost electronic biochips.

Batch fabrication is essential during research and optimization of the biochips. Such batch processing can make thousands of identical devices not subject to the variations present in individually constructed objects. Another application that well illustrates the advantages of batch processing is nucleic acid arrays fabricated by photolithography. For an $n \times n$ nucleic acid array, there are n^2 different oligonucleotides of length l. Synthesizing each oligonucleotide individually would require $n^2 l$ chemical steps. Lipshutz *et al.*'s method [11] uses selectively masked photochemistry to synthesize the oligonucleotides. It requires four chemical steps (one for each base) per unit length, or $4l$ steps irrespective of the number of different oligonucleotides. Thus, one can make a 4×4 array of octamers as easily as a 200×200 array. This dramatically decreases the difficulty of making large arrays.

Geometrical control can be very important for microstructures. Photolithography allows one to pattern largely varying geometries (1 μm to >1 cm) in the same space with micrometer dimensional accuracy. In addition, one can vary dimensions of the same feature on a mask, instantly making tens of different but similar structures. Often all one needs is a small constrained geometry, such as a small well. Constrained geometries can be used to confine either proteins or mechanical forces for preventing diffusion out, increasing a protein's local concentration. This is cleverly exploited by applications involving electrochemical or optical probing of proteins in small wells. The advantages of confining forces are well illustrated by the work of several investigators. Using cell-sized chambers microfabricated in glass cover slips, Holy *et al.* examined the assembly of microtubules (MTs) from artificial MT-organizing centers consisting of tubulin-covered beads. In these constrained geometries, results showed that MT polymerization alone could position the artificial MT-organizing centers in the middle of the well, suggesting that these forces are important when considering MT dynamics. Another study [110] used shallow channels with MTs attached to the bottom surface. By looking at MT bending as it polymerized and hit the wall of the channel, they could determine its force–velocity relationship. Both of these experiments would not work in free solution; microfabricated constrained geometries enable the experiments.

The electronic biochip is composed of arrayed pairs of working electrodes and counter electrodes. The detection scheme between florescent and electronic bioarray is totally different; an optical scanner can have images for all arrayed spots in minutes, but electronic detection should be conducted one by one through multiplexing. Thus, it is essential to make an electronically addressable bioarray, in which all arrayed working and counter electrodes are fabricated in such a way that all working and counter electrodes are electronically connected in row (x) and column (y), respectively, as shown in Fig. 11.27. During measurements, any pair of working/counter electrodes can

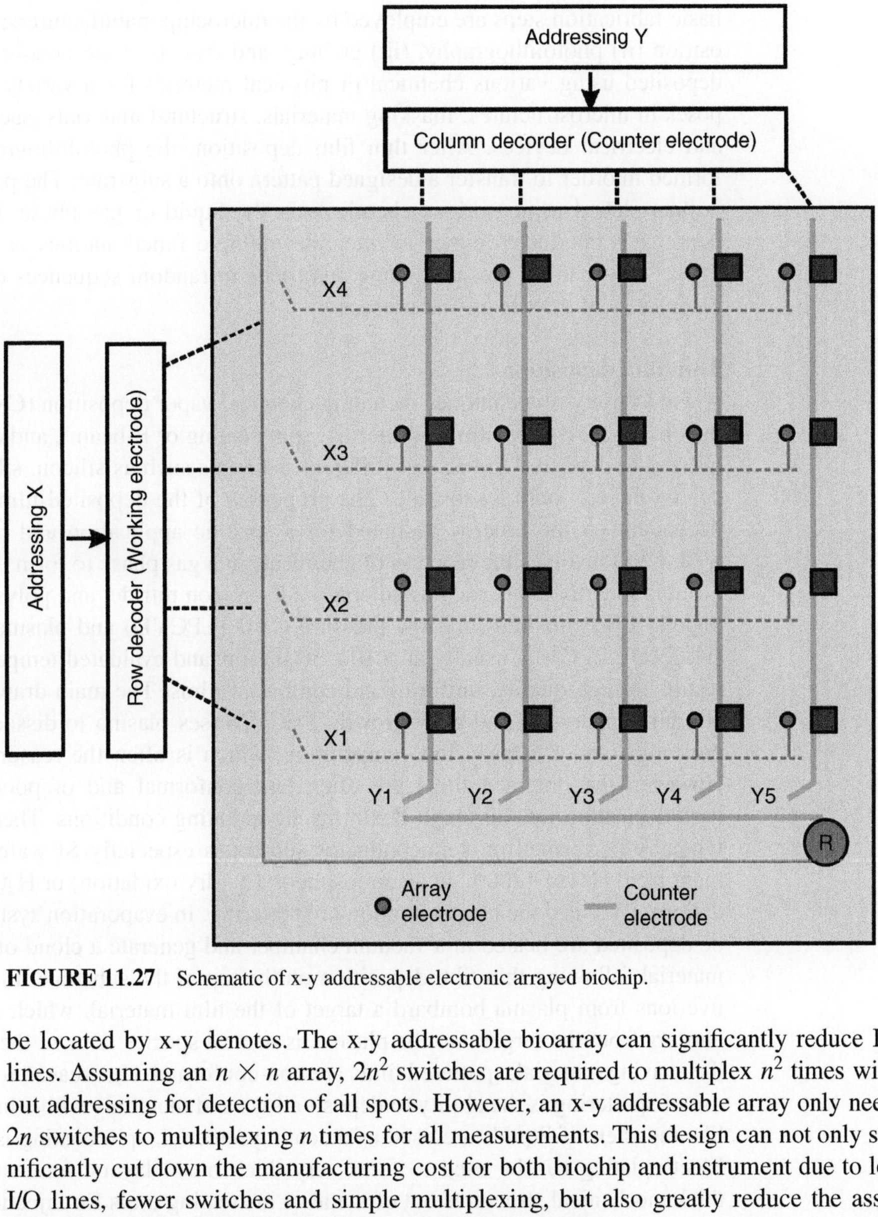

FIGURE 11.27 Schematic of x-y addressable electronic arrayed biochip.

be located by x-y denotes. The x-y addressable bioarray can significantly reduce I/O lines. Assuming an $n \times n$ array, $2n^2$ switches are required to multiplex n^2 times without addressing for detection of all spots. However, an x-y addressable array only needs $2n$ switches to multiplexing n times for all measurements. This design can not only significantly cut down the manufacturing cost for both biochip and instrument due to low I/O lines, fewer switches and simple multiplexing, but also greatly reduce the assay time that is particularly important in diagnostic applications.

11.4.2.2 Fabrication technology for silicon-based substrates

Most fabrication techniques for silicon-based biochips have their roots in the standard manufacturing methods developed for the semiconductor industry. Currently, four basic fabrication steps are employed by the microchip manufacturers: (i) thin film deposition (ii) photolithography, (iii) etching, and (iv) substrate bonding. Thin films are deposited using various chemical or physical methods for a variety of different purposes in microstructures: masking materials, structural materials, sacrificial materials, and electrical devices. After thin film deposition, the photolithography step is performed in order to transfer a designed pattern onto a substrate. The patterned substrate is then etched using various chemicals in the liquid or gas phase. Finally, substrate bonding is conducted either to integrate multiple functionalities or for packing use. These steps can be repeated numerous times in random sequences depending on the complexity of the design and process.

Thin film deposition

A wide variety of techniques including chemical vapor deposition (CVD), thermal oxidation, physical deposition (sputtering, spin coating or E-beam), and electroplating are utilized to deposit thin films of different materials such as silicon, silicon nitride, silicon oxide, etc. onto a substrate. The properties of the deposited film will be strongly dependent on the process designed for a specific application and materials used as well. CVD involves the reaction of chemicals in a gas phase to form the deposited film of inorganic materials such as silicon oxide, silicon nitride, and polysilicon. The most popular CVD methods are low pressure CVD (LPCVD) and plasma enhanced CVD (PECVD). LPCVD, usually at $\sim$100–300 mTorr and evaluated temperature, generally results in high quality, uniform, and conformal films. The main drawback of LPCVD is high temperature and slow growth. PECVD uses plasma to dissociate the reactive molecules under a fairly low temperature, which is often the reason for this choice. However, the obtained films are often less conformal and of poor quality despite high deposition rate and high flexibility in operating conditions. Thermal oxidation is typically performed on semiconductor substrates especially Si wafers by heat treatment from 800 to 1200°C in an atmospheric O_2 (dry oxidation) or H_2O vapor (wet oxidation). PVD is done by evaporation or sputtering. In evaporation systems, materials to be deposited are heated in a vacuum chamber and generate a cloud of the vapor of the material following thin film deposition on the top of the substrate. In sputtering, positive ions from plasma bombard a target of the film material, which remove atoms or clusters from the target into the plasma as neutral species. Due to the directionality of the arriving material, sputtered films are non-conformal. One reason for the widespread use of sputtering is the ability to deposit alloys and materials with a high melting point like tungsten. If liquids are used as a coating material, spin coating is often preferred. Electroplating can be used to form very thick metal layers, for instance in creating thick mechanical structures out of metal or in creating tools for injection molding.

Photolithography

This technique is used to transfer a computer-generated pattern onto a substrate. Here, a film of photoresist is spin-coated onto the substrate and exposed to UV light through a photolithographic mask; the light exposure transfers the desired pattern to the photoresist. Depending on whether the resist material is "positive" or "negative", the photoresist is developed by washing off the UV-exposed or -unexposed regions. When the substrate is subjected to chemicals, the photoresist would protect the surface below it and thus transfer the pattern to the substrate. A new pattern transfer method called microcontact printing has been recently reported. This method uses a conventionally microfabricated master to form a soft stamp by molding a polymeric substrate such as PDMS. The stamp is "inked" with alkanethiols or alkylsilanes and placed on a gold or silicon oxide coated surface, respectively. The printing process transfers the molecules from the stamp to the substrate, where patterned self-assembled monolayers can then be used as resists for etching or as passivation layers to prevent deposition. This method of pattern transfer is advantageous when handling non-cleanroom-compatible materials or chemicals, or non-planar substrates, although multilevel pattern registration is still unresolved.

Etching

As mentioned above, lithography is always followed by an etching step in order to obtain a patterned film or selective material removal from the substrate. Etching can be divided into wet or dry etching. In general, wet etchants are by and large isotropic in reactivity which etches in all directions equally, but shows superior material selectivity as compared to dry etching techniques. An extremely important exception to this rule is the anisotropic wet etching of monocrystalline substrates. The typical anisotropic silicon etchants are potassium hydroxide (KOH), ethylene diamine pyrochatechol (EDP), and tetramethyl ammonium hydroxide (TMAH). These etchants attack silicon along preferred crystallographic directions. Dry etching, especially reactive ion etching (RIE) is a combination of physical and chemical processes, where the active species react with the material only when the surfaces are activated by the collision of incident ions from the plasma. In some cases, sidewall passivation methods (such as DRIE) are used to further increase etch anisotropy. In this method, an already etched pit is coated with a passivation layer of polymer and followed by ion bombardment. Since ions are directly attacking the bottom of the pit, etching the bottom polymer and a thin layer of underlying substrate, the sidewall is kept intact and thus high aspect ratios would be achieved.

Substrate bonding

Substrate bonding refers to bonding two substrates such as silicon-silicon, silicon-glass, and glass-glass to fabricate complex 3D structures. A common example is the bonding of a glass capping wafer to a structured silicon wafer to form an optically accessible seal system. The two most popular bonding techniques are fusion bonding and anodic bonding. Fusion bonding is aimed primarily at bonding two silicon wafers via chemical reactions between the bonding surfaces, i.e. by contacting two ultraclean hydrophilic surfaces and annealing them at high temperature (800–1000ºC). Hydrogen bonds formed will join the substrate together. Other studies reveals that fusion bonding can also work with silicon oxide and nitride surfaces. Anodic bonding is a widespread method for microsensor packaging and device fabrication. This bond occurs between borosilicate glass (Pyrex 7740) and a silicon wafer through a cooperation of a high electric field, high pressure, and high temperature (400ºC). During this period, Na^+ in the borosilicate glass migrates towards the cathode under the function of electric field,

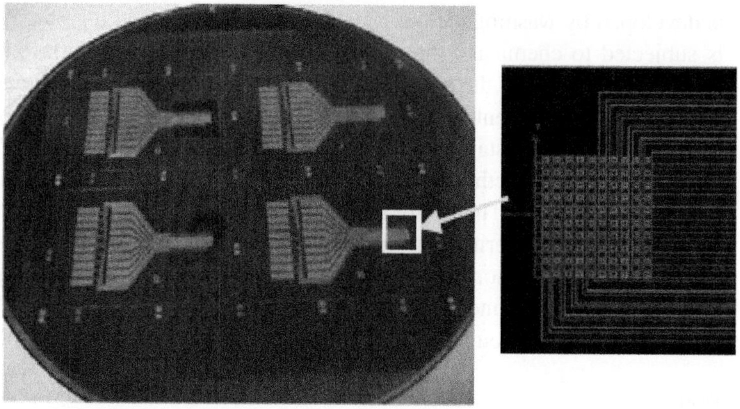

FIGURE 11.28 Silicon wafer-based electronic biochip (10 × 10 arrays) (see Plate 11 for color version).

leaving a negative charge region in the glass interface towards the silicon substrate, and then the two wafers are electrostatically attracted to each other. The presence of oxygen ions at the wafer interface will oxidize the bottom substrate and hence establish a bond in between, so the wafers won't fall apart even if the electric field disappears. The bond is so strong that any attempt to break the bond will lead to fracture of the wafers. In addition to the two methods mentioned above, several other alternatives like adhesive, eutectic, and glass frit are also investigated.

Figure 11.28 (see Plate 11 for color version) shows a silicon wafer-based x-y addressable bioarray that was made in the author's lab.

11.4.2.3 Fabrication technology for ceramic or plastic substrate

Large-scale biochip fabrication can be accomplished not only by lithographic techniques but also using screen-printing (thick film) processes [111] or popular PCB technology. The screen-printing technology relies on printing patterns of conductors and insulators onto the surface of planar (plastic or ceramic) substrate. The screen-printing process involves several steps including placement of the ink onto a pattern screen or stencil, followed by forcing it through the screen with the aid of a squeegee, and dying/curing the printed patterns [112]. Such a process yields mass-producible (uniform and disposable) electrodes of different shapes or sizes. The electrochemical reactivity and overall performance of screen-printed electrodes are dependent upon the composition of the ink employed and on the printing and curing conditions (e.g. pressure, temperature). Disposable potentiometric sensors can be fabricated by combination of ion-selective electrode slides require microliter (10–15 μL) sample volumes, and are ideally suited for various decentralized applications. Mass-produced potentiometric sensor arrays are also being developed in connection with future high speed clinical analyzers. These are being combined with advanced materials (e.g. hydrogels) that obviate the need for internal filling solutions. The screen-printing technology requires lower capital and production costs than the thin film lithographic approach, but is limited to

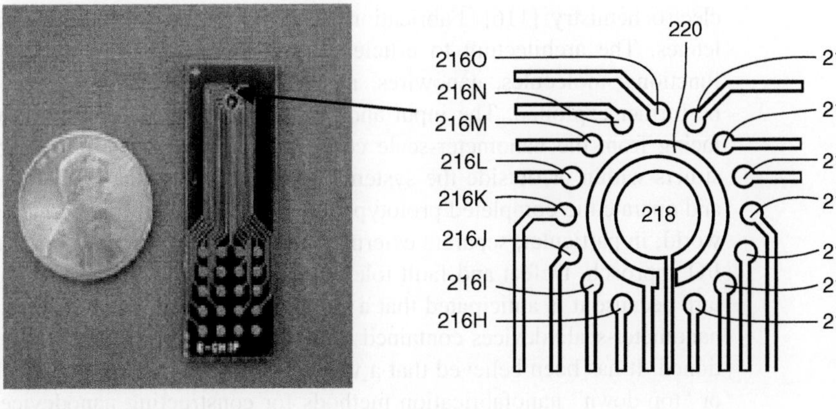

FIGURE 11.29 PCB biochip.

electrode structures larger than 100m. It is also possible to fabricate electrochemical devices combining the thin and thick film processes. Printed circuit boards (PCB) for electronic products have existed for many years. A mask is the picture that determines where the metal foil on a plastic substrate is either to be removed or left for a specific pattern. As circuit densities began to increase it was necessary to allow for more and more layers of interconnect to enable the complexity of design. The PCB technology can be used to fabricate low cost bioarray chips that can be integrated with detection circuits. Figure 11.29 shows a biochip made from PCB technology [113]. The biochip is composed of a large centered disk electrode outside a ring electrode, which are functionalized as counter and working electrode, respectively, and are surrounded by symmetric satellite UMEs as working electrodes. The design is particularly unique for eliminating the difference of any individual working electrode from the reference or counter electrode.

11.4.2.4 Fabrication of nanoarray biochips

The advances of nanoengineering and nanoscience are leading to fabrication of nanoarrays for biochips with extremely high density of sensors. A facile technique for fabrication of individually addressable, conducting polymer nanowire arrays of controlled dimension, high aspect ratio, and site-specific positioning using electrodeposition is reported [114]. A nanoarray was formed by a bottom-up approach to integrate multiwalled carbon nanotubes into multilevel interconnects on silicon substrate and demonstrated electrical properties consistent with their original structure [86]. A DNA nanaoarray detection method is reported in which the binding of oligonucleotides functionalized with gold nanoparticles leads to conductivity changes associated with target–probe binding events. The binding events localize gold nanoparticles in an electrode gap; silver deposition facilitated by these nanoparticles bridges the gap and leads to readily measurable conductivity changes [115]. Nanochannel arrays with diameters as small as 30 nm and aspect ratios up to 250 were prepared on silicon platforms by electrochemistry [116]. Fabrication of nanoarray biochips is still facing great challenges. The architecture to efficiently and inexpensively build a nanobiochip with functional molecules, nanowires, nanoparticles, and nanocarbontubes, etc. needs to be further exploited. The input and output strategy that specifies a method for interfacing from the nanometer-scale constructs of the systems to micrometer-scale constructs within or outside the system is not clear up to date. It can be possible to test and operate the completed prototype using an integrated interconnection to the outside world; in particular, separate external nanoprobe tips are not considered an acceptable I/O approach. Defect and fault tolerance are essential to the nanoscale array architecture because it is anticipated that a significant fraction of the hierarchically assembled nanometer-scale devices contained in the device may be defective or imprecisely positioned. It has been believed that a variety of "bottom-up" nanoassembly methods and/or "top-down" nanofabrication methods for constructing nanodevices will be hybridized for great performance.

11.4.3 Electrochemical detection

Although optical detection techniques are perhaps the most prevalent in biology and life sciences, electronic or electrochemical detection techniques have also been used in biochips due to their great sensitivity, high specificity, and low cost. These techniques can be amenable to portability and miniaturization, when compared to optical detection techniques [117]. Electrochemical detection include three basic types: (i) amperometry, which measures the electric current associated with the electron transfer involved in redox processes, (ii) potentiometry, which measures a change in potential at electrodes due to ions or chemical – biomolecular intereactions (such as an ion-sensitive FET), and (iii) impedimetry, which measures conductance or capacitance changes associated with changes in the overall ionic medium between the two electrodes. There are more reports on potentiometric and amperometric methods, particulary due to the established field of electrochemistry. Many of the sensors employing these two methods have been commercialized.

11.4.3.1 Amperometry

The most prevalent examples of the amperometric method employ an enzyme-catalyzed redox reaction, where the resulting redox electron current is measured at a working electrode for high sensitivity. A prominent advantage of amperometry for the biochip is the improvement of performance that accompanies the reduction of electrode size into the low micrometer range. Conventional macroelectrodes (typically millimeter in diameter) exhibit planar diffusion of electroactive species to the electrode surface for redox reactions. As the electrode diameter is decreased into the low micrometer range, a shift to non-planar diffusion occurs, causing an increase in the collection efficiency of the electroactive species at the surface. The practical result is an increase in the signal to noise ratio (S/N), which generally equals to a lower detection limit [118].

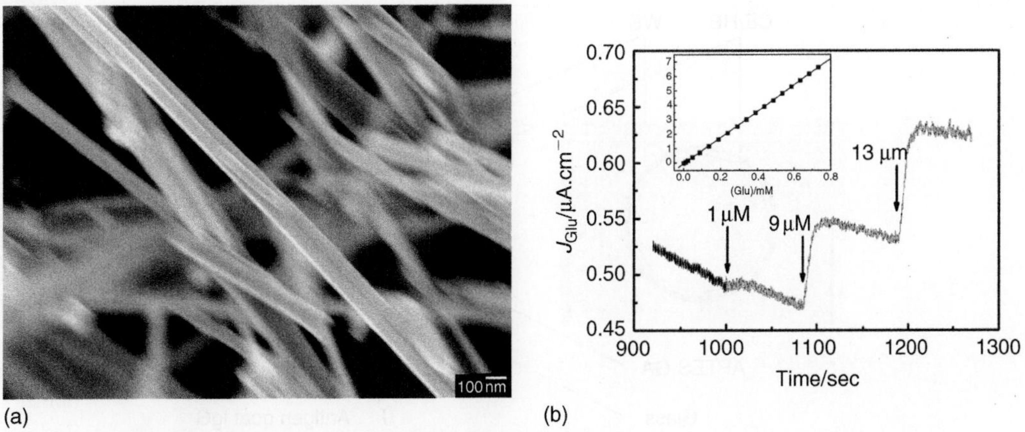

FIGURE 11.30 Zn Nanowire glucose sensor: (a) glucose oxidase impregnated Zn nanowire electrode; (b) amperimetric response of different glucose concentration.

The most widely used examples are that of detection of glucose, based on glucose oxidase, which generates hydrogen peroxide and gluconic acid in the presence of oxygen, glucose, and water [119]. These devices are designed either for monitoring formation of hydrogen peroxide or for consumption of oxygen for indirect determination of glucose concentration. Later advances of the sensors could detect the electron transfer of the glucose enzymatic oxidation by using an electron mediator [101]. Recently, Zang and Li *et al.* reported a glucose sensor based on detection of direct electron transfer by using nanostructructured zinc nanowires, which was impregnated by glucose oxidase as shown in Fig. 11.30. The detection of direct electron transfer for the enzymatic oxidation can significantly improve the sensing sensitivity. At the microscale, these sensors require the formation of the working and reference electrodes on a chip, and an enzymatic layer on the working electrode, as demonstrated for the detection of glucose, lactose, and urea [120]. More recently, hydrogels and conducting electroactive polymers have been integrated to develop electroactive hydrogels that physically entrap enzymes within their matrices for biosensor construction and chemically stimulated controlled release. Using these materials, the fabrication of glucose, cholesterol, and galactose amperometric biosensors has been demonstrated on a chip [121].

Another representative example is electrochemical immunoassays. Electrochemical immunoassays were introduced in the late 1970s and the methodology and applications matured rapidly. Important advantages of such microfabricated systems as compared with conventional systems include the ability to reproducibly handle very small amounts of samples, improved facilities for separations of analytes, minimized dilution of the products to be detected, and improved redox cycling efficiencies [122]. Wang and his coworkers [123] described microchip-based amperometric non-competitive and competitive immunoassays for detection down to pg ml^{-1}levels of mouse IgG and $\mu g\ ml^{-1}$ levels of 3,3′,5-triiodo-L-thyronine, respectively, using ferrocene redox tracers (antigen or antibodies) and capillary electrophoresis with amperometric detection. In another

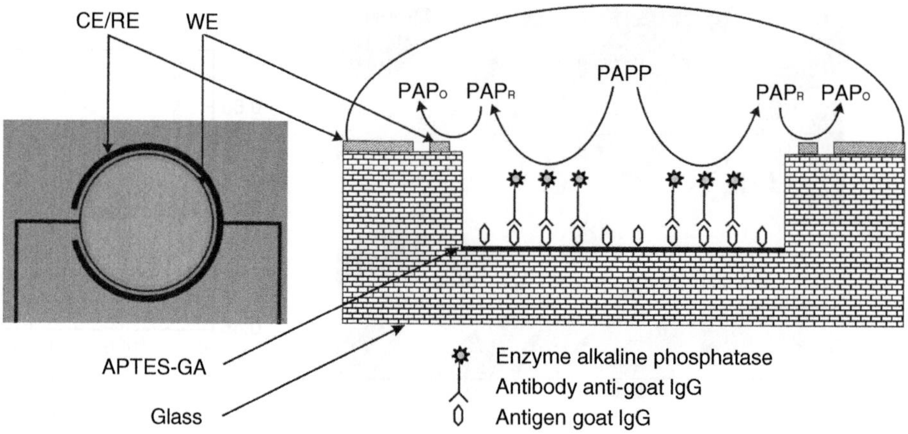

FIGURE 11.31 Schematic of electronic ELISA microchip.

study, Lim *et al.* [124] employed a microfabricated on-chip electrochemical flow immunoassay for measurements of down to 200 ng ml^{-1} concentrations of histamine in whole blood based on ferrocene-labeled IgG. Gabig-Ciminska *et al.* [125] described a silicon-based chip electric detector coupled to bead-based sandwich hybridization used for the measurement of 10^{10} molecules of 16S rRNA in an *E. coli* RNA extract. Paramagnetic beads with immobilized capture probes were used and the detection was accomplished employing redox cycling of generated p-aminophenol on an interdigitated array of gold microband electrodes in a flow system. Microelectrochemical immunoassays for very small sample volumes can also be realized by the use of antibodies attached to gold recessed microdisks at the bottom of microcavities. Aguilar *et al.* [126] measured pg ml^{-1} levels of mouse IgG by detecting the enzymatic generation of p-aminophenol on microband electrodes close to the recessed microdisks in the same microcavity. Based on a large number of such microcavities, the latter concept could be used to carry out a large number of stimultaneous immunoassays. Dong *et al.* [127] reported a microchip with novel electrochemical detection architecture as shown in Fig. 11.31 for enzyme immunoassay sensors. The microchip is composed of dual-ring working and counter electrodes, and a sensing cavity chamber was made on a glass slide. The glass surface of the microchip was coated by 3-aminopropyltriethoxysilane (APTES). Goat IgG, as an example, was covalently captured on APTES-modified glass surfaces through glutaraldehyde (GA) as a cross-linker. Enzyme substrate, p-aminophenyl phosphate (PAPP) was prepared by electrolysis. The enzyme conversion from home-synthetic PAPP to p-aminophenol (PAP) was detected by differential pulse voltammetry (DPV), demonstrating a good sensitivity of 100 fg/ml.

11.4.3.2 Potentiometry

The most common potentiometry involves an instrument called a "pH-Stat", in which a glass (pH) electrode follows reactions that either consume or produce protons. Since pH changes cause changes in enzyme activity, the pH is maintained at a constant value by the addition of acid or base. The rate of titrant addition is then proportional to the rate of the enzymatic reaction. Precise measurements using the pH-Stat require low buffer concentrations in the enzymatic assay mixture.

Besides, potentiometric sensors with ion-selective ionophores in modified poly(vinyl chloride) (PVC) have been used to detect analytes from human serum [128]. Cellular respiration and acidification due to the activity of the cells has been measured with CMOS ISFETS [129]. Some potentiometric methods employ gas-sensing electrodes for NH_3 (for deaminase reactions) and CO_2 (for decarboxylase reactions). Ion-selective electrodes have also been used to quantitate penicillin, since the penicillinase reaction may be mediated with I^-or CN^-.

In recent years, potentiometric sensors have been downscaled to nanometer dimension through the use of silicon nanowires [130] and carbon nanotubes as field-effect sensors [131], to take advantage of enhanced sensitivity due to higher surface area to volume ratio. The integration of these nanoscale sensors in lab-on-chips is somewhat difficult but recent advances in top-down fabrication techniques have been used to demonstrate such nanoscale structures [132]. Potentiometric sensors at the microscale have also been used to perform label-free detection of hybridization of DNA [133]. These sensors were incorporated within cantilevers so that they can be used within microfluidic channels. The DNA hybridization was detected by measuring the field effect in silicon by the intrinsic molecular charge on the DNA, using a buffer of poly-L-lysine later. The potentiometric methods often suffer from low sensitivity and sluggish response.

11.4.3.3 Impedimetry

Impedimetry measures the changes in the electrical impedance between two electrodes. The impedance changes at the electrode interface or in the bulk region can be used to identify biomolecular interreactions between DNAs, proteins, antigen–antibody bindings or excretion of cellular metabolic products. Impedimetric techniques are attractive due to their simplicity and ease of use since an electrochemical label into the target molecule is not needed, and have been used to detect a wide variety of entities such as agents of biothreat [134], biochemicals [135], toxins, and nucleic acids [94]. In addition, impedimetric sensors provide information on the ionic strength in electrolytes and can provide selectivity if coupled with enzyme membranes. These sensors have been used to detect different analytes, for example urea, glucose, etc. [136]. Measurement of impedance was also used to measure the metabolic activity of microorganisms within microfluidic biochips. As bacterial cells are grown within microfluidic channels and wells, the impedance changes in the medium can be detected using electrodes placed appropriately within the channels [137]. Electrical measurements of DNA hybridization using impedance techniques have been demonstrated where the binding of oligonucleotides functionalized with gold nanoparticles leads to conductivity changes associated with binding events [115].

In impedimetry, conductivity could simply provide a measure of the ionic concentration and mobility in a solution. However, the measurements are difficult due to the variable ionic background of clinical samples and the relatively small conductivity changes observed in high ionic strength solutions [138]. The capacitance measurement is limited to detect relatively large biomolecules, and requires more precision of capacitance measurements with commercial instruments [139]. Further, it is unlikely that individual electrodes in a biochip have identical electronic and physical properties, particularly in the case of film coated electrodes. This would cause significant variation. It is not practical to use simple subtraction of impedance changes before and after biomolecular interaction in biochip experiments. Actually, the simple subtraction method leads to a messy, scattering patterned plot of impedances vs target concentrations of the target molecule, due to variations from individual electrodes [95]. Li *et al.* reported impedance labelless detection of DNA and proteins [85, 94, 95] on polypyrrole deposited electrodes. In order to eliminate or reduce the variations from different single electrodes in multi-concentration analysis, a concept of normalized dimensionless impedance unit change was introduced to analyze the measured impedance data [95]. In this method, for example, the resistances measured at an Ab impregnated electrode before and after the target antigen incubation are assumed as R_1 and R_2, respectively. The normalized resistance unit change, ΔR_N, is

$$\Delta R_N = \frac{R_2 - R_1}{R_1} \tag{6}$$

The physical meaning of ΔR_N is the dimensionless unit resistance change. This normalization is different from the normalized change of intensity used in optical array biochips, which uses the responses at biomarked spots to be divided by the responses at the negative spots, and also is different from the normalized resistance change ($\Delta g/g_o$) used in literature [94] where $\Delta g = g - g_o$; g_o is the conductance of the sensor without any interest analyte and g is the conductance of the same sensor in the presence of the analyte. The normalized dimensionless unit resistance change introduced here is based on results from a single electrode. By using ΔR_N to process impedance data, the S/N ratio was significantly improved.

11.5 LAB-ON-CHIPS

The principle of the lab-on-chip (LOC) is to integrate all the necessary devices on a single small chip to perform complicated biological and chemical processes that are usually done with larger volumes in well-equipped laboratories. In the past ten years, the research and application in this area has been rapidly growing [140]. After an initial focus on electrokinetic separation techniques on the chip, the scope of the technology has widened to include topics like microfluidics, DNA analysis, cell analysis, microreactors, and mass spectrometer interfacing. Microfabrication and the drive to analyze thousands or hundreds of samples quickly and efficiently have led to the development of this new form of analytical technology. As well as the analytical chemistry community, synthetic chemists, chemical engineers, biochemists, and biomedical engineers are now becoming more and more interested in using new micro- and nanotechnological techniques in this area.

The idea of LOC is to have a bank of specialized modules that can be fitted together like Lego building blocks to create a system, custom-tailored toward a specific application. The LOC would combine tiny channels, pumps, and storage chambers with electronic and optical devices, actuators and sensors to perform multi-step tasks. These systems, fabricated using technologies adopted from the microelectronics industry, are enabling researchers from many disciplines to approach their activities in new ways. The term LOC is now intimately linked with the other buzzwords of the new micro-era: proteomics, genomics, combinatorial chemistry, drug discovery, ultra-high throughput screening, massively parallel synthesis, and even bioinformatics.

Microfluidics is the core technology that deals with the movement of small amounts of fluid. Subtechnologies would include electrophoresis, electrodynamics, semiconductor fabrication methods, labeling technology, laser fluorometry, and inkjet printing. A microfluidic device can be identified by the fact that it has one or more channels with at least one dimension less than 1 mm. Microfluidics can handle common fluids such as whole blood samples, bacterial cell suspensions, protein or antibody solutions and various buffers; it is the key toward the development of microsynthesis, microseparations and lab-on-chips for sample preparation, rinsing, mixing, reaction, and other fluid processing needs for small volumes that cannot be performed in traditional ways. It is expected that microfluidics will revolutionize the fields of chemistry and biology in many applications, such as proteomics and genomics research, high throughput and small sample analysis, on-site field and environmental analysis, clinical diagnostics, small-quantity syntheses and reactions, combinatorial synthesis, and on-line analysis in industrial processes. More detailed applications in a variety of interesting measurements include molecular diffusion coefficients [141], pH [142], chemical binding coefficients [141], and enzyme reaction kinetics [77, 143]. Other applications for microfluidic devices include capillary electrophoresis [144], isoelectric focusing [142, 145], immunoassays [146, 147], sample injection of proteins for analysis via mass spectrometry [148, 149], PCR amplification [150, 151], DNA analysis [152, 153], cell manipulation [154], cell separation [155], cell patterning [156], and chemical gradient formation [157, 158]. Many of these applications have utility for clinical diagnostics [159, 160].

The use of microfluidic devices to conduct biomedical research and create clinically useful technologies has a number of significant advantages. Because of the small volume, usually several nanoliters, the amount of reagents and analytes used is quite small. This is especially significant for expensive reagents. The microfluidic technologies for such devices are relatively inexpensive and are very amenable both to highly elaborate, multiplexed devices and also to mass production. One of the long-term goals in the field of microfluidics is to create integrated, portable clinical diagnostic devices for home and bedside use, thereby eliminating time-consuming laboratory analysis procedures. Additional possible benefits of devices based on microfluidics include automation, reduced waste, improved precision and accuracy, and disposability.

11.5.1 Theory of microfluidics

Microfluidics handles and analyzes fluids in structures of micrometer scale. At the microscale, different forces become dominant over those experienced in everyday life [161]. Inertia means nothing on these small sizes; the viscosity rears its head and becomes a very important player. The random and chaotic behavior of flows is reduced to much more "smooth" (laminar) flow in the smaller device. Typically, a fluid can be defined as a material that deforms continuously under shear stress. In other words, a fluid flows without three-dimensional structure. Three important parameters characterizing a fluid are its density, ρ, the pressure, P, and its viscosity, η. Since the pressure in a fluid is dependent only on the depth, pressure difference of a few μm to a few hundred μm in a microsystem can be neglected. However, any pressure difference induced externally at the openings of a microsystem is transmitted to every point in the fluid. Generally, the effects that become dominant in microfluidics include laminar flow, diffusion, fluidic resistance, surface area to volume ratio, and surface tension [162].

Laminar flow is the definitive characteristic of microfluidics. Fluids flowing in channels with dimensions on the order of 50 mm and at readily achievable flow speeds are characterized by low Reynolds number, Re, defined as

$$Re = \rho v^2 V/\eta v S = \rho v L/\eta \tag{7}$$

where v is the average flow speed of the fluid. The characteristic linear dimension L is the ratio of the volume V of the fluid to the surface area S of the walls that bound it [163]. Due to the small dimensions of microchannels, the Re is usually much less than 100, often less than 1. In this Reynolds number region, the flow is completely laminar. Laminar flow provides a means by which molecules can be transported in a relatively predictable manner through microchannels. To define the best dimension of the microchannels for calculating the Reynolds number, the hydraulic diameter, D_h, is introduced as

$$D_h = (4S/p_{wet}) \tag{8}$$

where p_{wet} is the wetted perimeter, which is all the perimeter that is in contact with the liquid. For a rectangular microchannel this corresponds to twice the width plus twice the height. For a circular cross-section, Eq. (8) is simplified to $D_h = d$, where d is its circular diameter. The theoretical framework to analyze fluid flow is based on Navier-Stokes equations [164]. However, the Navier-Stokes equations contain more unknown parameters than other equations, making complete analytical solution impossible. Typically, several boundary conditions and/or equations of state are used for solving the equation. An important solution to the equations, which can be used in the analysis of a microfluidic system, is the Poiseuille flow, which applies when a pressure gradient, ΔP, is used to drive a fluid through a capillary or channel. For a capillary with a cylindrical cross-section, the volume flow, Q, is found to be

$$Q = (\Delta V/t) = (\pi R^4/8\eta L)\Delta P \tag{9}$$

where R is the radius of the capillary. The velocity profile, i.e. the velocity, $v(r)$, at different radial positions between the center ($r = 0$) and the wall ($r = R$) are found to be

$$v(r) = (R^2 - r^2)(\Delta P/8\eta L) \tag{10}$$

TABLE 11.7
Terms for common geometries

Cross-section	Fluidic resistance
Circular	$8\eta L/\pi R^4$
Rectangular (low aspect)	$12\eta L/wh^4$ (h: height, w: width)
Square	$28.454\eta L/a^4$ (a: side length)
Regular triangle	$184.751\eta L/a^4$ (a: side length)

clearly describing a parabolic flow profile and also satisfying the no-slip condition. The term $8\eta L/[\pi\iota]\pi R^4$ in Eq. (9) is called the fluidic resistance. For channels with non-cylindrical cross-sections, the expressions are similar to those in Eq. (9), but with different terms for the resistance. Table 11.7 gives the terms for common geometries.

There are more important and useful relations that can be used in design and analysis of flow in microfluidic channel network with changing sections. Since

$$S_1 v_1 = S_2 v_2 = \text{constant} \tag{11}$$

the Bernoulli equation is a direct application of the law of energy conservation, relating pressure, kinetic energy, and potential energy in the following way,

$$P + 1/2\rho v^2 + \rho g h = \text{constant} \tag{12}$$

A knowledge of v can give an indication of the transit time of a plug of chemical or an ensemble of cells through a microfluidic channel network and thus to assess whether there is enough time for complete mixing or chemical reaction. Both Eq. (11) and Eq. (12) are strictly only valid under idealized conditions (i.e. incompressible and non-viscous fluids and steady flow), but can still be helpful for overall estimation and assessment.

The surface tension is of great importance when dealing with bubbles and particulate contaminations in microchannels and determining how strong the capillary forces are in a microchannel. For a cylindrical cross-section, the capillary force, F_{cap}, can be expressed quantitatively as shown in the following equation,

$$F_{\text{cap}} = 2\pi r\gamma \cos\Theta \tag{13}$$

where γ and Θ are the surface tension and the contact angle, respectively.

There are three types of mass transport processes within a microfluidic system: convection, diffusion, and immigration. Much more common are mixtures of three types of mass transport. It is essential to design a well-controlled transport scheme for the microsystem. Convection can be generated by different forces, such as capillary effect, thermal difference, gravity, a pressurized air bladder, the centripetal forces in a spinning disk, mechanical and electroosmotic pumps, in the microsystem. The mechanical and electroosmotic pumps are often used for transport in a microfluidic system due to their convenience, and will be further discussed in section 11.5.2. The migration is a direct transport of molecules in response to an electric field. In most cases, the moving molecules are ionized and driven by a coulombic force due to the electric field. The coulombic force, F, is given by

$$F = qE \tag{14}$$

where q is the charge on the molecule and E is the strength of the electric field. The velocity, v, of charged molecules first accelerates towards one of the electrodes, and then slows down to reach a terminal one, at which the coulombic force is balanced by a drag force, the Stokes force,

$$F = 6\pi\eta r_h v \tag{15}$$

where η is the viscosity of the fluid. The terminate v is calculated as

$$v = \mu E \tag{16}$$

where μ is the mobility of the molecules, given by

$$\varepsilon\ \theta]\mu = q/6\pi\eta r \tag{17}$$

Diffusion occurs when there is a concentration gradient of one kind of molecule within a fluid. In terms of random walk model, the average distance, x, after an elapsed time, t, between molecule collisions in a diffusion movement is characterized by the Einstein-Smoluchowski relation,

$$x = (2Dt)^{1/2} \tag{18}$$

where D is the diffusion coefficient. This allows us to estimate the time t_{cross} for taking a molecule to cross half the microchannel width W of the main channel of a T-junction,

$$t_{cross} = W^2/8D \tag{19}$$

More theoretical analysis of a diffusion process can be conducted by Fick's law as

$$j = D(\partial c/\partial x) \tag{20}$$

where j is the diffusion molecule flux and equal to cv. c is the molecule concentration.

In analysis of a mass transport process, the Peclet number, P_e, the ratio of the direct flow to the diffusion, is used and given as

$$P_e = vd/D \tag{21}$$

where d is a characteristic length of the microfluidic system. The Peclet number can be calculated in terms of Eq. (21), which is important in the design of a microfluidic system for control diffusion, such as for a chemical separation system or a mixer effect of a microreactor.

11.5.2 Components in lab-on-chip systems

Among the most important fluid handling components in an LOC are pumps and valves. There are two main methods by which fluid actuation through microchannels can be achieved: pressure driven and electroosmotic flow. In pressure driven flow, the fluid is pumped through the device via positive displacement pumps, such as syringe pumps, peristaltic pumps, etc. One of the basic laws of fluid mechanics for pressure driven laminar flow, the so-called no-slip boundary condition, states that the fluid velocity at the walls must be zero. This produces a parabolic velocity profile (Poiseuille flow profile) within the channel (Fig. 11.32a, see Plate 12 for color version), which has significant implications for the distribution of molecules or cells transported within a channel. Pressure driven flow can be a relatively inexpensive and reproducible approach. With the increasing efforts at developing functional micropumps, pressure driven flow is also amenable to miniaturization. However, pressure driven flow has two disadvantages. First, it is difficult to design and fabricate reliable mechanical pumps in the materials often used for a microfluidic device such as silicon and glass, due to multiple levels of fabrication and easy damages by particles of dust and contaminants in the fluid. Second, as illustrated in Fig. 11.32a (see Plate 12 for color version), Poiseuille flow is characterized by a parabolic velocity profile over the cross-section of the channel, with zero velocity at the walls and a maximum at the channel's center. This non-uniformity in flow velocity occurs because the imposed pressure exerts a uniform force over the cross-sectional area of the channel, but momentum leaves the flow, due to interactions with the solid boundary, only at the walls. The parabolic flow profile distorts a volume of fluid as it flows down the channel. When used to separate different molecules in solution, such a flow spatially broadens the bands of distinct species [165].

The electroosmotic pumping is executed when an electric field is applied across the channel. The moving force comes from the ion moves in the double layer at the wall towards the electrode of opposite polarity, which creates motion of the fluid near the walls and transfer of the bulk fluid in convection motion via viscous forces. The potential at the shear plane between the fixed Stern layer and Gouy-Champmon layer is called zeta potential, ξ, which is strongly dependent on the chemistry of the two phase system, i.e. the chemical composition of both solution and wall surface. The electroosmotic mobility, μ_{eo}, can be defined as follow,

$$\mu_{oe} = \xi_0 \varepsilon / 4\pi\eta \tag{22}$$

where ε is the dielectric coefficient. Thus, the electroosmotic velocity, v_{eo}, in an LOC can be calculated as

$$v_{eo} = \mu_{oe} E \tag{23}$$

The velocity profile is uniform across the entire width of the channel if the channel is open at the electrodes, as is most often the case. However, if the electric field is applied across a closed channel (or a backpressure exists that just counters that produced by the pump), a recirculation pattern forms in which fluid along the center of the channel moves in a direction opposite to that at the walls; further, the velocity along the centerline of the channel is 50% of that at the walls (Fig. 11.32a, see Plate 12 for color version). Figure 11.32b (see Plate 12 for color version) illustrates an electric field generating a net force on the fluid near the interface of the fluid/solid boundary, where a small separation of charge occurs due to the equilibrium between adsorption and desorption of ions. The charge region from excess cations localized near the interface by coulombic

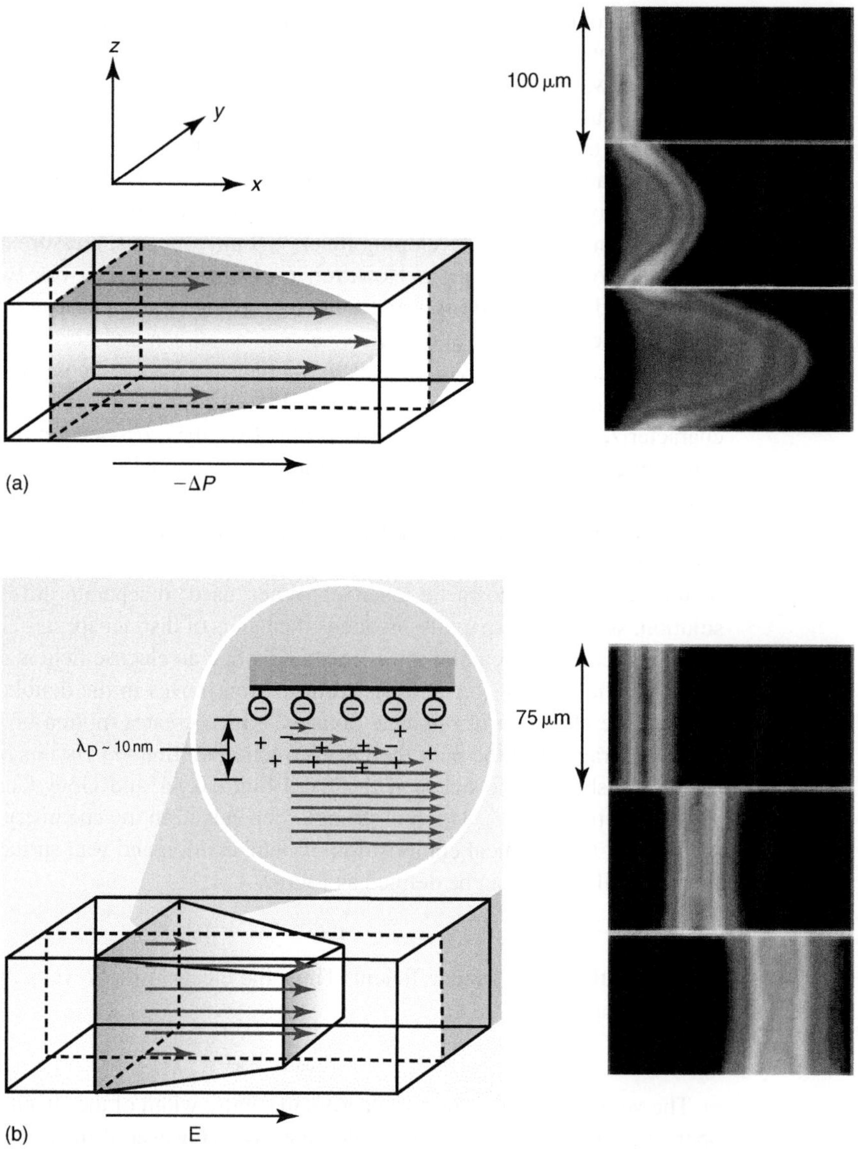

FIGURE 11.32 Flow profiles in microchannels. (a) A pressure gradient, $-\Delta P$, along a channel generates a parabolic or Poiseuille flow profile in the channel. The velocity of the flow varies across the entire cross-sectional area of the channel. On the right is an experimental measurement of the distortion of a volume of fluid in a Poiseuille flow. The frames show the state of the volume of fluid 0, 66, and 165 ms after the creation of a fluorescent molecule. (b) In electroosmotic flow in a channel, motion is induced by an applied electric field E. The flow speed only varies within the so-called Debye screening layer, of thickness λ_D. On the right is an experimental measurement of the distortion of a volume of fluid in an electroosmotic flow. The frames show the state of the fluorescent volume of fluid 0, 66, and 165 ms after the creation of a fluorescent molecule [165]. Source: http://www.niherst.gov.tt/scipop/sci-bits/microfluidics.htm (see Plate 12 for color version).

interactions, called a Debye screening layer, has a thickness, λ^D, typically less than 10 nm for aqueous buffers. As the fluid moves in the Debye layer, it carries the bulk liquid in the channel. As a result, the velocity profile is essentially flat across the channel [165]. This type of flow is ideal for separations based on the charge-to-size ratio of the molecules in biological samples, since broadening of the separated bands of differing species occurs only by diffusion, not as a result of the differences in flow velocity across the channel. Advantages of electroosmotic flow are that the blunt velocity profile avoids many of the diffusion non-uniformities that occur with pressure driven flow. However, electroosmotic flow has its own weaknesses: concerns on sample dispersion in the form of band broadening for the pumping, sensitivity to impurities that adsorb on the wall of the channel, ohmic generation of heat in the fluid, and the need for high voltages (order of kilovolts). The variability of surface properties can also affect the flow. Proteins, for example, can adsorb to the walls, substantially change the surface charge characteristics and, thereby, change the fluid velocity. This can result in unpredictable long-term time dependencies in the fluid flow. Thus, electroosmotic flow is usually not well suited for transporting multicomponent solutions when separation is not desired.

Valves are often classified by whether they work by themselves (passive or check valves) or if they need an external energy to work (active valves). Ideal valves are hoped to have characteristics of zero leakage, zero energy consumption, zero dead volume, and low cost. Figure 11.33a (see Plate 13 for color version) shows an example of a valve that exploits the electrometric character of polydimethylsiloxane (PDMS) [166]. This passive check valve is a fluidic rectifier: pressure in one side of the device prevents fluid from flowing through the device; pressure in the other side opens a flap and allows the fluid to flow. An active valve requires an actuator that mechanically moves a part to open or close the flow passage. The actuation principles are various including pneumatic (compressed air), thermopneumatic (heated fluids), piezoelectric (materials expansion when voltage applies), electrostatic (electric attraction), shape "memory" alloy (shape changes reversibly vs temperature), and electromagnetic actuation.

The importance of mixing in a chemical/biological microsystem is obvious, particularly in a microreactor. Passive mixing in a microsystem solely depends on diffusion. In terms of Eq. (18), the diffusion time can be reduced by 100 times if the diffusion distance, x, is reduced by ten times. We can design a narrow or high aspect ratio channel to reach a high efficient mixing design. A high aspect ratio, narrow channel would result in high fluidic resistance and an expensive manufacturing process. An alternative strategy is to design standard size microchannels followed by splitting each channel into an array of smaller channels, and then merging them again. In such a design, the fluidic resistance increase can be eliminated. The Peclet number in Eq. (21) is a good indicator for diffusion efficiency. The higher the Peclet number, the harder the diffusion mixing. Figure 11.33b (see Plate 13 for color version) shows a mixer designed by David Beebe's group at the University of Wisconsin (formerly at the University of Illinois) and illustrates the types of new devices that must be developed to perform familiar functions when turbulence is no longer available as an aid [167]. This 3D-serpentine channel acts as a passive mixer for laminarly flowing fluids based on a type of chaotic flow known as chaotic advection. Chaotic advection appears in certain steady

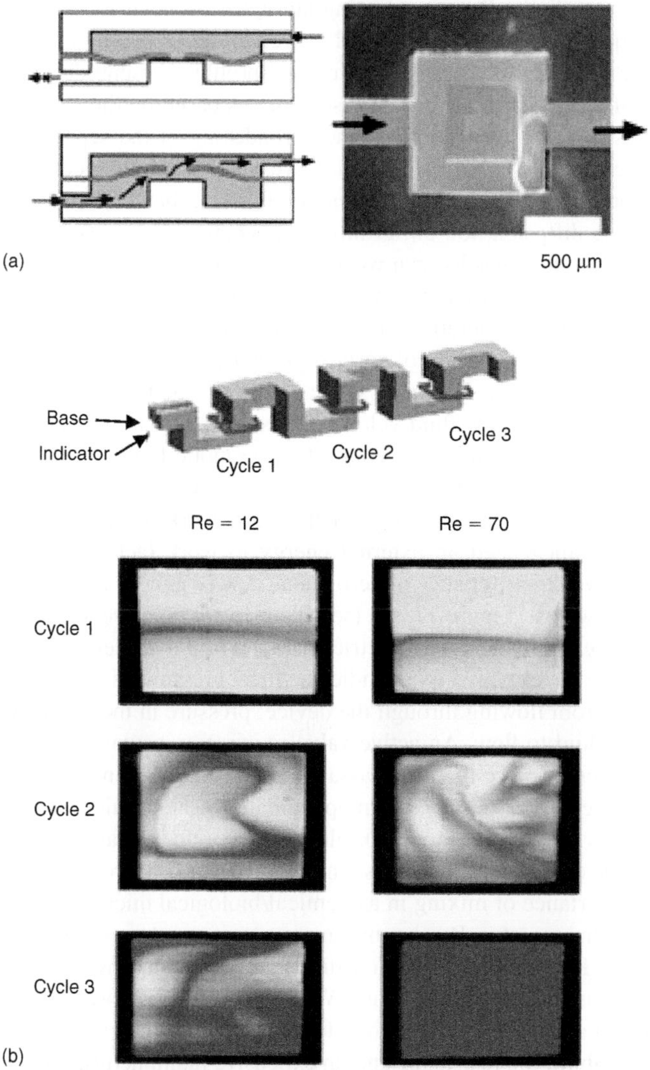

FIGURE 11.33 Microfluidic components. (a) A passive check valve in a multilayer structure of polydimethylsiloxane (PDMS). The upper schematic shows the valve in the closed position: pressure applied from the right presses the flexible diaphragm against a post, blocking flow through the hole in the center of the diaphragm. The lower schematic shows the valve in the open position: pressure applied from the left lifts the diaphragm off the post and allows flow through the diaphragm hole. The image on the right shows flow of a fluorescent fluid through the valve. (b) The design of this chaotic advection mixer exploits the character of laminar flow in a three-dimensional serpentine channel to mix adjacent streams in a single flow. A solution of pH indicator runs adjacent to a stream of basic solution and reveals the progress of mixing. As the two solutions mix, the indicator becomes red. Mixing is efficient even at low Reynolds numbers. Although the flow is laminar, there are weak eddies in the corners [163, 167]. Source: http://www.niherst.gov.tt/scipop/sci-bits/microfluidics.htm (see Plate 13 for color version).

3D flows and time-dependent 2D flows, and mixes the fluid by continuously stretching different volumes of the fluid and folding them into one another. In a qualitative sense, the path taken by a given fluid element in the flow depends in a sensitive way on its encounters with a series of weak secondary flows or eddies, present even at low *Re* in the corners of channels, which transport the element across the flow [167].

An LOC system mainly has four activities: sensing, actuation, heating, and processing. In analytical application, the sensors such as array biochips translate mechanical, thermodynamic or chemical information from their environments into electrical signals. That information is then processed, either by the LOC itself, by a nearby IC, or by a computer. The LOC device may then act on its environment by closing a valve, deflecting a light beam, pushing a fluid through a conduit, and moving a wire. These devices may be heated or cooled quite rapidly – a type of endogenous stimulation from which any of the other activities may ensue. Thus, except pumps and valves, other functional components, such as injection, dosing, metering, sensing, and temperature measurement, actuators, and control/sensing circuit components, are all or partially needed in the LOC. These are discussed in more detail in the literature [168].

11.5.3 Fabrication of BioMEMS

Rapid progress in microfabrication and assembly techniques has led to the development of extremely small-scale devices commonly referred to as MEMS (micro-electro-mechanical systems) or μTAS (micro-total analysis systems). Advances in MEMS are being applied to biomedical applications and has become a new field of research known as BioMEMS. In fact, BioMEMS is the technology to fabricate a microfluidic system containing functional components for a biological LOC system. Most of these devices are manufactured in silicon but recent developments have demonstrated that materials such as glass, quartz, ceramics, and polymers can also be used for MEMS. In early 1990, silicon and glass substrate-based microfluidic devices were first made by using conventional, planar fabrication techniques – photolithography and etching – to pattern microfluidic structures [169]. Relative high cost and limited material choice for the approach resulted in developing alternative techniques such as soft lithography and polymer molding for fabrication of BioMEMS microfluidic devices. These non-photolithographic microfabrication methods are based on printing and molding organic materials, which are much more straightforward, for making both prototype and special-purpose devices. They are also much simpler for building three-dimensional networks of channels and components as compared to photolithography. Precision machining techniques like milling, grinding, and turning can produce very accurate parts from a variety of engineering materials allowing to pattern three-dimensional structures with minimum feature size in the micrometer range when using special microcutting tools. The machining methods used include high speed cutting (HSC), micromilling as well as wire and electro discharge machining (EDM). The fabrication is relatively simple and suitable for rapid prototyping as it requires no optical mask, but its mass production with identical structures are cost prohibitive as each structure is machined individually (serial process). Generally, the design and microfabrication of BioMEMS is motivated by fast response times, well-controlled reaction conditions, small power and chemical consumption, and highly parallel screening and testing ability. Since each technique has its own advantages and disadvantages, the most suitable method of device fabrication often depends on the specific application of the device [170].

Silicon micromachining is based on a lithography process and is discussed in section 11.4.2.2. The use of glass instead of silicon in an LOC system is prompted by its unique properties such as chemical resistance, optical transparency, high dielectric coefficient, and biocompatibility. Borofloat glass is the most popular material in the fabrication of a LOC system. Fabrication of a glass microfluidic network involves the use of the traditional silicon-processing technique as discussed in section 11.4.2.2, such as photolithography and wet chemical etching. The wet etching of glass is mostly done with hydrofluoric acid (HF) or buffered hydrofluoric acid (BHF). The electronic ELISA chip in Fig. 11.28 (see Plate 11 for color version) made in the author's lab is based on glass substrate. A glass cover plate is often used to seal the etched microfluidic channel network. The three most frequently used glass-glass bonding methods are thermal fusion, anodic and adhesion bonding, and the most popular one among them is the fusion method due to its direct bonding without an intermediate layer. Many types of polymers show better chemical resistance and biocompatibility than silicon. These polymers can be mass produced at significantly lower cost than silicon microstructure. The methods to fabricate polymer microfluidic systems include thermal embossing, injection molding, casting, laser machining, milling, and x-ray/UV polymer lithography. The theory for polymer lithography is that some polymer materials are affected by energetic radiation such as UV and x-ray radiation, which may break chemical bonds or lead to chemical changes of the polymer and be removed by following solution treatment. There are quite a few well-developed polymer bonding techniques including gluing, laminating, thermal bonding, and ultrasonic and laser welding. One of the best methods for joining two plastic surfaces is gluing, due to its simplicity and low cost. Fabrication efforts include using optical lithography to make a silicon wafer mold master with complicated 3D structure, and then replicating thousands of polymer microfluidic devices by molding. This is an inexpensive method and the manufacturing process allows mass productions. Figure 11.34 shows a plastic flow-through ELISA microfluidic system made in the author's lab [171]. Taking advantage of microfabrication technology, the silicon wafer master (Fig. 11.34a) with 3D-structured multichannels was fabricated and served as a mold for plastic replica molding. The molded device is shown in Fig. 11.34b. The resulting images from different channels showed high sensitivity and specificity without cross-over interference from different detection sites. Furthermore, combining lithography approaches with electroplating will result in a robust metal structure suitable as a mold insert for a polymer replication process. The resist pattern is also used directly as a molding template to cast PDMS into the form for a replicating pattern. Recent research has started to focus on techniques like electron and ion beam lithography for patterning nanostructured fluidic systems. In all these individual or combining technologies, design and fabrication of 3D-structured LOC is essential. 3D structured systems can solve very complicated biochip tasks, such as separation, reaction, and sensing. As discussed above, it is crucial for addressable arrays and patterned electrodes to have inexpensive electronic biochips. However, most

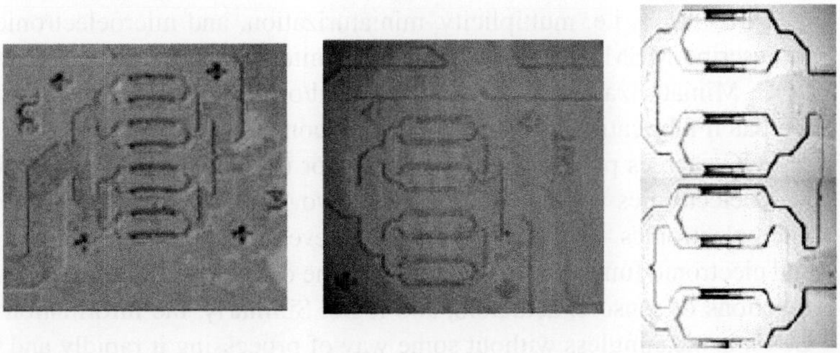

FIGURE 11.34 Photograph of fabricated silicon mask (a), molded PDMS ELISA flow-through biochip (b), and image of sample detection at different detection sites (c).

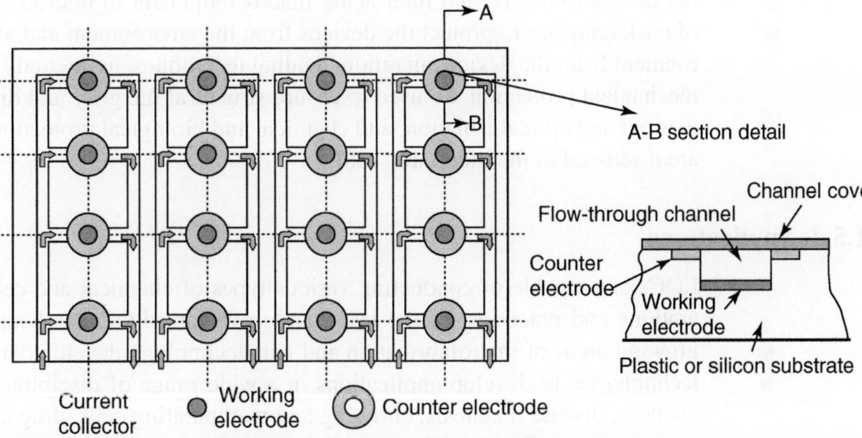

FIGURE 11.35 Flow-through electronic addressable bioarray devices.

of the biochips have to work in buffer solutions or biological fluids. Due to ionic conductivity, simple row-column (x-y) addressable array electrodes in a buffer are ionically shorted and cannot be detected for biological events occurring at individual electrodes. Li *et al.* disclosed a microfluidic design to fabricate a flow-through electronic array biochip for addressing individual electrodes without the ionic shortage [172]. The design is shown in Fig. 11.35, in which microchannels are designed to isolate every electronically connected column or row of electrodes for eliminating the ionic shortage during x-y addressing. Although these processes are expensive, slow and not commonly accessible to researchers, replication of these structures is becoming an emerging and established field summarized with the term nano-imprint-lithography (NIL). New lab-on-chip platforms often bring great advantages to their biomedical applications. Li *et al.* have reported a biotape device with an array biosensor associated with fluidic sample processing components that can be widely used in diagnostic applications [173].

Three "M"s, i.e. multiplicity, miniaturization, and microelectronics, are often used to describe MEMS technology. This definitely presents the prominent features of an LOC. Miniaturization refers to micron and/or nanoscale devices. Multiplicity refers to the batch fabrication on the scale of semiconductor manufacturing. This characteristic of LOC enables production of thousands or millions of identical devices concurrently. Microelectronics is what ties the other two "M"s together. LOC devices owe a lot to microelectronics fabrication R&D. However, without the integration of mechanical and electronic functions found in LOCs, the devices would have no capability to merge functions of sensors, actuators, and logic. Similarly, the information generated within LOCs is meaningless without some way of processing it rapidly and reproducibly. For both MEMS and LOCs, the relative positioning of electronics and mechanics is unimportant. There is no reason why electronics must be contained on the same chip as the other functions, although this is possible in both cases. Integration of an LOC involves critical, challenging, packaging issues including microfluidic interconnects (connecting device to device and interfacing macrocomponent to microdevice). The functions of packaging are to protect the devices from the environment and also protect the environment from the device operation. Sealing techniques, hermeticity measurements, and mechanical protection are used to ensure structural integrity and dimensional stability, thermal and optical isolation, and chemical and biological protection. These techniques are discussed in the literature [168].

11.5.4 Applications

LOCs are capable of conducting various types of chemical and cellular analysis, separations and reactions. In the last couple of years, LOC has been one of the fastest growing areas of microfabrication and nanotechnology development, integrating many technologies to develop applications in a wide range of disciplines including genetic analysis, disease diagnosis, culturing and manipulating cells, drug discovery, and materials chemistry. The LOC has enabled many types of sensor, actuator, and system to be reduced in size by orders of magnitude, while often even improving sensor performance (e.g. inertial sensors, optical switch arrays, biochemical analysis systems, etc.). It has demonstrated great potential in commercialization.

11.5.4.1 Cell sorting system

Cell sorting and counting are important methods for cell detection. Early detection of malignant cells is critical to the successful treatment of cancers. The difficulty in detecting these cells arises from the subtle onset of the disease in a single cell embedded in a host organ comprising billions of cells. Flow cytometry (FCM) is a very effective technique for tackling this challenge [174]. However, conventional flow cytometers tend to be large, expensive bench top systems that require specially trained personnel for sample tagging and instrument operation. A cytometry microsystem based on microfluidics could be a relatively cheap and portable cell interrogation tool with far-reaching applications in point of care diagnosis. Dielectrophoretic forces induced by high frequency alternating, rotating or traveling electric fields can be used to levitate and/or to move cells in weak electrolyte solutions. Counting red blood cells is a well-established method, but design of micromachined cell or particle counters needs to be further investigated. It is possible to count cells by the change in the electric resistance of a conducting fluid if a cell of different conductivity passes. A Coulter counter, where the change of the electrical resistance of a liquid filled capillary is measured when a cell flows through, has been made as a microsystem. Cell sorting and counting systems-based LOCs have good reproducibility and resolutions down to a cell size of 0.6 μM diameter [175]. Much work has focused on replacing the conventional flow chamber with microfabricated devices and it has been shown that cells, particles, and reagents can be manipulated by pressure, dielectrophoresis, and electroosmosis [176–180]. Several groups have demonstrated on-chip detection by means of impedance [178], fluorescence [179], and laser-based spectroscopy [180]. Optical forces have been reported in use for rapid (2–4 ms), active control of cell routing on a microfluidic chip [181]. The optical switch controls reduce the complexity of the chip and simplify connectivity. Using all-optical switching, a fluorescence-activated microfluidic cell sorted and evaluated its performance on live, stably transfected HeLa cells expressing a fused histone-green fluorescent protein.

11.5.4.2 Combinatorial synthesis for drug screening and materials discovery

The combinatorial synthesis approach is a powerful, yet simple, way to construct a large number of compounds in a very small area in a very short time. For example, by going through 32 iterations of oligonucleotide synthesis, scientists can produce all 65 536 oligos containing eight units in about one day. Each of these compounds is contained within a well-defined area on a microarray or biochip. Similarly, the combinatorial approach can be used with peptide synthesis to create an assortment of peptides of almost any length. Moreover, the combinatorial approach is not limited to oligonucleotides and peptides, and there is considerable potential for combining a limited array of combinatorial products with microsampling techniques and microseparations to create automated organic chemistry synthesizers, which will be very useful for drug discovery. Today, the drug industry has rethought its combinatorial strategies and focuses instead on limited libraries of up to several hundred compounds. This "rational design" of new drugs relies on techniques that use information about a molecule's shape, size, electronic topology, and lipophilicity or hydrophilicity to design a few, select, lead compounds. After testing these leads, chemists systematically alter the candidate molecules until the desired activity is optimized, generating thousands of candidates in a combinatorial library. Products have resulted from combichem/microsynthesis, and the list continues to grow. In the pharmaceutical/biotech arena, lead compounds for the treatment of Alzheimer's disease, tuberculosis, and inflammatory disorders are in the pipeline. Combinatorial drugs for pain, cancer, multidrug resistance in cancer cells, HIV, lupus, and asthma are in clinical trials.

Unfortunately, rational approaches do not work with all types of chemistry. The rational approach has achieved fewer successes in materials discovery than in pharmaceuticals.

This should come as no surprise, since the physical behavior of materials is non-linear and unpredictable, especially when materials are formulated or in combination. Two examples will suffice: high temperature ceramic superconductors and insulators above their critical temperatures or at non-ideal stoichiometries; composite structures may show several times the strength or impact resistance than would be expected from their component materials. Materials discovery will always require a good deal of trial and error, factors that may be mitigated by techniques that permit the simultaneous synthesis of large numbers of materials, followed by rapid or parallel screening for desired properties.

In material science, new phosphors have been discovered for eventual use in fluorescent lighting, flat-panel displays, and computer screens. An inkjet-based combinatorial chemistry synthesizer has been designed and built. A new, efficient blue photoluminescent composite has also been found. A library of electronic materials has been quickly tested in actual devices. A thin film material has been developed that has a higher dielectric constant than silicon dioxide, the insulator most commonly used in dynamic random-access memory (DRAM) computer chips. Studies suggest that combinatorial methods will be very useful in identifying superior materials for superconducting magnetoresistive applications as well, within classes of both known materials and entirely new ones. Numerous screening methods, ranging from NMR-on-chip to infrared thermography to activated resins, are being perfected. Moreover, the software that enables users to keep track of all these is commercially available.

11.5.4.3 Chemical and biological analysis

LOC-based chemical analysis systems consist of flow injection analysis as well as biological molecule analysis. Microfluidic injection analysis includes liquid/gas chromatography (LC, GC) or capillary electrophoresis (CE). The first micromachined flow injection analysis (FIA) system was a gas chromatograph developed by Terry [182], which consists of a long capillary column, active valve with magnetic actuation, and a detector element on a silicon wafer. Since then, different LOC chromatography devices have been reported [183, 184]. These miniaturized systems demonstrated good performance. The first capillary electrophoretic LOC was based on electroosmotic effect [185, 186], in which the capillaries are micromachined by silicon wet etching and sealed by anodic binding and separation. The LOC separation efficiency was improved by synchronized cyclic capillary electrophoresis [186]. More functions such as mixing, reaction [187], and sample detection with fluorescence [188] were incorporated onto the chip.

Almost all genetic tests today use one of two techniques: sequencing, which is more comprehensive and versatile, or hybridization, which is faster and more efficient for small pieces of genes. Rapid identification of genes or gene sequences has become one of the aims of the medical diagnostics business. Hybridization techniques lend themselves well to a combinatorial approach, and this has become the entry point into genetics for LOC developers. During the past few years, the HGP has provided the impetus for the evolution and maturation of DNA sequencing using capillary electrophoresis (CE), an important theme in LOC development. Recently, researchers have been looking ahead to the next generation of sequencer, a microfabricated chip-based electrophoresis system using channels cut in a glass or plastic chip, instead of capillary tubes. Microfabricated chip-based sequencers have many advantages over previous CE systems. The photolithography techniques used to create these microchannels facilitate the addition of numerous capillaries, making ultra-high lane densities – providing greater throughput capabilities than capillary sequencers. The intricate enclosed microchannels set in glass or fused silica substrates enable ultra-fast DNA separations because of simplified sample loading procedures, short separation distances, and optimal thermal characteristics. Precise optical positioning across an array of microchannels simplifies the detection process. And integration of sample handling and analysis reduces human interference, permitting automation and further contributing to low operating costs. Many other approaches have been tried, and new technologies are being developed, but none are likely to displace microfabricated chip-based electrophoresis systems as the most probable future technology for sequencing. Besides its promise for high throughput DNA sequencing, a microchip-based CE system is also being exploited for nucleic acid and protein analyses such as separation, sizing, quantifying, and identifying DNA and RNA samples. CE microchips have also been developed for rapid detection of single polynucleotide polymorphisms (SNPs) in the human genome. In this post-genomics era, the refinement of multiple channel microchip-based CE devices and integrated process automation will undoubtedly establish this technology as a powerful high throughput tool for invaluable genetic analyses. The possibility of building a complete DNA sequencing system on one integrated chip is very real. Such a chip can include the DNA preparation, loading the separation matrix, loading the sample, electrophoretic separation, and fluorescence detection – all on a single chip. Also, this chip can be mounted inside a computer like a PC board.

11.6 REFERENCES

1. B.F.C. Clark, Towards a total human protein map. *Nature* **292**, 491–492 (1981).
2. J.P. Jing, J. Reed, J. Huang, X. Hu, V. Clarke, J. Edington, D. Housman, T.S. Anantharaman, E.J. Huff, B. Mishra, B. Porter, A. Shenkeer, E. Wolfson, C. Hiort, R. Kantor, C. Aston, and D.C. Schwartz, Automated high resolution optical mapping using arrayed, fluid-fixed DNA molecules. *Proc. Nat. Acad. Sci.* **95**, 8046–8051 (1998).
3. C.M. Li, V.E. Choong, S. Shi, and G. Maracas, Method for enhanced bio-conjugation events. *US Pat* 6,602,400 (2003).
4. C.M. Li and B. Chui, P450 single nucleotide polymorphism biochip analysis. *US Pat* 20040229222 A1 (2004).
5. M. Schena, D. Schalon, and R.W. Davis, Quantitative monitoring of gene expression pattems with a complementary DNA microarray. *Science* **270**, 467–470 (1995).
6. L. Wodicka, H. Dong, M. Mittmann, M.H. Ho, and D.J. Lockhart, Genome-wide expression monitoring in saccharomyces cerevisiae. *Nat. Biotechnol.* **15**, 1359–1367 (1997).
7. N. Zhao, H. Hashida, N. Takahashi, Y. Misumi, and Y. Sakaki, High-density cDNA filter analysis: a novel approach for large-scale, quantitative analysis of gene expression. *Gene* **156**, 207–213 (1995).
8. M.J. Heller, A.H. Forster, and E. Tu, Active microelectronic chip devices which utilize controlled electrophoretic fields for multiplex DNA hybridization and other genomic applications. *Electrophoresis* **21**, 157–164 (2000).
9. V.G. Cheung, M. Morley, F. Aguilar, A. Massimi, R. Kucherlapati, and G. Childs, Making and reading microarrays. *Nat. Genet. (Suppl.)* **21**, 15–19 (1999).

10. P.A. Caviani, D. Solas, E.J. Sullivan, M.T. Cronin, C.P. Holmes, and S.P.A. Fodor, Light-generated oligonulceotide arrays for rapid DNA sequence analysis. *Proc. Natl. Acad. Sci.* **91**, 5022–5026 (1994).
11. R.J. Lipshutz, S.P.A. Fodor, T.R. Gingeras, and D.J. Lockhart, High density synthetic oligonucleotide arrays. *Nat. Genet. (Suppl.)* **21**, 20–24 (1999).
12. D. Wallace, Ink-jet based fluid microdispensing in biochemical applications. *Nucl. Med. Biol.* **21**, 6–9 (1996).
13. P. Doty, J. Marmur, J. Eigen, and C.E. Schildkraut, Strand separation and specific recombination in deoxyribonucleic acids: physical chemical studies. *Proc. Natl. Acad. Sci.* **46**, 461–476 (1960).
14. R. Drmanac, Z. Stresoska, I. Labat, S. Drmanac, and R. Crkvenjakov, Reliable hybridization of oligonucleotides as short as six nucleotides. *DNA Cell Biol.* **9**, 527–534 (1990).
15. R. Drmanac and S. Drmanac, cDNA screening by array hybridization. *Meth. Enzymol.* **303**, 165–178 (1990).
16. K.B. Mullis, The unusual origin of the polymerase chain reaction. *Sci. Am.* **262**, 56–61(1990).
17. R. Saiki, D. Gelfand, S. Stoffel, S. Scharf, R. Higuchi, G. Horn, K. Mullis, and H. Erlich, Primer-directed enzymatic amplification of DNA with a thermostable DNA polymerase. *Science* **239**, 487–491 (1988).
18. C.M. Li, J. Xia *et al.* Compositions and methods for rolling circle amplification. *US Pat* 20040014078 (2004).
19. A. Fire and S.Q. Xu, Rolling replication of short DNA circles. *Proc. Natl. Acad. Sci. U.S.A* **92**, 4641–4645 (1995).
20. D.Y. Liu, S.L. Daubendiek, M.A. Zillman, K. Ryan, and E.T. Kool, Rolling circle DNA synthesis: small circular oligonucleotides as efficient templates for DNA polymerases. *J. Am. Chem. Soc.* **118**, 1587–1594 (1996).
21. D.C. Thomas, G.A. Nardone, and S.K. Randall, Amplification of padlock probes for DNA diagnostics by cascade rolling circle amplification or the polymerase chain reaction. *Arch. Pathol. Lab. Med.* **123**, 1170–1176 (1999).
22. S. Wiltshire, S. O'Malley, J. Lambert, K. Kukanski, D. Edgar, S.F. Kingsmore, and B. Schweitzer, Detection of multiple allergen-specific IgEs on microarrays by immunoassay with rolling circle amplification. *Clin. Chem.* **46**, 1990–1993 (2000).
23. L. Alfonta, A.K. Singh, and I. Willner, Liposomes labeled with biotin and horseradish peroxidase: a probe for the enhanced amplification of antigen-antibody or oligonucleotide-DNA sensing processes by the precipitation of an insoluble product on the electrode. *Anal. Chem.* **73**, 91–102 (2001).
24. S.M. Hsu, L. Raine, and H. Fanger, A comparative study of the peroxidase-antiperoxidase method and an avidin-biotin complex method for studying polypeptide hormones with radioimmunoassay antibodies. *Am. J. Clin. Pathol.* **75**, 734–738 (1981).
25. M.J. Schermer, Confocal scanning microscopy in microarray detection, in *DNA Microarrays: A Practical Approach* (M. Schena, ed.), p. 17, Oxford University Press (1999).
26. T. Basarsky, Overview of a microarray scanner: design essentials for an integrated acquisition and analysis platform, in *Microarray Biochip Technology* (M. Schena, ed.), p. 265, Eaton Publishing (2000).
27. L. Ramdas, J. Wang, L. Hu, D. Cogdell, E. Taylor, and W. Zhang, Comparative evaluation of laser-based microarray scanners. *Biotechniques* **31**, 546–550 (2001).
28. Y.H. Yang, M.J. Buckley, and T.P. Speed, Analysis of cDNA microarray images. *Briefings in Bioinformatics* **2**, 341–349 (2001).
29. F.C. Tet, K.C. Chee, M. Chan-Park, W.C. Yew, S.R Conner, L.C.S Daniel, S.L.L. Shirley, P.L. Siew, L. Feng, K. Mao, U. Surana, S.K. Tan, S. Venkatraman, and F.M. Yeong, *Life Sciences – Engineering Applications in Biology*, 1st ed., McGraw Hill Education (2003).
30. D. Guschin, G. Yershov, A. Zaslavsky, A.Gemmell, V. Shick, D. Proudnikov, P. Arwnkov, and A. Mirzabekov, Manual manufacturing of oligonucleotide, DNA, and protein microchips. *Anal. Biochem.* **250**, 203–211 (1997).
31. V. Afanassiev, V. Hanemann, and S. Wölfl, Preparation of DNA and protein micro arrays on glass slides coated with an agarose film. *Nucl. Acids Res.* **28**, E66 (2000).
32. H. Zhu, J.F. Klemic, S. Chang, P. Bertone, A. Casamayor, K.G. Klemic, D. Smith, M. Gerstein, M.A. Reed, and M. Snyder, Analysis of yeast protein kinases using protein chips. *Nat. Genet.* **26**, 283–289 (2000).

33. T.O. Joos, M. Schrenk, P. Hopfl, K. Kroger, U. Chowdhury, D. Stoll, D. Shorner, M. Durr, K. Herick, and S. Rupp, A microarray enzyme-linked immunosorbent assay for autoimmune diagnostics. *Electrophoresis* **21**, 2641–2650 (2000).
34. G. MacBeath and S.L. Schreiber, Printing proteins as microarrays for high-throughput function determination. *Science* **289**, 1760–1763 (2000).
35. H. Zhu, M. Bilgin, R. Bangham, D. Hall, A. Casamayor, P. Bertone, and N. Lan, Global analysis of protein activities using proteome chips. *Science* **293**, 2101–2105 (2001).
36. B. Schweitzer and S.F. Kingsmore, Measuring proteins on microarrays. *Curr. Opin. Biotech*. **13**, 14–19 (2002).
37. L.O. Lomas, *Protein Microarray Technology*, Chap. 8, Wiley-VCH (2004).
38. V. Knezevic, C. Leethanakul, V.E. Bichsel, J.M. Worth, V.V. Prabhu, J.S. Gutkind, L.A. Liotta, P.J. Munson, E.F. Petricoin, and D.B. Krizman, Proteomic profiling of the cancer microenvironment by antibody arrays. *Proteomics* **1**, 1271–1278 (2001).
39. A. Sreekumar, M.K. Nyati, S. Varambally, T.R. Barrette, D. Ghosh, T.S. Lawrence, and A.M. Chinnaiyan, Profiling of cancer cells using protein microarrays: discovery of novel radiation-regulated proteins. *Cancer Res*. **61**, 7585–7593 (2001).
40. C.P. Paweletz, L. Charboneau, V.E. Bichsel, N.L. Simone, T. Chen, J.W. Gillespie, M.R. Emmert-Buck, M.J. Roth, E.F. Petricoin, and L.A. Liotta, Reverse phase protein microarrays which capture disease progression show activation of pro-survival pathways at the cancer invasion front. *Oncogene* **20**, 1981–1989 (2001).
41. L. Mezzasoma, T. Bacarese-Hamilton, M. Di Cristina, R. Rossi, F. Bistoni, and A. Crisanti, Antigen microarrays for serodiagnosis of infectious diseases. *Clin. Chem.* **48**, 121–130 (2002).
42. T. Bacarese-Hamilton, L. Mezzasoma, C. Ingham, A. Ardizzoni, R. Rossi, F. Bistoni, and A. Crisanti, Detection of allergen-specific IgE on microarrays by use of signal amplification techniques. *Clin. Chem.* **48**, 1367–1370 (2002).
43. P. Uetz, L. Giot, G. Cagney, T.A. Mansfield, R.S. Judson, J.R. Knight, D. Lockshon, V. Narayan, M. Srinivasan, P. Pochart, A. Qureshi-Emili, Y. Li, B. Godwin, D. Conover, T. Kalbfleisch, G. Vijayadamodar, M.J. Yang, M. Johnston, S. Fields, and J.M. Rothberg, A comprehensive analysis of protein-protein interactions in Saccharomyces cerevisiae. *Nature* **403**, 623–627 (2000).
44. A.C. Gavin, M. Bosche, R. Krause, P. Grandi, M. Marzioch, A. Bauer, J. Schultz, J.M. Rick, A.M. Michon, C.M. Cruciat, M. Remor, C. Hofert, M. Schelder, M. Brajenovic, H. Ruffner, A. Merino, K. Klein, M. Hudak, D. Dickson, T. Rudi, V. Gnau, A. Bauch, S. Bastuck, B. Huhse, C. Leutwein, M.A. Heurtier, R.R. Copley, A. Edelmann, E. Querfurth, V. Rybin, G. Drewes, M. Raida, T. Bouwmeester, P. Bork, B. Seraphin, B. Kuster, G. Neubauer, and G. Superti-Furga, Functional organization of the yeast proteome by systematic analysis of protein complexes *Nature* **415**, 141–147 (2002).
45. Y. Ho, A. Gruhler, A. Heilbut, G.D. Bader, L. Moore, S.L. Adams, A. Millar, P. Taylor, K. Bennett, K. Boutilier, L.Y. Yang, C. Wolting, I. Donaldson, S. Schandorff, J. Shewnarane, M. Vo, J. Taggart, M. Goudreault, B. Muskat, C. Alfarano, D. Dewar, Z. Lin, K. Michalickova, A.R. Willems, H. Sassi, P.A. Nielsen, K.J. Rasmussen, J.R. Andersen, L.E. Johansen, L.H. Hansen, H. Jespersen, A. Podtelejnikov, E. Nielsen, J. Crawford, V. Poulsen, B.D. Sorensen, J. Matthiesen, R.C. Hendrickson, F. Gleeson, T. Pawson, M.F. Moran, D. Durocher, M. Mann, C.W.V. Hogue, D. Figeys, and M. Tyers, Systematic identification of protein complexes in Saccharomyces cerevisiae by mass spectrometry. *Nature* **415**, 180–183 (2002).
46. H. Ge, UPA, a universal protein array system for quantitative detection of protein–protein, protein–DNA, protein–RNA and protein–ligand interactions. *Nucl. Acids Res*. **28**, e3 (2000).
47. A. Espejo, J. Cote, A. Bednarek, S. Richard, and M.T. Bedford, A protein-domain microarray identifies novel protein-protein interactions. *Biochem*. **367**, 697–702 (2002).
48. F. Emmrich, Reduction of non-specific human-IgM binding in solid-phase assays. *J. Immunol. Methods* **72**, 501–503 (1984).
49. P.W. Robertson, L.R. Whybin, and J. Cox, Reduction in non-specific binding in enzyme immunoassays using casein hydrolysate in serum diluents. *J. Immunol. Methods* **76**, 195–197 (1985).
50. H.C.B. Graves, The effect of surface-charge on non-specific binding of rabbit immunoglobulin-G in solid-phase immunoassays. *J. Immunol. Methods* **111**, 157–166 (1988).

51. J.G. Kenna, G.N. Major, and R.S. Williams, Methods for reducing nonspecific antibody-binding in enzyme-linked immunosorbent assays. *J. Immunol. Methods* **85**, 409–419 (1985).
52. C.A. Mauracher, L.A. Mitchell, and A.J. Tingle, Reduction of rubella ELISA background using heat denatured sample buffer. *J. Immunol. Methods* **145**, 251–254 (1991).
53. J.S. Albala and I. Humphery-Smith, *Protein Arrays, Biochips, and Proteomics*, Chap. 3, Marcel Dekker Inc. (2003).
54. L. Yu, C.M. Li, and Q. Zhou, Efficient probe immobilization on poly (dimethylsiloxane) for sensitive detection of proteins. *Front. Biosci.* **10**, 2848–2855 (2005).
55. W.H. Scouten, J.H. Luong, and R.S. Brown, Enzyme or protein immobilization techniques for applications in biosensor design. *Trends Biotechnol.* **13**, 178–185 (1995).
56. E.T. Vandenberg, L. Bertilsson, B. Liedberg, K. Uvdal, R. Erlandsson, H. Elwing, and I. Lundstrom, Structure of 3-aminopropyl triethoxy on silicon-oxide. *J. Colloid Interf. Sci.* **147**, 103–118 (1991).
57. A.A. Bergman, J. Buijs, J. Herbig, D.T. Mathes, J.J. Demarest, C.D. Wilson, C.T. Reimann, R.A. Baragiola, R. Hull, and S.O. Oscarsson, Nanometer-scale arrangement of human serum albumin by adsorption on defect arrays created with a finely focused ion beam. *Langmuir* **14**, 6785–6788 (1998).
58. K. Calkins, Putting chips down. *BioCentury* **9**, A1–A6 (2001).
59. K.B. Lee, S.J. Park, C.A. Mirkin, J.C. Smith, and M. Mrksich, Protein nanoarrays generated by dip-pen nanolithography. *Science* **295**, 1702–1705 (2002).
60. J. Boguslavsky, Microarray technology empowers proteomics. *Genomics Proteomics* **1**, 44–46 (2001).
61. R.P. Ekins, Ligand assays: from electrophoresis to miniaturized microarrays. *Clin. Chem.* **44**, 2015–2030 (1998).
62. A.S. Blawas and W.M. Reichert, Protein patterning. *Biomaterials* **19**, 595–609 (1998).
63. M. Mrksich, C.S. Chen, Y. Xia, L.E. Dike, D.E. Ingber, and G.M. Whitesides, Controlling cell attachment on contoured surfaces with self-assembled monolayers of alkanethiolates on gold. *Proc. Natl. Acad. Sci. U.S.A* **93**, 10775–10778 (1996).
64. V. Dolnik, S.R. Liu, and S. Jovanovich, Capillary electrophoresis on microchip. *Electrophoresis* **21**, 41–54 (2000).
65. A. Mirzabekov and A. Kolchinsky, Emerging array-based technologies in proteomics. *Curr. Opin. Chem. Biol.* **6**,70–75 (2002).
66. Z.P. Aguilar and I. Fritsch, Immobilized enzyme-linked DNA-hybridization assay with electrochemical detection for Cryptosporidium parvum hsp70 mRNA. *Anal. Chem.* **75**, 3890–3897 (2003).
67. E.F. Petricoin and L.A. Liotta, SELDI-TOF-based serum proteomic pattern diagnostics for early detection of cancer. *Curr. Opin. Biotech.* **15**, 24–30 (2004).
68. A.J. Sinskey, S.N. Finkelstein, and S.M. Cooper, The promise of protein microarrays. *PharmaGenomics* 20–22 (2002).
69. D.A. Hall, H. Zhu, X.W. Zhu, T. Royce, M. Gerstein, and M. Snyder, Regulation of gene expression by a metabolic enzyme. *Science* **306**, 482–484 (2004).
70. Y. Fang, A.G. Frutos, and J. Lahiri, Membrane protein microarrays. *J. Am. Chem. Soc.* **124**, 2394–2395 (2002).
71. G. Walter, K. Büssow, D. Cahill, A. Lueking, and H. Lehrach, Protein arrays for gene expression and molecular interaction screening. *Curr. Opin. Microbiol.* **3**, 298–302 (2000).
72. I.A. Darwish and D.A. Blake, One-step competitive immunoassay for cadmium ions: development and validation for environmental water samples. *Anal. Chem.* **73**, 1889–1895 (2001).
73. W. Huang, A. Feltus, A. Witkowski, and S. Daunert, Homogeneous bioluminescence competitive binding assay for folate based on a coupled glucose-6-phosphate dehydrogenase-bacterial luciferase enzyme system. *Anal. Chem.* **68**, 1646–1650 (1996).
74. X. Zhao and S.A. Shippy, Competitive immunoassay for microliter protein samples with magnetic beads and near-infrared fluorescence detection. *Anal. Chem.* **76**, 1871–1976 (2004).
75. E. Baldrich, J.L. Acero, G. Reekmans, W. Laureyn, and C.K. O'Sullivan, Displacement enzyme linked aptamer assay. *Anal. Chem.* **77**, 4774–4784 (2005).
76. U. Narang, P.R. Gauger, and F.S. Ligler, Capillary-based displacement flow immunosensor. *Anal. Chem.* **69**, 1961–1964 (1997).

77. D.C. Duffy, H.L. Gillis, J. Lin, N.F. Sheppard Jr, and G.J. Kellogg, Microfabricated centrifugal microfluidic systems: characterization and multiple enzymatic assays. *Anal. Chem.* **71**, 4669–4678 (1999).
78. J. Wang, M.P. Chatrathi, B. Tian, and R. Polsky, Microfabricated electrophoresis chips for simultaneous bioassays of glucose, uric acid, ascorbic acid, and acetaminophen *Anal. Chem.* **72**, 2514–2518 (2000).
79. K.A. Nemeth, A.V Singh., and T.B. Knudsen, Searching for biomarkers of developmental toxicity with microarrays: normal eye morphogenesis in rodent embryos. *Toxicol. Appl. Pharm.* **206**, 219–228 (2005).
80. C.M. Leys, S. Nomura, E. Montogomery, and J.R. Goldenring, Tissue microarray evaluation of prothymosin as a biomarker for human gastric metaplasia and neoplasia. *J. Surg. Res.* **121**, 327–328 (2004).
81. R. Galve, M. Nichkova, F. Camps, F. Sanchez-Baeza, and M.P. Marco, Development and evaluation of an immunoassay for biological monitoring chlorophenols in urine as potential indicators of occupational exposure. *Anal. Chem.* **74**, 468–478 (2002).
82. B. Wolf, M. Brischwein, W. Baumann, R. Ehret, and M. Kraus, Monitoring of cellular signalling and metabolism with modular sensor-technique: the PhysioControl-Microsystem (PCM®) *Biosens. Bioelectron*. **13**, 501–509 (1998).
83. R. Ehret, W. Baumann, M. Brischwein, M. Lehmann, T. Henning, I. Freund, S. Drechsler, U. Friedrich, M.-L. Hubert, E. Motrescu, A. Kob, H. Palzer, H. Grothe, and B. Wolf, Multiparametric microsensor chips for screening applications. *J. Fresenius, Anal. Chem.* **369**, 30–35 (2001).
84. C.M. Li, Addressable chem/bio chip array. *US Pat* 20050252777 (2005).
85. C.M. Li, J.R. Sawyer, Vi-En. Choong, G. Maracas, and P.M. Zhang, Protein and peptide sensors using electrical detection methods. *US Pat* 6,824,669 (2004).
86. J. Li, Q. Ye, A. Cassell, H.T. Ng, R. Stevens, J. Han, and M. Meyyappan, Bottom-up approach for carbon nanotube interconnects. *Appl. Phys. Lett.* **82**, 2491–2493 (2003).
87. C.M. Li, S. Shi, G. Maracas, and V.E. Choong, Reporterless genosensors using electrical detection methods. *US Pat* 20020051975 A1 (2002).
88. Buena Chui, C.M. Li *et al.*, P450 single nucleotide polymorphism biochip analysis. *US Pat* 6,986,992 (2006).
89. J. Xia, C.K. Brush, V. Gupta, H.S. Huang, C.M. Li, G. Maracas, R. Marrero, M.L. Ray, L. Sun, and P.M. Zhang, Compositions and methods for rolling circle amplification. *US Pat* 20040014078 (2004).
90. J. Wang, E. Palecek, P.E. Nielsen, G. Rivas, X.H. Cai, H. Shiraishi, N. Dontha, D. Luo, and P.A.M. Farias, Peptide nucleic acid probes for sequence-specific DNA biosensors. *J. Am. Chem. Soc.* **118**, 7667–7670 (1996).
91. M. Ozsoz, A. Erdem, P. Kara, K. Kerman, and D. Ozkan, Electrochemical biosensor for the detection of interaction between arsenic trioxide and DNA based on guanine signal. *Electroanal.* **15**, 613–619 (2003).
92. K. Kerman, M. Kobayashi, and E. Tamiya1, Recent trends in electrochemical DNA biosensor technology. *Meas. Sci. Technol.* **15**, R1–R11 (2004).
93. J. Wang, M. Jiang, A. Fortes, and B. Mukherjee, New label-free DNA recognition based on doping nucleic-acid probes within conducting polymer films. *Anal. Chim. Acta* **402**, 7–12 (1999).
94. C.M. Li, C.Q. Sun, S. Song, V.E. Choong, G. Maracas, and X.J. Zhang, Impedance labelless detection-based polypyrrole DNA biosensor. *Front. Biosci.* **10**, 180–186 (2005).
95. C.M. Li, W. Chen, and X. Yang, Impedance labelless detection-based polypyrrole protein biosensor. *Front. Biosci.* **10**, 2518–2526 (2005).
96. T.-Y. Lee and Y.-B. Shim, Direct DNA hybridization detection based on the oligonucleotide-functionalized conductive polymer. *Anal. Chem.* **73**, 5629–5632 (2001).
97. A.J. Bard and L.R. Faulkner, *Electrochemical Methods*, 2nd ed., John Wiley & Sons, Inc. (2001).
98. C.M. Li and C.S. Cha, The Tafel plots of thinyl chloride reduction. *Acta Physicochimica* **1**, 143 (1989).
99. C.M. Li and C.S. Cha, Studies of the reduction of thionyl chloride in dimethyl formamide using microdisk electrode. *J. Electroanal. Chem.* **260**, 91–99 (1989).
100. C.M. Li, The applications of microelectrodes in electrochemistry. *Chinese Chem. Lett.* **1**, 3 (1987).

101. C.M. Li and C.S. Cha, Porous carbon composite/enzyme glucose microsensor. *Front. Biosci.* **9**, 3324–3330 (2004).
102. C.M. Li and C.S. Cha, Article title. *Acta Chim. Sinica* **1**, 14 (1988).
103. C.M. Li and C.S. Cha, Powder microelectrodes I. Reversible electrode system. *Acta Physicochimica* **8**, 64 (1988).
104. C.M. Li and C.S. Cha, Powder microelectrodes II. Irreversible electrode system. *Acta Physicochimica* **4**, 273 (1988).
105. C.S. Cha, C.M. Li, H. Yang, and P.F. Liu, Powder microelectrodes. *J. Electroanal. Chem.* **368**, 47–54 (1994).
106. W.E. Morf and N.F. de Rooij, Performance of amperometric sensors based on multiple microelectrode arrays. *Sens. Actuators B.* **44**, 538–541 (1997).
107. C. Belmont, M.L. Tereier, J. Buffle, G. Fiaccabrino, and M. Koudeldahep, Mercury-plated iridium-based microelectrode arrays for trace metals detection by voltammetry: optimum conditions and reliability. *Anal. Chim. Acta* **329**, 203–214 (1996).
108. W.E. Morf and N.F. de Rooij, Micro-adaptation of chemical sensor materials. *Sens. Actuators B.* **51**, 89–95 (1995).
109. T. Gueshi, K. Tokudam, and H. Matsuda, Voltammetry at partially covered electrodes: Part I. Chronopotentiometry and chronoamperometry at model electrodes. *J. Electroanal. Chem.* **89**, 247–260 (1978).
110. M. Dogterom and B. Yurke, Measurement of the force-velocity relation for growing microtubules. *Science* **278**, 856–860 (1997).
111. S.A. Wring and J.P. Hart, Chemically modified, screen-printed carbon electrodes. *Analyst* **117**, 1281–1286 (1992).
112. J. Wang, *Analytical Electrochemistry*, 2nd ed., Wiley-VCH (2000).
113. C.M. Li, Addressable chem/bio chip array. *US Pat* 20050252777 (2005).
114. K. Ramanathan, M.A. Bangar, M. Yun, W. Chen, A. Mulchandani, and N.V. Myung, Individually addressable conducting polymer nanowires array. *Nano Lett.* **4**, 1237–1239 (2004).
115. S.J. Park, T.A. Taton, and C.A. Mirkin, Array-based electrical detection of DNA with nanoparticle probes. *Science* **295**, 1503–1506 (2002).
116. S.E. Létant, T.W. van Buuren, and L.J. Terminello, Nanochannel arrays on silicon platforms by electrochemistry. *Nano Lett.* **4**, 1705–1707 (2004).
117. R. Bashir, BioMEMS: state-of-the-art in detection, opportunities and prospects. *Adv. Drug Deliver. Rev.* **56**, 1565–1586 (2004).
118. A. Bange, H.B. Halsall, and W.R. Heineman, Microfluidic immunosensor systems. *Biosens. Bioelectron.* **20**, 2488–2503 (2005).
119. S. Poyard, N. Jaffrezic-Renault, C. Martelet, S. Cosnier, and P. Labbe, Optimization of an inorganic/bio-organic matrix for the development of new glucose biosensor membranes. *Anal. Chim. Acta* **364**, 165–172 (1998).
120. R. Hintsche, B. Moller, I. Dransfeld, U. Wollenberger, F. Scheller, and B. Hoffmann, Chip biosensors on thin-film metal-electrodes. *Sens. Actuators B.* **B4**, 287–291 (1991).
121. S. Brahim, D. Narinesingh, and A. Guiseppi-Elie, Polypyrrole-hydrogel composites for the construction of clinically important biosensors. *Biosens. Bioelectron.* **17**, 53–59 (2002).
122. L. Nyholm, Electrochemical techniques for lab-on-a-chip applications. *Analyst* **130**, 599–605 (2005).
123. J. Wang, A. Ibanez, and M.P. Chatrathi, Microchip-based amperometric immunoassays using redox tracers. *Electrophoresis* **23**, 3744–3749 (2002).
124. T.K. Lim, H. Ohta, and T. Matsunaga, Microfabricated on-chip-type electrochemical flow immunoassay system for the detection of histamine released in whole blood samples. *Anal. Chem.* **75**, 3316–3321 (2003).
125. M. Gabig-Ciminska, A. Holmgren, H. Andresen, K.B. Barken, M. Wumpelmann, J. Albers, R. Hintsche, A. Breitenstein, P. Neubauer, M. Los, A. Czyz, G. Wegrzyn, G. Silfversparre, B. Jurgen, T. Schweder, and S.O. Enfors, Electric chips for rapid detection and quantification of nucleic acids. *Biosens. Bioelectron.* **19**, 537–546 (2004).

126. Z.P. Aguilar, W.R. Vandaveer, and I. Fritsch, Self-contained microelectrochemical immunoassay for small volumes using mouse IgG as a model system. *Anal. Chem.*74, 3321–3329 (2002).
127. H. Dong, C.M. Li, Q. Zhou, J.B. Sun, and J.M. Miao, Sensitive electrochemical enzyme immunoassay microdevice based on architecture of dual ring electrodes with a sensing cavity chamber. *Biosen. Bioelectron.* **22**, 621–626 (2006).
128. R. Hintsche, C. Kruse, A. Uhlig, M. Paeschke, T. Lisec, U. Schnakenberg, and B. Wagner, Chemical microsensor systems for medical applications in catheters. *Sens. Actuators B.* **B27**, 471–473 (1995).
129. M. Lehmann, W. Baumann, M. Brischwein, H.J. Gahle, I. Freund, R. Ehret, S. Drechsler, H. Palzer, M. Kleintges, U. Sieben, and B. Wolf, Simultaneous measurement of cellular respiration and acidification with a single CMOS ISFET. *Biosens. Bioelectron.* **16**, 195–203 (2001).
130. Y. Cui, Q. Wei, H. Park, and C.M. Lieber, Nanowire nanosensors for highly sensitive and selective detection of biological and chemical species. *Science* **293**, 1289–1292 (2001).
131. K. Besteman, J.O. Lee, F.G.M. Wiertz, H.A. Heering, and C. Dekker, Enzyme-coated carbon nanotubes as single-molecule biosensors. *Nano Lett.* **3**, 727–730 (2003).
132. Y.K. Choi., T.J. King, and C. Hu, Nanoscale CMOS spacer FinFET for the terabit era. *IEEE. Electron. Dev. Lett.* **23**, 25–27 (2002).
133. J. Fritz, E.B. Copper, S. Gaudet, and P.K. Sorger, Electronic detection of DNA by its intrinsic molecular charge. *Proc. Natl. Acad. Sci. U.S.A.* **99**, 14142–14146 (2002).
134. T.Z. Muhammad and E.C. Alocilja, A conductometric biosensor for biosecurity. *Biosens. Bioelectron.* **18**, 813–819 (2003).
135. H. Suzuki and H. Arakawa, Fabrication of a sensing module using micromachined biosensors. *Biosens. Bioelectron.* **16**, 725–733 (2001).
136. A. Steinschaden, D. Adamovic, G. Jobst, R. Glatz, and G. Urban, Miniaturised thin film conductometric biosensors with high dynamic range and high sensitivity. *Sens. Actuators B.* **B44**, 365–369 (1997).
137. R. Gómez, R. Bashir, and A.K. Bhunia, Microscale electronic detection of bacterial metabolism. *Sens. Actuators B.* **86**, 198–208 (2002).
138. A.L. Ghindilis, P. Atanasov, M. Wilkins, and E. Wilkins, Immunosensors: electrochemical sensing and other engineering approaches. *Biosens. Bioelectron.* **13**, 113–131 (1998).
139. A. Gebbert, M. Alvarez-Icaza, W. Stocklein, and R.D. Schmid, Real-time monitoring of immunochemical interactions with a tantalum capacitance flow-through cell. *Anal. Chem.* **64**, 997–1003 (1992).
140. D. Figeys and D. Pinto, Lab-on-a-chip: a revolution in biological and medical sciences. *Anal. Chem.* **72A**, 330–335 (2000).
141. A.E. Kamholz, B.H. Weigl, B.A. Finlayson, and P. Yager, Quantitative analysis of molecular interaction in a micro fluidic channel: the T-sensor. *Anal. Chem.* **71**, 5340–5347 (1999).
142. K. Macounova, C.R. Cabrera, M.R. Holl, and P. Yager, Generation of natural PH gradients in micro fluidic channels for use in isoelectric focusing. *Anal. Chem.* **72**, 3745–3751 (2000).
143. A.G. Hadd, D.E. Raymond, J.W. Halliwell, S.C. Jacobson, and J.M. Ramsey, Microchip devices for performing enzyme assays. *Anal. Chem.* **69**, 3407–3412 (1997).
144. J. Kameoka, H.G. Craighead, H.W. Zhang, and J. Henion, A polymeric microfluidic chip for CE/MS determination of small molecules. *Anal. Chem.* **73**, 1935–1941 (2001).
145. K. Macounova, C.R. Cabrera, and P. Yager, Concentration and separation of proteins in microfluidic channels on the basis of transverse IEF. *Anal. Chem.* **73**, 1627–1633 (2001).
146. E. Eteshola and D. Leckband, Development and characterization of an ELISA assay in PDMS microfluidic channels. *Sens. Actuators B.* **72**, 129–133 (2001).
147. S.B. Cheng, C.D. Skinner, J. Taylor, S. Attiya, W.E. Lee, G. Picelli,. And D.J. Harrison, Development of a multichhannel microfluidic analysis system employing afinity capilary electrophoresis for immunoassay. *Anal. Chem.* **73**, 1472–1479 (2001).
148. D. Figeys, S.P. Gygi, G. McKinnon, and R. Aebersold, An integrated microfluidics tandem mass spectrometry system for automated protein analysis. *Anal. Chem.* **70**, 3728–3734 (1998).
149. J. Gao, J.D. Xu, L.E. Locascio, and C.S. Lee, Integrated microfluidic system enabling protein digestion, peptide separation, and protein identification. *Anal. Chem.* **73**, 2648–2655 (2001).

150. P. Belgrader, M. Okuzumi, F. Pourahmadi, D.A. Borkholder, and M.A. Northrup, A microfluidic cartridge to prepare spores for PCR analysis. *Biosens. Bioelectron.* **14**, 849–852 (2000).
151. J. Khandurina, T.E. McKnight, S.C. Jacobson, L.C. Waters, R.S. Foote, and J.M. Ramsey, Integrated system for rapid PCR-based DNA analysis in microfluidic devices. *Anal. Chem.* **72**, 2995–3000 (2000).
152. L. Koutny, D. Schmalzing, O. Salas-Solano, S. El-Difrawy, A. Adourian, S. Buonocore, K. Abbey, P. McEwan, P. Matsudaira, and D. Ehrlich, Eight hundred-base sequencing in a microfabricated electrophoretic device. *Anal. Chem.* **72**, 3388–3391 (2000).
153. G.B. Lee, S.H. Chen, G.R. Huang, W.C. Sung, and Y.H. Lin, Microfabricated plastic chips by hot embossing methods and their applications for DNA separation and detection. *Sens. Actuators B.* **75**, 142–148 (2001).
154. I.K. Glasgow, H.C. Zeringue, D.J. Beebe, S.J. Choi, J.T. Lyman, N.G. Chan, and M.B. Wheeler, Handling individual mammalian embryos using microfluidics. *IEEE. Trans. Biomed. Eng.* **48**, 570–578 (2001).
155. J. Yang, Y. Huang, X.B. Wang, F.F. Becker, and P.R.C. Gascoyne, Cell separation on microfabricated electrodes using dielectrophoretic/gravitational field-flow fractionation. *Anal. Chem.* **71**, 911–918 (1999).
156. A. Folch, B.H. Jo, O. Hurtado, D.J. Beebe, and M. Toner, Microfabricated elastomeric stencils for micropatterning cell cultures. *J. Biomed. Mater. Res.* **52**, 346–353 (2000).
157. N.L. Jeon, S.K.W. Dertinger, D.T. Chiu, I.S. Choi, A.D. Stroock, and G.M. Whitesides, Generation of solution and surface gradients using microfluidic systems. *Langmuir* **16**, 8311–8316 (2000).
158. S.K.W. Dertinger, D.T. Chiu, N.L. Jeon, and G.M. Whitesides, Generation of gradients having complex shapes using microfluidic networks. *Anal. Chem.* **73**, 1240–1246 (2001).
159. B.H. Weigl and P. Yager, Microfluidic diffusion-based separation and detection. *Science* **283**, 346–347 (1999).
160. D.D. Cunningham, Fluidics and sample handling in clinical chemical analysis. *Anal. Chim. Acta* **429**, 1–18 (2001).
161. J. Brody, P. Yager, R. Goldstein, and R. Austin, Biotechnology at low Reynolds numbers. *Biophys. J.* **71**, 3430–3441 (1996).
162. J.B. David, A.M. Glennys, and M.W. Glenn, Physics and applications of microfluidics in biology. *Annu. Rev. Biomed. Eng.* **4**, 261–286 (2002).
163. G. Whitesides and A. Stroock, Flexible methods for microfluidics. *Phys. Today* **54**, 42–48 (2001).
164. R.F. Probstein, *Physicochemical Hydrodynamics: An Introduction*, 2nd ed., Wiley, New York (1994).
165. P.H. Paul, M.G. Garguilo, and D.J. Rakestraw, Imaging of pressure- and electrokinetically driven flows through open capillaries. *Anal. Chem.* **70**, 2459–2467 (1998).
166. R.H. Liu, M.A. Stremler, K.V. Sharp, M.G. Olsen, J.G. Santiago, R.J. Adrian, H. Aref and D.J. Beebe, Passive mixing in a three-dimensional serpentine microchannel. *J. Microelectromech. Syst.* **9**, 190–197 (2000).
167. J.M. Ottino, *The Kinetics of Mixing: Stretching, Chaos, and Transport*, Cambridge University Press, Cambridge (1989).
168. O. Geschke, H. Klank, and P. Telleman, *Microsystem Engineering of Lab-on-a-chip System*, Wiley-VCH (2004).
169. P. Gravesen, J. Branebjerg, and O. Jensen, Microfluidics – a review. *J. Micromech. Microeng.* **3**, 168–182 (1993).
170. H. Becker and C. Gartner, Polymer microfabrication methods for microfludic analytical applications. *Electrophoresis* **21**, 12–26 (2000).
171. C.M. Li, Microchip and method for detecting molecules and molecular interactions. *Pat WO*2005095262 (2005).
172. C.M. Li, Addressable transistor chip for conducting assays. *Pat WO*2005095938 (2005).
173. C.M. Li, L.K. Pan, and J.H.T. Luong, Capacitance immunosensors based on an array biotape. *Analyst* **131**, 788–790 (2006).
174. J. Kruger, K. Singh, A. O'Neill, C. Jackson, A. Morrison, and P. O'Brien, Development of a microfluidic device for fluorescence activated cell sorting. *J. Micromech. Microeng.* **12**, 486–494 (2002).

175. D. Koutsouris, R. Guillet, J.C. Lelievre, M.T. Guilemin, B.P. Beuzard, Y. Beuzard, and M. Bounard, Determination of erythrocyte transit times through micropores: I – Basic operational principles. *Biorheology* **25**, 763–772 (1988).
176. P. Telleman, U.D. Larsen, J. Philip, and G. Blankenstein, μ-Total Analysis Systems '98, Vol. 39, (1998).
177. A.Y. Fu, C. Spence, A. Scherer, F.H. Arnold, and S.R. Quake, A microfabricated fluorescence-activated cell sorter. *Nat. Biotechnol.* **17**, 1109–1111 (1999).
178. S. Gaward, L. Schild, and P. Renaud, Micromachined impedance spectroscopy flow cytometer for cell analysis and particle sizing. *Lab on a Chip* **1**, 76–82 (2001).
179. K.D. Kramer, K.W. Oh, C.H. Ahn, J.J. Bao, and K.R. Wehmeyer, An optical MEMS-based fluorescence detection. *Spie-Microfluidic Devices and Systems '98* **3515**, 76–85 (1998).
180. L. Nihlen and H. Capps, Nanolaser/microfluidic biochip for realtime tumor pathology. *Biomedical Microdevices* **2**, 111–122 (1999).
181. M.M. Wang, E. Tu, D.E. Raymond, J.M. Yang, H. Zhang, N. Hagen, B. Dees, E.M. Mercer, A.H. Forster, I. Kariv, P.J. Marchand, and W.F. Butler, Microfluidic sorting of mammalian cells by optical force switching. *Nat. Biotechnol.* **23**, 83–87 (2005).
182. A. Manz, Y. Miyahara, J. Miura, Y. Watanabe, H. Miyagi, and K. Sato, Design of an open-tubular column liquid chromatograph using silicon chip technology. *Sens. Actuators* **B1**, 249–255 (1990).
183. S.C. Terry, J.H. Jerman, and J.B. Angell, A gas chromatographic air analyzer fabricated on a silicon wafer. *IEEE. Trans. Electron Dev.* **ED-26**, 1880–1886 (1979).
184. G. Ocivirk, E. Verpoorte, A. Manz, and H.M. Widmer, Integration of a micro liquid chromatograph onto a silicon chip. *Proceedings Transducers (Stockholm, Sweden)* 756–759 (1995).
185. D.J. Harrison, A. Manz, and P.G. Glavina, Electroosmotic pumping within a chemical sensor system integratedon silicon. *Proceedings Transducers (San Francisco, USA)* 792–795 (1991).
186. A. Manz, D.J. Harrison, J.C. Verpoorte, H. Ludi, and H.M. Widmer, Integrated electroosmotic pumps and flow manifolds for total chemical analysis systems. *Proceedings Transducers (San Francisco, USA)* 939–941 (1991).
187. C.S. Effenhauser, A. Manz, and M. Widmer, Glass chips for high-speed capillary electrophoresis separations with submicrometer plate heights. *Anal. Chem.* **65**, 2637–2642 (1993).
188. D.J. Harrison, K. Fluri, N. Chiem, T. Tang, and Z. Fan, Micromachinng chemical and biochemical analysis and reaction systems on glass substrates. *Proceedings Transducers (Stockholm, Sweden)* 752–755 (1995).

CHAPTER 12

Powering fuel cells through biocatalysis

Dónal Leech, Marie Pellissier, and Frédéric Barrière

12.1 INTRODUCTION

Monitoring and control of a range of medical conditions will increasingly be performed by sophisticated, miniaturized, integrated, implanted medical devices [1]. Provision of miniaturized, implantable power sources to drive these devices is therefore of significant importance. Current miniaturized battery technology uses highly reactive lithium or corrosive alkaline electrolytes. This necessitates use of protective cases and seals to prevent leakage, making miniaturization expensive and difficult. Biocatalytic fuel cells have the potential to deliver a simple, inexpensive, miniaturized, implanted power supply [1–3].

Fuel cells generate electricity through an electrochemical process in which the energy stored in a fuel is converted directly into electricity. Fuel cells chemically combine the molecules of a fuel and oxidant, without burning, dispensing with the inefficiencies and pollution of traditional combustion.

In principle, a fuel cell operates like a battery. Unlike a battery, a fuel cell does not run down or require recharging: it will produce electricity as long as fuel and oxidant are supplied. The electrochemical reactions of a fuel cell consist of two separate reactions: an oxidation half-reaction occurring at the anode and a reduction half-reaction occurring at the cathode. The anode and the cathode are separated from each other by the electrolyte and an ion-exchange membrane. In the oxidation half-reaction, the input fuel passes over the anode and is catalytically split, producing ions, which travel through the electrolyte to the cathode, and electrons, which travel through an external circuit to serve an electric load, which consumes the power generated, to the cathode. In the reduction half-reaction, an oxidant, supplied from air or fluid flowing past the cathode, combines with the ions and electrons to complete the circuit. The power output of the fuel cell is the product of the cell voltage and the cell current. The theoretical thermodynamic cell voltage is the difference in standard reduction potentials of the oxidant and fuel. System losses, however, in the form of kinetic and mass transport limitations and IR drop can severely lower the power output of fuel cells. For example, in commercial fuel cells, catalysts are used on both the anode and cathode to increase the kinetics of each half-reaction. The catalyst that works the best on each electrode is platinum, a non-selective catalyst, and a very expensive material.

Biocatalytic fuel cells are fuel cells which rely upon biocatalytic reactions at the electrodes to convert chemical fuels and oxidants into electrical power. The biocatalytic reactions used to produce power range from reactions of fermentation broths containing whole cells, to isolated enzyme biocatalysts. Fermentation broths containing microbial cells can be used to produce chemical fuels, such as sugars or hydrogen, in the anodic compartment of a fuel cell [2]. The production of the fuels from microbial cells by fermentation may alternatively be decoupled from the fuel cell, with the fuel being fed into a conventional fuel cell [2]. Extraction of electrical power from microbial fermentation processes can also be by addition of small redox molecules that can mediate electron transfer from the microbial electron transport pathway to the electrode surface [2]. An advantage of using whole cells to produce power is that the biocatalysts and micro-organisms can be maintained in their natural environment, while efficiently producing power over long periods. The power densities that can be extracted from these fuel cells are, however, typically low, and they are thus expected to find limited application in implanted electronic devices. Renewed research in development of these types of fuel cells has instead been driven by the goal of large-scale clean power production. Given the low power densities of these cells, however, it is doubtful if this technology will ever compete with conventional fuel cells [3].

Biocatalytic fuel cells using isolated redox enzymes were first investigated in 1964 [4]. These fuel cells represent a more realistic opportunity for provision of implantable power, given the exquisite selectivity of enzyme catalysts, their activity under physiological conditions, and the relative ease of immobilization of isolated enzymes, compared to the microbial fuel cells. Implantable biocatalytic fuel cells have thus been proposed, where the body's own chemicals are used to produce power *in vivo*.

In this chapter, progress on biocatalytic fuel cell research is reviewed. Particularly, reported biocatalytic fuel cell prototypes are critically assessed against the long-term goal of the design of a small, implantable, and long-lived, low power source for biomedical applications. In this context, the substrates of choice to power such a device are free oxygen as the oxidant and glucose as the fuel, both present in significant concentration in physiological media. Although other recently reported biocatalytic fuel cells not exclusively relying on glucose or oxygen are of interest to the field, we will mainly focus on the glucose–oxygen system, with eventual implantation in mind.

The first oxygen–glucose biocatalytic fuel cell working at neutral pH was reported in 1964 [4]. The overambitious goal of the time was to power energy-demanding devices like the artificial heart. The development of the field occurred concomitantly with the development of bioelectrochemistry [5]. The low current densities obtained in these early, and to some extent current, prototypes gradually reoriented the biocatalytic fuel cell research towards more modest and sensible aims [1–3]. The recent upsurge in interest in biocatalytic fuel cell research is driven in part by the convergence of advances, on the one hand, in biosensor design and enzyme electrochemistry, particularly in terms of increasing stable current densities at modified electrodes, and on the other hand, in microelectronics, where ever smaller and lower energy consuming devices are being manufactured. These advances now possibly make the implantable low power biocatalytic fuel cell a realistic goal, although many issues still need to be resolved. Several interesting reviews on aspects of biocatalytic fuel cell research have appeared in the last decade [1–3, 6, 7]. Here, we review the development of this exciting research area, concentrating initially on some basic principles, then on the cathode compartment of the cell, then on the anode, and finally summarizing research on combining the two electrodes to provide prototype biocatalytic fuel cells.

12.2 BIOCATALYTIC FUEL CELL DESIGN

Contrary to traditional fuel cells, biocatalytic fuel cells are in principle very simple in design [1]. Fuel cells are usually made of two half-cell electrodes, the anode and cathode, separated by an electrolyte and a membrane that should avoid mixing of the fuel and oxidant at both electrodes, while allowing the diffusion of ions to/from the electrodes. The electrodes and membrane assembly needs to be sealed and mounted in a case from which plumbing allows the fuel and oxidant delivery to the anode and cathode, respectively, and exhaustion of the reaction products. In contrast, the simplicity of the biocatalytic fuel cell design rests on the specificity of the catalyst brought upon by the use of enzymes.

Provided that the required enzymes can be immobilized at, and electrically communicated with, the surface of an electrode, with retention of their high catalytic properties and there is no electrolysis of fuel at the cathode or oxidant at the anode, or a solution redox reaction between fuel and oxidant, the biocatalytic fuel cell then simply

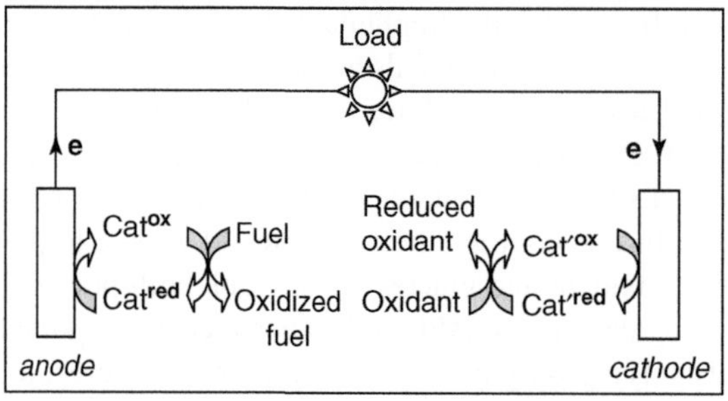

FIGURE 12.1 Schematic depiction of a biocatalytic fuel cell, with fuel oxidation by a biocatalyst (Cat) at the anode and oxidant reduction by a biocatalyst (Cat′) at the cathode, in a membraneless assembly, providing power to the load.

consists of the modified cathode and anode separated by an electrolyte containing both fuel and oxidant, connected to a load, as depicted in the schematic in Fig. 12.1. The design simplicity allows miniaturization of the cell. Realistic reachable power outputs, however, should restrict future devices to powering low energy demanding systems.

12.3 ELECTRON TRANSFER REACTIONS

Electronic communication between electrode surfaces and biocatalysts can be achieved by direct electron transfer if the active site of the biocatalyst is not located too remote from the protein surface, as discussed elsewhere in this book (Chapter 17). Direct electron transfer is an attractive process for fuel cells as no other molecules except the substrate and the enzyme are involved in the electrocatalytic reaction, as depicted in the schematic in Fig. 12.2. The enzyme is the relay for the electron transfer between the substrate and the electrode surface. Recent advances in tailoring surface nanostructural features to match the size of co-substrate channels in biocatalysts, and in reconstituting active prosthetic groups tethered to, and communicating electronically with, surfaces, with apo-enzymes, are elegant demonstrations of direct electron transfer to biocatalyst active sites that were previously considered inaccessible to electrode surfaces [8–17].

Current and power densities achieved with electrodes using the direct electron transfer approach will be limited, however, because of the need to have intimate contact between the two-dimensional electrode surface and a coating monolayer of correctly oriented biocatalyst. The use of small redox molecules that can mediate electron transfer between the biocatalyst and the electrode surface offers an opportunity to improve output from biocatalytic electrodes, as three-dimensional films of biocatalysts may now be used. In addition the distance between the active site of the enzyme and the electrode surface is often too great to allow efficient direct electron transfer. In these cases the electron transfer rate is not effective because of the insulation of the redox active site by the surrounding protein. A redox mediator can shuttle electrons between the enzyme and the surface. In the example of redox mediated biocatalytic oxidation of a fuel, depicted in Fig. 12.3, the enzyme catalyzes the oxidation of the mediator

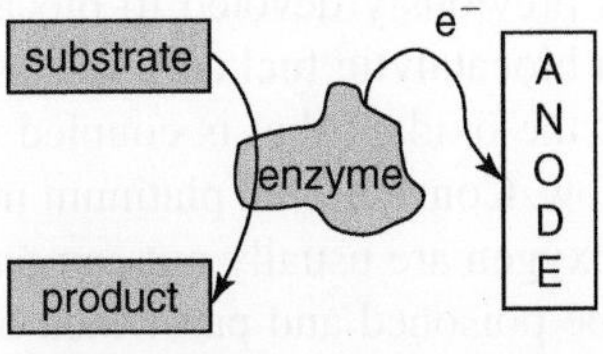

FIGURE 12.2 Schematic depicting a direct electron transfer process between an enzyme and the electrode, acting as an anode in this case.

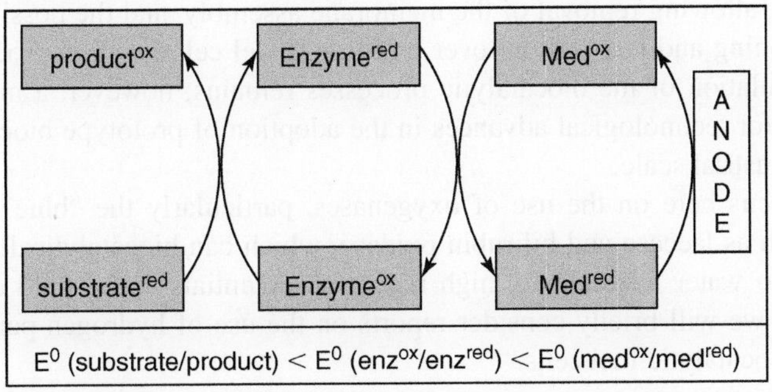

FIGURE 12.3 Schematic depicting the electron flow in an enzyme-catalyzed mediated electron transfer oxidation of substrate. The relative magnitudes of the standard reduction potentials of each element for efficient mediation are shown beneath the scheme.

and the mediator is considered to be the second substrate (co-substrate) of the catalytic process. The use of mediators can increase the rate of electron transfer, sometimes by several orders of magnitude.

In redox mediation, to have an effective electron exchange, the thermodynamic redox potentials of the enzyme and the mediator have to be accurately matched. For biocatalytic electrodes, efficient mediators must have redox potentials downhill from the redox potential of the enzyme: a 50 mV difference is proposed to be optimal [1, 18]. The tuning of these potentials is a compromise between the need to have a high cell voltage and a high catalytic current. Furthermore, an obvious requirement is that the mediator must be stable in the reduced and oxidized states. Finally, for operation of a membraneless miniaturized biocatalytic fuel cell, the mediators for both the anode and the cathode must be immobilized to prevent power dissipation by solution redox reactions between them.

12.4 BIOCATALYTIC CATHODES

12.4.1 Enzymes and substrates

In this section we review research on biocatalytic cathodes for oxidant reduction. The biocatalytic reduction of oxidants has only recently attracted renewed attention, with most research on biocatalytic fuels cells previously devoted to biocatalytic oxidation of fuels. However, in order to construct a biocatalytic fuel cell, it is essential to design a functional cathode for the reduction of the oxidant that is coupled to the anode and allows the electrically balanced current flow. Conventional platinum metal cathode catalysts used in fuel cells for reduction of oxygen are usually not compatible with oxidation of biocatalytic fuels since they can be poisoned and passivated by components in the electrolyte. In addition, in the absence of a membrane assembly in miniaturized membraneless fuel cells, oxidation of the fuel can occur at the cathode catalyst [1–3]. Biocatalytic processes at the cathode offer the advantages of selectivity for the oxidant over fuel, allowing removal of the membrane assembly, and the possibility of decreasing poisoning and passivation, over traditional fuel cell catalytic processes. Inhibition and modulation of the biocatalytic processes remains, however, a major, unresolved, problem for technological advances in the adoption of prototype biocatalytic cathodes on an industrial scale.

We focus here on the use of oxygenases, particularly the "blue" copper oxygenases, such as laccase and bilirubin oxidase, which can biocatalytically reduce oxygen directly to water at relatively high reduction potentials under mild conditions. First, however, we will briefly consider reports on the use of hydrogen peroxide as an oxidant in biocatalytic fuel cells.

12.4.2 Peroxidases

The use of hydrogen peroxide as an oxidant is not compatible with the operation of a biocatalytic fuel cell *in vivo*, because of low levels of peroxide available, and the toxicity associated with this reactive oxygen species. In addition peroxide reduction cannot be used in a membraneless system as it could well be oxidized at the anode. Nevertheless, some elegant approaches to biocatalytic fuel cell electrode configuration have been demonstrated using peroxidases as the biocatalyst and will be briefly reviewed here.

Peroxide is a strong oxidant, with a standard reduction potential of +1.78 V vs NHE, and is thus a good candidate for an oxidant [2b]. The direct reduction of peroxide at electrodes, however, has a high overpotential, thus necessitating the use of catalysts. A recent interesting development for the design of peroxide-reducing cathodes is the use of ferrocene-mediated peroxide reduction by the enzyme horseradish peroxidase at electrodes prepared by spray-painting of graphite, enzyme, mediator, and binder onto a polymeric support [19]. These electrodes demonstrated peroxide reduction occurring close to the ferrocene/ferricenium redox potential of ~ +0.25 V vs SCE (~ +0.50 V vs NHE). Willner and co-authors have developed an impressive protocol for the tethering and immobilization of microperoxidase-11 at gold electrodes, to yield electrodes that reduce peroxide by direct electron transfer to the microperoxidase [15, 20, 21]. Microperoxidases are produced via proteolytic digestion of cytochrome *c* (Cyt. *c*) to yield a heme-bound peptide of six, eight, nine or 11 amino acids. Microperoxidase-11 thus consists of 11 amino acids and a covalently linked heme site. Microperoxidase-11 was covalently linked via carbodiimide coupling chemistry of carboxylic acid functions of the microperoxidase, to the terminal amine group of a self-assembled monolayer of cystamine on a gold electrode, yielding electrodes with redox potentials for microperoxidase-11 heme of ~ −0.4 V vs SCE [20]. There is some doubt about whether a single monolayer of microperoxidase-11 forms using this coupling approach, or if multi-layers are formed, based upon examination of the rate of heterogeneous electron transfer to the microperoxidase-11 using this and other immobilization approaches [22]. The microperoxidase-11 modified electrodes were capable of biocatalytic reduction of peroxide at potentials as positive as +0.3 V vs SCE, with the reason for this potential shift postulated to be a result of the formation of an Fe(IV) intermediate species in the presence of peroxide [20, 23].

Recently, direct electron transfer to microperoxidases adsorbed on carbon nanotube-modified platinum electrodes has been observed [24]. The redox potential for this direct electron transfer is ~ −0.4 V vs SCE, the same as that for the microperoxidase-11 on the cystamine-modified gold. However, curiously, biocatalytic reduction of peroxide proceeds at this redox potential, −0.4 V vs SCE, at the carbon nanotube-modified electrodes, and not shifted positively, as was reported for the cystamine-modified gold [20].

Conversion of a peroxide-reducing cathode into a cathode that reduces dissolved oxygen is also possible, as recently demonstrated by Ramanavicius *et al.* [25]. In this study, microperoxidase-8 was co-immobilized with glucose oxidase to provide a cathode that couples glucose oxidation, producing peroxide from the oxygen co-substrate, with peroxide reduction by the microperoxidase, and subsequent direct electron transfer from the electrode to the microperoxidase. While this is an elegant demonstration of a novel combination of biocatalyst to provide high potential (~ +0.15 V vs SCE) reduction of oxygen, the fact that glucose is depleted, as it effectively acts as a co-substrate, would mitigate against adoption of this approach for an implantable biocatalytic fuel cell using glucose as a fuel.

12.4.3 Oxygenases

The theoretical thermodynamic reduction potential for oxygen is +1.23 V vs NHE at pH 0, or +0.82 V vs NHE at pH 7, and it is thus, like peroxide, a strong oxidant [1–3]. The reduction of oxygen at electrodes is, again like peroxide, hampered by large overpotentials, with direct electrochemical reduction occurring only at ~ −0.1 V vs NHE at gold and carbon electrodes at neutral pH. Catalysts, such as platinum, are therefore used to decrease this overpotential in fuel cell cathodes. As stated previously, however, the use of expensive, and non-selective, platinum catalysts is not compatible with operation of a putative miniaturized membraneless fuel cell *in vivo*. An additional disadvantage of oxygen reduction at platinum catalysts is that it occurs, at neutral pH, via a two-electron reduction, to produce peroxide, a toxic reactive oxygen species *in vivo*.

Catalytic reduction of oxygen directly to water, while not as yet possible with traditional catalyst technology at neutral pH, is achieved with some biocatalysts, particularly by enzymes with multi-copper active sites such as the laccases, ceruloplasmins, ascorbate oxidase and bilirubin oxidases. The first report on the use of a biocatalyst in the cathode of a fuel cell for reduction of oxygen was by Palmore and Kim [26], who investigated the reduction of oxygen to water by a solution-phase laccase from *Pyricularia oryzae* using 2,2′-azinobis(3-ethylbenzothiazoline-6-sulfonate) (ABTS) as a diffusional mediator.

The laccases, classed as polyphenol oxidases, catalyze the oxidation of diphenols, polyamines, as well as some inorganic ions, coupled to the four-electron reduction of oxygen to water: see Fig. 12.4 for the proposed catalytic cycle. Due to this broad specificity, and the recognition that this specificity can be extended by the use of redox mediators [27], laccases show promise in a range of applications [28], from biosensors [29–32], biobleaching [27, 33–35] or biodegradation [36], to biocatalytic fuel cells [1–3, 18, 26, 37–42].

Laccase was first isolated by Yoshida in 1883 [43] from tree lacquer of *Rhus vernicifera*. Laccases can thus be classified according to their source: plant, fungal or, more recently, bacterial or insect [44]. The laccase enzyme active site contains four copper ions classified into three types based upon their geometry and coordinating ligands, denoted

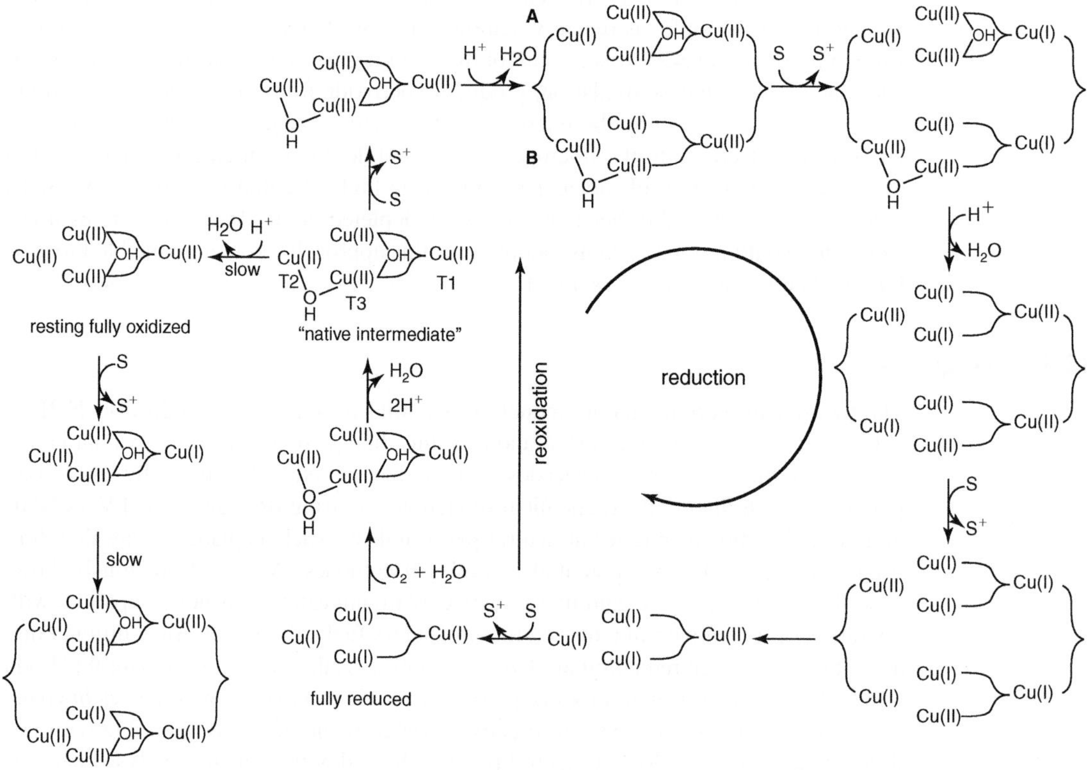

FIGURE 12.4 Proposed catalytic cycle for laccase, where S represents substrate. (From [44], with permission from the American Chemical Society.)

type 1 (T1), type 2 (T2), and type 3 (T3) [44]. The T2 and T3 coppers form a tri-nuclear cluster for reduction of oxygen, whereas successive one-electron oxidation of the substrates occur at the so-called "blue" T1 site, approximately 1.3 nm distal to the T2/T3 cluster. The key characteristic of the "blue" multi-copper oxygenases is the standard reduction potential of the T1 site. The catalytic efficiency of the laccase reaction with its substrate has been shown to depend on the thermodynamic driving force for electron transfer: the potential difference between substrate and T1 copper site [34, 44–48]. Laccases possessing T1 sites of relatively high reduction potentials can drive oxidation of otherwise recalcitrant organic or biopolymeric substrates, finding application in bioremediation and dye and pulp bleaching. In addition, the oxidation of substrates coupled to intra-molecular electron transfer to the T2/T3 cluster can result in the reduction of oxygen at relatively high potentials. The reduction potential of the T1 site can be determined by redox titrimetry [44–49].

The copper-containing redox enzymes have also been shown to transfer electrons directly with electrode materials, allowing determination of the reduction potentials of the active site using voltammetry, and possible correlation with structure and activity. Direct electron transfer to a laccase was first reported by Yaropolov's group at carbon electrodes [50]. Subsequent studies [8, 9, 51, 52] have investigated direct electron transfer to the copper active sites of the multi-copper oxidases as a means to classify the oxidases. The laccases can thus be classified, as suggested recently by Shleev *et al.* [9, 10], into three separate groups, based upon the reduction potential of the T1 copper site. The plant laccases have a low T1 potential of ~ +0.43 V vs NHE, while fungal laccases possess T1 sites of, either middle potential of +0.47 to +0.71 V vs NHE, or high potential of ~ +0.78 V vs NHE.

While direct electron transfer to laccases may help elucidate the mechanism of action of these enzymes it is unlikely that this process will supply sufficient power for a viable implantable biocatalytic fuel cell, because of difficulties associated with the correct orientation of the laccase and the two-dimensional nature of the biocatalytic layer on the surface. However, a recent attempt to immobilize laccase in a carbon dispersion, to provide electrodes with correctly oriented laccase for direct electron transfer, and a higher density of electrode material shows promise [53].

Mediated reduction of oxygen by laccase, particularly from fungal sources with high T1 potentials, as demonstrated by Palmore and Kim [26], does show great promise for the development of biocatalytic fuel cell cathodes. Immobilization of both mediator and laccase provides a biocatalytic cathode for oxygen reduction that may be used in a miniaturized membraneless biocatalytic fuel cell. Trudeau *et al.* [30] were the first to report on oxygen reduction by films of immobilized mediator and laccase, formed by cross-linking laccase to an osmium-based redox polymer film on carbon electrodes. The redox polymer structure, shown in Fig. 12.5, was prepared by substitution of one of the chloride ligands of an Os(2,2′-bipyridine)$_2$Cl$_2$ complex, with every tenth imidazole monomeric unit of the polyvinylimidazole polymer backbone.

Co-immobilization of this redox polymer with a fungal laccase from *Trametes versicolor,* possessing a T1 copper site reduction potential of ~ +0.57 V vs Ag/AgCl (~ +0.77 vs NHE), was achieved using a diepoxide cross-linker, in an approach

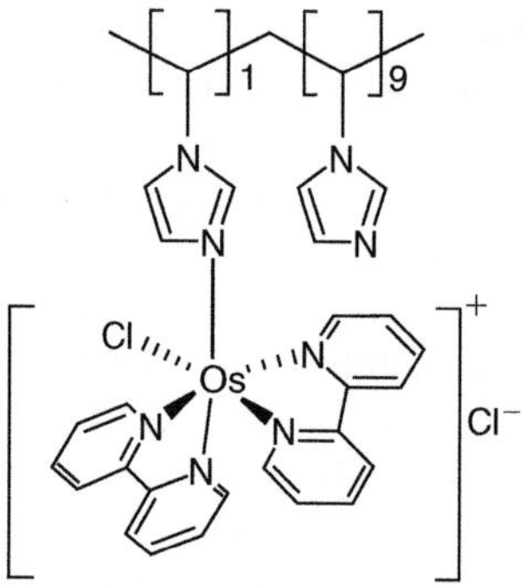

FIGURE 12.5 Structure of the osmium redox polymer, OsPVI, formed by coordination of an [Os(2,2′-bipyridine)$_2$Cl]$^+$ complex to polyvinylimidazole in a usually 1:9 ratio.

pioneered by the Heller group [54]. Although the laccase biocatalytic electrode was developed for the reagentless detection of modulators of enzyme activity, steady-state current densities of greater the 125 μAcm^{-2} were achieved at potentials of ~ +0.15 V vs Ag/AgCl in oxygen sparged acetate buffered solutions, pH 4.5, as shown in the cyclic voltammograms in Fig. 12.6 [55]. Subsequent to this report, this group and others have investigated mediated laccase catalyzed reduction of oxygen in films of redox polymers on electrode surfaces for application as biocatalytic cathodes. For example, substitution of the chloride ligand of a [Os(4,4′-dimethyl, 2,2′-bipyridine)(2,2′:6′,2″-terpyridine)Cl]$^+$ complex with imidazole units of PVI yields a redox polymer that may be co-immobilized with laccase from *Coriolus hirsutus* on carbon cloth fiber. The resulting biocatalytic oxygen cathodes operate at mAcm^{-2} current densities at a potential of +0.7 V vs NHE in pH 5 buffer, 37°C, when rotated at 4000 rpm [37].

Further refinement of this cathode is also possible, by judicious choice of biocatalyst. The high potential fungal laccases are reportedly sensitive to chloride inhibition and have optimal acidic pH maxima, for example, seemingly precluding their use in physiological solutions. Co-immobilization of the redox polymer described above with a laccase from *Pleurotus ostreatus*, which has been reported to retain a high level of substrate oxidation activity at pH 7, yielded biocatalytic oxygen cathodes capable of operating at +0.62 V vs NHE in pH 7, 0.1 M NaCl solution at 37°C, with 6% of their pH 5, chloride-free, current density [40].

The chloride and pH sensitivity of the laccase catalyzed oxygen reduction has led some groups to focus on another class of "blue" copper enzymes, that are active under physiological conditions of pH and NaCl, the bilirubin oxidases. Bilirubin oxidase catalyzes the oxidation of bilirubin to biliverdin coupled to the four-electron reduction of oxygen to water. The catalytic site of BOD, like laccase, consists of a tri-nuclear T2/T3 oxygen-reducing copper site and a T1 substrate oxidizing copper site [44, 45]. Unfortunately, the reduction potential of the T1 site of the bilirubin oxidases is of the medium potential classification of ~ +0.3 V vs Ag/AgCl [9, 10, 45]. The first report on a BOD-based biocatalytic oxygen cathode focused on homogeneous ABTS mediated reduction of oxygen at carbon felt electrodes using a BOD from *Myrothecium*

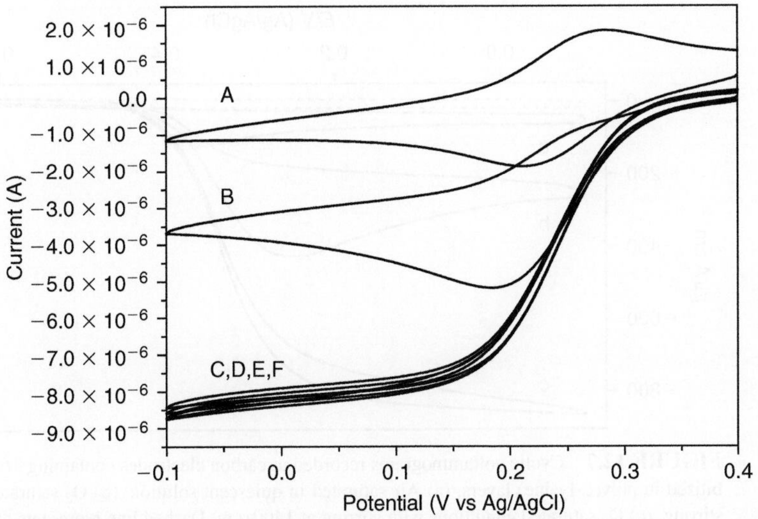

FIGURE 12.6 Cyclic voltammograms at a laccase biosensor in the absence (a) and presence (b) of O_2. Sparging of O_2 through the electrolyte between scans (2 min) yields reproducible and increased catalytic reduction currents (C–F) compared to ambient O_2 levels. Scan rate $10\,mVs^{-1}$ in 0.05 M acetate buffer of pH 4.5. (From [55], with permission from Wiley.)

verrucaria (*Mv*) in phosphate buffer, pH 7.0 [56]. Subsequent optimization of the BOD cathode focused on immobilization of both mediator and enzyme, and matching of mediator redox potential to that of the enzyme. The Ikeda group have achieved current densities of $17\,mAcm^{-2}$ at +0.25 V vs Ag/AgCl for mediated oxygen reduction by electrostatically entrapping *Mv*BOD with $[W(CN)_8]^{3-/4-}$ in poly(L-lysine) at carbon felt electrodes rotated at 4000 rpm [57]. The Heller group have co-immobilized *Mv*BOD and a redox polymer prepared by substitution of one of the chloride ligands of an Os(4,4′-dichloro-2,2′-bipyridine)$_2$Cl$_2$ complex with imidazole units of a copolymer of poly(vinylimidazole) and polyacrylamide on carbon cloth fibers to yield biocatalytic oxygen cathodes that provide current densities of $0.7\,mAcm^{-2}$ at a potential of +0.3 V vs Ag/AgCl in non-stirred phosphate buffered saline at 37ºC [58]. Further improvements in the BOD cathode were obtained by replacement of the *Mv*BOD by a BOD from *Trachyderma tsunodae*, which is claimed to have a T1 redox potential of +0.44 V vs Ag/AgCl. Current densities of $6.25\,mAcm^{-2}$ at potentials of +0.3 V vs Ag/AgCl were obtained at carbon cloth electrodes rotated at 4000 rpm in oxygenated phosphate buffered saline at 37ºC [59]. Purification of the BOD yielded biocatalytic films that provided higher current densities of $9.5\,mAcm^{-2}$ under these conditions [60]. More recently, direct electron transfer "type" reactions of the *Mv*BOD enzyme immobilized in poly(L-lysine) layers at carbon electrodes containing a high density of crystal edges have been reported [61]. Using this approach, from the cyclic voltammograms (CVs) in Fig. 12.7, steady-state current densities of $0.85\,mAcm^{-2}$ for oxygen reduction at potentials of ~ +0.2 V vs Ag/AgCl in oxygen saturated phosphate buffer at pH 7.0 with rotation at 1400 rpm can be estimated.

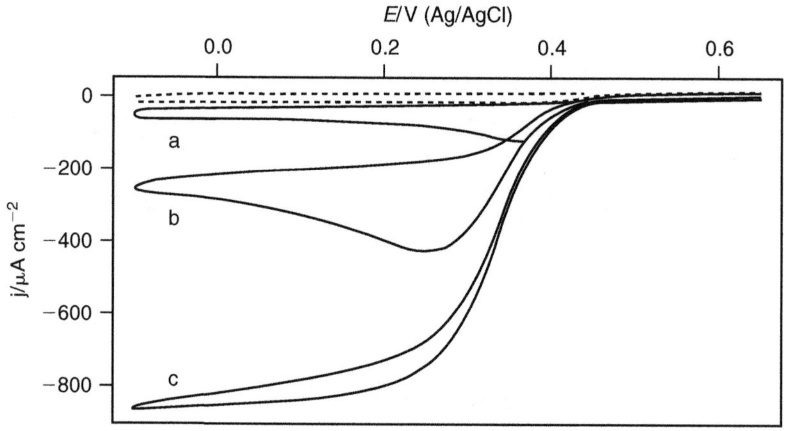

FIGURE 12.7 Cyclic voltammograms recorded at carbon electrodes containing *Mv*BOD enzyme immobilized in poly(L-lysine) layers. (a) Air saturated in quiescent solution. (b) O_2 saturated conditions without stirring. (c) O_2 saturated conditions with stirring at 1400 rpm. Dashed line represents the CV recorded under Argon. (From [61], with permission from Elsevier.)

12.5 BIOCATALYTIC ANODES

12.5.1 Enzymes and substrates

In this section, the enzymes, and associated substrates, used as biocatalysts in anodes are presented. For the development of biocatalytic anodes, there is a wide range of fuels available for use as substrates, such as alcohols, lactate, hydrogen, fructose, sucrose, all of which can be oxidized by biocatalysts. The fuel that is the most widely considered, however, in the context of an implantable biocatalytic fuel cell is glucose. We shall focus our attention on this fuel, but will mention briefly research on the use of some other fuels in biocatalytic anodes.

12.5.2 Glucose oxidase

Glucose oxidase (GOx) is an FAD (Flavin Adenine Dinucleotide, see Fig. 12.8) enzyme which very specifically oxidises β-D-glucose, reducing dioxygen to hydrogen peroxide in the process. Glucose is an attractive substrate, for future implantable devices, because of its relatively high concentration in blood, of around 9 mM, and also because GOx is an extremely stable enzyme with high substrate specificity.

As referred to previously, if the active site of a biocatalyst is close enough to the electrode surface, direct electron transfer to/from an electrode can result. It has been shown in recent years that direct electron transfer from the GOx active site is possible using appropriate electrode preparation procedures. These preparation procedures usually aim to provide nano-structured features on the electrode surface that can penetrate sufficiently the GOx active site to allow for direct electron transfer. The direct electron

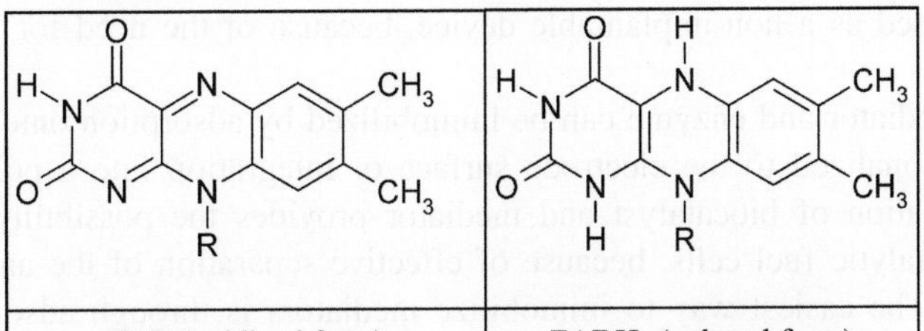

FIGURE 12.8 Structure of the $FAD/FADH_2$ active site of glucose oxidase.

transfer occurs between the $FAD/FADH_2$ active site of the GOx and the electrode surface, at redox potentials close to the −0.41 V vs SCE reduction potential of FAD [62] at neutral pH. Examples of direct electron transfer to GOx have been reported for GOx adsorbed or attached to carbon paste [63], carbon nanotube [64–66] and colloidal gold nanoparticle [67, 68] modified electrodes. However, in these cases the electrode replaces the glucose substrate, thus resulting in the biocatalytic reduction of oxygen to hydrogen peroxide, in the presence of oxygen.

$$\text{GOD(FAD)} + 2e^- + 2H^+ \rightleftharpoons \text{GOD(FADH}_2\text{)} \quad (1)$$

$$\text{GOD(FADH}_2\text{)} + O_2 \rightarrow \text{GOD(FAD)} + H_2O_2 \quad (2)$$

In the presence of glucose and oxygen the bioelectrocatalytic reduction of oxygen is diminished due to competition for the FAD active site between glucose and the electrode surface. A method for the determination of glucose levels, based on this decrease in reduction current in the presence of glucose, has been proposed [67]. In the presence of anaerobic glucose solutions, only the redox process for $FAD/FADH_2$ is observed, with no biocatalytic oxidation of glucose. In conclusion, the direct electron transfer to GOx at nanostructured electrode surfaces yields biocatalytic oxygen reduction currents. GOx direct electron transfer processes cannot therefore be usefully used as yet to develop an anode for biocatalytic fuel cells.

Since the first report on the ferrocene mediated oxidation of glucose by GOx [69], extensive solution-phase studies have been undertaken in an attempt to elucidate the factors controlling the mediator–enzyme interaction. Although the use of solution-phase mediators is not compatible with a membraneless biocatalytic fuel cell, such studies can help elucidate the relationship between enzyme structure, mediator size, structure and mobility, and mediation thermodynamics and kinetics. For example, comprehensive studies on ferrocene and its derivatives [70] and polypyridyl complexes of ruthenium and osmium [71, 72] as mediators of GOx have been undertaken. Ferrocenes have come to the fore as mediators to GOx, surpassing many others, because of factors such as their mediation efficiency, stability in the reduced form, pH independent redox potentials, ease of synthesis, and substitutional versatility. Ferrocenes are also of sufficiently small size to diffuse easily to the active site of GOx. However, solution phase mediation can only be used if the future biocatalytic fuel cell is to be used as a non-implantable device, because of the need for a membrane and casing.

The mediator and enzyme can be immobilized by adsorption onto electrodes, linkage functionalized to the electrode surface or integration into a polymer layer. The immobilization of biocatalyst and mediator provides the possibility for membraneless biocatalytic fuel cells, because of effective separation of the anode and cathode reagents. The easiest way to immobilize mediators is through adsorption onto electrode surfaces. Cass *et al.* [69] deposited solutions of ferrocene derivatives onto pyrolytic graphite electrodes and allowed them to air-dry. In this approach, the solubility of ferrocenes in aqueous solution must be low to aid entrapment within the electrode. Covalent attachment of GOx (through carbodiimide activation) to the electrode and covering with polycarbonate membranes provided biocatalytic surfaces for mediated oxidation of glucose. All the ferrocene derivatives investigated act as rapid oxidants for the enzyme glucose oxidase [69, 70].

Increased micro- and nanoscopic surface areas can lead to improved signal output for mediated biocatalytic electrodes. For example, Liu *et al.* [73] used porous carbon as a matrix to load glucose oxidase to provide an anode in a biocatalytic fuel cell. Not surprisingly, given the increased loading of enzyme, this biocatalytic anode displayed higher oxidation currents than that observed for GOx on a glassy carbon electrode in the presence of glucose and solution-phase ferrocene monocarboxylic acid as a mediator. The formal potential of the mediator of +0.34 V vs Ag/AgCl is, however, remote from the FAD/$FADH_2$ reduction potential and further refinement of this system is focused on selecting more appropriate mediators and the co-immobilization of the mediator with enzyme.

Co-immobilization of mediator and enzyme may be achieved using a novel reconstitution approach. For example, Katz *et al.* [15] have developed a biocatalytic anode functionalized by a surface reconstitution of apo-GOx onto FAD that was previously coupled to a pyrrolo-quinoline quinine (PQQ) relay conjugated to a self-assembled monolayer of cysteamine on gold. The CV study of this assembly in the presence of glucose yields an electrocatalytic current for glucose oxidation commencing at −0.12 V vs SCE at pH 7. The chronoamperometric output from this assembly can achieve current densities of 300 $mAcm^{-2}$ at +0.2 V vs SCE in 80 mM of glucose [2b]. Further modification, to what the authors term the electrical contacting of the enzyme, involved the use of alternate conjugation strategies [74], and of nanostructured surfaces [75, 76]. Xiao *et al.* [76] have reconstituted apo-GOX onto FAD-functionalized gold nanoparticles that are then tethered via a dithiol spacer to a bulk gold electrode. This gold nanoparticle assembly is an efficient relay of electrons between the FAD active site and the bulk electrode. Unfortunately, the biocatalytic oxidation of glucose using this assembly only commences at potentials greater than +0.4 V vs SCE, with the overpotential proposed to result from a tunnelling barrier introduced by the dithiol spacer.

An alternative strategy for co-immobilization of mediator and GOx is based on adsorption of enzyme, cross-linked, as was described for the laccase-based biocatalytic cathodes [30, 37–42], to an osmium-based redox polymer film, on carbon electrodes [1–3, 54].

In the case of these redox–polymer mediators a key parameter to take into account is usually the polymer–enzyme ratio [30, 77, 78]. An excessive enzyme weight fraction can decrease the current density because the enzyme is an electronic insulator. When the weight fraction of redox polymer is excessive, the flux of electrons is reduced because of the smaller number of enzyme molecules. The initial studies on these redox polymer GOx films for mediated oxidation of glucose focused on co-immobilization with the OsPVI polymer shown in Fig. 12.5 on carbon electrodes [42, 54]. Refinement of the redox potential of the polymer was achieved by replacing the 2,2′-bipyridine ligands of osmium with, initially, 4,4-dimethyl-2,′-bipyridine, resulting in biocatalytic oxidation of glucose at ~ +0.1 V vs Ag/AgCl compared to +0.25 V vs Ag/AgCl [79]. Replacement of this ligand with 4,4′-dimethoxy-2,2′-bipyridine, subsequent to a report by Zakeerudin *et al.* [71] on increased electron transfer rates from GOx to osmium complexed to such ligands, yielded a redox polymer with a redox potential of −0.07 V vs Ag/AgCl [54]. Finally, further refinement of the redox polymer is possible by replacement of the 2,2′-bipyridine ligands of osmium with 4,4′-diamino-2,2′-bipyridine, to yield a redox potential of ~ −0.15 V vs Ag/AgCl [77]. A current density of ~65 μA/cm^2 was observed for oxidation of 10 mM of glucose in room temperature pH 7.4 phosphate buffered saline solution at films of GOx co-immobilized with the latter redox polymer on graphite electrodes. This output may be improved upon by the use of electrode surfaces of higher micro- and nanoscopic areas. For example, a current density of ~150 μA/cm^2 was observed for similar films immobilized on a 7 μm diameter carbon fiber electrode, in solutions containing 15 mM glucose, pH 7.4 in phosphate buffered saline at 37ºC [77, 80].

A significant limiting factor to current density output in these redox polymer films is the ease of physical displacement of the redox center in the polymer film to allow intimate contact, and electron transfer, between the redox complex and the biocatalyst [81]. To address the mobility of the redox complex sites in the redox polymer, represented by an apparent diffusion coefficient *D*, Mao *et al.* [82] have grafted a 13-atom-long flexible spacer between a poly(vinylpyridine) polymer backbone and an alkyl functional group of an [Os(N,N′-dialkylated-2,2′-biimidazole)$_3$]$^{2+/3+}$ redox center. This strategy allows both improved electron transfer between GOx-FADH$_2$ centers and the redox polymer, and charge transport within the redox polymer. Carbon fibers coated with cross-linked films of GOx and this polymer have a redox potential of −0.19 V vs Ag/AgCl, and provide a current density of 1.15 mAcm^{-2} in solutions containing 15 mM glucose, pH 7.4 in phosphate buffered saline at 37ºC. A further improvement in current density, to yield 1.5 mAcm^{-2} at a potential of −0.1 V vs Ag/AgCl, is obtained when the ratio of osmium complex to polymer monomeric unit is optimized [83].

It should be noted that biocatalytic fuel cell anodes based on the GOx enzyme face a significant problem: O_2 is the natural electron acceptor of GOx. GOx therefore catalyzes the oxidation of glucose to gluconolactone, producing also hydrogen peroxide when dioxygen is the electron acceptor. The mediators described previously attempt to compete with the oxygen reduction in order to avoid oxygen depletion, because it is required for the cathode, and production of peroxide, a highly toxic product. It may, therefore, be interesting to focus on enzymes which oxidize fuels yet don't produce hydrogen peroxide, such as dehydrogenases.

12.5.3 Dehydrogenases

Dehydrogenases, which represent a majority of redox enzymes, are mostly NAD (Nicotinamide Adenine Dinucleotide, see Fig. 12.9) dependent. This cofactor is not directly bound to the enzyme but its presence in the medium is necessary because it acts as a carrier of two electrons and one proton, and it activates the biocatalytic function of the enzyme.

The thermodynamic redox potential of NAD^+/NADH is −0.56 V vs SCE at neutral pH. The NADH cofactor itself is not a useful redox mediator because of the high overpotential and lack of electrochemical reversibility for the NADH/NAD^+ redox process, and the interfering adsorption of the cofactor at electrode surfaces.

As dehydrogenases (DH) are widely distributed enzymes, a number of studies have been carried out with these biocatalysts. For example, Willner *et al.* [20] have used a PQQ-monolayer functionalized gold electrode for the catalytic oxidation of NADH in the presence of Ca^{2+}. In this scheme, the pyrrolo-quinoline quinine co-factor, PQQ, was covalently linked, as before for the GOx system [15, 20, 21], to the Au electrode, and was capable of oxidizing NADH at ∼ −0.15 V vs SCE at pH 8.0. A theoretical current density of 185 μA/cm^2 is estimated for this monolayer-modified system. This method is attractive as it may be applied to oxidative reactions of the dehydrogenases for which NAD is the cofactor.

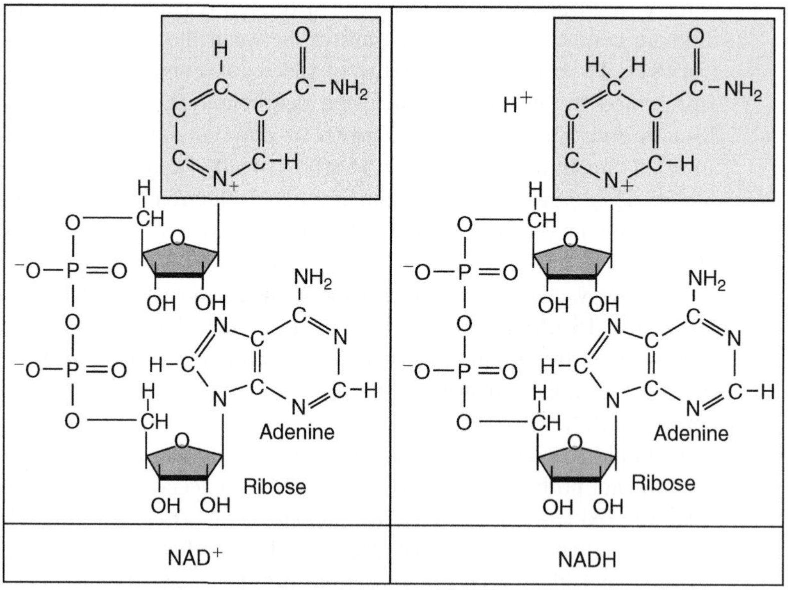

FIGURE 12.9 Structure of the NAD^+/NADH cofactor.

Glucose dehydrogenase (GDH) is a class of enzyme that is gaining increased attention as a catalyst for the oxidation of glucose in biosensing and biocatalytic fuel cell applications. GDH is either a NAD^+- or PQQ-dependent enzyme. The pyrrolo-quinoline quinine (PQQ) cofactor (prosthetic group) has a thermodynamic redox potential of −0.125 V vs SCE at pH 7 [20], which is less negative than that for the NAD^+/NADH couple. In addition, the application of the PQQ-dependent GDH to biocatalytic fuel cells may be limited because of its relative instability with respect to GOx [84]. Nevertheless, the application of electrodes modified with the PQQ-dependent GDH as glucose sensors and biocatalytic anodes has been explored owing to its oxygen insensitivity, the fact that the cofactor is bound to the enzyme, unlike the NAD^+-dependent GDH, and the high catalytic efficiency of this PQQ-dependent GDH [84–87].

For example, Ye *et al.* [85] have used carbon electrodes coated with cross-linked films of soluble PQQ-dependent GDH, from *Acinetobacter calcoaceticus*, and an osmium redox polymer to achieve glucose oxidation current densities of 1.8 $mA cm^{-2}$ at +0.4 V vs SCE in solutions containing glucose concentrations above 40 mM. Tsujimura *et al.* [87] have recently utilized this approach to devise a biocatalytic fuel cell anode for glucose. Zayats *et al.* [74], in an approach similar to that devised for the apo-GOx system [15, 20, 21], have reconstituted PQQ-dependent apo-GDH on PQQ-functionalized gold nanoparticles assembled onto a gold surface. Unfortunately, while the onset of bioelectrocatalytic oxidation of glucose is close to the redox potential of the PQQ, −0.125 V vs SCE at pH 7, appreciable glucose oxidation currents are only observed at an overpotential of approximately 0.35 V vs the PQQ redox potential, at +0.2 V, precluding its use as a biocatalytic anode for glucose oxidation. This overpotential is similar to that observed for reconstituted apo-GOx on gold nanoparticles (see above) and is again attributed to a tunneling barrier introduced by the dithiol spacer.

Yuhashi *et al.* [84] have described a biocatalytic fuel cell anode for glucose based on PQQ-dependent GDH carbon paste electrodes. The enzyme mixed with carbon paste is lyophilized and then packed into the end of a carbon electrode and treated with glutaraldehyde for immobilization. The research was extended to investigate improving the stability of the GDH using a Ser415Cys mutant GDH, previously developed for biosensor applications [88]. For the wild type GDH, after 24 hours of operation only 40% of the initial catalytic response remained. Meanwhile, the Ser415Cys mutant GDH showed improved stability, with 80% of the initial response remaining after the same period [84].

The use of the NAD^+-dependent GDH as a glucose biocatalytic anode has been investigated recently by Sato *et al.* [89]. This research focused on evaluating the use of glassy carbon electrodes coated with diaphorase/GDH and a polymeric vitamin K_3-based mediator with a redox potential of −0.25 V vs Ag/AgCl at pH 7, as a glucose oxidizing anode. Vitamin K_3 had previously been identified as a promising mediator of the diaphorase oxidation of NADH to NAD^+ [90]. The anode operation is thus based on vitamin K_3 mediation of diaphorase oxidation of NADH to NAD^+ and GDH oxidation of glucose coupled to NAD^+ reduction.

Palmore *et al.* [91] have reported on a graphite plate biocatalytic anode that uses solution-phase dehydrogenases to catalyse the successive oxidation of methanol to CO_2.

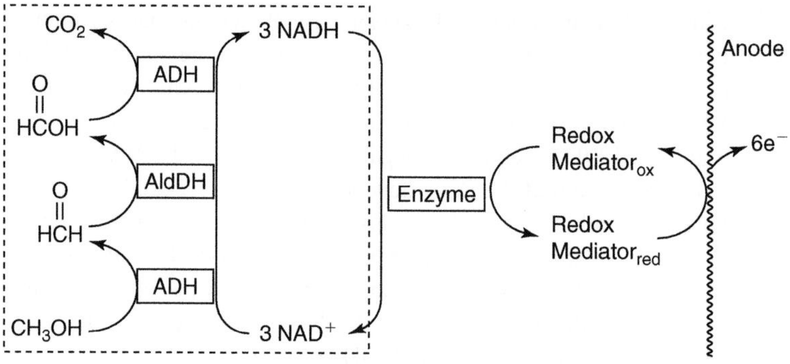

FIGURE 12.10 Oxidation of methanol to CO_2, catalyzed by NAD^+-dependent alcohol (ADH), aldehyde (AldDH), and formate (FDH) dehydrogenase, with regeneration of NAD^+ via redox mediation to diaphorase. (From [91], with permission from Elsevier.)

While this anode is not useful in the context of implantable fuel cells, it is of interest because methanol is an attractive anodic fuel due to its availability and ease of transport and storage. The oxidation of one equivalent of methanol requires the reduction of three equivalents of NAD^+ to NADH. As the NADH cofactor itself is not a useful redox mediator, a benzylviologen/diaphorase redox cycle, with a redox potential of $-0.55\,V$ vs SCE at pH 7, was used to regenerate NAD^+ for use by the dehydrogenases, as depicted in Fig. 12.10.

Simon *et al.* [92] investigated a biocatalytic anode based on lactate oxidation by lactate dehydrogenase (LDH). The anodic current is generated by the oxidation of NADH (produced by NAD^+ and substrate) while LDH catalyzes the electro-oxidation of lactate into pyruvate. As previously mentioned, the oxidation of NADH at bare electrodes requires a large overpotential, so these authors used poly(aniline) films doped with polyanions to catalyze NADH oxidation. Subsequent research by this group focused on targeting mutants of LDH that are amenable to immobilization on the polyaniline surface [93].

12.6 BIOCATALYTIC FUEL CELLS

12.6.1 Physiological conditions

In this section, recent progress on assembled biocatalytic fuel cells is reviewed. One long-term goal of biocatalytic fuel cell research is the development of a low powering device implanted in the human body and extracting electrical energy through the oxidation of a biocatalytic fuel (e.g. glucose) and reduction of a bio-available oxidant (e.g. O_2). Therefore it is important to keep in mind the basic physiological conditions such as, among others, concentration of substrates or inhibitors, fluid pH, temperature and velocity, which will eventually become the immediate working environment of the implanted biocatalytic fuel cell. Although the list focuses on blood parameters, it is not exhaustive and other physiological tissues may be considered as well for an eventual implantation.

The dioxygen partial pressure in blood vessels is about 95 mmHg in arteries and 40 mmHg in veins, which corresponds to a concentration of respectively ~2.14×10^{-4} M and 5.4×10^{-5} M in free dioxygen. The concentration of glucose in blood vessels is ca. 9 mM in healthy adults. Chloride is the anion present in the highest concentration in physiological media with chloride concentration in blood about 136–145 mM. It is an important plasmolyte to take into account as it can be an inhibitor of some enzymes, for instance via binding to copper sites in some oxidases. Other anions present in appreciable concentration are bicarbonate (~10 mM), phosphate (~0.6 mM), and sulfate (~0.2 mM). Sodium is the main cation in blood (132 to 144 mM) followed by potassium 4 mM, calcium 2 mM, and magnesium 1 mM. Other elements such as aluminum, copper, fluorine, iron, and zinc are present in low concentration (~5 to 30 μM), while manganese, lead, tin, bromine, and iodine are present in very low concentration (~0.1 to 5 μM). Blood is a buffered solution with a pH of 7.4. Other body fluids are also close to neutral pH, like, for example, saliva at the pH of 7.2. A notable exception is acidic gastric juices (pH 2). The human body is also thermostated at 37ºC, a peculiarity common to all mammals where constant body temperature lies in the range 34 to 40ºC depending on the species. The velocity of blood in blood vessels is of the order of 1 to 10 cm/s. Other body fluids may be considered as well, for instance the pH of tears is ca. 7.5 and contains, not surprisingly, low levels of glucose (50–500 μM).

12.6.2 Assembled glucose–oxygen biocatalytic fuel cells

Many assembled glucose–O_2 biocatalytic fuel cells have been reported in recent years. This section aims at giving an overview of the methods that have been used to build these biocatalytic fuel cells and of the performances obtained by these devices depending on their different designs.

Katz *et al.* [17] reported on a biocatalytic fuel cell based on the surface reconstitution of apo-GOx onto FAD that was previously coupled to a pyrrolo-quinoline quinine (PQQ) relay conjugated to a self-assembled monolayer of cysteamine on gold. The cathode consisted of a cross-linked affinity complex between cytochrome *c* (Cyt. *c*) and cytochrome *c* oxidase (COx) assembled on gold through a thiol modified Cyt. *c* promoter. The promoter is a maleimide monolayer that is specifically reacted with the cysteine102 residue of Cyt. *c* from *Saccharomyces cerevisae* ensuring proper alignment between the electrode surface and the heme site of the protein, and consequently electrical communication between the electrode and the protein. Electron transfer between the biocatalyst and the electrode was particularly efficient at the anode and thought not to be perturbed by molecular oxygen, the natural electron donor, or by typical interfering species like ascorbic acid or uric acid. Comparatively, interfacial electron transfer was much less efficient at the cathode. The reduction of O_2 occurred at 0 V vs SCE at the cathode while oxidation of glucose at the anode occurred at the potential of PQQ (−0.12 V vs SCE at pH 7). The maximum reported power for this cell was 4 μW at a load of 0.9 kΩ and an electromotive force of 50 mV (power density 5 μWcm^{-2}).

The cell was studied for 48 hours in a flow system under constant glucose concentration (1 mM) under air and showed no sign of current or cell potential decrease.

A biocatalytic fuel cell reported by Tsujimura *et al.* [87] is based on polished glassy carbon electrodes modified by *Mv*BOD for the anode and PQQ-GDH from *Acenitobacter calcoaceticus* for the cathode. Redox mediation is effected by $[Os(2,2'\text{-bipyridine})_2Cl]^+$ conjugated to poly(4-vinylpyridine) quaternized with bromoethylamine (redox potential of 0.35 V vs Ag/AgCl) for the cathode, while the anode contains $[Os(4,4'\text{-dimethyl-}2,2'\text{-bipyridine})_2Cl]^+$ conjugated to poly(vinylimidazole) (redox potential of +0.15 V vs Ag/AgCl). The cell, assembled and studied in MOPS buffer, pH 7, containing 3 mM $CaCl_2$, yielded an open circuit voltage of 0.44 V (25°C), with maximum current density of $430\,\mu A cm^{-2}$ and maximum power density of $58\,\mu W cm^{-2}$ at a cell potential of 0.19 V. The cathode was identified as limiting the cell performance, as a three times lower current density than the anode was obtained under the same conditions. The advantage of using the GDH enzyme at the anode rather than GOx, in addition to insensitivity to O_2, is that GDH is more tolerant to redox mediators of redox potential close to that of the enzyme. Indeed, the authors showed that despite a higher E^0 (−0.18 V for GDH and −0.35 V for GOx at pH 7) the bio-anode modified with GDH and the 0.15 V redox polymer yielded more than twice the current density than that modified with GOx and the same redox polymer. Finally GDH is less substrate selective than GOx. This is a drawback for glucose determination but may be an advantage for some fuel cell applications. The fuel cell stability was not reported but the need to stabilize the GDH enzyme was stressed.

Recently, Yuhashi *et al.* [84] attempted to address the GDH instability by using a mutated GDH (Ser415Cys). Their glucose–O_2 glucose dehydrogenase/bilirubin oxidase biocatalytic fuel cell was studied in a two-compartment cell linked by a salt bridge and containing ABTS and phenazine methosulfonate (PMS) redox mediators in the cathode and anode compartment, respectively. The electrodes were made by packing a lyophilized mixture of the enzyme and carbon paste at the surface of a carbon electrode, followed by treatment by a 1% glutaradehyde solution. The anode was allowed to undergo holoformation in a solution containing $CaCl_2$ (1 mM) and PQQ (5 μM) before use. The open circuit voltage of the cell was 0.577 V (25°C, pH 7). The maximum current density was $61.4\,\mu A cm^{-2}$ under stirring (250 rpm) and the maximum power density was $17.6\,\mu W cm^{-2}$ at a cell potential of 0.5 V with an external load of 200 kΩ. The half-life of the cell was under a week, while that of the wild type GDH was only 2 days. The eventual inactivation of the cell is ascribed to mediator instability. Sato *et al.* [89] used an NAD^+-dependent GDH anode that was coupled to a platinum cathode, not modified by any bio-catalyst, in their fuel cell. Diaphorase from *Bacillus stearothermophilus*, was co-immobilized with polyallylamine functionalized with a derivative of vitamin K_3 as the redox mediator to yield an NADH oxidation layer. A second layer of an electrostatic adducts between GDH and polylysine effected glucose oxidation. Crucial to the efficient electrical conduction between the vitamin K_3 mediator and the electrode was the addition of carbon black in the diaphorase layer. Indeed the current density was two orders of magnitude higher in the presence of the conducting material. The "double-layer" design of the anode implies that the optimized diaphorase layer may be used with other NADH dependent enzymes such as alcohol dehydrogenase. The platinum cathode was coated with polydimethylsiloxane to yield an O_2-selective electrode. The assembled biocatalytic fuel cell was studied at 37°C in air saturated phosphate buffer solution at pH 7 containing glucose (10 mM) and NADH (0.5 mM). The maximum power density was 14.5 μWcm^{-2} (40.3 μAcm^{-2} current density) at a potential of 0.36 V with an external load of 200 kΩ. The cell power dropped steadily for five days to 30% of its initial value and was then stable at about 4 μWcm^{-2} for more than 2 weeks. The use of a non-biocatalytic cathode is interesting in terms of electrode stability, while obvious bio-compatibility of vitamin K_3 is attractive for the prospect of an implantable device. However, the reduction product of oxygen at pH 7 is inevitably hydrogen peroxide.

Despite the known loss of activity of fungal laccase enzymes at neutral pH, and its reported inhibition by chloride, we were curious to investigate what cell performance could be reached with a *Trametes versicolor* laccase in a glucose–O_2 biocatalytic fuel cell working in pseudo-physiological conditions [18]. Our pseudo-physiological media was a pH 7.4 phosphate buffer containing 0.1 M sodium chloride and 10 mM glucose. The solution was thermostated under air at body temperature (37°C). Graphite electrodes were modified by a mixture of enzyme, redox polymer, and diepoxide cross-linker. The GOx-based anode redox polymer was [Os(4,4′-diamino-2,2′bipyridine)$_2$Cl]$^+$ conjugated to polyvinylimidazole, of redox potential −0.11 V vs Ag/AgCl. An [Os(1,10-phenanthroline)$_2$]$^{2+}$ complex conjugated to two units of imidazole in a polyvinylimidazole polymer, of redox potential 0.49 V vs Ag/AgCl, was used at the cathode. At pH 7.4 the maximum power density for the cell was 16 μWcm^{-2} at a cell voltage of 0.25 V. The fact that the cathode was limiting the device in this case is exemplified by the higher maximum power density of 40 μWcm^{-2} at pH 5.5, where laccase is much more active. Liu *et al.* [73] have exploited the pH dependence of laccase to build a biocatalytic fuel cell with a tunable power ouput. To increase power density porous carbon was chosen as the electrode material and GOx and laccase enzymes were entrapped in suspensions of carbon nanotube/chitosan and then cast on the porous carbon matrix. Solution redox mediators ferrocene monocarboxylic acid and ABTS were dissolved, respectively, in the anolyte and catholyte which were separated by a Nafion membrane. The maximum power density dropped from 100 μWcm^{-2} at pH 4 to 2 $\mu W/cm^{-2}$ at pH 7.

The biocatalytic fuel cell prototypes presented have not as yet been refined in a rational manner. The situation is different for the cells designed by the Heller group, where an extensive literature is available. The publication of the first oxygen–glucose biocatalytic fuel cell from this group appeared in 2001 [38], after several years of refining each individual electrode [3]. One interesting aspect of their study is the miniaturization of the cell by the use of small carbon fibers (~3 cm long, 7 μm diameter and 0.44 mm^2 active area) as the electrode material. The fibers were first rendered hydrophilic by plasma oxidation, and modified by a mixture of enzyme, an osmium-based redox polymer and a diepoxide cross-linker. The GOx anode was modified with [Os(4,4′-dimethyl-2,2′-bipyridine)$_2$Cl]$^+$ complex conjugated to a polyvinylimidazole *co*-acrylamide polymer, redox potential of +0.10 V vs Ag/AgCl, while the laccase

(*Coriolus hirsitus*) cathode had [Os(4,4′-dimethyl-2,2′-bipyridine)$_2$(2,2′,6′,2″-terpyridine)]$^{2+}$ conjugated to a similar polymer backbone, redox potential of +0.55 V vs Ag/AgCl, as mediators. The cell was studied under conditions appropriate for high activity of the laccase enzyme: aerated pH 5 citrate buffer and chloride-free medium containing 15 mM glucose. Maximum power density reached 137 μWcm^{-2} at 37°C. The power dropped by less than 10% after a day of continuous operation and by 25% after 3 days. The cell power output was then refined [41] by the use of an anodic redox polymer with a more appropriate redox potential of −0.29 V and a flexible tether to improve charge transport diffusion. In this case an [Os(N,N′-alkylated-2,2′-biimidazole)$_3$]$^{2+}$ redox active moiety was tethered to the polymer backbone via one of its alkylated biimidazole ligands through a 13 atom long flexible spacer. At 37°C the maximum power density for the cell was 268 μWcm^{-2} at 0.78 V. Under continuous operation for a week at 0.78 V and 37°C the cell power dropped by 10% a day. The same refinement procedure was followed for the cathode [94], where an [Os(4,4′-dimethyl-2,2′-bipyridine)$_2$(4-aminomethyl-4′-methyl-2,2′-bipyridine)]$^{2+}$ complex was reacted with partially quaternized polyvinylpyridine to yield an 8 atom long flexible tether. The power density of this miniature biocatalytic fuel cell peaked (350 μWcm^{-2}) at 0.88 V, the highest cell voltage to date for a working miniature glucose–O_2 biocatalytic fuel cell, and just 300 mV below the thermodynamic potential of the glucose–O_2 cell reaction.

Along the principles just delineated above and almost concomitantly, the same group developed their own GOx-BOD fuel cell, for reasons already discussed. The refinement of the biocatalytic fuel cell was conducted in pseudo-physiological conditions of pH 7.4, 20 mM phosphate buffer containing 0.15 M NaCl and 15 mM glucose, air saturated solution with oxygen concentration ~0.2 mM at 37°C. In all cases the cathode redox polymer consisted of a polyvinylimidazole–polyacrylamide copolymer functionalized with [Os(4,4′-dichloro-2,2′-bipyridine)$_2$Cl]$^+$, redox potential of +0.35 V vs Ag/AgCl. The anode redox polymer was initially [Os(4,4′-diamino-2,2′-bipyridine)$_2$Cl]$^+$, redox potential of −0.16 V vs Ag/AgCl, complexed to a partially quaternized polyvinylimidazole [81]. The assembled biocatalytic fuel cell power density was 50 μWcm^{-2} at 0.5 V. Two days of continuous operation resulted in 40% loss of the initial power density. Using a redox polymer containing [Os(4,4′-dimethoxy-2,2′-bipyridine)$_2$Cl], with a slightly more anodic potential of −0.07 V vs Ag/AgCl, for the anode [95], the maximum power density increased significantly to 244 μWcm^{-2} at the lower voltage of 0.36 V. This illustrates the limiting effect of the anode on the biocatalytic fuel cell performance under these conditions. One day of continuous operation of this cell resulted in less than 10% loss at 0.36 V, and 45% loss after 4 days. Further refinement of the anode involved, as was the case for the GOx-laccase fuel cell, introduction of a long flexible spacer between the polymer and [Os(N,N′-alkylated-2,2′-biimidazole)$_3$]$^{2+/3+}$ redox active moiety to improve diffusional charge transport between the enzyme active site and the electrode [96]. Despite the lowering of the redox potential to −0.19 V vs Ag/AgCl, the cell produced a power density of 430 μWcm^{-2} at the relatively high voltage of 0.52 V. One week of continuous operation at 0.52 V resulted in an ~6% loss in power density per day. The operation of this cell in a living plant has been demonstrated [97]. A grape sap was chosen for its high glucose content (>30 mM). The biocatalytic fuel cell performance depended on O_2 transport. In the O_2-deficient grape center the power density was 47 μWcm^{-2} whereas near the better oxygenated grape skin it reached 240 μWcm^{-2}, with an operating voltage of 0.52 V for both cases. Further improvement in output for this biofuel cell was achieved by increasing the redox site density of the anode redox polymer, resulting in a glucose flux-limited current density increase of 20% and an overpotential decrease of 50 mV [98]. This optimized biocatalytic fuel cell operated at +0.60 V with a 480 μWcm^{-2} power density in pH 7.2 phosphate buffer containing 0.1 M NaCl, 15 mM glucose at 37.5ºC, losing about 8% of its power each day of operation.

12.7 CONCLUSIONS

It seems that using and developing the different strategies discussed above, biocatalytic fuel cells with power densities and operating voltage high enough to power low energy microdevices are now at hand. Indeed, a tremendous improvement has been shown from the initial units of μWcm^{-2} for the cell power density and tens of millivolts for the operating voltage, to achieve outputs approaching mWcm^{-2} and operating voltages of ~0.5 V. Biocatalytic fuel cell optimization to date has mostly dealt with attempts to achieve direct electron transfer to the biocatalyst, and with matching the redox potential of the mediator with the biocatalytic elements. An important factor to consider, addressed by some, in the biocatalytic fuel cell refinement process is the nature of the electrode. The introduction of solid nanoparticles and fibers, for example of gold or carbon, in the biocatalytic electrodes can further improve current and power densities, while opening up the possibility of direct electron transfer with the biocatalytic active site.

A remaining crucial technological milestone to pass for an implanted device remains the stability of the biocatalytic fuel cell, which should be expressed in months or years rather than days or weeks. Recent reports on the use of BOD biocatalytic electrodes in serum have, for example, highlighted instabilities associated with the presence of O_2, urate or metal ions [99, 100], and enzyme deactivation in its oxidized state [101]. Strategies to be considered include the use of new biocatalysts with improved thermal properties, or stability towards interferences and inhibitors, the use of nanostructured electrode surfaces and chemical coupling of films to such surfaces, to improve film stability, and the design of redox mediator libraries tailored towards both mediation and immobilization.

12.8 REFERENCES

1. A. Heller, *AIChE Journal* **51**, 1054 (2005).
2. Useful reviews for an introduction to this area are: (a) S.C. Barton, J. Gallaway, and P. Atanassov, *Chem. Rev.* **104**, 4867 (2004). (b) E. Katz, A.N. Shipway, and I. Willner, in *Handbook of Fuel Cells – Fundamentals, Technology and Applications* (W. Vielstich, H.A. Gasteiger, and A. Lamm, eds), Vol. 1,

Chap. 21, Biochemical fuel cells, John Wiley & Sons, Ltd 2003. (c) K. Rabaey and W. Verstraete *Tr. Biotechnol.* **23**, 91 (2005).

3. A. Heller, *Phys. Chem. Chem. Phys.* **6**, 209 (2004).
4. A.T. Yahiro, S.M. Lee, and D.O. Kimble, *Biochim. Biophys. Acta* **88**, 375 (1964).
5. H.A.O. Hill, *Coord. Chem. Rev.* **151**, 115 (1996).
6. J. Kim, H. Jia, and P. Wang, *Biotechnol. Advances* **24**, 296 (2006).
7. R.A. Bullen, T.C. Arnot, J.B. Lakeman, and F.C. Walsh, *Biosens. Bioelectron.* **21**, 2015 (2006).
8. S. Shleev, J. Tkac, A. Christenson, T. Ruzgas, A.I. Yaropolov, J.W. Whittaker, and L. Gorton, *Biosens. Bioelectron.* **20**, 2517 (2005).
9. S. Shleev, A. Jarosz-Wilkolazka, A. Khalunina, O. Morozova, A. Yaropolov, T. Ruzgas, and L. Gorton, *Bioelectrochem.* **67**, 115 (2005).
10. S. Tsujimura, K. Kano, and T. Ikeda, *J. Electroanal. Chem.* **576**, 113 (2005).
11. W. Zheng, Q.F. Li, L. Su, Y.M. Yan, J. Zhang, and L.Q. Mao, *Electroanal.* **18**, 587 (2006).
12. Y. Liu, M.K. Wang, F. Zhao, Z.A. Xu, and S. Dong, *Biosens. Bioelectron.* **21**, 984 (2005).
13. C. Cai and J. Chen, *Anal. Biochem.* **332**, 75 (2004).
14. I. Willner, V. Heleg-Shabtai, R. Blonder, E. Katz, G. Tao, A.F. Buckmann, and A. Heller, *J. Am. Chem. Soc.* **118**, 10321 (1996).
15. E. Katz, A. Riklin, V. Heleg-Shabtai, I. Willner, and A.F. Buckmann, *Anal. Chim. Acta* **385**, 45 (1999).
16. E. Katz, B. Filanovsky, and I. Willner, *New J. Chem.* **23**, 481 (1999).
17. E. Katz, I. Willner, and A.B. Kotlyar, *J. Electroanal. Chem.* **479**, 64 (1999).
18. F. Barrière, P. Kavanagh, and D. Leech, *Electrochim. Acta* **51**, 5187 (2006).
19. A. Pizzariello, M. Stred'ansky, and S. Miertuš, *Bioelectrochem.* **56**, 99 (2002).
20. I. Willner, G. Arad, and E. Katz, *Bioelectroch. Bioener.* **44**, 209 (1998).
21. I. Willner, E. Katz, F. Patolsky, and A.F. Bückmann, *J. Chem. Soc., Perkin Trans.* **2** 1817 (1998).
22. T. Ruzgas, A. Gaigalas, and L. Gorton, *J. Electroanal. Chem.* **469**, 123 (1999).
23. T. Lotzbeyer, W. Schuhmann, and H.-L. Schmidt, *Bioelectroch. Bioener.* **42**, 1 (1997).
24. Z. Xu, N. Gao, H. Chen, and S. Dong, *Langmuir* **21**, 10808 (2005).
25. A. Ramanavicius, A. Kausaite, and A. Ramanaviciene, *Biosens. Bioelectron.* **20**, 1962 (2005).
26. G.T.R. Palmore and H.-H. Kim, *J. Electroanal. Chem.* **464**, 110 (1999).
27. R. Bourbonnais and M.G. Paice, *FEBS Letters* **267**, 99 (1990).
28. AI. Yaropolov, O.V. Skorobogat'ko, S.S. Vartanov, and S.D. Varfolomeyev, *Appl. Biochem. Biotechnol.* **49**, 257 (1994).
29. N. Duran, M.A. Rosa, A. D'Annibale, and L. Gianfreda, *Enz. Microb. Technol.* **31**, 907 (2002).
30. F. Trudeau, F. Daigle, and D. Leech, *Anal. Chem.* **69**, 882 (1997).
31. B.A. Kuznetsov, G.P. Shumakovich, O.V. Koroleva, and A.I. Yaropolov, *Biosens. Bioelectron.* **16**, 73 (2001).
32. Y. Ferry and D. Leech, *Electroanal.* **17,** 113 (2005).
33. R. Bourbonnais, M.G. Paice, and D. Leech, *Biochim. Biophys. Acta* **1379**, 381 (1998).
34. D. Rochefort, D. Leech, and R. Bourbonnais, *Green Chem.* **6**, 14 (2004).
35. H.P. Call and I. Mucke, *J. Biotech.* **53**, 163 (1997).
36. E. Torres, I. Bustos-Jaimes, and S. le Borgne, *Appl. Catal. B* **46**, 1 (2003).
37. S.C. Barton, H.-H. Kim, G. Binyamin, Y. Zhang, and A. Heller, *J. Am. Chem. Soc.* **123**, 5802 (2001).
38. T. Chen, S.C. Barton, G. Binyamin, Z.Q. Gao, Y.C. Zhang, H.-H. Kim, and A. Heller, *J. Am. Chem. Soc.* **123**, 8630 (2001).
39. S.C. Barton, H.-H. Kim, G. Binyamin, Y. Zhang, and A. Heller, *J. Phys. Chem. B* **105**, 11917 (2001).
40. S.C. Barton, M. Pickard, R. Vazquez-Duhalt, and A. Heller *Biosens. Bioelectron.* **17**, 1071 (2002).
41. N. Mano, F. Mao, W. Shin, T. Chen, and A. Heller, *Chem. Comm.* 518 (2003).
42. F. Barrière, Y. Ferry, D. Rochefort, and D. Leech, *Electrochem. Comm.* **6**, 237 (2004).
43. H. Yoshida, *J. Chem. Soc.* **43**, 472 (1883).
44. E.I. Solomon, U.M. Sundaram, and T.E. Machonkin, *Chem. Rev.* **96**, 2563 (1996).
45. F. Xu, W. Shin, S.H. Brown, J.A. Wahleithner, U.M. Sundaram, and E.I. Solomon, *Biochim. Biophys. Acta* **1292**, 303 (1996).

46. F. Xu, H.J.W. Deussen, B. Lopez, L. Lam, and K.C. Li, *Eur. J. Biochem.* **268**, 4169 (2001).
47. F. Xu, J.J. Kulys, K. Duke, K.C. Li, K. Krikstopaitis, H.J.W. Deussen, E. Abbate, V. Galinyte, and P. Schneider, *Appl. Env. Microbiol.* **66**, 2052 (2000).
48. F. Xu, *J. Biol. Chem.* **272**, 924 (1997).
49. B. Reinhammer, *Biochim. Biophys. Acta* **205**, 35 (1970).
50. I.V. Berezin, V.A. Bogdanovskaya, S.D. Varfolomeev, M.R. Tarasevich, and A.I. Yaropolov, *Dokl. Akad. Nauk SSSR* **240**, 615 (1978).
51. C.W. Lee, H.B. Gray, F.C. Anson, and B.G. Malmstrom, *J. Electroanal. Chem.* **172**, 289 (1984).
52. M.H. Thuesen, O. Farver, B. Reinhammar, and J. Ulstrup, *Acta Chem. Scand.* **52**, 555 (1998).
53. M.R. Tarasevich, V.A. Bogdanovskaya, and A.V. Kapustin, *Electrochem. Comm.* **5**, 491 (2003).
54. C. Taylor, G. Kenausis, I. Katakis, and A. Heller, *J. Electroanal. Chem.* **396**, 511 (1995).
55. D. Leech and K.O. Feerick, *Electroanal.* **12**, 1339 (2000).
56. S. Tsujimura, H. Tatsumi, J. Ogawa, S. Shimizu, K. Kano, and T. Ikeda, *J. Electroanal. Chem.* **496**, 69 (2001).
57. S. Tsujimura, M. Kawaharada, T. Nakagawa, K. Kano, and T. Ikeda, *Electrochem. Comm.* **5**, 138 (2003).
58. N. Mano, H.-H. Kim, Y. Zhang, and A. Heller, *J. Am. Chem. Soc.* **124**, 6480 (2002).
59. N. Mano, H.-H. Kim, and A. Heller, *J. Phys. Chem., B* **106**, 8842 (2002).
60. N. Mano, J.L. Fernandez, Y. Kim, W. Shin, A.J. Bard, and A. Heller, *J. Am. Chem. Soc.* **125**, 15290 (2003).
61. S. Tsujimura, K. Kano, and T. Ikeda, *J. Electroanal. Chem.* **576**, 113 (2005).
62. M.T. Stankovich, L.M. Schopfer, and V. Massey, *J. Biol. Chem.* **253**, 4971 (1978).
63. C. Godet, M. Boujtita, and N. El Murr, *New J. Chem.* **23**, 795 (1999).
64. C. Cai and J. Chen, *Anal. Biochem.* **332**, 75 (2004).
65. J.H.T. Luong, S. Hrapovic, D. Wang, F. Benseba, and B. Simard, *Electroanal.* **16**, 132 (2004).
66. J. Liu, A. Chou, W. Rahmat, M.N. Paddon-Row, and J.J. Gooding, *Electroanal.* **17**, 38 (2005).
67. S. Liu and H. Ju, *Biosens. Bioelectron.* **19**, 177 (2003).
68. F. Zhang, S.S. Cho, S.H. Yang, S.S. Seo, G.S. Cha, and H. Nam, *Electroanal.* **18**, 217 (2006).
69. A.E.G. Cass, G. Davis, GD. Francis, H.A.O. Hill, W.J. Aston, I.J. Higgins, E.V. Plotkin, L.D. Scott, and A.P.F. Turner, *Anal. Chem.* **56**, 667 (1984).
70. N.J. Forrow, G.S. Sanghera, and S.J. Walters, *J. Chem. Soc. Dalton.* 3187 (2002).
71. S.M. Zakeeruddin, D.M. Fraser, M.-K. Nazeeruddin, and M. Grätzel, *J. Electroanal. Chem.* **337**, 253 (1992).
72. D.M. Fraser, S.M. Zakeeruddin, and M. Gratzel, *J. Electroanal. Chem.* **359**, 125 (1993).
73. Y. Liu, M. Wang, F. Zhao, B. Liu, and S. Dong, *Chem. Eur. J.* **11**, 4970 (2005).
74. M. Zayats, E. Katz, and I. Willner, *J. Am. Chem. Soc.* **124**, 2120 (2002).
75. F. Patolsky, Y. Weizmann, and I. Willner, *Angew. Chem. Int. Ed.* **43**, 2113 (2004).
76. Y. Xiao, F. Patolsky, E. Katz, J.F. Hainfeld, and I. Willner, *Science* **299**, 1877 (2003).
77. N. Mano, F. Mao, and A. Heller, *J. Electroanal. Chem.* **574**, 347 (2005).
78. D. Leech and F. Daigle, *Analyst* **123**, 1971 (1998).
79. T. Delumleywoodyear, P. Rocca, J. Lindsay, Y. Dror, A. Freeman, and A. Heller, *Anal. Chem.* **67**, 1332 (1995).
80. H.-H. Kim, N. Mano, Y. Zhang, and A. Heller, *J. Electrochem. Soc.* **150**, A209 (2003).
81. D.N. Blauch and J.-M. Savéant, *J. Phys. Chem.* **97**, 6444 (1993) and *J. Am. Chem. Soc.* **114**, 3323 (1992).
82. F. Mao, N. Mano, and A. Heller, *J. Am. Chem. Soc.* **125**, 4951 (2003).
83. N. Mano, F. Mao, and A. Heller, *Chem. Commun.* 2116 (2004).
84. N. Yuhashi, M. Tomiyama, J. Okuda, S. Igarashi, K. Ikebukuro, and K. Sode, *Biosens. Bioelectron.* **20**, 2145 (2005).
85. L. Ye, M. Hiimmerle, A.J.J. Olsthoorn, W. Schuhmann, H.-L. Schmidt, J.A. Duine, and A. Heller, *Anal. Chem.* **65**, 238 (1993).
86. T. Ikeda and K. Kano, *Biochim. Biophys. Acta* **1647**, 121 (2003).
87. S. Tsujimura, K. Kano, and T. Ikeda, *Electrochemistry* **70**, 940 (2002).

88. S. Igarashi and K. Sode, *Mol. Biotechnol.* **24**, 97 (2003).
89. F. Sato, M. Togo, M.K. Islam, T. Matsue, J. Kosuge, N. Fukasaku, S. Kurosawa, and M. Nishizawa, *Electrochem. Comm.* **7**, 643 (2005).
90. Y. Ogino, K. Takagi, K. Kano, and T. Ikeda, *J. Electroanal. Chem.* **396**, 517 (1995).
91. G.T.R. Palmore, H. Bertschy, S.H. Bergens, and G.M. Whitesides, *J. Electroanal. Chem.* **443**, 155 (1998).
92. E. Simon, C.M. Halliwell, C.S. Toh, A.E.G. Cass, and P.N. Bartlett, *Bioelectrochem.* **55**, 13 (2002).
93. C.M. Halliwell, E. Simon, C.S. Toh, A.E.G. Cass, and P.N. Bartlett, *Bioelectrochem.* **55**, 21 (2002).
94. V. Soukharev, N. Mano, and A. Heller, *J. Am. Chem. Soc.* **126**, 8368 (2004).
95. N. Mano and A. Heller, *J. Electrochem. Soc.* **150**, A1136 (2003).
96. N. Mano, F. Mao, and A. Heller, *J. Am. Chem. Soc.* **124**, 12962 (2002).
97. N. Mano, F. Mao, and A. Heller, *J. Am. Chem. Soc.* **125**, 6588 (2003).
98. N. Mano, F. Mao, and A. Heller, *ChemBioChem* **5**, 1703 (2004).
99. G. Binyamin, T. Chen, and A. Heller, *J. Electroanal. Chem.* **500**, 604 (2001).
100. C. Kang, H. Shin, Y. Zhang, and A. Heller, *Bioelectrochem.* **65**, 83 (2004).
101. C. Kang, H. Shin, and A. Heller, *Bioelectrochem.* **68**, 22 (2006).

CHAPTER 13

Chemical and biological sensors based on electroactive inorganic polycrystals

Arkady Karyakin

13.1 INTRODUCTION

Accurate, rapid, cheap, and selective analysis is required nowadays for clinical and industrial laboratories. Electrochemical biosensors seem to accomplish this function.

Except for their low cost and simplicity, sensors allow continuous monitoring of key analytes, which is important in certain cases of clinical, industrial, and environmental analysis. Besides that, modern clinical diagnostics requires analysis of blood or tissue liquids directly in the target organ, because a conventional procedure, which includes cutting a piece of tissue and delivering it to an analytical instrument, in some cases becomes non-informational due to decomposition of certain key metabolites. Obviously, such clinical analysis can be realized only with chemical or biological sensors.

To provide its successful operation in real samples one has to address sufficient selectivity of chemical or biological sensors. That's why the use of common electrocatalysts like platinum, which requires membrane technologies to shield selectively electrode surface, is not convenient. On the contrary, the use of inert electrode supports modified with certain monomers or polymers may result in highly selective and even more sensitive chemical sensors. The use of a biological recognition element, which distinguishes biological sensors, already provides high selectivity. However, to provide biosensor operation a successful coupling of biological and electrode reactions is required, and some of the selective chemical sensors can serve as suitable transducers.

Among the variety of materials used for electrode modification the electroactive organic and inorganic polymers seem to be the most prominant ones. In this chapter the electroactive polycrystals of transition metals, hexacyanoferrates, will be discussed for the development of chemical and biological sensors.

13.2 PROPERTIES OF TRANSITION METAL HEXACYANOFERRATES

Prussian blue, or ferric hexacyanoferrate, is definitely one of the most ancient coordination materials. The earliest announcements were from the very beginning of the eighteenth century [1, 2]. However, a quite recent investigation by Neff [3], that Prussian blue forms electroactive layers after electrochemical or chemical deposition onto the electrode surface, has opened a new area in fundamental investigation of this unique inorganic polycrystal.

13.2.1 Structure of transition metal hexacyanoferrates

The fact that Prussian blue is indeed ferric ferrocyanide ($Fe_4^{III}[Fe^{II}(CN)_6]_3$) with iron(III) atom coordinated to nitrogen and iron(II) atom coordinated to carbon has been established by spectroscopic investigations [4]. Prussian blue can be synthesized chemically by the mixing of ferric (ferrous) and hexacyanoferrate ions with different oxidation state of iron atoms: either $Fe^{3+} + [Fe^{II}(CN)_6]^{4-}$ or $Fe^{2+} + [Fe^{III}(CN)_6]^{3-}$. After mixing, an immediate formation of the dark blue colloid is observed. However, the mixed solutions of ferric (ferrous) and hexacyanoferrate ions with the same oxidation state of iron atoms are apparently stable.

The crystalline structure of Prussian blue was first discussed by Keggin and Miles on the basis of powder diffraction patterns [5] and then has been determined more precisely by Ludi and coworkers from single crystals by electron and neutron diffraction measurements [6]. Prussian blue has a basic cubic structure consisting of alternating iron(II) and iron(III) located on a face centered cubic lattice (Fig. 13.1), in such a way that the iron(III) ions are surrounded octahedrically by nitrogen atoms, and iron(II) ions are surrounded by carbon atoms. The cubic unit cell dimensions are of 10.2Å.

13.2.2 Electrochemistry of transition metal hexacyanoferrates

Commonly, deposition of Prussian blue on various conductive surfaces is carried out from the aqueous solutions containing a mixture of ferric (Fe^{3+}) and ferricyanide ($[Fe^{III}(CN)_6]^{3-}$) ions, either spontaneously in open-circuit regime or by applying a reductive electrochemical driving force. Chronopotentiometric investigations in equimolar ferric–ferricyanide mixtures has shown the two basic plateaus: at 0.7 V and at 0.4 V (SCE) [7]. These plateaus have been attributed to reduction of the one-to-one complex of $Fe^{III}[Fe^{III}(CN)_6]$, discovered earlier [8], and of Fe^{3+} ions, respectively. At 0.7 V Prussian blue is deposited according to reduction of the ferric–ferricyanide complex ($Fe^{III}[Fe^{III}(CN)_6]$). Around 0.4 V the bulk precipitation of Prussian blue occurs due to reduction of Fe^{3+} to Fe^{2+}, the latter reacts with ($[Fe^{III}(CN)_6]^{3-}$). The open-circuit deposition is highly dependent on the electrode support. Its mechanism is probably the oxidation of the conductive material with the ($Fe^{III}[Fe^{III}(CN)_6]$) complex, which forms Prussian blue after one-electron reduction. Posing of the electrodes to the potentials lower than 0.2 V is, according to our experience, not plausible for deposition of Prussian blue, because both Fe^{3+} and $[Fe^{III}(CN)_6]^{3-}$ ions are reduced, and the structure of the resulting polycrystal is less regular.

A cyclic voltammogram of a Prussian blue-modified electrode is shown in Fig. 13.2. In between the observed two sets of peaks the oxidation state, which is correspondent to the Prussian blue itself, occurs. Its reduction is accompanied with loss of

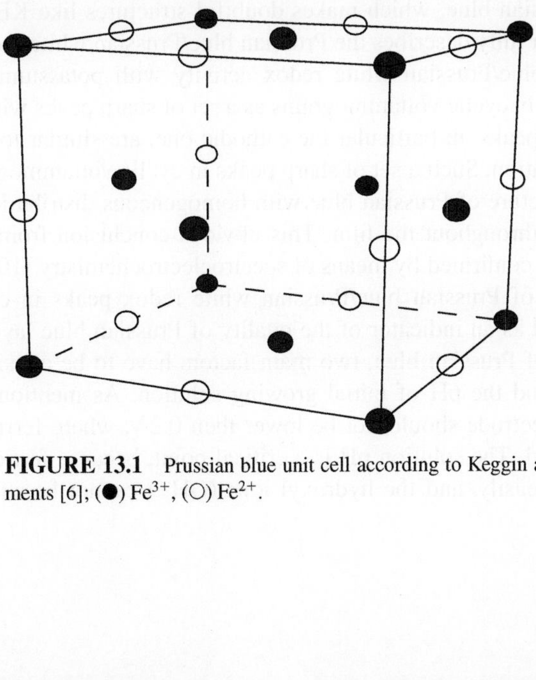

FIGURE 13.1 Prussian blue unit cell according to Keggin and Miles, plotted using data from measurements [6]; (●) Fe^{3+}, (○) Fe^{2+}.

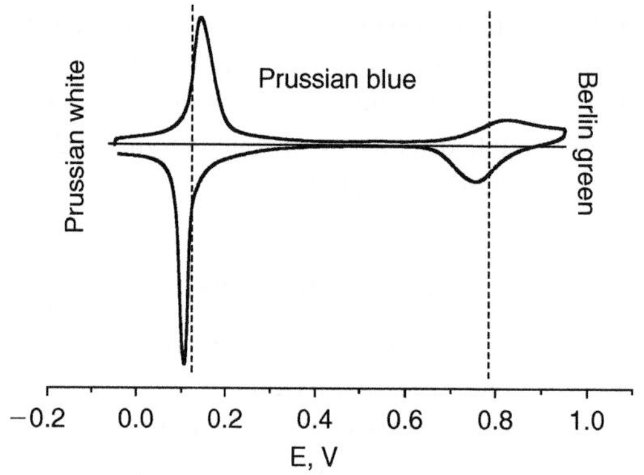

FIGURE 13.2 Typical cyclic voltammogram of Prussian blue-modified smooth (mirrored glassy carbon) electrode; 0.1 M KCl, 40 mV s^{-1}.

the color, and the reduced form of the polycrystal is denoted as Prussian white. The transfer of electrons is compensated by the entrapment of cations in the film according to the equation:

$$Fe_4^{III}[Fe^{II}(CN)_6]_3 + 4\bar{e} + 4K^+ \longleftrightarrow K_4Fe_4^{II}[Fe^{II}(CN)_6]_3 \quad (1)$$

In the literature the term "soluble Prussian blue" introduced by Keggin and Miles [5] to determine the $KFeFe(CN)_6$ compound is still widely used. However, it is important to note, that the term "soluble" refers to the ease with which the potassium ion can be "peptized" rather than to the real solubility of Prussian blue. Indeed, it can be easily shown by means of cyclic voltammetry that the stability of Prussian blue films on electrode supports is nearly independent of their saturation by potassium cations. Moreover, Itaya and coworkers [9] have not found any appreciable amount of potassium ions in Prussian blue, which makes doubtful structures like $KFeFe(CN)_6$. Thus, the above equation fully describes the Prussian blue/Prussian white redox reaction.

The Prussian blue/Prussian white redox activity with potassium as the countercation is observed in cyclic voltammograms as a set of sharp peaks with a separation of 15–30 mV. These peaks, in particular the cathodic one, are similar to the peaks of the anodic demetallization. Such a set of sharp peaks in cyclic voltammograms correspond to the regular structure of Prussian blue with homogeneous distribution of charge and ion transfer rates throughout the film. This obvious conclusion from electrochemical investigations was confirmed by means of spectroelectrochemistry [10].

The sharpness of Prussian blue/Prussian white redox peaks in cyclic voltammograms can be used as an indicator of the quality of Prussian blue layers. To achieve a regular structure of Prussian blue, two main factors have to be considered: the deposition potentials and the pH of initial growing solution. As mentioned, the potential of the working electrode should not be lower then 0.2 V, where ferricyanide ions are intensively reduced. The solution pH is a critical point, because ferric ions are known to be hydrolyzed easily, and the hydroxyl ions (OH^-) cannot be substituted in their coordination sphere in the course of Prussian blue crystallization. According to our findings initial solution with pH 1 is optimal for deposition [11, 12].

It is important to note that not all cations promote Prussian blue/Prussian white electroactivity. Except for potassium, only ammonium (NH_4^+), cesium (Cs^+), and rubidium (Rb^+) were found able to penetrate the Prussian blue lattice. Other mono- and divalent cations are considered as blocking ones.

At high anodic potentials Prussian blue converts to its fully oxidized form as is clearly seen in cyclic voltammograms due to the presence of the corresponding set of peaks (Fig. 13.2). The fully oxidized redox state is denoted as Berlin green or in some cases as "Prussian yellow". Since the presence of alkali metal ions is doubtful in the Prussian blue redox state, the most probable mechanism for charge compensation in Berlin green/Prussian blue redox activity is the entrapment of anions in the course of oxidative reaction. The complete equation is:

$$Fe_4^{III}[Fe^{II}(CN)_6]_3 - 3\overline{e} + 3A^- \longleftrightarrow Fe_4^{III}[Fe^{III}(CN)_6A]_3 \qquad (2)$$

Except for deposition of Prussian blue from the mixture of ferric and ferricyanide ions, its electrosynthesis from the single ferricyanide solution is reported [13]. Ferricyanide ions are not extremely stable even in aqueous solution, which is noticed in the change of color after a few days of storage. Thus, the coordination sphere can be destroyed also in the course of electrochemical reactions. The mentioned processes may lead to formation of ferric–ferricyanide complex or "free" ferric ions. The reduction of the resulting mixture leads to the formation of Prussian blue.

Just a few years after the discovery of the deposition and electroactivity of Prussian blue, other metal hexacyanoferrates were deposited on various electrode surfaces. However, except for ruthenium and osmium, the electroplating of the metal or its anodizing was required for the deposition of nickel [14], copper [15, 16], and silver [9] hexacyanoferrates. Later studies have shown the possibilities of the synthesis of nickel, cobalt, indium hexacyanoferrates similar to the deposition of Prussian blue [17–19], as well as palladium [20–22], zinc [23, 24], lanthanum [25–27], vanadium [28], silver [29], and thallium [30] hexacyanoferrates.

As mentioned, potassium ion promotes the redox activity of Prussian blue, whereas sodium ion blocks it. However, indium, cobalt, and nickel hexacyanoferrates were successfully grown and then cycled in the presence of sodium as the counter-cation [18]. There are two possible explanations of redox activity in the presence of sodium. All these hexacyanoferrates give the single set of peaks in their cyclic voltammograms. This set of peaks may be attributed to the redox reaction with the charge compensation due to entrapment of anions rather than cations, similar to the Berlin green/Prussian blue redox reaction. Alternatively, sodium ions may penetrate the lattice of these hexacyanoferrates. Transition metal hexacyanoferrates can be deposited on a variety of materials. The only requirement to support this is its stability at high anodic potentials. Most commonly Prussian blue and its analogs are deposited on carbon materials and gold, and in some cases on platinum. For optical measurements ITO can be used as a substrate. More recently the reports on the deposition of nickel [31], cobalt [32], manganese [33], and palladium [21] hexacyanoferrates on aluminum have been reported.

13.3 AMPEROMETRIC SENSORS FOR REDOX-INACTIVE CATIONS AND ELECTROACTIVE COMPOUNDS

13.3.1 Sensors for redox-inactive cations

Redox-inactive cations attract a particular interest for analytical chemists because of their importance in environmental control, industry, and medicine. For instance, in clinical diagnostics, tests for blood electrolytes (Na^+, K^+) are routine, because deviation of cation content from their normal values indicates a number of pathologies.

The participation of cations in redox reactions of metal hexacyanoferrates provides a unique opportunity for the development of chemical sensors for non-electroactive ions. The development of sensors for thallium (Tl^+) [15], cesium (Cs^+) [34], and potassium (K^+) [35, 36] pioneered analytical applications of metal hexacyanoferrates (Table 13.1). Later, a number of cationic analytes were enlarged, including ammonium ($NH_4{}^+$) [37], rubidium (Rb^+) [38], and even other mono- and divalent cations [39]. In most cases the electrochemical techniques used were potentiometry and amperometry either under constant potential or in cyclic voltammetric regime. More recently, sensors for silver [29] and arsenite [40] on the basis of transition metal hexacyanoferrates were proposed. An apparent list of sensors for non-electroactive ions is presented in Table 13.1.

Some monovalent ions promote the electroactivity of metal hexacyanoferrates rather similarly, which affects selectivity of the corresponding sensors. In particular, it

TABLE 13.1
Sensors for redox inactive cations

Analyte	Me-hexacyanoferate, Me:	Ref.
Cs^+	Cu	[34, 47]
K^+	Cu, $Ag^{I-}[Mo(CN)_8]^{4-}$, Fe, Co, Ni, Zn	[36, 48], [35], [49–51], [41], [42], [24]
$NH_4{}^+$	Cu	[52]
Tl^+	Cu, Fe	[15, 53], [54, 55]
Ag^+	Ag	[29]
As^{3+}	Fe	[40]
K^+, Cs^+	Fe, Cu, Ag, Ni, Cd	[56]
K^+, $NH_4{}^+$	Cu	[37]
K^+, Rb^+, Cs^+, $NH_4{}^+$	Cu, Ni, Fe	[38], [57, 58]
Cs^+, K^+, Na^+, Li^+	Ni	[59]
Li^+, Na^+, K^+, Rb^+, Cs^+, H^+, Tl^+	Tl^+	[30]
Mono- and divalent cations	Cu, Ni	[39]

is rather hard to distinguish between the alkali metal ions and ammonium ion. Indeed, the same metal hexacyanoferrates were used in some cases as potassium, cesium or rubidium sensors, and in other cases as ammonium sensors (Table 13.1).

Particular cases are potassium selective potentiometric sensors based on cobalt [41] and nickel [38, 42] hexacyanoferrates. As mentioned, these hexacyanoferrates possess quite satisfactory redox activity with sodium as counter-cation [18]. According to the two possible mechanisms of such redox activity (either sodium ions penetrate the lattice or charge compensation occurs due to entrapment of anions) there is no thermodynamic background for selectivity of these sensors. In these cases electroactive films seem to operate as "smart materials" similar to conductive polymers in electronic noses.

Except for sensor applications, the intercalation of alkali metal ions in metal hexacyanoferrates was used for adsorption and separation of cesium ions from different aqueous solutions with Prussian blue [43, 44] and cupric hexacyanoferrate [45, 46].

13.3.2 Amperometric sensors for electroactive compounds

Whereas detection of electroinactive ions was principally worked out at the end of last century, the use of transition metal hexacyanoferrates as sensors for various electroactive compounds still attracts particular interest of scientists. Although the cross-selectivity of such compounds must be low, a number of them have been successfully used for analysis of real objects.

Electrocatalysis in oxidation has apparently first been shown for ascorbic acid oxidation by Prussian blue [60] and later by nickel hexacyanoferrate [61]. More valuable for analytical applications was the discovery in the early 1990s of the oxidation of sulfite [62] and thiosulfate [18, 63] at nickel [62, 63] and also ferric, indium, and cobalt [18] hexacyanoferrates. More recently electrocatalytic activity in thiosulfate oxidation was shown also for zinc [23] hexacyanoferrate. Prussian blue-modified electrodes allowed sulfite determination in wine products [64], which is important for the wine industry.

A particular interest for clinical applications was a possibility for detection of dopamine by its oxidation on nickel [19], cobalt [65], and osmium [66] hexacyanoferrates. Except for oxidation of dopamine, cobalt and osmium hexacyanoferrates were active in oxidation of epinephrine and norepinephrine. For clinical analysis it is also important to carry out the detection of morphine on cobalt [67] and ferric [68] hexacyanoferrates, as well as the detection of oxidizable amino acids (cystein, methionine) by manganous [69] and ruthenium [70] hexacyanoferrate-modified electrodes. In general, oxidation of thiols was first shown for Prussian blue [71] and nickel hexacyanoferrate [72]. This approach has been used for the detection of thiols in rat striatum microdialysate [73]. Alternatively, the detection of thiocholine with Prussian blue was employed for pesticide determination in acetylcholine-esterase test [74].

Nitric oxide (NO) and nitrite were found to be oxidized by Prussian blue and indium hexacyanoferrate-modified electrodes [75–77]. For pharmaceutical application oxidation of isoprenaline [78] and vitamin B-6 [79] at cupric hexacyanoferrate-modified electrodes was shown.

Cobalt hexacyanoferrates and Prussian blue have shown high activity in oxidations of hydrazine and hydroxylamine [80–82]. Electrocatalytic activity in this reaction has also been found for nickel [83] and manganese [69] hexacyanoferrates.

Ten years ago oxidation of NADH [84, 85] seems not to be important because of the other more powerful electrocatalysts [86–89]. A possibility for oxidation of guanine even in DNA [90] with cobalt hexacyanoferrate, on the contrary, seems to be more apposite.

In contrast to a variety of oxidizable compounds, only a few examples for the detection of strong oxidants with transition metal hexacyanoferrates were shown. Among them, hydrogen peroxide is discussed in the following section. Except for H_2O_2, the reduction of carbon dioxide [91] and persulfate [92] by Prussian blue-modified electrode was shown. The detection of the latter is important in cosmetics. It should be noted that the reduction of Prussian blue to Prussian white occurs at the lowest redox potential as can be found in transition metal hexacyanoferrates.

13.4 ADVANCED SENSOR FOR HYDROGEN PEROXIDE

13.4.1 H_2O_2 as important analyte for medicine, biology, environmental control, and industry

Monitoring of hydrogen peroxide is of great importance for modern medicine, environmental control, and various branches of industry. H_2O_2 is a chemical threat agent; its excessive concentration as a product of industry and from atomic power stations affects the environment [93, 94]. In addition, hydrogen peroxide is used for the disinfection of water pools, food, and beverage packages [95, 96], which makes it important to measure its residual concentration. On the other hand, H_2O_2 is the most valuable marker for oxidative stress, recognized as one of the major risk factors in the progression of disease-related pathophysiological complications in diabetes, atherosclerosis, renal disease, cancer, aging, and other conditions [97–101]. Except for oxidative stress, H_2O_2 is the most valuable marker for inflammatory processes [102], and a mediator for apoptotic cell death [103, 104].

Selective detection of hydrogen peroxide is also of great importance for biosensors. More, than 90% of enzyme-based biosensors and analytical kits use the enzyme oxidase as a terminal (signal generation) one. Operation of biosensors requires successive coupling of the enzyme and electrochemical reactions. As stated, *first-generation* biosensors are based on direct electrochemical detection of substrate or product of the enzyme reaction. In the case of oxidases, these are oxygen and hydrogen peroxide, respectively. Amperometric detection of these substances was usually undertaken using platinum or platinized electrodes [105–108]. Detection of oxygen consumption at negative potentials (−0.6 V vs Ag/AgCl) [105, 106] was the most simple procedure. Nevertheless, such biosensors were not able to detect low analyte concentrations. The reasons were: (i) great excess of oxygen, (ii) variation of oxygen concentrations in biological liquids, and (iii) reduction of hydrogen peroxide at similar potentials.

Electrochemical biosensors based on detection of hydrogen peroxide at platinized electrodes were found to be more versatile allowing a decrease in detection limit down to 1 μmol L^{-1} [109]. However, all biological liquids contain a variety of electrochemically easily oxidizable reductants, e.g. ascorbate, urate, bilirubin, catecholamines, etc., which are oxidized at similar potentials and dramatically affect biosensor selectivity producing parasitic anodic current [110].

Hence, for successful operation of first-generation oxidase-based biosensors the selective low potential detection of hydrogen peroxide is required. As will be shown, this is possible only with transition metal hexacyanoferrates. Using a peroxidase enzyme electrode (peroxidases are responsible in nature for the reduction of hydrogen peroxide) [111, 112], one can develop a selective method for the detection of H_2O_2. However, enzymes being biological macromolecules obviously cannot provide long-term sensor operation due to their inherent instability. Moreover, some commonly found reducing agents in "real" samples may compete with the electrode as the source of electrons for the oxidized form of the peroxidase leading to erratic signals.

13.4.2 Advanced electrocatalyst for hydrogen peroxide reduction

Through optimization of the deposition procedure for Prussian blue, a selective electrocatalyst for H_2O_2 reduction able to operate in the presence of oxygen in a wide potential range has been synthesized [11]. Selectivity of Prussian blue in relation to oxygen is illustrated in hydrodynamic steady-state current potential curves (Fig. 13.3). As seen in air saturated solution (oxygen concentration is approximately 0.2 mmol L^{-1}), only a minor current is observed. However, when the twofold lower concentration of hydrogen peroxide is added, a well-defined polarographic wave with the half-wave potential coincided with the Prussian blue/Prussian white redox potential appears. At the current plateau region, the current of the hydrogen peroxide reduction is two orders of magnitude

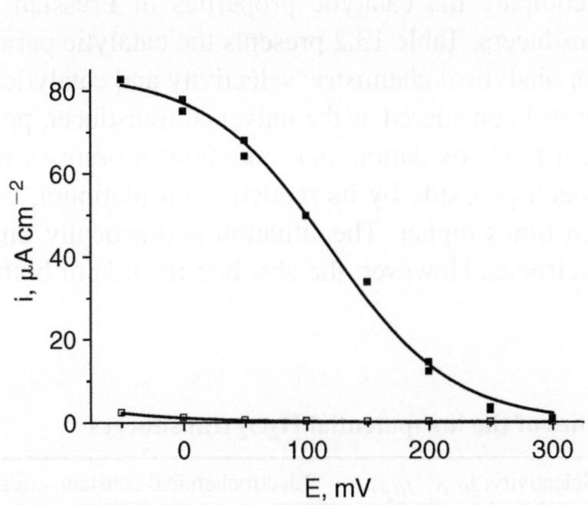

FIGURE 13.3 Hydrodynamic voltammograms of Prussian blue-modified electrodes in a wall-jet cell with continuous flow of 0.8 ml/min: (■) background in air saturated solution (0.1 M KCl + 0.01 M phosphate, pH 6.0), (!) 0.1 mM H_2O_2.

higher than of oxygen reduction. At more positive potentials the selectivity of Prussian blue in H_2O_2 reduction is even higher.

Stability of the Prussian blue-based hydrogen peroxide transducer is a crucial point commonly raised as an objection against its practical applications. Indeed Prussian white (the redox state of Prussian blue at 0.0 V) is thermodynamically unstable on electrode surfaces. In addition, hydroxyl ions being the products of hydrogen peroxide reduction in neutral media [113] are able to solubilize the inorganic polycrystal. However, through continuous efforts in improving the crystalline structure of the deposited Prussian blue and its additional post-treatment, excellent operational stability, which even exceeds the stability of the known H_2O_2 transducers, has been achieved [114, 115].

The kinetics of hydrogen peroxide reduction catalyzed by Prussian blue has been investigated [12, 113]. In neutral media the reaction scheme of H_2O_2 reduction has been found to be the following:

$$H_2O_2 + 2\bar{e} \xrightarrow{k_{cat.}} 2OH^- \quad (3)$$

The electrochemical rate constants for hydrogen peroxide reduction have been found to be dependent on the amount of Prussian blue deposited, confirming that H_2O_2 penetrates the films, and the inner layers of the polycrystal take part in the catalysis. For 4–6 nmol cm^{-2} of Prussian blue the electrochemical rate constant exceeds 0.01 cm s^{-1} [12], which corresponds to the bi-molecular rate constant of $kcat = 3 \times 10^3$ L $mol^{-1}s^{-1}$ [114]. The rate constant of hydrogen peroxide reduction by ferrocyanide catalyzed by enzyme peroxidase was 2×10^4 L $mol^{-1}s^{-1}$ [116]. Thus, the activity of the natural enzyme peroxidase is of a similar order of magnitude as the catalytic activity of our Prussian blue-based electrocatalyst. Due to the high catalytic activity and selectivity, which are comparable with biocatalysis, we were able to denote the specially deposited Prussian blue as an *artificial peroxidase* [114, 117].

It is important to compare the catalytic properties of Prussian blue with known hydrogen peroxide transducers. Table 13.2 presents the catalytic parameters, which are of major importance for analytical chemistry: selectivity and catalytic activity. It is seen that platinum, which is still considered as the universal transducer, possesses rather low catalytic activity in both H_2O_2 oxidation and reduction. Moreover, it is nearly impossible to measure hydrogen peroxide by its reduction on platinum, because the rate of oxygen reduction is ten times higher. The situation is drastically improved in case of enzyme peroxidase electrodes. However, the absolute records of both catalytic activity

TABLE 13.2
Electrocatalytic properties of the low-potential H_2O_2 transducers

Transducer	Selectivity: $j_{H_2O_2}/j_{H_2O}$	Electrochemical constant, cm/s
Pt	0.1	4×10^{-6}
peroxidase electrodes	30–40	1×10^{-3}
Prussian blue	400–600	1×10^{-2}

and selectivity are reached in the case of Prussian blue. Compared with platinum, still the most widely used transducer, Prussian Blue is *three* orders of *magnitude* more "active", and *three* orders of *magnitude* more "selective". Moreover, the cost of raw materials, which is highly important for mass production, is also about *three* orders of *magnitude* less in the case of Prussian blue. The catalytic parameters show promise that on the basis of Prussian blue a highly advantageous H_2O_2 transducer can be structured.

13.4.3 An advanced sensor for hydrogen peroxide based on Prussian blue

The likelihood of selective detection of hydrogen peroxide by its reduction in the presence of oxygen using Prussian blue-modified electrodes was first announced by our group in 1994 [118]. Indeed, analytical performances of Prussian blue-modified electrodes in hydrogen peroxide detection investigated in flow-injection system equipped with a wall-jet cell, show the linear calibration range prolonged over four orders of magnitude of hydrogen peroxide concentration. The lower limit of the linear calibration range referred to as the detection limit is 10^{-7} mol L^{-1}. Using Prussian blue as transducer for hydrogen peroxide it was possible to achieve the sensitivity in flow-injection mode of 0.6 A L mol^{-1} cm^{-2} [114, 119], which taking into account the dispersion coefficient [120] corresponds to the sensitivity of 1–2 A L mol^{-1} cm^{-2} in either batch regime or under continuous flow. The latter is at the upper sensitivity level limited by hydrogen peroxide diffusion [112].

13.4.4 Non-conductive polymers on the surface of Prussian blue modified electrodes

Further improvement of the Prussian blue-based transducer presents two principal problems. First, Prussian blue layers are not mechanically stable, especially on smooth electrode surfaces because of their polycrystalline nature. Second, despite the low electrode potential used, the most powerful reductants like ascorbic acid still interfere with sensor response if present in excessive concentrations.

To improve the stability and selectivity, additional electrode coverings were considered. Conducting and non-conducting polymers were known to reduce the interference effect. Covering the electrode with Nafion [37, 121], sol-gel [122, 123] or other thick films [124] facilitated an increase in sensor selectivity by approximately ten times.

Electrosynthesis of polymers compares favorably with the thick film method providing addressable and controlled deposition. In terms of selectivity to hydrogen peroxide in the presence of interferents, the most promising results were obtained with poly-1,2-diaminobenzene (poly-1,2-DAB)-modified electrodes [125].

The possibility for electropolymerization on the top surface of Prussian blue films was probably first shown in [126] describing the high oxidizing ability of Berlin green, the fully oxidized form of Prussian blue. Afterwards non-conducting polymers were synthesized on the top surface of transition metal hexacyanoferrate-modified electrodes for immobilization of the enzyme [127].

In contrast to metal surfaces, the growth of non-conducting polymers on the top surface of Prussian blue-modified electrodes can be independently monitored due to

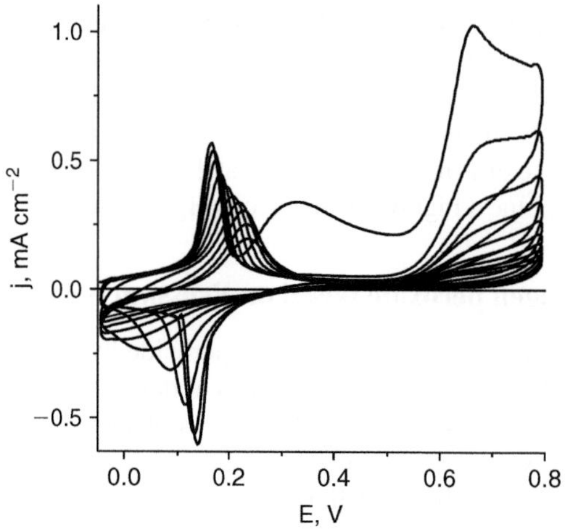

FIGURE 13.4 Electropolymerization of 1,2-DAB on the top of Prussian blue-modified electrode: 5 mM DAB, 0.02 M phosphate pH 6.0 with 0.1 M KCl, 20 mV s^{-1}.

the electroactivity of the inorganic polycrystal, which is dependent on the presence of counter-cations. As seen in Fig. 13.4, in the course of polymer growth, which is noticed by the decrease of the current of monomer irreversible oxidation, the set of peaks peculiar to Prussian blue redox activity is also changed: the peaks become wider and lower indicating that a barrier to cations appears. Depending on the counter-cation, the permeability of a non-conducting film can thus be monitored during the deposition procedure.

Concerning *analytical characteristics* of the resulting Prussian blue (PB)-based hydrogen peroxide transducer, the best performance characteristics were obtained for PB electrodes covered with electropolymerized *o*-phenylenediamine. The reported transducer remained at a 100% response state for more than 20 hours under continuous flow of 0.1 mmol L^{-1} hydrogen peroxide (flow rate 1 ml min^{-1}), which improves the stability level among selective H_2O_2 sensors by one order of magnitude. The selectivity factor of the PB-poly (1,2-diaminobenzene)-based transducer relative to ascorbate is nominally 600. The PB-poly(1,2-diaminobenzene)-modified electrode allows the detection of hydrogen peroxide in flow-injection mode down to 10^{-7} mol L^{-1} with a sensitivity of 0.3 A L $mol^{-1}cm^{-r2}$, which is two times lower compared with the uncoated Prussian Blue-based transducer [128].

Hence, non-conducting polymers deposited on the top surface of Prussian blue-modified electrodes only slightly decrease sensor response, but dramatically improve both stability and selectivity of the transducer.

13.4.5 Nano-electrode arrays: towards the sensor with the record analytical performances

Despite the possibility for detection of hydrogen peroxide down to 10^{-7} mol L^{-1} achieved, both clinical diagnostics and environmental control in certain cases require monitoring of lower H_2O_2 levels. The decrease in detection limits is possible by electrode miniaturization. Miniaturization usually creates new scientific and technological fields, and electroanalysis is not an exception. Nano-technology is expected to play a key role in accessing limiting performance characteristics of chemical and biological sensors.

Diffusion of electroactive species to the surface of conventional disk (macro-) electrodes is mainly planar. When the electrode diameter is decreased the edge effects of hemi-spherical diffusion become significant. In 1964 Lingane derived the corrective term bearing in mind the edge effects for the Cotrell equation [129, 130], confirmed later on analytically and by numerical calculation [131, 132]. In the case of ultramicroelectrodes this term becomes dominant, which makes steady-state current proportional to the electrode radius [133–135]. Since capacitive and other diffusion-unrelated currents are proportional to the square of electrode radius, the signal-to-noise ratio is increased as the electrode radius is decreased.

In the case of a single electrode, however, the decrease of its dimensions requires the measurement of very low currents. To overcome this problem it is convenient to use microelectrode arrays [136, 137]. Despite the fact that in such arrays microelectrodes are electronically connected to each other, analytical properties of such assemblies are advantageous over those of a conventional macro-electrode [138, 139].

Micro- (and even nano-) electrode arrays are commonly produced with photolithography and electronic beam techniques by insulating of macro-electrode surface with subsequent drilling micro-holes in an insulating layer [136, 137]. Physical methods are, however, expensive and, besides that, unsuitable for sensor development in certain cases (for instance, for modification of the lateral surface of needle electrodes). That's why an increasing interest is being applied to chemical approaches of material nano-structuring on solid supports [140, 141].

Nano-electrode arrays can be formed through nano-structuring of the electrocatalyst on an inert electrode support. Indeed, if the current of the analyte reduction (oxidation) on a blank electrode is negligible compared to the activity of the electrocatalyst, the former can be considered as an insulator surface. Hence, for the synthesis of nano-electrode arrays one has to carry out material nano-structuring. Recently, an elegant approach [140] for the electrosynthesis of mesoporous nano-structured surfaces by depositioning different metals (Pt, Pd, Co, Sn) through lyotropic liquid crystalline phases has been proposed [141–143].

Prussian blue-based nano-electrode arrays were formed by deposition of the electrocatalyst through lyotropic liquid crystalline [144] or sol templates onto inert electrode supports. Alternatively, nucleation and growth of Prussian blue at early stages results in nano-structured film [145]. Whereas Prussian blue is known to be a superior electrocatalyst in hydrogen peroxide reduction, carbon materials used as an electrode support demonstrate only a minor activity. Since the electrochemical reaction on the blank electrode is negligible, the nano-structured electrocatalyst can be considered as a nano-electrode array.

The morphology of Prussian blue electrodeposited onto a mono-crystalline graphite surface was investigated by atomic force microscopy (AFM) and is presented in

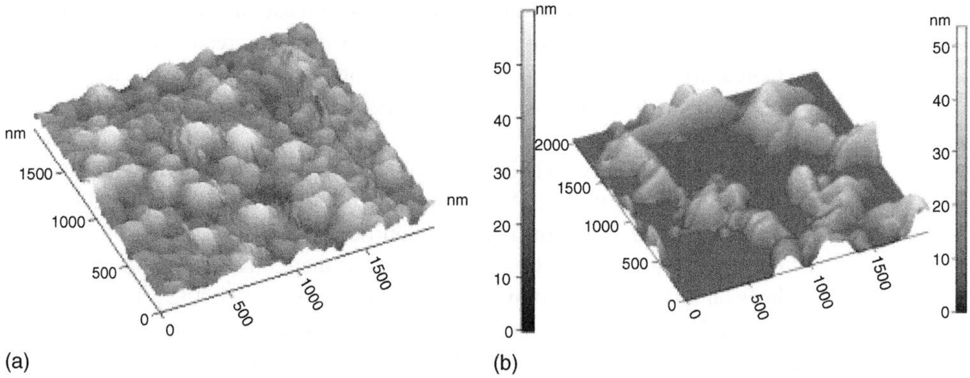

FIGURE 13.5 AFM images of Prussian blue-modified monocrystalline graphite: (a) conventional Prussian blue deposited without surfactants, (b) Prussian blue electrochemically deposited through liquid crystalline phase of non-ionic surfactant Brij-56.

Fig. 13.5. As seen, a conventional Prussian blue film is of polycrystalline structure, however, the layer covers the surface completely.

When Prussian blue has been growing through the liquid crystalline template of non-ionic surfactant, the corresponding AFM-image (Fig. 13.5b) displays an archipelago of new structures, which may be attributed only to ferric hexacyanoferrate [144]. These structures of submicron dimensions are surrounded by very smooth areas, which, taking into account polycrystalline Prussian blue morphology (Fig. 13.5a), displays an obviously unmodified graphite surface.

The analytical performance of Prussian blue-modified electrodes in hydrogen peroxide detection were investigated in a flow-injection system equipped with a wall-jet cell. Nano-structured Prussian blue-modified electrodes demonstrate a significantly decreased background, which results in improved signal-to-noise ratio.

Nano-structuring also results in a decreased detection limit. Since the latter has different explanations in analytical literature, we define it as the lower limit of the linear calibration range. For nano-structured Prussian blue the detection limit was found to be of 1×10^{-9} mol L^{-1} (Fig. 13.6).

Since among the main disadvantages of peroxidase-based H_2O_2 sensors is the saturation of the enzyme with substrate, linear calibration range never reaches the four orders of magnitude of H_2O_2 concentrations. It was essential therefore to investigate the upper detection limit of nano-structured Prussian blue. It is seen that the linear calibration range is extended over *seven* orders of magnitude for H_2O_2 concentration (1×10^{-9}–1×10^{-2} mol L^{-1}). The significant deviation from the linearity is observed only at decimolar (0.1 mol L^{-1}) hydrogen peroxide content.

The resulting Prussian blue-based nano-electrode arrays in FIA demonstrate a sub-ppb detection limit (1×10^{-9} mol L^{-1}) and a linear calibration range starting from the detection limit and extending over seven orders of magnitude of H_2O_2 concentrations ($1 \times 10^{-9} \div 1 \times 10^{-2}$ mol L^{-1}), which is the most advantageous analytical performance in electroanalysis. As a conclusion from the evidence in this chapter, Prussian

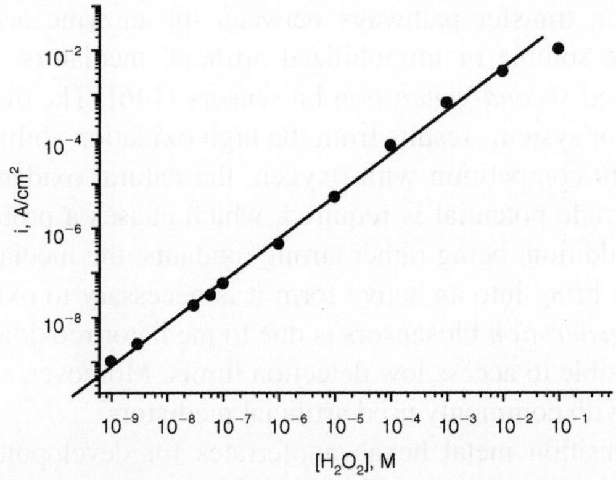

FIGURE 13.6 Calibration plot for hydrogen peroxide detection in flow-injection mode with nano-structured Prussian blue as a detector; Prussian blue electrodeposited through sol template based on the vinyltriethoxysilane, operating potential 50 mV, phosphate buffer pH 6.0 + 0.1 M KCl, flow rate 0.7 ml/min.

blue can be considered to be an ideal electrocatalyst for hydrogen peroxide reduction. The above examples illustrate the performance characteristics of Prussian blue-based transducers, which offer exceptional benefits to electroanalysis.

13.5 BIOSENSORS BASED ON TRANSITION METAL HEXACYANOFERRATES

13.5.1 Transducing principles for oxidase-based biosensors

More than 90% of commercially available enzyme-based biosensors and analytical kits contain oxidases as terminal enzymes responsible for generation of analytical signal. These enzymes catalyze oxidation of specific analyte with molecular oxygen producing hydrogen peroxide according to the reaction:

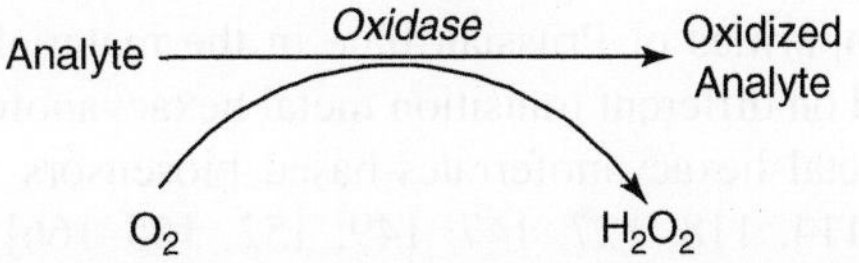

(4)

Among different approaches providing operation of the oxidase-based biosensors, the detection of hydrogen peroxide production was found to be the most progressive one, allowing detection of low levels of analytes [107]. However, the detection of H_2O_2 has to be carried out at low potentials in order to reduce the interference of easily oxidizable compounds [110].

To provide electron transfer pathways between the enzyme active site and the electrode one can use soluble or immobilized artificial mediators. This approach is realized in the so-called *second-generation* biosensors [146]. The disadvantage of the oxidase-based mediator systems results from the high oxidation ability of the mediator required for successful competition with oxygen, the natural oxidant. To oxidize the mediator a high electrode potential is required, which causes a problem with reductants (see above). In addition, being rather strong oxidants, the mediators exist in their reduced states, and to bring into an active form it is necessary to oxidize them. Thus, the signal of *second-generation* biosensors is due to mediator reoxidation, which obviously makes it impossible to access low detection limits. Moreover, some oxidases are reported not to react with commonly used artificial mediators.

Application of transition metal hexacyanoferrates for development of biosensors was first announced by our group in 1994 [118]. The goal was to substitute platinum as the most commonly used hydrogen peroxide transducer for Prussian blue-modified electrode. The enzyme glucose oxidase was immobilized on the top of the transducer in the polymer (Nafion) membrane. The resulting biosensor showed advantageous characteristics of both sensitivity and selectivity in the presence of commonly tested reductants, such as ascorbate and paracetamol.

Another approach for development of Prussian blue-based biosensors was published in 1995 and involved enzyme immobilization by entrapment into Prussian blue films during its deposition [147]. However, as was mentioned, the best media for deposition of Prussian blue is 0.1 M HCl, which is not acceptable for enzymes, in particular for glucose oxidase [148]. Moreover, the entrapment of the enzyme in metal hexacyanoferrates during their deposition does not provide enough enzyme activity of the resulting film, which results in a rather low sensitivity of the resulting biosensor [149] compared to the sensitivity of the corresponding H_2O_2 transducer [150].

13.5.2 Biosensors based on transition metal hexacyanoferrates

Except for Prussian blue activity in hydrogen peroxide, reduction has been shown for a number of transition metal hexacyanoferrates. The latter were cobalt [151], nickel [152], chromium [150], titanium [153], copper [154], manganese [33], and vanadium [28] hexacyanoferrates. However, as was shown in review [117], catalytic activity of the mentioned inorganic materials in H_2O_2 reduction is either very low, or is provided by impurities of Prussian blue in the material. Nevertheless, a number of biosensors based on different transition metal hexacyanoferrates have been developed.

Metal hexacyanoferrates-based biosensors were developed for analysis of glucose [11, 114, 118, 127, 147, 149, 152, 155–166], ethanol [11], D-alanine [147], oxalate [167–169], cholesterol [170, 171], glutamate [114, 119], sucrose [172], and choline [163]. Among the transducers used Prussian blue undoubtedly dominates especially if one takes into account that instead of both chromium and cobalt hexacyanoferrates the activity of the transducers in publications [149, 159, 167, 168] was most probably provided by Prussian blue [117]. The sensitivity of cupric hexacyanoferrate is several orders of magnitude lower compared to Prussian blue. However, chemically synthesized cupric hexacyanoferrate is useful for carbon paste biosensors [162]. An apparently first publication on Prussian blue-based immuno-sensors appeared recently [173].

13.5.3 Immobilization of the enzymes using non-conventional media

A necessary part of biosensor construction is the biorecognition element immobilized on the top surface of the transducer. Thus, immobilization of biocomponents (enzymes in the case of the present review) is an important and crucial point of biosensor construction. Certainly, enzyme immobilization is a widely investigated area, which can even be recognized as a precursor to modern biotechnology. This undoubtedly makes it impossible to review in a single chapter. Hence, only the original protocol for enzyme immobilization providing substantial improvement of analytical performances of the resulting biosensors will be presented.

Among various enzyme immobilization protocols, entrapment in polymer membranes is a general one for a variety of transducers. Formation of a membrane from a solution of already synthesized polymer is simpler and reproducible compared to chemical polymerization. The simplicity of this immobilization procedure should provide reproducibility for the resulting biosensors; the latter is strongly required for mass production.

Since the polymer has to be water-insoluble, its complete solution may occur only in an organic solvent. Casting the polymer from such a solution obviously can improve the properties of the resulting membrane. Thus, to prepare an enzyme containing casting solution the protein has to be exposed to organic solvent.

The investigation of enzymes in water-miscible organic solvents trivially called "non-aqueous enzymology" about 20 years ago became an independent part of modern biochemistry and enzymology [174–176]. In concentrated organic solvents, with the water content less than 10–15%, the enzymes are rather stable and can even retain their activity [176, 177]. Recent studies even demonstrated improvement of the enzyme activity in concentrated organic solvents [178].

The methodology of the proposed immobilization approach [164, 179] is, however, quite different from "non-aqueous enzymology". Though during an immobilization procedure, enzymes have to be exposed to organic solvents, their activity is required only in aqueous solution in which resulting biosensors are operated. Hence, it is only important that the enzymes are able to retain their catalytic properties after exposure to organic solvents.

13.5.3.1 Tolerance of the enzymes to organic solvents

It was thus necessary to investigate what happens with the enzyme activity after exposure to an organic solvent. Figure 13.7 illustrates the remaining activity of alcohol dehydrogenase and glucose oxidase after 30 minutes in concentrated ethanol, isopropanol, and acetonitrile, related to initial enzyme activity in aqueous solution. As the water content in the water–organic mixture is increased, the activity of the enzymes is also initially increased, but further additions of water result in its decrease. The general conclusion made is that after exposure to organic solvents, the enzymes may retain up to 100% of their initial activity in water.

The relative remaining activity of alcohol dehydrogenase in an optimal water–ethanol mixture exceeded 100%. The observation of the improved activity of some of the enzyme samples after this procedure [179] was a precursor to later achievements on enzyme activation in organic media [178].

13.5.3.2 Enzyme-containing perfluorosulfonated membranes

The procedure for the preparation of enzyme containing polymer membranes according to the proposed protocol is simple: casting of the mixture containing the enzyme suspension and polymer solution in organic solvent over the target surface and allowing the volatile organic solvent to evaporate. The casting solution is prepared by (i) a suspension of enzyme in organic solvent and (ii) mixing of the enzyme suspension with the polymer solution.

On the basis of the knowledge that different polyelectrolytes can stabilize proteins [180], even in water–organic mixtures [181], it was important to check the membrane-forming polyelectrolyte itself as a potential stabilizer. Indeed, as was found from spectrophotometric investigations, Nafion stabilizes glucose oxidase suspensions in organic solvents.

Suspending enzyme in polymer solution instead of in pure organic solvent not only simplifies preparation of the casting solution, the enzyme suspensions became more uniform and stable. It was also found that at certain concentrations (enzyme, polyelectrolyte, and water) the resulting membranes exhibited extremes in both stability and

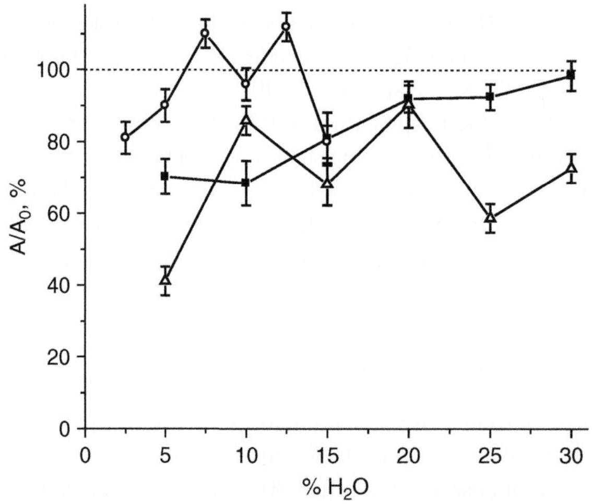

FIGURE 13.7 The remaining activity of alcoholdehydrogenase (○) and glucose oxidase (■, △) after 30 minutes exposing to concentrated ethanol (○), isopropanol (△), and acetonitrile (■) as a function of water content in water–organic mixture.

specific activity of the immobilized glucose oxidase. Since molar concentrations of the enzyme and the polyelectrolyte in the casting solution are of a similar order of magnitude, the observed extremes indicate the formation of an enzyme–polyelectrolyte complex at a molecular recognition level in concentrated organic solvents [164].

Enzyme containing Nafion membranes prepared according to the proposed protocol have shown high specific activity and stability of immobilized glucose oxidase. As expected, the simplicity of preparation provided high reproducibility. When the same casting solution is used, the maximum deviation in membrane activity is <2%. This, however, is also the precision limit for kinetic investigations.

The advanced protocol for enzyme immobilization allowed a glucose biosensor to be produced with good analytical performances.

13.5.4 Towards the biosensors with the best analytical performance characteristics

The glucose biosensor, except for its highly important practical applications, is considered as a test to demonstrate the achievements in biosensorics. Indeed, having more than 40 years of history, glucose biosensors were categorized according to the majority of the principles available.

To make the most advantageous glucose biosensor, it is important to combine the best transduction principle with the best immobilization protocol. As mentioned, the most progressive way to couple the oxidase and the electrode reaction is a low potential detection of hydrogen peroxide. Among available H_2O_2 transducers, Prussian blue is the most advantageous one.

A glucose biosensor was made by immobilization of glucose oxidase over a Prussian blue-modified electrode (Fig. 13.8). The scheme of biosensor operation included glucose oxidation by molecular oxygen producing hydrogen peroxide. The latter is reduced on Prussian blue-modified electrodes producing an analytical signal.

Analytical performances of the glucose biosensor in flow-injection mode are as follows. The biosensor allows detection of glucose down to the 0.1 μM level. The

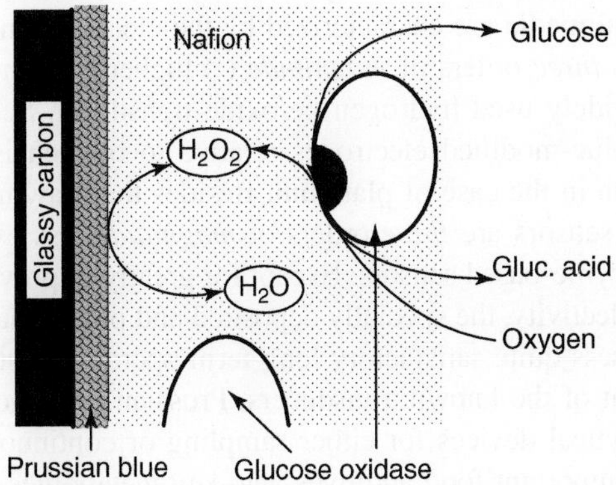

FIGURE 13.8 Principal scheme of the Prussian blue-based glucose biosensor.

calibration plot is linear over four orders of magnitude of glucose concentration from 0.1 μM to 1 mM. The sensitivity of FIA glucose detection, which has been calculated in the linear range up to 1 mM, is of 0.05 A M^{-1} cm^{-2}. This is only ten times lower than the sensitivity of the transducer to hydrogen peroxide used in this biosensor [164].

Let us compare the elaborated biosensor with the glucose sensitive electrodes using a similar transducer. Considering the immobilization of glucose oxidase in poly(vinyl alcohol) grafted 4-vinylpyridine [182], in an additional metal hexacyanoferrate layer during electrodeposition [149], covering by Nafion membrane [160], in gelatin gel [159], and in poly(o-phenylenediamine) during electropolymerization [127], one concludes that the proposed approach displays a dramatic (100-fold) improvement in sensitivity of the resulting biosensor. Moreover, a comparison with the other known biosensors for glucose [183–187] has shown that the proposed sensor is characterized by the best analytical performances [164].

Combined with the attractive performance of a Prussian blue-based hydrogen peroxide transducer, the proposed immobilization protocol provides elaboration of the most advantageous first-generation glucose biosensor concerning its sensitivity and detection limit.

13.6 CONCLUSIONS

In conclusion, the unique properties of Prussian blue and other transition metal hexacyanoferrates, which are advantageous over existing materials concerning their analytical applications, should be mentioned. First, metal hexacyanoferrates provide the possibility to develop amperometric sensors for non-electroactive cations. In contrast to common "smart materials", the sensitivity and selectivity of metal hexacyanoferrates to such ions is provided by thermodynamic background: non-electroactive cations are entrapped in the films for charge compensation upon redox reactions.

Second, Prussian blue is considered the most advantageous low potential transducer for hydrogen peroxide not only among hexacyanoferrates, but over all known systems. The specific activity and, thus, the sensitivity of Prussian blue-modified electrodes to H_2O_2 reduction in neutral media are characterized by the electrochemical rate constant of 0.01 cm s^{-1}, which is *three* orders of *magnitude* (!) higher than in the case of platinum, being the most widely used hydrogen peroxide transducer. Selectivity in H_2O_2 reduction on Prussian blue-modified electrodes relative to oxygen is *three* orders of *magnitude* (!) higher than in the case of platinum. Besides that, raw materials required for Prussian blue-based sensors are *three* orders of *magnitude* less expensive than for noble metal-based or enzyme-based sensors. Except for greatly improved sensitivity (up to 1 A $M^{-1}cm^{-2}$) and selectivity, the specially deposited and post-treated Prussian blue-modified electrodes possess quite satisfactory long-term operational stability compared to or even exceeding that of the known transducers. Prussian blue-modified electrodes are ready to use in analytical devices for either sampling or continuous monitoring of chemical threat agents, important food additives, and key metabolites of life pathways. The particular importance of application of Prussian blue-based biosensors is expected in certain areas of clinical diagnostics, where high sensitivity and selectivity as well as the possibility of miniaturization are required, e.g. brain research and non-invasive monitoring of blood chemistry.

13.7 ACKNOWLEDGMENTS

Financial support through Science Support Foundation, FASIE-Start-05 grant (Russia), is greatly acknowledged.

13.8 REFERENCES

1. *Miscellanea Berolinensia ad Incrementium Scientiarum*, Berlin (1710), p. 377.
2. D. Brown, *J. Philos. Trans.* **33**, 17 (1724).
3. V.D. Neff, Electrochemical oxidation and reduction of thin films of Prussian blue. *J. Electrochem. Soc.* **128**, 886–887 (1978).
4. J.F. Duncan and P.W.R. Wrigley, The electronic structure of the iron atoms in complex iron cyanides. *J. Chem. Soc.* 1120–1125 (1963).
5. J.F. Keggin and F.D. Miles, Structure and formulae of the Prussian blue and related compounds. *Nature* **137**, 577–578 (1936).
6. F. Herren, P. Fisher, A. Ludi, and W. Halg, Neutron difraction study of Prussian blue, $Fe_4[Fe(CN)_6]_3$ xH_2O. Location of water molecules and long-range magnetic order. *Inorgan. Chem.* **19**, 956–959 (1980).
7. K. Itaya, T. Ataka, and S. Toshima, Spectroelectrochemistry and electrochemical preparation method of Prussian blue modified electrodes. *J. Am. Chem. Soc.* **104**, 4767–4772 (1982).
8. J.A. Ibers and N. Davidson, On the interaction between hexacyanatoferrate (III) ions and hexacyanoferrate (II) or iron (III) ions. *J. Am. Chem. Soc.* **73**, 476–478 (1951).
9. K. Itaya, I. Uchida, and V.D. Neff, Electrochemistry of polynuclear transition metal cyanides: Prussian blue and its analogues. *Acc. Chem. Res.* **19**, 162–168 (1986).
10. D. Ellis, M. Eckhoff, and V.D. Neff, Electrochromism in the mixed-valence hexacyanides. 1. Voltammetric and spectral studies of the oxidation and reduction of thin films of Prussian Blue. *J. Phys. Chem.* **85**, 1225–1231 (1981).
11. A.A. Karyakin, E.E. Karyakina, and L. Gorton, Prussian Blue based amperometric biosensors in flow-injection analysis. *Talanta* **43**, 1597–1606 (1996).
12. A.A. Karyakin, E.E. Karyakina, and L. Gorton, The electrocatalytic activity of Prussian blue in hydrogen peroxide reduction studied using a wall-jet electrode with continuous flow. *J. Electroanal. Chem.* **456**, 97–104 (1998).
13. R. Yang, Z.B. Qian, and J.Q. Deng, Electrochemical deposition of Prussian blue from a single ferricyanide solution. *J. Electrochem. Soc.* **145**, 2231–2236 (1998).
14. A.B. Bocarsly and S. Sinha, Chemically derivatized nickel surfaces: synthesis of a new class of stable electrode surfaces. *J. Electroanal. Chem.* **137**, 157–162 (1982).
15. A.K. Jain, R.P. Singh, and C. Bala, Solid membranes of copper hexacyanoferrate(III) as thallium(I)-sensitive electrode. *Anal. Lett.* **15**, 1557–1563 (1982).
16. L.M. Siperko and T. Kuwana, Electrochemical and spectroscopic studies of metal hexacyanometalate films. I. Cupric hexacyanoferrate. *J. Electrochem. Soc.* **130**, 396–402 (1983).
17. P.J. Kulesza and Z. Galus, Polynuclear transition metal hexacyanoferrate films. In-situ electrochemical determination of their composition. *J. Electroanal. Chem.* **267**, 117–127 (1989).
18. S.-M. Chen, Electrocatalytic oxidation of thiosulfate by metal hexacyanoferrate film modified electrodes. *J. Electroanal. Chem.* **417**, 145–153 (1996).

19. D.M. Zhou, H.X. Ju, and H.Y. Chen, Catalytic oxidation of dopamine at a microdisc platinum electrode modified by electrodeposition of nickel hexacyanoferrate and Nafion. *J. Electroanal. Chem.* **408**, 219–223 (1996).
20. P.J. Kulesza, M.A. Malik, R. Schmidt, A. Smolinska, K. Miecznikowski, S. Zamponi, A. Czerwinski, M. Berrettoni, and R. Marassi, Electrochemical preparation and characterization of electrodes modified with mixed hexacyanoferrates of nickel and palladium. *J. Electroanal. Chem.* **487**, 57–65 (2000).
21. M.H. Pournaghi-Azar and H. Dastangoo, Electrochemical characteristics of an aluminum electrode modified by a palladium hexacyanoferrate film, synthesized by a simple electroless procedure. *J. Electroanal. Chem.* **523**, 26–33 (2002).
22. H. Razmi and A. Azadbakht, Electrochemical characteristics of dopamine oxidation at palladium hexacyanoferrate film, electroless plated on aluminum electrode. *Electrochim. Acta* **50**, 2193 (2005).
23. A. Eftekhari, Electrochemical behavior and electrocatalytic activity of a zinc hexacyanoferrate film directly modified electrode. *J. Electroanal. Chem.* **537**, 59–66 (2002).
24. H.A. Arida and R.F. Aglan, A solid-state potassium selective electrode based on potassium zinc ferrocyanide ion exchanger. *Anal. Lett.* **36**, 895–907 (2003).
25. S.Q. Liu and H.Y. Chen, Spectroscopic and voltammetric studies on a lanthanum hexacyanoferrate modified electrode. *J. Electroanal. Chem.* **528**, 190–195 (2002).
26. A. Eftekhari, Electrochemical properties of lanthanum hexacyanoferrate particles immobilized onto electrode surface by Au-codeposition method. *Electroanalysis* **16**, 1324 (2004).
27. P. Wu and C.X. Cai, Electrochemical preparation and characterization of lanthanum hexacyanoferrate modified electrode. *Chin. J. Chem.* **23**, 127 (2005).
28. C.G. Tsiafoulis, P.N. Trikalitis, and M.I. Prodromidis, Synthesis, characterization and performance of vanadium hexacyanoferrate as electrocatalyst of H_2O_2. *Electrochem. Commun.* **7**, 1398 (2005).
29. A. Eftekhari, Silver-selective electrode based on a direct modified electrode silver hexacyanoferrate(II) film. *Anal. Lett.* **33**, 2873–2882 (2000).
30. S.M. Chen, K.T. Peng, and K.C. Lin, Preparation of thallium hexacyanoferrate film and mixed-film modified electrodes with cobalt(II) hexacyanoferrate. *Electroanalysis* **17**, 319 (2005).
31. M.H. Pournaghi-Azar and H. Razmi-Nerbin, Voltammetric behaviour and electrocatalytic activity of the aluminum electrode modified with nickel and nickel hexacyanoferrate films, prepared by electroless deposition. *J. Electroanal. Chem.* **456**, 83–90 (1998).
32. M.H. Pournaghi-Azar and R. Sabzi, Preparation of a cobalt hexacyanoferrate film-modified aluminum electrode by chemical and electrochemical methods: enhanced stability of the electrode in the presence of phosphate and ruthenium(III). *J. Solid State Electrochem.* **6**, 553–559 (2002).
33. A. Eftekhari, Aluminum electrode modified with manganese hexacyanoferrate as a chemical sensor for hydrogen peroxide. *Talanta* **55**, 395–402 (2001).
34. A.K. Jain, R.P. Singh, and C. Bala, Studies on an Araldite-based membrane of copper hexacyanoferrate(III) as a caesium-ion-sensitive electrode. *J. Chem. Technol. Biotechnol., Chem. Technol.* **34A**, 363–366 (1984).
35. J.A. Cox and B.K. Das, Voltammetric determination of nonelectroactive ions at a modified electrode. *Anal. Chem.* **57**, 239–240 (1985).
36. D. Engel and E.W. Grabner, Copper hexacyanoferrate-modified glassy carbon: a novel type of potassium-selective electrode. *Ber. Bunsenges. Phys. Chem.* **89**, 982–986 (1985).
37. K.N. Thomsen and R.P. Baldwin, Amperometric detection of non-electroactive cations in flow systems at a cupric hexacyanoferrate electrode. *Anal. Chem.* **61**, 2594–2598 (1989).
38. K.N. Thomsen and R.P. Baldwin, Evaluation of electrodes coated with metal hexacyanoferrate as amperometric sensors for non-electroactive cations in flow systems. *Electroanalysis* **2**, 263–271 (1990).
39. Y. Tani, H. Eun, and Y. Umezawa, A cation selective electrode based on copper(II) and nickel(II) hexacyanoferrates: dual response mechanisms, selective uptake or adsorption of analyte cations. *Electrochim. Acta* **43**, 3431–3441 (1998).
40. J.M. Zen, P.Y. Chen, and A.S. Kumar, Flow injection analysis of an ultratrace amount of arsenite using a Prussian blue-modified screen-printed electrode. *Anal. Chem.* **75**, 6017–6022 (2003).
41. Z. Gao, X. Zhou, G. Wang, P. Li, and Z. Zhao, Potassium ion-selective electrode based on a cobalt(II) – hexacyanoferrate film-modified electrode. *Anal. Chim. Acta* **244**, 39–48 (1991).

42. R.J. Mortimer, P.J.S. Barbeira, A.F.B. Sene, and N.R. Stradiotto, Potentiometric determination of potassium cations using a nickel(II) hexacyanoferrate-modified electrode. *Talanta* **49**, 271–275 (1999).
43. T. Ikeshoji, Separation of alkali-metal ions by intercalation into a Prussian blue electrode. *J. Electrochem. Soc.* **133**, 2108–2109 (1986).
44. N. Kobayashi, Y. Yamamoto, and M. Akashi, Prussian blue as an agent for decontamination of 137Cs in radiation accidents. *Hoken Butsuri* **33**, 323–330 (1998).
45. M.T. Ganzerli-Valentini, R. Stella, L. Maggi, and G. Ciceri, Copper hexacyanoferrate-(II) and -(III) as trace caesium adsorbers from natural waters. *J. Radioanal. Nucl. Chem.* **114**, 105–112 (1987).
46. L. Johansson, C. Samuelsson, and E. Holm, Adsorption of cesium in urine on copper hexacyanoferrate(II) – a contamination control kit for large scale in situ use. *Radiat. Prot. Dosim.* **81**, 147–152 (1999).
47. C.Y. Huang, J.D. Lee, C.L. Tseng, and J.M. Lo, Rapid method for the determination of caesium-137 in environmental water samples. *Anal. Chim. Acta* **294**, 221–226 (1994).
48. J.-J. Xu, H.-Q. Fang, and H.-Y. Chen, The electrochemical characteristics of an inorganic monolayer film modified gold electrode and its molecular recognition of alkali metal cation. *J. Electroanal. Chem.* **426**, 139–143 (1997).
49. V. Krishnan, A.L. Xidis, and V.D. Neff, Prussian blue solid-state films and membranes as potassium ion-selective electrodes. *Anal. Chim. Acta* **239**, 7–12 (1990).
50. K.C. Ho and C.L. Lin, A novel potassium ion sensing based on Prussian blue thin films. *Sens. Actuators B-chem.* **76**, 512–518 (2001).
51. C. Gabrielli, P. Hemery, P. Liatsi, M. Masure, and H. Perrot, An electrogravimetric study of an all-solid-state potassium selective electrode with Prussian blue as the electroactive solid internal contact. *J. Electrochem. Soc.* **152**, H219 (2005).
52. R.M. Liu, B. Sun, D.J. Liu, and A.L. Sun, Flow injection gas-diffusion amperometric determination of trace amounts of ammonium ions with a cupric hexacyanoferrate. *Talanta* **43**, 1049–1054 (1996).
53. M. Zadronecki, I.A. Linek, J. Stroka, P.K. Wrona, and Z. Gajus, High affinity of thallium ions to copper hexacyanoferrate films. *J. Electrochem. Soc.* **148**, E348–E355 (2001).
54. H. Kahlert, S. Komorsky-Lovric, M. Hermes, and F. Scholz, Prussian blue-based reactive electrode (reactrode) for the determination of thallium ions. *Fresenius' J. Anal. Chem.* **356**, 204–208 (1996).
55. J.M. Zen, H. Ho, and P.Y. Chen, Voltammetric determination of thallium(I) on a Prussian blue/cinder paste electrode. *Indian J. Chem. Sect. A Inorgan. Bio-Inorgan. Phys. Theoret. Anal. Chem.* **42**, 839–842 (2003).
56. H. Duessel, A. Dostal, and F. Scholz, Hexacyanoferrate-based composite ion-sensitive electrodes for voltammetry. *Fresenius' J. Anal. Chem.* **355**, 21–28 (1996).
57. M. Hartmann, E.W. Grabner, and P. Bergveld, Prussian blue-coated inter-digitated array electrodes for possible analytical application. *Anal. Chim. Acta* **242**, 249–257 (1991).
58. M. Hartmann, E.W. Grabner, and P. Bergveld, Alkali ion sensor based on Prussian blue-covered inter-digitated array electrodes. *Sens. Actuators B* **B4**, 333–336 (1991).
59. M. Giorgetti, E. Scavetta, M. Berrettoni, and D. Tonelli, Nickel hexacyanoferrate membrane as a coated wire cation-selective electrode. *Analyst* **126**, 2168–2171 (2001).
60. F. Li and S. Dong, The electrocatalytic oxidation of ascorbic acid on Prussian blue film modified electrodes. *Electrochim. Acta* **32**, 1511–1513 (1987).
61. S.F. Wang, M.A. Jiang, and X.Y. Zhou, Electrocatalytic oxidation of ascorbic acid on nickel hexacyanoferrate film modified electrode. *Gaodeng Xuexiao Huaxue Xuebao* **13**, 325–327 (1992).
62. Z.P. Wang, S.F. Wang, M. Jiang, and X.Y. Zhou, Electrocatalytic oxidation of sulfite at nickel hexacyanoferrate-film modified electrode and its application. *Fenxi-Shiyanshi* **12**, 91–93 (1993).
63. X.Y. Zhou, S.F. Wang, Z.P. Wang, and M. Jiang, Electrocatalytic oxidation of thiosulfate on a modified nickel hexacyanoferrate-film electrode. *Fresenius' J. Anal. Chem.* **345**, 424–427 (1993).
64. T. Garcia, E. Casero, E. Lorenzo, and F. Pariente, Electrochemical sensor for sulfite determination based on iron hexacyanoferrate film modified electrodes. *Sens. Actuators B chem.* **106**, 803 (2005).
65. S.M. Chen and K.T. Peng, The electrochemical properties of dopamine, epinephrine, norepinephrine, and their electrocatalytic reactions on cobalt(II) hexacyanoferrate films. *J. Electroanal. Chem.* **547**, 179–189 (2003).

66. S.M. Chen and C.J. Liao, Preparation and characterization of osmium hexacyanoferrate films and their electrocatalytic properties. *Electrochim. Acta* **50**, 115–125 (2004).
67. F. Xu, M.N. Gao, L. Wang, T.S. Zhou, L.T. Jin, and J.Y. Jin, Amperometric determination of morphine on cobalt hexacyanoferrate modified electrode in rat brain microdialysates. *Talanta* **58**, 427–432 (2002).
68. K.C. Ho, C.Y. Chen, H.C. Hsu, L.C. Chen, S.C. Shiesh, and X.Z. Lin, Amperometric detection of morphine at a Prussian blue-modified indium tin oxide electrode. *Biosens. and Bioelectron.* **20**, 3 (2004).
69. P. Wang, X.Y. Jing, W.Y. Zhang, and G.Y. Zhu, Renewable manganous hexacyanoferrate-modified graphite organosilicate composite electrode and its electrocatalytic oxidation of L-cysteine. *J. Solid State Electrochem.* **5**, 369–374 (2001).
70. L.G. Shaidarova, S.A. Ziganshina, L.N. Tikhonova, and G.K. Budnikov, Electrocatalytic oxidation and flow-injection determination of sulfur-containing amino acids at graphite electrodes modified with a ruthenium hexacyanoferrate film. *J. Anal. Chem.* **58**, 1144–1150 (2003).
71. P.N. Deepa and S.S. Narayanan, Sol-gel coated Prussian blue modified electrode for electrocatalytic oxidation and amperometric determination of thiols. *Bull. Electrochem.* **17**, 259–264 (2001).
72. D.R. Shankaran and S.S. Narayanan, Amperometric sensor for thiols based on mechanically immobilised nickel hexacyanoferrate modified electrode. *Bull. Electrochem.* **17**, 277–280 (2001).
73. M.C. Liu, P. Li, Y.X. Cheng, Y.Z. Xian, C.L. Zhang, and L.T. Jin, Determination of thiol compounds in rat striatum microdialysate by HPLC with a nanosized CoHCF-modified electrode. *Anal. Bioanal. Chem.* **380**, 742 (2004).
74. F. Ricci, F. Arduini, A. Amine, D. Moscone, and G. Palleschi, Characterisation of Prussian blue modified screen-printed electrodes for thiol detection. *J. Electroanal. Chem.* **563**, 229–237 (2004).
75. E. Casero, F. Pariente, and E. Lorenzo, Electrocatalytic oxidation of nitric oxide at indium hexacyanoferrate film-modified electrodes. *Anal. Bioanal. Chem.* **375**, 294–299 (2003).
76. J.M. Zen, A.S. Kumar, and H.W. Chen, Electrochemical behavior of stable cinder/Prussian blue analogue and its mediated nitrite oxidation. *Electroanalysis* **13**, 1171–1178 (2001).
77. K.C. Pan, C.S. Chuang, S.H. Cheng, and Y.O. Su, Electrocatalytic reactions of nitric oxide on Prussian blue film modified electrodes. *J. Electroanal. Chem.* **501**, 160–165 (2001).
78. V.G. Bonifacio, L.H. Marcolino, M.F.S. Teixeira, and O. Fatibello-Filho, Voltammetric determination of isoprenaline in pharmaceutical preparations using a copper(II) hexacyanoferrate(III) modified carbon paste electrode. *Microchem. J.* **78**, 55–59 (2004).
79. M.F.S. Teixeira, A. Segnini, F.C. Moraes, L.H. Marcolino, O. Fatibello, and E.T.G. Cavalheiro, Determination of vitamin B-6 (pyridoxine) in pharmaceutical preparations by cyclic voltammetry at a copper(II) hexacyanoferrate(III) modified carbon paste electrode. *J. Brazilian Chem. Soc.* **14**, 316–321 (2003).
80. U. Scharf and E.W. Grabner, Electrocatalytic oxidation of hydrazine at a Prussian Blue-modified glassy carbon electrode. *Electrochim. Acta* **41**, 233–239 (1996).
81. S.M. Golabi and F. Noor-Mohammadi, Electrocatalytic oxidation of hydrazine at cobalt hexacyanoferrate-modified glassy carbon, Pt and Au electrodes. *J. Solid State Electrochem.* **2**, 30–37 (1998).
82. S.-M. Chen, Characterization and electrocatalytic properties of cobalt hexacyanoferrate films. *Electrochim. Acta* **43**, 3359–3369 (1998).
83. D.R. Shankaran and S.S. Narayanan, Amperometric sensor for hydrazine determination based on mechanically immobilized nickel hexacyanoferrate modified electrode. *Russian J. Electrochem.* **37**, 1149–1153 (2001).
84. C.X. Cai, H.X. Ju, and H.Y. Chen, Cobalt hexacyanoferrate-modified microband gold electrode and its electrocatalytic activity for oxidation of NADH. *J. Electroanal. Chem.* **397**, 185–190 (1995).
85. C.X. Cai, H.X. Ju, and H.Y. Chen, Catalytic oxidation of reduced nicotinamide adenine dinucleotide at a microband gold electrode modified with nickel hexacyanoferrate. *Anal. Chim. Acta* **310**, 145--151 (1995).
86. A.A. Karyakin, Y.N. Ivanova, K.V. Revunova, and E.E. Karyakina, Electropolymerized flavin adenine dinucleotide as an advanced NADH transducer. *Anal. Chem.* **76**, 2004–2009 (2004).
87. A. Ciszewski and G. Milczarek, Electrocatalysis of NADH oxidation with an electropolymerized film of 1,4-bis(3,4-dihydroxyphenyl)-2,3-dimethylbutane. *Anal. Chem.* **72**, 3203–3209 (2000).

88. N. Mano and A. Kuhn, Immobilized nitro-fluorenone derivatives as electrocatalysts for NADH oxidation. *J. Electroanal. Chem.* **477**, 79–88 (1999).
89. A.A. Karyakin, E.E. Karyakina, W. Schuhmann, and H.L. Schmidt, Electropolymerized azines: Part II. In search of the best electrocatalyst of NADH oxidation. *Electroanalysis* **11**, 553–557 (1999).
90. A. Abbaspour and M.A. Mehrgardi, Electrocatalytic oxidation of guanine and DNA on a carbon paste electrode modified by cobalt hexacyanoferrate films. *Anal. Chem.* **76**, 5690–5696 (2004).
91. K. Ogura, N. Endo, M. Nakayama, and H. Ootsuka, Mediated activation and electroreduction of CO_2 on modified electrodes with conducting polymer and inorganic conductor films. *J. Electrochem. Soc.* **142**, 4026–4032 (1995).
92. M.F. de Oliveira, R.J. Mortimer, and N.R. Stradiotto, Voltammetric determination of persulfate anions using an electrode modified with a Prussian blue film. *Microchem. J.* **64**, 155–159 (2000).
93. Y. Wang, J. Huang, C. Zhang, J. Wei, and X. Zhou, Determination of hydrogen peroxide in rainwater by using a polyaniline film and platinum particles co-modified carbon fiber microelectrode. *Electroanalysis* **10**, 776–778 (1998).
94. W.B. Nowall and W.G. Kuhr, Detection of hydrogen peroxide and other molecules of biological importance at an electrocatalytic surface on a carbon fiber microelectrode. *Electroanalysis* **9**, 102–109 (1997).
95. A. Schwake, B. Ross, and K. Cammann, Chrono amperometric determination of hydrogen peroxide in swimming pool water using ultramicroelectrode array. *Sens. Actuators, B* **B 46**, 242–248 (1998).
96. B. Strausak and W. Schoch, in *European Patent Application*, Vol. bulletine 85/15. EP 0136973, European Patent Application (1985).
97. P.A. MacCarthy and A.M. Shah, Oxidative stress and heart failure. *Coron. Artery Dis.* **14**, 109–113 (2003).
98. R. Rodrigo and G. Rivera, Renal damage mediated by oxidative stress: a hypothesis of protective effects of red wine. *Free Rad. Biol. Med.* **33**, 409–422 (2002).
99. R.S. Sohal, R.J. Mockett, and W.C. Orr, Mechanisms of aging: an appraisal of the oxidative stress hypothesis. *Free Rad. Biol. Med.* **33**, 575–586 (2002).
100. T.T.C. Yang, S. Devaraj, and I. Jialal, Oxidative stress and atherosclerosis. *J. Clin. Ligand Assay* **24**, 13–24 (2001).
101. M.A. Yorek, The role of oxidative stress in diabetic vascular and neural disease. *Free Rad. Res.* **37**, 471–480 (2003).
102. A.W. Boots, G. Haenen, and A. Bast, Oxidant metabolism in chronic obstructive pulmonary disease. *Eur. Respir. J.* **22**, 14S–27S (2003).
103. H. Chang, W. Oehrl, P. Elsner, and J.J. Thiele, The role of H_2O_2 as a mediator of UVB-induced apoptosis in keratinocytes. *Free Rad. Res.* **37**, 655–663 (2003).
104. S. Tada-Oikawa, Y. Hiraku, M. Kawanishi, and S. Kawanishi, Mechanism for generation of hydrogen peroxide and change of mitochondrial membrane potential during rotenone-induced apoptosis. *Life Sci.* **73**, 3277–3288 (2003).
105. L.C. Clark and C. Lyons, Electrode systems for continuous monitoring in cardiovascular surgery. *Ann. NY Acad. Sci.* **102**, 29–45 (1962).
106. S.J. Updike and J.P. Hiks, The enzyme electrode. *Nature* **214**, 986–988 (1967).
107. G.G. Guilbault, G.J. Lubrano, and D.N. Gray, Glass-metal composite electrodes. *Anal. Chem.* **45**, 2255–2259 (1973).
108. G.G. Guilbault and G.J. Lubrano, Amperometric enzyme electrodes. Amino acid oxidase. *Anal. Chim. Acta* **69**, 183–185 (1974).
109. G.G. Guilbault and G.J. Lubrano, An enzyme electrode for amperometric determination of glucose. *Anal. Chim. Acta* **64**, 439–455 (1973).
110. F.W. Scheller, D. Pfeifer, F. Schubert, R. Reneberg, and D. Kirstein, in Biosensors: Fundamental and Applications (A.P.F. Turner, I. Karube, and J.S. Wilson, eds). Oxford University Press, Oxford (1987).
111. A.I. Yaropolov, V. Malovik, S.D. Varfolomeev, and I.V. Berezin, Electroreduction of hydrogen peroxide on an electrode with immobilized peroxidase. *Dokl. Akad. Nauk SSSR* **249**, 1399–1401 (1979).
112. T. Ruzgas, E. Csщregi, J. Emnйus, L. Gorton, and G. Marko-Varga, Peroxidase-modified electrodes. fundamentals and applications. *Anal. Chim. Acta* **330**, 123–138 (1996).

113. A.A. Karyakin, E.E. Karyakina, and L. Gorton, On the mechanism of H_2O_2 reduction at Prussian blue modified electrodes. *Electrochem. Commun.* **1**, 78–82 (1999).
114. A.A. Karyakin and E.E. Karyakina, Prussian Blue-based "artificial peroxidase" as a transducer for hydrogen peroxide detection. Application to biosensors. *Sens. Actuators, B* **B57**, 268–273 (1999).
115. I.L. Mattos, L. Gorton, T. Ruzgas, and A.A. Karyakin, Sensor for hydrogen peroxide based on Prussian blue modified electrode: improvement of the operational stability. *Anal. Sci.* **16**, 1–5 (2000).
116. B.B. Hasinoff and H.B. Dunford, Kinetics of the oxidation of ferrocyanide by hosradish peroxidase compounds I and II. *Biochemistry* **9**, 4930 (1970).
117. A.A. Karyakin, Prussian Blue and its analogues: electrochemistry and analytical applications. *Electroanalysis* **13**, 813–819 (2001).
118. A.A. Karyakin, O.V. Gitelmacher, and E.E. Karyakina, A high-sensitive glucose amperometric biosensor based on Prussian blue modified electrodes. *Anal. Lett.* **27**, 2861–2869 (1994).
119. A.A. Karyakin, E.E. Karyakina, and L. Gorton, Amperometric biosensor for glutamate using Prussian Blue-based "artificial peroxldase" as a transducer for hydrogen peroxide. *Anal. Chem.* **72**, 1720–1723 (2000).
120. J. Ruzicka and E.H. Hansen, *Flow Injection Analysis*. John Willey & Sons, New York, Toronto (1988).
121. W.Y. Hou and E. Wang, Flow-injection amperometric detection of hydrazine by electrocatalytic oxidation at a Prussian blue film-modified electrode. *Anal. Chim. Acta* **257**, 275–280 (1992).
122. J. Wang, P.V.A. Pamidi, and D.S. Park, Sol-gel-derived metal-dispersed carbon composite amperometric biosensors. *Electroanalysis* **9**, 52–55 (1997).
123. Y. Guo, A.R. Guadalupe, O. Resto, L.F. Fonseca, and S.Z. Weisz, Chemically derived Prussian blue sol-gel composite thin films. *Chem. Mater.* **11**, 135–140 (1999).
124. Y. Sato, T. Sawaguchi, Hirata, Y., F. Mizutani, and S. Yabuki, Glucose xidse/polyion complex-bilayer membrane for elimination of electroactive interferents in amperometric glucose sensor. *Anal. Chim. Acta* **364**, 173–179 (1998).
125. J.P. Lowry and R.D. O'Neill, Partial characterization in vitro of glucose oxidase modified poly(phenylenediamine)-coated electrodes for neurochemical analysis in vivo. *Electroanalysis* **6**, 369–379 (1994).
126. A.A. Karyakin and M.F. Chaplin, Polypyrrole-Prussian Blue films with controlled level of doping: codeposition of polypyrrole and Prussian Blue. *J. Electroanal. Chem.* **370**, 301–303 (1994).
127. R. Garjonyte and A. Malinauskas, Amperometric glucose biosensor based on glucose oxidase immobilized in poly(o-phenylenediamine) layer. *Sens. Actuators, B* **B56**, 85–92 (1999).
128. L.V. Lukachova, E.A. Kotel'nikova, D. D'Ottavi, E.A. Shkerin, E.E. Karyakina, D. Moscone, G. Palleschi, A. Curulli, and A.A. Karyakin, Electrosynthesis of poly-o-diaminobenzene on the Prussian Blue modified electrodes for improvement of hydrogen peroxide transducer characteristics. *Bioelectrochemistry* **55**, 145–148 (2002).
129. P.J. Lingane, Chronopotentiometry and chronoamperometry with unshielded planar electrodes. *Anal. Chem.* **36**, 1723–1726 (1964).
130. Z.G. Soos and P.J. Lingane, Derivation of the chronoamperometric constant for unshielded, circular, planar electrodes. *J. Phys. Chem.* **68**, 3821–3828 (1964).
131. K. Aoki and J. Osteryoung, Diffusion-controlled current at the stationary finite disk electrode – theory. *J. Electroanal. Chem.* **122**, 19–35 (1981).
132. J. Heinze, Diffusion processes at finite (micro) disk electrodes solved by digital simulation. *J. Electroanal. Chem.* **124**, 73–86 (1981).
133. K. Aoki, Theory Of ultramicroelectrodes. *Electroanalysis* **5**, 627–639 (1993).
134. J. Newman, Resistance for flow of current to a disk. *J. Electrochem. Soc.* **113**, 501–502 (1966).
135. K.B. Oldham, Edge effects in semiinfinite diffusion. *J. Electroanal. Chem.* **122**, 1–17 (1981).
136. R. Feeney and S.P. Kounaves, Microfabricated ultramicroelectrode arrays: developments, advances, and applications in environmental analysis. *Electroanalysis* **12**, 677–684 (2000).
137. G.C. Fiaccabrino and M. Koudelka-Hep, Thin-film microfabrication of electrochemical transducers. *Electroanalysis* **10**, 217–222 (1998).
138. C. Amatore, J.M. Saveant, and D. Tessier, Charge transfer at partially blocked surfaces. A model for the case of microscopic active and inactive sites. *J. Electroanal. Chem.* **147**, 39–51 (1983).

139. T. Gueshi, K. Tokuda, and H. Matsuda, Voltammetry at partially covered electrodes. Part I. Chronopotentiometry and chronoamperometry at model electrodes. *J. Electroanal. Chem.* **89**, 247–260 (1978).
140. G.S. Attard, P.N. Bartlett, N.R.B. Coleman, J.M. Elliott, J.R. Owen, and J.H. Wang, Mesoporous platinum films from lyotropic liquid crystalline phases. *Science* **278**, 838–840 (1997).
141. P.N. Bartlett, P.N. Birkin, M.A. Ghanem, P. de Groot, and M. Sawickib, The electrochemical deposition of nanostructured cobalt films from lyotropic liquid crystalline media. *J. Electrochem. Soc.* **148**, C119–C123 (2001).
142. G.S. Attard, P.N. Bartlett, N.R.B. Coleman, J.M. Elliott, and J.R. Owen, Lyotropic liquid crystalline properties of nonionic surfactant/H_2O/hexachloroplatinic acid ternary mixtures used for the production of nanostructured platinum. *Langmuir* **14**, 7340–7342 (1998).
143. A.H. Whitehead, J.M. Elliott, J.R. Owen, and G.S. Attard, Electrodeposition of mesoporous tin films. *Chem. Commun.*, 331–332 (1999).
144. A.A. Karyakin, E.A. Puganova, I.A. Budashov, I.N. Kurochkin, E.E. Karyakina, V.A. Levchenko, V.N. Matveyenko, and S.D. Varfolomeyev, Prussian Blue based nanoelectrode arrays for H_2O_2 detection. *Anal. Chem.* **76**, 474–478 (2004).
145. E.A. Puganova and A.A. Karyakin, New materials based on nanostructured Prussian blue for development of hydrogen peroxide sensors. *Sens. Actuators B chem.* **109**, 167–170 (2005).
146. A.E.G. Cass, G. Davis, G.D. Francis, H.A.O. Hill, W.G. Aston, I.J. Higgins, E.V. Plotkin, L.D.L. Scott, and A.P.F. Turner, Ferrocene-mediated enzyme electrode for amperometric detection of glucose. *Anal. Chem.* **56**, 667–671 (1984).
147. Q.J. Chi and S.J. Dong, Amperometric biosensors based on the immobilization of oxidases in a Prussian blue film by electrochemical codeposition. *Anal. Chim. Acta* **310**, 429–436 (1995).
148. R. Wilson and A.P.F. Turner, Glucose oxidase: an ideal enzyme. *Biosens. Bioelectr.* **7**, 165–185 (1992).
149. M.S. Lin and W.C. Shih, Chromium hexacyanoferrate based glucose biosensor. *Anal. Chim. Acta* **381**, 183–189 (1999).
150. M.S. Lin, T.F. Tseng, and W.C. Shih, Chromium(III) hexacyanoferrate(II)-based chemical sensor for the cathodic determination of hydrogen peroxide. *Analyst* **123**, 159–163 (1998).
151. M.S. Lin and B.I. Jan, Determination of hydrogen peroxide by utilizing a cobalt(II)hexacyanoferrate-modified glassy carbon electrode as a chemical sensor. *Electroanalysis* **9**, 340–344 (1997).
152. S. Milardovic, I. Kruhak, D. Ivekovic, V. Rumenjak, M. Tkalcec, and B.S. Grabaric, Glucose determination in blood samples using flow injection analysis and an amperometric biosensor based on glucose oxidase immobilized on hexacyanoferrate modified nickel electrode. *Anal. Chim. Acta* **350**, 91–96 (1997).
153. Y. Mishima, J. Motonaka, K. Maruyama, and S. Ikeda, Determination of hydrogen peroxide using a potassium hexacyanoferrate(III) modified titanium dioxide electrode. *Anal. Chim. Acta* **358**, 291–296 (1998).
154. R. Garjonyte and A. Malinauskas, Operational stability of amperometric hydrogen peroxide sensors, based on ferrous and copper hexacyanoferrates. *Sens. Actuators, B* **B56**, 93–97 (1999).
155. A.A. Karyakin, O.V. Gitelmacher, and E.E. Karyakina, Prussian blue-based first-generation biosensor. A sensitive amperometric electrode for glucose. *Anal. Chem.* **67**, 2419–2423 (1995).
156. X. Zhang, J. Wang, B. Ogorevc, and U.E. Spichiger, Glucose nanosensor based on Prussian-blue modified carbon-fiber cone nanoelectrode and an integrated reference electrode. *Electroanalysis* **11**, 945–949 (1999).
157. S.A. Jaffari and J.C. Pickup, Novel hexacyanoferrate(III)-modified carbon electrodes: application in miniaturized biosensors with potential for in vivo glucose sensing. *Biosens. Bioelect.* **11**, 1167–1175 (1996).
158. S.A. Jaffari and A.P.F. Turner, Novel hexacyanoferrate(III) modified graphite disc electrodes and their application in enzyme electrodes.1. *Biosens. Bioelectr.* **12**, 1–9 (1997).
159. J.Z. Zhang and S.J. Dong, Cobalt(II)hexacyanoferrate film modified glassy carbon electrode for construction of a glucose biosensor. *Anal. Lett.* **32**, 2925–2936 (1999).
160. I.L. Mattos, L. Gorton, T. Laurell, A. Malinauskas, and A.A. Karyakin, Development of biosensors based on hexacyanoferrates. *Talanta* **52**, 791–799 (2000).

161. J. Wang and X. Zhang, Screen printed cupric-hexacyanoferrate modified carbon enzyme electrode for single-use glucose measurements. *Anal. Lett.* **32**, 1739–1749 (1999).
162. J. Wang, X.J. Zhang, and M. Prakash, Glucose microsensors based on carbon paste enzyme electrodes modified with cupric hexacyanoferrate. *Anal. Chim. Acta* **395**, 11–16 (1999).
163. D. Moscone, D. D'Ottavi, D. Compagnone, G. Palleschi, and A. Amine, Construction and analytical characterization of Prussian Blue-based carbon paste electrodes and their assembly as oxidase enzyme sensors. *Anal. Chem.* **73**, 2529–2535 (2001).
164. A.A. Karyakin, E.A. Kotel'nikova, L.V. Lukachova, E.E. Karyakina, and J. Wang, Optimal environment for glucose oxidase in perfluorosulfonated ionomer membranes: improvement of first-generation biosensors. *Anal. Chem.* **74**, 1597–1603 (2002).
165. J. Wang and A.S. Arribas, Biocatalytically induced formation of cupric ferrocyanide nanoparticles and their application for electrochemical and optical biosensing of glucose. *Small* **2**, 129 (2006).
166. X.C. Tan, Y.X. Tian, P.X. Cai, and X.Y. Zou, Glucose biosensor based on glucose oxidase immobilized in sol-gel chitosan/silica hybrid composite film on Prussian blue modified glass carbon electrode. *Anal. Bioanal. Chem.* **381**, 500 (2005).
167. S. Milardovic, Z. Grabaric, V. Rumenjak, and M. Jukic, Rapid determination of oxalate by an amperometric oxalate oxidase-based electrode. *Electroanalysis* **12**, 1051–1058 (2000).
168. S. Milardovic, Z. Grabaric, M. Tkalcec, and V. Rumenjak, Determination of oxalate in urine; using an amperometric biosensor with oxalate oxidase immobilized on the surface of a chromium hexacyanoferrate-modified graphite electrode. *J. AOAC Int.* **83**, 1212–1217 (2000).
169. P.A. Fiorito and S.I.C. de Torresi, Hybrid nickel hexacyanoferrate/polypyrrole composite as mediator for hydrogen peroxide detection and its application in oxidase-based biosensors. *J. Electroanal. Chem.* **581**, 31 (2005).
170. J.C. Vidal, J. Espuelas, E. Garcia-Ruiz, and J.R. Castillo, Amperometric cholesterol biosensors based on the electropolymerization of pyrrole and the electrocatalytic effect of Prussian-Blue layers helped with self-assembled monolayers. *Talanta* **64**, 655 (2004).
171. J.P. Li, T.Z. Peng, and Y.Q. Peng, A cholesterol biosensor based on entrapment of cholesterol oxidase in a silicic sol-gel matrix at a Prussian blue modified electrode. *Electroanalysis* **15**, 1031–1037 (2003).
172. B. Haghighi, S. Varma, F.M. Alizadeh, Y. Yigzaw, and L. Gorton, Prussian blue modified glassy carbon electrodes – study on operational stability and its application as a sucrose biosensor. *Talanta* **64**, 3–12 (2004).
173. J.G. Guan, Y.Q. Miao, and J.R. Chen, Prussian blue modified amperometric FIA biosensor: one-step immunoassay for alpha-fetoprotein. *Biosens. Bioelectr.* **19**, 789–794 (2004).
174. Y.L. Khmelnitsky, A.V. Levashov, N.L. Klyachko, and K. Martinek, Engineering biocatalytic systems in organic media with low water content. *Enzyme Microb. Technol.* **10**, 710–724 (1988).
175. H. Kise and H. Shirato, Enzymatic reactions in aqueous-organic media. V. Medium effect on the esterification of aromatic amino acids by a-chymotrypsin. *Enzyme Microb. Technol.* **10**, 582–585 (1988).
176. A. Zaks and A.M. Klibanov, Enzymatic catalysis in nonaqueous solvents. *J. Biol. Chem.* **263**, 3194–3201 (1988).
177. K. Griebenow and A.M. Klibanov, On protein denaturation in aqueous-organic mixtures but not in pure organic solvents. *J. Am. Chem. Soc.* **118**, 11 695–11 700 (1996).
178. L.Z. Dai and A.M. Klibanov, Striking activation of oxidative enzymes suspended in nonaqueous media. *Proc. Nat. Acad. Sci. U.S.A* **96**, 9475–9478 (1999).
179. A.A. Karyakin, E.E. Karyakina, L. Gorton, O.A. Bobrova, L.V. Lukachova, A.K. Gladilin, and A.V. Levashov, The improvement of electrochemical biosensors using enzyme immobilisation from water-organic mixtures with the high content of organic solvent. *Anal. Chem.* **68**, 4335–4341 (1996).
180. B. Appleton, T.D. Gibson, and J.R. Woodward, High temperature stabilisation of immobilised glucose oxidase: potential applications in biosensors. *Sens. Actuators, B* **B43**, 65–69 (1997).
181. E.V. Kudryashova, A.K. Gladilin, A.V. Vakurov, F. Heitz, A.V. Levashov, and V.V. Mozhaev, Enzyme-polyelectrolyte complexes in water-ethanol mixtures: negatively charged groups artificially introduced into alpha-chymotrypsin provide additional activation and stabilization effects. *Biotechnol. Bioeng.* **55**, 267–277 (1997).

182. Q. Deng, B. Li, and S. Dong, Self-gelatinizable copolymer immobilized glucose biosensor based on Prussian Blue modified graphite electrode. *Analyst* **123**, 1995–1999 (1998).
183. W. Sung and Y. Bae, A glucose oxidase electrode based on electropolymerized conducting polymer with polyanion-enzyme conjugated dopant. *Anal. Chem.* **72**, 2177–2181 (2000).
184. S. Kruger, S. Setford, and A. Turner, Assessment of glucose oxidase behaviour in alcoholic solutions using disposable electrodes. *Anal. Chim. Acta* **368**, 219–231 (1998).
185. J. Pei and X. Li, Amperometric glucose enzyme sensor prepared by immobilizing glucose oxidase on CuPtCl6 chemically modified electrode. *Electroanalysis* **11**, 1266–1272 (1999).
186. J. Wang, G. Rivas, and M. Chicharro, Iridium-dispersed carbon paste enzyme electrodes. *Electroanalysis* **8**, 434–437 (1995).
187. Z. Zhang, H. Liu, and J. Deng, A glucose biosensor based on immobilization of glucose oxidase in electropolymerized o-aminophenol film on platinized glassy carbon electrode. *Anal. Chem.* **68**, 1632–1638 (1996).

CHAPTER 14

Nanoparticle-based biosensors and bioassays

Guodong Liu, Jun Wang, Yuehe Lin, and Joseph Wang

14.1 INTRODUCTION

The emergence of nanotechnology is opening new horizons for the application of nanoparticles in biosensors and bioassays. In particular, nanoparticles are of considerable interest in the world of nanoscience due to their unique physical and chemical properties. Such properties offer excellent prospects for chemical and biological sensing [1–3]. Nanoparticles with different compositions and dimensions have been widely used in recent years as versatile and sensitive tracers for the electronic, optical, and microgravimetric transduction of different biomolecular recognition events [4–8]. Colloidal gold and semiconductor quantum dot nanoparticles are particularly attractive for numerous bioanalytical applications. The power and scope of such nanoparticles can be greatly enhanced by coupling them with biological recognition reactions and electrical processes (i.e. nanobioelectronics). The enormous signal enhancement associated with the use of nanoparticle amplifying labels and with the formation of nanoparticle–biomolecule assemblies provides the basis for ultrasensitive optical and electrical detection with polymerase chain reaction (PCR)-like sensitivity. Such protocols couple the amplification features of nanoparticle–biomolecule assemblies with highly sensitive optical or electrochemical transduction schemes. Multi-amplification protocols, combining several nanomaterial-based amplification units and processes, can also be designed for addressing further the high sensitivity demands of modern bioassays. There has also been a substantial interest recently in using biomolecules to construct nanostructured architectures [1, 9] and in tailoring and functionalizing the surfaces of nanoparticles [2, 9]. Nanoparticle-based biosensors thus offer great potential for DNA and protein diagnostics and can have a profound impact upon bioanalytical chemistry.

The applications of nanoparticles in biosensors can be classified into two categories according to their functions: (1) nanoparticle-modified transducers for bioanalytical applications and (2) biomolecule–nanoparticle conjugates as labels for biosensing and bioassays. We intend to review some of the major advances and milestones in biosensor development based upon nanoparticle labels and their roles in biosensors and bioassays for nucleic acids and proteins. Moreover, we focus on some of the key fundamental properties of certain nanoparticles that make them ideal for different biosensing applications.

14.2 WHY NANOPARTICLES?

The unique properties of nanoscale materials offer excellent prospects for designing highly sensitive and selective bioassays of nucleic acids and proteins. The creation of such designer nanomaterials for specific biosensing and bioassay applications greatly benefits from being able to vary the size, composition, and shape of the materials and hence tailor their physical and chemical properties. Owing to the tiny size of nanomaterials, their properties are strongly influenced by the binding of target biomolecules. Nanoparticles of different compositions and dimensions have been widely used in recent years as versatile and sensitive tracers for the electronic, optical, and microgravimetric transduction of different biomolecular recognition events [4–8]. The enormous signal enhancement associated with using nanoparticle amplifying labels and with forming nanoparticle–biomolecule assemblies provides the basis for ultrasensitive optical and electrical detection with PCR-like sensitivity. Such protocols couple the amplification features of nanoparticle–biomolecule assemblies with highly sensitive optical or electrochemical transduction schemes. Multi-amplification protocols, combining several nanomaterial-based amplification units and processes, can also be designed for addressing further the high sensitivity demands of modern bioassays. The unique catalytic properties of metal nanoparticles stimulate their enlargement by the same metal or another one to offer substantial signal amplification. It is possible also to dramatically increase the number of tags per binding event (and achieve an enormous signal amplification) by encapsulating numerous signal-generating molecules within a nanoparticle host. These nanomaterial-based biosensing and bioassays can be combined with additional amplification processes, such as surface preconcentration or enzymatic recycling. The following sections discuss nanoparticle labels for ultrasensitive optical and electronic biosensing, and bioassays.

14.3 NANOPARTICLE-BASED OPTICAL BIOSENSORS AND BIOASSAY

The driving force behind the use of nanoparticle labels in optical biosensors (or bioassays) has been to address the significant chemical and spectra limitations of organic fluorophores. Early work by Mirkin and coworkers demonstrated that the aggregation of gold nanoparticles, induced by DNA hybridization, leads to materials with remarkable optical properties [10, 11]. For example, the distance-dependent optical properties of aggregated gold nanoparticles were exploited for developing a simple and fast colorimetric protocol for detecting polynucleotides [11]. Such hybridization-induced aggregation of nanoparticle-modified DNA led to a rapid change of the solution color from red to blue. The resulting gold-nanoparticle–DNA bioassemblies displayed remarkably sharp DNA melting curves that allowed convenient differentiation of oligonucleotides with single-base imperfections [10, 12]. Taton *et al.* described a highly sensitive scanometric DNA array detection based on the use of oligonucleotide targets, labeled with gold nanoparticles, for recognizing DNA segments on a chip (Fig. 14.1) [13]. A nanoparticle-promoted reduction of silver(1) led to a dramatic (105-fold) signal amplification and to

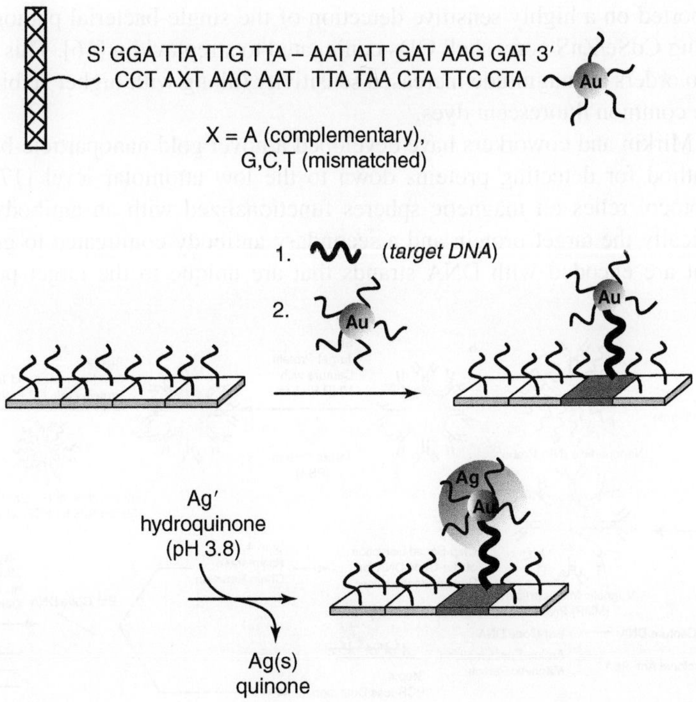

FIGURE 14.1 Scanometric DNA array detection with enlarged nanoparticle probes. Use of oligonucleotide targets, labeled with gold nanoparticles, for recognizing DNA segments on a chip (reproduced from [13] with permission).

a 100-fold increase in sensitivity compared to analogous fluorescence-based assays. The silver enhancement relies on the chemical reduction of silver ions by hydroquinone to silver metal on the surface of the gold nanoparticles. Such silver precipitation facilitated visualization of the nanoparticle label and enabled quantitation of the hybridized target based on the imaged grayscale values. In addition, the use of nanoparticle labels altered the melting profiles to allow effective discrimination against single-base mismatches. Pavlov *et al.* reported on the use of gold nanoparticles for amplified optical transduction of aptamer–protein interactions [14]. The gold nanoparticles were functionalized with the thiolated aptamer (80 aptamers per particle). The aptamer binding to the thrombin protein analyte caused the gold nanoparticles to aggregate and their plasmon absorbance spectra to decrease.

Surface-enhanced Raman scattering (SERS) is another spectroscopic transduction mode that can greatly benefit from the use of gold nanoparticles. Cao *et al.* used nanoparticles functionalized with oligonucleotides and Ramanactive dyes for detecting DNA hybridization [15]. The gold nanoparticles facilitated the formation of a silver coating that acted as a promoter for the Raman scattering of the dyes. High sensitivity down to the 20 fM DNA level was reported. Multiplexed detection was accomplished by using different Raman dyes. The high fluorescence intensity of semiconductor quantum dots (QDs) can also lead to remarkably sensitive bioassays. Hahn *et al.* reported on a highly sensitive detection of the single-bacterial pathogen *E. coli* O157 using CdSe/ZnS core–shell QDs conjugated to streptavidin [16]. This system exhibited two orders of magnitude increased sensitivity (along with higher stability) compared to the common fluorescent dyes.

Mirkin and coworkers have developed a novel gold nanoparticle-based bio-barcode method for detecting proteins down to the low attomolar level [17]. This powerful protocol relies on magnetic spheres functionalized with an antibody that binds specifically the target protein and a secondary antibody conjugated to gold nanoparticles that are encoded with DNA strands that are unique to the target protein (Fig. 14.2,

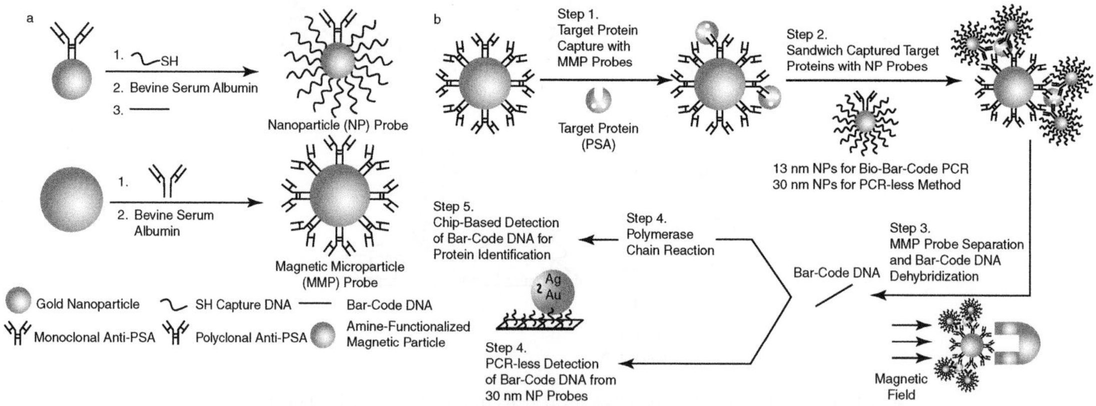

FIGURE 14.2 The bio-bar-code assay method. (a) Probe design and preparation. (b) PSA detection and bar-code DNA amplification and identification (reproduced from [17] with permission) (see Plate 14 for color version).

see Plate 14 for color version). Magnetic separation of the complexed probes and target, followed by thermal dehybridization of the oligonucleotides on the nanoparticle surface, allowed highly sensitive scanometric measurements of the target protein by identifying the oligonucleotide sequence released from the nanoparticle probe. Substantial amplification was achieved because each nanoparticle probe carried a large number of oligonucleotides per protein binding event. A multianalyte immunoassay could be accomplished by using different oligonucleotide sequences for encoding different target antigens. The nanoparticle-based bio-barcode assay was applied for detecting the prostate-specific antigen (PSA) at the low attomolar (aM = 10^{-18} M) level, that is, a sensitivity that is six orders of magnitude greater than the corresponding ELISA assay. In addition, the same assay was used for measuring the pathogenic Alzheimer's disease marker ADDL in the cerebral spinal fluid at picomolar levels [18]. The bio-barcode concept was extended also to PCR-less amplified detection of DNA hybridization down to the 500 zeptomolar level (i.e. 30 copies in 30 μL samples; zM = 10^{-21} M) [19]. A recent contribution from Niemeyer and coworkers demonstrated a sensitive optical detection of proteins using difunctional DNA–gold nanoparticles [20]. In this case, the sandwich immunoassay caused multilayers of DNA-linked gold nanoparticles to form that were detected by ultraviolet/visible (UV/Vis) spectroscopy.

Another attractive route for optical transduction of biorecognition events involves the encapsulation of a huge amount of a fluorescent marker within nanoparticle carriers [21–26]. Harma *et al.* reported a europium-entrapped fluorescence nanoparticle label for an ultrasensitive prostate-specific antigen (PSA) assay. PSA was detected in microtiter wells coated with a PSA-specific antibody using biotinylated antibody and streptavidin-coated, highly fluorescent 107 nm nanoparticles that contained more than 30 000 europium ions entrapped by beta-diketones. PSA was monitored directly on the surface of a well without any additional enhancement step. The sensitivity of the assay was 1.6 ng/L, corresponding to 50 fmol L^{-1} fM or 250 zeptomoles (250×10^{-21} mol^{-1}) of PSA. The high specific activity and low non-specific binding of the streptavidin coated nanoparticles improved the sensitivity of the PSA assay 100-fold compared to the conventional europium-labeled streptavidin tracer in the same assay format. This nanoparticle label has been successfully used for free PSA detection in a serum sample [22] and thyroid-stimulating hormone (a marker of thyroid function) detection [23].

Trau and coworkers reported on a highly sensitive immunoassay of proteins based on polyelectrolyte encapsulated microcrystalline fluorescent material interfaced to the antibody [24]. A dramatically (~2000-fold) amplified immunoassay was reported in connection with the release of the fluorescent molecules to the detection medium following the antibody–antigen interaction. An analogous route for maximizing the number of fluorescent molecules per binding event was reported by Tan and coworkers [25]. For this purpose, the fluorescent dye was encapsulated within silica nanoparticles functionalized with an oligonucleotide probe. The method offered an extremely low detection limit of 0.8 fM with effective discrimination against mismatched DNA. The silica matrix also provided good protection against bleaching of the fluorophore. The fluorescence-bioconjugated silica nanoparticles were also applied for ultrasensitive bacterial detection, down to a single *E. coli* O157 cell [26].

14.4 NANOPARTICLE-BASED ELECTROCHEMICAL BIOSENSORS AND BIOASSAY

Electrochemical devices have recently received considerable attention in the development of immunosensors and sequence-specific DNA hybridization biosensors [27–29]. Electrochemical biosensors are devices that intimately couple a biological recognition element to an electrode transducer that relies on the conversion of the antibody–antigen and Watson–Crick base-pair recognition event into a useful electrical signal. Electrochemical devices offer elegant routes for interfacing – at the molecular level – the DNA (or antibody–antigen) recognition and signal transduction elements and are uniquely qualified for meeting the size, cost, low volume, and power requirements of decentralized DNA and protein diagnostics [27–29]. The high sensitivity of electrochemical biosensors, coupled with their inherent miniaturization, compatibility with modern microfabrication technologies, low-cost and power requirements, and independence of sample turbidity make such devices excellent candidates for centralized and decentralized genetic and protein testing. Although the use of electrochemical biosensors or chips is at an early stage, easy-to-use hand-held electrical DNA and protein analyzers are already approaching the marketplace [30] and are expected to have a considerable impact on future DNA and protein diagnostics. Electrochemical transduction of DNA hybridization events has commonly been achieved in connection with electroactive indicators/intercalators or enzyme tags. Electrochemical transduction of antibody–antigen recognition events has been realized in connection with metal ion and enzyme labels. The use of nanoparticle tracers is relatively new in electrical detection and offers unique opportunities for electrochemical transduction of protein and DNA sensing events.

14.4.1 Nanoparticle-based electrochemical DNA biosensors and bioassays

Nanoparticle-based electrical routes for gene detection have attracted much interest. Such new protocols are based on the use of colloidal gold [31, 32], semiconductor quantum dot tracers [33, 34], iron/gold alloy nanoparticles [35], copper/gold alloy nanoparticles [36], and silver nanoparticles [37]. These nanoparticle materials offer elegant ways for interfacing DNA recognition events with electrochemical signal transduction and for amplifying the resulting electrical response. Most of these schemes have commonly relied on a highly sensitive electrochemical stripping transduction/measurement of the metal tracers. Stripping voltammetry is a powerful electroanalytical technique for trace metal measurements [38]. Its remarkable sensitivity is attributed to the "built-in" preconcentration step during which the target metals are accumulated (deposited) onto the working electrode. The detection limits are thus lowered by three to four orders of magnitude, compared to pulse-voltammetric techniques used earlier for monitoring DNA hybridization. Such ultrasensitive electrical detection of metal tags has been accomplished in connection with a variety of new and novel DNA-linked particle nanostructure networks.

Several groups, including ours, have developed powerful nanoparticle-based electrochemical DNA hybridization assays [31–37]. Such protocols have relied on capturing the DNA coded nanoparticle probes to the hybridized target, followed by acid dissolution and anodic-stripping electrochemical measurement of the metal tracers (Fig. 14.3, see Plate 15 for color version). The probe or target immobilization is accomplished in connection with streptavidin coated magnetic beads [31], through the use of chitosan or polypyrrole surface layers [37], or via adsorption onto the walls of polystyrene microwells [32]. Picomolar and subnanomolar levels of the DNA target have thus been detected. Further sensitivity enhancement can be obtained by catalytic enlargement of the gold tracer in connection with nanoparticle-promoted precipitation of gold [31] or silver [39–41] (Fig. 14.3b, see Plate 15 for color version). Combining such enlargement of the metal particle tags, with the effective "built-in" amplification of electrochemical stripping analysis, paved the way to subpicomolar detection limits. The silver-enhancement electrical route has been applied recently for detection in "lab-on-chip" systems in connection with

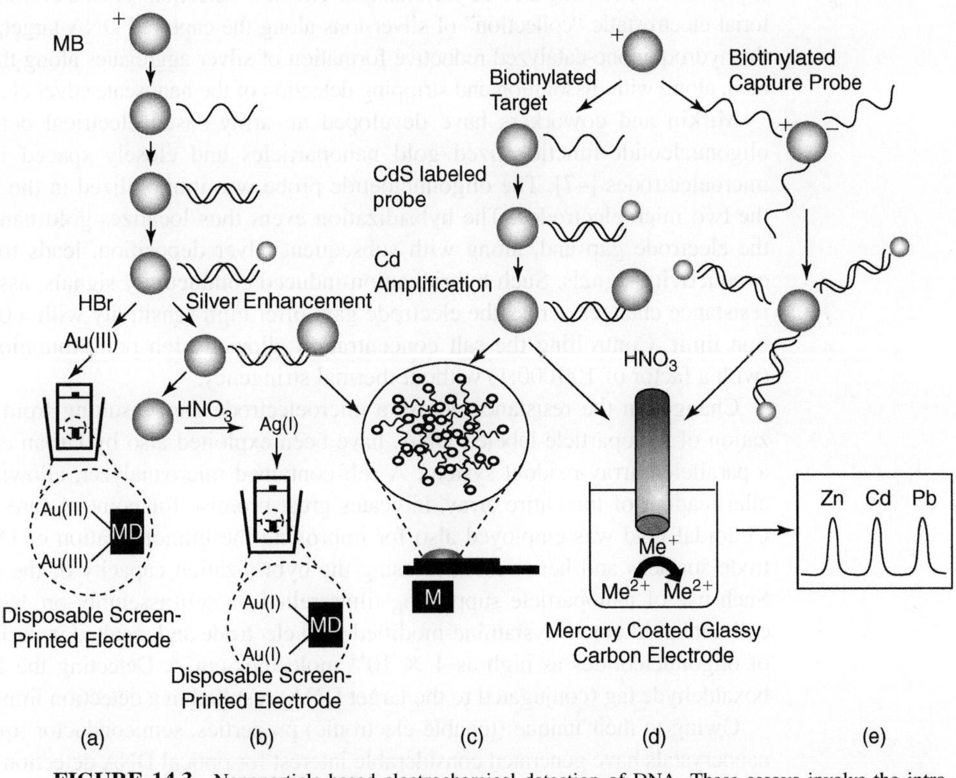

FIGURE 14.3 Nanoparticle-based electrochemical detection of DNA. These assays involve the introduction of the probe coated magnetic beads, the addition of the target/hybridization event, magnetic removal of unwanted materials, binding of the metal, and amplified electrochemical detection of the dissolved gold (a), silver (b), and cadmium sulfide (d) nanoparticles. Me: metal tag. Also shown are solid-state stripping (c) and multi-target (e) detection protocols. See text for details (see Plate 15 for color version).

on-chip PCR amplification [42]. The hybridization of probe coated magnetic beads with the gold-tagged targets results in three-dimensional network structures of "large" (micrometer) magnetic beads, cross-linked together through the DNA and gold nanoparticles. In these aggregates, the DNA duplex "bridges" the magnetic beads with the metal nanoparticles. No such aggregation was observed in the presence of non-complementary or mismatched oligonucleotides. Similar DNA-induced aggregation has been exploited by Mirkin and coworkers to detect the hybridization in connection with distance-dependent color changes [11]. Very recently, we described an electrochemical protocol for detecting DNA hybridization based on preparing the metal marker along the DNA backbone (instead of capturing it at the end of the duplex) [43]. The new protocol relies on DNA template-induced generation of conducting nanowires as a mode of capturing the metal tag. The use of DNA as a metallization template has evoked substantial research activity directed at the generation of conductive nanowires and the construction of functional circuits [44–46]. Such an approach was applied to grow silver [44], palladium [45], or platinum [46] clusters on DNA templates. Yet, the DNA-templated assembly of metal wires has not been exploited for detecting DNA hybridization. The new detection scheme consists of the vectorial electrostatic "collection" of silver ions along the captured DNA target, followed by the hydroquinone-catalyzed reductive formation of silver aggregates along the DNA skeleton, along with dissolution and stripping detection of the nanoscale silver cluster.

Mirkin and coworkers have developed an array-based electrical detection using oligonucleotide-functionalized gold nanoparticles and closely spaced interdigitated microelectrodes [47]. The oligonucloetide probe was immobilized in the gap between the two microelectrodes. The hybridization event thus localizes gold nanoparticles in the electrode gap and, along with subsequent silver deposition, leads to measurable conductivity signals. Such hybridization-induced conductivity signals, associated with resistance changes across the electrode gap, offer high sensitivity with a 0.5 pM detection limit. Controlling the salt concentration allowed high point-mutation selectivity (with a factor of 100 000:1) without thermal stringency.

Changes in the resistance across a microelectrode gap, resulting from the hybridization of nanoparticle-labeled DNA, have been exploited also by Urban *et al.* [48] for a paralleled array-readout system. A self-contained microanalyzer, allowing such parallel readout of the entire array, indicates great promise for point-of-care applications. Colloidal gold was employed also for improving the immobilization of DNA on electrode surfaces and hence for increasing the hybridization capacity of the surface [49]. Such use of nanoparticle supporting films relied on self-assembly on 16 nm diameter colloidal gold onto a cystamine-modified gold electrode and resulted in surface densities of oligonucleotides as high as 4×10^{14} molecules cm^{-2}. Detecting the ferrocenecarboxaldehyde tag (conjugated to the target DNA) resulted in a detection limit of 500 pM.

Owing to their unique (tunable-electronic) properties, semiconductor (quantum dots) nanocrystals have generated considerable interest for optical DNA detection [12]. Recent activity has demonstrated the utility of quantum dot nanoparticles for enhanced electrical DNA detection [33, 34, 50]. Willner *et al.* reported on a photoelectrochemical transduction of DNA sensing events in connection with DNA cross-linked CdS nanoparticle arrays [50]. The electrostatic binding of the $Ru(NH_3)_6^{3+}$ electron acceptor to the dsDNA units provided a tunneling route for the electron-band electrons and thus led to increased photocurrents. We reported on the detection of DNA hybridization in connection with cadmium sulfide nanoparticle tracers and electrochemical stripping measurements of the cadmium [33]. A nanoparticle-promoted cadmium precipitation was used to enlarge the nanoparticle tag and amplify the stripping DNA hybridization signal (Fig. 14.3, see Plate 15 for color version). In addition to measurements of the dissolved cadmium ion, we demonstrated solid-state measurements following a "magnetic" collection of the magnetic bead/DNA hybrid/CdS tracer assembly onto a thick film electrode transducer. Such a protocol combines the amplification features of nanoparticle/polynucleotide assemblies and highly sensitive potentiometric stripping detection of cadmium with an effective magnetic isolation of the duplex. The low detection limit (100 fmol) was coupled to good reproducibility (RSD = 6%). Such a protocol was recently extended to other inorganic colloids (e.g. ZnS or PbS) which can be similarly synthesized in reversed micelles. Such extension has paved the way for an electrochemical coding technology for the simultaneous detection of multiple DNA targets based on nanocrystal tags with diverse redox potentials [34]. Functionalizing the nanocrystal tags with thiolated oligonucleotide probes thus offered a voltammetric signature with distinct electrical hybridization signals for the corresponding DNA targets (Fig. 14.3e, see Plate 15 for color version). The position and size of the resulting stripping peaks provided the desired identification and quantitative information, respectively, on a given target DNA. The multi-target DNA detection capability was coupled to the amplification feature of stripping voltammetry (to yield femtomolar detection limits) and with an efficient magnetic removal of non-hybridized nucleic acids to offer high sensitivity and selectivity. Up to five to six targets can thus be measured simultaneously in a single run in connection with ZnS, PbS, CdS, InAs, and GaAs semiconductor particles. Conducting massively parallel assays (in microwells of microtiter plates or using multi-channel microchips, with each microwell or channel carrying out multiple measurements) could thus lead to a high throughput operation.

14.4.2 Nanoparticle-based electrochemical immunosensors and immunoassays

On the basis of a specific reaction of the antibody and antigen, immunosensors provide a sensitive and selective tool for determining immunoreagents. Here, the immunologic material is immobilized on a transducer; the analyte is measured through a label species conjugated with one of the immunoreagents. Quantification is generally achieved by measuring the specific activity of a label, i.e. its radioactivity, enzyme activity, fluorescence, chemiluminescence or bioluminescence. There is no ideal label, and each of them has its own advantages and disadvantages. Metalloimmunoassays, i.e. immunoassays involving metal-based labels, were developed in the 1970s [51] to overcome problems associated with the common radioisotopic, fluorescent or enzyme labels. Metalloimmunoassays with electrochemical detection can offer several advantages; e.g. measurement can be performed in very low volumes of liquid (a few microliters), eventually in turbid media, with the possibility of having good sensitivity for a relatively inexpensive instrumentation (field portable). Although the electrochemical techniques allow organometallic compounds [52, 53] or metal ions [54, 55] to be detected at nanomolar concentrations, their sensitivities remain insufficient compared with fluorescent europium chelate labels for which picomolar levels can be determined.

A promising route for maximizing the number of metal markers per binding event is the use of a metal nanoparticle label, which is composed of thousands of metal atoms. Colloidal gold is one of the best candidates for electrochemical immunosensing and immunoassay because of its excellent redox activity. Murielle *et al.* first reported a sensitive electrochemical immunoassay based on a colloidal gold label that, after oxidative gold metal dissolution in an acidic solution, was indirectly determined by anodic stripping voltammetry (ASV) at a single-use carbon-based screen-printed electrode (SPE) [56]. The method was evaluated for a non-competitive heterogeneous immunoassay of an immunoglobulin G (IgG) and a concentration as low as 3×10^{-12}M was determined, which was competitive with colorimetric enzyme-linked immunosorbent assay (ELISA) or with immunoassays based on fluorescent europium chelate labels. Chu *et al.* reported a similar electrochemical immunoassay based on silver on colloidal gold labels which, after silver metal dissolution in an acidic solution, was indirectly determined by ASV at a glassy-carbon electrode [57]. The method was evaluated for a non-competitive heterogeneous immunoassay of an immunoglobulin G (IgG) as a model. The anodic stripping peak current of silver depended linearly on the IgG concentration over the range of 1.66 ng mL^{-1} to 27.25 ng mL^{-1} in a logarithmic plot. A detection limit as low as 1 ng mL^{-1} (i.e. 6×10^{-12}M) human IgG was achieved.

A new method based on cyclic accumulation of gold nanoparticles has been proposed for detecting human immunoglobulin G (IgG) by anodic stripping voltammetry [58]. The dissociation reaction between dethiobiotin and avidin in the presence of biotin provides an efficient means for the cyclic accumulation of gold nanoparticles used for the final analytical quantification. The anodic peak current increases gradually with the increasing accumulation cycles. Five cycles of accumulation are sufficient for the assay. The low background of the proposed method is a distinct advantage providing a possibility for determination of at least 0.1 ng/mL^{-1} human IgG.

An electrochemical immunoassay method based on Au nanoparticle-labeled immunocomplex enlargement used to detect complement C-3 was reported by Zhou *et al.* [59]. When the aggregates formed from nano-Au-labeled goat-anti-human C-3 and nano-Au-labeled rabbit-anti-goat IgG were immobilized on the electrode surface by the sandwich method (antibody–antigen–aggregate), the electrochemical signal of the electrode was enlarged greatly. The reported immunosensor could quantitatively determine complement C-3 in the range of 0.12, similar to ~117.3 ng mL^{-1}, and the detection limit was 0.02 ng mL^{-1}.

Liao *et al.* reported an amplified electrochemical immunoassay by autocatalytic deposition of Au^{3+} onto gold nanoparticle labels [60]. By coupling the autocatalytic deposition with square-wave stripping voltammetry, enlarged gold nanoparticles labeled on goat anti-rabbit immunoglobulin G (GaRIgG-Au) and, thus, the rabbit immunoglobulin G (RIgG) analyte, could be determined quantitatively. From a calibration graph over a broad dynamic range of concentrations (1–500 pg mL^{-1}; $R^2 = 0.9975$), a very low detection limit was obtained. The limit was 0.25 pg mL^{-1} (1.6 fM), which is three orders of magnitude lower than that obtained by a conventional immunoassay using the same gold nanoparticle labels.

Recently, a particle-based renewable electrochemical magnetic immunosensor was developed in our group by using magnetic beads and gold nanoparticle labels [61]. Anti-IgG antibody-modified magnetic beads were attached to a renewable carbon paste transducer surface by a magnet that was fixed inside the sensor. Gold nanoparticle labels were capsulated to the surface of magnetic beads with a sandwich immunoassay (Fig. 14.4). Highly sensitive electrochemical stripping analysis offers a simple and fast

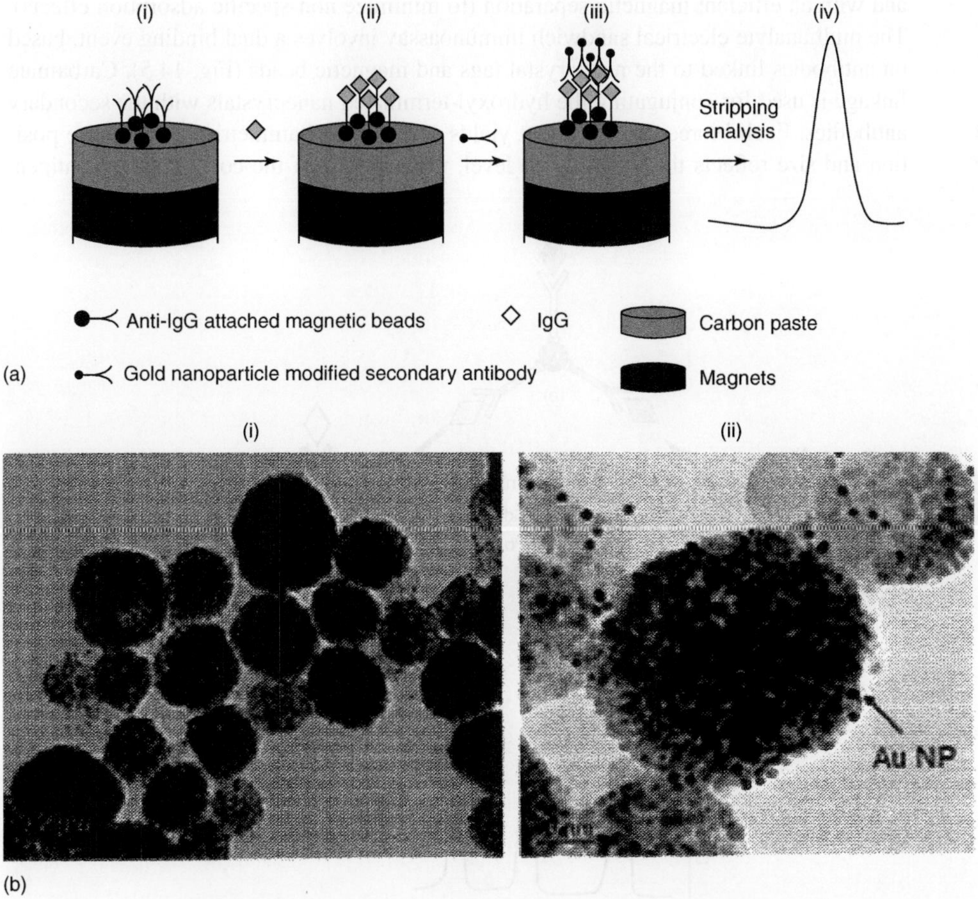

FIGURE 14.4 (a) Particle-based electrochemical immunoassay protocol: (i) introduction of antibody-modified magnetic beads to magnet/carbon paste electrochemical transducer surface; (ii) binding of the IgG antigen to the antibodies on the magnetic beads; (iii) capture of the gold nanoparticle-labeled secondary antibodies; (iv) electrochemical stripping detection of gold nanoparticles. (b) TEM images of magnetic beads–gold nanoparticles assembly resulting from the sandwich immunoreaction of $0.2\,\mu g\ mL^{-1}$ IgG analyte (i) and an expanded view of the particle assembly (ii) (reproduced from [61] with permission).

method to quantify the captured gold nanoparticle tracers and avoid the dissolution step and the use of an enzyme label and substrate. The stripping signal of gold nanoparticles is related to the concentration of target IgG in the sample solution. The detection limit of $0.02\,\mu g\ mL^{-1}$ of IgG was obtained under optimum experimental conditions.

The above immunosensors and immunoassays based on a gold nanoparticle label are used to detect one target analyte and cannot be used for multiple target assays. An electrochemical immunoassay protocol for the simultaneous measurements of proteins, based on the use of different inorganic nanocrystal tracers, was reported by Liu *et al.* [29]. The multi-protein electrical detection capability is coupled to the amplification feature of electrochemical stripping transduction (to yield fmol detection limits) and with an efficient magnetic separation (to minimize non-specific adsorption effects). The multianalyte electrical sandwich immunoassay involves a dual binding event, based on antibodies linked to the nanocrystal tags and magnetic beads (Fig. 14.5). Carbamate linkage is used for conjugating the hydroxyl-terminated nanocrystals with the secondary antibodies. Each biorecognition event yields a distinct voltammetric peak, whose position and size reflects the identity and level, respectively, of the corresponding antigen.

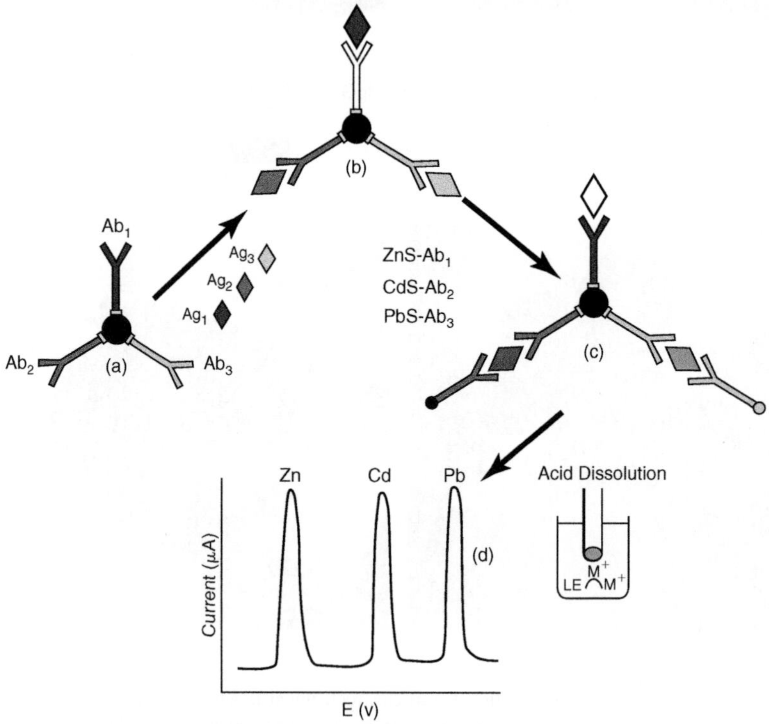

FIGURE 14.5 Multiprotein electrical detection protocol based on different inorganic colloid nanocrystal tracers. (a) Introduction of antibody-modified magnetic beads; (b) binding of the antigens to the antibodies on the magnetic beads; (c) capture of the nanocrystal-labeled secondary antibodies; (d) dissolution of nanocrystals and electrochemical stripping detection (reproduced from [29] with permission).

The concept is demonstrated for a simultaneous immunoassay of β2-microglobulin, IgG, bovine serum albumin, and C-reactive protein in connection with ZnS, CdS, PbS, and CuS colloidal crystals, respectively (Fig. 14.6). These nanocrystal labels exhibit similar sensitivity. Such electrochemical coding could be readily multiplexed and scaled up in multiwell microtiter plates to allow simultaneous parallel detection of numerous proteins or samples and is expected to open new opportunities for protein diagnostics and biosecurity.

To enhance the sensitivity of the nanoparticle label-based electrochemical immunosensors and immunoassays, we recently developed a novel electrochemical immunosensor based on poly(guanine)-functionalized silica nanoparticle labels and mediator-generated catalytic reaction [62]. Figure 14.7 (see Plate 16 for color version) schematically illustrates the principle of electrochemical immunosensing based on poly[G]-covered silica NPs. Biotinylated primary antibodies are first immobilized on an avidin-modified electrode and mouse IgG then bound onto the antibody, followed by interaction with mouse IgG specific antibody–silica NPs covered with poly[G], which introduces a large amount of guanine residues on the electrode surface. Guanines on silica NPs catalyze the oxidation of $Ru(bpy)_3^{2+}$. The amplitude of the oxidation current depends on the amount of guanine, which is related to the concentrations of sample solutions. The amplification of the catalytic signals is attributed to the attachment of a large number of guanine markers

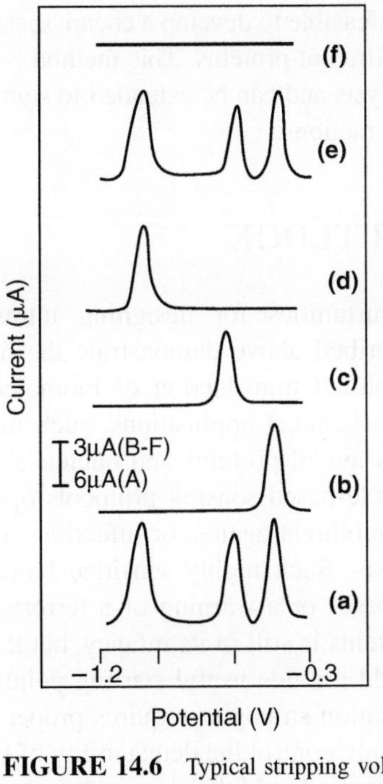

FIGURE 14.6 Typical stripping voltammograms for (a) nanocrystal-labeled antibodies and (b–f) magnetic bead–Ab–Ag–Ab-nanocrystal complexes. (b) Response for a solution containing dissolved ZnS anti-β2-microglobulin, PbS–anti-BSA, and CdS–anti-IgG conjugates (reproduced from [29] with permission).

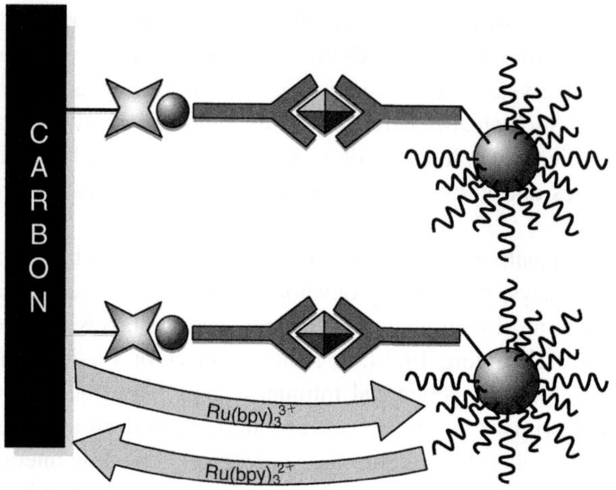

FIGURE 14.7 Schematic illustration of an electrochemical immunoassay based on the poly[G]-functionalized silica NPs on carbon electrode. Diamond stands for the mouse IgG; "Y" shape for mouse antibody; red strings for poly[G] (reproduced from [62] with permission) (see Plate 16 for color version).

per antibody–antigen–antibody complex formed. This immunobiosensor is very sensitive, and the limit of detection was found to be down to 0.02 ng mL^{-1}. An attractive feature of this approach is that it makes it feasible to develop a cheap, sensitive, and portable device for multiplexed diagnoses of different proteins. This method is simple, selective, and reproducible for trace protein analysis and can be extended to study protein–protein, peptide–protein, and DNA–protein interactions.

14.5 CONCLUSION AND OUTLOOK

Nanotechnology offers unique opportunities for designing ultrasensitive biosensors and bioassays. The studies described above demonstrate the broad potential of bioconjugated nanoparticles for amplified transduction of biomolecular recognition events. Given the optical and electrochemical applications, such nanoparticle labels provide the basis for ultrasensitive assays of proteins and nucleic acids. The remarkable sensitivity of the new nanoparticle-based sensing protocols opens up the possibility for detecting disease markers, biothreat agents, or infectious agents that cannot be measured by conventional methods. Such highly sensitive biodetection schemes could provide early detection of diseases or a warning of a terrorist attack. The use of nanoparticle tags for detecting proteins is still in its infancy, but the lessons learned in ultrasensitive DNA detection should provide useful starting points. The successful realization of the new signal-amplification strategies requires proper attention to non-specific adsorption issues that commonly control the detectability of bioaffinity assays. Proper washing and surface blocking steps should thus be employed to avoid amplification of background signals (associated with non-specific adsorption of the nanoparticle amplifiers). A wide range of newly introduced nanomaterials is expected to further expand the realm of nanomaterial-based biosensors.

14.6 ACKNOWLEDGMENTS

The work performed at Pacific Northwest National Laboratory (PNNL) was supported by the Laboratory Directed Research and Development Program, US Department of Energy-Environmental Management Science Program (DOE-EMSP), Strategic Environmental Research and Development Program (SERDP) (Project ID 1297), and NIH/1R01 ES010976-01A2. The authors' research described in this chapter was performed in part at the Environmental Molecular Sciences Laboratory, a national scientific user facility sponsored by the DOE Office of Biological and Environmental Research and located at PNNL. PNNL is operated for DOE by Battelle under Contract DE-AC05-76RL01830.

14.7 REFERENCES

1. F. Carusu, Nanoengineering of particle surfaces. *Adv. Mater.* **13**, 11–22 (2001).
2. J.J. Storhoff and C.A. Mirkin, Programmed materials synthesis with DNA. *Chem. Rev.* **99**, 1849–1862 (1999).
3. I. Willner and B. Willner, Functional nanoparticle architectures for sensoric, optoelectronic, and bioelectronic applications. *Pure Appl. Chem.* **74**, 1773–1783 (2002).
4. C.M. Niemeyer, Nanopartikel, Proteine und Nucleinsäuren: Die Biotechnologie begegnet den Materialwissenschaften. *Angew. Chem.* **113**, 4254–4287 (2001).
5. P. Alivisatos, The use of nanocrystals in biological detection. *Nat. Biotechnol.* **22**, 47–52 (2004).
6. E. Katz, I. Willner, and J. Wang, Electroanalytical and bioelectroanalytical systems based on metal and semiconductor nanoparticles. *Electroanal.* **16**, 19–44 (2004).
7. I. Willner and E. Katz, Integrated nanoparticle-biomolecule hybrid systems: synthesis, properties, and applications. *Angew. Chem. Int. Ed.* **43**, 6042–6108 (2004).
8. N.L. Rosi and C.A. Mirkin, Nanostructures in biodiagnostics. *Chem. Rev.* **105**, 1547–1562 (2005).
9. C.M. Niemeyer, Nanoparticles, proteins, and nucleic acids: biotechnology meets materials science. *Angew. Chem. Int. Ed.* **40**, 4129–4158 (2001).
10. R.L. Letsinger, R.C. Mucic, and J.J. Storhoff, A DNA-based method for rationally assembling nanoparticles into macroscopic materials.*Nature* **382**, 607–609 (1996).
11. J.J. Storhoff, R. Elghanian, R.C. Mucic, C.A. Mirkin, and R.L. Letsinger, One-pot colorimetric differentiation of polynucleotides with single base imperfections using gold nanoparticle probes. *J. Am. Chem. Soc.* **120**, 1959–1964 (1998).
12. R. Elghanian, J.J. Storhoff, R.C. Mucic, R.L. Letsinger, and C.A. Mirkin, Selective colorimetric detection of polynucleotides based on the distance-dependent optical properties of gold nanoparticles. *Science* **277**, 1078–1081 (1997).
13. TA. Taton, C.A. Mirkin, and R.L. Letsinger, Scanometric DNA array detection with nanoparticle probes. *Science* **289**, 1757–1760 (2000).
14. V. Pavlov, B. Shyahovsky, and I. Willner, Aptamer-functionalized Au nanoparticles for the amplified optical detection of thrombin. *J. Am. Chem. Soc.* **126**, 11 768–11 769 (2004).
15. Y.C. Cao, R. Jin, and C.A. Mirkin, Nanoparticles with raman spectroscopic fingerprints for DNA and RNA detection. *Science* **297**, 1536–1540 (2002).

16. M.A. Hahn, J.S. Tabb, and T.D. Krauss, Detection of single bacterial pathogens with semiconductor quantum dots. *Anal. Chem.* **77**, 4861–4869 (2005).
17. J.M. Nam, C.S. Thaxton, and C.A. Mirkin, Nanoparticle-based bio-bar codes for the ultrasensitive detection of proteins. *Science* **301**, 1884–1886 (2003).
18. D.G. Georganopoulou, L. Chang, J.M. Nam, C.S. Thaxton, E.J. Mufson, W.L. Klein, and C.A. Mirkin, Nanoparticle-based detection in cerebral spinal fluid of a soluble pathogenic biomarker for Alzheimer's disease. *Proc. Natl. Acad. Sci. U.S.A.* **102**, 2273–2276 (2005).
19. J.M. Nam, S.I. Stoeva, and C.A. Mirkin, Bio-bar-code-based DNA detection with PCR-like sensitivity. *J. Am. Chem. Soc.* **126**, 5932–5933 (2004).
20. P. Hazarika, B. Ceyhan, and C.M. Niemeyer, Sensitive detection of proteins using difunctional DNA-gold nanoparticles. *Small* **1**, 844–848 (2005).
21. H. Harma, T. Soukka, S.Lonnberg, J. Paukkunen, P. Tarkkinen, and T. Lovgren, Zeptomole detection sensitivity of prostate-specific antigen in a rapid microtitre plate assay using time-resolved fluorescence. *Luminescence* **15**, 351–355 (2000).
22. T. Soukka, K. Antonen, H. Harma, A.M. Pelkkikangas, P. Huhtinen, and T. Lovgren, Highly sensitive immunoassay of free prostate-specific antigen in serum using europium(III) nanoparticle label technology. *Clin. Chim. Acta* **328**, 45–58 (2003).
23. A.M. Pelkkikangas, S. Jaakohuhta, T. Lovgren, and H. Harma, Simple, rapid, and sensitive thyroid-stimulating hormone immunoassay using europium(III) nanoparticle label. *Anal. Chim. Acta* **517**, 169–176 (2004).
24. D. Trau, W. Yang, M. Seydack, F. Caruso, N.T. Yu, and R. Renneberg, Nanoencapsulated microcrystalline particles for superamplified biochemical assays. *Anal. Chem.* **74**, 5480–5486 (2002).
25. X. Zhao, R. Tapec-Dytioco, and W. Tan, Ultrasensitive DNA detection using highly fluorescent bioconjugated nanoparticles. *J. Am. Chem. Soc.* **125**, 11474–11475 (2003).
26. X. Zhao, L.R. Hilliard, S.J. Mechery, Y. Wang, R.P. Bagwe, S. Jin, and W. Tan, From the cover: a rapid bioassay for single bacterial cell quantitation using bioconjugated nanoparticles. *Proc. Natl. Acad. Sci. U.S.A.* **101**, 15027–15032 (2004).
27. E. Palecek and M. Fojta, DNA hybridization and damage, *Anal. Chem.* **73**, 75A–83A (2001).
28. J. Wang, Electrochemical nucleic acid biosensors. *Anal. Chim. Acta* **469**, 63–71 (2002).
29. G. Liu, J. Wang, J. Kim, M.R. Jan, and G.E. Collins, Electrochemical coding for multiplexed immunoassays of proteins. *Anal. Chem.* **76**, 7126–7130 (2004).
30. V. Chan, Y. Chong, L. Cheung, J. Vielmetter, and D.H. Farkas, Bioelectronic detection of point mutations using discrimination of the H63D polymorphism of the Hfe gene as a model. *Mol. Diagn.* **5**, 321–328 (2000).
31. J. Wang, D. Xu, A.N. Kawde, and R. Polsky, Metal nanoparticle-based electrochemical stripping potentiometric detection of DNA hybridization. *Anal. Chem.* **73**, 5576–5581 (2001).
32. L. Authier, C. Grossiord, P. Brossier, and B. Limoges, Gold nanoparticle-based quantitative electrochemical detection of amplified human cytomegalovirus DNA using disposable microband electrodes. *Anal. Chem.* **73**, 4450–4456 (2001).
33. J. Wang, G. Liu, and R. Polsky, Electrochemical stripping detection of DNA hybridization based on cadmium sulfide nanoparticle tags. *Electrochem. Commun.* **4**, 722–726 (2002).
34. J. Wang, G. Liu, and A. Merkoçi, Electrochemical coding technology for simultaneous detection of multiple DNA targets. *J. Am. Chem. Soc.* **125**, 3214–3215 (2003).
35. J. Wang, G. Liu, and A. Merkoci, Particle-based detection of DNA hybridization using electrochemical stripping measurements of an iron tracer. *Anal. Chim. Acta* **482**, 149–155 (2003).
36. H. Cai, N.N. Zhu, Y. Jiang, P.G. He, and Y.Z Fang, Cu–Au alloy nanoparticle as oligonucleotide labels for electrochemical stripping detection of DNA hybridization. *Biosens. Bioelectron.* **18**, 1311–1319 (2003).
37. H. Cai, Y. Xu, N. Zhu, P. He, and Y. Fang, An electrochemical DNA hybridization detection assay based on a silver nanoparticle label. *Analyst* **127**, 803–808 (2002).
38. J. Wang, *Stripping Analysis*, VCH Publishers, New York (1985).
39. J. Wang, R. Polsky, and X. Danke, Silver-enhanced colloidal gold electrochemical stripping detection of DNA hybridization. *Langmuir* **17**, 5739–5741 (2001).

40. H. Cai, Y. Wang, P. He, and Y. Fang, Electrochemical detection of DNA hybridization based on silver-enhanced gold nanoparticle label. *Anal. Chim. Acta* **469**, 165–172 (2002).
41. T.M.H. Lee, L.L. Li, and I.M. Hsing, Enhanced electrochemical detection of DNA hybridization based on electrode-surface modification. *Langmuir* **19**, 4338–4343 (2003).
42. T.M.H. Lee, M. Carles, and I.M. Hsing, Microfabricated PCR-electrochemical device for simultaneous DNA amplification and detection. *Lab. Chip* **3**, 100–105 (2003).
43. J. Wang, O. Rincon, R. Polsky, and E. Dominguez, Electrochemical detection of DNA hybridization based on DNA-templated assembly of silver cluster. *Electrochem. Commun.* **5**, 83–86 (2003).
44. E. Braun, Y. Eichen, U. Sivan, and G. Ben-Yoseph, DNA-templated assembly and electrode attachment of a conducting silver wire. *Nature* **391**, 775–778 (1998).
45. J. Richter, R. Seidel, R. Kirsch, M. Mertig, W. Pompe, J. Plashke, and H. Schackert, Nanoscale palladium metallization of DNA. *Adv. Mater.* **12**, 507–510 (2000).
46. M. Mertig, L.C. Ciacchi, R. Siedel, W. Pompe, and A. de Vita, DNA as a selective metallization template. *Nano Lett.* **2**, 841–844 (2002).
47. S. Park, T.A. Taton, and C.A. Mirkin, Array-based electrical detection of DNA with nanoparticle probes. *Science* **295**, 1503–1506 (2002).
48. M. Urban, R. Moller, and W. Fritzsche, A paralleled readout system for an electrical DNA-hybridization assay based on a microstructured electrode array. *Rev. Sci. Instrum.* **74**, 1077–1081 (2003).
49. H. Cai, C. Xu, P. He, and Y. Fang, Colloid Au-enhanced DNA immobilization for the electrochemical detection of sequence-specific DNA. *J. Electroanal. Chem.* **510**, 78–85(2001).
50. I. Willner, P. Patolsky, and J. Wasserman, Photoelectrochemistry with controlled DNA-cross-linked CdS nanoparticle arrays. *Angew. Chem. Int. Ed.* **40**, 1861–1864 (2001).
51. M. Cais, S. Dani, Y. Eden, O. Gandolfi, M. Horn, E.E. Isaacs, Y. Joseph, Y. Saar, E. Slovin, and L. Snarskt, Metalloimmunoassay. *Nature* 270, 534–535 (1977).
52. M.J. Doyle, H.B. Halsall, and W.R. Heineman, Heterogeneous immunoassay for serum proteins by differential pulse anodic stripping voltammetry. *Anal. Chem.* **54**, 2318–2322 (1982).
53. B. Limoges, C. Degrand, and P.J. Brossier, Redox cationic or procationic labeled drugs detected at a perfluorosulfonated ionomer film-coated electrode. *J. Electroanal. Chem.* **402**, 175–187 (1996).
54. F.J. Hayes, H.B. Halsall, and W.R. Heineman, Simultaneous immunoassay using electrochemical detection of metal ion labels. *Anal. Chem.* **66**, 1860–1865 (1994).
55. J. Wang, B. Tian, and K.R. Rogers, Thick-film electrochemical immunosensor based on stripping potentiometric detection of a metal ion label. *Anal. Chem.* 70, 1682–1685 (1998).
56. M. Dequaire, C. Degrand, and B. Limoges, An electrochemical metalloimmunoassay based on a colloidal gold label. *Anal. Chem.* **72**, 5521–5528 (2000).
57. X. Chu, X. Fu, K. Chen, G. Shen, and R. Yu, An electrochemical stripping metalloimmunoassay based on silver-enhanced gold nanoparticle label. *Biosens. Bioelectron.* **20**, 1805–1812 (2005).
58. X. Mao, J.H. Jiang, J.W. Chen, Y. Huang, G.L. Shen, and R.Q. Yu, Cyclic accumulation of nanoparticles: a new strategy for electrochemical immunoassay based on the reversible reaction between dethiobiotin and avidin. *Anal. Chim. Acta* **557**, 159–163 (2006).
59. G.Z. Zhou, J.S. Li, J.H. Jiang, G.L. Shen, and R. Yu, Chronopotentiometry based on nano-Au labeled aggregate enlargement used for the immunoassay of complement C_3. *Acta Chimi. Sinica* **63**, 2093–2097 (2005).
60. K.T. Liao and H.J. Huang, Femtomolar immunoassay based on coupling gold nanoparticle enlargement with square wave stripping voltammetry. *Anal. Chim. Acta* **538**, 159–164 (2005).
61. G. Liu and Y. Lin, Electrochemical magnetic immunosensor based on gold nanoparticle labels. *J. Nanosci. Nanotech.* **5**, 1060–1065 (2005).
62. J. Wang, G. Liu, and Y. Lin, Bioassay label based on electroactive silica beads. *Small* **2**, 1134–1138 (2006).

CHAPTER 15

Electrochemical sensors based on carbon nanotubes

Manliang Feng, Heyou Han, Jingdong Zhang, and Hiroyasu Tachikawa

15.1 INTRODUCTION

Since the discovery of carbon nanotubes (CNTs) in 1991 [1], there has been growing interest in using CNTs in chemical and biochemical sensing [2–5] and nanoscale electronic devices due to their remarkable electronic and mechanical properties. CNTs behave as a metal or semiconductor depending on their structures. CNT-modified electrodes have better conductivity than graphite [6, 7] and show a superior performance compared with such electrodes as Au, Pt, and other carbon electrodes. CNTs have a hollow core, which is suitable for storing guest molecules. Proteins and enzymes can be immobilized in the hollow core as well as to the surface of CNT without losing biological activity. The electrical conductivity of nanotubes can be improved by modifying the original CNT structure. CNTs possess fascinating mechanical strength and are the strongest and stiffest material known [8]. CNTs, especially the side walls, are relatively inert. However, the ends of nanotubes are more reactive than the cylindrical parts [8, 9]. For the application of CNTs to electrochemical sensing, CNTs show the enhanced electrochemical response to some important biomolecules [5, 10] and promote the electron transfer reactions of proteins [4]. These characteristics demonstrate clearly that nanotubes have significant potential for the design of electrochemical sensors. Indeed, in the past several years, there have been extensive studies on the applications of CNTs in electrochemical sensors [11]. Other applications include hydrogen storage [12], catalysis, micro-chemical and micro-biological detectors [13–15], biological cell electrodes [16], nanoscale electronic and mechanical systems [17], scanning probe microscope, and electron field emission tips [18, 19].

In this chapter, we will discuss electrochemical sensors based on CNTs. First, the properties and structures of CNTs, the preparation and purification of CNTs, and the advantages of electrochemical sensors based on CNTs are described, then, the fabrication of electrochemical sensors based on CNTs, applications of electrochemical sensors based on CNTs, and the spectroscopic characterization of CNT sensors are described. In conclusion, we will look into some aspects of the future direction for CNT sensors in clinical and biomedical research.

15.2 THE STRUCTURE AND PROPERTIES OF CNTS

15.2.1 The structure of CNTs

CNTs are unique tubular structures of nanometer diameter and have large length/diameter ratio. CNTs are divided into two main groups: single-walled carbon nanotube (SWNT) and multi-walled carbon nanotube (MWNT). SWNT can be considered as a long wrapped graphene sheet by rolling it in certain directions. The properties of the nanotubes are mainly dictated by the rolling direction as well as the diameter. SWNT consists of two separate regions (the side wall and the end cap) with very different physical and chemical properties. The end-cap structure is similar to or derived from a small fullerene in which the carbon atoms are in both pentagon and hexagon rings. The side wall only consists of hexagon rings. MWNT can be considered as a

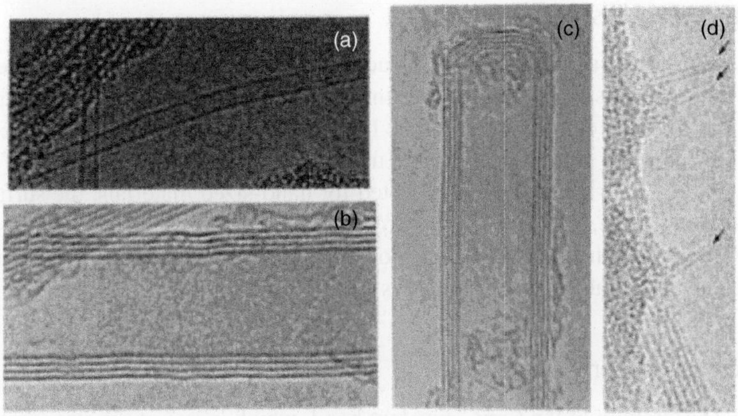

FIGURE 15.1 High-resolution transmission electron microscopy images of CNTs. (a) SWNT; (b) MWNT; (c) closed MWNT tips (MWNT tips); and (d) closed SWNT tip. The separation between the closely spaced fringes in the MWNT (b, c) is 0.34 nm, close to the spacing between graphite planes. The diameter of the SWNT (a, d) is ~1.2 nm. (Reprinted with permission from [8]. Copyright (1999) American Chemical Society.)

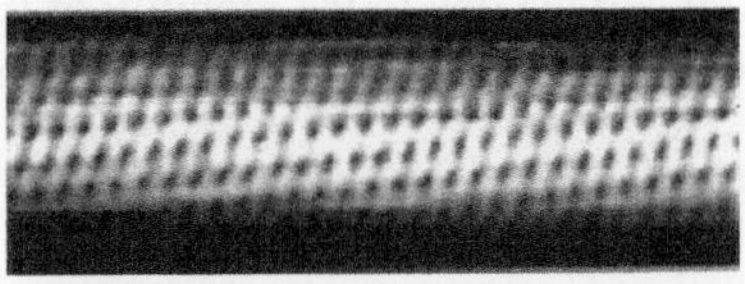

FIGURE 15.2 High resolution STM image of the lattice structure of a helical semi-conducting SWNT. (Reprinted with permission from [8]. Copyright (1999) American Chemical Society.)

collection of concentric SWNTs with different diameters. MWNTs may consist of one up to tens and hundreds of concentric shells of carbons with adjacent shell separations of ~0.34 nm. The carbon network of shells is closely related to the honeycomb arrangement of the carbon atoms in the graphite sheets.

The diameters of CNTs range from 0.2 to 2 nm for SWNTs and from 2 to 100 nm for MWNTs, while the lengths of CNTs range from several hundred nanometers to several micrometers [13, 16]. The structural information of CNTs has been provided by both transmission electron microscope (TEM) and scanning tunneling microscopy/spectroscopy (STM/STS) at the atomic level. Figure 15.1 shows TEM images of both SWNTs and MWNTs. In these images obtained by high resolution TEM, the diameter of the SWNTs of ~1.2 nm is clearly seen. The spacing (0.34 nm) between graphite planes of MWNTs as well as the closed nanotube tips is also seen in the TEM images. Atomically resolved lattices of SWNTs have been observed by STM as shown in Fig. 15.2. This STM image also shows the helicity of SWNTs. For the formation of curved structures such as the tips of CNTs and fullerenes from a planar hexagonal graphite lattice, certain topological defects and pentagons (12 pentagons are needed to close the hexagonal lattice) need to be introduced.

15.2.2 Properties of CNTs

The carbon nanotubes possess such properties as high conductivity, excellent strength and stiffness, and chemical inertness. CNTs also show unusual electronic characteristics that are dependent on lattice helicity and elasticity. The density of SWNTs is estimated to be smaller (0.6 g/cm^3) than graphite due to the presence of hollow channels in the center of CNTs. As expected for nano-sized materials, the surface area of CNTs is very large, e.g. ~10–20 m^2/g for MWNTs and the value of SWNTs is expected to be an order of magnitude higher. Some detailed discussion of the mechanical, electronic, and chemical properties of CNTs can be found in the following sections.

15.2.2.1 Mechanical properties

The strength of the C—C covalent bond, which is one of the strongest in nature, makes CNTs one of the strongest and stiffest materials. Treacy *et al.* estimated the elastic modulus of CNTs to be in the terapascal range by measuring their thermal vibrational amplitudes using TEM [20]. Because of the hollow structure and closed topology, CNTs can sustain extreme strains (40%) in tension without showing plastic deformation of bond rupture [21]. Under strain, some local bonds are broken, but this local defect is redistributed over the entire surface due to the mobility of these defects. This process changes the helicity of CNTs and eventually affects its electronic property.

15.2.2.2 Electronic properties

The electronic properties of CNTs can be related to their structures. In particular, both Wildöer *et al.* and Odom *et al.* explained the relationship between the structure and electronic conductivity of SWNTs using STM/STS images and current voltage curves obtained by tunneling spectroscopy on individual CNTs. Their studies indicate that so-called armchair tubes are metallic, and the zigzag tubes and chiral tubes are either metallic or semiconducting depending on the wrapping angle and the length of CNTs. Nevertheless, the SWNT samples exhibit many different structures with no one species dominating. Four-probe measurements of MWNTs reveal that the electrical conductivity of individual MWNTs is metallic, semiconducting or semimetallic [8]. The electrical conductivity of MWNTs becomes metallic by doping with boron and nitrogen [22], and the conductivity of SWNTs becomes an order of magnitude higher by intercalating the CNT tubes with alkali and halogen dopants.

15.2.2.3 Chemical properties

It has been known that the basal graphite plane (graphene hexagon) is chemically inert. However, CNTs are susceptive to some chemical reactions due to the π-orbital mismatch in the curvature structures. Oxidation studies have revealed that the tips (caps) of CNTs are more reactive than the cylindrical parts [8, 20]. *Ab initio* calculations indicate that the average charge density of a pentagon (at the tips) is 3–4 times larger than that at the graphene hexagon in the basal graphite plane and would act as an electrophilic reaction site. There have been numerous reports on the enhanced electron transfer of analytes when CNTs are used as electrode materials. In carbon nanotubes, it has been suggested that the presence of defects creating overall topological change may inherently increase reactivity compared with their graphite counterparts [8]. The moderate reactivity made it easy to introduce functional groups to the CNTs (both side wall and end cap) which is essential for sensor design in many cases. For example, during the purification of CNTs with a strong acid, the carboxyl functional groups are introduced to the surface of CNTs, especially at the tips. The carboxy functional groups are involved in nanotube chemical modification with amide-linked groups at the tip ends of the CNTs [23, 24].

15.2.3 Preparation of CNTs

The preparation of CNTs is a prerequisite step for the further study and application of CNTs. Considerable efforts have been made to synthesize high quality CNTs since their discovery in 1991. Numerous methods have been developed for the preparation of CNTs such as arc discharge, laser vaporization, pyrolysis, and plasma-enhanced or thermal chemical vapor deposition (CVD). Among these methods, arc discharge, laser vaporization, and chemical vapor deposition are the main techniques used to produce CNTs.

Arc discharge [25] is initially used for producing C_{60} fullerenes. Nanotubes are produced by arc vaporization of two carbon rods placed in a chamber that is filled with low pressure inert gas (helium, argon). The composition of the graphite anode determines the type of CNTs produced. A pure graphite anode produce preferably MWNT while catalyst (Fe, Co, Ni, Y or Mo) doped graphite anode produces mainly SWNT. This technique normally produces a complex mixture of components, and requires further purification to separate the CNTs from the soot and the residual catalytic metals present in the crude product.

The laser ablation technique was developed in 1995 by Smalley's group [26] at Rice University. Samples were prepared by laser vaporization of graphite rods with a catalyst mixture of Co and Ni (particle size $\sim 1\,\mu m$) at 1200ºC in flowing argon followed by heat treatment in a vacuum at 1000ºC to remove the C_{60} and other fullerenes. The material produced by this method appears as a mat of "ropes", 10–20 nm in diameter and up to 100 μm or more in length. The average nanotube diameter and size distribution can be tuned by varying the growth temperature, the catalyst composition, and other process parameters.

Despite the frequent use of arc-discharge and laser ablation techniques, both of these two methods suffer from some drawbacks. The first is that both methods involve evaporating the carbon source, which makes it difficult to scale up production to the industrial level using these approaches. Second, vaporization methods grow CNTs in highly tangled forms, mixed with unwanted forms of carbon and/or metal species. The CNTs thus produced are difficult to purify, manipulate, and assemble for building nanotube-device architectures in practical applications.

Chemical vapor deposition (CVD) [27] of hydrocarbons over a metal catalyst is a method that has been used to synthesize carbon fibers, filaments, etc. for over 20 years. Large amounts of CNTs can be formed by catalytic CVD of acetylene over Co and Fe catalysts on silica or zeolite.

CNTs can also be produced by diffusion flame synthesis, electrolysis, use of solar energy, heat treatment of a polymer, and low temperature solid pyrolysis. In flame synthesis, combustion of a portion of the hydrocarbon gas provides the elevated temperature required, with the remaining fuel conveniently serving as the required hydrocarbon reagent. Hence, the flame constitutes an efficient source of both energy and hydrocarbon raw material. Combustion synthesis has been shown to be scalable for a high volume commercial production.

15.2.4 Purification of carbon nanotubes

In order to obtain the optimal performance of CNTs in various applications, high purity CNTs will be required. Purification of CNTs generally refers to the separation of CNTs from other entities, such as carbon nanoparticles, amorphous carbon, residual catalyst, and other unwanted species. A number of purification methods including acid oxidation, gas oxidation, filtration, and chromatography have been developed to date. In many cases, various combinations of these methods are used to obtain high quality CNTs.

The acid reflux procedure was first described by Rinzler *et al.* [28], in which raw nanotube materials are refluxed in nitric acid to oxidize the metals and carbon impurities. Acid-treated CNTs are considered to have carboxylic acid groups at the tube ends and, possibly, at defects on the side walls. The functionalized SWNTs have considerably different properties from those of the pristine tubes.

Gas phase oxidation is commonly used for the purification of CNTs. The method proposed by Ebbesen *et al.* [20] uses heat treatment of crude CNT products under a gas containing oxygen. This method has been explored extensively because it can yield the most highly graphitized tubes without being contaminated by the metal catalyst.

Filtration is also used to purify CNTs. Bandow [29] have reported a procedure for a one-step SWNT purification by microfiltration in an aqueous solution with a cationic surfactant. Shelimov [30] developed an ultrasonically assisted filtration method which allows the purity of nanotubes to reach $>90\%$.

Many chromatographic methods such as permeation chromatography, column chromatography, and size exclusion chromatography have been used to purify CNTs. The size exclusion chromatography (SEC) is the only carbon nanotube purification method in the literature that is not subjected to the acid treatments which tend to create the carboxylic functionality on CNTs.

Sample purity is documented with SEM, TEM, and electron microprobe elemental analysis. Raman and UV-vis-near-IR spectra are also useful techniques that can be used to examine the quality of CNTs at the different stages of the purification procedure.

15.2.5 Advantages of electrochemical sensors based on CNTs

CNTs have been one of the most actively studied electrode materials in the past few years due to their unique electronic and mechanical properties. From a chemistry point of view, CNTs are expected to exhibit inherent electrochemical properties similar to other carbon electrodes widely used in various electrochemical applications. Unlike other carbon-based nanomaterials such as C_{60} and C_{70} [31], CNTs show very different electrochemical properties. The subtle electronic properties suggest that carbon nanotubes will have the ability to mediate electron transfer reactions with electroactive species in solution when used as the electrode material. Up to now, carbon nanotube-based electrodes have been widely used in electrochemical sensing [32–35]. CNT-modified electrodes show many advantages which are described in the following paragraphs.

First, the CNT-modified electrodes catalyze the redox reactions of analytes. It has been reported that the oxidation of such analytes as dopamine, H_2O_2, and NADH are catalyzed at the various types of CNT-modified electrodes. Second, the biomacromolecules such as enzymes and DNA can be immobilized on the CNT-modified electrodes and maintain their biological activities. Third, the CNTs are a good material for constructing the electrodes. CNTs are small, straight, and strong, and they also have chemical stability. These characteristics of CNTs are advantageous for constructing CNT-modified electrodes. A couple of examples are vertically aligned CNT modified electrodes and the electrodes based on an individual nanotube. Fourth, CNTs can be functionalized mostly through the carboxyl group on the tips which helps to immobilize the enzymes, etc. for the development of various types of sensors. The fifth advantage of CNTs is their good electronic conductivity. The conductivity of CNTs is also affected by the structural changes such as twisting and bending of CNTs which may be applied for the sensing purpose.

These advantages combined with others such as the porous structure may contribute to having good wetting property for the solvents, a better electrode–electrolyte interface, and a large surface area. The central hollow cores and outside walls are a superior material to adsorb and store gases such as oxygen, hydrogen, and nitrogen oxide [4, 36, 37]. The investigation of gas sensors using the adsorptive properties of carbon nanotubes to detect oxygen and carbon oxide has been reported [4, 38]. The CNTs in many cases can serve as molecular wires that connect the electrode surface to the active site of enzymes. The direct or enhanced electrochemistry of several proteins and enzymes has been observed without the needs of mediators. Many enzyme-based electrochemical biosensors have been reported using CNT-modified electrodes.

15.3 FABRICATION AND APPLICATION OF ELECTROCHEMICAL SENSORS BASED ON CARBON NANOTUBES

A large number of papers on CNT-based sensors have been published in the last several years mainly because CNTs have the following advantages for the electrochemical sensor applications: (1) small size with large surface area, (2) high sensitivity, (3) fast response time, (4) enhanced electron transfer, (5) easy protein immobilization with retention of activity, and (6) alleviating surface fouling effects. For the fabrication of CNT-modified electrodes, several methods have been developed. Most of these methods involve the immobilization of enzyme molecules on the electrode surface. In the following sections, the preparation methods and some representative results using such electrodes as (1) CNT-composite electrodes, (2) vertically aligned nanotube electrode arrays, (3) layer-by-layer electrodes, and (4) CNT coated electrodes will be described. In addition, advantages as well as the applications of each type of these CNT-modified electrodes are discussed. Since the efficient charge transfer between the electrode surface and macrobiomolecules such as proteins and DNA is important for the development of biosensors, the direct electrochemistry of proteins and DNA is discussed in this section. The discussion in this section also includes CNT enzyme-based biosensors.

15.3.1 Preparation of carbon nanotube electrodes and their electrochemical characteristics

15.3.1.1 CNT-composite electrodes

The first CNT-modified electrode was reported by Britto *et al.* in 1996 to study the oxidation of dopamine [16]. The CNT-composite electrode was constructed with bromoform as the binder. The cyclic voltammetry showed a high degree of reversibility in the redox reaction of dopamine (see Fig. 15.3). Valentini and Rubianes have reported another type of CNT paste electrode by mixing CNTs with mineral oil. This kind of electrode shows excellent electrocatalytic activity toward many materials such as dopamine, ascorbic acid, uric acid, 3,4-dihydroxyphenylacetic acid [39], hydrogen peroxide, and NADH [7]. Wang and Musameh have fabricated the CNT/Teflon composite electrodes with attractive electrochemical performance, based on the dispersion of CNTs within a Teflon binder. It has been demonstrated that the electrocatalytic properties of CNTs are not impaired by their association with the Teflon binder [15].

Other CNT-composite electrodes are prepared by using conducting polymers such as polypyrrole [40] and polyaniline [41]. The PPy/CNT composite can be prepared electrochemically by repeated scanning the potential in the oxidation range in an aqueous solution containing CNTs, surfactant SDS, and pyrrole. A CNT-composite electrode has also been prepared using the sol-gel matrix as a glucose sensor [42]. The preparation of binderless biocomposite with CNTs has been reported for the preparation of glucose sensor [15]. After mixing CNTs with glucose oxidase (GOx), the composite was packed inside of a needle and was used for the detection of glucose.

15.3.1.2 Vertically aligned CNT-modified electrode

The electrodes modified with vertically aligned CNTs have recently received much interest for the purpose of designing CNT-based electrodes and electrochemical sensors.

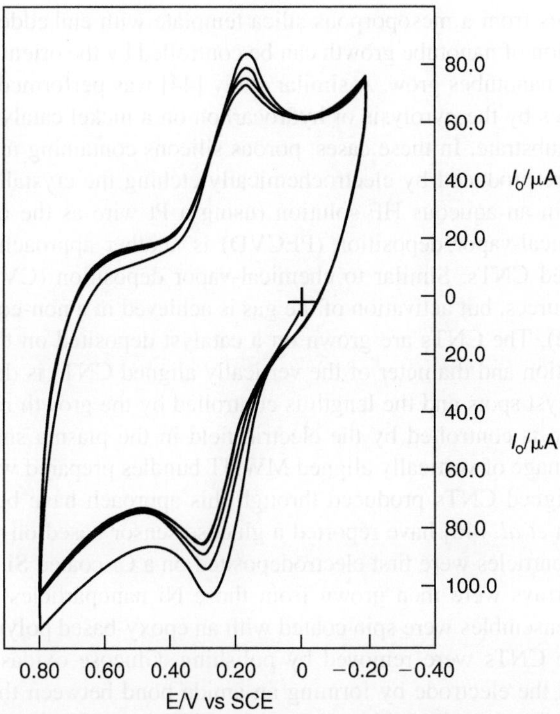

FIGURE 15.3 Cyclic voltammetric curve of 5 mM dopamine in PBS (pH 7.4) at a CNT electrode at 20 mV s^{-1}. (Reprinted with permission from [16]. Copyright (1996) Elsevier.)

The aligned CNT-modified electrode has added advantages to electrodes modified with a random tangle of CNTs (normally formed by drop coating) and CNT-composite electrodes. For the preparation of this type of electrode, CNTs are vertically assembled on the substrate surface; proteins and other electroactive species are then attached to the end tip of the CNTs. The vertically aligned CNT-modified electrode is helpful for understanding the electron transfer mechanism since proteins or other electroactive species attached to the CNTs through a known and definite way (at the end tip). While in other types of CNT-based electrodes, the electroactive species can either be attached to the end tip or to the side wall which yields an unknown spatial geometry. Second, the vertically aligned CNTs serve as molecular wires that facilitate the electron transfer between the electrodes and redox centers of enzymes without the mediator. This feature simplifies the fabrication of electrochemical sensors using redox enzymes as sensing agents. Third, vertically aligned CNTs can be used to make nano-electrode arrays.

Vertically aligned CNTs can be obtained through guided growth (physical method) or self-assembly of modified CNTs on certain substrates. In the first case, CNTs are synthesized in a controlled manner. One approach to achieve guided growth of CNTs is by using a porous template (such as mesoporous silica, alumina nanoholes) in the CVD method. For example, Li *et al.* [43] report the large-scale CVD growth of aligned carbon nanotubes from a mesoporpous silica template with embedded iron nanoparticles. The direction of nanotube growth can be controlled by the orientation of the pores from which the nanotubes grow. A similar study [44] was performed to make aligned carbon nanotubes by the pyrolysis of hydrocarbon on a nickel catalyst embedded in a porous silicon substrate. In these cases, porous silicons containing micro-, meso-, and macropores were produced by electrochemically etching the crystalline silicon wafer (as the anode) in an aqueous HF solution (using a Pt wire as the cathode). Plasma-enhanced chemical-vapor deposition (PECVD) is another approach used to prepare vertically aligned CNTs. Similar to chemical-vapor deposition (CVD), PECVD also uses gaseous sources, but activation of the gas is achieved in a non-equilibrium plasma (glow discharge). The CNTs are grown on a catalyst deposited on the substrate as in CVD. The location and diameter of the vertically aligned CNTs is defined by the pattern of the catalyst spots and the length is controlled by the growth rate and time. The growth direction is controlled by the electric field in the plasma sheath. Figure 15.4 shows a SEM image of vertically aligned MWNT bundles prepared with PECVD [45].

Vertically aligned CNTs produced through this approach have been used for sensor designs. Lin *et al.* [46] have reported a glucose sensor based on vertically aligned CNTs. Ni nanoparticles were first electrodeposited on a Cr coated Si. Low site density aligned CNT arrays were then grown from those Ni nanoparticles by PECVD. The resulting CNT ensembles were spin coated with an epoxy-based polymer. The protruding parts of the CNTs were removed by polishing. Glucose oxidase was covalently immobilized on the electrode by forming an amide bond between the amine residues of the protein and carboxylic acid groups on the CNT tips. With this electrode, glucose is indirectly sensed by measuring the catalytic reduction current of hydrogen peroxide generated by the reaction of glucose oxidase catalyzed oxidation of glucose. The electrode prepared in this way has also been used for DNA detection [45] by attaching oligonucleotide probes to the open ends of the CNTs. This electrode can detect subattomole of DNA by combining with the $Ru(byp)_3^{2+}$-mediated oxidation of guanine.

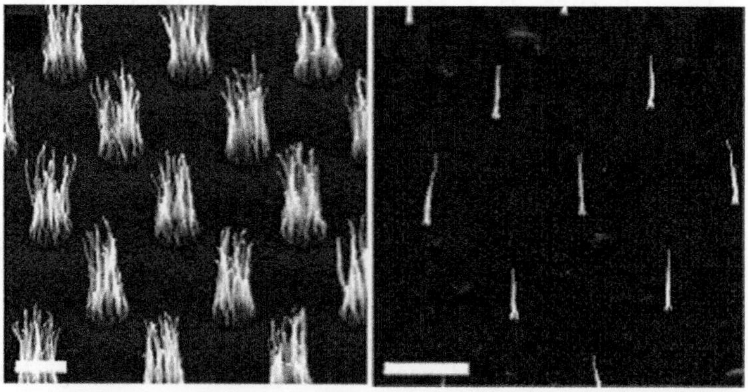

FIGURE 15.4 SEM images of vertically aligned MWNTs at (left) UV lithography and at (right) e-beam patterned Ni spots. (Reprinted with permission from [45]. Copyright (2003) American Chemical Society.)

Vertically aligned CNTs can also be obtained by using the self-assembly strategy. To perform self-assembling, the end tips of shortened CNTs first need to be modified with functional groups (such as —COOH, —SH); then the nanotubes are attached to a substrate (gold, polymer, etc.) through a chemical reaction. AFM images of electrodes prepared in this way normally show vertically aligned CNTs in bundles. In one example, Gooding *et al.* [47] and Yu *et al.* [48] modified the gold electrode with vertically aligned SWNTs through a self-assembling approach. SWNTs were first shortened by sonicating in a 3:1 v/v solution of concentrated sulfuric acid (98%) and concentrated nitric acid (70%) for 4 h. The shortened tubes were filtered and washed with Milli Q water to bring the pH to greater than 5. Then the shortened SWNT was dispersed with a concentration of 0.2 mg ml^{-1} in dimethylformamide containing 0.5 mg/mL of dicyclohexyl carbodiimide (DCC) to convert the carboxyl groups at the ends of the shortened SWNT into active carbodiimide esters. The substrate gold electrode which was previously modified with cysteamine was placed in the nanotube solution for 4 h. The —COOH group at the end tip of SWNTs then forms amide bonds with the amines group of the cysteamine and results in SWNTs vertically aligned to the electrode surface. Microperoxidase MP-11 is then attached to the end tip of CNTs by incubating in the enzyme solution. The resulting electrode shows a CV with $E_{1/2}$ at −390 mV (vs Ag/AgCl) for the redox reaction of MP-11. In a similar manner, Diao *et al.* [49] also prepared vertically aligned CNTs on a gold substrate and examined the electrochemical properties of the resulting electrode. The procedure is schematically shown in Fig. 15.5, and Fig. 15.6 is the AFM image of the modified electrode.

Liu *et al.* [50] have prepared aligned SWNTs by self-assembling nanotubes end functionalized with thiol groups on a gold substrate. The procedure is schematically shown in Fig. 15.7. The shortened SWNTs with carboxyl-end terminals formed by acid oxidation were first functionalized with thiol-containing alkyl amines through the amide linkage. The self-assembly of aligned SWNTs was obtained by incubating and ultrasonicating the gold substrate with the thiol-functionalized SWNT suspension.

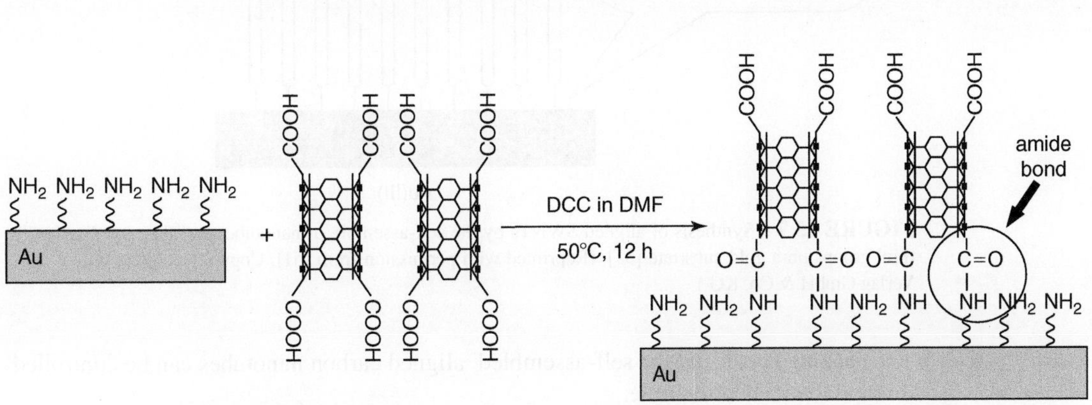

FIGURE 15.5 Schematic representation of the formation of SWNT assemblies. (Reprinted with permission from [49]. Copyright (2002) Wiley VCH Verlag GmbH & Co. KG.)

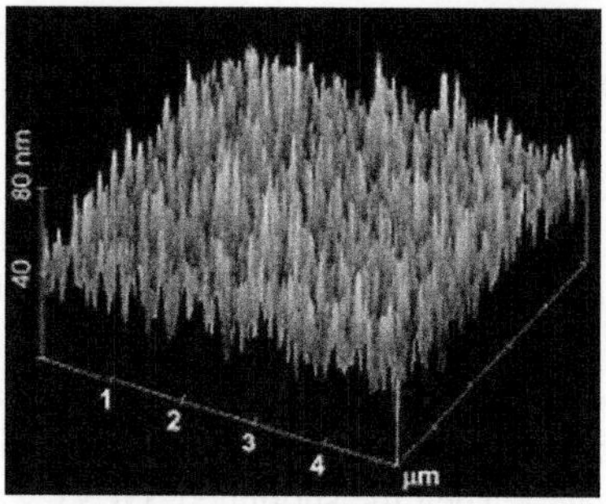

FIGURE 15.6 AFM image of self-assembled CNT. (Reprinted with permission from [49]. Copyright (2002) Wiley VCH Verlag GmbH & Co. KG.)

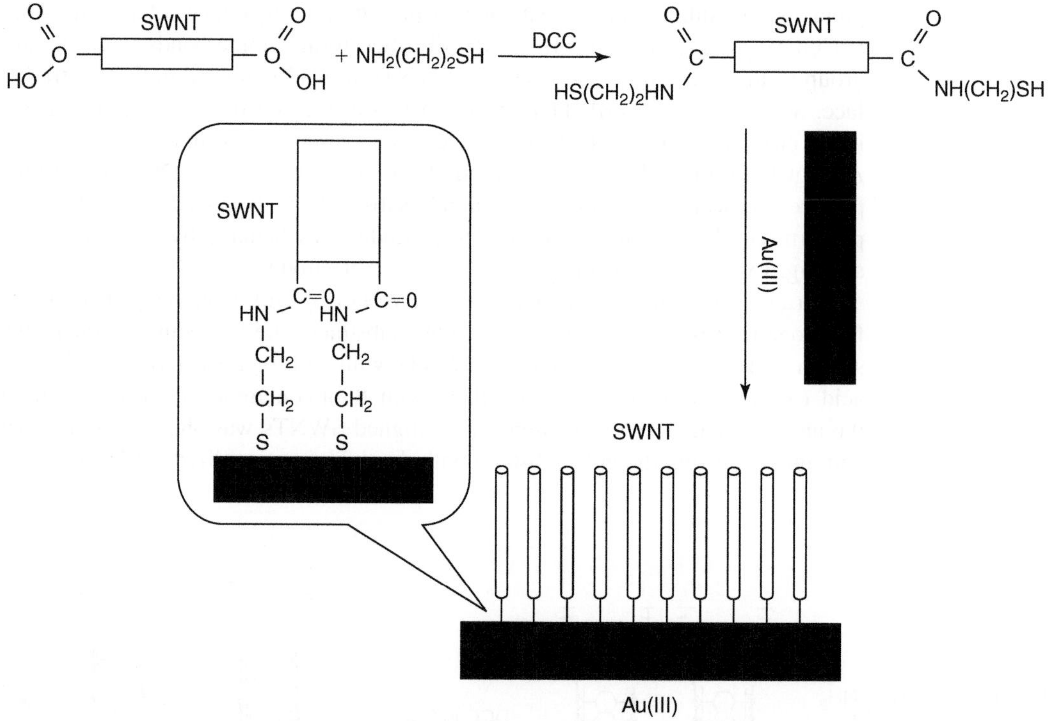

FIGURE 15.7 Synthesis of aligned SWNTs by the self-assembly of nanotubes end-functionalized with thiol groups on a gold substrate [51]. (Reprinted with permission from [51]. Copyright (2003) Wiley VCH Verlag GmbH & Co. KG.)

The packing density of the self-assembled, aligned carbon nanotubes can be controlled through the incubation time.

Lan *et al.* [52] have reported an approach for vertically aligning multi-wall carbon nanotubes on diverse substrates treated with polyelectrolyte. The shortened multi-wall carbon nanotubes were first functionalized with acyl chloride in thionyl chloride ($SOCl_2$).

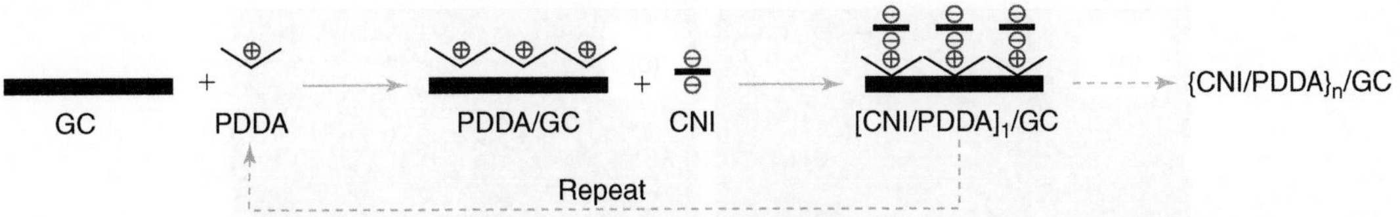

FIGURE 15.8 Schematic description of layer-by-layer assembly of CNTs with PDDA on the GC surface.

The vertically aligned monolayer of MWNTs was obtained by dipping the polyelectrolyte-modified substrate into a tetrahydrofuran suspension of the functionalized MWNTs.

15.3.1.3 Layer-by-layer fabrication of CNT electrode

The carboxylic acid group of CNTs is negatively charged, and an electrostatic assembly of CNTs with a positively charged surface helps to form CNT multilayer films [53]. Zhang *et al.* have adopted this layer-by-layer technique and have prepared an assembled CNT multilayer film on a GC electrode surface. Figure 15.8 describes the layer-by-layer assembly process of CNTs with poly(diallyldimethylammonium chloride) (PDDA) on the surface of a GC electrode. The electrostatic attraction between negatively charged CNTs and positively charged PDDA molecules leads to the formation of stable CNT multiple layers on the GC electrode surfaces (Fig. 15.9) by the layer-by-layer method, which is different from other preparation methods including cast coating, mixing the CNTs with carbon paste, and confining the CNTs with polymer matrixes, in which the CNTs are mainly in the form of big bundles. CNT multilayer films assembled by the layer-by-layer method possess many striking features such as having an adsorbed form of small CNT bundles or single tubes on the substrate, uniformity, good stability, and remarkable electrocatalytic activity toward O_2 reduction in alkaline media [54]. Moreover, this layer-by-layer technique has promising applications in CNT-based biosensor fabrication and enzyme immobilization. With layer-by-layer fabrication of DNA-wrapped CNTs, a DNA electrochemical sensor has been developed [55]. Munge *et al.* have reported a novel method for dramatically amplifying the electrochemical detection of proteins or DNA based on a stepwise layer-by-layer assembly of multilayer enzyme films on a CNT template [56].

15.3.1.4 CNT-coated electrodes

The coating of CNT solution or suspension on the electrode surface followed by evaporating the solvent is a simple but effective strategy to prepare CNT-modified electrodes. However, due to the strong van der Waals forces between nanotubes, CNTs tend to aggregate, thus leading to the poor solubility of CNTs in water and other solvents, which limits the preparation and application of CNT solutions. Ausman *et al.* have demonstrated the ability of organic solvents to form stable CNT dispersion. CNTs dispersed in several organic solvents such as dimethylformamide (DMF) [35, 57], acetone [58], acetonitrile [59], and ethanol [60] have been coated on the electrode surface. On these CNT-modified electrodes, the electrocatalytic behavior of ferricyanide,

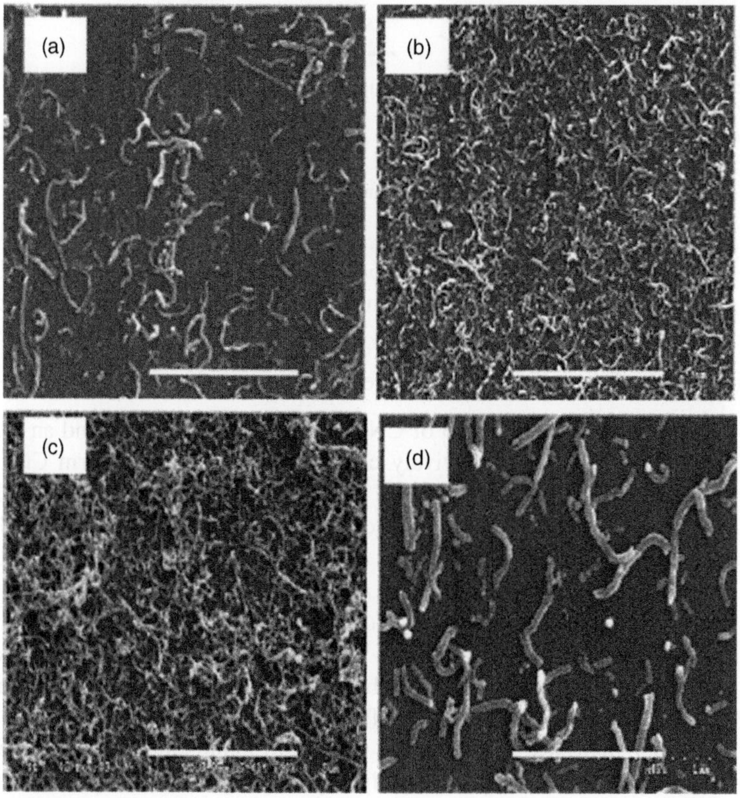

FIGURE 15.9 Representative SEM images of [PDDA/MWNT] (a), [PDDA/MWNT]3 (b), and [PDDA/MWNT]5 (c) assembled on a silicon wafer. The scale bar in (a)–(c) was 5 μm. (d) represents the SEM image of [PDDA/MWNT]1 with a high amplification with a scale bar of 1 μm. (Reprinted with permission from [54]. Copyright (2004) American Chemical Society.)

NADH, epinephrine or norepinephrine has been demonstrated with enhanced currents and reduced peak-to-peak separations in the cyclic voltammetry in comparison with unmodified electrodes. However, the dispersion of CNTs in organic solvents is not suitable in such cases as fabricating biosensors with biomolecules and CNTs, because the denaturalization of biomolecules caused by organic solvents may lead to a decrease or loss of activity. Thus, the preparation of aqueous dispersions and solubilization of CNTs have attracted much attention [61]. As described in 15.2.4, the acid treatment of CNTs form carboxyl sites on the surface of CNTs and cause tube shortening. A typical acid treatment procedure for CNTs is as follows [62]: a purified CNT "bucky paper" is suspended in 40 ml of a 3:1 mixture of concentrated H_2SO_4/HNO_3 in a 100 ml test tube and sonicated in a water bath for 24 h at 35 to 40°C. The resultant suspension was then diluted with 200 ml of water, and the larger cut CNTs are collected on a 100 nm pore filter membrane and washed with 10 mM NaOH solution. The cut tubes are then further treated with a 4:1 mixture of concentrated H_2SO_4 and 30% aqueous H_2O_2 under

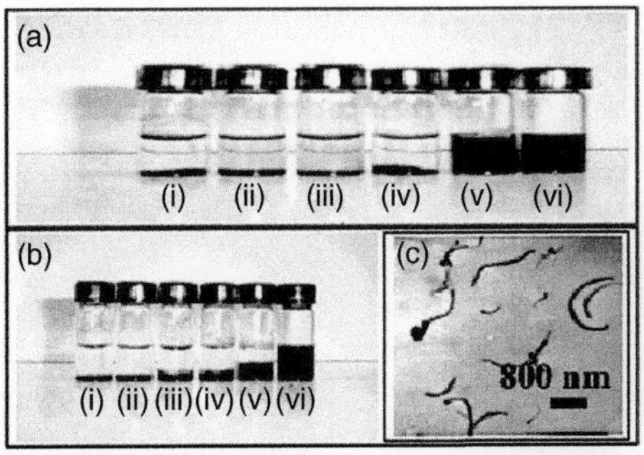

FIGURE 15.10 Photographs of vials containing 0.5 mg mL^{-1} of SWCNT (a) and MWCNT (b) in different solutions: phosphate buffer (0.05 M, pH 7.4) (i), 98% ethanol (ii), 10% ethanol in phosphate buffer (iii), 0.1% Nafion in phosphate buffer (iv), 0.5% Nafion in phosphate buffer (v), and 5% Nafion in ethanol (vi). Also shown (c) is a TEM image of a 0.5% Nafion solution containing 0.3 mg mL^{-1} of MWCNT. (Reprinted with permission from [64]. Copyright (2003) American Chemical Society.)

stirring conditions at 70ºC for 30 min. After filtering and washing again on a 100 nm pore filter, the cut nanotubes are suspended at a density of 0.1 mg ml^{-1} in water with the aid of 0.5% weight Triton X-100 surfactant. Such an acid treatment shortens the CNTs and functionizes CNTs with carboxylic acid groups. Cyclic voltammetry shows well-behaved peak currents for the redox reaction of the carboxylic acid group at a CNT-modified GC electrode in buffer solutions [35].

The oxidative procedure of CNTs alone may generate a stable CNT suspension which can be used for the preparation of CNT coated electrodes. Other procedures to prepare CNT suspensions are based on non-covalent stabilization of CNTs by using either surfactants or polymers. An excellent review on this subject has been published [61]. For the preparation of polymer-assisted CNT suspensions, a simple procedure for dispersing CNTs in aqueous solutions with a natural polymer gum arabic has been described [63]. Due to the physical adsorption of the polymer, a stable dispersion of full-length, well-separated, individual CNTs is formed in water. Wang *et al.* have demonstrated the ability of a perfluorosulfonated polymer Nafion to solubilize CNTs in water (see Fig. 15.10). The CNT coated electrodes formed with this suspension dramatically enhanced the redox activity of hydrogen peroxide at CNT/Nafion coated electrodes and helped to prepare oxidase-based amperometric biosensors [64]. Zhang *et al.* have solubilized CNTs in aqueous solution of a biopolymer chitosan (CHIT) which was used to modify electrodes for the development of sensors and biosensors [65]. Such a CNT/CHIT dispersion can easily attach the redox mediators such as toluidine blue [66] and azure dye [67] to form composite films that facilitate the electrooxidation of NADH that can be applied to a large group of NAD^+-dependent dehydrogenase enzymes for designing bioelectrochemical devices.

Surfactants are a low cost but effective additive for suspending CNTs in water. There have been numerous reports regarding the dispersion of CNTs in aqueous solution with the help of surfactants. It has been demonstrated that various surfactants have different influences on the suspendability of CNTs due to distinct CNT–surfactant

FIGURE 15.11 Photographs of vials containing $1\,mg\,ml^{-1}$ CNT dispersed in water (a) and in 0.1% CTAB solution by ultrasonication (b). Photograph (a) was taken 0.5 h after CNT dispersed in water, photograph (b) taken 1 week after CNT dispersed in CTAB solution. (Reprinted with permission from [72]. Copyright (2004) Elsevier.)

interactions [68, 69]. Islam *et al.* [70] have reported that the solubilization of high weight fraction CNTs in water is assisted by appropriate surfactants. The surfactant-dispersed CNT film electrodes have been widely applied in electrochemical and electroanalytical studies. Chen *et al.* [71] have developed a carbon fiber nanoelectrode modified by CNTs dispersed in a sodium dodecylsulfate (SDS) solution, which displays excellent electrochemical behaviors such as very high sensitivity toward neurotransmitters including dopamine, epinephrine, and norepinephrine. A stable suspension of CNTs was also obtained by dispersing CNTs in a solution of a cationic surfactant cetyltrimethylammonium bromide (CTAB) as demonstrated in Fig. 15.11. The direct electron transfer of glucose oxidase (GOx) immobilized on the CNT/CTAB-modified electrode has been reported [72]. A hydrophobic surfactant such as dihexadecyl hydrogen phosphate (DHP) was used to prepare a CNT film coated GC electrode, which showed an enhanced-reduction peak current of metronidazole and improved the sensitivity for the voltammetric determination of metronidazole [73].

The electrochemical response of analytes at the CNT-modified electrodes is influenced by the surfactants which are used as dispersants. CNT-modified electrodes using cationic surfactant CTAB as a dispersant showed an improved catalytic effect for negatively charged small molecular analytes, such as potassium ferricyanide and ascorbic acid, whereas anionic surfactants such as SDS showed a better catalytic activity for a positively charged analyte such as dopamine. This effect, which is ascribed mainly to the electrostatic interactions, is also observed for the electrochemical response of a negatively charged macromolecule such as DNA on the CNT (surfactant)-modified electrodes (see Fig. 15.12). An oxidation peak current near +1.0 V was observed only at the CNT/CTAB-modified electrode in the DNA solution (curve (ii) in Fig. 15.12a). The differential pulse voltammetry of DNA at the CNT/CTAB-modified electrode also showed a sharp peak current, which is due to the oxidation of the adenine residue in DNA (curve (ii) in Fig. 15.12b). The different effects of surfactants for CNTs to promote the electron transfer of DNA are in agreement with the electrostatic interactions

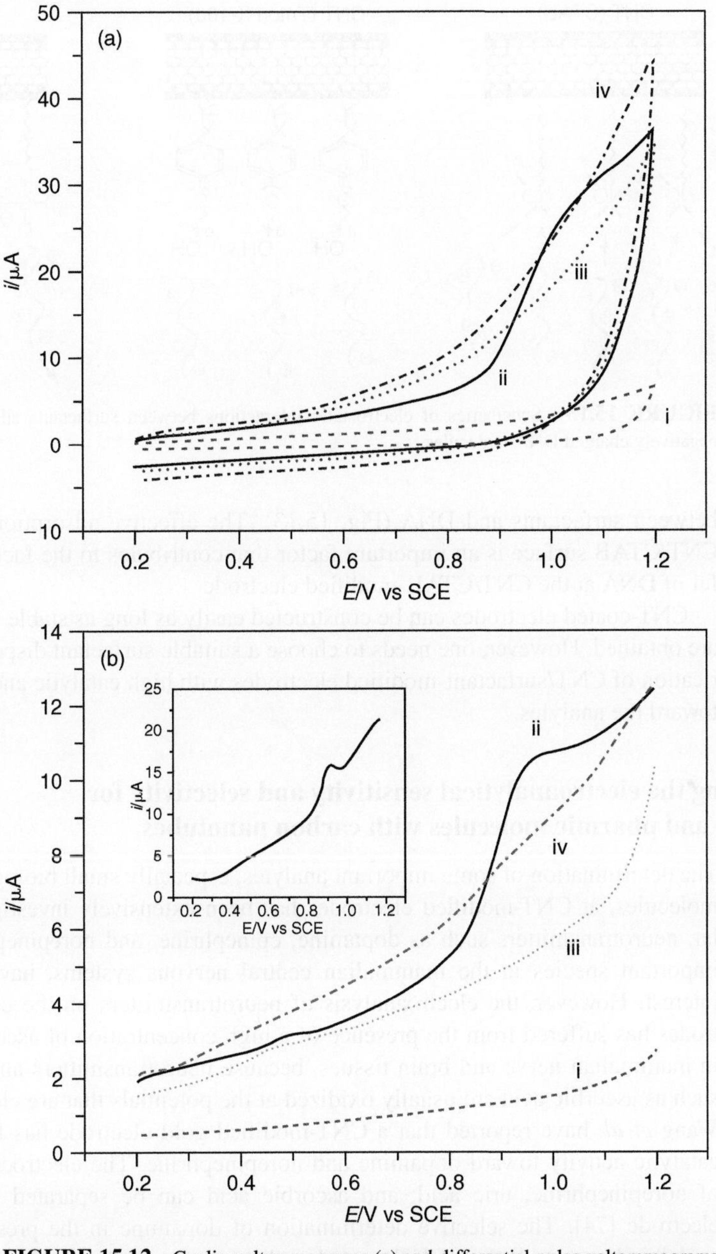

FIGURE 15.12 Cyclic voltammograms (a) and differential pulse voltammograms (b) of 1 mg m^{-1} DNA in a 0.1 M phosphate buffer (pH 7.4) at a bare GC electrode (i) and CNT-modified GC electrodes using CTAB (ii), Triton X-100 (iii), and SDS (iv) as dispersants. The inset in (b) shows a differential pulse voltammogram of 1 μg ml^{-1} adenine in a 0.1 M phosphate buffer (pH 7.4) at the CNT/CTAB-modified GC electrode. Accumulation time: 5 min. The scan rate of cyclic voltammetry is 0.1 V s^{-1}. The pulse amplitude of differential pulse voltammetry is 50 mV, and the pulse width is 50 ms (unpublished results).

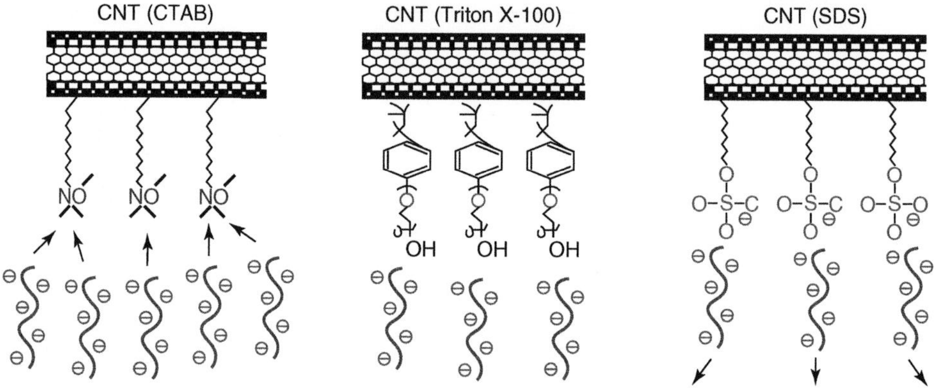

FIGURE 15.13 Schematics of electrostatic interactions between surfactants adsorbed on CNTs and negatively charged DNA molecules.

between surfactants and DNA (Fig. 15.13). The effective adsorption of DNA on the CNT/CTAB surface is an important factor that contributes to the facile electron transfer of DNA at the CNT/CTAB-modified electrode.

CNT-coated electrodes can be constructed easily as long as stable CNT suspensions are obtained. However, one needs to choose a suitable surfactant dispersant for the fabrication of CNT/surfactant-modified electrodes with high catalytic and good selectivity toward the analytes.

15.3.2 Improving the electroanalytical sensitivity and selectivity for small biological and pharmic molecules with carbon nanotubes

The determination of some important analytes, especially small biological and pharmic molecules, at CNT-modified electrodes has been extensively investigated. In particular, neurotransmitters such as dopamine, epinephrine, and norepinephrine, which are important species in the mammalian central nervous systems, have received much interest. However, the electroanalysis of neurotransmitters on the conventional electrodes has suffered from the presence of a high concentration of ascorbic acid present in mammalian nerve and brain tissues, because neurotransmitters and the interferents such as ascorbic acid are usually oxidized at the potentials that are close to each other. Wang *et al.* have reported that a CNT-modified gold electrode has favorable electrocatalytic activity toward dopamine and norepinephrine. The electrochemical response of norepinephrine, uric acid, and ascorbic acid can be separated on this modified electrode [74]. The selective determination of dopamine in the presence of ascorbic acid has been studied at CNT coated and intercalated electrodes [75], CNT/Nafion coated GC electrodes [76], CNT-modified gold electrodes [77], single carbon fiber microelectrodes modified with CNTs and Nafion [78], layer-by-layer assembled CNTs [79], DNA-functionalized CNTs [80], and water-soluble CNTs prepared via noncovalent functionalization by Congo red through a physical grinding treatment [81].

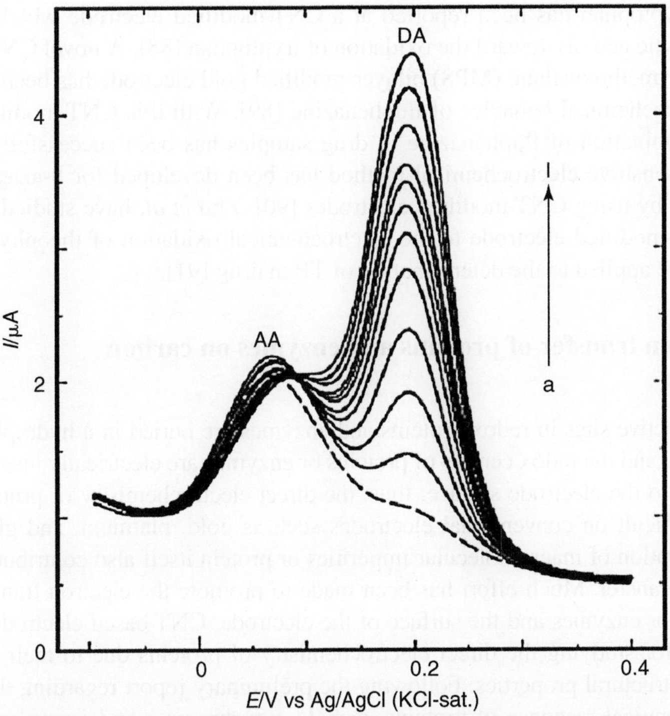

FIGURE 15.14 Osteryoung square wave voltammograms at the (PDDA/MWNT)5/GC electrode for ascorbic acid (0.10 mM) in the presence of different concentrations of dopamine: (a) 0, (b) 2, (c) 4, (d) 6, (e) 8, (f) 10, (g) 12, (h) 17, (i) 20, (j) 24, (k) 28, and (l) 32 μM. OSWV conditions were 4 mV step height, 20 mV pulse height, 2 Hz frequency, and 2 s quite time. (Reprinted with permission from [79]. Copyright (2005) Elsevier.)

Figure 15.14 illustrates a typical voltammetric result for the determination of dopamine in the presence of ascorbic acid with a CNT-modified electrode. The selective voltammetric detection of uric acid [82] or norepinephrine [83] in the presence of ascorbic acid has been demonstrated with a β-cyclodextrin-modified electrodes incorporating CNTs. Ye *et al.* [84] have studied the electrocatalytic oxidation of uric acid and ascorbic acid at a well-aligned CNT electrode, which can be used for the selective determination of uric acid in the presence of ascorbic acid. The simultaneous determination of dopamine and serotonin on a CNT-modified GC electrode has also been described [85].

It has been demonstrated that the presence of CNTs greatly increases the oxidation peak current of 6-benzylaminopurine. The CNT-modified electrode is suitable for the determination of trace amounts of benzylaminopurine and has the advantages of high sensitivity, quick response, and good stability [86]. Wang *et al.* have studied the electrocatalytic oxidation of thymine at a α-cyclodextrin incorporated CNT coated electrode in an alkaline media. A sensitive detection scheme for thymine has been further developed by using differential pulse voltammetry [87]. The electrochemical determination of tryptophan has been reported at a CNT-modified electrode which indicates a high catalytic activity toward the oxidation of tryptophan [88]. A novel CNT/(3-mercaptopro pyl)trimethoxysilane (MPS) bilayer-modified gold electrode has been used to study the electrochemical behavior of fluphenazine [89]. With this CNT-modified electrode, the determination of fluphenazine in drug samples has been successfully carried out. An ultrasensitive electrochemical method has been developed for 8-azaguanine in human urine by using CNT-modified electrodes [90]. Zhu *et al.* have studied the promotion of CNT-modified electrode to the electrochemical oxidation of theophylline (TP), which can be applied to the determination of TP in drug [91].

15.3.3 Direct electron transfer of proteins and enzymes on carbon nanotube electrodes

The active sites in redox proteins and enzymes are buried in a hydrophobic polypeptide chain, and the redox centers of proteins or enzymes are electrically insulated and inaccessible to the electrode surface; thus, the direct electrochemistry of proteins and enzymes is difficult on conventional electrodes such as gold, platinum, and glassy carbon. The adsorption of macromolecular impurities or protein itself also contributes to a slow electron transfer. Much effort has been made to promote the electron transfer between proteins or enzymes and the surface of the electrode. CNT-based electrodes have paved the way for studying the direct electrochemistry of proteins due to their unique electronic and structural properties. Following the preliminary report regarding the promoted electrochemical response of proteins, namely cytochrome *c* and azurin, at CNT electrodes [92], Wang *et al.* have displayed a pair of well-defined redox peaks for a cytochrome *c* aqueous solution at a GC electrode modified with CNTs. The peak current increases linearly with the concentration of cytochrome *c* [93]. The direct electrochemistry of cytochrome *c* adsorbed on the surface of CNTs has also been investigated, and a reagentless biosensor has been constructed for the determination of H_2O_2 using CNT/Cyt. *c*-modified electrode [94]. Voltammetric studies of myoglobin (Mb) and horseradish peroxidase (HRP) covalently attached onto the ends of vertically oriented CNT forest arrays show quasi-reversible redox behavior for Fe(III)/Fe(II) and demonstrate the electrochemically manifested peroxidase activity of Mb and HRP attached to CNTs [48]. CNT-modified electrodes also improved the direct electron transfer of Mb [95–97] and hemoglobin (Hb) [98–100]. Based on the electrocatalytic activity of the CNT-modified electrode, reagentless biosensors for H_2O_2 and NO have been constructed.

Moreover, it has been demonstrated that CNTs promote the direct electrochemistry of enzymes. Dong and coworkers have reported the direct electrochemistry of microperoxidase 11 (MP-11) using CNT-modified GC electrodes [101] and layer-by-layer self-assembled films of chitosan and CNTs [102]. The immobilized MP-11 has retained its bioelectrocatalytic activity for the reduction of H_2O_2 and O_2, which can be used in biosensors or biofuel cells. The direct electrochemistry of catalase at the CNT-modified gold and GC electrodes has also been reported [103–104]. The electron transfer rate involving the heme Fe(III)/Fe(II) redox couple for catalase on the CNT-modified electrode is much faster than that on an unmodified electrode or other

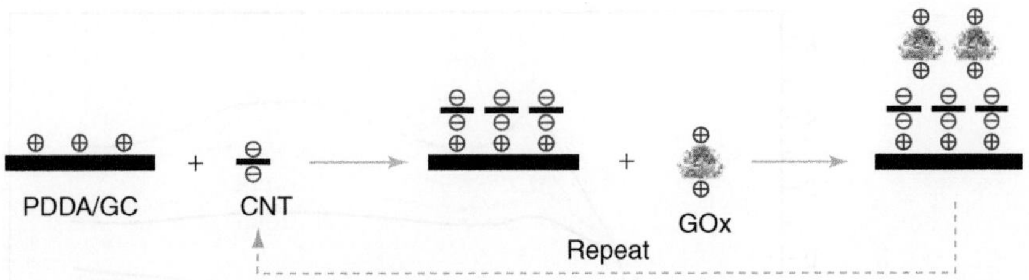

FIGURE 15.15 Schematic description of the fabrication of multilayer films of GOx onto CNTs.

types of carbon electrodes. Zhao *et al.* have reported the direct electrochemistry of HRP [105] and GOx [106] at carbon nanotube powder microelectrodes. Liu *et al.* have probed the direct electron transfer between the redox active center of GOx, flavin adenine dinucleotide (FAD), and the CNT-modified gold electrodes [107]. These results are helpful in understanding the intrinsic redox behaviors of enzymes. Furthermore, the immobilized GOx retains its bioelectrocatalytic activity for the oxidation of glucose [72, 108], which can be used to fabricate a glucose biosensor.

Negatively charged species such as carboxylic acid group in acid-treated CNTs can attract positively charged enzymes from solution as long as the pH value of the enzyme solution is controlled to be lower than the iso-electric point of the enzyme; thus, multilayer films of the enzyme can be formed by the layer-by-layer technique. For example, five layers of GOx can be immobilized on the electrode surface by alternatively dipping a poly(diallyldimethylammonium chloride (PDDA))-functionalized GC into a CNT solution and a GOx solution (pH 3.8). Figure 15.15 illustrates the preparation process for the formation of a multilayer film of GOx on the electrode.

The cyclic voltammograms of the GOx/CNT-modified GC electrodes in phosphate buffer solution (pH 7.4) show two pairs of redox peak currents. The first pair of peaks ($E_{1/2} = -0.09$ V vs Ag|AgCl) is attributed to the carboxylic acid groups in CNTs, while the second pair of peak currents ($E_{1/2} = -0.46$ V vs Ag|AgCl) is assigned to the direct electron tranfer of GOx. These peak currents increase as the number of layers increase indicating that the effective immobilization of GOx on CNTs using the layer-by-layer technique (Fig. 15.16, see Plate 17 for color version).

15.3.4 Electrochemical biosensors based on carbon nanotubes

CNTs offer an exciting possibility for developing ultrasensitive electrochemical biosensors because of their unique electrical properties and biocompatible nanostructures. Luong *et al.* have fabricated a glucose biosensor based on the immobilization of GOx on CNTs solubilized in 3-aminopropyltriethoxysilane (APTES). The as-prepared CNT-based biosensor using a carbon fiber has achieved a picoamperometric response current with the response time of less than 5 s and a detection limit of 5–10 μM [109]. When Nafion is used to solubilize CNTs and combine with platinum nanoparticles, it displays strong interactions with Pt nanoparticles to form a network that connects Pt nanoparticles to the electrode surface. The Pt-CNT nanohybrid-based glucose biosensor

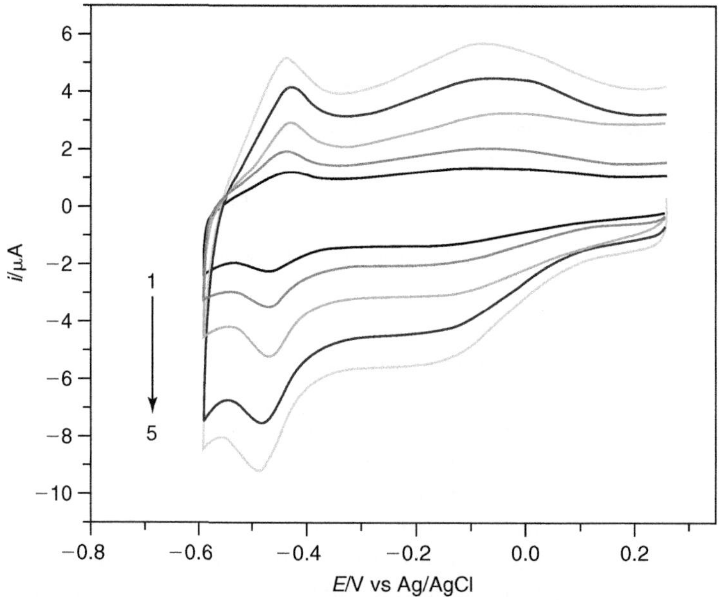

FIGURE 15.16 Cyclic voltammograms of $[GOx/CNT]_n$ immobilized on PDDA/GC electrodes in 0.1 M phosphate buffer solution (pH 7.4) at $0.1\,V\,s^{-1}$, indicating an increasing response of $[GOx/CNT]_n$ with layer number from 1 to 5 (unpublished results, J. Zhang, M. Feng, and H. Tachikawa) (see Plate 17 for color version).

constructed using a GC or carbon fiber microelectrode responds even more sensitively to glucose than the GC/GOx electrode modified with Pt nanoparticles or CNTs alone. The response time and detection limit of this biosensor is 3 s and 0.5 μM, respectively [110]. A glucose biosensor based on the adsorption of GOx on Pt nanoparticle-modified CNT electrode that is fabricated by growing CNTs directly on the graphite substrate has also been reported [111]. Joshi *et al.* [112] have constructed a glucose biosensor by incubating the enzyme in a CNT solution and cross-linking with a redox polymer poly[(vinylpyridine)Os(bipyridyl)$_2$Cl$^{2+/3+}$] film. This glucose sensor showed a three-fold increase of the glucose oxidation current ($1\,mA\,cm^{-2}$). Moreover, new CNT-based glucose electrochemical sensors are still attracting much interest as a model enzyme biosensor. Salimi *et al.* have fabricated a glucose biosensor prepared by immobilizing GOx into a sol-gel composite at the surface of a graphite electrode modified with CNTs [42]. A disposable glucose biosensor based on CNTs and screen-printed carbon electrodes has been developed [113]. Li *et al.* formed covalently linked composites of CNTs and GOx with a high function density in order to use as a biosensing interface [114]. These sensors show some excellent properties for glucose determination, such as high sensitivity, quick response, good reproducibility, and long-term stability.

Other enzymes have also been immobilized on CNTs for the construction of electrochemical biosensors. Deo *et al.* [115] have described an amperometric biosensor for organophosphorus (OP) pesticides based on a CNT-modified transducer and OP hydrolase, which is used to measure as low as 0.15 μM paraoxon and 0.8 μM parathion with sensitivities of 25 and 6 nA μM^{-1}, respectively. Xu *et al.* have explored the pH-sensitive property of the CNT-modified electrode and immobilized the enzymes with sol-gel hybrid material on the CNT-modified electrode to construct voltammetric urease and acetylcholinesterase biosensors based on pH detection [116]. Luong *et al.* [117] have constructed a putrescine biosensor based on putrescine oxidase (POx) immobilized on CNTs modified by APTES, which is able to efficiently monitor the direct electroactivity of POx at the electrode surface. Rubianes and Rivas [118] have demonstrated that the strong electrocatalytic activity of CNTs towards the reduction of hydrogen peroxide and quinones and the oxidation of NADH has allowed an effective low potential amperometric determination of lactate, phenols, catechols and ethanol, with the incorporation of lactate oxidase, polyphenol oxidase, and alcohol dehydrogenase/NAD^+, respectively, within the composite matrix.

CNT-based DNA sensors are also important in the application of CNTs in bioelectrochemistry. Due to the electrocatalytic activity of CNTs and interfacial accumulation, the direct electrochemical oxidation of natural DNA with an enhanced signal has been demonstrated on either multi-wall [119] or single-wall [120] CNT-modified electrodes. Cai *et al.* have described a sensitive electrochemical DNA sensor based on CNTs with a carboxylic acid group for covalent DNA immobilization and enhanced hybridization. The CNT-based assay with its large surface area and good charge transfer characteristics dramatically improves DNA attachment and complementary DNA detection sensitivity compared with the previous DNA sensors which directly incorporate oligononucleotides on carbon electrodes [121]. A nanoelectrode array based on vertically aligned multi-wall CNTs embedded in SiO_2 as an ultrasensitive DNA detection device has been developed. The hybridization of subattomole DNA target can be detected by combining with $Ru(bpy)_3{}^{2+}$ which mediates guanine oxidation [45]. Kerman *et al.* [122] have demonstrated that the combination of sidewall- and end-functionalization of CNTs provides a significant enhancement in the voltammetric signal of guanine oxidation and creates a large surface area for DNA immobilization. Wang *et al.* have described that DNA biosensors based on self-assembled CNTs have a higher hybridization efficiency compared with those based on random CNTs [123]. In addition, combining CNTs with other nanoparticles such as CdS [124], Pt [125], and magnetite nanoparticles [126] has offered great promise for constructing ultrasensitive DNA electrochemical biosensors.

15.4 SPECTROSCOPIC CHARACTERIZATION OF CARBON NANOTUBE SENSORS

15.4.1 Raman spectroscopy of carbon nanotubes

15.4.1.1 General features of Raman spectra from carbon nanotubes

Raman scattering is one of the most useful and powerful techniques to characterize carbon nanotube samples. Figure 15.17 shows the Raman spectrum of a single SWNT [127]. The spectrum shows four major bands which are labeled RBM, D, G, and G′.

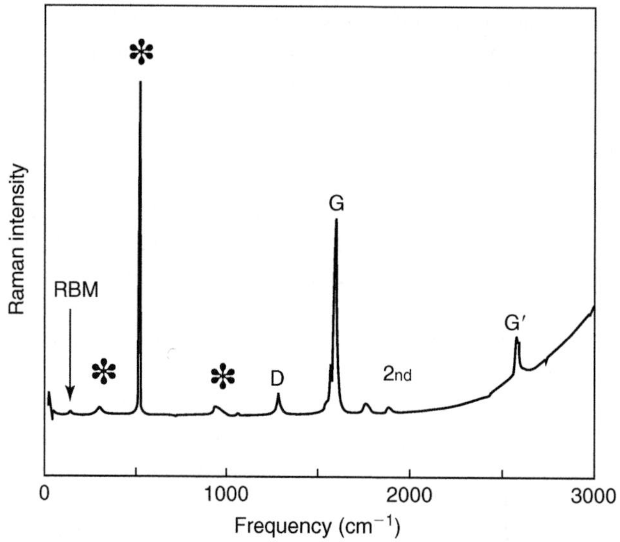

FIGURE 15.17 Raman spectrum of SWNT. (Reprinted with permission from [134]. Copyright (2004) The Royal Society.)

The radial breathing mode (RBM) appearing in the 160 to 300 cm^{-1} region is associated with a symmetric movement of all carbon atoms in the radial direction (Fig. 15.18a) [128]. The frequency of the radial breathing mode is inversely proportional to the diameter of the individual nanotube through the relation $\nu_{RBM} = A/d_t + B$, where the A and B parameters are determined experimentally. For isolated SWNTs, $A = 248\,\text{cm}^{-1}\text{nm}$ and $B = 0$ have been found [129]. In nanotube aggregates, the weak intertubule coupling via van der Waals interactions normally results in a 6–20 cm^{-1} shift to higher frequencies due to space restrictions imposed by neighboring tubes [130]. For typical SWNT bundles in the diameter range $d_t = 1.5 \pm 0.2\,\text{nm}$, $A = 234\,\text{cm}^{-1}\text{nm}$ and $B = 10\,\text{cm}^{-1}$ (where B reflects the tube–tube interaction) have been found [131]. For large diameter tubes ($d_t > 2\,\text{nm}$) the intensity of the RBM feature is weak and is hardly observable.

The tangential G-band (or high energy mode) around 1580 cm^{-1} also provides a signature of carbon nanotube. The G mode in SWNTs gives rise to a multi-peak feature which originates from the symmetry breathing of the tangential vibration when the graphite sheet is rolled to make a cylindrically shaped tube (see Fig. 15.18b). The two most intense G peaks are labeled G^+, for modes with atomic displacements along the tube axis, and G^-, for modes with atomic displacement along the circumferential direction (see Fig. 15.18b). The G^- bands for semiconductive and metallic SWNTs are different in the line shape and frequencies (Fig. 15.19), i.e. broadened for metallic SWNTs while sharp and weak for semiconductive SWNTs [128]. This feature can be used to distinguish between the two types of SWNTs. The G^- mode shifts to lower frequencies with decreasing diameter, and this decrease becomes larger as the curvature of the sheet increases [132, 133].

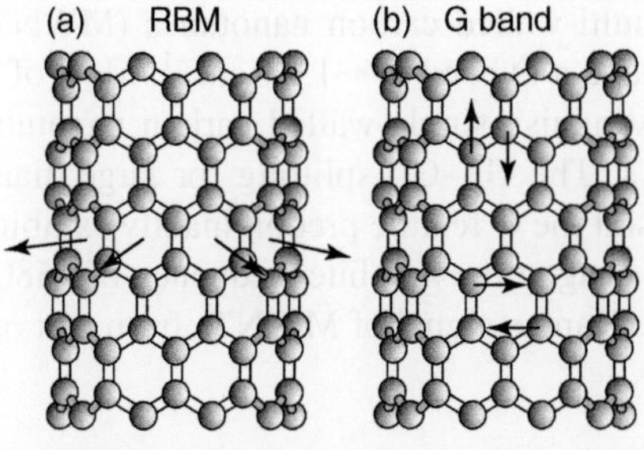

FIGURE 15.18 Schematic picture showing the atomic vibrations for (a) the RBM and (b) the G band modes. (Reprinted with permission from [128]. Copyright (2003) IOP Publishing Ltd.)

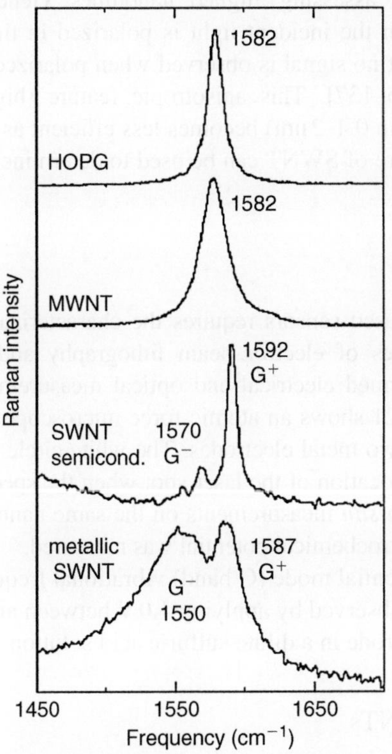

FIGURE 15.19 G band for highly ordered pyrolytic graphite (HOPG), MWNT bundles, one isolated semiconducting SWNT and one isolated metallic SWNT. (Reprinted with permission from [128]. Copyright (2003) IOP Publishing Ltd.)

The D band, the disorder induced mode, normally appears between 1250 and $1450\,cm^{-1}$. This band is activated in the first-order scattering process of sp^2 carbons by the presence of in-plane substitutional hetero-atoms, vacancies, grain boundaries or other defects and by finite-size effects [134]. The G′ band is the second-order overtone of the D band.

The Raman spectrum of multi-walled carbon nanotunes (MWNTs) only shows a single G band at $\sim 1580\,\mathrm{cm}^{-1}$ and a D band at $\sim 1360\,\mathrm{cm}^{-1}$. Most of the spectral characteristics that are used to distinguish single-walled carbon nanotubes from graphite are not so obvious in MWNTs. The G^{+}–G^{-} splitting for large diameter MWNTs is small and smeared out; as a result the G feature predominantly exhibits a weakly asymmetric band, with a peak appearing at the graphite frequency of $1580\,\mathrm{cm}^{-1}$. Therefore it is difficult to differentiate the Raman signal of MWNTs from that of graphite.

15.4.1.2 Anisotropy of SWNT

The isolated SWNT acts as a dipolar antenna, polarized along the tube axis. Polarization effects are very important for assessing aligned nanotubes. Generally the highest Raman signal is observed when the incident light is polarized in the direction parallel to the tube axis while almost no signal is observed when polarized perpendicular to the axis (see Fig. 15.20) [135–137]. This anisotropic feature (highly effective for nanotubes with a diameter within 0.4–2 nm) becomes less efficient as the tube diameter increases. The anisotropic feature of SWNT can be used to determine the orientation of the tubes.

15.4.1.3 Single nanotube characterization

Designing nano-sized SWNT-based sensors requires the characterization of individual nanotubes. Using the techniques of electron-beam lithography and nano-fabrication Cronin *et al.* [138] have performed electrical and optical measurements on the same individual nanotube. Figure 15.21 shows an atomic force microscope image of a single carbon nanotube contacted by two metal electrodes. The white circle in the figure indicates the approximate size and location of the laser spot when the spectrum on the right was taken. By performing the *in-situ* measurements on the same nanotube the response of a nanotube to changes in electrochemical potential was measured.

A $9\,\mathrm{cm}^{-1}$ upshift of the tangential mode (G band) vibrational frequency as well as a 90% decrease in intensity was observed by applying 1.0 V between an individual nanotube and a silver reference electrode in a dilute sulfuric acid solution.

15.4.1.4 Raman spectroscopy of modified CNTs

Chemical modification of CNTs is an essential step towards the fabrication of CNT-based electrochemical sensors. Raman spectroscopy provides an effective way to monitor the modification process and to characterize the functionalized CNTs.

Modified CNTs feature various spectral changes depending on the methods and the location of modifications. These changes include variations in band frequencies, width, and intensities. For example, aryldiazonium salts [139] were used to modify individual sodium dodecyl sulphate (SDS) coated SWNTs with aryl group. The Raman spectrum of functionalized (SDS-free) SWNTs shows a disorder mode much higher than pristine SWNT; the radial breathing modes are nearly unobservable.

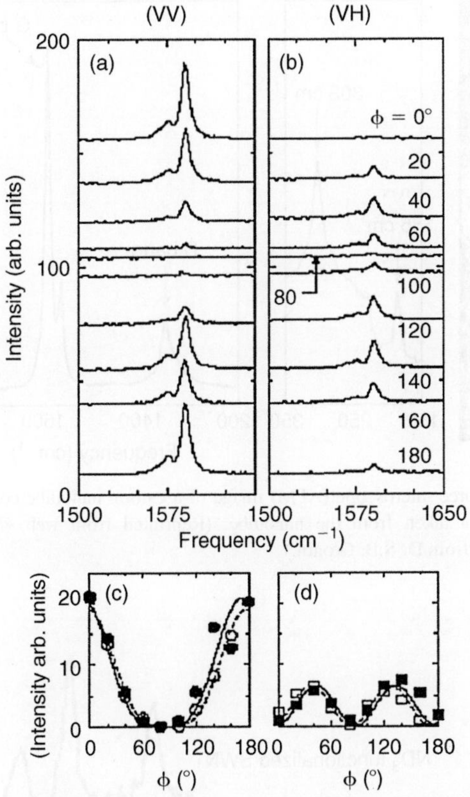

FIGURE 15.20 Polarization dependence of the G band from one isolated semiconducting SWNT sitting on an Si/SiO_2 substrate with incident and scattered light parallel (a) and crossed (b) to each other. $\phi = 0°$ stands for incident light polarized along the nanotube axis. Points on (c) and (d) plot the G band intensity dependence on ϕ for (a) and (b), respectively. The solid curves fit the data points with the functions $\cos^4(\phi)$ and $\cos^2(\phi)\sin^2(\phi)$. (Reprinted with permission from [137]. Copyright (2002) American Physical Society.)

Khare *et al.* [141] functionalized single-walled carbon nanotubes through a microwave discharge of ammonia. SWNTs exposed to both NH_3 and ND_3 discharges display RBM frequency shifts from that of the pristine nanotubes (Fig. 15.22) since sidewall functionalization of the nanotubes results in a general increase of CNT diameter. The frequency of the G band (1594 cm^{-1} in pristine SWNT) also shifts toward the lower energy (1583 cm^{-1}) in the NH_3 functionalized SWNT, and the line shape is broadened. This change in G band frequency further suggests the sidewall interactions with NH species.

In another example [56] SWNT was modified with peroxytrifluoroacetic acid (PTFAA). Raman spectrum of the carbon nanotubes after the PTFAA treatment shows a D-line substantially increased indicating the formation of "defect" sites with sp^3-hybridized carbon atoms on the sidewalls due to the addition of the functional groups. The RBM bands in the region of 170–270 cm^{-1} decreased and shifted to higher

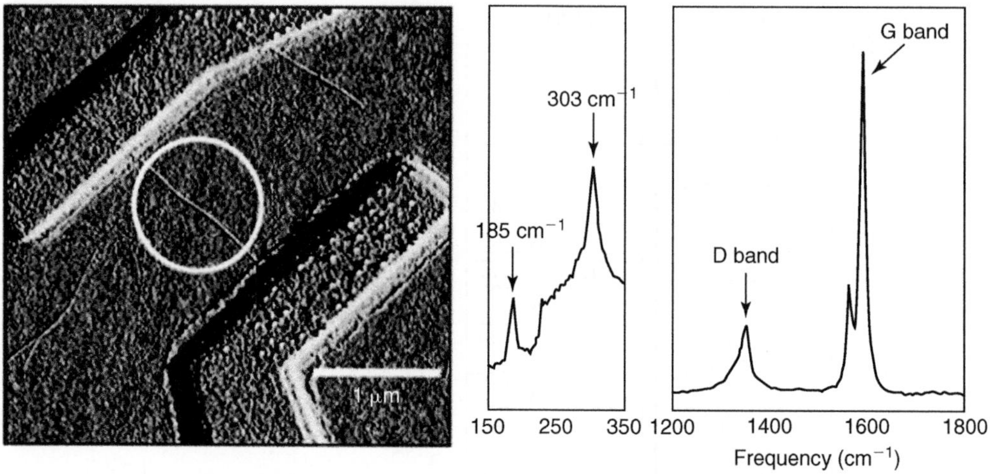

FIGURE 15.21 Atomic force microscope (AFM) image of a carbon nanotube contacted by two metal electrodes and Raman spectra taken from the nanotube. (Reprinted from web @ http://www-rcf.usc.edu/~scronin with permission from Dr S.B. Cronin.)

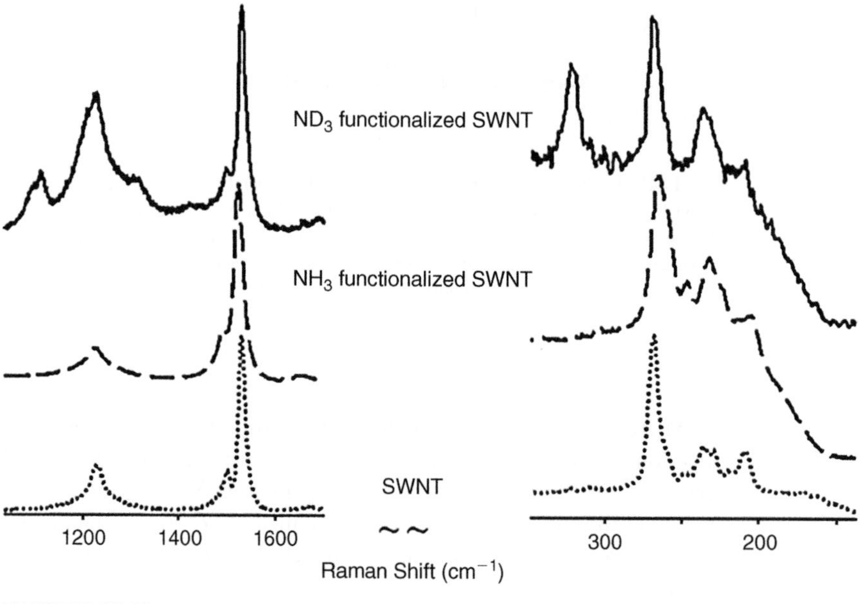

FIGURE 15.22 Raman spectra of NH_3 and ND_3 functionalized SWNTs by glow-discharge. (Reprinted with permission from [140]. Copyright (2004) American Chemical Society.)

wavenumbers after the treatment. The change in the radial breathing modes indicates the covalent functionalization of SWNTs on the sidewall [141].

15.4.1.5 Raman spectroscopy of self-assembled carbon nanotubes

One interesting development in the carbon nanotube-based electrochemical sensor is the ability to self-assemble the CNT to other types of nano materials such as gold and silver nanoparticles or to a polymer surface. The enhancement of Raman signals of carbon nanotubes through the adsorption on gold or silver substrate has been also reported [142–146].

With the aid of a bi-functionalized reagent (terminated with pyrenyl unit at one end and thiol group at the other end), gold nanoparticles were self-assembled onto the surface of solubilized carbon nanotubes [147]. Raman spectrum of the gold nanoparticle bearing CNTs is enhanced possibly due to charge transfer interactions between nanotubes and gold nanoparticles.

Self-assembling of CNTs normally results in vertically aligned CNTs on the substrate surface. The self-aligned SWNT can be confirmed by polarized Raman spectra. Diao *et al.* [49] reported the fabrication and characterization of chemically assembled SWNTs (ca-SWNTs), which are constructed by the combination of a self-assembling procedure and a surface condensation reaction between the —NH_2 group on the gold substrate and the —COOH group on SWNT. Polarized Raman spectra at the cross-section of Au substrate with aligned SWNTs show much stronger signals when the incident light is parallel to the CNT axis than when perpendicular to the nanotube axis.

15.4.1.6 Raman spectroscopy of CNT composites

CNT composites have also been used to develop CNT-based sensors. Raman spectroscopy has been used to study the CNT–polymer composite concerning the orientation of nanotubes in polymers and CNT–polymer interactions. Poulin [148] and Frogley [149] have performed a thorough study of nanotube alignment in polymers using polarized Raman spectroscopy, and have compared a large amount of experimental results with existing models. Generally, the interaction between nanotubes and polymers is featured by peak shifts or broadenings [134, 150].

Conducting polymers have been extensively used to make CNT composite modified electrode because of their unique electrochemical properties. Zhang *et al.* [40] studied the doping level of CNT in PPy using Raman spectra. The Raman spectrum only shows PPy bands since CNTs were completely coated with the polymer. A strong peak at 1583 cm^{-1} is assigned to the backbone stretching mode of C=C bonds, and the peaks at 1377 and 1083 cm^{-1} to the ring stretching and the N–H in-plane deformation of the oxidized (doped) species of the PPy, respectively. The corresponding band of the neutral species at 1048 cm^{-1} is relatively weak indicating a relatively high level of doping of PPy.

Recently, a layer-by-layer (LBL) technique has been introduced into the fabrication of CNT-modified electrode and received a great deal of interest. He *et al.* [55] reported a fabrication of DNA-wrapped carbon nanotubes using the LBL technique.

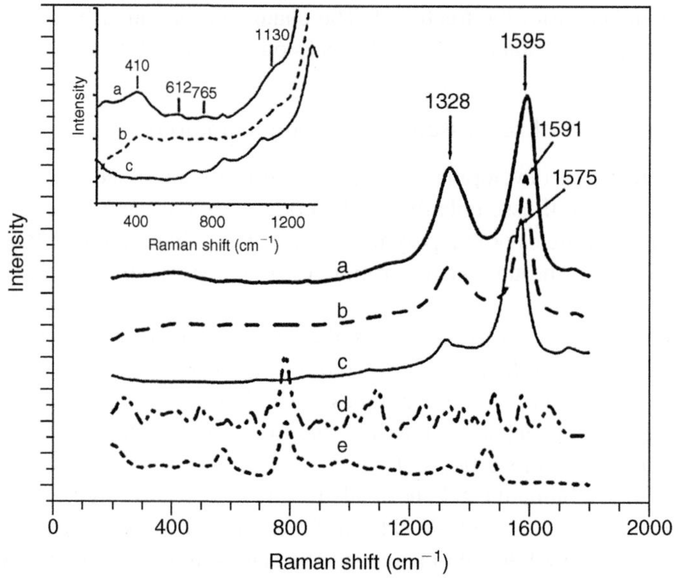

FIGURE 15.23 Raman spectra of (a) ds-calf thymus DNA/PDDA/SWNTs, (b) PDDA/SWNTs, (c) oxidized SWNTs, (d) calf thymus DNA, and (e) PDDA. Inset: Magnification of the Raman spectra of DNA/PDDA/SWNTs, PDDA/SWNTs, and oxidized SWNTs. (Reprinted with permission from [55]. Copyright (2005) American Chemical Society.)

Poly(diallyldimethylammonium) (PDDA), a positively charged polyelectrolyte, and DNA, a negatively charged counterpart, are alternatively deposited on the water-soluble oxidized SWNTs. A purified DNA/PDDA/SWNT composite-modified electrode is used as a DNA sensor. Raman spectroscopy was used to monitor the individual steps of the LBL fabrication process and to characterize the resulting DNA/PDDA/SWNT samples [55]. Figure 15.23 shows typical Raman spectra at different steps of the LBL fabrication process. The disorder-induced D band at 1328 cm^{-1} and the tangential stretch G band at 1580 cm^{-1} are observed for oxidized SWNTs, PDDA/SWNTs, and DNA/PDDA/SWNTs. The PDDA/SWNTs and DNA/PDDA/SWNTs show tangential G bands upshifted 16 and 22 cm^{-1}, respectively, relative to that of bare oxidized SWNTs at 1575 cm^{-1}. Similar upshifts in the tangential mode have also been reported in SWNTs/polymer composites which indicate the presence of polymer molecules wrapped around the sidewall of SWNTs [151–152]. The intensities of the typical disorder-induced D band at 1328 cm^{-1} for both PDDA/SWNTs and DNA/PDDA/SWNTs are markedly increased compared to the intensity of the same band in oxidized SWNTs owing to the overlapping Raman features from added layers of PDDA and/or DNA on oxidized SWNTs. Raman spectra in the range from 200 to 1350 cm^{-1} also have four very weak Raman scattering peaks at 1130, 765, 615, and 410 cm^{-1} for DNA/PDDA/SWNTs (inset). These peaks originated from Raman features of both PDDA and DNA confirmed the stepwise layer-by-layer preparation of the DNA/PDDA/SWNT composite.

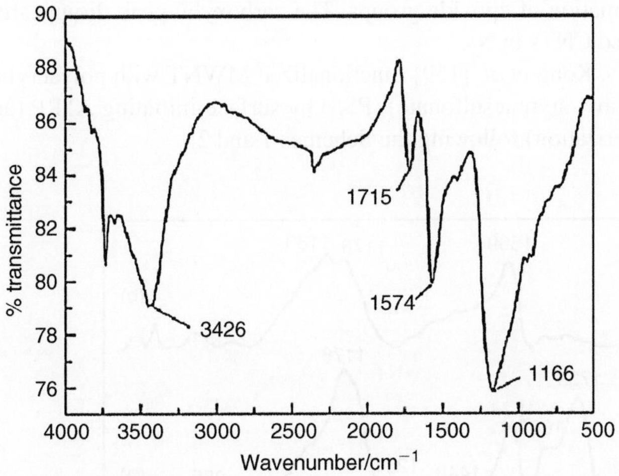

FIGURE 15.24 FTIR spectrum of acid-treated MWNT. (Reprinted with permission from [72]. Copyright (2004) Elsevier.)

15.4.2 FTIR of CNT-based sensors

Since FTIR is a powerful tool to identify functional groups, it has been extensively used to characterize CNT-based sensors which normally require the modification of nanotubes. For instance, acid (HNO_3/H_2SO_4) and ultrasonic treatments, which are common procedures to purify CNTs, are known to shorten and introduce oxygen-containing groups (such as the carboxylic group) to CNTs. This can be easily confirmed by the FTIR spectrum [72, 154, 155]. The FTIR spectrum of acid-treated MWNTs shows the peaks at 1715 cm^{-1} (carboxylic) and 1574 cm^{-1} (carboxylate). The peaks at 3426 and 1166 cm^{-1} are attributed to stretching vibrations of —OH and C—OH, respectively (Fig. 15.24).

As mentioned earlier in this section, peroxytrifluoroacetic acid (PTFAA) can be used to modify SWNTs with carboxylic groups [56]. FTIR was used to characterize the modified SWNTs and the results are compared with that from thermal oxidation (used to remove amorphous carbon) by heating the catalyst-free SWNTs in air. The thermally oxidized SWNTs show a band at 1744 cm^{-1} (originated from the formed carboxylic groups) in addition to the modes at 1570 cm^{-1} (in-plane E_{1u} mode of SWNTs [156, 157]) and at 1165 cm^{-1}. Figure 15.25 shows the IR spectra of the SWNTs after the PTFAA treatment. The band at 1733 cm^{-1} in the PTFAA-treated sample (Fig. 15.25b) is from carboxylic groups and/or ester groups. The shift of the carboxylic band from 1744 cm^{-1} for the air-oxidized sample to 1733 cm^{-1} for the PTFAA-treated sample (Fig. 15.25b) is due to the formation of a hydrogen bond between the carboxylic groups (—COOH) suggesting abundance in —COOH groups after the PTFAA oxidation [158]. The bands at 1660 cm^{-1} and 1448 cm^{-1} are assigned to the C═O stretching mode in quinone groups, while the bands at 1281 cm^{-1} and 855 cm^{-1} are the evidence for the formation of epoxide groups. The carboxylic peak disappeared after annealing the modified CNTs in N_2.

Recently, Kong *et al.* [159] functionalized MWNT with polyacrylic acid (PAA) and poly(sodium 4-styrenesulfonate) (PSS) by surface-initiating ATRP (atom transfer radical polymerization) following the Schemes 1 and 2:

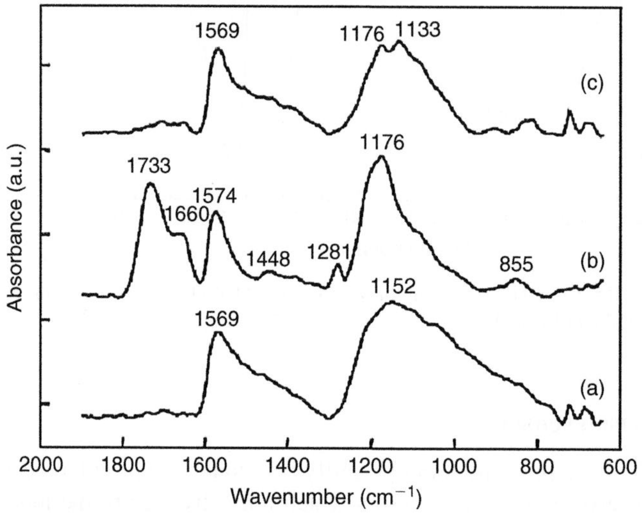

FIGURE 15.25 IR spectra of SWNTs: (a) p-SWNTs, (b) after peroxytrifluoroacetic acid treatment, and (c) N_2 annealing after treatment. (Reprinted with permission from [158]. Copyright (2005) Elsevier.)

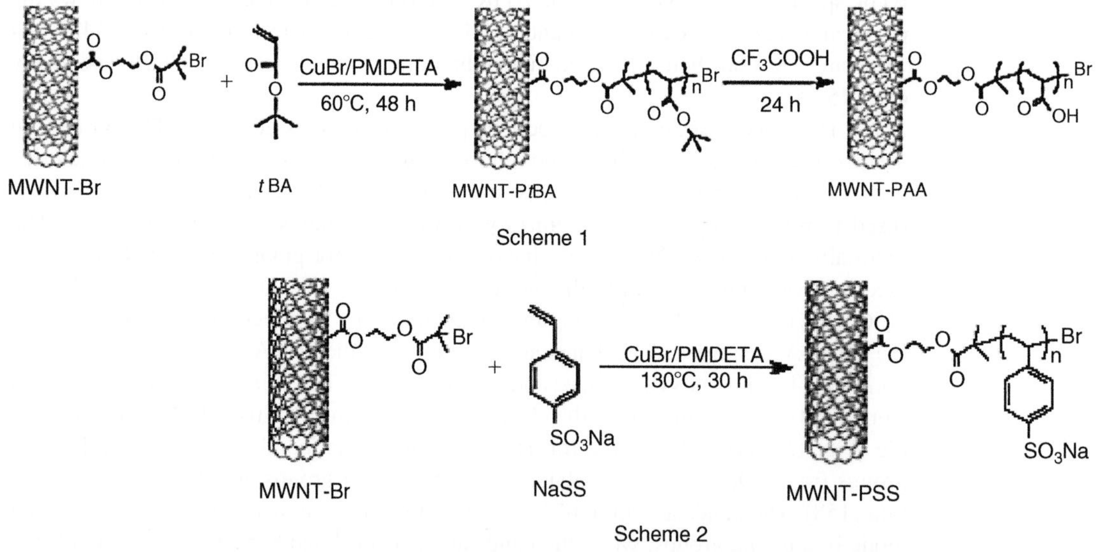

SCHEMES 1 AND 2 (Reprinted with permission from [159]. Copyright (2005) Elsevier.)

FTIR spectroscopy was used to monitor the functionalization process. In the first scheme, MWNT-P*t*BA shows a strong C=O stretch (ca. $1730\,cm^{-1}$), characteristic of carbonyl groups, while only a weak absorption peak at $1730\,cm^{-1}$ was found in the spectrum of MWNT-Br. When PtBA was hydrolyzed to PAA, a wide absorption band assigned to the hydroxyl groups around $3440\,cm^{-1}$ was observed in the FTIR spectrum. FTIR spectra of MWNTs modified according to the second scheme show the aromatic C—H stretch at 3098 and $1580–1602\,cm^{-1}$, the aliphatic C—H stretch of the polymer backbones at 2921 and $2848\,cm^{-1}$, and the absorption peaks of the O=S=O stretch at 1208 and $1156\,cm^{-1}$. The S—O and C—S stretches were observed at 663 and $641\,cm^{-1}$, respectively. A strong O—H stretch was also observed at $3436\,cm^{-1}$ for MWNT-PSS due to H-bonding with water. The resulting polyelectrolyte-functionalized MWNT as well as MWNT-COOH were used as templates for further surface functionalization by the LBL self-assembly approach on the basis of their high surface charge densities. Linear poly[2-(N,N-dimethylaminoethyl) methacrylate] (PDMAEMA) and hyperbranched poly(sulfone amine) (HPSA) were selected as polycations. A strong band at $1100\,cm^{-1}$ from the C—O (PDMAEMA) or O=S=O (HPSA) absorption, confirmed the assemblage of polyelectrolytes.

In another study, cationic polyethyleneamine (PEI) and anionic citric acid (CA) as well as heat treatment in NH_3 are also found to modify the nanotube surface and change the nanotube properties [160]. FTIR spectrum of CA coated CNTs shows absorption peaks corresponding to stretching vibration of C=O (1724 and $1650\,cm^{-1}$), and C—O (1392 and $1218\,cm^{-1}$). The NH_3-treated and PEI coated CNTs show basic nitrogen-containing groups (e.g. amine) on the nanotubes, as suggested by the bending vibration of N—H (1633, 1581, 1639, and $1581\,cm^{-1}$), and stretching vibration of C—N ($1093\,cm^{-1}$, $1164\,cm^{-1}$, and $1114\,cm^{-1}$).

Amide bond is an effective anchor to connect CNTs to substrate surfaces. Lan *et al.* [52] covalently assembled shortened multi-walled carbon nanotubes (s-MWNT) on polyelectrolyte films. The shortened MWNT is functionalized with acyl chloride in thionyl chloride ($SOCl_2$) before self-assembling. The FTIR spectrum of self-assembled MWNT (SA-MWNT) adsorbed on a CaF_2 plate modified with PEI/$(PSS/PEI)_2$ shows two characteristic absorption peaks at $1646\,cm^{-1}$ (amide I bond) and $1524\,cm^{-1}$ (amide II bond) resulting from the amide bond formed between the polyelectrolyte films and s-MWNTs.

FTIR has also been used to study how the functional group is attached to the CNTs. Khara *et al.* [140] functionalized single-walled carbon nanotubes through a microwave discharge with ammonia. As it is shown in Fig. 15.26, the N—H absorption bands at 3343 and $3198\,cm^{-1}$ are much lower in frequencies than the end-functionalized open metallic SWNT with CONH-4-$C_6H_4(CH_2)_{13}CH_3$ showing an N—H stretching frequency of $3450\,cm^{-1}$ [161]. The lowered N—H frequency is indicative of an N—H directly attached to the side walls.

CNT-doped conducting polymers possess improved mechanical, chemical, and optical properties. They also provide a simple strategy for making aligned CNTs. The disappearance of the characteristic peaks of carbon nanotubes in the FTIR spectrum of polymer/CNT composite films is normally an indication of perfect enwrapping of CNTs with the deposited conducting polymer [162, 163]. Zhang *et al.* [40] have studied the

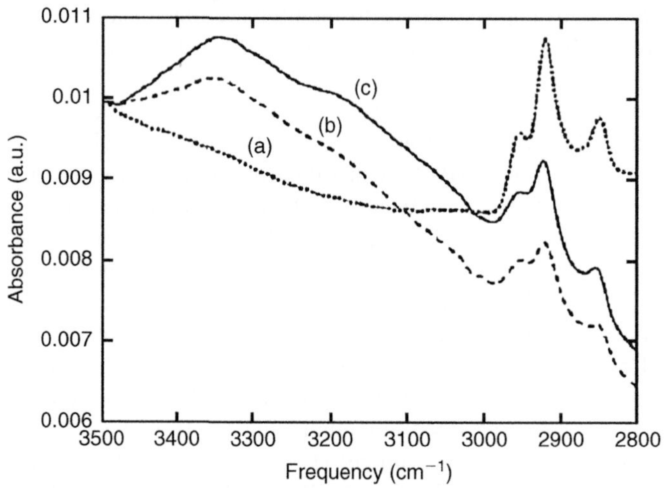

FIGURE 15.26 FTIR spectra of SWNT after exposure to NH_3 microwave plasma: (a) SWNT on CaF_2 substrate after heating; (b) after 20 min of plasma exposure; (c) after 40 min of exposure. (Reprinted with permission from [140]. Copyright (2004) American Chemical Society.)

PPy/CNT composite film using both FTIR and Raman spectroscopy. The films were obtained by repeatedly cycling the potential in the range of −0.2 to +0.8 V in an aqueous solution of 0.1 mol L^{-1} SDS, 25 mmol L^{-1} pyrrole, and 0.17 g L^{-1} single-walled carbon nanotubes. The FTIR spectrum shows characteristic absorption bands of PPy and doped SDS but no corresponding band from CNT. For example, the strong peaks at 1549 and 1469 cm^{-1} are due to the antisymmetric and symmetric ring-stretching modes of the PPy, respectively. The peak at 1176 cm^{-1} indicates the doping state of polypyrrole, and the peaks at 1045 cm^{-1} and 1315 cm^{-1} are attributed to C—H deformation vibrations and C—N stretching vibrations of the PPy, respectively. Two very strong peaks around 2914 cm^{-1} and 2846 cm^{-1} were attributed to the stretching vibration mode of methylene, indicating that surfactant SDS had been doped into the chains of PPy.

FTIR is also an effective way to assess the orientation or alignment of CNTs since the absorption is much stronger when the electric field of the incident light wave is parallel to the axis of aligned CNTs. The anisotropic feature has been found in SWNT-PAN composite [164]. The infrared spectra of PAN fibers for the two polarization directions (parallel and perpendicular) are comparable, while the composite fibers show significant differences in the two polarization directions.

Zengin *et al.* [41] characterized a polyaniline (PANI)/MWNT composite. The FTIR spectra of the composite film show benzoid and quinoid ring vibrations at 1500 cm^{-1} and 1600 cm^{-1}, respectively, which indicate the presence of emeraldine salt (ES) of polyaniline. A weak broad band near 3400 cm^{-1} is assigned to the N—H stretching mode. The strong band at 1150 cm^{-1} is characteristic of PANI conductivity. The FTIR spectrum of PANI/MWNT composite in the ES form exhibits several clear differences from the spectrum of neat ES PANI: (1) the composite spectrum shows an inverse

1600/1500 cm^{-1} intensity ratio indicating that PANI in the composite is richer in quinoid units than the pure ES PANI possibly because the nanotube–PANI interactions promote and (or) stabilize the quinoid ring structure; (2) the N—H stretching region near 3400 cm^{-1} is broad and strong in the composite but weak in the pure ES PANI since the sp^2 carbon of CNTs competes with the chloride ion, and thus perturbs the H-bonding environment and increases the N—H stretching intensity.

15.5 CONCLUSIONS

This chapter described the general aspects of carbon nanotubes (CNTs) and electrochemical sensors based on carbon nanotubes. There are several methods for forming CNT-modified electrodes that can be used for detecting analytes. Both advantages and examples of application of each method are described. This chapter also included examples of a fairly large number of analytes which have been determined by CNT-modified electrodes as effective and sensitive transducers. Since there have been a large number of publications in this exciting field of analytical chemistry in the last ten years, some of the important papers, particularly those published recently, might not be included.

Among those several different types of transducers based on CNTs, the CNT-composite electrode, which was the first CNT electrode tested in 1996, is still widely used with different composite materials such as conducting polymers, nanoparticles, sol-gel, etc. The usefulness of these electrodes is based on their high sensitivity, quick response, good reproducibility, and particularly long-term stability. We expect to see continued research activities using CNT-composite electrodes.

In recent years, there are more applications based on the layer-by-layer fabrication techniques for CNT-modified electrodes. This technique clearly provides thinner and more isolated CNTs compared with other methods such as CNT-composite and CNT coated electrodes in which CNTs are in the form of big bundles. This method should help biomolecules such as enzymes and DNA to interact more effectively with CNTs than other methods, and sensors based on this technique are expected to be more sensitive. Important biosensors such as glucose sensors have been developed using this technique, and further development of other sensors based on the layer-by-layer technique is expected.

Vertically aligned CNT-modified electrodes are based on a more elaborated technique than other methods, and microscopic images are used to characterize the integrity of this type of electrode. The technique has been applied for the immobilization of enzymes and DNA, and the sensors based on this technique have shown a lower detection limit than those based on other methods. More research activities using this technique, particularly with low density CNT arrays, are expected in the near future because of its sensitivity and versatility.

Even though the number of reports on the electrochemistry based on an individual nanotube is very limited, mainly due to technical difficulty, the development of sensors using such electrodes may be imminent for the recognition of species in cells as well as other biomedical applications.

It is generally agreed that the tip portion of CNTs particularly those shortened with oxidation treatments, is more reactive than the side wall of CNTs. Yet, the reason for the presence of high catalytic activity of CNTs is still unclear, and more research activities in this area using advanced spectroscopic and microscopic techniques are expected.

15.6 REFERENCES

1. S. Iijima, Helical microtubules of graphitic carbon. *Nature* **354**, 56–58 (1991).
2. K. Gong, Y. Yan, M. Zhang, L. Su, S. Xiong, and L. Mao, Electrochemistry and electroanalytical applications of carbon nanotubes: a review. *Anal. Sci.* **21**, 1383–1393 (2005).
3. A. Merkoci, Carbon nanotubes in analytical sciences. *Microchim. Acta* **152**,157–174 (2006).
4. J. Wang, Carbon-nanotube based electrochemical biosensors: a review. *Electroanalysis* **17**, 7–14 (2005).
5. Q. Zhao, Z. Gan, and Q. Zhuang, Electrochemical sensors based on carbon nanotubes. *Electroanalysis* **14**, 1609–1613 (2002).
6. J. Li, A. Cassell, L. Delzeit, J. Han, and M. Meyyappan, Novel three-dimensional electrodes: electrochemical properties of carbon nanotube ensembles. *J. Phys. Chem. B* **106**, 9299–9305 (2002).
7. F. Valentini, A. Amine, S. Orlanducci, M.L. Terranova, and G. Palleschi, Carbon nanotube purification: preparation and characterization of carbon nanotube paste electrodes. *Anal. Chem.* **75**, 5413–5421 (2003).
8. P.M. Ajayan, Nanotubes from carbon. *Chem. Rev.* **99**, 1787–1799 (1999).
9. S.C. Tsang, P.J.F. Harris, and M.L.H. Green, Thinning and opening of carbon nanotubes by oxidation using carbon dioxide. *Nature* **362**, 520–522 (1993).
10. M. Musameh, J. Wang, A. Merkoci, and Y. Lin, Low-potential stable NADH detection at carbon-nanotube-modified glassy carbon electrode. *Electrochem. Commun.* **4**, 743–746 (2002).
11. S. Sotiropoulou and N.A. Chaniotakis, Carbon nanotube array-based biosensor. *Anal. Bioanal. Chem.* **375**, 103–105 (2003).
12. A.C. Dillon, K.M. Jones, T.A. Bekkedahl, C.H. Kiang, D.S. Bethune, and M.J. Heben, Storage of hydrogen in single-walled carbon nanotubes. *Nature* **386**, 377–379 (1997).
13. P.J.F. Harris, *Carbon Nanotubes and Related Structures*. Cambridge University Press, Cambridge (1999).
14. M.D. Rubianes and G.A. Rivas, Carbon nanotubes paste electrode. *Electrochem. Commun.* **5**, 689–694 (2003).
15. J. Wang and M. Musameh, Carbon nanotube/Teflon composite electrochemical sensors and biosensors. *Anal. Chem.* **75**, 2075–2079 (2003).
16. P.J. Britto, K.S.V. Santhanam, and P.M. Ajayan, Carbon nanotube electrode for oxidation of dopamine. *Bioelectrochem. Bioenerg.* **41**, 121–125 (1996).
17. P.G. Collins, A. Zettl, H. Bando, A. Thess, and R.E. Smalley, Nanotube nanodevice. *Science* **278**, 100–102 (1997).
18. H. Dai, J.H. Hafner, A.G. Rinzler, D.T. Colbert, and R.E. Smalley, Nanotubes as nanoprobes in scanning probe microscopy. *Nature* **384**, 147–150 (1996).
19. A.G. Rinzler, J.H. Hafner, P. Nikolaev, P. Nordlander, D.T. Colbert, R.E. Smalley, L. Lou, S.G. Kim, and D. Tomanek, Unraveling nanotubes: field emission from an atomic wire. *Science* **269**, 1550–1553 (1995).
20. M.M.J. Treacy, T.W. Ebbesen, and J.M. Gibson, Exceptionally high Young's modulus observed for individual carbon nanotubes. *Nature* **381**, 678–680 (1996).
21. B.I. Yakobson, C.J. Brabec, and J. Bernholc, Nanomechanics of carbon tubes: instabilities beyond linear response. *Phys. Rev. Lett.* **76**, 2511–2514 (1996).

22. D.L. Carroll, X. Blase, J.C. Charlier, P. Redlich, P.M. Ajayan, S. Roth, and M. Ruhle, Effects of nanodomain formation on the electronic structure of doped carbon nanotubes. *Phys. Rev. Lett.* **81**, 2332–2335 (1998).
23. E.W. Wong, P.E. Sheehan, and C.M. Lieber, Nanobeam mechanics: elasticity, strength, and toughness of nanorods and nanotubes. *Science* **277**, 1971–1975 (1997).
24. S.S. Wong, E. Joselevich, A.T. Woolley, C.L. Cheung, and C.M. Lieber, Covalently functionalized nanotubes as nanometre-sized probes in chemistry and biology. *Nature* **394**, 52–55 (1998).
25. Y. Liu, X. Song, T. Zhao, J. Zhu, H. Michael, and P. Fritz, Amorphous carbon nanotubes produced by a temperature controlled DC arc discharge. *Carbon* **42**, 1852–1855 (2004).
26. T. Guo, P. Nikolaev, A. Thess, D.T. Colbert, and R.E. Smalley, Catalytic growth of single-walled nanotubes by laser vaporization. *Chem. Phys. Lett.* **243**, 49–54 (1995).
27. P. Coquay, E. Flahaut, E. de Grave, A. Peigney, R.E. Vandenberghe, and C. Laurent, Fe/Co alloys for the catalytic chemical vapor deposition synthesis of single- and double-walled carbon nanotubes (CNTs). 2. The CNT-Fe/Co-$MgAl_2O_4$ system. *J. Phys. Chem. B* **109**, 17 825–17 830 (2005).
28. A. Rinzler, J. Liu, H. Dai, P. Nikolaev, C. Huffman, F. Rodriguez-Macias, P. Boul, A. Lu, D. Heymann, D.T. Colbert, R.S. Lee, J. Fischer, A. Rao, P.C. Eklund, and R.E. Smalley, Large-scale purification of single-wall carbon nanotubes: process, product, and characterization. *Appl. Phys. A* **67**, 29–37 (1998).
29. S. Bandow, A.M. Rao, K.A. Williams, A. Thess, R.E. Smalley, and P.C. Eklund, Purification of single-wall carbon nanotubes by microfiltration. *J. Phys. Chem. B* **101**, 8839–8842 (1997).
30. K.B. Shelimov, R.O. Esenaliev, A.G. Rinzler, C.B. Huffman, and R.E. Smalley, Purification of single-wall carbon nanotubes by ultrasonically assisted filtration. *Chem. Phys. Lett.* **282**, 429–434 (1998).
31. Q. Xie, F. Arias, and L. Echegoyen, Electrochemically-reversible, single-electron oxidation of C60 and C70. *J. Am. Chem. Soc.* **115**, 9818–9819 (1993).
32. P.J. Britto, K.S.V. Santhanam, A. Rubio, J.A. Alonso, and P.M. Ajayan, Improved charge transfer at carbon nanotube electrodes. *Adv. Mater.* **11**, 154–157 (1999).
33. J.K. Campbell, L. Sun, and R.M. Crooks, Electrochemistry using single carbon nanotubes. *J. Am. Chem. Soc.* **121**, 3779–3780 (1999).
34. G. Che, B.B. Lakshmi, E.R. Fisher, and C.R. Martin, Carbon nanotubule membranes for electrochemical energy storage and production. *Nature* **393**, 346–349 (1998).
35. H. Luo, Z. Shi, N. Li, Z. Gu, and Q. Zhuang, Investigation of the electrochemical and electrocatalytic behavior of single-wall carbon nanotube film on a glassy carbon electrode. *Anal. Chem.* **73**, 915–920 (2001).
36. J.J. Davis, K.S. Coleman, B.R. Azamian, C.B. Bagshaw, and M.L.H. Green, Chemical and biochemical sensing with modified single walled carbon nanotubes. *Chem. Eur. J.* **9**, 3732–3739 (2003).
37. M. Valcarcel, B.M. Simonet, S. Cardenas, and B. Suarez, Present and future applications of carbon nanotubes to analytical science. *Anal. Bioanal. Chem.* **382**, 1783–1790 (2005).
38. C.K.W. Adu, G.U. Sumanasekera, B.K. Pradhan, H.E. Romero, and P.C. Eklund, Carbon nanotubes: a thermoelectric nano-nose. *Chem. Phys. Lett.* **337**, 31–35 (2001).
39. J. Wang, M. Li, Z. Shi, N. Li, and Z. Gu, Electrocatalytic oxidation of 3,4-dihydroxyphenylacetic acid at a glassy carbon electrode modified with single-wall carbon nanotubes. *Electrochim. Acta* **47**, 651–657 (2001).
40. X. Zhang, J. Zhang, and Z. Liu, Conducting polymer/carbon nanotube composite films made by in situ electropolymerization using an ionic surfactant as the supporting electrolyte. *Carbon* **43**, 2186–2191 (2005).
41. H. Zengin, W. Zhou, J. Jin, R. Czerw, D.W.S. Jr, L. Echegoyen, D.L. Carroll, S.H. Foulger, and J. Ballato, Carbon nanotube doped polyaniline. *Adv. Mater.* **14**, 1480–1483 (2002).
42. A. Salimi, R.G. Compton, and R. Hallaj, Glucose biosensor prepared by glucose oxidase encapsulated sol-gel and carbon-nanotube-modified basal plane pyrolytic graphite electrode. *Anal. Biochem.* **333**, 49–56 (2004).
43. W. Li, S. Xie, L. Qian, B. Chang, B. Zou, W. Zhou, R. Zhao, and G. Wang, Large-scale synthesis of aligned carbon nanotubes. *Science* **274**, 1701–1703 (1996).

44. J. Li, C. Papadopoulos, and J. Xu, Highly-ordered carbon nanotube arrays for electronics applications. *Appl. Phys. Lett.* **75**, 367–369 (1999).
45. J. Li, H.T. Ng, A. Cassell, W. Fan, H. Chen, Q. Ye, J. Koehne, J. Han, M. Meyyappan, Carbon nanotube nanoelectrode array for ultrasensitive DNA detection. *Nano Lett.* **3**, 597–602 (2003).
46. Y. Lin, F. Lu, Y. Yu, and Z. Ren, Glucose biosensors based on carbon nanotube nanoelectrode ensembles. *Nano Lett.* **4**, 191–195 (2004).
47. J.J. Gooding, R. Wibowo, J. Liu, W. Yang, D. Losic, S. Orbons, F.J. Mearns, J.G. Shapter, and D.B. Hibbert, Protein electrochemistry using aligned carbon nanotube arrays. *J. Am. Chem. Soc.* **125**, 9006–9007 (2003).
48. X. Yu, D. Chattopadhyay, I. Galeska, F. Papadimitrakopoulos, and J.F. Rusling, Peroxidase activity of enzymes bound to the ends of single-wall carbon nanotube forest electrodes. *Electrochem. Commun.* **5**, 408–411 (2003).
49. P. Diao, Z. Liu, B. Wu, X. Nan, J. Zhang, and Z. Wei, Chemically assembled single-wall carbon nanotubes and their electrochemistry. *ChemPhysChem* **3**, 898–901 (2002).
50. Z. Liu, Z. Shen, T. Zhu, S. Hou, L. Ying, Z. Shi, and Z. Gu, Organizing single-walled carbon nanotubes on gold using a wet chemical self-assembling technique. *Langmuir* **16**, 3569–3573 (2000).
51. L. Dai, A. Patil, X. Gong, Z. Guo, L. Liu, and Y. Liu, Aligned nanotubes. *ChemPhysChem* **4**, 1150–1169 (2003).
52. Y. Lan, E. Wang, Y. Song, Z. Kang, M. Jiang, L. Gao, S. Lian, D. Wu, L. Xu, and Z. Li, Covalent assembly of shortened multiwall carbon nanotubes on polyelectrolyte films and relevant electrochemistry study. *J. Colloid Interf. Sci.* **284**, 216–221 (2005).
53. J.H. Rouse and P.T. Lillehei, Electrostatic assembly of polymer/single walled carbon nanotube multilayer films. *Nano Lett.* **3**, 59–62 (2003).
54. M. Zhang, Y. Yan, K. Gong, L. Mao, Z. Guo, and Y. Chen, Electrostatic layer-by-layer assembled carbon canotube multilayer film and its electrocatalytic activity for O_2 reduction. *Langmuir* **20**, 8781–8785 (2004).
55. P. He and M. Bayachou, Layer-by-layer fabrication and characterization of DNA-wrapped single-walled carbon nanotube particles. *Langmuir* **21**, 6086–6092 (2005).
56. B. Munge, G. Liu, G. Collins, and J. Wang, Multiple enzyme layers on carbon nanotubes for electrochemical detection down to 80 DNA copies. *Anal. Chem.* **77**, 4662–4666 (2005).
57. M. Musameh, N.S. Lawrence, and J. Wang, Electrochemical activation of carbon nanotubes. *Electrochem. Commun.* **7**, 14–18 (2005).
58. C.Y. Liu, A.J. Bard, F. Wudl, I. Weitz, and J.R. Heath, Electrochemical characterization of films of single-walled carbon nanotubes and their possible application in supercapacitors. *Electrochem. Solid State Lett.* **2**, 577–578 (1999).
59. R.R. Moore, C.E. Banks, and R.G. Compton, Basal plane pyrolytic graphite modified electrodes: comparison of carbon nanotubes and graphite powder as electrocatalysts. *Anal. Chem.* **76**, 2677–2682 (2004).
60. J. Qu, Y. Shen, X. Qu, and S. Dong, Preparation of hybrid thin film modified carbon nanotubes on glassy carbon electrode and its electrocatalysis for oxygen reduction. *Chem. Commun.* **1**, 34–35 (2004).
61. Y. Lin, S. Taylor, H. Li, K.A.S. Fernando, L. Qu, W. Wang, L. Gu, B. Zhou, and Y.P. Sun, Advances toward bioapplications of carbon nanotubes. *J. Mater. Chem.* **14**, 527–541 (2004).
62. J. Liu, A.G. Rinzler, H. Dai, J.H. Hafner, R.K. Bradley, P.J. Boul, A. Lu, T. Iverson, K. Shelimov, C.B. Huffman, F. Rodriguez-Macias, Y.S. Shon, T.R. Lee, D.T. Colbert, and R.E. Smalley, Fullerene pipes. *Science* **280**, 1253–1256 (1998).
63. R. Bandyopadhyaya, E. Nativ-Roth, O. Regev, and R. Yerushalmi-Rozen, Stabilization of individual carbon nanotubes in aqueous solutions. *Nano Lett.* **1**, 25–28 (2002).
64. J. Wang, M. Musameh, and Y. Lin, Solubilization of carbon nanotubes by Nafion toward the preparation of amperometric biosensors. *J. Am. Chem. Soc.* **125**, 2408–2409 (2003).
65. M. Zhang, A. Smith, and W. Gorski, Carbon nanotube-chitosan system for electrochemical sensing based on dehydrogenase enzymes. *Anal. Chem.* **76**, 5045–5050 (2004).

66. M. Zhang and W. Gorski, Electrochemical sensing platform based on the carbon nanotubes/redox mediators-biopolymer system. *J. Am. Chem. Soc.* **127**, 2058–2059 (2005).
67. M. Zhang and W. Gorski, Electrochemical sensing based on redox mediation at carbon nanotubes. *Anal. Chem.* **77**, 3960–3965 (2005).
68. O. Matarredona, H. Rhoads, Z. Li, J.H. Harwell, L. Balzano, and D.E. Resasco, Dispersion of single-walled carbon nanotubes in aqueous solutions of the anionic surfactant NaDDBS. *J. Phys. Chem. B* **107**, 13357–13367 (2003).
69. K. Shen, S. Curran, H. Xu, S. Rogelj, Y. Jiang, J. Dewald, and T. Pietrass, Single-walled carbon nanotube purification, pelletization, and surfactant-assisted dispersion: a combined term and resonant micro-Raman spectroscopy study. *J. Phys. Chem. B* **109**, 4455–4463 (2005).
70. M.F. Islam, E. Rojas, D.M. Bergey, A.T. Johnson, and A.G. Yodh, High weight fraction surfactant solubilization of single-wall carbon nanotubes in water. *Nano Lett.* **3**, 269–273 (2003).
71. R.S. Chen, W.H. Huang, H. Tong, Z.L. Wang, and J.K. Cheng, Carbon fiber nanoelectrodes modified by single-walled carbon nanotubes. *Anal. Chem.* **75**, 6341–6345 (2003).
72. C. Cai and J. Chen, Direct electron transfer of glucose oxidase promoted by carbon nanotubes. *Anal. Biochem.* **332**, 75–83 (2004).
73. S. Lu, K. Wu, X. Dang, and S. Hu, Electrochemical reduction and voltammetric determination of metronidazole at a nanomaterial thin film coated glassy carbon electrode. *Talanta* **63**, 653–657 (2004).
74. J. Wang, M. Li, Z. Shi, N. Li, and Z. Gu, Investigation of the electrocatalytic behavior of single-wall carbon nanotube films on an Au electrode. *Microchem. J.* **73**, 325–333 (2002).
75. Z. Wang, J. Liu, Q. Liang, Y. Wang, and G. Luo, Carbon nanotube-modified electrodes for the simultaneous determination of dopamine and ascorbic acid. *Analyst* **127**, 653–658 (2002).
76. K. Wu and S. Hu, Electrochemical study and selective detn. of dopamine at multi-wall carbon nanotube-Nafion film coated glassy carbon electrode. *Microchim. Acta* **144**, 131–137 (2004).
77. P. Zhang, F.H. Wu, G.C. Zhao, and X.W. Wei, Selective response of dopamine in the presence of ascorbic acid at multi-walled carbon nanotube modified gold electrode. *Bioelectrochem.* **67**, 109–114 (2005).
78. S.B. Hocevar, J. Wang, R.P. Deo, M. Musameh, and B. Ogorevc, Carbon nanotube modified microelectrode for enhanced voltammetric detection of dopamine in the presence of ascorbate. *Electroanalysis* **17**, 417–422 (2005).
79. M. Zhang, K. Gong, H. Zhang, and L. Mao, Layer-by-layer assembled carbon nanotubes for selective determination of dopamine in the presence of ascorbic acid. *Biosens. Bioelectron.* **20**, 1270–1276 (2005).
80. C. Hu, Y. Zhang, G. Bao, Y. Zhang, M. Liu, and Z.L. Wang, DNA functionalized single-walled carbon nanotubes for electrochemical detection. *J. Phys. Chem. B* **109**, 20072–20076 (2005).
81. C. Hu, X. Chen, and S. Hu, Water-soluble single-walled carbon nanotubes films: preparation, characterization and applications as electrochemical sensing films. *J. Electroanal. Chem.* **586**, 77–85 (2006).
82. Z. Wang, Y. Wang, and G. Luo, A selective voltammetric method for uric acid detection at β-cyclodextrin modified electrode incorporating carbon nanotubes. *Analyst* **127**, 1353–1358 (2002).
83. G. Wang, X. Liu, B. Yu, and G. Luo, Electrocatalytic response of norepinephrine at a β-cyclodextrin incorporated carbon nanotube modified electrode. *J. Electroanal. Chem.* **567**, 227–231 (2004).
84. J.S. Ye, Y. Wen, W.D. Zhang, L.M. Gan, G.Q. Xu, and F.S. Sheu, Selective voltammetric detection of uric acid in the presence of ascorbic acid at well-aligned carbon nanotube electrode. *Electroanalysis* **15**, 1693–1698 (2003).
85. K. Wu, J. Fei, and S. Hu, Simultaneous determination of dopamine and serotonin on a glassy carbon electrode coated with a film of carbon nanotubes. *Anal. Biochem.* **318**, 100–106 (2003).
86. G. Zhao, K. Liu, S. Lin, J. Liang, X. Guo, and Z. Zhang, Application of a carbon nanotube modified electrode in anodic stripping voltammetry for determination of trace amounts of 6-benzylaminopurine. *Microchim. Acta* **143**, 255–260 (2003).
87. Z. Wang, Y. Wang, and G. Luo, The electrocatalytic oxidation of thymine at β-cyclodextrin incorporated carbon nanotube-coated electrode. *Electroanalysis* **15**, 1129–1133 (2003).

88. F.H. Wu, G.C. Zhao, X.W. Wei, and Z.S. Yang, Electrocatalysis of tryptophan at multi-walled carbon nanotube modified electrode. *Microchim. Acta* **144**, 243–247 (2004).
89. B. Zeng and F. Huang, Electrochemical behavior and determination of fluphenazine at multi-walled carbon nanotubes/(3-mercaptopropyl)trimethoxysilane bilayer modified gold electrodes. *Talanta* **64**, 380–386 (2004).
90. S. Lu, Electrochemical determination of 8-azaguanine in human urine at a multi-carbon nanotubes modified electrode. *Microchem. J.* **77**, 37–42 (2004).
91. Y.H. Zhu, Z.L. Zhang, and D.W. Pang, Electrochemical oxidation of theophylline at multi-wall carbon nanotube modified glassy carbon electrodes. *J. Electroanal. Chem.* **581**, 303–309 (2005).
92. J.J. Davis, R.J. Coles, and H.A.O. Hill, Protein electrochemistry at carbon nanotube electrodes. *J. Electroanal. Chem.* **440**, 279–282 (1997).
93. J. Wang, M. Li, Z. Shi, N. Li, and Z. Gu, Direct electrochemistry of cytochrome *c* at a glassy carbon electrode modified with single-wall carbon nanotubes. *Anal. Chem.* **74**, 1993–1997 (2002).
94. G.C. Zhao, Z.Z. Yin, L. Zhang, and X.W. Wei, Direct electrochemistry of cytochrome *c* on a multi-walled carbon nanotube modified electrode and its electrocatalytic activity for the reduction of H_2O_2. *Electrochem. Commun.* **7**, 256–260 (2005).
95. G.C. Zhao, X.W. Wei, and Z.S. Yang, A nitric oxide biosensor based on myoglobin adsorbed on multi-walled carbon nanotubes. *Electroanalysis* **17**, 630–634 (2005).
96. G.C. Zhao, L. Zhang, and X.W. Wei, An unmediated H_2O_2 biosensor based on the enzyme-like activity of myoglobin on multi-walled carbon nanotubes. *Anal. Biochem.* **329**, 160–161 2004).
97. G.C. Zhao, L. Zhang, X.W. Wei, Z.S. Yang, Myoglobin on multi-walled carbon nanotubes modified electrode: direct electrochemistry and electrocatalysis. *Electrochem. Commun.* **5**, 825–829 (2003).
98. C. Cai and J. Chen, Direct electron transfer and bioelectrocatalysis of hemoglobin at a carbon nanotube electrode. *Anal. Biochem.* **325**, 285–292 (2004).
99. P. Yang, Q. Zhao, Z. Gu, and Q. Zhuang, The electrochemical behavior of hemoglobin on SWNTs/DDAB film modified glassy carbon electrode. *Electroanalysis* **16**, 97–100 (2004).
100. Y.D. Zhao, Y.H. Bi, W.D. Zhang, and Q.M. Luo, The interface behavior of hemoglobin at carbon nanotube and the detection for H_2O_2. *Talanta* **65**, 489–494 (2005).
101. M. Wang, Y. Shen, Y. Liu, T. Wang, F. Zhao, B. Liu, and S. Dong, Direct electrochemistry of microperoxidase 11 using carbon nanotube modified electrodes. *J. Electroanal. Chem.* **578**, 121–127 (2005).
102. Z. Xu, N. Gao, H. Chen, and S. Dong, Biopolymer and carbon nanotubes interface prepared by self-assembly for studying the electrochemistry of microperoxidase-11. *Langmuir* **21**, 10808–10813 (2005).
103. A. Salimi, A. Noorbakhsh, and M. Ghadermarz, Direct electrochemistry and electrocatalytic activity of catalase incorporated onto multiwall carbon nanotubes-modified glassy carbon electrode. *Anal. Biochem.* **344**, 16–24 (2005).
104. L. Wang, J. Wang, and F. Zhou, Direct electrochemistry of catalase at a gold electrode modified with single-wall carbon nanotubes. *Electroanalysis* **16**, 627–632 (2004).
105. Y.D. Zhao, W.D. Zhang, H. Chen, Q.M. Luo, and S.F.Y. Li, Direct electrochemistry of horseradish peroxidase at carbon nanotube powder microelectrode. *Sens. Actuators B* **87**, 168–172 (2002).
106. Y.D. Zhao, W.D. Zhang, H. Chen, and Q.M. Luo, Direct electron transfer of glucose oxidase molecules adsorbed onto carbon nanotube powder microelectrode. *Anal. Sci.* **18**, 939–941 (2002).
107. J. Liu, A. Chou, W. Rahmat, M.N. Paddon-Row, and J.J. Gooding, Achieving direct electrical connection to glucose oxidase using aligned single walled carbon nanotube arrays. *Electroanalysis* **17**, 38–46 (2005).
108. B.R. Azamian, J.J. Davis, K.S. Coleman, C.B. Bagshaw, and M.L.H. Green, Bioelectrochemical single-walled carbon nanotubes. *J. Am. Chem. Soc.* **124**, 12664–12665 (2002).
109. J.H.T. Luong, S. Hrapovic, D. Wang, F. Bensebaa, and B. Simard, Solubilization of multiwall carbon nanotubes by 3-aminopropyltriethoxysilane towards the fabrication of electrochemical biosensors with promoted electron transfer. *Electroanalysis* **16**, 132–139 (2004).
110. S. Hrapovic, Y. Liu, K.B. Male, and J.H.T. Luong, Electrochemical biosensing platforms using platinum nanoparticles and carbon nanotubes. *Anal. Chem.* **76**, 1083–1088 (2004).

111. H. Tang, J. Chen, S. Yao, L. Nie, G. Deng, and Y. Kuang, Amperometric glucose biosensor based on adsorption of glucose oxidase at platinum nanoparticle-modified carbon nanotube electrode. *Anal. Biochem.* **331**, 89–97 (2004).
112. P.P. Joshi, S.A. Merchant, Y. Wang, and D.W. Schmidtke, Amperometric biosensors based on redox polymer-carbon nanotube-enzyme composites. *Anal. Chem.* **77**, 3183–3188 (2005).
113. W.J. Guan, Y. Li, Y.Q. Chen, X.B. Zhang, and G.Q. Hu, Glucose biosensor based on multi-wall carbon nanotubes and screen printed carbon electrodes. *Biosens. Bioelectron.* **21**, 508–512 (2005).
114. J. Li, Y.B. Wang, J.D. Qiu, D.C. Sun, and X.H. Xia, Biocomposites of covalently linked glucose oxidase on carbon nanotubes for glucose biosensor. *Anal. Bioanal. Chem.* **383**, 918–922 (2005).
115. R.P. Deo, J. Wang, I. Block, A. Mulchandani, K.A. Joshi, M. Trojaowicz, F. Scholz, W. Chen, and Y. Lin, Determination of organophosphate pesticides at a carbon nanotube/organophosphorus hydrolase electrochemical biosensor. *Anal. Chim. Acta* **530**, 185–189 (2005).
116. Z. Xu, X. Chen, X. Qu, J. Jia, and S. Dong, Single-wall carbon nanotube-based voltammetric sensor and biosensor. *Biosens. Bioelectron.* **20**, 579–584 (2004).
117. J.H.T. Luong, S. Hrapovic, and D. Wang, Multiwall carbon nanotube (MWCNT) based electrochemical biosensors for mediatorless detection of putrescine. *Electroanalysis* **17**, 47–53 (2005).
118. M.D. Rubianes and G.A. Rivas, Enzymatic biosensors based on carbon nanotubes paste electrodes. *Electroanalysis* **17**, 73–78 (2005).
119. K. Wu, J. Fei, W. Bai, and S. Hu, Direct electrochemistry of DNA, guanine and adenine at a nanostructured film-modified electrode. *Anal. Bioanal. Chem.* **376**, 205–209 (2003).
120. J. Wang, M. Li, Z. Shi, N. Li, and Z. Gu, Electrochemistry of DNA at single-wall carbon nanotubes. *Electroanalysis* **16**, 140–144 (2004).
121. H. Cai, X. Cao, Y. Jiang, P. He, and Y. Fang, Carbon nanotube-enhanced electrochemical DNA biosensor for DNA hybridization detection. *Anal. Bioanal. Chem.* **375**, 287–293 (2003).
122. K. Kerman, Y. Morita, Y. Takamura, M. Ozsoz, and E. Tamiya, DNA-directed attachment of carbon nanotubes for the enhanced electrochemical label-free detection of DNA hybridization. *Electroanalysis* **16**, 1667–1672 (2004).
123. S.G. Wang, R. Wang, P.J. Sellin, and Q. Zhang, DNA biosensors based on self-assembled carbon nanotubes. *Biochem. Biophys. Res. Commun.* **325**, 1433–1437 (2004).
124. J. Wang, G. Liu, M.R. Jan, and Q. Zhu, Electrochemical detection of DNA hybridization based on carbon-nanotubes loaded with CdS tags. *Electrochem. Commun.* **5**, 1000–1004 (2003).
125. N. Zhu, P. He, and Y. Fang, Electrochemical DNA biosensors based on platinum nanoparticles combined carbon nanotubes. *Anal. Chim. Acta* **545**, 21–26 (2005).
126. G. Cheng, J. Zhao, Y. Tu, P. He, and Y. Fang, A sensitive DNA electrochemical biosensor based on magnetite with a glassy carbon electrode modified by multi-walled carbon nanotubes in polypyrrole. *Anal. Chim. Acta* **533**, 11–16 (2005).
127. M.S. Dresselhaus, G. Dresselhaus, A. Jorio, A.G.S. Filho, and R. Saito, Raman spectroscopy on isolated single wall carbon nanotubes. *Carbon* **40**, 2043–2061 (2002).
128. A. Jorio, M.A. Pimenta, A.G.S. Filho, R. Saito, G. Dresselhaus, and M.S. Dresselhaus, Characterizing carbon nanotube samples with. resonance Raman scattering. *New J. Phys.* **5**, 139.1–139.17 (2003).
129. A. Jorio, R. Saito, J.H. Hafner, C.M. Lieber, M. Hunter, T. McClure, G. Dresselhaus, and M.S. Dresselhaus, Structural (n, m) determination of isolated single-wall carbon nanotubes by resonant Raman scattering. *Phys. Rev. Lett.* **86**, 1118–1121 (2001).
130. U.D. Venkateswaran, A.M. Rao, E. Richter, M. Menon, A. Rinzler, R.E. Smalley, P.C. Eklund, Probing the single-wall carbon nanotube bundle: Raman scattering under high pressure. *Phys. Rev. B* **59**, 10928–10934 (1999).
131. M. Milnera, J. Kurti, M. Hulman, and H. Kuzmany, Periodic resonance excitation and intertube interaction from quasicontinuous distributed helicities in single-wall carbon nanotubes. *Phys. Rev. Lett.* **84**, 1324–1327 (2000).
132. A. Jorio, A.G.S. Filho, G. Dresselhaus, M.S. Dresselhaus, A.K. Swan, M.S. Unlu, B.B. Goldberg, M.A. Pimenta, J.H. Hafner, C.M. Lieber, and R. Saito, G-band resonant Raman study of 62 isolated single-wall carbon nanotubes. *Phys. Rev. B* **65**, 155412.1–155412.9 (2002).

133. A. Jorio, M.A. Pimenta, A.G.S. Filho, G.G. Samsonidze, A.K. Swan, M.S. Unlu, B.B. Goldberg, R. Saito, G. Dresselhaus, and M.S. Dresselhaus, Resonance Raman spectra of carbon nanotubes by cross-polarized light. *Phys. Rev. Lett.* **90**, 107403.1–107403.4 (2003).
134. Q. Zhao and H.D. Wagner, Raman spectroscopy of carbon-nanotube-based composites. *Phil. Trans. R. Soc. Lond. A* **362**, 2407–2424 (2004).
135. G.S. Duesberg, I. Loa, M. Burghard, K. Syassen, and S. Roth, Polarized Raman spectroscopy on isolated single-wall carbon nanotubes. *Phys. Rev. Lett.* **85**, 5436–5439 (2000).
136. H.H. Gommans, J.W. Alldredge, H. Tashiro, J. Park, J. Magnuson, and A.G. Rinzler, Fibers of aligned single-walled carbon. nanotubes: polarized Raman spectroscopy. *J. Appl. Phys.* **88**, 2509–2514 (2000).
137. A. Jorio, A.G.S. Filho, V.W. Brar, A.K. Swan, M.S. Unlu, B.B. Goldberg, A. Righi, J.H. Hafner, C.M. Lieber, R. Saito, G. Dresselhaus, and M.S. Dresselhaus, Polarized resonant Raman study of isolated single-wall carbon nanotubes: symmetry selection rules, dipolar and multipolar antenna effects. *Phys. Rev. B* **65**, 121402.1–121402.4 (2002).
138. S.B. Cronin, R. Barnett, M. Tinkham, S.G. Chou, O. Rabin, M.S. Dresselhaus, A.K. Swan, M.S. Unlu, and B.B. Goldberg, Electrochemical gating of individual single-wall carbon nanotubes observed by electron transport measurements and resonant Raman spectroscopy. *Appl. Phys. Lett.* **84**, 2052–2054 (2004).
139. C.A. Dyke and J.M. Tour, Unbundled and highly functionalized carbon nanotubes from aqueous reactions. *Nano Lett.* **3**, 1215–1218 (2003).
140. B.N. Khare, P. Wilhite, R.C. Quinn, B. Chen, R.H. Schingler, B. Tran, H. Imanaka, C.R. So, J.C.W. Bauschlicher, and M. Meyyappan, Functionalization of carbon nanotubes by ammonia glow-discharge: experiments and modeling. *J. Phys. Chem. B* **108**, 8166–8172 (2004).
141. M.S. Strano, C.A. Dyke, M.L. Usrey, P.W. Barone, M.J. Allen, H. Shan, C. Kittrell, R.H. Hauge, J.M. Tour, R.E. Smalley, Electronic structure control of single-walled carbon nanotube functionalization. *Science* **301**, 1519–1522 (2003).
142. P. Corio, S.D.M. Brown, A. Marucci, M.A. Pimenta, K. Kneipp, and G. Dresselhaus, M.S. Dresselhaus, Surface-enhanced resonant Raman spectroscopy of single-wall carbon nanotubes adsorbed on silver and gold surfaces. *Phys. Rev. B* **61**, 13202–13211 (2000).
143. H. Grebel, Z. Iqbal, and A. Lan, Detecting single-wall carbon nanotubes with surface-enhanced Raman scattering from metal-coated periodic structures. *Chem. Phys. Lett.* **348**, 203–208 (2001).
144. K. Kneipp, A. Jorio, H. Kneipp, S.D.M. Brown, K. Shafer, J. Motz, R. Saito, G. Dresselhaus, and M.S. Dresslhaus, Polarization effects in surface-enhanced resonant Raman scattering of single-wall carbon nanotubes on colloidal silver clusters. *Phys. Rev. B* **63**, 081401.1–081401.4 (2001).
145. K. Kneipp, H. Kneipp, P. Corio, S.D.M. Brown, K. Shafer, J. Motz, L.T. Perelman, E.B. Hanlon, A. Marucci, G. Dresselhaus, and M.S. Dresslhaus, Surface-enhanced and normal stokes and anti-stokes Raman spectroscopy of single-walled carbon nanotubes. *Phys. Rev. Lett.* **84**, 3470–3473 (2000).
146. S. Lefrant, I. Baltog, M. Baibarac, J. Schreiber, and O. Chauver, Modification of surface-enhanced Raman scattering spectra of single-walled carbon nanotubes as a function of nanotube film thickness. *Phys. Rev. B* **65**, 235401.1– 235401.9 (2002).
147. L. Liu, T. Wang, J. Li, Z.-X. Guo, L. Dai, D. Zhang, and D. Zhu, Self-assembly of gold nanoparticles to carbon nanotubes using a thiol-terminated pyrene as interlinker. *Chem. Phys. Lett.* **367**, 747–752 (2003).
148. P. Poulin, B. Vigolo, and P. Launois, Films and fibers of oriented single wall nanotubes. *Carbon* **40**, 1741–1749 (2002).
149. M.D. Frogley, Q. Zhao, and H.D. Wagner, Polarized resonance Raman spectroscopy of single-wall carbon nanotubes within a polymer under strain. *Phys. Rev. B* **65**, 113413.1–113413.4 (2002).
150. H.J. Barraza, F. Pompeo, E.A. O'Rear, and D.E. Resasco, SWNT-filled thermoplastic and elastomeric composites prepared by miniemulsion polymerization. *Nano Lett.* **2**, 797–802 (2002).
151. B.A. Bhattacharyya, V.T. Sreekumar, T. Lui, S. Kumar, M.L. Ericson, H.R. Hauge, and E.R. Smalley, Crystallization and orientation studies in polypropylene/single wall carbon nanotube composite. *Polymer* **44**, 2373-–2377 (2003).
152. G. Chambers, C. Carroll, G.F. Farrell, A.B. Dalton, M. McNamara, M.I.H. Panhuis, and H.J. Byrne, Characterization of the interaction of gamma cyclodextrin with single-walled carbon nanotubes. *Nano Lett.* **3**, 843–846 (2003).

153. V.G. Hadjiev, M.N. Lliev, S. Arepalli, P. Nikolaev, and B.S. Files, Raman scattering test of single-wall carbon nanotube composites. *App. Phys. Lett.* **78**, 3193–3195 (2001).
154. J. Chen, M.A. Hamon, H. Hu, Y. Chen, A.M. Rao, P.C. Eklund, and R.C. Haddon, Solution properties of single-walled carbon nanotubes. *Science* **282**, 95–98 (1998).
155. B. Kim and W.M. Sigmund, Functionalized multiwall carbon nanotube/gold nanoparticle composites. *Langmuir* **20**, 8239–8242 (2004).
156. J. Kastner, T. Pichler, H. Kuzmany, S. Curran, W. Blau, D.N. Weldon, M. Delamesiere, S. Draper, and H. Zandbergen, Resonance Raman and infrared spectroscopy of carbon nanotubes. *Chem. Phys. Lett.* **221**, 53–58 (1994).
157. D.S. Knight and W.B. White, Characterization of diamond films by Raman spectroscopy. *J. Mater. Res.* **4**, 385–393 (1989).
158. M. Liu, Y. Yang, T. Zhu, and Z. Liu, Chemical modification of single-walled carbon nanotubes with peroxytrifluoroacetic acid. *Carbon* **43**, 1470–1478 (2005).
159. H. Kong, P. Luo, C. Gao, and D. Yan, Polyelectrolyte-functionalized multiwalled carbon nanotubes: preparation, characterization and layer-by-layer self-assembly. *Polymer* **46**, 2472–2485 (2005).
160. L. Jiang and L. Gao, Modified carbon nanotubes: an effective way to selective attachment of gold nanoparticles. *Carbon* **41**, 2923–2929 (2003).
161. M.A. Hamon, J. Chen, H. Hu, Y. Chen, M.E. Itkis, A.M. Rao, P.C. Eklund, and R.C. Haddon, Dissolution of single-walled carbon nanotubes. *Adv. Mater.* **11**, 834–840 (1999).
162. X. Zhang, J. Zhang, R. Wang, and Z. Liu, Cationic surfactant directed polyaniline/CNT nanocables: synthesis, characterization, and enhanced electrical properties. *Carbon* **42**, 1455–1461 (2004).
163. X. Zhang, J. Zhang, R. Wang, T. Zhu, and Z. Liu, Surfactant-directed polypyrrole/CNT nanocables: synthesis, characterization, and enhanced electrical properties. *ChemPhysChem* **5**, 998–1002 (2004).
164. T.V. Sreekumar, T. Liu, B.G. Min, H. Guo, S. Kumar, R.H. Hauge, and R.E. Smalley, Polyacrylonitrile single-walled carbon nanotube composite fibers. *Adv. Mater.* **16**, 58–61 (2004).

CHAPTER 16

Biosensors based on immobilization of biomolecules in sol-gel matrices

Vivek Babu Kandimalla, Vijay Shyam Tripathi, and Huangxian Ju

16.1 INTRODUCTION

In recent years a lot of effort and research have been carried out towards the development of biosensors for environmental and biomedical monitoring. A biosensor is an analytical device composed of a biological sensing element (enzyme, antibody, DNA) in intimate contact with a physical transducer (optical, mass or electrochemical), which together relates the concentration of an analyte to a measurable electrical signal [1–3]. The stability of biomolecules and signal transfer to transducer surface are crucial factors in the stability and sensitivity of biosensors. In aqueous solutions, biomolecules such as enzymes lose their catalytic activity rather rapidly, because enzymes can suffer oxidation reactions or their tertiary structure can be destroyed at the air–water interface, hence making the use of enzymes and reagents both expensive and complex [4]. This problem can be resolved by immobilization of biomolecules. By attachment to an inert support material, bioactive molecules may be rendered, retaining catalytic activity and thereby extending their useful life [5, 6]. In view of the above, a variety of techniques have been developed to immobilize biomolecules, including adsorption, covalent attachment, and entrapment in various polymers.

In general the immobilization method should have the following characteristics: it should be simple and fast, inert, biocompatible, have high retention capacity, controllable porosity and the film should be stable at different environmental and experimental conditions such as pH and temperature for the development of biosensors. Sol-gel films have most of the above-mentioned properties; hence they have been widely employed in biosensor design in recent years. The sol-gel process can be carried out at low temperature, be chemically inert, have tunable porosity, optical transparency, mechanical stability, and negligible swelling behavior [7–11]. Several studies have been carried out to examine the properties of the porous sol-gel matrix, such as pore size distribution, surface area, pore geometry, morphology, and polarity [12–14]. Using the sol-gel technique different biorecognition agents such as enzymes [15, 16], antibodies [17], and whole cells [18, 19] have been successfully immobilized and employed in multifarious applications, e.g. biosensors [1, 3], solid phase extraction sorbents [20, 21], etc. In recent years several reviews on sol-gel technology have appeared in specific applied areas [3, 7, 8, 16, 22–24]. The present chapter highlights the advantages, recent developments, biosensing applications, and future perspectives of sol-gel entrapped biomolecules.

16.2 SOL-GEL

16.2.1 Sol-gel chemistry and matrix characteristics

The sol-gel process involves hydrolysis of alkoxide precursors under acidic or basic conditions, followed by condensation and polycondensation of the hydroxylated units, which lead to the formation of porous gel. Typically a low molecular weight metal alkoxide precursor molecule such as tetramethoxy silane (TMOS) or tetra ethoxysilane (TEOS) is hydrolyzed first in the presence of water, acid catalyst, and mutual solvent

[7, 22]. The hydrolysis of metal alkoxide (e.g. TEOS or TMOS) precursors results in the formation of silanol groups (Si-OH), through condensation these silanol moieties react further and form siloxanes (—Si-O-Si—), finally through polycondensation of silanol and siloxanes SiO_2 matrices are formed after aging and drying processes as shown in Eqs (1)–(3) (Fig. 16.1) and Fig. 16.2. The resulting sol-gel is an interconnected rigid network with pores of submicrometer dimensions and polymeric chains whose average length is greater than a micrometer. HCl and ammonia are the most generally used catalysts for the hydrolysis, however, other catalysts such as acetic acid, KOH, amines, KF, and HF are also used. The rate and extent of the hydrolysis are mostly influenced by the strength and concentration of the acid or base catalyst [25]. When the liquid in the pore is removed at or near ambient pressure by thermal evaporation, drying and shrinkage occurs, the resulting monolith is termed xerogel. If the liquid is primarily alcohol, the monolith is termed alcogel.

In recent years, silica sol-gel-based inorganic–organic hybrid materials have also been reported. The introduction of various functional groups into organic alkoxide has led to organically modified sol-gel glasses (ormosils). Some of the ormosil monomers and ormosil formations can be found in Fig. 16.3. Redox molecules can be coupled with the ormosil monomer functional group. The immobilized redox molecules can

$$(RO)_3SiOR + H_2O \longrightarrow (RO)_3SiOH + ROH \quad (1)$$

$$2(RO)_3SiOH \longrightarrow (RO)_3Si\text{–}O\text{–}Si\,(OR)_3 + H_2O \quad (2)$$

$$(RO)_3SiOH + ROSi(OR)_3 \longrightarrow (RO)_3Si\text{–}O\text{–}Si\,(OR)_3 + ROH \quad (3)$$

FIGURE 16.1 Reaction scheme for formation of sol-gel.

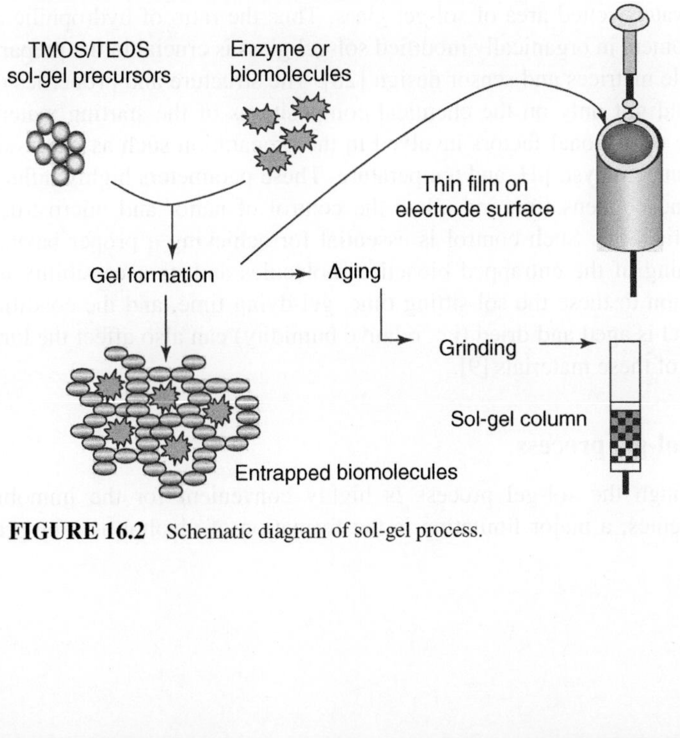

FIGURE 16.2 Schematic diagram of sol-gel process.

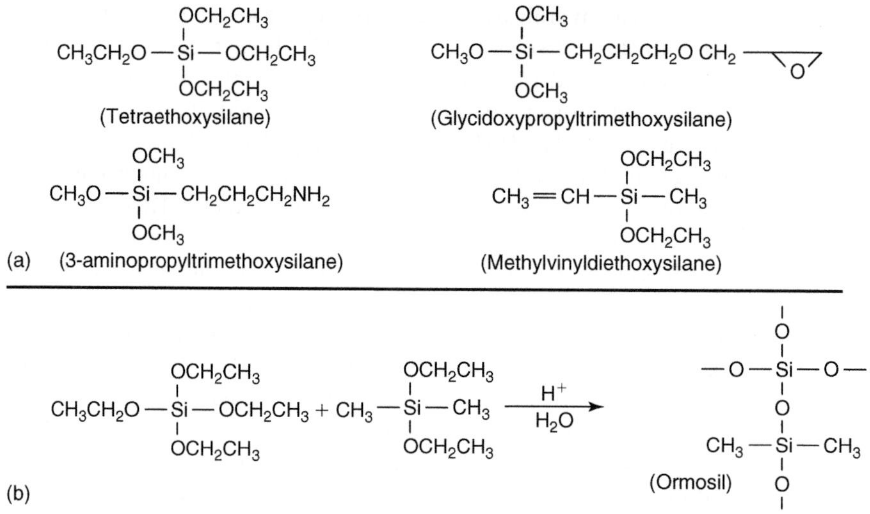

FIGURE 16.3 (a) TEOS and other commonly used ormosil monomers and (b) reaction scheme for ormosil formation.

act as electron shuttles to transfer the electrons between electrode surface and redox enzymes. The use of ormosils in bioencapsulation provides interesting properties to the host matrixes from hydrophobic to hydrophilic (hydrogels) [26]. When hydrophobic silica forming monomers are used, the resulting electrodes reject water, leaving only segregated islands of carbon at the outermost surface in contact with electrolyte [27].

On the other hand, when hydrophilic monomers are used, there is an increase in the water-wetted area of sol-gel glass. Thus the ratio of hydrophilic and hydrophobic monomers in organically modified sol-gel glass is crucial in the preparation of biocompatible matrices and sensor design [28]. The structure and properties of doped sol-gels depend not only on the chemical compositions of the starting materials, but also on many operational factors involved in the preparation such as water–silica molar ratio, solvent, catalyst, pH, and temperature. These parameters highly influence the hydrolysis and condensation and allow the control of nano- and microstructure of the final materials [8]. Such control is essential for achieving a proper balance between non-leaching of the entrapped bioactive molecules and its accessibility to the analyte. In addition to these the sol-sitting time, gel-dying time, and the conditions under which the gel is aged and dried (i.e. relative humidity) can also affect the long-term performance of these materials [9].

16.2.2 Progress in sol-gel process

Although the sol-gel process is highly convenient for the immobilization of biomolecules, a major limitation is the formation of alcohol as a by-product during the hydrolysis and condensation of the alkoxide precursors. Alcohol can have a detrimental effect on the activity of the biomolecules. Gill and Ballesteros [16] introduced poly(glyceryl silicate) (PGS) sol-gel precursors. The PGS can be rapidly hydrolyzed and gelled in aqueous, buffered milieu without the need for any catalyst, to form silica hydrogels, which produced transparent, mesoporous, and physically stable silica xerogels after aging, washing to remove glycerol, and drying. These sol-gel materials showed good porosity, less shrinkage and high percentage of bioencapsulation, and the entrapped biomolecules were retained almost complete activity (98%). Later Liu and Chen [29] reported an alcohol-free aqueous colloidal sol-gel process and encapsulated cytochrome *c*, catalase, myoglobin, and hemoglobin with good retained activities. In another alcohol-free sol-gel approach, sodium silicate was employed as a starting precursor, in which proteins showed preserved activity [30]. Recently *Moraxella* spp. cells engineered to express recombinant organophosphorus hydrolase (OPH) were encapsulated by the sodium silicate method and displayed higher activity retention compared to those by the traditional alkoxide process [31]. Perhaps in the sodium silicate route the precursors release a high Na^+ concentration, which must be eliminated through an acidic cation-exchange resin. Ferrer *et al.* [32] reported another alcohol-free and simple aqueous sol-gel method. In this approach the alcohol formed during the hydrolysis was removed through the rotavapor method. The horseradish peroxidase (HRP) immobilized in this alcohol-free route exhibited completely preserved activity and showed higher specific activity compared with the regular sol-gel method.

Some non-silica sol-gel materials have also been developed to immobilize bioactive molecules for the construction of biosensors and to synthesize new catalysts for the functional devices. Liu *et al.* [33] proved that alumina sol-gel was a suitable matrix to improve the immobilization of tyrosinase for detection of trace phenols. Titania is another kind of non-silica material easily obtained from the sol-gel process [34, 35]. Luckarift *et al.* [36] introduced a new method for enzyme immobilization in a biomimetic silica support. In this biosilicification process precipitation was catalyzed by the R5 peptide, the repeat unit of the silaffin, which was identified from the diatom *Cylindrotheca fusiformis*. During the enzyme immobilization in biosilicification the reaction mixture consisted of silicic acid (hydrolyzed tetramethyl orthosilicate) and R5 peptide and enzyme. In the process of precipitation the reaction enzyme was entrapped and nm-sized biosilica-immobilized spheres were formed. Carturan *et al.* [11] developed a biosil method for the encapsulation of plant and animal cells.

16.2.3 Advantages and disadvantages

Different sol-gel matrices such as inorganic, organically modified (ormosils), hybrid sol-gels and interpenetrating polymer networks have been used for encapsulation. Perhaps each type of sol-gel has its own advantages and disadvantages [1]. Inorganic sol-gels are good in optical transparency; chemical robustness but brittleness and low porosity in xerogels are major limitations. Similarly organically modified sol-gels have good tunable porosity and electrochemical activities, but are relatively fragile and have limited optical transparency [8, 9]. Hybrid sol-gels can be prepared with flexible rigidity, controlled porosity, and balance hydrophobicity and hydrophilicity, but poor optical transparency and structural collapse on drying are somewhat limiting factors. Interpenetrating polymer networks are combined matrices of sol-gel with water-soluble polymers such as carrageenan, alginate, agar, polyvinyl alcohol (PVA), polyethylene glycol (PEG), etc. These matrices are highly biocompatible for the fragile molecules such as organelles and living cells. Compared with alginate and carrageenan beads sol-gel layered beads are stable against chelating agents and physicochemical perturbations due to the supporting action of the outer sol-gel layer [11, 37].

16.2.4 Porosity and dynamics of proteins in sol-gel

The pores of the immobilization matrix should be large enough to allow unrestricted transport of molecules including buffer ions, substrates, products of the reaction, and analytes. Similarly the pores should also be small enough to prevent leakage of encapsulated macromolecules [7, 8]. The entrapped enzymes show an increase in K_m, which means high substrate concentration compared to the native enzyme [30]. It is mainly due to the diffusion resistance to the transport of substrate to the enzyme. If the diffusion rate of the substrate is sufficiently slow compared to enzymatic catalysis, the enzyme molecules close to the surface can use up most of the substrate molecules entering the matrix, effectively making the substrate concentration zero in the interior of the matrix. Similar transport or diffusional problems arise in the case of antigen–antibody reaction. To overcome these problems pore size and density should be controlled for better performance. Different agents including surfactants and non-surfactants have been used as pore-improving agents. Various alcohols and mixed solvent systems have also been used for the improvement of pore size [38]. To increase the pore size, PVA [39], PEG [40], and Triton-X100 [2, 41] have been employed. Proteins such as cytochrome *c*, RNAse [42], and antibodies [43, 44] can reversibly immerse into the sol-gel and selectively bind with the doped molecules.

The vicinity of entrapped proteins is completely different from the native environment. Hence the conformational, rotational and translational dynamics and the accessibility of the entrapped proteins should be closely monitored. The conformational and dynamic motions of the entrapped proteins have been examined widely using absorbance, fluorescence [12], resonance Raman [45], dipolar relaxation [13], and time resolved fluorescence anisotropy [12] measurements. After encapsulation, proteins such as bovine serum albumin (BSA), human serum albumin (HSA), and monellin retain their conformation [46–49], but small protein molecules such as myoglobin (Mb) [46] undergo substantial conformational changes during the entrapment. Sol-gel cage restricts the conformational change of macromolecules, which leads to partial unfolding, but small protein molecules can easily be obtained by more conformational changes/denaturation [10]. Edmiston *et al.* [46] studied the behaviors of myoglobin- and acrylodon-labeled bovine serum albumin (BSA-Ac) entrapped in TMOS-derived xerogels. Jordan *et al.* [12] observed the nanosecond and picosecond dynamics of BSA-Ac and acrylodan-labeled human serum albumin (HSA-Ac) when they were sequestered within sol-gel-derived xerogel glasses. These experiments indicated that the "global" protein rotational motion was not arrested within a xerogel. Brennan and coworkers [50] studied the real-time behavior of monellin sequestered within thin TEOS-derived xerogels as the protein was challenged by the quencher acrylamide and the chemical denaturant guanidine hydrochloride. The rotational mobility of glucose odixase (GOD) and its active site flavin adenine dinucleotide (FAD) have been investigated by Hartnett *et al.* [51]. The rotational mobility of GOD caged in sol-gel is reduced twice compared to that in solution, while the active site pocket is similar to that in solution. Hence while optimizing the influencing parameter for encapsulated biomolecules, dynamics of the proteins inside the "cage" should also be taken into consideration for their better performance.

16.2.5 Interactions and stability of biomolecules in sol-gel

Even though pore size is ideal for diffusion in reaching the enzyme active site, the interaction between sol-gel matrices and analytes/substrate also plays a key role in accessibility. This interaction might be electrostatic, hydrogen bonding and/or hydrophobic. Badjic and Kostic [52, 53] studied the interaction between polar silica and organic compounds. Silica monoliths immersed in solution containing styrene were evenly dispersed, as styrene could not form hydrogen bonding with silica. After soaking of the silica matrix in electrolyte solutions at a pH value at which pore walls were negatively charged, anions such as $[Fe(CN)_6]^{3-}$ were only partially taken up, whereas cations such as $[Ru(NH_3)_6]^{3+}$ were excessively taken up by the sol-gel matrix from the surrounding solution. In either case the internal and external concentrations of the ions were unequal even after the equilibrium was reached [54].

Several bioactive proteins retained their activity and conformation in sol-gel matrices. The sol-gel entrapped heme proteins such as cytochrome *c* and Mb showed good stability against pH and thermal perturbations compared to protein in solution [29, 55]. The sol-gel caged cytochrome *c* (cyt *c*) showed high thermal stability due to the exact fitting of the protein inside the cage, which was controlled by the protein size [56]. Sol-gel encapsulated acid phosphatase [57] and bovine carbonic anhydrase II (BCA II) [58] showed improved thermal stability. This enhanced stability was due to the protective nature of the cage and the rigidity of the SiO_2 matrix, which reduced the freedom of peptide-chain refolding molecular motions [7]. Trypsin and acid phosphatase entrapped in silicate sol-gel along with PEG had a half-life 100-fold higher than that of enzyme in solution at 70ºC [58]. More interestingly the creatine kinase encapsulated in TMOS sol-gel exhibited fourfold-improved activity upon short exposure to the elevated temperatures [59]. Through resonance Raman spectra, Das *et al.* [45] proved that Mb could be preserved in native form even at low pH by encapsulating them in sol-gel glasses. Different types of additives have been employed as stabilizers to the entrapped proteins, including ligand-based stabilizers (Cod III parvalbumin [47], oncomodulin [60]), methyltrimethoxysilane-based materials (to stabilize atrazine chlorohydrolase) [61], the incorporation of organosilanes and polymers into lipase-doped silica [62], poly-(ethylene glycol) (to stabilize acetylcholinesterase and butyrylcholinesterase) [63], and graft copolymers of polyvinylimidazole and polyvinylpyridine (to stabilize entrapped glucose oxidase and horseradish peroxidase) [64, 65]. Brennan *et al.* [66] reported the addition of sugar (sorbitol) and amino acids (*N*-methylglycine) increased the thermal stability and improved the α-chymostriosin and RNAse T1 activity, because the added osmolytes (sorbitol, *N*-methylglycine) altered the hydration effects, protein–silica interaction, and pore morphology.

16.2.6 Improvement of biocompatibility and conductivity of sol-gels

During the gel formation, alcohol liberation will take place. The retained water content is also low in aged sol-gel matrices. In such a condition most of the proteins are not stable or lose their activity. Hence there is a great need to stabilize the protein molecules by the addition of protein stabilizing agents or employment of biocompatible sol-gel monomers. As discussed above several new methods have been introduced for the reduction of the negative affect of alcohol. Some of the studies reported that the doping of BSA, cellulose, and chitosan into the sol-gel matrices can improve the hydrophilic nature of the matrices, so that the enzyme is stabile for quite longer time [67–69]. In view of the fact that most of the sol-gel matrices are not conductive materials, the doping of highly conductive nanoparticles such as carbon nanotubes, palladium, graphite, etc. into the sol-gel matrices/ormosil greatly enhances the conductive and catalytic properties [69–71].

16.3 APPLICATIONS OF SOL-GEL ENTRAPPED BIOACTIVE MOLECULES

16.3.1 Enzyme-based biosensors

Several enzymes have been immobilized in sol-gel matrices effectively and employed in diverse applications. Urease, catalase, and adenylic acid deaminase were first encapsulated in sol-gel matrices [72]. The encapsulated urease and catalase retained partial activity but adenylic acid deaminase completely lost its activity. After three decades considerable attention has been paid again towards the bioencapsulation using sol-gel glasses. Braun *et al.* [73] successfully encapsulated alkaline phosphatase in silica gel, which retained its activity up to 2 months (30% of initial) with improved thermal stability. Further Shtelzer *et al.* [58] sequestered trypsin within a binary sol-gel-derived composite using TEOS and PEG. Ellerby *et al.* [74] entrapped other proteins such as cytochrome *c* and Mb in TEOS sol-gel. Later several proteins such as Mb [8], hemoglobin (Hb) [56], cyt *c* [55, 75], bacteriorhodopsin (bR) [76], lactate oxidase [77], alkaline phosphatase (AP) [78], GOD [51], HRP [79], urease [80], superoxide dismutase [8], tyrosinase [81], acetylcholinesterase [82], etc. have been immobilized into different sol-gel matrices. Hitherto some reports have described the various aspects of sol-gel entrapped biomolecules such as conformation [50, 60], dynamics [12, 83], accessibility [46], reaction kinetics [50, 54], activity [7, 84], and stability [1, 80].

16.3.1.1 Biosensor applications of enzymes

The inherent features of the sol-gel matrices, such as optical transparency, high surface area, tunable porosity, chemical and photochemical inertness, and the ability to obtain any desired shape (monoliths, thin films, powders, fibers), enable the design of biosensors [27, 85, 86]. The slow diffusion of the electroactive species inside the sol-gel matrix causes long response times. To dissolve this limitation sol-gel matrices are doped with metal particles, such as graphite, carbon nanotubes, palladium, etc. in the construction of electrochemical biosensors [70, 71, 87]. Very recently carbon nanotubes were decorated with platinum (CNT-Pt) by the chemical reduction method [88]. This new kind of CNT-Pt was intercalated with graphite to prepare a modified electrode covered with cholineoxidase-doped TEOS film. The CNT-Pt-doped electrodes showed higher catalytic activity than the CNT electrode for the reduction of hydrogen peroxide. The linear range for cholesterol measurement was 4.0×10^{-6} to 1.0×10^{-4} M with a detection limit of 1.4×10^{-6} M. Mediators such as ferrocene and its derivatives can also be entrapped in the sol-gel/ormosil matrices in two ways, either by direct conjugation with sol-gel precursors [27] or conjugation with active biomolecule and then entrapment [69, 89]. In such a case the mediator is in close proximity to the active protein and transducer or electrode surface. Further coimmobilization of mediator and redox enzyme is a highly convenient way for the development of reagentless biosensors. Sol-gel-derived electrochemical biosensors mainly rely on two basic configurations: conductive ceramic composites and electrode surface coatings.

16.3.1.2 Carbon–ceramic composite electrodes (CCEs)

Since the pioneering work of Lev and coworkers [90] sol-gel derived composite carbon electrodes have been widely used to develop various kinds of amperometric biosensors. Carbon–ceramic composite electrodes (CCEs) are comprised of a dispersion of carbon powder in organically modified or non-modified silica matrixes. Usually the electrodes are prepared by mixing an appropriate amount of carbon black or graphite powder with the sol-gel precursors. A porous, brittle composite matrix can be formed after gelation and drying. The composite electrodes benefit from the mechanical properties of the silicate backbone, from the electron percolation conductivity through the interconnected carbon powder and from the ability to manipulate the physicochemical characteristics of the matrix easily by incorporation of suitable monomer precursors or sol-gel dopants. It is also possible to cast silica–carbon matrixes in virtually any desired geometrical configuration, including flat layers spread on insulating or conductive matrixes, monolithic disks or rods and even in the form of a miniature. The choice of carbon powder significantly affects the properties of the CCEs [91]. Wang and coworkers reported that the pore size distribution could be increased by increasing the H_2O: Si ratio used for the preparation of the CCEs [92]. For the long-term use of enzyme biosensors one of the hurdles is fouling and contamination of the surface during operation; however, an advantage is the use of polishable biosensors/renewable biosensors. The renewable amperometric biosensors are commonly comprised of either carbon paste or carbon-epoxy materials. Sampath and Lev [93, 94] reported a renewable GOD-entrapped glucose biosensor by the addition of hydrophilic PEG into the hydrophobic ormosil (present on the electrode surface). The same group [95] reported another sol-gel-derived ferrocenyl modified silicate–graphite composite electrode employed in glucose biosensor design. Li *et al.* [96] reported MTMOS-derived carbon composite-based renewable glucose biosensor using vinyl ferrocene as a mediator. To circumvent cracking and swelling Lev *et al.* [97] reported some composite ceramic–carbon materials along with surfactants, but these materials were less biocompatible and the surfactants were detrimental to the enzyme and needed high amounts of enzyme.

16.3.1.3 Electrode surface coatings

Mainly, three approaches have been used to immobilize the enzyme on transducer or electrode surface, single layer, bilayer, and sandwich configurations [69, 98]. In some studies enzymes are covalently linked with sol-gel thin films [99]. Sol-gel thin films are highly convenient for fast, large, and homogeneous electron transfer [17]. With an increase in gel thickness the signal decays and diffusion of analytes to biomolecule active site becomes difficult; eventually these factors lead to poor response. By employing thin films various biosensors such as optical and electrochemical biosensors have been reported.

16.3.1.4 Optical biosensors

Brun *et al.* [100] reported an optical biosensor based on the xerogel disk doped with GOD, peroxidase, and dye for the detection of glucose. Tatsu *et al.* [101] prepared tetraethyl orthosilicate-derived silica gel doped with GOD and used it as a glucose recognition element in a flow-injection analytical system. Gulcev [102] reported an optical biosensor by coimmobilizing sensitive fluorophore and enzyme into the silica sol-gel matrix. The changes in the enzyme reaction mixtures affect the fluorophore response. The coimmobilization of dextran conjugates of fluorescein or carboxy-seminaphtharhodafluor-1 (SNARF-1) and enzyme showed less leaching and quantifiable pH response of fluorescence in reaction mixtures of lipase and urease. Another pH sensitive fluorescent biosensor was reported for the detection of acetylcholine and praoxon [103], in which fluorescene isothiocyanate (FITC)-Dextran conjugate and acethylcholinesterase (AChE) were coimmobilized in TMOS sol-gel. This biosensor showed a linear range from 0.5 to 20 mM for achetylcholine and 30% inhibition of AChE activity at 152 mg ml^{-1} of paraoxon. Hydroxyethyl carboxymethyl cellulose and TEOS hybrid polymer could improve the stability of entrapped GOD (up to 3 years) [68]. The aging of sol-gel matrix at 4°C resulted in reduction of shrinkage and good porosity. When integrating this encapsulated enzyme with optical biosensor, it showed good linearity for glucose quantification between 50 and 200 μM in urine. A glass capillary-based optical biosensor for retinol was reported in which the sensing element was retinol-binding protein (RBP) [104]. The binding events of retinoic-acid-horseradish peroxidase (conjugate) and retinol to RBP were measured using a simple photomultiplier tube, chemiluminescent signal generated upon reaction of hydrogen peroxide, and luminol with the conjugate bound to RBP. Tsai and Doong [105] reported an array-based optical biosensor for the determination of urea, acetylcholine, and heavy metals. Urease and acetyleholinesterase (AChE) were coentrapped with FITC-dextran in sol-gel and fixed on a sensing probe. The linear ranges of detection were 2.5 to 50 μM for urea and 10 to 100 nM for Cd(II), Cu(II), and Hg(II).

16.3.1.5 Electrochemical biosensors

The most widely used enzymes in biosensor design are GOD and HRP due to their well-known structure and mechanism of action. Metal alkoxides are more attractive matrixes because they provide good conductivity and the possibility of manipulation of polarity, rigidity, pore size, distribution, and electronic conductivity. Owing to the advantages of metal alkoxides Glezer and Lev [106] prepared GOD-doped vanadium pentoxide sol-gel film-modified platinum electrodes. Liu *et al.* [29] reported a glucose biosensor based on immobilization of enzyme in alumina (aluminum iso-propoxide) sol-gel film on a platinized glassy carbon electrode. The low operating potential greatly minimized the interference from coexisting electroactive species. Ju and coworkers [35, 107] proposed a simple and mild vapor deposition method to prepare titania sol-gel for immobilization of HRP, GOD, and other biomolecules, which retained their catalytic activity and exhibited good analytical performance on electrode surface. The titania sol-gel matrix had uniform porous structure and very low mass transport barrier (Fig. 16.4).

As shown in Fig. 16.5a, no response of the enzyme electrode was observed in the absence of catechol and H_2O_2. The electrode displayed a low background current. When 1.0 mM catechol was added to PBS, the cyclic voltammogram showed a couple of oxidation and reduction peaks for catechol (curve (ii) in Fig. 16.5a). Upon addition of

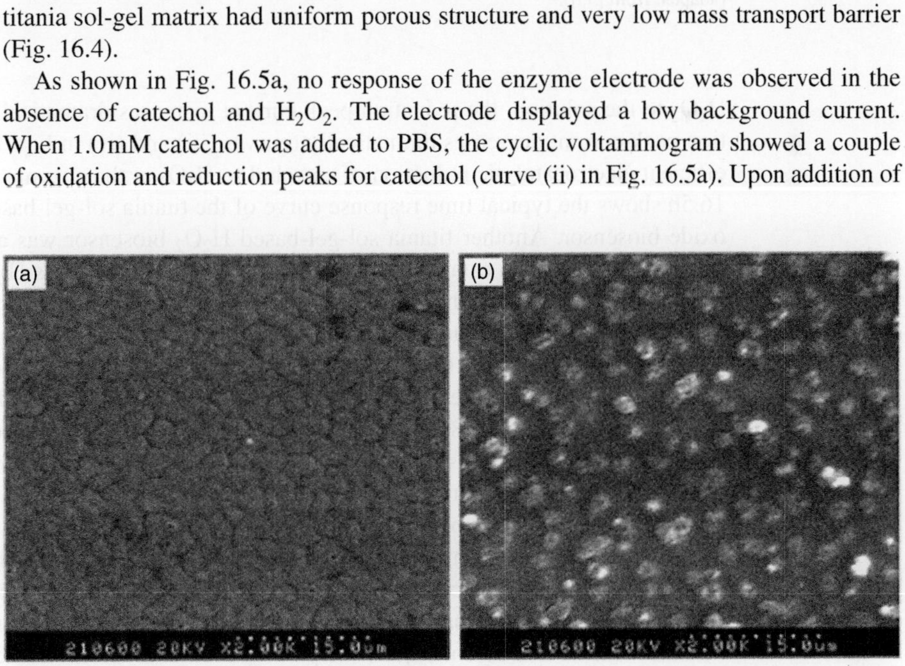

FIGURE 16.4 Scanning electron micrographs of glassy carbon electrodes coated with (a) titania sol-gel and (b) titania sol-gel doped with HRP (adapted from [35]).

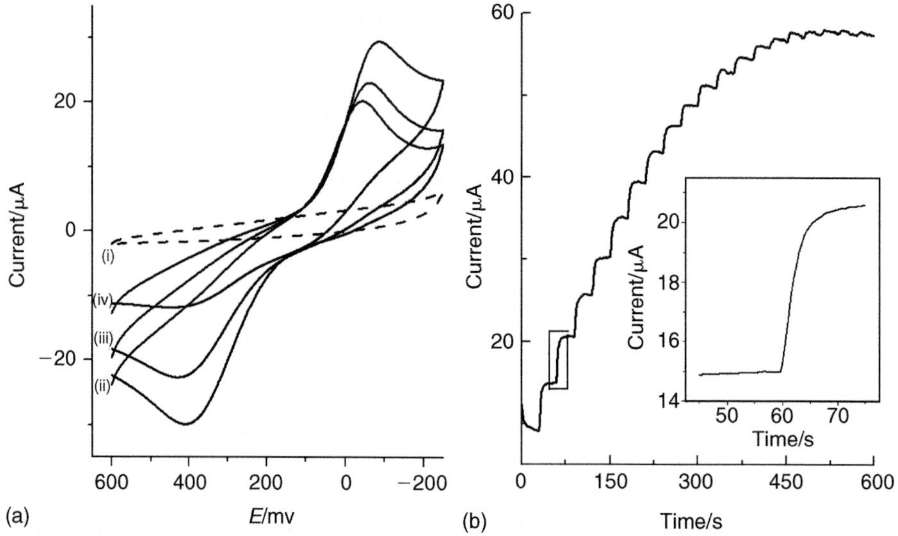

FIGURE 16.5 (a) Cyclic voltammograms of HRP enzyme electrode in (i) 0.02 M pH 7.0 PBS, (ii) (i) + 1.0 mM catehol, (iii) (ii) + 0.3 mM H_2O_2, and (iv) (ii) + 1.0 mM H_2O_2 at 100 mV s^{-1} and (b) typical current–time curve for the sensor upon successive addition of 0.08 mM H_2O_2 to 0.02 M pH 7.0 PBS containing 1.0 mM catehol at an applied potential of −150 mV. Inset: a magnification of the second addition of H_2O_2 (adapted from [35]).

H_2O_2 to the solution, the reduction peak current increases dramatically and the oxidation peak current decreases dramatically (curves (iii) and (iv) in Fig. 16.5a), due to the efficient electrocatalytic reduction of catechol to H_2O_2 at the enzyme electrode. Figure 16.5b shows the typical time response curve of the titania sol-gel-based hydrogen peroxide biosensor. Another titania sol-gel-based H_2O_2 biosensor was also presented by immobilizing Hb as a catalyst in titania sol-gel matrix [107]. This H_2O_2 sensor exhibited a fast response (less than 5 s) and sensitivity as high as 1.29 mA mM^{-1} cm^{-2}. The linear range for H_2O_2 determination was from 5.0×10^{-7} to 5.4×10^{-4} M with a detection limit of 1.2×10^{-7} M. Narang *et al.* [80] reported a glucose biosensor by immobilizing GOD between two sol-gel layers (sandwich). Using the advantages of ormosils Pandey *et al.* [28] reported a glucose biosensor based on the immobilization of GOD in sol-gel glasses derived from 3-aminopropyltrimethoxy silane and 2-(3,4-epoxycyclohexyl)-ethyltrimethoxy silane. An organophosphorus pesticide amperometric biosensor has been reported by immobilizing the AChE in TEOS [108]. Similarly in recent years several amperometric biosensors have been reported based on sol-gel-modified surfaces, which can be found in reviews [3, 16, 22, 69, 98].

Electrochemical biosensors can be divided into mediated and non-mediated biosensors. Biosensors that do not require the participation of redox molecules with reversible electrochemistry are referred to as non-mediated biosensors, whereas the participation of redox mediators in signal transduction generates a category of mediated biosensors.

The mediators such as ferrocene and its derivatives, osmium complexes, quinone salts and dyes, and ruthenium hexamine have been employed in biosensors. Some studies have added a mediator directly to the reaction mixture [109]. Sensors in which the mediator is coimmobilized with enzyme on electrode do not require the addition of the mediator to the test solution thereby eliminating the many practical limitations. The immobilized mediator catalyzes the redox reactions of biomolecules and shuttles the electrons more efficiently between enzyme and electrode to greatly enhance the sensing performance. Recently a sensitive mediated biosensor has been reported, in which glucose oxidase was immobilized in chitosan/methyltrimethoxysilane (MTOS) sol-gel and casted on Prussian blue electrodeposited GCE [110]. The formed sol-gel/chitosan film was covered with a Nafion coating to eliminate the effect of interfering compounds such as ascorbic acid, uric acid, etc. The linear response of the obtained glucose biosensor was 50 μM to 20 mM with a detection limit of 8 μM (S/N = 3). Tan *et al.* [111] reported a cholesterol amperometric biosensor based on hybrid sol-gel film. MTOS sol-gel was doped with MWNTs, chitosan, and cholineoxidase and overlaid on Prussian blue electrodeposited carbon glassy electrode. The MWNTs act as both nanometer conducting wire and catalyst. Recently Ju and coworkers [69] reported a highly conductive, biocompatible and mediator-entrapped ormosil composite matrix. The composite matrix was porous, stable, and crack free (Fig. 16.6). The electrodes modified with ormosil composite showed well-defined redox peaks and the presence of MWNTs decreased the background current and increased the peak currents (Fig. 16.7a). In the conjugated form of ferrocene monocarboxylic acid and bovine serum albumin (FMC-BSA), the mediator was well retained inside the matrix. In batch analysis the K_M^{app} and linear detection range were 6.6 mM and 0.06–18 mM, respectively (Fig. 16.7b). The FIA-based glucose biosensor showed a wide linear range from 0.05 to 20 mM. In another study HRP was doped in ormosil composite matrix for preparation of an H_2O_2 biosensor [112]. This H_2O_2 biosensor exhibited a linear range of 0.02 to 4.0 mM with a detection limit of 5.0 μM (S/N = 3) and a K_M^{app} value of 2.0 mM.

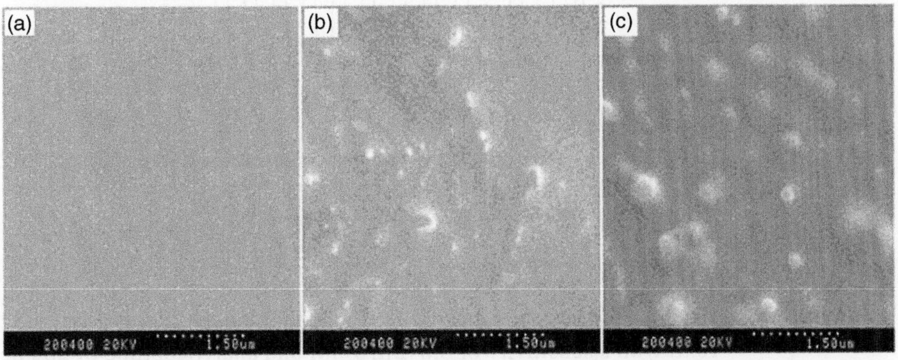

FIGURE 16.6 Scanning electron microphotos of (a) ormosil, (b) MWNTs/ormosil composite, and (c) FMC-BSA conjugate/ormosil films (adapted from [69]).

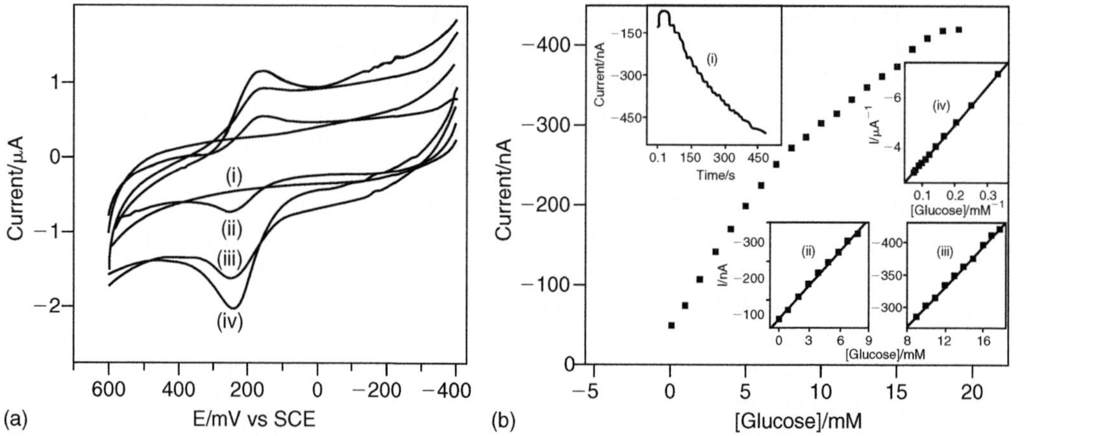

FIGURE 16.7 (a) Cyclic voltammograms of (i) ormosil, (ii) FMC-BSA conjugate/ormosil, (iii) MWNTs covered by FMC-BSA conjugate/ormosil, and (iv) FMC-BSA conjugate/MWNTs/ormosil composite film-modified electrodes in 0.1 M pH 7.4 PBS at 100 mV s^{-1} and (b) amperometric response of the biosensor to glucose in 0.1 M pH 7.0 PBS at +300 mV. Inset: (i) typical steady-state response curve, (ii) and (iii) linear plots, and (iv) Lineweaver-Burk plot (adapted from [69]).

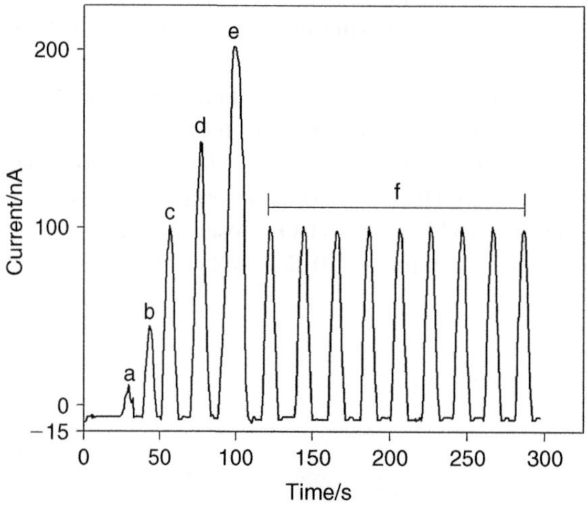

FIGURE 16.8 Flow-injection analysis with successive injection of 0.02, 0.5, 2.0, 3.0 and 4.5 mM H_2O_2 (from a to e) and eight successive additions of 2.0 mM H_2O_2 (f) to 0.1 M pH 6.8 PBS at flow rate of 1.0 ml min^{-1} at an applied potential of +220 mV (adapted from [112]).

The linear detection range in flow injection analysis was from 0.02 to 4.5 mM (Fig. 16.8) with a sensitivity of 0.042 μA/mM and analytical time of 20 s per sample.

On the other hand, the inclusion of electrocatalytically active metal nanocrystals (graphite-supported palladium, platinum, and ruthenium or gold nanocrystals) allowed the construction of mediator-free electrodes that are highly sensitive, selective, and stable [93, 94, 113]. Mediatorless amperometric biosensors have been prepared by coupling GOD with HRP or soybean peroxidases, which mediate the reduction of hydrogen peroxide [93, 94, 114]. The fundamental problem arising in the construction of an amperometric glucose biosensor is the selectivity of the substrate detection. The often-used detection approach is electrochemical oxidation of the liberated hydrogen peroxide through the enzymatic reaction and requires a relatively high working potential. At such a potential, endogenous or exogenous compounds present in biological samples (e.g. urate, ascorbate or paracetamol) can be electrochemically oxidized, leading to a high level of interference and false results in the quantification of glucose concentration. To overcome this limitation Wang *et al.* [115] and Sampath and Lev [93, 94] took the advantage of the catalytic properties of palladium-modified carbon particles to detect glucose at lower potentials (+0.3, +0.5 V) via a screen-printing process or by molding the porous organically modified silica in a glass capillary. One quick and stable biosensor performance was reported by the doping of grafting copolymer poly(vinyl alcohol) grafting 4-vinylpyridine into TEOS sol-gel [116]. The response of the glucose amperometric biosensor was less than 20 s with a linear range up to 9 mM and a sensitivity of 405 nA/mM. This polymer retained well the activity of entrapped enzyme and was able to firmly adhere to the electrode surface and enhanced hydrophilic nature [2, 116]. Similarly the GOD entrapped in cellulose-doped titanium oxide composite showed improved stability and lifetime of electrode, but the main drawback of this biosensor was the long time (>0.5 h) required to reach a steady state, which limited its application in analysis of a practical sample online [117].

Organic phase enzyme biosensors have been received remarkable importance in pharmaceutical and other sectors. Such organic phase biosensors provide some distinct advantages, such as an extended analyte range due to increased analyte solubility of certain reactants, monitoring of many hydrophobic substrates, prevention of undesirable side reactions and decreased microbial contamination. Enzymes usually lose their protein conformation in pure organic solvents. To avoid this some of the studies added trace amounts of water to the sides of the electrode [118]. Silica sol-gel-entrapped tyrosinase-based biosensors have also been reported for the phenol determination in non-aqueous phase [119]. This sensor has taken about 18 seconds to reach steady-state current. Recently Yu and Ju [120] reported a titania sol-gel-based tyrosinase biosensor for the phenol determination in organic phase. At −100 mV vs SCE this biosensor showed a good amperometric response to phenols in pure chloroform without any mediator and rehydration of the enzyme. For catechol determination the sensor exhibited a fast response of less than 5 s. The sensitivity of different phenols was as follows: catechol > phenol > p-cresol. The biosensor showed good reproducibility and stability (70 days at 4β), attributed to the titania sol-gel membrane which effectively retained the essential water layer around the enzyme molecule needed for maintaining its activity in organic phase.

The fouling and microbial contamination of the surface during operation are main problems in the long-term use of enzyme biosensors. Wang *et al.* [121] reported for the first time a screen-printed electrode modified with GOD/HRP that was stable up to 3 months. The renewable or screen-printed sensors were highly convenient for the filed applications and low cost. In order to circumvent limitations of microbial adhesion and growth on the surface, recently nitric oxide (NO) releasing sol-gel particles have been reported [122]. Nitric oxide is a potential antimicrobial agent [123]; if it releases slowly from the matrix adhesion the growth of microbes can be controlled. NO-releasing glucose biosensors prepared by doping diazeniumdiolate-modified sol-gel particles in a polyurethane membrane exhibited high sensitivity (mention details), reproducibility, and fast response up to 18 days [122].

16.3.2 Photoactive proteins-based biosensors

Photoactive proteins bR and phycoerythrin (PE) retain their optical activity when encapsulated within sol-gel glasses, with enhanced stability against photodegradation [124]. Bacteriorhodopsin is another naturally occurring transmembrane protein that converts light energy into metabolic energy. The absorption of light initiates a photocycle in the bR molecule, which accompanies the transportation of protons. The characteristics and effects of this photocycle make it a potentially useful material for development as an optically sensitive film that is self-developing and erasable. The encapsulation of light sensitive proteins in transparent matrices is of interest because of the potential application to photovoltaic devices, photoimaging, molecular computing, and chemical sensing. Wu *et al.* [84] and Weetall [125, 126] entrapped bR in wet sol-gel glasses. The D96N mutant bR retained its activity in a dried sol-gel glass. Encapsulation of photochromatic proteins in transparent films can be employed in optoelectronic sol-gel devices, which seems to be technically feasible. Phycobiliproteins are biomolecular assemblies located on the outer thylakoid membranes of marine algae. Although PE was stable under ambient light it denatured at intense light illumination. The sol-gel-entrapped PE retained its conformation, exhibited improved photodegradation capacity and was more stable than in solution [124]. Such stabilized photoactive proteins could be used in potential applications in biomolecular sensing, imaging, and information processing and storage. The bioluminescence from aequorin, a bioluminescent protein found in the jellyfish *Aequorea* sp., is specifically triggered by the presence of calcium ions. Based on this triggering effect of calcium ion, the concentration of calcium ion has been determined using aequorin immobilized in a porous sol-gel matrix [127].

16.3.3 Immunosensors

Compared with enzymes fewer reports are available on immobilization of antibody (Ab) in sol-gels and their applications in immunosensing. Immobilization of Abs on a solid support was first reported in 1967 [128] and the technology has widespread application in affinity chromatography and other areas. However, the major problem associated with covalent immobilization of antibody on solid surface is partial loss of biological activity due to the random orientation of the asymmetric macromolecules, sometimes steric hindrance caused by the neighboring antibody molecules [129]. The Abs encapsulated in sol-gel-derived glasses can interact with target molecules, with a same degree of specificity as in solution, and the signal can be detected using an appropriate sensing scheme [17]. Wang *et al.* [130] encapsulated first antifluorescein antibodies in TMOS sol-gel. The entrapped antibodies retained activity significantly and could bind with fluorescein molecules, which led to the decrease in the fluorescence. The entrapped antibodies stored at 4°C in water was stable up to 4 weeks. In another report [131] antifluorescin antibody was entrapped in a sandwich configuration, in aerosol-generated sol-gel-derived thin films ($0.62 \pm 0.05\,\mu m$). The fluorescent hapten 5-(and 6-)carboxy-4′5 dimethylfluorescein (Me_2F) was employed to determine the accessibility and viability of the entrapped antifluorescein antibody. The antibody was efficient in recognition and binding of the Me_2F up to 13 weeks when stored in 0.1 M pH 8 PBS at 4°C [83]. The addition of PEG into the sol-gel matrix improved the binding activity of the encapsulated antibody. The binding efficiency of sol-gel-entrapped antigentamicin antibody was improved from 42% to 95% in the presence of PEG and prevented the shrinking of the sol-gel glass matrix and denaturation of the immobilized antibody. This improved binding activity was probably due to the high diffusion rate of the gentamicin [132]. Hence while designing the matrices for immobilization of biomolecules, it is imperative to choose dimensions of the support matrices so as to minimize diffusional resistance and make the entire population of encapsulated biomolecules participate in the reaction.

Like other immunoassays sol-gel-based immunosensors can also entrap either antigen or antibody. Sol-gel-encapsulated anti-TNT antibody has been used as a detector for TNT [133]. The entrapped antibody is more stable than that immobilized through surface attachment. It is able to differentiate TNT and analog trinitrobenzene (TNB), and can be employed in different immunoassay formats such as competitive and displacement assays. The aged gels exhibit significantly faster response than xerogels due to the larger pore diameter. Tofiño *et al.* [134] developed a flow-through fluoroimmunosensor for the determination of isoproturon. The encapsulated antibodies could completely retain their activity and exhibited good performance in competitive immunoassay. The antigentamicin Mab entrapped in mesoporous TMOS so-gel monolith has been employed for the development of flow-injection fluorescence immunoassay for the quantitative analysis of the gentamicn [132]. The entrapped antigentamicin MAbs were able to recognize antibiotic up to $200\,ng\,ml^{-1}$ in the serum with a working range of $250–5000\,ng\,ml^{-1}$. The immunoreactor column was stable up to one month at room temperature and 3 months at 4°C. Anticortisol antibody has been encapsulated in optically transparent sol-gel silica matrices for cortisol detection of $1–100\,\mu g\,dl^{-1}$ with competitive immunoassays [17]. The antibody-doped thin film showed good accessibility for antigen and significant reduction in assay time. Wang *et al.* [64] reported a sol-gel-derived thick film amperometric immunosensor by encapsulation of antigen (RIgG) in TEOS sol-gels. This thick film immunosensor showed very fast response (20 s) due to the effective contact between electrode surface and reaction centers though the doped graphite powder, retained antigenic properties of the sol-gel-entrapped antigen, and effective binding efficiency between antigen and anti-IgG-HRP conjugate.

In competitive immunoassays antigen has also been immobilized in sol-gel matrices for the development of immunosensors. Gong *et al.* [135] immobilized *Schistosoma japonicum* antigens (SjAg) in TMOS along with BSA and graphite powder, finally squeezing the mixture into a PVC tube with a screw thread at one end. Through the screwing up and polishing, the electrode surface could be easily regenerated. Using competitive immunoassay between HRP-SjAb, SjAb, and SjAg entrapped on the electrode surface, the immunosensor was able to detect SjAb up to 4.5 ng ml^{-1} through the fluorescence detection method. They then designed an amperometric immunosensor by encapsulating the new castle disease antigen (NDAg) in TEOS sol-gel [136]. In this investigation antigen was entrapped in two different matrices, i.e. paraffin-ND Ag-BSA and sol-gel-ND Ag-BSA. Sol-gel-NDAg-BSA had good sensitivity in determination of NDAb present in the rabbit serum, and the antigen retained antigenic properties very effectively in sol-gel. Du *et al.* [137] reported an immunosensor for the detection of carcinoma antigen (CA) 19-9 by immobilizing the antigen in titania sol-gel matrix. The incubation of the immunosensor in a solution containing horseradish peroxidase (HRP)-labeled CA19-9 antibody led to the binding of HRP-labeled antibody with the immobilized antigen. The immobilized HRP catalyzed the oxidation of catechol by H_2O_2 and provided a competitive method for the measurement of serum CA19-9. The current decrease of the immunosensor was proportional to CA19-9 concentration in the range 3–20 U/ml with a detection limit of 2.68 U/ml. By taking advantage of the direct electrochemistry of HRP, Ju and coworkers [138, 139] developed two reagentless immunosensors for CA125, and human serum chorionic gonadotrophin (HCG). These immunosensors were prepared by immobilizing CA125 or HCG with titania sol-gel on glassy carbon electrodes by the vapor deposition method. The incubation of the immunosensors in PBS including HRP-labeled CA125 or HRP-labeled HCG antibody led to the formation of HRP-modified surface (Fig. 16.9). The immobilized HRP displayed its direct electrochemistry. Under optimal conditions, the current decrease was proportional to CA125 or HCG concentration ranging from 2 to 14 units ml^{-1} or 2.5 to 12.5 mIU ml^{-1} with a detection limit of 1.29 units ml^{-1} or 1.4 mIU ml^{-1} using a competition mechanism and differential pulse voltammetric determination method (Fig. 16.10). The CA125 immunosensor showed good accuracy and acceptable precision and

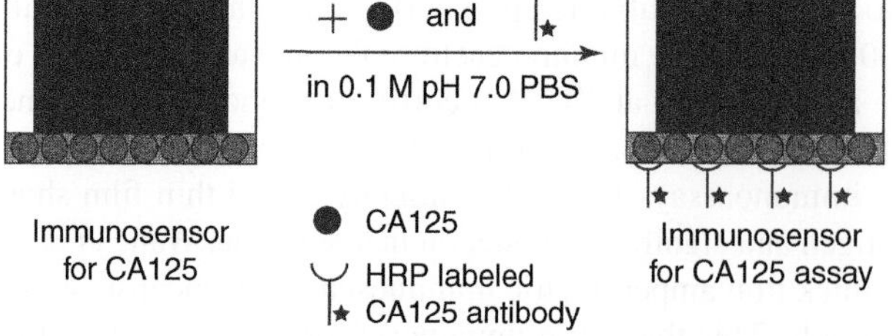

FIGURE 16.9 Principle of reagentless amperometric immunosensor based on immobilized antigen, competitive immunological reaction, and direct electrochemistry of HRP label (adapted from [138]).

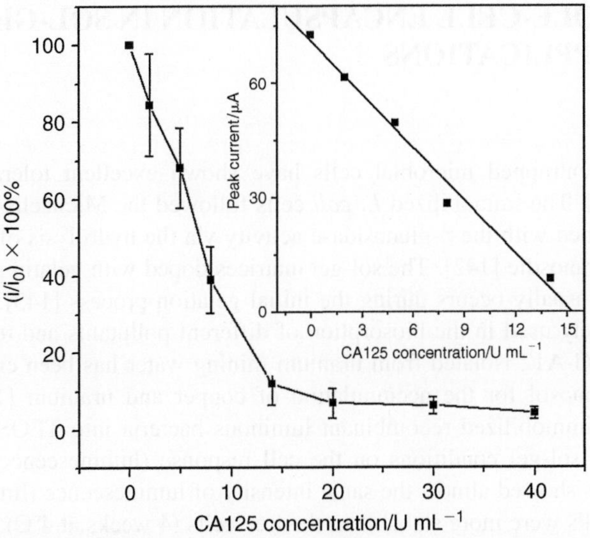

FIGURE 16.10 Calibration for CA125 determination. Inset: plot of the decrease in DPV peak current vs CA125 concentration (adapted from [138]).

fabrication reproducibility. These kinds of reagentless immunosensors are highly useful for medical applications, because the use of fewer reagents in clinical assays leads to less interference and more accuracy, especially in real samples such as serum, urine, and other biological fluids.

16.3.4 Immunoaffinity columns

The sol-gel-entrapped Abs have also been employed in affinity purification by packing into column (immunosorbent). The successful entrapment of dopants requires substantial screening of the sol-gel preparation procedure parameters, including examination of the effects of sol-gel format and composition on binding [40, 63]. Altstein *et al.* [140] immobilized TNT IgG antibodies in TMOS monoliths. The resulting wet gels were thoroughly crushed and packed into columns for the separation and quantification of TNT. Lira *et al.* [141] reported 2,4-dichlorophenoxyacetic acid (2,4-D) immunosorbent for the extraction of 2,4-D ester. The immunosorbent showed good precision, no leakage of the antibody, and a binding capacity of 130 ng of 2,4-D ester per mg of immobilized antibody, corresponding to 42% of the free antibody activity. Similarly Bronshtein *et al.* [20] reported an immunosorbent column for atrazine using monoclonal antibodies without purification. These promising results illustrated that sol-gel entrapped antibodies could be efficiently employed in affinity separation and online detection.

16.4 WHOLE-CELL ENCAPSULATION IN SOL-GELS AND THEIR APPLICATIONS

16.4.1 Microbial cells

The sol-gel-entrapped microbial cells have shown excellent tolerance to different alcohols [99]. The immobilized *E. coli* cells followed the Michaelis-Menten equation when quantified with the β-glucosidase activity via the hydrolysis of 4-nitrophenyl-β-D-galactopyranosdie [142]. The sol-gel matrices doped with gelatin prevented the cell lysis, which usually occurs during the initial gelation process [143]. Microorganisms are now widely used in the biosorption of different pollutants and toxicants. *Bacillus sphaericus* JG-A12 isolated from uranium mining water has been entrapped in aqueous silica nanosol for the accumulation of copper and uranium [144]. Premkumar *et al.* [145] immobilized recombinant luminous bacteria into TEOS sol-gel to study the effect of sol-gel conditions on the cell response (luminescence). The entrapped and free cells showed almost the same intensity of luminescence (little lower), but the entrapped cells were more stable than the free cells (4 weeks at 4°C). This kind of stable cell could be employed in biosensors in the near future.

16.4.2 Plant and animal cells

The cell-based biosensors are highly useful for the functional characterization and detection of various compounds such as drugs and toxins, and in drug testing. The sol-gel immobilized whole cells can also be employed in biotechnological processes for the production of various primary and secondary metabolites. However, the entrapment of plant and animal whole cell in sol-gel matrices is still at an early stage. A new sol-gel method for biosil immobilization has been introduced, in which sol-gel siliceous layers were deposited through airflow on to the cell surface. This method is employed for entrapment of plant and animal cells [19]. Pressi *et al.* [37] immobilized *Ajuga reptans* AYN-IRB1 plant cells by the biosil method for invertase production. Carturan *et al.* [146] immobilized *Catharanthus roseus* plant cells though the biosil method for production of the alkaloids vincristin and vinblastin. The immobilized cells were stable up to 100 days. Human hepatoblastoma (HepG2) and human T leukaemia Jurkat cells have been encapsulated in alginate silica microspores for the production of albumin (HSA) and human interleukin 2 (IL2), respectively [23]. The main limitation in the development of sol-gel-entrapped whole cell-based biosensors is signal transfer. The process of the signal to the transducer surface has to cross two main barriers, cell walls and sol-gel layer, leading to signal decay or sometimes signal hindrance.

16.5 CONCLUSIONS

The functionality of immobilized biomolecules is mainly governed by their nature, preparation method as well as the nature and structure of the immobilization matrix.

Although various sol-gel precursors and sol-gel preparation methods have been introduced, there is still a great need to explore more mild biocompatible matrices. The long-term use and reduction of film fouling though microbial growth or adhesion of new matrices, which can inhibit the growth, have to be explored. This is possible by developing new sol-gel precursor molecules, which are modified with microbial inhibition compounds or metal ions. The immobilized antibodies are showing promising results in immunosensor design, but their stability and regeneration are limited, hence more focus is needed in these areas to design robust immunosensors.

Relatively less work has been done on immobilization of plant and animal cells and spores of microbes in silica matrixes. The main drawback is less viability of the cells in sol-gel matrices. Thus more refined methods are required to utilize harness of the whole cells entrapped in sol-gel matrices and biosensing applications. At the same time studies such as interactions between sol-gel matrices and whole cells and metabolic changes during immobilization have to be closely monitored for the exploration of new matrices and methods.

16.6 ACKNOWLEDGMENTS

We gratefully acknowledge the financial support of the National Science Funds for Distinguished Young Scholars (20325518) and Creative Research Groups (20521503), the Key (20535010) and General (20275017) Programs from the National Natural Science Foundation of China, the Specialized Research Fund for Excellent Young Teachers from Ministry of Education of China and the Science Foundation of Jiangsu (BS2006006, BS2006074). V.S. Tripathi and V.B. Kandimalla are extremely thankful to Nanjing University for providing postdoctoral fellowships.

16.7 REFERENCES

1. I. Gill, Bio-doped nanocomposite polymers: sol-gel bioencapsulates. *Chem. Mater.* **13**, 3404–3421 (2001).
2. B. Wang, B. Li, Q. Deng, and S. Dong, Amperometric glucose biosensor based on sol-gel organic-inorganic hybrid material. *Anal. Chem.* **70**, 3170–3174 (1998).
3. V.B. Kandimalla, V.S. Tripathi, and H.X. Ju, Immobilization of biomolecules in sol-gels: biological and analytical applications. *Crit. Rev. Anal. Chem.* **36**, 73–106 (2006).
4. N.A. Chaniotakis, Enzyme stabilization strategies based on electrolytes and polyelectrolytes for biosensor applications. *Anal. Bioanal. Chem.* **378**, 89–95 (2002).
5. R. Fernandez–Lafuente, V. Rodriguez, and J.M. Guisan, The coimmobilization of D-amino acid oxidase and catalase enables the quantitative transformation of D-amino acids (D-phenylalanine) into alpha-keto acids (phenylpyruvic acid). *Enz. Microb. Technol.* **23**, 28–33 (1998).
6. M.N. Gupta, Thermostabilization of proteins. *Biotechnol. Appl. Biochem.* **14**, 1–11 (1991).
7. D. Avnir, S. Braun, L. Ovadia, and M. Ottolengthi, Enzymes and other proteins entrapped in sol-gel materials. *Chem. Mat.* **6**, 1605–1614 (1994).
8. B.C. Dave, B. Dunn, J.S. Valentine, and J.I. Zink, Sol-gel encapsulation methods for biosensors. *Anal. Chem.* **66**, 1120A–1126A (1994).

9. M.M. Collinson, Recent trends in analytical applications of organically modified silicate materials. *Trends Anal. Chem.* **21**, 31–39 (2002).
10. W. Jin and J.D. Brenan, Properties and applications of proteins encapsulated within sol-gel derived materials. *Anal. Chim. Acta* **461**, 1–36 (2002).
11. G. Carturan, R.D. Toso, S. Boninsegna, and R.D. Monte, Encapsulation of functional cells by sol-gel silica: actual progress and perspectives for cell therapy. *J. Mater. Chem.* **14**, 2087–2098 (2004).
12. J.D. Jordan, R.A. Dunbar, and F.V. Briht, Dynamics of scrylodan-labeled bovine and human serum albumin entrapped in a sol-gel-derived biogel. *Anal. Chem.* **67**, 2436–2443 (1995).
13. B.C. Dave, H. Soyez, J.M. Miller, B. Dunn, J.S. Valentine, and J.I. Zink, Synthesis of protein-doped sol-gel SiO_2 thin films: evidence for rotational mobility of encapsulated cytochrome c. *Chem. Mater.* **7**, 1431–1434 (1995).
14. M. Kanungo and M.M. Collinson, Diffusion of redox probes in hydrated sol-gel-derived glasses. Effect of gel structure. *Anal. Chem.* **75**, 6555–6559 (2003).
15. P. Audebert, C. Demaille, and C. Sanchez, Electrochemical probing of the activity of glucose oxidase embedded sol-gel matrixes. *Chem. Mater.* **5**, 911–913 (1993).
16. I. Gill and A. Ballesteros, Encapsulation of biologicals within silicate, siloxane, and hybrid sol-gel polymers: an efficient and generic approach. *J. Am. Chem. Soc.* **120**, 8587–8589 (1998).
17. J.C. Zhou, M.H. Chuang, E.H. Lan, and B. Dunn, Immunoassays for cortisol using antibody-doped sol-gel silica. *Mater. Chem.* **14**, 2311–2316 (2004).
18. S. Chia, J. Urano, F. Tamanoi, B. Dunn, and J.I. Zink, Patterned hexagonal arrays of living cells in sol-gel silica films. *J. Am. Chem. Soc.* **122**, 6488–6489 (2000).
19. S. Boninsegna, P. Bosetti, G. Carturan, G. Dellagiacoma, R.D. Monte, and M. Rossi, Encapsulation of individual pancreatic islets by sol-gel SiO_2: a novel procedure for perspective cellular grafts. *J. Biotech.* **100**, 277–286 (2003).
20. A. Bronshtein, N. Aharonson, D. Avnir, A. Turniansky, and M. Altstein, Sol-gel matrixes doped with atrazine antibodies: atrazine binding properties. *Chem. Mater.* **9**, 2632–2639 (1997).
21. A. Bronshtein, N. Aharonson, A. Turniansky, and M. Altstein, Sol-gel-based immunoaffinity chromatography: application to nitroaromatic compounds. *Chem. Mater.* **12**, 2050–2058 (2000).
22. I. Gill and A. Ballesteros, Bioencapsulation within synthetic polymers (Part 1): sol-gel encapsulated biologicals. *Trends Biotech.* **18**, 282–296 (2000).
23. T. Coradin, N. Nassif, and J. Livage, Silica-alginate composites for microencapsulation. *Appl. Microb. Biotech.* **61**, 429–434 (2003).
24. A.C. Pierre, The sol-gel encapsulation of enzymes. *Biocatal. Biotrans.* **22**, 145–170 (2004).
25. R. Aelion, A. Loebel, and F. Eirich, Hydrolysis of ethyl silicate. *J. Am. Chem. Soc.* **72**, 5705–5712 (1950).
26. M.S. Rao, I.S. Dubenko, S. Roy, N. Ali, and B.C. Dave, Matrix-assisted biomimetic assembly of ferritin core analogues in organosilica sol-gels. *J. Am. Chem. Soc.* **123**, 1511–1512 (2001).
27. O. Lev, M. Tsionsky, L. Ravinovich, V. Glezer, S. Sampath, L. Pankratov, and J. Gun, Organically modified sol-gel sensors. *Anal. Chem.* **67**, 22A–30A (1995).
28. P.C. Pandey, S. Upadhayay, H.C., Pathak, and C.M.D. Pandey, Studies on ferrocene immobilized sol-gel glasses and its application in the construction of a novel solid-state ion sensor. *Electroanalysis* **11**, 950–956 (1999).
29. M.D. Liu and I.W. Chen, Encapsulation of protein molecules in transparent porous silica matrices via an aqueous colloidal sol-gel process. *Acta Mater.* **47**, 4535–4544 (1999).
30. R.B. Bhatia, C.J. Brinker, A.K. Gupta, and A.P. Singh, Aqueous sol-gel process for protein encapsulation. *Chem. Mater.* **12**, 2434–2441 (2000).
31. D. Yu, J. Volponi, S. Chhabra, C.J. Brinker, A. Mulchandani, and A.K. Singh, Aqueous sol-gel encapsulation of genetically engineered Moraxella spp. cells for the detection of organophosphates. *Biosens. Bioelectron.* **20**, 1433–1437 (2005).
32. M.L. Ferrer, F. del Monte, and D. Levy, A novel and simple alcohol-free sol-gel route for encapsulation of labile proteins. *Chem. Mater.* **14**, 3619–3621 (2002).
33. Z.J. Liu, B.H. Liu, J.L. Kong, and J.Q. Deng, Probing trace phenols based on mediator-free alumina sol-gel-derived tyrosinase biosensor. *Anal. Chem.* **72**, 4707–4712 (2000).

34. T. Akiyama, A. Miyazaki, M. Sutoh, I. Ichinose, T. Kunitake, and S. Yamada, Fabrication of porphyrin-titaninm oxide-fullerene assemblies on an ITO electrode and their photocurrent responses. *Colloids Surf. A* **169**, 137–141 (2000).
35. J. Yu and H.X. Ju, Preparation of porous titania sol-gel matrix for immobilization of horseradish peroxidase by a vapor deposition method. *Anal. Chem.* **74**, 3579–3583 (2002).
36. H.R. Luckarift, J.C. Spain, R.J. Naik, and M.O. Stone, Enzyme immobilization in a biomimetic silica support. *Nat. Biotech.* **22**, 211–213 (2004).
37. G. Pressi, R.D. Toso, R.D. Monte, and G. Carturan, Production of enzymes by plant cells immobilized by sol-gel silica. *J. Sol-Gel Sci. Tech.* **26**, 1189–1193 (2003).
38. J.H. Harreld, T. Ebina, N. Tsubo, and G. Stucky, Manipulation of pore size distributions in silica and ormosil gels dried under ambient pressure conditions. *J. Non-Cryst. Solids* **298**, 241–251 (2002).
39. Q. Jie, K. Lin, J. Zhong, Y. Shi, Q. Li, J. Chang, and R. Wang, Preparation of macroporous sol-gel bioglass using PVA particles as pore former. *J. Sol-Gel Sci. Tech.* **30**, 49–61 (2004).
40. C.M.F. Soares, O.A. dos Santos, J.E. Olivo, H.F. de Castro, F.F. de Moraes, and G.M. Zanin, Influence of the alkyl-substituted silane precursor on sol-gel encapsulated lipase activity. *J. Mol. Catal. B.* **2969**, 69–79 (2004).
41. O. Lev, Diagnostic applications of organically doped sol-gel porous glass. *Analysis* **20**, 543–553 (1992).
42. U. Narang, M.H. Rahman, J.H.Wang, P.N. Prasad, and F.V. Bright, Removal of ribonucleases from solution using an inhibitor-based sol-gel-derived biogel. *Anal. Chem.* **67**, 1935–1939 (1995).
43. D.L. Venton, K.L. Cheesman, R.T. Chattertan, and T.L. Anderson, Entrapment of a highly specific antiprogesterone antiserum using polysiloxane copolymers. *Biochim. Biophy. Acta* **797**, 343–347 (1984).
44. C. Roux, J. Livage, K. Farhati, and L. Monjour, Antibody-antigen reactions in porous sol-gel matrices. *J. Sol-Gel Sci. Tech.* **8**, 663–666 (1997).
45. T.K. Das, I. Khan, D.L. Rousseau, and J.M. Friedman, Preservation of the native structure in myoglobin at low pH by sol-gel encapsulation. *J. Am. Chem. Soc.* **120**, 10268–10269 (1998).
46. P.L. Edmiston, C.L. Wambolt, M.K. Smith, and S.S. Saavedra, Spectroscopic characterization of albumin and myoglobin entrapped in bulk sol-gel glasses. *J. Colloid Inter. Sci.* **163**, 395–406 (1994).
47. K. Flora and J.D. Brennan, Fluorometric detection of Ca^{2+} based on an induced change in the conformation of sol-gel entrapped parvalbumin. *Anal. Chem.* **70**, 4505–4513 (1998).
48. L. Zheng and J.D. Brennan, Measurement of intrinsic fluorescence to probe the conformational flexibility and thermodynamic stability of a single tryptophan protein entrapped in a sol-gel derived glass matrix. *Analyst* **123**, 1735–1744 (1998).
49. H. Frenkel-Mullerad and D. Avnir, Sol-gel materials as efficient enzyme protectors: preserving the activity of phosphatases under extreme pH conditions. *J. Am. Chem. Soc.* **127**, 8077–8081 (2005).
50. L. Zheng, W.R. Reid, and J.D. Brennan, Measurement of fluorescence from tryptophan to probe the environment and reaction kinetics within protein-doped sol-gel-derived glass monoliths. *Anal. Chem.* **69**, 3940–3949 (1997).
51. A.M. Hartnett, C.M. Ingersoll, G.A. Baker, and F.V. Bright, Kinetics and thermodynamics of free flavins and the flavin-based redox active site within glucose oxidase dissolved in solution or sequestered within a sol-gel-derived glass. *Anal. Chem.* **71**, 1215–1224 (1999).
52. J.D. Badjic and N.M. Kostic, Unexpected interactions between sol-gel silica glass and guest molecules. Extraction of aromatic hydrocarbons into polar silica from hydrophobic solvents. *J. Phy. Chem. B* **4**, 11081–11087 (2000)
53. D. Badjic and N.M. Kostic, Behavior of organic compounds confined in monoliths of sol-gel silica glass. Effects of guest–host hydrogen bonding on uptake, release, and isomerization of the guest compounds. *J. Mater. Chem.* **11**, 408–418 (2001).
54. C. Shen and N.M. Kostic. Kinetics of photoinduced electron-transfer reactions within sol-gel silica glass doped with zinc cytochrome *c*. Study of electrostatic effects in confined liquids. *J. Am. Chem. Soc.* **119**, 1304–1312 (1997).
55. B. Dunn, J.M. Miller, B.C. Dave, J.S. Valentine, and J.I. Zink, Initiation toughness of silicon/glass anodic bonds. *Acta Mater.* **46**, 737–741 (1998).
56. E.H. Lan, B.C. Dave, J.M. Fukuto, B. Dunn, J.I. Zink, and J.S. Valentine, Synthesis of sol-gel encapsulated heme proteins with chemical sensing properties. *J. Mater. Chem.* **9**, 45–53 (1999).

57. J.D. Badjic and N.M. Kostic, Effects of encapsulation in sol-gel silica glass on esterase activity, conformational stability, and unfolding of bovine carbonic anhydrase II. *Chem. Mater.* **11**, 3671–3679 (1999).
58. S. Shtelzer, S. Rappoport, D. Avnir, M. Ottolenghi, and S. Braun, Properties of trypsin and of acid phosphatase immobilized in sol-gel glass matrixes. *Biotechnol. Appl. Biochem.* **15**, 227–235 (1992).
59. D.T. Nguyen, M. Smit, B. Dunn, and J.I. Zink, Stabilization of creatine kinase encapsulated in silicate sol-gel materials and unusual temperature effects on its activity. *Chem. Mater.* **14**, 4300–4308 (2002).
60. L. Zheng, K. Flora, and J.D. Brennan, Improving the performance of a sol-gel-entrapped metal-binding protein by maximizing protein thermal stability before entrapment. *Chem. Mater.* **10**, 3974–3983 (1998).
61. C. Kauffmann and R.T. Mandelbaum, Entrapment of atrazine chlorohydrolase in sol-gel glass matrix. *J. Biotech.* **62**, 169–176 (1998).
62. M.T. Reetz, A. Zonta, and J. Simpelkamp, Efficient heterogeneous biocatalysts by entrapment of lipases in hydrophobic sol-gel materials. *Angew. Chem. Int. Ed.* **34**, 301–303 (1995).
63. M. Altstein, G. Segev, N. Aharonson, O. Ben-Aziz, A. Turniansky, and D. Avnir, Sol-gel-entrapped cholinesterases: a microtiter plate method for monitoring anti-cholinesterase compounds. *J. Agric. Food Chem.* **46**, 3318–3324 (1998).
64. J. Wang, DNA biosensors based on peptide nucleic acid (PNA) recognition layers. A review. *Biosens. Bioelectron.* **13**, 757–762 (1998).
65. B.Q. Wang and S.J. Dong, Sol-gel-derived amperometric biosensor for hydrogen peroxide based on methylene green incorporated in Nafion film. *Talanta* **51**, 565–572 (2000).
66. J.D. Brennan, D. Benjamin, E. Dibattista, and M.D. Gulcev, Using sugar and amino acid additives to stabilize enzymes within sol-gel derived silica. *Chem. Mater.* **15**, 737–745 (2003).
67. Y. Miao and S.N. Tan, Amperometric hydrogen peroxide biosensor with silica sol-gel/chitosan film as immobilization matrix. *Anal. Chim. Acta* **437**, 87–93 (2001).
68. X.J. Wu and M.M.F. Choi, An optical glucose biosensor based on entrapped-glucose oxidase in silicate xerogel hybridised with hydroxyethyl carboxymethyl cellulose. *Anal. Chim. Acta* **514**, 219–226 (2004).
69. V.B. Kandimalla, V.S. Tripathi, and H.X. Ju, A conductive ormosil encapsulated with ferrocene conjugate and multiwall carbon nanotubes for biosensing application. *Biomaterials* **27**, 1167–1174 (2006).
70. P.C. Pandey, S. Upadhyay, N.K. Shukla, and S. Sharma, Studies on the electrochemical performance of glucose biosensor based on ferrocene encapsulated ORMOSIL and glucose oxidase modified graphite paste electrode. *Biosens. Bioelectron.* **18**, 1257–1268 (2003).
71. V.G. Gavalas, S.A. Law, J.C. Ball, R. Andrews, and L.G. Bachasa, Carbon nanotube aqueous sol-gel composites: enzyme-friendly platforms for the development of stable biosensors. *Anal. Biochem.* **329**, 247–252 (2004).
72. F.H. Dickey, Specific adsorption. *J. Phys. Chem.* **58**, 695–707 (1955).
73. S. Braun, S. Rappoport, R. Zusman, D. Avnir, and M. Ottolenghi, Biochemically active sol-gel glasses: the trapping of enzymes. *Mater. Lett.* **10**, 1–5 (1990).
74. L.M. Ellerby, C.R. Nishida, F. Nishida, S.A. Yamanaka, B. Dunn, J.S. Valentine, and J.I. Zink, Encapsulation of proteins in transparent porous silicate glasses prepared by the sol-gel method. *Science* **255**, 1113–1115 (1992).
75. G. Fiandaca, E. Vitrano, and A. Cupane, Ferricytochrome c encapsulated in silica nanoparticles: structural stability and functional properties. *Biopolymers* **74**, 55–59 (2004).
76. L.M. Shamansky, K.M. Luong, D. Han, and E.L. Chronister, Photoinduced kinetics of bacteriorhodopsin in a dried xerogel glass. *Biosens. Bioelectron.* **17**, 227–231 (2002).
77. B. Lillis, C. Grogan, H. Berney, and W.A. Lane, Investigation into immobilisation of lactate oxidase to improve stability. *Sens. Actuat. B* **68**, 109–114 (2000).
78. A.N. Diaz, F.G. Sanchez, M.C. Ramos, and M.C. Torijas, Horseradish peroxidase sol-gel immobilized for chemiluminescence measurements of alkaline-phosphatase. *Sens. Actuat. B* **82**, 176–179 (2002).
79. G. Wang, J.J. Xu, H.Y. Chen, and Z.H. Lu, Amperometric hydrogen peroxide biosensor with sol-gel/chitosan network-like film as immobilization matrix. *Biosens. Bioelectron.* **18**, 335–343 (2003).
80. U. Narang, P.N. Prasad, and F.V. Bright, A novel protocol to entrap active urease in a tetraethoxysilane-derived sol-gel thin-film architecture. *Chem. Mater.* **6**, 1596–1598 (1994).

81. J. Yu, S. Liu, and H.X. Ju, Mediator-free phenol sensor based on titania sol-gel encapsulation matrix for immobilization of tyrosinase by a vapor deposition method. *Biosens. Bioelectron.* **19**, 509–514 (2003).
82. A.K. Singh, A.W. Flounders, J.V. Volponi, C.S. Ashley, K. Wally, and J.S. Schoeniger, Development of sensors for direct detection of organophosphates. Part I: immobilization, characterization and stabilization of acetylcholinesterase and organophosphate hydrolase on silica supports. *Biosens. Bioelectron.* **14**, 703–713 (1999).
83. M.A. Doody, G.A. Baker, S. Pandey, and V. Bright, Affinity and mobility of polyclonal anti-dansyl antibodies sequestered within sol-gel-derived biogels. *Chem. Mater.* **12**, 1142–1147 (2000).
84. S. Wu, L.M. Ellerby, J.S. Cohan, B. Dunn, M.A. El-Sayed, J.S. Valentine, and J.I. Zink, Bacteriorhodopsin encapsulated in transparent sol-gel glass: a new biomaterial. *Chem. Mater.* **5**, 115–120 (1993).
85. M.M. Collinson and A.R. Howells, Sol-gel and electrochemistry: research at the intersection. *Anal. Chem.* **72**, 702A–709A (2000).
86. E. Mei, A.M. Bardo, M.M. Collinson, and D.A. Higgins, Single-molecule studies of sol-gel-derived silicate films. Microenvironments and film-drying conditions *J. Phys. Chem., B* **104**, 9973–9980 (2000).
87. S. Bharathi and O. Lev, Sol-gel-derived nanocrystalline gold-silicate composite biosensor. *Anal. Commun.* **35**, 29–31 (1998).
88. Q.C. Shi, T.Z. Peng, Y.N. Zhu, and C.F. Yang, An electrochemical biosensor with cholesterol oxidase/sol-gel film on a nanoplatinum/carbon nanotube electrode. *Electroanalysis* **17**, 857–861 (2005).
89. Y. Degani and A. Heller, Direct electrical communication between chemically modified enzymes and metal electrodes. I. Electron transfer from glucose oxidase to metal electrodes via electron relays, bound covalently to the enzyme. *J. Phys. Chem.* **91**, 1285–1289 (1987).
90. L. Rabinovich and O. Lev, Sol-gel derived composite ceramic carbon electrodes. *Electroanalysis*, **13**, 265–275 (2001).
91. J. Gun and O. Lev, Voltammetric studies of composite ceramic carbon working electrodes. *Anal. Chim. Acta*, **294**, 261–270 (1994).
92. J. Wang, D.S. Park, and P.V.A. Pamidi, Tailoring the macroporosity and performance of sol-gel derived carbon composite glucose sensors. *J. Electroanal. Chem.* **434**, 185–189 (1997).
93. S. Sampath and O. Lev, Inert metal-modified, composite ceramic-carbon, amperometric biosensors: renewable, controlled reactive layer. *Anal. Chem.* **68**, 2015–2021 (1996).
94. S. Sampath and O. Lev, Renewable, reagentless glucose sensor based on a redox-modified enzyme and carbon–silica composite. *Electroanalysis* **8**, 1112–1116 (1996).
95. J. Gun and O. Lev, Sol-gel derived, ferrocenyl-modified silicate-graphite composite electrode: wiring of glucose oxidase. *Anal. Chim. Acta* **336**, 95–106 (1996).
96. J. Li, L.S. Chia, N.K. Goh, and S.N. Tan, Renewable silica sol-gel derived carbon composite based glucose biosensor. *J. Electroanal. Chem.* **460**, 234–241 (1999).
97. M. Tsionsky, G. Gun, V. Giezer, and O. Lev, Sol-gel-derived ceramic-carbon composite electrodes: introduction and scope of applications. *Anal. Chem.* **66**, 1747–1753 (1994).
98. V.S. Tripathi, V.B. Kandimalla, and H.X. Ju, Preparation of ormosil and its applications in the immobilizing biomolecules. *Sens. Actuat. B* **114**, 1071–1082 (2006).
99. X. Yang, L. Huaa, H. Gonga, and S.N. Tan, Covalent immobilization of an enzyme (glucose oxidase) onto a carbon sol-gel silicate composite surface as a biosensing platform. *Anal. Chim. Acta* **478**, 67–75 (2003).
100. S. Braun, S. Shtelzer, S. Rappoport, D. Avnir, and M. Ottolenghi, Biocatalysis by sol-gel entrapped enzymes. *J. Non-Cryst. Solids* **147/148**, 739–743 (1992).
101. Y. Tatsu, K. Yamashita, M. Yamaguchi, S. Yamamura, H. Yamammoto, and S. Yoshikawa, Entrapment of glucose oxidase in silica gel by the sol-gel method and its application to glucose sensor. *Chem. Lett.* **36**, 1615–1618 (1992).
102. M.D. Gulcev, G.LG. Goring, M. Rakic, and J.D. Brennan, Reagentless pH-based biosensing using a fluorescently-labelled dextran co-entrapped with a hydrolytic enzyme in sol-gel derived nanocomposite films. *Anal. Chim. Acta* **457**, 47–59 (2002).
103. R.A. Doong and H.C. Tsai, Immobilization and characterization of sol-gel-encapsulated acetylcholinesterase fiber-optic biosensor. *Anal. Chim. Acta* **434**, 239–246 (2001).

104. K. Ramanathan, J. Svitel, A. Dzgoev, P.V. Sundaram, and B. Danielsson, Biomaterials for molecular electronics: development of an optical biosensor for retinol. *Appl. Biochem. Biotech.* **96**, 277–291 (2001).
105. H.C. Tsai and R.A. Doong, Simultaneous determination of pH, urea, acetylcholine and heavy metals using array-based enzymatic optical biosensor. *Biosens. Bioelectron.* **20**, 1796–1804 (2005).
106. V. Glezer and O. Lev, Sol-gel vanadium pentaoxide glucose biosensor. *J. Am. Chem. Soc.*, **115**, 2533 (1993).
107. J. Yu and H.X. Ju, Amperometric biosensor for hydrogen peroxide based on hemoglobin entrapped in titania sol-gel film. *Anal. Chim. Acta* **486**, 209–216 (2003).
108. K. Anitha, S.V. Mohan, S.J. Reddy, Development of acetylcholinesterase silica sol-gel immobilized biosensor – an application towards oxydemeton methyl detection. *Biosens. Bioelectron.* **20**, 848–856 (2004).
109. J. Yu, S. Liu, and H.X. Ju, Glucose sensor for flow injection analysis of serum glucose based on immobilization of glucose oxidase in titania sol-gel membrane. *Biosens. Bioelectron.* **19**, 401–409 (2003).
110. X.C. Tan, Y.X. Tian, P.X. Cai, and X.Y. Zou, Glucose biosensor based on glucose oxidase immobilized in sol-gel chitosan/silica hybrid composite film on Prussian blue modified glass carbon electrode. *Anal. Bioanal. Chem.* **381**, 500–507 (2005).
111. X.C. Tan, M.J. Li, P.X. Cai, L.J. Luo, and X.Y. Zou, An amperometric cholesterol biosensor based on multiwalled carbon nanotubes and organically modified sol-gel/chitosan hybrid composite film. *Anal. Biochem.* **337**, 111–120 (2005).
112. V.S. Tripathi, V.B. Kandimalla, and H.X. Ju, Amperometric biosensor for hydrogen peroxide based on ferrocene-bovine serum albumin and multiwall carbon nanotube modified ormosil composite. *Biosens. Bioelectron.* **21**, 1529–1535 (2006).
113. S. Sampath and O. Lev, Membrane-free, rhodium-modified, methyl silicate-graphite amperometric biosensor. *J. Electroanal. Chem.* **426**, 131–137 (1997).
114. L. Coche-Guerente, S. Cosnier, and P. Labbe, Sol-gel derived composite materials for the construction of oxidase/peroxidase mediatorless biosensors. *Chem. Mater.* **9**, 1348–1352 (1997).
115. J. Wang, P.V.A. Pamidi, and D.S. Park, Screen-printable sol-gel enzyme-containing carbon inks. *Anal. Chem.* **68**, 2705–2708 (1996).
116. X. Chen and S. Dong, Sol-gel-derived titanium oxide/copolymer composite based glucose biosensor. *Biosens. Bioelectron.* **18**, 999–1004 (2003).
117. Y. Kurokawa, T. Sano, H. Ohta, and Y. Nakagawa, Immobilization of enzyme onto cellulose-titanium oxide composite fiber. *Biotech. Bioenerg.* **42**, 394–397 (1993).
118. G.F. Hall, D.J. Best, and A.P.F. Turner, The determination of p-cresol in chloroform with an enzyme electrode used in the organic phase. *Anal. Chim. Acta* **213**, 113–119 (1988).
119. B. Wang and S.J. Dong, Organic-phase enzyme electrode for phenolic determination based on a functionalized sol-gel composite. *J. Electroanal. Chem.* **487**, 45–50 (2000).
120. J. Yu and H.X. Ju, Pure organic phase phenol biosensor based on tyrosinase entrapped in a vapor deposited titania sol-gel membrane. *Electroanalysis* **16**, 1305–1310 (2004).
121. B. Wang, B. Li, Z. Wang, G. Xu, Q. Wang, and S. Dong, Sol-gel thin-film immobilized soybean peroxidase biosensor for the amperometric determination of hydrogen peroxide in acid medium. *Anal. Chem.* **71**, 1935–1939 (1999).
122. J.H. Shin, S.M. Marxer, and M.H. Schoenfisch, Nitric oxide-releasing sol-gel particle/polyurethane glucose biosensors. *Anal. Chem.* **76**, 4543–4549 (2004).
123. B.J. Nablo, T.Y. Chen, and M.H. Schoenfisch, Sol-gel derived nitric-oxide releasing materials that reduce bacterial adhesion. *J. Am. Chem. Soc.* **123**, 9712–9713 (2001).
124. Z. Chen, L.A. Samuelson, J. Akkara, D.L. Kaplan, H. Gao, J. Kumar, K.A. Marx, and S.K. Tripathy, Sol-gel encapsulated light-transducing protein phycoerythrin: a new biomaterial. *Chem. Mater.* **7**, 1779–1783 (1995).
125. H. Weetall, D96N mutant bacteriorhodopsin immobilized in sol-gel glass characterization. *Appl. Biochem. Biotech.* **49**, 241–256 (1994).

126. H. Weetall, Retention of bacteriorhodopsin activity in dried sol-gel glass. *Biosens. Bioelectron.* **11**, 327–333 (1996).
127. D.J. Blyth, S.J. Poynter, and D.A. Russell, Calcium biosensing with a sol-gel immobilized photoprotein. *Analyst* **121**, 1975–1978 (1996).
128. K. Catt, H.D. Niall, and G.W. Tregear, Solid phase radioimmunoassay. *Nature* **213**, 825–827 (1967).
129. R. Walters, Affinity chromatography. *Anal. Chem.* **57**, 1099A–1114A (1985).
130. R. Wang, U. Narang, P.N. Prasad, and F.V. Bright, Affinity of antifluorescein antibodies encapsulated within a transparent sol-gel glass. *Anal. Chem.* **65**, 2671–2675 (1993).
131. J.D. Jordan, R.A. Dunbar, and F.V. Bright, Aerosol-generated sol-gel-derived thin films as biosensing platforms. *Anal. Chim. Acta* **332**, 83–91 (1996).
132. H.H. Yang, Q.Z. Zhu, H.Y. Qu, X.L. Chen, M.T. Ding, and J.G. Xua, Flow injection fluorescence immunoassay for gentamicin using sol-gel-derived mesoporous biomaterial. *Anal. Biochem.* **308**, 71–76 (2002).
133. E.H. Lan, B. Dunn, and J.I. Zink, Sol-gel encapsulated anti-trinitrotoluene antibodies in immunoassays for TNT. *Chem. Mater.* **12**, 1874–1878 (2000).
134. P.P. Tofiño, J.M.B. Morenol, and M.C.P. Conde, Sol-gel glass doped with isoproturon antibody as selective support for the development of a flow-through fluoroimmunosensor. *Anal. Chim. Acta* **429**, 337–345 (2001).
135. F.C. Gong, Z.J. Zhou, G.L. Shen, and R.Q. Yu, Schistosoma japonicum antibody assay by immunosensing with fluorescence detection using 3,3′,5,5′-tetramethylbenzidine as substrate. *Talanta* **58**, 611–618 (2002).
136. J.L. Gong, F.C. Gong, G.M. Zeng, G.L. Shen, and R.Q. Yu, An amperometric immunosensor for the Newcastle disease antibody assay. *Anal. Lett.* **36**, 287–302 (2003).
137. D. Du, F. Yan, S. Liu, and H.X. Ju, Immunological assay for carbohydrate antigen 19-9 using an electrochemical immunosensor and antigen immobilization in titania sol-gel matrix. *J. Immu. Methods* **283**, 67–75 (2003).
138. Z. Dai, F. Yan, Z. Chen, and H.X. Ju, Reagentless amperometric immunosensors based on direct electrochemistry of horseradish peroxidase for determination of carcinoma antigen-125. *Anal. Chem.* **75**, 5429–5434 (2003).
139. J. Chen, F. Yan, Z. Dai, and H.X. Ju, Reagentless amperometric immunosensor for human chorionic gonadotrophin based on direct electrochemistry of horseradish peroxidase. *Biosens. Bioelectron.* **21**, 330–336 (2005).
140. M. Altstein, A. Bronshtein, B. Glattstein, A. Zeichner, T. Tamiri, and J. Almog, Immunochemical approaches for purification and detection of TNT traces by antibodies entrapped in a sol-gel matrix. *Anal. Chem.* **73**, 2461–2467 (2001).
141. J.C.V. Lira, E.C. Frias, A.P. Alvarez, and L.E.V. Avila, Preparation and characterization of a sol-gel immunosorbent doped with 2,4-D antibodies. *Chem. Mater.* **15**, 154–161 (2003).
142. S. Fennouh, S. Guyon, C. Jourdant, J. Livage, and C. Roux, eds, Encapsulation of bacteria, in *Silica Gels*, Vol. 2 (Translated). Comptes Rendus Academy of Sciences, Paris, p. 625 (1999).
143. N. Nassif, A. Coiffier, T. Coradin, C. Roux, and J. Livage, Viability of bacteria in hybrid aqueous silica gels. *J. Sol-Gel Sci. Tech.* **26**, 1141–1144 (2003).
144. J. Raff, U. Soltmann, S. Matys, S. Selenska-Pobell, H. Bottcher, and W. Pompe, Biosorption of uranium and copper by biocers. *Chem. Mater.* **15**, 240–244 (2003).
145. J. R. Premkumar, O. Lev, R. Rosen, and S. Belkin, Encapsulation of luminous recombinant E-coli in sol-gel silicate films. *Adv. Mater.* **13**, 1773–1775 (2001).
146. G. Carturan, R.D. Monte, G. Pressi, S. Secondin, and P. Verza, Production of valuable drugs from plant cells immobilized by hybrid sol-gel SiO_2. *J. Sol-Gel. Sci. Technol.* **13**, 273–276 (1998).

CHAPTER 17

Biosensors based on direct electron transfer of protein

Shengshui Hu, Qing Lu, and Yanxia Xu

17.1 INTRODUCTION

17.1.1 Introduction of biosensors on direct electron transfer of protein

Electrochemical biosensors are analytical devices in which an electrochemical device serves as a transduction element. They are of particular interest because of practical advantages, such as operation simplicity, low expense of fabrication, and suitability for real-time detection. Since the first proposal of the concept of an enzyme-based biosensor by Clark, Jr [1], significant progress in this field has been achieved with the inherited sensitivity and selectivity of enzymes for analytical purposes.

The electron-transfer rate between large redox protein and electrode surface is usually prohibitively slow, which is the major barricade of the electrochemical system. The way to achieve efficient electrical communication between redox protein and electrode has been among the most challenging objects in the field of bioelectrochemistry. In summary, two ways have been proposed. One is based on the so-called "electrochemical" mediators, both natural enzyme substrates and products, and artificial redox mediators, mostly dye molecules and conducted polymers. The other approach is based on the direct electron transfer of protein. With its inherited simplicity in either theoretical calculations or practical applications, the latter has received far greater interest despite its limited applications at the present stage.

17.1.2 Advantage of biosensors on direct electron transfer of protein

For biosensors based on direct electron transfer of protein, the absence of mediator is the main advantage, providing them with superior selectivity, both because they should operate in a potential window closer to the redox potential of the enzyme and are therefore less prone to interfering reaction, but also because of the lack of yet another reagent in the reaction sequence, which simplifies the reaction system. Another attractive feature of the system, based on direct electron transfer of protein, is the possibility of modulating the desired properties of an analytical device using protein modification with genetic or chemical engineering techniques on one hand and novel interfacial technologies on the other.

17.2 DIRECT ELECTRON TRANSFER OF PROTEIN

17.2.1 Methods of protein immobilization

A biosensor is an analytical device comprising a biological recognition element directly interfaced to a signal transducer, which together relates the concentration of an analyte, or group of analytes, to a measurable response. Thus, for electrochemical biosensors based on direct electron transfer of protein, how to immobilize the biological recognition elements, such as redox proteins, onto the signal transducers, here the electrodes, are an important process.

17.2.1.1 Adsorption of protein

The physical adsorption of protein onto the surface of an electrode is a simple immobilization method. The adsorption is obtained by volatilizing the buffers containing proteins. The physical adsorption needs no chemical reagent, seldom activation and rinse, so that the bioactivities of the immobilized proteins can be retained well. However, the immobilized proteins are easy to break off from the electrode, which restrict broad applications of this method. Below are some examples of the physical adsorption of proteins immobilized on electrodes.

Ikariyama [2] described a unique method for the preparation of a glucose oxidase (GOD) electrode in their work. The method is based on two electrochemical processes, i.e. electrochemical adsorption of GOD molecules and electrochemical growth of porous electrode. GOD immobilized in the growing matrix of platinum black particles is employed for the microfabrication of the enzyme electrode. It demonstrated high performance with high sensitivity and fast responsiveness.

Suaud-Chagny and Gonon [3] presented a new procedure for protein immobilization adapted to carbon microelectrode characteristics. The principle of this method of immobilization is based on the association of the protein with an inert porous film immobilized around the active tip of the electrode. For this purpose the carbon was coated with an inert, electrochemically obtained protein sheath (bovine serum albumin, BSA) a few micrometers thick. Then the sheath around the fiber was impregnated with lactate dehydrogenase (LDH), which could be immobilized onto the electrode and resulted in an electrode sensitive to pyruvate.

17.2.1.2 Covalent bonding of protein

Protein is immobilized by combining with the surface of the electrode through a covalent bond, which is called covalent bonding of protein. The process requires low temperature (0ºC), low ion intensity, and physiological pH conditions. Although covalent bonding onto the surface of an electrode is more difficult than adsorption, it can provide a more stable immobilized protein.

Moody [4] described in his paper a miniature enzyme electrode consisting of GOD covalently attached to a silanized and anodized platinum wire surface via the bifunctional glutaraldehyde enzyme-immobilizing reagent. Its response characteristics were determined in a three-electrode amperometric mode by monitoring the anodic decomposition of hydrogen. The system exhibited good linearity (for glucose concentrations of 0.1–20 mM), where $\log (I/A) = 0.992 \log ([\text{glucose}]/M) - 3.94$, with a correlation coefficient of 0.999. Response times (<25 s) and wash times (<30 s) were short, and a lifetime of 9 h was obtained for continuous exposure to 2.5 and 10 mM glucose.

GOD and L-amino acid oxidase have been covalently bonded to chemically modified graphite electrodes via the cyanuric chloride linkage to yield glucose and L-phenylalanine sensors, respectively, by Ianniello [5]. These enzymes catalyze the oxidation of their respective substrates in the presence of O_2 to yield H_2O_2 as one of the products. The H_2O_2 produced during the enzymatic reaction is electrochemically consumed resulting in a measurable, steady-state current. These electrodes displayed relatively rapid response, expanded linear response range, and unique catalytic properties as compared to previously reported amperometric enzyme electrodes. The electrode remained active over a 20–30 day period, provided the proper storage conditions were maintained.

17.2.1.3 Sol-gel/polymer embedment of protein

So far, the prevalent method to immobilize protein is sol-gel/polymer embedment of protein. It usually embeds and immobilizes protein in a three-dimensional netlike structure of macromolecule polymer. This technology has some characteristics such as mild conditions, multifarious sol-gels, and controllable membrane aperture and figure. It can also retain the bioactivity of protein well when applied to a high concentration of protein. However, the aggregation process is difficult to control.

Fortier [6] found that AQ polymer from Eastman was not deleterious for the activity of a variety of enzymes such as L-amino acid oxidase, choline oxidase, galactose oxidase, and GOD. Following mixing of the enzyme with the AQ polymer, the mixture was cast and dried onto the surface of a platinum electrode. The film was then coated with a thin layer of Nafion to avoid dissolution of the AQ polymer film in the aqueous solution when the electrode was used as a biosensor. These easy-to-make amperometric biosensors, which were based on the amperometric detection of H_2O_2, showed high catalytic activity.

Tor [7] developed a new method for the preparation of thin, uniform, self-mounted enzyme membrane, directly coating the surface of glass pH electrodes. The enzyme was dissolved in a solution containing synthetic prepolymers. The electrode was dipped in the solution, dried, and drained carefully. The backbone polymer was then cross-linked under controlled conditions to generate a thin enzyme membrane. The method was demonstrated and characterized by the determination of acetylcholine by an acetylcholine esterase electrode, urea by a urease electrode, and penicillin G by a penicillinase electrode. Linear response in a wide range of substrate concentrations and high storage and operational stability were recorded for all the enzymes tested.

17.2.1.4 Surfactant embedment of protein

Surfactant has a similar amphoteric structure as lipid, which makes it possible to form a stable membrane the same as a lipid membrane and can be used to embed proteins. A surfactant membrane has many characteristics similar to those of a biomembrane, so that it can retain the bioactivities of proteins well. The process of preparing a surfactant/protein-modified electrode is simple and viable. There are usually two methods to incorporate proteins into films. In the first, a solution of surfactant in chloroform is spread onto the electrode surface, and solvent is evaporated. The coated electrode is then placed into an electrochemical cell containing a solution of protein, which is taken up into the film. This works best with films in the liquid crystal phase. An alternative preparation controls the amount of protein in the film. An aqueous vesicle dispersion of surfactant is made and then mixed with a solution containing protein. A precise volume of this mixture is spread onto an electrode and dried. The vesicles flatten as they dry, resulting in multiple stacks of bilayers.

Didodecyldimethylammonium bromide (DDAB) is a kind of surfactant, which can form stable film. DDAB film is lamellar liquid crystals at room temperature and this fluid state facilitates good mass and charge transport necessary for catalytic applications. Rusling and Nasser [8] reported the preparation of stable myoglobin (Mb)-DDAB films on pyrolytic graphite (PG) electrodes, which was by spontaneous insertion of Mb from solution into water-insoluble cast films of DDAB. The heterogeneous electron-transfer rate on pyrolytic graphite for the Mb Fe(III)/Fe(II) redox couple in these films was enhanced up to 1000-fold over those in aqueous solution. Electron-transfer rates of Mb were also enhanced in films of soluble cationic and anionic surfactants adsorbed on PG. The resulting films were stable for a month in pH 5.5–7.5 buffers containing 50 mM NaBr. Spectroscopic, thermal, and electrochemical characterizations suggested that the films consist of lamellar liquid crystal DDAB containing preferentially oriented myoglobin with the iron heme in a high spin state. Mb-DDAB films showed good charge-transport rate, which allowed Mb to be used as a redox catalyst. Reductions of the organohalide acid and ethylene dibromide were catalyzed by Mb-DDAB films on PG electrodes at voltages 1.0–1.3 V less negative than direct reductions.

17.2.1.5 Nanoparticles embedment of protein

Nanoparticles of metals and metal oxides have been investigated extensively in recent years due to their novel material properties, which differ greatly from the bulk substances [9–10]. Very small clusters (<50 metal atoms) act like large molecules, whereas large ones (>300 atoms) exhibit characteristics of a bulk sample of those atoms [11]. Between these extremes, the materials (usually nanomaterial) have largely unknown chemical and physical properties, which is also the reason why a glut of research activities has been focused on nanoparticles. Materials in the nanometric size category display size-dependent optical, electronic, and chemical properties. Nanoparticles can be applied to many fields, such as optical devices, electronic devices, catalysis, sensor technology, biomolecular labeling [12–13], etc. Considering the virtues of nanoparticles, such as large surface area, high powered catalysis and excellent affinity, nowadays, varied material nanoparticles have been used to embed proteins. Carbon nanotubes (CNTs) [14], Au colloid [15–16], quantum dots [17–18], and nano TiO_2 [19] are generally used.

Since their discovery in 1991 [20], CNTs have generated a frenzy of excitement [21]. CNTs, consisting of cylindrical graphitic sheets with nanometer diameters, possess superb electrical conductivity, high chemical stability, and extremely remarkable mechanical strength and modulus [22]. Cai and Chen [14] obtained a stable suspension of CNTs by dispersing the CNTs in the solution of the surfactant cetyltrimethylammonium bromide (CTAB). When haemoglobin (Hb) was immobilized onto the surface of CNTs, its direct electron-transfer rate was greatly enhanced. Cyclic voltammetric (CV) results showed a pair of well-defined redox peaks, which corresponded to the direct electron transfer of Hb. The electrochemical parameters such as apparent heterogeneous electron-transfer rate constant (k_s) and the value of formal potential ($E^{0\prime}$) were estimated. The experimental results also demonstrated that the immobilized Hb retained its bioelectrocatalytic activity to the reduction of H_2O_2. Other proteins, such as cyt *c*, catalyase, xanthine oxidase (XOD) and GOD, can also be immobilized onto the surface of CNTs, and achieve the direct electrochemistry as reported.

17.2.1.6 Other methods of protein immobilization

Some physical mechanisms that might assure the function of DNA as a molecular wire have been considered on the basis of recent progress in understanding charge transfer of biological molecules [23–26]. So, the advent of molecular electronics has stimulated an interest in the possibility of exploiting this molecule in functional mesoscopic electronic devices [27–28] and considerable interest has focused on the application of DNA based on its p-stacked base pairs [29–30]. This key structure feature is similar to the conductive one-dimensional aromatic crystals, which suggests that in the interior of the double helix, the stack of base pairs can provide a one-dimensional pathway for charge migration and acts as a "π way" for the efficient transfer of electrons [31–33]. Using DNA to immobilize proteins is a new embedment method developed recently. Rusling and Nassar [34] reported that stable films of calf thymus (CT) double stranded (ds) DNA and proteins on PG electrodes could be obtained. They also achieved direct electron transfer involving heme protein Fe(III)/Fe(II) couples in DNA/protein films. In their work, DNA films on PG electrodes also extracted heme proteins from solution. Mb diffused into pure DNA films much faster than haemoglobin.

Layer-by-layer (LbL) assembly of proteins or enzymes with polyelectrolytes is a novel general method for protein film fabrication that emerged over the past decade [35], and establishes a new procedure for studying redox proteins with electrochemical technology [36]. The principle of the LbL assembly is based on alternate adsorption of oppositely charged species from their solutions by electrostatic interaction between them. Compared with cast method, the LbL assembly technology develops a "molecular architecture" with precise control of the composition, the number of layers, and the thickness of films at a molecular or nanometer level. Moreover, the LbL method is simple and suitable to a variety of substrate matrices with different shapes. Recently, direct electrochemistry of proteins in LbL films assembled with oppositely charged polyions was studied [37–39]. In general, proteins in these films retained their native structures and electroactivities, and were used for electrocatalysis. The LbL assembly technique was recently extended to fabricate ultrathin protein films with inorganic nanoparticles, since proteins and nanoparticles could carry opposite surface charges under appropriate conditions and might be assembled layer by layer by electrostatic interaction [40–42]. The direct electrochemistry of proteins in LbL films with nanoparticles assembled on electrodes was also studied [43–45]. In these films, inorganic nanoparticles, with their unique and excellent properties such as large surface area and good biocompatibility, provided a favorable microenvironment for redox proteins to transfer electrons directly with underlying electrodes. For example, Lvov [44] assembled LbL films of Mb with MnO_2 or SiO_2 nanoparticles at PG electrodes. A pair of reversible, symmetric, reduction-oxidation CV peaks of Mb heme Fe(III)/Fe(II) couple were observed.

17.2.2 Direct electron transfer of proteins

Nowadays, studies of direct electrochemistry of redox proteins at the electrode–solution interface have held more and more scientists' interest. Those studies are a convenient and informative means for understanding the kinetics and thermodynamics of biological redox processes. And they may provide a model for the study of the mechanism of electron transfer between enzymes in biological systems, and establish a foundation for fabricating new kinds of biosensors or enzymatic bioreactors.

Historically, several factors have plagued direct electron transfer between electrodes and proteins [46]. These factors include (i) electroactive prosthetic groups deep within the protein structure, (ii) adsorptive denaturation of proteins onto electrodes, and (iii) unfavorable orientations of proteins at electrodes. Remarkable recent progress provides several strategies for achieving direct electron exchange between electrodes and proteins. With few exceptions [47], special electrode preparations are required. One approach employs highly purified protein solution and specially cleaned electrodes [48]. Another coats electrodes with promoter molecules, which facilitate electron transfer by blocking adsorptive denaturation and favorably orienting the protein [46].

The first reports on direct electrochemistry of a redox active protein were published in 1977 by Hill [49] and Kuwana [50]. They independently reported that cytochrome *c* (cyt *c*) exhibited virtually reversible electrochemistry on gold and tin doped indium oxide (ITO) electrodes as revealed by cyclic voltammetry, respectively. Unlike using specific promoters to realize direct electrochemistry of protein in the earlier studies, recently a novel approach that only employed specific modifications of the electrode surface without promoters was developed. From then on, achieving reversible, direct electron transfer between redox proteins and electrodes without using any mediators and promoters had made great accomplishments.

17.2.2.1 Direct electron transfer of cytochrome *c*

Cyt *c* is one of most important and extensively studied electron-transfer proteins, partly because of its high solubility in water compared with other redox-active proteins. *In vivo*, cyt *c* transfers an electron from complex III to complex IV, membrane-bound components of the mitochondrial electron-transfer chain. The electrochemical interrogation of cyt *c* has, however, been hindered because the redox-active heme center is buried beneath the surface of the protein. This difficulty has been overcome by the introduction of modified electrode surfaces, with monolayers able to interact with both the heme center and the underlying electrode. Since the direct electron transfer of cyt *c* was first observed in 1977 [49–50], its direct electrochemistry and voltammetric measurements have been extensively described at various chemically modified electrodes [51–59]. Direct electron transfer of cyt *c* can also be achieved by using electrode materials such as oxides [50]. Otherwise, direct electrochemical studies of cyt *c* can be made by using the protein-film voltammetry approach, pioneered by Armstrong and coworkers, where proteins are adsorbed on rough hydrophilic surfaces such as edge-plane pyrolytic graphite [60]. These works have led to a good understanding of the electron-transfer mechanism between cyt *c* and chemically modified electrodes.

The modifiers for preparation of those chemically modified electrodes are generally organic compounds [61]. Compared with organic compounds, inorganic materials are intrinsically more stable catalysts because of their layered oxide structure. Recently, a series of inorganic porous materials such as clay [62], montmorillonite [63–64], porous alumina [65], and sol-gel matrix [66] have been proven to be promising as the immobilization matrices. They have the advantages of high mechanical, thermal, and chemical stability, good adsorption and penetrability due to their regular structure and appreciable surface area. Also, the unique structural and catalytic properties of zeolites for structuring an electrochemical/electron transfer environment and resistance to biodegradation have also attracted considerable attention [67]. NaY zeolite possesses a microporous diameter of 0.81 nm [67]. The electrochemical behaviors of redox substances such as GOD [68–69] and horseradish peroxidase [70] incorporated in NaY modified matrixes have been studied. Ju [71] used polyvinyl alcohol (PVA) as a supporting medium for NaY immobilization on electrode surface and reported the direct electrochemistry of cyt *c* incorporated in NaY–PVA mixed media. An interaction between cyt *c* and NaY zeolite particles was observed with UV-Vis spectroscopy and CV. NaY zeolite particles effectively retained the activity of the immobilized cyt *c* and facilitated the electron exchange between the cyt *c* and the electrode.

It is well known that cyt *c* adsorbs strongly on Pt, Hg, Au, Ag, and other electrodes, which results in large changes in its conformation and often in denaturation of the protein. The denaturation of cyt *c* will hamper its electrochemistry [48, 72]. Conversely, direct adsorption of proteins onto the uncoated, nanometer-sized colloidal Au particles would not denature the proteins, for it has been demonstrated that electrostatically bound colloidal Au and protein conjugates typically retain biological activity [73–74]. Indeed, Crumbliss and coworkers have shown that several enzymes could maintain their enzymatic and electrochemical activity when immobilized on gold nanoparticles (GNPs) [75–76]. Recently, Au nanoparticles were self-assembled onto two-dimensional and three-dimensional superstructures on a variety of substrates such as glass, alumina, and so on by using amino/thiolsiloxanes and dithiols/diamines/bipyridinium as cross-linkers [77]. These GNPs can be strongly bound to the surface through covalent bonds to the polymer functional groups, such as —CN, —NH_2, or —SH, and a GNP monolayer can be prepared by self-assembly on the polymer coated substrate [78]. These nanoparticles can act as tiny conduction centers and facilitate the transfer of electrons.

Such functionalized superstructures have been shown to have sensoric and photoelectrochemical applications [79–81]. These metal colloidal films have also been used as a basic interface to construct amperometric biosensors to realize the direct electron transfer between the electrode and redox proteins or enzymes [82–83]. So far, the immobilization of a redox enzyme on colloidal gold is thought either to help the protein to assume a favored orientation or to make possible conducting channels between the prosthetic groups and the electrode surface, and they will both reduce the effective electron transfer distance, thereby facilitating charge transfer between electrode and enzyme [76]. Sol-gel technology has been the subject of many studies, because it provides a unique low temperature methodology to prepare a three-dimensional network suited for the encapsulation of many different molecules [84–86]. Wang [16] used (3-mercaptopropyl)-trimethoxysilane (MPTMS) for designing a three-dimensional interfacial structure of silica gel on a gold electrode via the direct coupling of sol-gel and self-assembled technology. And quasi-reversible and direct electrochemistry of cyt *c* was obtained at a novel electrochemical interface constructed by self-assembling GNPs onto a three-dimensional silica gel network, without polishing or any modification of the surface. These nanoparticles inhibited the adsorption of cyt *c* onto bare electrode and act as a bridge of electron transfer between protein and electrode.

It is known that CNTs could self-organize with DNA molecules [87–88]. The DNA/CNTs layer could be used as new electronic materials, which was based on theoretical prediction [14–89] and experimental confirmation [89–91]. Since DNA solution can gelatinize, the mixed DNA/CNTs layer can keep stability on the surface of metal electrode, and can be used to investigate some electrochemical phenomena or probe electrochemical properties of some proteins. Furthermore, proteins containing positive charges could be immobilized on the modified layer due to the massive negative charges of DNA. In Wang's work [92], multi-walled carbon nanotube (MWNT) was successfully immobilized on the surface of platinum electrode by mixing with DNA. Further research indicated that cyt *c* could be strongly adsorbed on the surface of the modified electrode, and formed an approximate monolayer. The immobilized MWNTs could promote the redox of horse heart cyt *c*, which gave reversible redox peaks with a formal potential of 81 mV.

17.2.2.2 Direct electron transfer of myoglobin

Myoglobin (Mb) is a kind of heme protein containing a single polypeptide chain with an iron heme as its prosthetic group. The physiological function of Mb is to store dioxygen and increase the diffusion rate of dioxygen in the cell. Although Mb does not function physiologically as an electron carrier, it undergoes the oxidation and reduction process in the respiratory system. Thus, its electron-transfer reactions play essential roles in biological processes. It is an ideal model molecule for the study of electron-transfer reactions of heme proteins, biosensing, and electrocatalysis.

The electrochemistry of Mb has been achieved by using mercury electrodes [93], methyl-viologen-modified gold electrodes [94], and ultraclean and hydrophilic indium oxide electrodes [95–96]. The electrochemical behavior is unstable and extremely sensitive to the sample purity and the conditions of the electrode surface [97]. Great efforts have been made to enhance its electron transfer by using mediators, promoters, or some special modified materials [98–102]. Among these, surfactant micelles have been shown to be efficient in promoting the electron transfer of Mb. Extensive studies on electrochemistry of Mb using various surfactants such as DDAB, lipids, LB films, etc. [99, 102–103] demonstrate that the surfactant film can effectively enhance the rate of electron transfer between the protein and the electrode.

Polyelectrolyte–surfactant complex films can be viewed as a new type of biomembrane-like films, which promote the electron transfer of proteins. Such complexes combine in unique ways the properties of amphiphilic surfactants with those of polymers. The polymeric components can provide mechanical strength and good stability, while the surfactants retain their tendency to assemble in bilayered structures [104–105]. Proteins in this kind of film show well-behaved electrochemistry and good stability [106–107]. Since these films are amenable to a variety of electrochemical and other experiments for a longer time than surfactant films, they may have more practical applications as biosensors or bioreactors. Wang and Hu [108] prepared the polyelectrolyte–surfactant complex DHP–PDDA by reacting the anionic surfactant dihexadecylphosphate (DHP) with polycationic poly(diallyldimethylammonium) (PDDA). Thin films made from DHP–PDDA on solid substrates demonstrated an ordered multibilayer structure by XRD and DSC. Mb incorporated in DHP–PDDA films on PG electrodes showed a pair of well-defined and nearly reversible CV peaks for the Mb Fe(III)/Fe(II) couple, which reflected that the electron transfer between Mb and PG electrodes was greatly facilitated in the film microenvironment. Mb could act as an enzyme-like catalyst in DHP–PDDA films as demonstrated by catalytic reduction of trichloroacetic acid, nitrite, and oxygen with a decrease in the electrode potentials required.

A series of inorganic porous materials, such as clay, montmorillonite, etc., have been shown to be promising as immobilization matrices. They have the advantages of high mechanical, thermal, and chemical stability and good adsorption and penetrability due to their regular structures and appreciable surface areas. In addition, the unique structural and catalytic properties of molecular sieves for structuring an electrochemical/electron-transfer environment and resistance to biodegradation have attracted considerable attention. Among those protein immobilization matrices, molecular sieves can combine with proteins through physical or chemical action. Ju and Dai [109] used a kind of mesoporous silica material, hexagonal mesoporous silica (HMS), which processes a porous size of nanoscale dimension to make it more suitable for enzyme intercalation and loading. When it is immersed into aqueous solution, the vacancy of the hexagonal mesopores is difficult to be saturated by water due to its hydrophobic surface. Thus, it can be used for the intercalation of protein. The direct electrochemistry of Mb immobilized on an HMS-modified glassy carbon electrode was described in their work. Two couples of redox peaks corresponding to Fe(III) to Fe(II) conversion of Mb intercalated in the mesopores and adsorbed on the surface of the HMS were observed with the formal potentials of -0.167 and $-0.029\,V$ in 0.1 M, pH 7.0, phosphate buffer solution, respectively. The electrode reaction showed a surface-controlled process with one proton transfer. The immobilized Mb displayed good electrocatalytic responses to the reduction of both H_2O_2 and nitrite (NO_2^-), which were used to develop novel sensors for H_2O_2 and NO_2^-. The HMS provided a novel matrix for protein immobilization and the construction of biosensors via the direct electron transfer of immobilized protein.

17.2.2.3 Direct electron transfer of hemoglobin

Hemoglobin (Hb) is a heme protein that can store and transport oxygen in blood in vertebrates. It has a molar mass of approximately 67 000 g mol^{-1} and comprises four polypeptide subunits (two α and two β subunits). The electron-transfer reactivity of Hb is physiologically hampered, though it contains four iron-bearing hemes within molecularly accessible crevices, which are known to act as electron-transfer centers in other proteins. Although Hb does not function biologically or physiologically as an electron carrier, it can be used as an ideal model molecule for study of electron transfer of heme enzymes due to its commercial availability and a known structure. Unlike some other small heme proteins such as cyt *c*, it, however, is difficult for Hb to exhibit heterogeneous electron-transfer process in most cases, which means that the rate of electron transfer of Hb is very slow. As a result, no useful currents appear at the conventional electrodes, even when rather large overvoltages are applied, due to its extended three-dimensional structure and resulting inaccessibility of the electroactive centers as well as its strong adsorption onto the electrode surface for subsequent passivation. Great efforts have been made to enhance the electron transfer of Hb by using mediators and promoters [110–111] and some interesting results were obtained. Recently, Rusling [112], Lu [113], and Wang [114] developed a technique of protein-film to incorporate Hb into films, such as composite films, protein–polyion layer-by-layer assembly films–polymer and natural lipid films on PG electrodes. For example, Wang [114] observed the direct electrochemistry of Hb in stable thin film composed of a natural lipid (egg–phosphatidylcholine) and Hb on PG electrode. Hb in lipid films showed thin layer electrochemistry behavior and exhibited elegant catalytic activity for electrochemical reduction of H_2O_2, based on which an unmediated biosensor for H_2O_2 was developed.

Some efforts have been taken to obtain the electrochemical response of Hb at solid electrode surfaces. Fan's electrochemical researches revealed that the electron-transfer reactivity of Hb could be greatly enhanced, simply by treating it with an organic solvent, dimethyl sulfoxide (DMSO) [115]. Hb can also achieve its direct electron transfer in *N,N*-dimethylformamide (DMF) film, as Xu [116] reported. These, therefore, suggested that there are many different factors that regulate electron-transfer reactivity of proteins. It also pointed out the complicated and precise regulation mechanisms of proteins *in vivo*.

Nanotechnology has provided a novel way to enhance the electron-transfer rates between Hb and the electrode. As in the case of cyt *c* and Mb, nanocrystalline TiO_2 film has been proposed to be a promising interface for the immobilization of Hb. GNPs are renowned for their good biocompatibility. With the help of these GNPs, Hb can exhibit a direct electron-transfer reaction without being denatured. To improve the stability of these particles, a kind of gold nanoparticle protected by lipid (DDAB) was invented [15]. The electron transfer of Hb at electrodes modified with colloidal clay nanoparticles has also been well studied due to its simplicity, effectiveness, and low cost. A new electrochemical sensing system for direct electron transfer of heme protein was developed that relied on the virtues of excellent biocompatibility, nano-dimensions and semiconductor properties of quantum dots in Lu's work [17]. To demonstrate the conception, Hb was immobilized in a water-soluble quantum dots (QDs, mercapto-coated CdSe-ZnS) film on glassy carbon (GC) electrode. From the results of cyclic voltammetry (CV), a pair of well-defined and quasi-reversible peaks for Fe(III)/Fe(II) redox couple of Hb was obtained, which reflected direct electron transfer of the heme protein. Scanning electron microscopy (SEM), fluorescence (FL), and electrochemical impedance spectroscopy (EIS) demonstrated that electrostatic attractions existed between Hb and QDs. These interactions led to the arrangement of Hb in the film in a favorable orientation for exchanging electrons with the electrode, the inhibition of Hb adsorption onto bare electrode, and the offering of an electron transfer bridge between Hb and electrode, which were believed to be responsible for the direct electrochemistry of Hb.

Cellulose is one of the naturally occurring biopolymers and widely existent in wood and other plants. Cellulose in its native form is not soluble in water. It can be rendered water soluble by chemical reaction of its hydroxyl groups with hydrophilic substituents. Carboxymethyl cellulose (CMC) is one of the water-soluble cellulose derivatives. CMC contains a hydrophobic polysaccharide backbone and many hydrophilic carboxyl groups, and hence shows amphiphilic characteristics. Due to its desirable properties such as non-toxicity, biocompatibility, biodegradability, high hydrophilicity, and good film forming ability, CMC has been used in various practical fields. CMC has also been used to study interactions with proteins. For example, Cark and Glatz described the formation of complex of CMC with casein by electrostatic interaction [117]. Lii *et al.* [118] reported the formation of CMC–casein complex by covalent bonds with electrosynthesis. These complexes were very stable to pH and ionic strength changes and exhibited very good emulsifying properties and increased thermal stability. Thus, protein–CMC films were made by casting a solution of Hb and CMC on PG electrodes [119]. In pH 7.0 buffers, Hb incorporated in CMC films gave a pair of well-defined and quasi-reversible CV peaks at about -0.34 V (vs SCE), which was characteristic of heme Fe(III)/Fe(II) redox couples of the protein. The electrochemical parameters such as apparent standard rate constants (k_s) of heterogeneous electron transfer and formal potentials ($E^{0'}$) could be estimated by square wave voltammetry with non-linear regression analysis. In aqueous solution, stable CMC films absorbed large amounts of water and formed hydrogel. The more loosening structure of CMC in its hydrogel form would provide a more suitable microenvironment for Hb to transfer electrons with underlying PG electrodes.

Recently, a novel method of Hb immobilization was achieved by Lu [120]. The direct electrochemistry of Hb was successfully achieved by adsorbed Hb onto the surface of a yeast cell through electrostatic attractions on a GC electrode. The bioactivity of Hb immobilized in yeast cell film was retained, and the catalytic reduction of NO and H_2O_2 was estimated.

17.2.3 Direct electron transfer of enzymes

Study of electron-transfer reactions of enzymes at the electrode–solution interface is a convenient and informative means for understanding the kinetics and thermodynamics of biological redox processes [121–123]. It can also establish a foundation for fabricating the third-generation biosensors without using redox mediators. Work on direct electrochemistry of biomolecules has largely focused on relatively small proteins [121]. However, few studies have been reported on the direct electrochemistry of redox enzymes. Part of the difficulty may have stemmed from the large spatial separation between the prosthetic group(s) of catalytically active proteins/enzymes and the electrode surface. Many types of metal electrodes or chemically modified electrodes may not be suitable for aligning the redox center(s) of the enzymes close to the surface or could even lead to denaturation of the enzymes upon adsorption [124].

17.2.3.1 Direct electron transfer of HRP

Horseradish peroxidase (HRP) is a member of the large class of peroxidases, which are enzymes defined as oxidoreductases using hydroperoxide as electron acceptor. Due to its commercial availability in high purity, HRP has long been a representative system for investigating the structure, dynamic, and thermodynamic properties of peroxidases, especially for understanding their biological behaviors of catalyzing oxidation of substrates by H_2O_2 [125]. HRP can react with H_2O_2 to form a powerful enzymatic oxidizing agent known as Compound I, which is a two-equivalent oxidized form containing an oxyferryl heme ($Fe^{4+}{=}O$) and a porphyrin π cation radical. Compound I is catalytically active and can abstract one electron from the substrate to form a second intermediate, called Compound II, which is subsequently reduced to the resting state of the native enzyme, HRP–Fe(III), by accepting an additional electron from the substrate. HRP–Fe(III) can also be further reduced to HRP–Fe(II). Efficient electron transfer between HRP and electrodes has been reported for many years [126]. However, in most cases, direct electrochemistry was proven in the presence of H_2O_2 or other peroxides by amperometry, and attributed to electrochemical reduction of Compound I or Compound II. Only a few examples of independent quasi-reversible CVs of HRP were reported in the absence of peroxides, probably because of the large molecular mass and extended structure of HRP, and the inaccessibility of its redox centers [127–130]. Ferri [127–128] reported a pair of quasi-reversible CV peaks for HRP Fe(III)/Fe(II) redox couple by entrapping HRP in the film of tributylmethyl phosphonium chloride (TBMPC) polymer bound to an anionic exchange resin at PG electrodes. Chen [129–130] then explored the direct electrochemistry of HRP Fe(III)/Fe(II) couple by CV in DDAB and DNA films at PG electrodes. These films provided a suitable microenvironment for HRP, which greatly facilitated the electron exchange between HRP and electrodes. Electrochemical catalytic reduction of H_2O_2 by HRP in these films was also described. Recently, a novel Nafion–cysteine functional membrane was constructed [131]. Rapid and direct electron transfer of HRP was carried out on the functional membrane-modified gold electrode with good stability and repeatability.

Poly(ester sulfonic acid), with the trade name of Eastman AQ, is a kind of anionic ionomer. Unlike another more popular ionomer, Nafion, this thermoplastic, amorphous polymer gives translucent, low viscosity dispersions in water without adding organic solvent. AQ films cast onto electrodes do not dissolve in water. Similar to Nafion, AQ films bind hydrophobic cations preferentially and exclude negatively charged species [132]. Enzymes can be added directly to aqueous dispersions of AQ ionomers and deposited onto solid surfaces to form films without denaturation or significant loss of activity [133]. Direct electrochemistry of small redox proteins such as cyt *c* [134], Mb [135], and Hb [136] in AQ films at PG electrodes was studied previously. Thus, Huang Rong and Hu Naifei [137] made stable films from ionomer poly(ester sulfonic acid) or Eastman AQ29 on PG electrodes and achieved direct electrochemistry for incorporated enzyme HRP. Cyclic voltammetry of HRP–AQ films showed a pair of well-defined, nearly reversible peaks at about -0.33 V (vs SCE) at pH 7.0 in blank buffers, characteristic of HRP heme Fe(III)/Fe(II) redox couple. The electron transfer between HRP and PG electrode was greatly facilitated in AQ films. Reflectance absorption infrared (RAIR) and ultraviolet visible (UV-Vis) absorption spectra demonstrated that HRP retained a near native conformation in AQ films. The embedded HRP in AQ films retained the electrocatalytic activity for oxygen, nitrite, and H_2O_2.

Agarose is a polysaccharide with an average molecular weight of 120 000 Da, consisting of 1,3-l-dgalactopyranoseand 1,4-linked 3,6-anhydro-k-l-galactose units [138]. In a hot solution, agarose chains exist in a stiff and disordered configuration. Upon cooling below 40°C, the coils form orderly helices which subsequently aggregate into thick bundles in which there exist large pores of water. Agarose gel matrix exhibits strong elasticity, high turbidity, an aqueous microenvironment, and bioaffinity, which make it an ideal biopolymer to immobilize proteins on solid substrates. Agarose gel was proven to be the best one for constructing the bioreactor with cytochrome P450, compared with other gels, such as PAM, calcium alginate, and prepolymerized polyacrylamide hydrazide [139]. Liu [140] immobilized HRP on edge-plane PG electrodes modified by agarose hydrogel and obtained the direct electrochemistry of HRP. The protein entrapped in the agarose film underwent fast direct electron transfer reactions, corresponding to

$$Fe^{III} + e^- \rightarrow Fe^{II} \tag{1}$$

The $E^{0'}$ was linearly dependent on solution pH (redox Bohr effect), indicating that the electron transfer was proton coupled. UV-Vis absorption spectra and RAIR spectra suggested that the conformation of HRP in the agarose film was little different from that protein alone, and the conformation changed reversibly in the range pH 3.0–10.0. Atomic force microscopy images of the agarose film indicated a stable and crystal-like structure formed possibly due to the synergistic interaction of hydrogen bonding between DMF, agarose hydrogel, and HRP. This suggested a strong interaction between the heme protein and the agarose hydrogel. DMF played an important role in immobilizing protein and enhancing the rate of electron transfer between protein and electrode.

It is well known that TiO_2 is widely used in cosmetics, solar cells, batteries, additives in toothpaste and white paint, and others. Recently, there is a considerable interest in using TiO_2 nanoparticles as a film-forming material since they have high surface area, optical transparency, good biocompatibility, and relatively good conductivity. Various TiO_2 films were also used to immobilize proteins or enzymes on the electrode surface for either mechanistic study of the proteins or fabricating electrochemical biosensors. For example, Durrant and coworkers immobilized a range of proteins into nanoporous TiO_2 film-modified electrodes and successfully used this strategy to develop electrochemical and optical biosensors [141–142]. Luo and coworkers used nanocrystalline TiO_2 films on electrodes to entrap heme proteins such as cyt *c*, Mb, and Hb, and observed the direct electrochemistry of these proteins [143]. Accumulation and electroactivity of cyt *c* in mesoporous layer-by-layer films of TiO_2 and phytate at ITO electrodes was also studied [144]. Titania sol-gel matrix films were used to immobilize HRP, and with the aid of mediators, HRP–TiO_2 film electrodes were used to detect H_2O_2 by amperometry [66, 145]. Hu's group [146] also incorporated HRP in TiO_2 nanoparticle films modified on electrodes to achieve the direct electron transfer of HRP. HRP–TiO_2 film electrodes were fabricated by casting the mixture of HRP solution and aqueous titania nanoparticle dispersion onto PG electrodes and letting the solvent evaporate. The HRP incorporated in TiO_2 films exhibited a pair of well-defined and quasi-reversible CV peaks at about $-0.35\,V$ in pH 7.0 buffer, which reflected that the rate of electron exchange between the enzyme and PG electrodes was greatly enhanced in the TiO_2 nanoparticle film microenvironment. The HRP–TiO_2 film electrodes were quite stable and amenable to long-time voltammetric experiments. The UV-Vis spectroscopy showed that the position and shape of the Soret absorption band of HRP in TiO_2 films kept nearly unchanged and were different from those of hemin or hemin–TiO_2 films, suggesting that HRP retains its native-like tertiary structure in TiO_2 films. The electrocatalytic activity of HRP embedded in TiO_2 films toward O_2 and H_2O_2 was retained.

17.2.3.2 Direct electron transfer of catalase

Catalase (cat) is also a heme enzyme, which is present in almost all aerobic organisms [147]. Cat has a molecular weight of approximately 240 000, and is composed of four identical subunits, each containing a single heme prosthetic group. The heme group consists of a protoporphyrin ring and a central Fe atom, where iron is usually in the ferric oxidation state as its stable resting state [148]. As a catalyst, cat functions either in the catabolism of H_2O_2 or in the peroxidatic oxidation of small molecule substrates by H_2O_2 [149–150]. Under normal physiological conditions, cat controls the H_2O_2 concentration so that it does not reach toxic levels that could bring about oxidative damage in cells. The mechanism of disproportionation of H_2O_2 catalyzed by cat can be expressed as [149–150]:

$$H_2O_2 + \text{CatFe(III)} \rightarrow \text{Compound I} + H_2O \tag{2}$$

$$H_2O_2 + \text{Compound I} \rightarrow \text{CatFe(III)} + O_2 + H_2O \tag{3}$$

where CatFe(III) is the resting state of catalase, Compound I is a two-equivalent oxidized form of CatFe(III) containing an oxyferryl heme ($Fe^{IV} = O$) and a porphyrin π-cation radical. H_2O_2 first oxidizes CatFe(III) to form Compound I and H_2O, and then reduces Compound I to CatFe(III) and produces O_2. As a well-known efficient catalyst, cat acts either as a reductant or as an oxidant in the reactions, and returns to its resting state after one catalytic cycle, while H_2O_2 undergoes dismutation to produce H_2O and O_2. Although the heme groups are the electroactive center of cat, it is usually difficult to observe the direct electron exchange of cat in solution with electrodes, probably because the hemes are buried deeply inside the polypeptide chains of relatively large cat molecules. Films modified on solid electrodes provided an approach to realize its direct electrochemistry. A recent report by Kong [151] showed that in DDAB films cast on PG electrodes, cat gave a pair of direct CV peaks in blank buffers. Rusling and coworkers [152] used dimyristoylphosphatidylcholine (DMPC) as a film-forming material to incorporate catalase. The Cat–DMPC films cast on PG electrodes showed a pair of well-defined, quasi-reversible peaks, characteristic of heme Fe(III)/Fe(II) redox couples of cat. Both DDAB and DMPC are water-insoluble, double-chain surfactants, and can form multibilayer films from their organic solution or aqueous dispersion. It is these biomembrane-like surfactant films that provided a suitable microenvironment for cat to transfer electrons with PG electrodes.

Polyacrylamide (PAM) is a kind of widely used polymer. PAM has a long hydrophobic hydrocarbon backbone with hydrophilic amide groups, showing the amphiphilic property. The electrochemistry of heme proteins in PAM polymer environment has been studied. Murray and coworkers has reported "solid-state" voltammetry of cyt *c* in PAM gel solvent [153]. In the previous study, it was found that PAM could form stable films on PG surface and absorb considerable amounts of water in aqueous solution. Heme proteins such as Hb [154] and Mb [155] in the PAM hydrogel films-modified PG electrodes demonstrated a reversible CV response for the heme Fe(III)/Fe(II) redox couple. As Lu reported [156], PAM hydrogel films could also provide a favorable microenvironment for incorporated cat, and cat in PAM films gave a direct electrochemical response at the electrode surface. Cat–PAM film electrode showed a pair of well-defined and nearly reversible cyclic voltammetry peaks for Cat Fe(III)/Fe(II) redox couples at approximately $-0.46\,V$ (vs SCE) in pH 7.0 buffers. This suggested that the electron transfer between cat and PG electrode was greatly facilitated in the microenvironment of PAM films. The apparent heterogeneous electron-transfer rate constant (k_s) and formal potential ($E^{0\prime}$) were estimated by fitting square wave voltammograms with non-linear regression analysis. The formal potential of Cat Fe(III)/Fe(II) couples in PAM films have a linear relationship with pH between pH 4.0 and 9.0 with a slope of $-56\,mV\ pH^{-1}$, suggesting that one proton is coupled with single-electron transfer for each heme group of cat in the electrode reaction. UV-Vis absorption spectroscopy demonstrated that cat retained a nearly native conformation in PAM films at medium pH. The embedded cat in PAM films showed the electrocatalytic activity toward O_2 and H_2O_2.

Several studies have shown that the direct electron transfer between cat and graphite or carbon soot electrodes in deoxygenated solutions was sluggish, with large peak separations [157–159]. Based on the consideration that CNTs might be the best candidate among the various carbonaceous substrates for promoting the electron transfer reaction of cat and for biomolecular attachment [160–163], Wang [164] accomplished the direct electrochemistry of cat at a gold electrode modified with single-wall carbon nanotubes (SWNTs). A pair of well-defined redox peaks was obtained for cat with the reduction peak potential at −0.414 V and a peak potential separation of 32 mV at pH 5.9. Both reflectance FT-IR spectra and the dependence of the reduction peak current on the scan rate revealed that cat adsorbed onto the SWNTs surfaces. The redox wave corresponds to the Fe(III)/Fe(II) redox center of the heme group of the cat adsorbate. Compared to other types of carbonaceous electrode materials (e.g. graphite and carbon soot), the electron-transfer rate of cat redox reaction was greatly enhanced at the SWNTs-modified electrode. The catalytic activity of cat adsorbate at the SWNTs appeared to be retained, as the addition of H_2O_2 produced a characteristic catalytic redox wave. The facile electron-transfer reaction of cat could be attributed to the unique properties of SWNTs (e.g. the excellent electrical conductivity of SWNTs, the enhanced surface area arising from the high aspect ratio of the nanotubes, and the amenability of SWNTs for the attachment of biomolecules).

17.2.3.3 Direct electron transfer of GOD

Glucose oxidase (GOD) is a typical flavin enzyme with flavin adenine dinucleotide (FAD) as redox prosthetic group. Its biological function is to catalyze glucose to form gluconolaction, while the enzyme itself is turned from GOD(FAD) to GOD($FADH_2$). The GOD molecule is a structurally rigid glycoprotein with a molecular weight of 152 000–186 000 Da, and consists of two identical polypeptide chains, each containing a FAD redox center. Since the FAD moiety is deeply embedded within a protective protein shell, well-defined direct electrochemical behavior of GOD is rather difficult. In the literature, only a few examples of quasi-reversible voltammograms for direct electron transfer between the GOD-active site and the electrode surface were reported. Ianniello [165] reported the direct electron transfer of adsorbed GOD at a graphite electrode and a cyanuric chloride-modified graphite electrode using differential pulse voltammetry. The direct electrochemistry of GOD, immobilized at a self-assembled monolayer of 3,3-dithiobis-sulfosuccinimidyl propionate, was obtained by Jiang [166]. Direct electron transfer of GOD realized by CNTs was also studied by various groups in recent years [167–169]. Wang [170] found that the direct electrochemistry of GOD could be conducted at SWNT-modified Au electrodes. A reversible redox wave could be observed with the cathodic peak potential at −0.465 V and the separation of the peak potentials of 23 mV at pH 7.0. The peak potential was pH dependent and shifted to the cathodic direction by 48 mV per unit of pH. The cathodic peak current was found to be proportional to the scan rate, suggesting that the redox wave was given rise by the GOD adsorbate at the SWNTs. The specific enzyme activity of the GOD adsorbates at the SWNTs was found to be retained, suggesting that SWNT-modified electrodes covered with redox-active enzymes could retain the enzyme activities and provide an attractive route for the development of biosensors and nanobiosensors.

The metal colloidal films have been used to construct the interface for direct electron transfer of redox-active proteins and retaining their bioactivity [171]. Colloidal gold is an extensively used metal colloid, which has been used for the study of direct electrochemistry of proteins [172–173]. It provides an environment similar to that of redox proteins in native systems and gives the protein molecules more freedom in orientation, thus reducing the insulating property of the protein shell for the direct electron transfer and facilitating the electron transfer through the conducting tunnels of colloidal gold. Ju [174] combined the advantageous features of colloidal gold and carbon paste technology to achieve the direct electron transfer of GOD. GOD adsorbed on a colloidal gold-modified carbon paste electrode displayed a pair of redox peaks with a formal potential of $-(449 \pm 1)$ mV in 0.1 M pH 5.0 phosphate buffer solution. The response showed a surface-controlled electrode process with an electron-transfer rate constant of (38.99 ± 5.3)/s determined in the scan rate range from 10 to 100 mV/s. GOD adsorbed on gold colloid nanoparticles maintained its bioactivity and stability. The immobilized GOD could electrocatalyze the reduction of dissolved oxygen and resulted in a great increase of the reduction peak current.

17.2.3.4 Direct electron transfer of other active enzymes

Several dehydrogenases harboring pyrroloquinoline quinone (PQQ) as their prosthetic group have been reported, such as glucose dehydrogenases, ethanol dehydrogenases, and methanol dehydrogenases [175–179]. They are divided into two categories, quinoprotein (PQQGDHs, PQQMDH, and type I PQQADH) and quinohemoprotein (types II and III PQQADH), the latter containing an additional heme *c* prosthetic group together with PQQ. These quino- and quinohemoproteins form protein complexes composed of catalytic subunits, with PQQ as their redox center, and electron acceptors, such as cyt *c*, which transfer electrons from reduced PQQ to the respiratory chain. The 3D structure of the quinohemoprotein ethanol dehydrogenase (QH-EDH) from *Comamonas testosteroni* was recently determined [180]. This enzyme has two domains separated with a peptide linker region, an eight-bladed b propeller fold catalytic domain containing PQQ, and a cyt *c* domain that is located at the C-terminal region. The heme is located on top of the catalytic site of the first domain, thereby allowing smooth electron transfer from the catalytic site via heme to the external electron acceptor. Due to their superior electron transferability, such heme-containing, multicofactor dehydrogenases, consisting of an FAD- or PQQ-harboring catalytic subunit and a heme-containing electron transfer subunit/domain, were recently shown to display direct electron transfer with the electrode [181–182]. The direct electron transfer mechanism has been investigated as the ultimate enzyme sensor format, allowing the direct monitoring of the catalytic reaction by the electrode. Unfortunately, only a limited number of enzymes, such as multicofactor enzymes, are capable of direct electron transfer to electrode. Among the numerous dehydrogenases, the water-soluble PQQGDH from *Acinetobacter calcoaceticus* (GDH-B) is one of the most industrially attractive enzymes, as a sensor constituent for glucose sensing, because of its high catalytic activity and insensitivity to oxygen. Okuda [183] attempted to engineer GDH-B to enable electron transfer to the electrode in the absence of artificial electron mediator by mimicking the domain structure of the quinohemoprotein ethanol dehydrogenase (QH-EDH) from *Comamonas testosteroni*, which is composed of a PQQ-containing catalytic domain and a cyt *c* domain. They genetically fused the cyto *c* domain of QH-EDH to the C-terminal of GDH-B. The constructed fusion protein showed not only intramolecular electron transfer, between PQQ and heme of the cyt *c* domain, but also electron transfer from heme to the electrode, thereby allowing the construction of a direct electron-transfer-type glucose sensor.

The large redox enzyme xanthine oxidase (XOD, mw 286000), being a unimolecular, multicomponent electron-transport biomacromolecule, features one olybdenum center, two Fe_2S_2 centers, and one flavin adenine dinucleotide (FAD), and has been the target of extensive study, especially in the study of biosensors. XOD is implicated as a key oxidative enzyme in many physiological processes. The enzyme can catalytically oxidize many substrates, including purines, pteridines, heterocyclic molecules, and aldehydes [184]. The oxidation of xanthine takes place at the molybdenum center and the electrons distributed to other redox centers. The reoxidation of the reduced enzyme by the oxidant substrate, either NAD^+ or molecular oxygen, occurs through FAD [185]. Furthermore, the electron transfers of O_2, NAD^+, methylene blue, quinines, and nitrate are also associated with XOD. So, electrochemical researchers have studied the electron-transfer property of XOD [186–188], and used this enzyme to detect xanthine, hypoxanthine, and other biological molecules [189–191]. Nonetheless, the direct electron-transfer reaction of XOD is still very difficult to achieve. Although various methods have been employed, there is no finding that electrochemical signals both of FAD and molybdenum centers can be observed simultaneously. And no response is observed unless denaturation of the enzyme has occurred, in which case one finds a surface-confined chemically reversible process for free FAD. Li and coworkers [192] used DNA as a matrix to embed XOD, and obtained the electrochemical responds of FAD and molybdenum center of XOD. Meanwhile, XOD kept its enzymatic activity to hypoxanthine. In DNA, because p-conjugated nucleic acid bases are stacked, it is naturally considered to act as an effective molecular wire for electron transfer. The experimental results indicated that DNA could simultaneously activate both the FAD and the molybdenum centers of XOD. Wang [193] also obtained the direct electrochemistry of XOD at a gold electrode modified with SWNTs.

17.3 APPLICATION OF BIOSENSORS BASED ON DIRECT ELECTRON TRANSFER OF PROTEIN

17.3.1 Biosensors based on direct electron transfer of proteins

One of the important purposes for the study of the direct electron transfer of protein is to construct the mediator-free protein-based biosensors. These biosensors can determine many small molecules like H_2O_2, O_2, NO, nitrite, small organic peroxide, and so on. They also can determine glucose, alcohol, and amino acids by coimmobilization of the corresponding oxidase on the electrode surface. A biosensor is a detection system composed of a biological sensing element (e.g. enzyme) and a transducer (e.g. electrode). The biosensor measures the signal caused by change in the concentration of a co-reactant which reacts with the analyte or a co-product which is produced with the analyte of a biological reaction (e.g. enzyme reaction). When an electrode is used as a transducer in a biosensor, the electrode converts the change in concentration of a product or a reactant of a biological reaction into an electric signal.

The first enzyme biosensor was a glucose sensor reported by Clark in 1962 [194]. This biosensor measured the product of glucose oxidation by GOD using an electrode which was a remarkable achievement even though the enzyme was not immobilized on the electrode. Updark and Hicks have developed an improved enzyme sensor using enzyme immobilization [194]. The sensor combined the membrane-immobilized GOD with an oxygen electrode, and oxygen measurements were carried out before and after the enzyme reaction. Their report showed the importance of biomaterial immobilization to enhance the stability of a biosensor.

The enzyme-based biosensor has come through three steps: (1) with oxygen for the media; (2) with artificial intermediate for media; and (3) without media and based for the direct electron transfer of redox proteins. The following is an example:

1. With oxygen for media:

$$\text{In solution: } GOD_{ox} + \text{glucose} \rightarrow \text{gluconolactone} + GOD_{re} \tag{4}$$

$$GOD_{red} + O_2 \rightarrow GOD_{ox} + H_2O_2 \tag{5}$$

$$\text{On electrode: } H_2O_2 \rightarrow O_2 + 2H^+ + 2e \tag{6}$$

2. With artificial intermediate for media:

$$\text{In solution: } GOD_{ox} + \text{glucose} \rightarrow \text{gluconolactone} + GOD_{red} \tag{7}$$

$$GOD_{red} + M_{ox} \rightarrow GOD_{ox} + M_{red} \tag{8}$$

$$\text{On electrode: } M_{red} \rightarrow M_{ox} + ne \tag{9}$$

3. Without media:

$$\text{In solution: } GOD_{ox} + \text{glucose} \rightarrow \text{gluconolactone} + GOD_{red} \tag{10}$$

$$\text{On electrode: } GOD_{red} \rightarrow GOD_{ox} + 2e \tag{11}$$

In this chapter, we mainly describe the third-generation biosensor: mediator-free protein-based biosensor. This biosensor has many advantages compared with other two-generation biosensors. Many redox proteins and redox enzymes including cytochrome *c* [195], myglobin [196], hemoglobin [197], horseradish peroxidase [198], cytochrome P450 [199], catalase [200], glucose oxidase [201], and other active enzymes have been used to make this type of biosensor in recent years.

17.3.1.1 Biosensors based on direct electron transfer of proteins cytochrome *c*

It has been said above that cyt *c* was one of the most important and extensively studied electron-transfer proteins with active heme centers. Thus, cyt *c* was widely used in enzyme-based biosensors and to study the mechanism of the catalytic process between redox enzyme and substrate.

The determination of H_2O_2 is very important in many different fields, such as in clinical, food, pharmaceutical, and environmental analyses [202]. Many techniques such as spectrophotometry, chemiluminesence, fluorimetry, acoustic emission, and electrochemistry methods have been employed to determine H_2O_2. Electrochemical methods are often used because of their advantages. Among these electrochemical methods, the construction of the mediator-free enzyme-based biosensors based on the direct electrochemistry of redox proteins has been reported over the past decade [203–204]. The enzyme-based biosensors, which use cyt *c* as biocatalyzer to catalyze H_2O_2, were widely studied.

Many materials were used to immobilize cyt *c*, and multi-walled carbon nanotubes (MWNTs) were one of these materials. CNTs consist of cylindrical graphitic sheets with nanometer diameters, superb electrical conductivity, high chemical stability, and extremely remarkable mechanical strength and modulus [22], as described above. Zhao [195] developed a hydrogen peroxide biosensor based on the direct electrochemistry of cyt *c* on an MWNTs-modified GC electrode to study the electrochemical reduction of hydrogen peroxide. As shown in Fig. 17.1, no redox response can be observed in the potential range from 0.4 to -1.0 V at the bare MWNTs-modified GC electrode. However, at the cyt *c*/MWNTs-modified glass carbon electrode, an obvious catalytic reduction peak appears at the potential of -0.277 V. The cathodic peak current of cyt *c* increased but its anodic peak current decreased with an increase in the concentration of H_2O_2, indicating the typical electrocatalytic reduction process. Cyt *c* immobilized with other materials has the same behavior when it catalyzes H_2O_2 [205]. The catalytic currents increased linearly with the concentration of H_2O_2 in different ranges with different immobilized materials, and this is the base of the determination of hydrogen peroxide. The calculated apparent Michaelis-Menten constant (K_m^{app}), which can indicate the catalytic activity of enzyme to its substrate, can be obtained from the Lineweaver-Burk equation:

$$1/I_{ss} = 1/I_{max} + K_m^{app}/I_{max}C \qquad (12)$$

where I_{ss} is the steady-state current after the addition of a substrate, which can be obtained from amperometric experiments. I_{max} is the maximum current under saturated substrate condition and C is the bulk concentrate of the substrate. The value of the apparent Michaelis-Menten constant (K_m^{app}) can be calculated from the slope (K_m^{app}/I_{max}) and the intercept ($1/I_{max}$) of the plot of the reciprocals of the I_{ss} vs $C_{H_2O_2}$. The concentration of H_2O_2 can also be determined by amperometric method. The amperometric response of the cyt *c*/MWNTs-modified electrode to H_2O_2 was recorded through successively adding H_2O_2 to a continuous stirring PBS solution. The amperometric response has linear relationship with the concentration of H_2O_2 (Fig. 17.2).

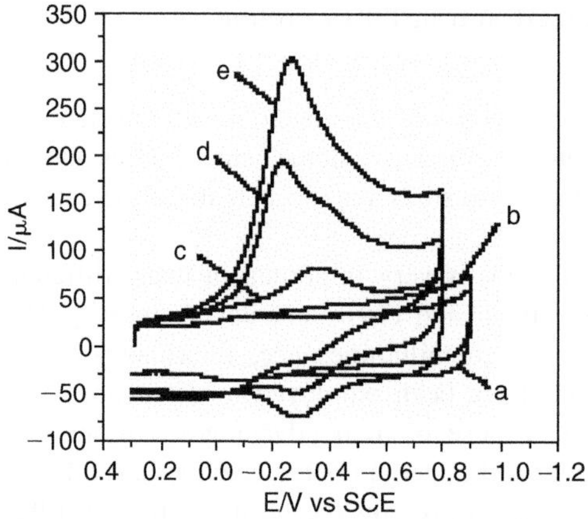

FIGURE 17.1 Cyclic voltammograms of cyt *c*/MWNT modified electrode (c, d, and e) and MWNTs modified electrode (a and b) in 0.1 mol/l PBS containing no H_2O_2 (a and c), 0.8 mM H_2O_2 (d), and 1.6 mM H_2O_2 (b and e). The scan rate is 200 mV/s. (From [195], with permission.)

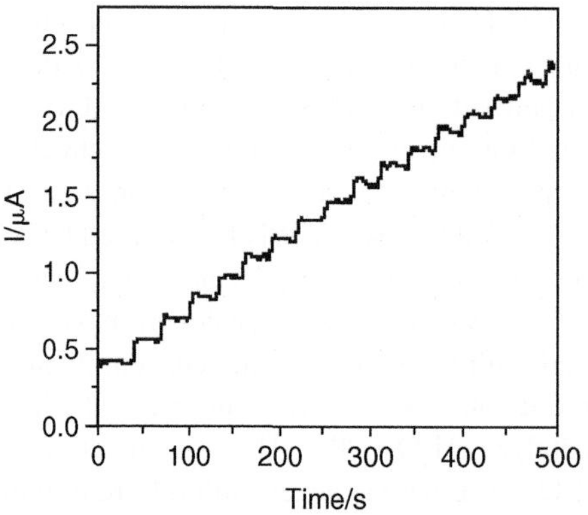

FIGURE 17.2 Amperometric response of cyt *c*/MWNTs electrode to H_2O_2. Conditions: a −0.20 V constant potential modulated with 50 mV pulse in the time intervals of 0.5 s, successive additions of 10 μL 1.8 mM H_2O_2 to 10 mL 0.1 M pH 7.0 PBS and the stirring rate of solution is 400 rpm. (From [205], with permission.)

Oxygen is an essential component of living organisms, so the study of reactive oxygen intermediates is an important field. About 5% of oxygen consumed by biosystems is used to produce superoxide [206]. The main sources of reactive oxygen species (ROS) are electron-transport chains, active phagocytosis, and the activity of some enzymes [206–208]. ROS are able to trigger transcription [209] and to protect macroorganisms from bacteria and viruses [210]. ROS have been linked to several toxicity processes, including damage of proteins, membranes, DNA, and enzymes [211]. It has also been linked to various pathological situations, including cancerogenesis, heart disease, reperfusion injuries, rheumatoid arthritis, inflammation, and aging [212].

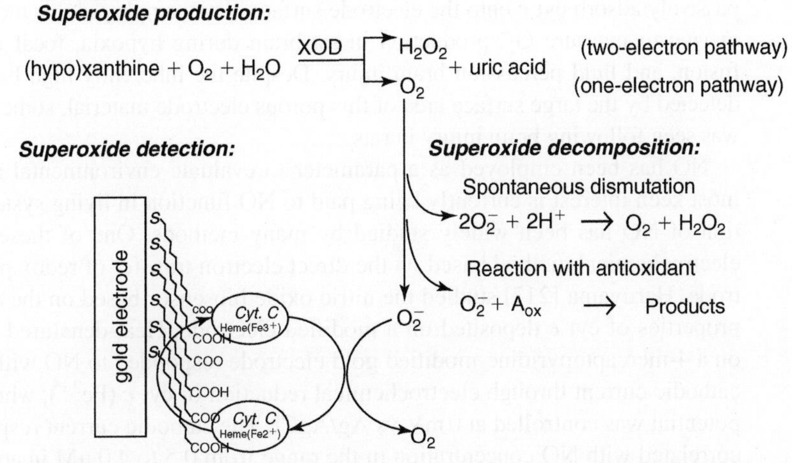

FIGURE 17.3 Scheme of the detection principle of antioxidant activity using a cytochrome *c* functionalized gold electrode. A_{ox} is the antioxidant under investigation. (From [213], with permission.)

Overproduction of ROS can be harmful. Thus, cells have developed a complex antioxidant defense system to counteract the biological potential of ROS formation [213].

Many methods including photometric, fluorimetric, chromatographic, and electrochemical methods have been used to detect the antioxidants so far. Recently, electrochemical methods have intensively been used for antioxidant detection. Among the electrochemical methods, the detection of antioxidant based on the direct redox transformation of cyt *c* has been studied over the decade. Since cyt *c* can act as an oxidant of superoxide, the superoxide level in solution can be detected as an oxidation current at the sensor electrode due to electron transfer from the radical via cyt *c* to the electrode.

The principle of antioxidant detection is shown in Fig. 17.3. Superoxide was enzymatically produced and dismutated spontaneously to oxygen and H_2O_2. Under controlled conditions of superoxide generation such as air saturation of the buffer, optimal hypoxanthine concentration (100 μM) and XOD activity (50 mU ml^{-1}) a steady-state superoxide level could be obtained for several min (580–680 s). Since these steady-state superoxide concentrations can be detected by the cyt *c*-modified gold electrode, the antioxidate activity can be quantified from the response of the sensor electrode by the percentage of the current decrease.

O_2^- also can be detected by cyt *c*-modified electrode [214]. When cyt *c* was covalently attached to the modified electrode, the immobilized cyt *c* was used as an integral part of an amperometric O_2^- sensor. Superoxide generated by xanthine/XOD caused the one-electron reduction of cytochrome c^{3+} to cytochrome c^{2+}. The reduced protein was then reoxidized at the electrode surface. McNeil used this immobilization procedure to detect O_2^- production by stimulated human neutrophils [215]. The neutrophils that stimulated with phorbol-12 myristate-13 acetate (PMA) produced current changes that were cell number dependent. Fabian [216] used a platinized activated carbon electrode (PACE) to passively adsorb cyt *c* onto the electrode surface. They employed this modified electrode *in vivo* to measure O_2^- production in rat brain during hypoxia, focal ischemia, reperfusion, and fluid percussion brain injury. Despite the inherently high background noise detected by the large surface area of this porous electrode material, some O_2^- generation was seen following brain injury in rats.

NO has been employed as a parameter to evaluate environmental air pollution. A most keen interest is currently being paid to NO function in living systems. The detection of NO has been widely studied by many methods. One of these methods is an electrochemical method based on the direct electron transfer of redox proteins on electrode. Haruyama [217] studied the nitric oxide biosensor based on the electrochemical properties of cyt *c* deposited on a modified electrode. Heat-denatured cyt *c* deposited on a 4-mercaptopyridine-modified gold electrode responded to NO with an increase of cathodic current through electrochemical reduction of cyt *c* (Fe^{3+}), when the electrode potential was controlled at 0 mV vs Ag/AgCl. The cathodic current response is linearly correlated with NO concentration in the range from 0.5 to 4.0 μM in aqueous solution. The NO sensing, which used heat-denatured cyt *c*, was performed at a lower potential of 0 mV, where no serious interference might be associated in living systems. A notable NO sensing system has been accomplished by implementing heat-denatured cyt *c* on a mercaptopyridine-modified gold electrode. Besides these normal small molecules, many other molecules like phenol, chromate, and histamine can also be detected by cyt *c*-based biosensors.

17.3.1.2 Biosensors based on direct electron transfer of proteins cytochrome p450 [218]

Cytochrome P450 (CYP) is a large family of enzymes containing heme as the active site. It forms a large family of heme enzymes that catalyze a diversity of chemical reactions such as epoxidation, hydroxylation, and heteroatom oxidation. The enzymes are involved in the metabolism of many drugs and xenobiotics. They are also responsible for bioactivation. Many of these compounds are even inducers for CYP expression in different organs [219]. The catalytic abilities of the CYP family attracted the interest of enzyme engineers in the 1970s [220]. However, these studies had to face complications due to the limited stability of the labile multi-enzyme system and the need of the regeneration of the cofactor NADPH or NADH. Many biosensers are based on the direct electrochemistry of CYP [221].

The electrochemistry of CYP has been investigated using a variety of metal electrodes such as Au, Pt, and Tin oxide, as well as non-metal electrodes such as glassy carbon, pyrolytic graphite, edge-plane graphite, and carbon cloth. Although direct electron transfer has been observed on bare solid electrodes, modifying the electrode with an appropriate medium like a polymer or a polyelectrolyte has been very popular in recent years. These modifying materials can attain native structure and appropriate orientation of enzyme, which increases electron transfer between the enzyme and the electrode. In the first biosensor based on the direct electron transfer of CYP [221], solubilized CYP from rabbit liver showed a polarographic reduction step at a mercury electrode of −580 mV and was partially reduced in constrast to the microsomal CYP which was not detectable. Besides the mercury electrode, the solid electrodes are widely used to study the electrochemical behavior of CYP. But the rather low electron transfer between the protein and the electrode at bare solid electrodes limits their use for the construction of efficient CYP biosensors. Thus, electrodes modified with compounds that can facilitate electron transfer, prevent denaturation of protein, and cause appropriate orientation of the protein have been widely used.

A biosensor based on mediator-free CYP2B4 catalysis by immobilizing monomerized CYP2B4 in montmorillonite was studied by Shumyantseva [222]. When substrates were added to air saturated buffer solution, there was an increase in the reduction current. A typical concentration dependence measured in chronamperometry is shown for aminopyrine and benzphetamine (Fig. 17.4). The reaction was inhibited by metyrapone. This indicates the catalytic activity of CYP2B4 in the presence of substrate.

The majority of CYP enzymes are located in a hydrophobic environment in the endoplasmic reticulum of cells, although cytosolic enzymes also exist, such as CYP101. In order to mimic the physiological environment of CYP enzymes, a number of groups have used phospholipids to construct biosensors such as DDAB, dimeristoyl-L-α-phosphatidylcholine (DMPC), dilauroylphosphatidylethanolamine (DLPE) and distearoylphosphatidylethanolamine (DSPE). Phospholipid layers form stable vesicular dispersions that bear structural relationship with the phospholipid components of biologically important membranes. By this way a membranous environment is created that facilitates electron transfer between the enzyme's redox center and the electrode.

A CYP101 biosensor was created for monitoring drug conversion by this approach [223]. The biosensor comprised a GC electrode modified with CYP101 contained in DDAB vesicle dispersion. CVs of the CYP electrode in air-free buffer showed direct electron exchange between the heme group of CYP101 and the electrode. The E^0 was found to be -260 ± 10 mV at a scan rate of 500 mV s^{-1} and the peak separation was 36 mV. When 3 mM of ethanolic solution of camphor was added to the measuring buffer, a catalytic response was observed. The cathodic peak potential and peak

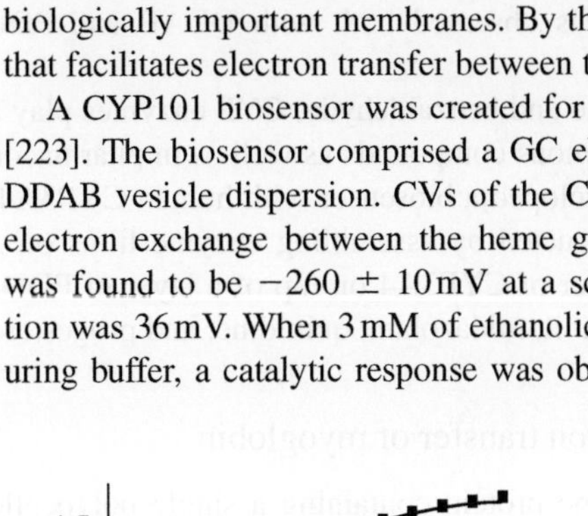

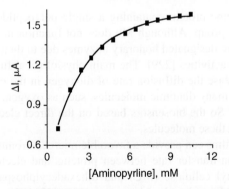

FIGURE 17.4 Relationship of catalytic current, obtained with CYP2B4 biosensor after addition of aminopyrine, with increasing concentrations of aminopyrine. (From [222], with permission.)

current were −375 mV and 44 μA in the absence of camphor and −350 mV and 50 μA in presence of camphor, respectively. Under anaerobic conditions addition of 3 mM ethanolic solution of camphor produced a catalytic response with a cathodic peak potential of −430 mV and a cathodic peak current of 14 μA. This indicated a typical fast reversible electrochemistry of heme $Fe^{3+/2+}$ redox species coupled to a subsequent process, such as hydroxylation of camphor or H_2O_2 production by uncoupled turnover. The authors therefore claimed that even in degassed buffer the oxygen in the aerobic ethanolic camphor solution was sufficient for the reaction to take place.

Cytochrome CYP101 immobilized with polyion can successfully catalyze the oxidation of styrene [224–225]. GC–MS product analysis (styrene oxide) showed a turnover number of $9.3\,h^{-1}$, which is larger than the turnover number when CYP101 was in solution, which is $0.35\,h^{-1}$. Catalysis of styrene as well as benzaldehyde oxidation has also been observed by protein polyion/CYP101-modified carbon cloth (CC) electrodes [226]. The formal potential was the same as that reported with CYP101 multilayer-modified Au electrode [224], but the turnover number for the catalysis of styrene was slightly lower at $7.2\,h^{-1}$. Conversion of styrene-to-styrene oxide has also been studied with a biosensor based on CYP1A2 [227]. Cytochrome CYP1A2 is the main enzyme that metabolizes caffeine but it is also relatively active in converting styrene, although not as active as CYP2E1 and CYP2B6. In addition to CYP101 and CYP1A2, CYP2E1 has also been studied with phospholipid-modified GC and Au electrodes. Reversible electrochemical response was observed both with DDAB and PDDA-modified GC electrodes.

Besides catalyzing styrene and benzaldehyde, CYP enzymes play an important role in the metabolism of endogenous compounds as well as in pharmacokinetics and toxicokinetics. Joseph [228] developed a biosensor with human CYP3A4 as a novel drug-screening tool. It was constructed by assembling enzyme films on Au electrodes by alternate adsorption of a layer of CYP3A4 on top of a layer of PDDA. The biosensor was applied to detect verapamil, midazolam, quinidine, and progesterone.

17.3.1.3 Biosensors based on direct electron transfer of myoglobin

Myoglobin is a kind of heme protein containing a single polypeptide chain with an iron heme as its prosthetic group. Although Mb does not function in biological electron-transfer chains, they are designated honorary enzymes due to their peroxidase and cytochrome P450 catalytic activities [229]. The main physiological function of Mb is to store dioxygen and increase the diffusion rate of dioxygen in the cell. The ferrous heme iron in Mb can bind many diatomic molecules, such as oxygen, carbon monoxide, NO, H_2O_2, and so on. So the biosensors based on the direct electron transfer of Mb are often used to detect these molecules.

The biomembrane-like films can provide a favorable microenvironment for proteins and enhance direct electron-transfer rate between proteins and electrodes, so many biopolymers such as methyl cellulose (MC) and dihexadecylphospate (DHP) have been used to immobilize Mb and make biosensors [230–232]. The electrochemical catalytic reduction of oxygen by the Mb–MC/EPG was examined by CVs. Mb–MC presents a pair of redox peaks within the potential range from 0.2 to −0.8 V in deoxygenated PBS (pH 7.0). When a volume of air was injected into the solution, a significant increase in the reduction peak at −0.292 V, accompanied by the disappearance of the oxidation peak of the $MbFe^{III}/Fe^{II}$ redox couple, was observed. The increase of the reduction peak current and the shift of peak potential with an increase in the amount of oxygen in solution indicated that Mb entrapped in MC film had reacted with oxygen. Oxygen can bind reversibly to heme in Mb, forming $MbFe^{II}—O_2$ which can then undergo electrochemical reduction at the potential of $MbFe^{III}$ reduction, producing $MbFe^{II}$ again. The mechanisms can be expressed as the following:

$$MbFe^{II} + O_2 \rightarrow MbFe^{II} - O_2 \quad (13)$$

$$MbFe^{II}\,O_2 + 2e^- + 2H^+ \rightarrow MbFe^{II} + H_2O_2 \quad (14)$$

The Mb–MC/EPG can also determine H_2O_2 by CVs. When the Mb–MC/EPG was placed in pH 7.0 PBS containing H_2O_2, a reduction peak at about −0.30 V was increased and the $MbFe^{II}$ oxidation peak was decreased at the same time. However, direct reduction of H_2O_2 was not observed at EPG or EPG coated MC in the potential range investigated. These results are characteristic of catalytic reduction of H_2O_2 by Mb entrapped in MC film, because Mb has close structural similarity to peroxidase with an intrinsic catalytic activity toward peroxide compounds. The peak reduction current of Mb–MC/EPG was linearly proportional to H_2O_2 concentrations within a range of 16.8–120.7 μM.

Besides MC, agarose is another kind of biopolymer material often used to immobilize proteins. Electrochemical catalytic reduction of H_2O_2 using Mb–agarose/EPG was studied by voltammetry. The reduction peak at about −0.30 was increased and the $MbFe^{II}$ oxidation peak was decreased with the addition of H_2O_2 in pH 7.0 PBS for Mb–agarose/EPG. However, direct reduction of H_2O_2 could not be obtained at EPG or EPG coated with agarose hydrogel in the potential range scanned, indicating that the reduction of H_2O_2 was catalyzed by Mb entrapped in agarose film. The reduction peak current increased with increasing concentration of H_2O_2. The reduction peak currents were linearly proportional to H_2O_2 concentration for Mb–agarose/EPG in the range of 4.2–53.6 μM. The reduction peak potential shifted negatively with increasing pH value, suggesting that the reduction of H_2O_2 is a proton-coupled reaction. The peak potential depended linearly on pH in the range of 3.0–10.0 with the slope value of 28 mV pH^{-1}, which indicated a two-electron redox process involving one proton in the rate-controlled step. The proposed mechanism for the reduction of H_2O_2 by Mb can be described as follows:

$$\text{Ferric Mb} + H_2O_2 \rightarrow \text{Compound I} + H_2O \quad (15)$$

$$\text{Compound I} + H^+ + e^- \rightarrow \text{Compound II} \quad (16)$$

$$\text{Compound II} + H^+ + e^- \rightarrow \text{Ferric Mb} \quad (17)$$

NO is found *in vivo* and regulates various physiological functions including blood pressure, platelet aggregation, and neurotransmission. NO binding to heme proteins such as Mb and Hb has been linked to a number of important physiological processes. NO is one of the products of the disproportionation reaction of NO_2^- when it occurs in acidic solution.

Cyclic voltammograms of an Mb–agarose/EPG in pH 7.0 PBS show a pair of reversible redox peaks at approximately −0.30 V. When NO was added to the solution, an irreversible cathodic peak at −0.84 V was observed over a potential range of +0.2 to −1.2 V. The peak could be assigned to reduction of the NO–myoglobin adduct ($MbFe^{II}$—NO) [233]. At the same time, the reduction peak of $MbFe^{III}$ widened, and two cathodic peaks were observed at lower scan rates, suggesting NO coordinated with $MbFe^{II}$. However, the oxidation peak of $MbFe^{II}$ kept its original shape, indicating that NO interacts with $MbFe^{II}$ more strongly than $MbFe^{III}$. Both the peak potential and current were dependent on pH. The peak potential shifted negatively with increasing pH, suggesting a proton-coupled reaction. The peak potential was linearly dependent on pH in the range of 3.0–10.0 with the slope value of 29 mV pH^{-1}, indicating that one proton transferred in the redox reaction. In addition, large peak currents were obtained in neutral or weakly acidic solution.

Biocompatible nanosized polyamidoamine (PAMAM) dendrimer films provided a suitable microenvironment for heme proteins to transfer electron directly with underlying pyrolytic graphite electrodes. The Mb–PAMAM film can catalytically reduced oxygen, hydrogen peroxide, and nitrite, indicating that the potential applicability of the film can be used to fabricate a new type of biosensor or bioreactor based on the direct electron transfer of Mb [234].

Mb–CMC film was made by casting solution of myoglobin and carboxymethyl cellulose (CMC) on pyrolytic graphite electrode. Trichloroacetic acid (TCA), nitrite, oxygen, and hydrogen peroxide can be catalytically reduced at the Mb–CMC film electrode. When TCA was added to a pH 3.0 buffer, an increase in the $MbFe^{III}$ reduction peak at about −0.2 V was observed, accompanied by a decrease of the $MbFe^{II}$ oxidation peak. The reduction peak current increased as the TCA concentration increased. The catalytic reduction peak current showed a linear relationship with TCA concentration in the range of $7.5 \times 10^{-4} - 2 \times 10^{-3}$ M. Compared with the direct reduction of TCA on Mb-free CMC film at the potential more negative than −1.1 V, Mb–CMC films lowered the reduction overpotential of TCA by at least 0.9 V. The $MbFe^{III}$ in CMC films was electrochemically reduced at electrodes forming $MbFe^{II}$. The electrode reaction product $MbFe^{II}$ was then chemically oxidized by TCA and returned to $MbFe^{III}$. This formed a catalytic cycle, which presumably resulted in the reductive dechlorination of TCA.

Nitrite can also be catalytically reduced by the Mb–CMC film electrode. When an Mb–CMC film electrode was placed in a pH 5.5 buffer containing NO_2^-, a new reduction peak appeared at about −0.8 V while the original Mb Fe^{III}/Fe^{II} peak pair at about −0.25 V was intact. This new peak increased with the concentration of NO_2^- and the catalytic reduction peak of nitrite increased linearly with nitrite concentration in the range of 0.6–8 mM with a detection limit of 0.32 mM. Oxygen and hydrogen peroxide are also often detected by biosensors based on the direct electrochemistry of Mb in the Mb–CMC film electrode.

Recently, many inorganic porous materials such as clay [62], montmorillonite [63–64], porous alumina [65], sol-gel matrix [66], etc. have been shown to be promising as immobilization matrices. They have many significant advantages of high mechanical, thermal, and chemical stability and good adsorption and penetrability due to their regular structures and appreciable surface areas.

Hexagonal mesoporous silica (HMS) is a mesoporous silica material which processes a porous size of nanoscale dimension to make it more suitable for enzyme intercalation and loading. The Mb–HMS/GCE electrode exhibits a good electrocatalytic behavior with respect to the pseudo-peroxidase activity to H_2O_2 reduction and a fast amperometric response to $NaNO_2$ and the prepared sensors could be used for the determinations of H_2O_2 and NO_2^- [235]. From the amperometric response of Mb–HMS/GCE with successive additions of H_2O_2 to 0.1 M, pH 7.0, PBS at an applied potential of −400 mV, we can see that the reduction current increased steeply to reach a stable value upon addition of an aliquot of H_2O_2 to the buffer solution. Under optimal conditions the linear response range of the sensor to H_2O_2 concentration was from 4.0 to 124 μM and the detection limit was estimated to be 6.2×10^{-8} M at a signal-to-noise ratio of 3. The K_m^{app} value for the electrocatalytic activity of Mb–HMS/GCE to H_2O_2 was determined to be 0.065 ± 0.005 mM, indicating that the Mb entrapped in the HMS matrix was of a high affinity to H_2O_2. The cyclic voltammograms of Mb–HMS/GCE in 0.1 M, pH 6.0, PBS upon addition of $NaNO_2$ showed that there was a reduction peak of NO at about −0.96 V. NO is produced from the nitrite disproportionation reaction:

$$NO_2^- + 2H^+ + e^- \rightarrow NO + H_2O \qquad (18)$$

Thus −1.0 V was applied to amperometric detection of NO_2^-. The linear response range of the sensor to NO_2^- concentration is from 8.0 to 216 μM and the detection limit of NO_2^- concentration was 8.0×10^{-7} M at a signal-to-noise ratio of 3. The $K_m{}^{app}$ value of the Mb–HMS/GCE to $NaNO_2$ was estimated to be 0.72 mM, indicating a high affinity of the Mb–HMS/GCE to NO_2^-.

Clays are usually cation-exchangeable aluminosilicates, and exfoliated clay particles have a platelet shape with nanoscopic size. Cast protein–clay films on electrodes have been used to immobilize proteins. The $\{Clay/Mb\}_n$ electrode has good electrocatalytic properties for the reduction of oxygen and hydrogen peroxide [236] and the biosensors can also be made based on these properties.

17.3.1.4 Biosensors based on direct electron transfer of hemoglobin

Hemoglobin (Hb) is a heme protein that can store and transport oxygen in blood in vertebrates. It has a molar mass of approximately 67 000 g mol^{-1} and comprises four polypeptide subunits (two α and two β subunits). The structure of each subunit is similar to that of Mb. The materials that are used to immobilize Mb for biosensors are also suitable for Hb. The substrates that can be catalyzed by Mb also can be catalyzed by Hb. Kieselgubr is a kind of inorganic material with a porous structure, which is promising as an immobilization matrix because of its good mechanical, thermal, and chemical stability. Moreover, kieselgubr is of special interest because of its effectiveness in the preservation of enzyme activity, and its fine catalytic ability. The Hb–kieselgubr–PG electrode can catalytically reduce H_2O_2 and make an H_2O_2 biosensor [237]. With the addition of H_2O_2 to the electrochemical cell with an Hb–kieselgubr-modified PG working electrode, an obivious increase in the cathodic peak and a decrease in the anodic peak are observed, which is characteristic of an electrochemically catalytic reaction. The catalytic peak is located at about −250 mV. On the other hand, no electrochemical signal corresponding to H_2O_2 can be observed in the potential range at a bare PG or an electrode modified with kieselgubr alone. Therefore, the catalytic peak should come from the interaction between Hb and H_2O_2. This result shows that Hb incorporated in the kieselgubr film can act as an effective catalyst to the reduction of H_2O_2. Meanwhile, the height of the reduction peak current apparently increases with the concentration of H_2O_2 and reaches a plateau at a higher concentration of 500 μM in a stirring solution, suggesting a typical electrocatalysis process that coincides with a Michaelis-Menten kinetics model. The decrease of the anodic peak current with the addition of H_2O_2 implies that the oxidized form of Hb can be quickly reduced to its reduced form by acquiring an electron from the PG electrode. There is a linear relation between the catalytic peak current and H_2O_2 concentration in the range of 5.0×10^{-6} to 3.0×10^{-4} mol L^{-1}, while it reaches a plateau at a concentration of 6.0×10^{-4} mol L^{-1} and the detection limit is estimated to be 2.1×10^{-6} mol L^{-1} (from three times S.D.). The apparent Michaelis-Menten constant for the Hb–kieselgubr-modified H_2O_2 sensor is found to be 975 μM.

The porous inorganic sol-gel matrix is particularly attractive for the development of electrochemical biosensors because the matrix possesses physical rigidity, chemical inertness, high photochemical, biodegradational, and thermal stabilities, and experiences negligible swelling in aqueous solutions. When hemoglobin was immobilized by a thin silica sol-gel film which derived from tetraethylorthosilicate (TEOS) on a carbon paste electrode (CPE), the modified electrode can electrocatalytically reduce O_2, NO_2^-, and H_2O_2 [197]. Electrochemical catalytic reduction of oxygen by Hb is observed through CV. When a volume of air is passed through a pH 7.0 phosphate buffer, a significant increase in reduction peak at about −0.3 V is observed for the Hb/sol-gel film-modified CPE, as compared to the reduction peak of Hb Fe(III) of the Hb/sol-gel film-modified CPE without oxygen. This increase in the reduction peak is accompanied by the decrease of the oxidation peak of HbFe(II), which suggests that HbFe(II) partially reacted with oxygen. For sol-gel-modified CPE without any Hb, the reduction peak of oxygen is observed at about 0.83 V, far more negative than the catalytic peak potential. Electrochemical catalytic reduction of NO_2^- was also tested with Hb immobilized on CPE by sol-gel film. The addition of $NaNO_2$ in a pH 5.5 buffer results in a new reduction peak at about −0.6 V. Further addition of $NaNO_2$ caused an increase of the peak. It has been demonstrated that this new reduction peak does not come from NO_2^- but from the $[\text{Hb Fe(II) (NO)}^+]^+$ nitrosyl adduct [101]. The Hb/sol-gel film-modified CPE was also used to reduce H_2O_2. When H_2O_2 was added to a pH 5.5 buffer solution, an increase in the reduction peak at about −0.3 V was observed. The reduction peak current increases with the concentration of H_2O_2 in the solution. The concentration between 5×10^{-6} and 7×10^{-4} M, the electrocatalytic reduction peak current and the H_2O_2 concentration behave in a linear relationship. The linear regression equation is I (μA) = 6.60[H_2O_2] (μM) − 2.29 with a correlation coefficient of 0.9972. When the concentration of H_2O_2 is higher than 7.0×10^{-4} M, a response plateau is observed, showing the characteristics of the Michaelis-Menten kinetic mechanism. The $K_m{}^{app}$ value was found to be 8.98×10^{-4} M for the Hb/sol-gel film-modified CPE.

Direct electron transfer of haemoglobin can also be achieved in Hb/montmorillonite (MMT)/polyvinyl alcohol multi-assembly at a pyrolytic graphite (PG) electrode. Accordingly, a novel nitric oxide (NO) biosensor is proposed [238]. A pair of well-defined peaks can be observed for the CV response of Hb in an MMT/PVA multi-assembly at a PG electrode surface at pH 5.5 (anodic peak at −0.382 V and cathodic peak at −0.333 V). In the presence of NO, a new peak appears at the potential of −783 mV, in the negative direction of the redox pair of Hb. Notably, no corresponding signal is visible at a bare electrode or an electrode modified with MMT alone under the same condition. Thus, this peak can be attributed to the reduction of NO catalyzed by the heme of Hb. Moreover, the peak current apparently increases along with NO concentrations which will be employed in the following NO measurements. The peak current of NO reduction is observed to be linearly proportional to the NO concentration in the range from 1.0×10^{-6} to 2.5×10^{-4} M. The detection limit is estimated to be 5.0×10^{-7} M, defined from a signal-to-noise ratio of 3. The K_m^{app} is estimated to be 81.4 μM.

Gold nanoparticles are the most intensively studied and applied metal nanoparticles in electrochemistry owing to their stable physical and chemical properties, useful catalytic activities, and small dimensional size. A novel Hb-based H_2O_2 sensor has been constructed on a gold nanoparticles-modified ITO [239]. Although the direct electrochemical response of Hb can be observed on some electrodes, no characteristic peak for Hb appears in the cyclic voltammogram recorded with the Hb/Au/ITO electrode in a PBS solution at pH 7.0. While 2 mM H_2O_2 was present in the solution, noticeable cathodic currents could be observed starting at −0.10 V, indicating the reduction of H_2O_2 on this electrode. Gold nanoparticles can improve electron transfer kinetics of Hb and enhance the reduction current for H_2O_2 because of its larger conductive area, good biocompatibility, and useful catalytic activities. The catalytic current is linearly proportional to H_2O_2 concentration from 1×10^{-5} to 7×10^{-3} M with a correlation coefficient of 0.9995. The linear regression equation is expressed as I(A) = 0.2977 [H_2O_2] (mM) + 0.0642. The detection limit (S/N = 3) on the Hb/Au/ITO electrode is estimated to be 4.5×10^{-6} M. Besides gold nanoparticles, Hb can also be absorbed by a compound composited with gold nanoparticles and other materials. Haemoglobin has been adsorbed onto a chitosan-stabilized gold nanoparticles (Chit-Aus)-modified Au electrode via a molecule bridge like cysteine. The resultant electrode displayed an excellent electrocatalytic response to the reduction of H_2O_2, long-term stability, and good reproducibility [240]. The Hb/Chit–Aus/Cys/Au electrode showed increasing amperometric responses to H_2O_2 with a linear range from 13 to 0.74 mM. The $K_m{}^{app}$ value for Hb activity of the Hb/Chit-Aus/Cys/Au electrode to H_2O_2 was determined to be about 1.4 mM. Besides these, metal oxide nanoparticles are also usually used to enhance the direct electron-transfer rate of Hb. Samili [241] used a nickel oxide nanoparticle to immobilize Hb on a carbon electrode and used the modified electrode to catalyze H_2O_2 and O_2. Hb in the modified electrode showed catalytic activity for the reduction of H_2O_2 with a K_m^{app} 1.37 mM.

Some non-ionic surfactant, such as Triton X-100, is usually chosen to preserve the functional state of proteins and to accelerate electron transfer in the supramolecular complex. A third-generation biosensor for H_2O_2 can be constructed based on the direct electrochemistry of Hb by incorporating Hb with Triton-100 at the PG electrode surface [242]. When an aliquot of H_2O_2 was added into a buffer solution of 0.1 mol L^{-1} NaAc–HAc (pH 6.0), the current of cathodic peak increased gradually. The catalytic peak currents were proportional to the concentration of H_2O_2 in the range from 1.0×10^{-6} to 1.0×10^{-4} M. The linear regression equation was I (A) = 38.59 [H_2O_2] (mM) + 1.12, $r = 0.999$. The detection limit is estimated to be 3.0×10^{-7} mol L^{-1} when the signal-to-noise ratio is 3. The K_m^{app} was 4.27 mM for this biosensor.

Some biomembrane-like films can provide a favorable microenvironment for proteins and enhance direct electron transfer between proteins and electrodes. Poly-3-hydroxybutyrate (PHB), a linear polymer of betahydroxylate, is produced within bacterial cytoplasm as an energy reserve by a range of prokaryotic cells. Due to its good property of biodegradability PHB has been widely used as degradable plastics, and has an extensive application in medicine, membrane technology, and other biotechnologies. Thus, PBS might be a suitable material to incorporate proteins. The Hb–PHB/PG electrode can show direct, reversible electrochemistry for heme FeIII/FeII redox couples. The electrochemical catalytic reductions of hydrogen peroxide (H_2O_2), nitric oxide (NO), and trichloroacetic acid (TCA) have been observed, showing the potential applicability of the films as biosensor [243]. A linear dependence between the catalytic current and the concentration of H_2O_2 is obtained in the range 6.0×10^{-7} to 8.0×10^{-4} M for Hb–PHB/PG electrode. The linear regression equation is $y = 5.95528 + 0.03541x$, with a correlation coefficient of 0.999. Its detection limit is 2.0×10^{-7} M with sensitivity of 0.03541 μA μM^{-1} H_2O_2. The K_m^{app} of Hb–PHB film is calculated to be 1076 μM for H_2O_2. Electrocatalytic reduction of TCA can also be tested by the Hb in PHB films. When TCA is added to a pH 5.0 buffer, the HbFe(III) reduction peak of Hb–PHB film electrodes at about −0.28 V increased in height, accompanied by a decrease of HbFe(II) oxidation peak. The reduction peak current is linearly proportional to the concentration of TCA. When the concentration of TCA is larger than 0.04 M, a new reduction peak located at −0.45 V (vs SCE) is observed, and the peak current increased with the concentration of TCA. Lu [244] also developed a hydrogen peroxide by immobilizing Hb to a water-soluble polymer, poly-α, βN-(2-hydroxyethyl)-l-aspartamide] (PHEA) film.

Direct electrochemistry of hemoglobin was observed in stable thin film composed of a natural lipid (egg–phosphatidylcholine) and hemoglobin on a PG electrode. Hemoglobin in lipid films shows thin layer electrochemistry behavior. Hemoglobin in the lipid film exhibited elegant catalytic activity for electrochemical reduction of H_2O_2, from which a mediator-free biosensor for H_2O_2 could be developed [245]. There was a linear relation of the current with concentration of H_2O_2 between 10 and 100 μM for the biosensor. The detection limit is 3.9 μM with the signal-to-noise ratio of 3.

17.3.2 Biosensors based on direct electron transfer of enzymes

17.3.2.1 Biosensors based on direct electron transfer of horseradish peroxidase

Horseradish peroxidase (HRP) is a member of the large class of peroxidases, which are enzymes defined as oxidoreductases using hydroperoxide as electron acceptor. HRP has been widely used for the construction of amperometric biosensor for the determination of H_2O_2 and small organic and inorganic substrates.

Poly(ester sulfonic acid), with the trade name of Eastman AQ, is a kind of anionic ionomer. Enzymes can be added directly to aqueous dispersions of AQ ionomers and deposited onto solid surfaces to form films without denaturation or significant loss of activity. The embedded HRP in AQ films retained the electrocatalytic activity for oxygen, nitrite, and hydrogen peroxide [246]. A certain volume of air can increase the reduction peak of HRP–AQ films in CVs. This increase in reduction peak was accompanied by the disappearance of the oxidation peak for HRP–Fe (II) because HRP–Fe (II) reacted with oxygen. An increase in the amount of oxygen in solution increased the reduction peak current. NO_2^- can also be reduced by the HRP–AQ/PG electrode. When NO_2^- was added into a pH 6.0 buffer, a catalytic peak was observed at about -0.84 V. The peak current increased with the concentration of NO_2^-. Electrochemical catalytic response of hydrogen peroxide was also observed by HRP–AQ films. When H_2O_2 was added to a pH 7.0 buffer, an increase in reduction peak at about -0.36 V was seen with the disappearance of oxidation peak for HRP–Fe(II). The reduction peak current increased with the concentration of H_2O_2 in solution. However, there was no reduction current observed at AQ film electrodes in the presence of H_2O_2.

Biomembranes in living organisms are composed of lipid, protein, and carbohydrate. The biomembranes can be prepared by casting aqueous vesicle of lipid onto the surface of electrodes. The lipids were arranged in multiple bilayers by self-assembling and retain their bilayer properties similar to biomembrane after drying. When we immobilize HRP in the DDAB films by casting the mixture of the aqueous vesicle of DDAB and HRP on GC electrodes, the enzyme electrode showed catalytic activity toward H_2O_2 [204]. The electrocatalytic reduction of H_2O_2 by HRP immobilized in the DDAB films was tested by CVs. When H_2O_2 was added to the buffer solution, the cathodic peak increases dramatically and the anodic peak almost disappears for the HRP–DDAB/GC electrode. These results indicate that HRP incorporated into the DDAB films can catalyze the reduction of H_2O_2 efficiently and still retains its biological activity. The amperometric response of the enzyme electrode resulted from increasing concentrations of H_2O_2 was investigated in the stirred buffer solution (pH 5.5). When an aliquot of H_2O_2 is added into the buffer solution, the reductive current rises steeply to reach a stable value. There is a linear relationship between the current response and H_2O_2 concentration in the range from 1×10^{-3} to 4×10^{-3} M. The regression equation is $y = 0.04007 + 0.05186x$, with a correlation coefficient of 0.998.

Biopolymers such as agarose hydrogel can also be used to immobilize HRP and to reduce oxygen, hydrogen peroxide and nitric oxide [231]. For an HRP–agarose/EPG in a pH 7.0 PBS, a reduction peak at -0.30 V was observed after addition of H_2O_2.

At the same time the HRP–Fe(II) oxidation peak disappeared. However, at the EPG or EPG coated with agarose hydrogen in the potential range scanned, the direct reduction of H_2O_2 could not be observed. This indicated that the reduction of H_2O_2 was catalyzed by HRP entrapped in agarose film. The reduction peak current increased with increasing concentration of H_2O_2. The calibration curve gradually tended to a plateau and then dropped down with adding H_2O_2, implying a progressive enzyme inactivation in the presence of higher concentration of H_2O_2.

Biogenic amines, including catecholamines such as dopamine, norepinephrine, epinephrine, and indoleamines such as serotonin, are neurotransmitters that have special roles in neuroscience. An amperometric biosensor using a carbon paste electrode modified with horseradish peroxidase (HRP) enzyme can be used for total biogenic amine determinations [247]. The HRP immobilization on graphite was made using bovine serum albumin, carbodiimide, and glutaraldehyde. The biosensor response was optimized using serotonin and it presented the best performance in 0.1 M phosphate buffer (pH × 7.0) containing 10 μM of hydrogen peroxide. Under optimized operational conditions at −50 mV, a linear response range from 40 to 470 ng ml^{-1} was obtained. The detection limit was 17 ng ml^{-1} and the response time was 0.5 s. The proposed sensor presented a stable response during 4 h under continuous monitoring. The difference of the response between six sensor preparations was <2%. The sensor was applied in the determination of total biogenic amines (neurotransmitters) in rat blood samples with success, obtaining a recovery average of 102%.

17.3.2.2 Biosensors based on direct electron transfer of catalase

Catalase is a heme protein belonging to the class of oxidoreductases with ferriprotoporphyrin-IX at the redox center, and it catalyzes the disproportionation of hydrogen peroxide into oxygen and water without the formation of free radicals.

Catalase was immobilized with gelatin by means of glutaraldehyde and fixed on a pretreated Teflon membrane served as enzyme electrode to determine hydrogen peroxide [248]. The electrode response reached a maximum when 50 mM phosphate buffer was used at pH 7.0 and at 35°C. Catalase enzyme electrode response depends linearly on hydrogen peroxide concentration between 1.0×10^{-5} and 3.0×10^{-3} M with response time 30 s.

Catalase, which was adsorbed on the surface of multi-wall carbon nanotubes (MWNTs), could show the direct voltammetry and electrocatalytic properties [249]. The catalase MWNTs-modified GC electrode has excellent and strong mediation properties and facilitates the low potential amperometric measurement of hydrogen peroxide. An obvious catalytic reduction peak appears at the potential of −0.25 V for the catalase MWNTs-modified GC electrode when H_2O_2 was added to the solution. By increasing the concentration of H_2O_2, the cathodic peak current of catalase increased while its anodic peak current decreased. The catalytic currents increased linearly with the H_2O_2 concentration in the range of 10 μM to 1 mM. The linear regression equation of catalytic currents vs hydrogen peroxide concentration can be obtained from the experimental data: $I_p(\mu A) = 6.563[H_2O_2]\ (\mu M) + 18.3$ with a correlation coefficient of 0.9889 and a detection limit of 50 μM at a signal-to-noise ratio of 3.

Collagen, an electrochemically inert protein, formed films on pyrolytic graphite (PG) electrodes, which provided a suitable microenvironment for heme proteins to transfer electrons directly with the underlying electrodes. When catalase was incorporated into the PG electrode by collagen, it can exhibit electrochemical behavior well and often was used to catalyze the reduction of nitrite, oxygen, and hydrogen peroxide [250]. For Cat–collagen films, when a certain amount of air was injected into a pH 7.0 buffer by a syringe, a significant increase in reduction peak at about −0.5 V was observed, accompanied by the decrease or even disappearance of the oxidation peak of CatFe(II), suggesting that CatFe(II) had reacted with oxygen. The reduction peak current increased with the amount of oxygen in the solution. The electrochemical catalytic reduction of hydrogen peroxide also can be observed on the protein–collagen films by CVs. When H_2O_2 was added into the pH 7.0 buffer solution, an increase in the reduction peak was observed for heme Fe(III), accompanied by the decrease or disappearance of the oxidation peak for heme Fe(II). The reduction peak current increased linearly with the concentration of H_2O_2. When NO_2^- was injected into the pH 5.5 buffer solution, a new reduction peak was observed at about −0.8 V and the reduction current increased with the concentration of NO_2^-. The detection limit of NO_2^- for the Cat–collagen/PG electrode is 3.15 mM.

PAM can absorb large amounts of water and form hydrogel, which is widely used in the field of life science. When catalase was incorporated into the PG electrode by PAM, the modified electrode showed good electrocatalytic activity toward dioxygen and hydrogen peroxide [251]. At Cat–PAM film electrodes, the position of catalytic reduction peak potential of hydrogen peroxide was almost the same as that of oxygen, indicating that the reaction mechanism between the two systems was similar. When a certain amount of oxygen was passed through a pH 7.0 buffer solution by a syringe, an increase in the reduction peak at approximately −0.5 V was observed for Cat–PAM films, accompanied by a disappearance of the oxidation peak for CatFe(II). The reduction peak current increased with the amount of oxygen in solution. Reduction of H_2O_2 was also electrochemically catalyzed by Cat–PAM films. When H_2O_2 was added to the pH 7.0 buffer solution, an increase in the reduction peak at approximately −0.5 V was observed, accompanied by a disappearance of the oxidation peak. The reduction peak current increased with the concentration of H_2O_2 in the solution. Catalytic efficiency, expressed as a ratio of the reduction peak current of Cat–PAM films in the presence (I_c) and absence (I_d) of H_2O_2, I_c/I_d, decreased with the increase in scan rate.

17.3.2.3 Biosensors based on direct electron transfer of GOD

Glucose oxidase (GOD) is a typical flavin enzyme with flavin adenine dinucleotide (FAD) as redox prosthetic group. Its biological function is to catalyze glucose to form gluconolaction, while the enzyme itself is turned from GOD(FAD) to GOD($FADH_2$). GOD was used to prepare biosensors in extensive fields. Many materials that can be used to immobilize other proteins can be suitable for GOD. GOD adsorbed on CdS nanoparticles maintained its bioactivity and structure, and could electrocatalyze the reduction of dissolved oxygen and glucose [252]. In the presence of oxygen, the reduced enzyme is oxidized very quickly at the surface of the electrode:

$$\text{GOD–FADH}_2 + O_2 \rightarrow \text{GOD–FAD} + H_2O_2 \quad (19)$$

In the O_2-saturated buffer solution, an increase in reduction peak was observed, accompanied by a decrease in the oxidation peak current, which demonstrates that GOD in the film nicely catalyzes the oxygen reduction. When glucose is added to the O_2-saturated buffer solution, the reduction peak current decreases with the increase of the glucose concentration. β-(D)-glucose is the substrate of GOD and it will react with the enzyme and decrease the concentration of the oxidized form of GOD on the electrode surface:

$$\text{Glucose} + \text{GOD(FAD)} \rightarrow \text{gluconolactone} + \text{GOD(FADH}_2\text{)} \quad (20)$$

The decrease of the reduction current has been employed to determine the glucose concentration and probe the enzyme activity of GOD. The peak current responds linearly to the glucose concentration ranging from 0.5 to 11.1 mM ($r = 0.998$), with a sensitivity of 7.0 μA mM^{-1} and the estimated detection limit of 0.05 mM at a signal-to-noise ratio of 3. The $K_m{}^{app}$ is equal to 5.1 mM.

Liu [253] studied the direct electron transfer of glucose oxidase and glucose biosensor based on carbon nanotubes–chitosan matrix. The characteristics of the biosensor are investigated by chronoamperometric measurement under the optimum conditions. From the current–time responses, we can know the linear range of the substrate. For glucose, the calibration curve with linear range spans the concentration of glucose from 0 to 7.8 mM (Fig. 17.5). And it deviates from linearity at higher concentration representing a typical characteristic of Michaelis-Menten kinetics. $K_m{}^{app}$ is evaluated to be 8.2 mM.

GOD adsorbed on a colloidal gold-modified carbon paste electrode can transfer electrons directly between the protein and the electrode [254]. The immobilized GOD could electrocatalyze the reduction of dissolved oxygen and resulted in a great increase in the reduction peak current. Upon addition of β-D(+)-glucose to air-saturated PBS, the reduction current response of GOD/Au/CPE decreased. The decrease increased with the increase of β-D(+)-glucose concentration and the decrease has a linear relationship with the concentration of β-D(+)-glucose from 0.04 to 0.28 mM with a correlation coefficient of 0.997. The detection limit is 0.01 mM at a signal-to-noise ratio of 3. The sensitivity of GOD/Au/CPE to β-D(+)-glucose was found to be 8.4 μA mM^{-1}.

Wang [255] developed a novel biosensor which comprised an electrode of gold/multi-walled carbon nanotubes–glucose oxidase (Au/MWNTs–GOD) to detect glucose. The biosensor exhibits excellent bioelectrocatalytic oxidation activity for glucose at a static applied potential (+0.45 V vs Ag/AgCl). Addition of glucose to the buffer solution results in a remarkable increase in the oxidation current, whereas the time required to reach the 95% steady-state response is within 16 s. The current increased linearly with the concentration of glucose in the range from 0.05 to 13 mM. The biosensor also has a very low detection limit of less than 0.01 mM for glucose. Wu [256] used colloid Au–dihexadecylphosphate composite to immobilize GOD and develop a

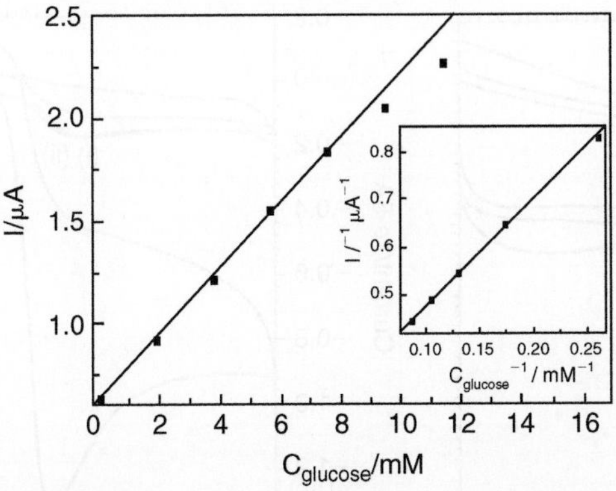

FIGURE 17.5 The calibration curve of the electrocatalytic current on the concentration of glucose. Inset is the plot of Lineweaver-Burk equation. (From [253], with permission.)

glucose biosensor. Metal oxide nanoparticles are also used to make glucose biosensor by immobilizing GOD [257]. The K_m^{app} was 7.5 mM for glucose by GOD with TiO_2 nanostructured films.

17.3.2.4 Biosensors based on direct electron transfer of other active enzymes

Besides the enzymes in common use like HRP, Cat and GOD, many other uncommon enzymes have also been studied to develop biosensors based on their direct electrochemistry. Uricase is an enzyme participating in the final step of purine degradation. Uric acid represents the major catabolite of purine breakdown in humans. For this reason it remains an important marker molecule for disorders associated with purine metabolism, most notably gout, hyperuric aemia, and the Lesch-Nyhan syndrome [258]. Zhang [259] developed a reagentless uric acid biosensor by immobilizing uricase on ZnO nanorods. This sensor showed a high thermal stability up to 85ºC and an electrocatalytic activity to the oxidation of uric acid without the presence of an electron mediator. Figure 17.6a shows the CV of the uricase/GCE (i), ZnO/GCE (ii), and uricase/ZnO/GCE (iii) in PBS (pH 6.9); Fig. 17.6b shows the CVs the uricase/GCE (i), ZnO/GCE (ii), and uricase/ZnO/GCE (iii) in PBS (pH 6.9) containing 5.0×10^{-4} mol L^{-1} uric acid. Comparing the three voltammograms, a remarkable electrocatalytic oxidation of sensor (iii) was observed. The peak current of sensor (iii) reached to 12.85 µA, after calibration by ZnO/GCE (ii) as a blank, which was 11.55 µA and about 8.37 times that (1.38 µA) obtained from sensor (i). DPV was used to study the biosensor. The biosensor has good linear relationship with the uric acid concentration from the DPV response. The calibrated response to uric acid was linear in the range of 5.0×10^{-6} to 1.0×10^{-3} mol L^{-1} ($r = 0.9983$). The detection limit was 2.0×10^{-6} mol L^{-1} at a signal-to-noise ratio of 3. When the concentration of uric acid was higher than 1.0×10^{-3} mol L^{-1}, a plateau was observed, showing a characteristic of the Michaelis-Menten kinetic mechanism. The K_m^{app} is estimated to be 0.238 mM.

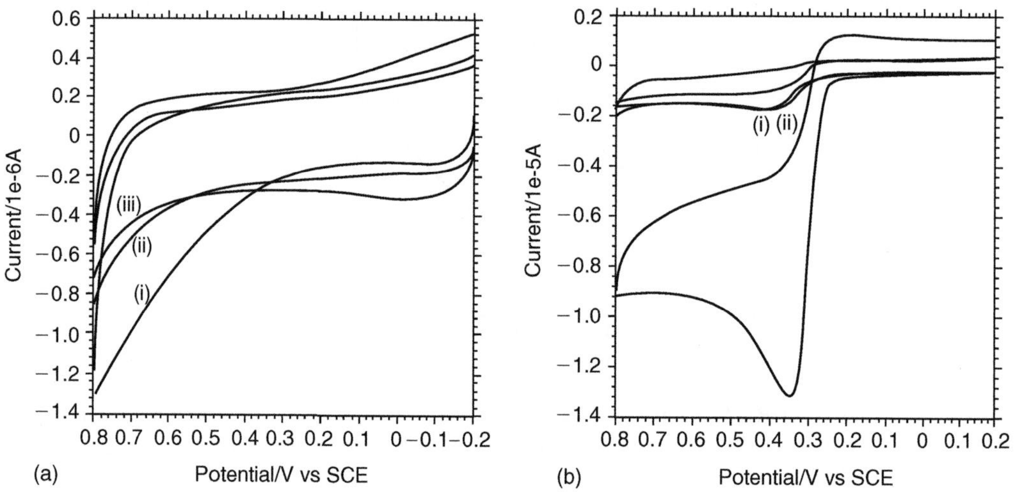

FIGURE 17.6 (a) Cyclic voltammograms of the uricase/GCE (i), ZnO/GCE (ii), and uricase/ZnO/GCE (iii) in PBS (pH 6.9), scan rate: $100\,mVs^{-1}$. (b) Cyclic voltammograms of the uricase/GCE (i), ZnO/GCE (ii), and uricase/ZnO/GCE (iii) in PBS (pH 6.9) containing 5.0×10^{-4} mol L^{-1} uric acid, scan rate: $100\,mVs^{-1}$. (From [259], with permission.)

An amperometric urate sensor based on uricase-immobilized silk fibroin membrane was developed by Zhang [256]. The biosensor can be used to measure the urate level in human serum or urine and standard additions of uric acid. F or this biosensor, the recoveries of uric acid in human serum and urine are in the range of 94.2 ± 102.6% to 92.5 ± 97.9%, respectively. The relative standard deviations for repeatedly monitoring standard urate solution, human serum, and urine are 2.37, 3.72, and 2.95%, respectively, based on 100 measurements.

XOD is one of the most complex flavoproteins and is composed of two identical and catalytically independent subunits; each subunit contains one molybdenium center, two iron sulfur centers, and flavine adenine dinucleotide. The enzyme activity is due to a complicated interaction of FAD, molybdenium, iron, and labile sulfur moieties at or near the active site [260]. It can be used to detect xanthine and hypoxanthine by immobilizing xanthine oxidase on a glassy carbon paste electrode [261]. The elements are based on the chronoamperometric monitoring of the current that occurs due to the oxidation of the hydrogen peroxide which liberates during the enzymatic reaction. The biosensor showed linear dependence in the concentration range between 5.0×10^{-7} and 4.0×10^{-5}M for xanthine and 2.0×10^{-5} and 8.0×10^{-5}M for hypoxanthine, respectively. The detection limit values were estimated as 1.0×10^{-7}M for xanthine and 5.3×10^{-6}M for hypoxanthine, respectively. Li used DNA to embed xanthine oxidase and obtained the electrochemical response of FAD and molybdenum center of xanthine oxidase [262]. Moreover, the enzyme keeps its native catalytic activity to hypoxanthine in the DNA film. So the biosensor for hypoxanthine can be based on the electron transfer of xanthine oxidase embedded on DNA. When hypoxanthine was added to the pH 5.0 phosphate buffer solution, a catalytic peak appears at −0.500 V, and the anodic peak at −0.420 V disappears by increasing hypoxanthine.

Ghamouss [263] developed a screen-printed carbon electrode modified with both HRP and LOD (SPCE–HRP/LOD) to determine l-lactate. The sensitivity of the optimized SPCE–HRP/LOD to l-lactate was 0.84 nALμM^{-1} in a detection range between 10 and 180 μM.

17.4 CONCLUSIONS

The direct electron transfer of redox proteins and enzymes is very difficult on many bare solid electrodes. Many methods and materials are used to immobilize the proteins and enzymes in order to enhance the direct electron-transfer rate and provide a suitable microenvironment for them. Proteins and enzymes prepared by these methods can retain their catalytic activity and be used to make the mediator-free biosensors. These biosensors are widely used in environmental, food, industrial, and other fields by detection of hydrogen peroxide, glucose, NO, uric acid, xanthine, hypoxanthine, and so on. Because of the specificity of proteins and enzymes these mediator-free biosensors have more selectivity than those biosensors without enzymes. The absence of mediator can also simplify the preparation of the biosensor and reduce its charge. Many scientists work in this field and achieve much progress. However, only a few of the proteins and enzymes can achieve direct electrochemistry. There is no common method to immobilize the proteins and enzymes to achieve their direct electrochemistry. At the same time, we are not very clear about the mechanism of direct electron transfer between proteins and electrodes. Future work will focus on finding efficient methods to achieve direct electron transfer of most proteins and enzymes, pointing out the mechanism of the electron-transfer process in biologic systems, and preparing excellent and steady enzyme-based mediator-free biosensors.

17.5 ACKNOWLEDGMENTS

We are grateful to all the authors and coworkers cited in the references for their efforts in carrying out the work described in this chapter. We are also grateful to the National Natural Science Foundation of China (Nos 60571042/30370397 and 60171023) and the State Key Laboratory of Transducer Technology, Chinese Academy of Sciences, for financial support.

17.6 REFERENCES

1. L.C. Clark and C. Lyons, Electrode systems for continuous monitoring in cardiovascular surgery. *Ann. N Y. Acad. Sci.* **102**, 29–45 (1962).
2. Y. Ikariyama, S. Yamauchi, T. Yukiashi, and H. Vshioda, One step fabrication of microbiosensor prepared by the codeposition of enzyme and platinum particles. *Anal. Lett.* **20**, 1791–1801 (1987).

3. M.F. Suaud-Chagny and F.G. Gonon, Immobilization of lactate dehydrogenase on a pyrolytic carbon fiber microelectrode. *Anal. Chem.* **58**, 412–415 (1986).
4. G.J. Moody, G.S. Sanghera, and J.D.R. Thomas, Modified platinum wire glucose oxidase amperometric electrode. *Analyst* **111**, 1235–1238 (1986).
5. R.M. Ianniello and A.M. Yacynych, Immobilized enzyme chemically modified electrode as an amperometric sensor. *Anal. Chem.* **53**, 2090–2095 (1981).
6. G. Fortier, R. Beliveau, E. Leblond, and D. Belanger, Development of biosensors based on immobilization of enzymes in Eastman AQ polymer coated with a layer of Nafion. *Anal. Lett.* **23**, 1607–1619 (1990).
7. R. Tor and A. Freeman, New enzyme membrane for enzyme electrodes. *Anal. Chem.* **58**, 1042–1046 (1986).
8. J.F. Rusling and A.E.F. Nassar, Enhanced electron transfer for myoglobin in surfactant films on electrodes. *J. Am. Chem. Soc.* **115**, 11 891–11 897 (1993).
9. H. Kamaya, I. Ueda, and H. Eyring, Progress in anesthesiology, in *Molecular Mechanisms of Anesthesia* (B.R. Fink, ed.), vol. 2, p. 429, Raven Press, New York (1980).
10. T. Yoshida, H. Okabayashi, K. Takahashi, and I.Ueda, A proton nuclear magnetic resonance study on the release of bound water by inhalation anesthetic in water-in-oil emulsion. *Biochim. Biophys. Acta*, **772**, 102–107 (1984).
11. M.J. Hostetler, J.E. Wingate, C.J. Zhong, J.E. Harris, R.W. Vachet, M.R. Clark, J.D. Londono, S.J. Green, J.J. Stokes, G.D. Wignall, G.L. Glish, M.D. Porter, N.D. Evans, and R.W. Murray, Alkanethiolate gold cluster molecules with core diameters from 1.5 to 5.2 nm: core and monolayer properties as a function of core size. *Langmuir*, **14**, 17–30 (1998).
12. S. Ribrioux, G. Kleymann, W. Haase, K. Heitmann, C. Ostermeier, and H. Michel, Use of nanogold- and fluorescent-labeled antibody Fv fragments in immunocytochemistry. *J. Histochem, Cytochem.* **44**, 207–213 (1996).
13. J.F. Hainfeld and F.R. Furuya, A 1.4-nm gold cluster covalently attached to antibodies improves immunolabeling. *J. Histochem, Cytochem.* **40**, 177–184 (1992).
14. C.X. Cai and J. Chen, Direct electron transfer and bioelectrocatalysis of haemoglobin at a carbon nanotube electrode. *Anal. Biochem.* **325**, 285–292 (2004).
15. X.J. Han, W.L. Cheng, Z.L. Zhang, S.J. Dong, and E.K. Wang, Direct electron transfer between haemoglobin and a glassy carbon electrode facilitated by lipid-protected gold nanoparticles. *Biochim. Biophys. Acta* **1556**, 273–277 (2002).
16. L. Wang and E.K. Wang, Direct electron transfer between cytochrome c and a gold nanoparticles modified electrode. *Electrochem. Commun.* **6**, 49–54 (2004).
17. Q. Lu, S.S. Hu, D.W. Pang, and Z.K. He, Direct electrochemistry and electrocatalysis with haemoglobin in water-soluble quantum dots film on glassy carbon electrode. *Chem. Commun.* **20**, 2584–2585 (2005).
18. Y.X. Xu, J.G. Liang, C.G. Hu, S.S. Hu, and Z.K. He, A hydrogen peroxide biosensor based on the direct electrochemistry of haemoglobin modified with quantum dots. *J. Biolog. Inorg. Chem.* Online.
19. E. Topoglidis, A.E.G. Cass, B. O'Regan and J.R. Durrant, Immobilization and bioelectrochemistry of proteins on nanoporous TiO_2 and ZnO films. *J. Electroanal. Chem.* **517**, 20–27 (2001).
20. S. Iijima, Helical microtubules of graphitic carbon. *Nature* **354**, 56–58 (1991).
21. P.M. Ajayan, Nanotubes from carbon. *Chem. Rev.* **99**, 1787–1800 (1999).
22. A. Guiseppi-Elie, C.H. Lei, and R.H. Baughman, Direct electron transfer of glucose oxidase on carbon nanotubes. *Nanotechnology* **13**, 559–564 (2002).
23. D.B. Hall, R.E. Holmlin, and J.K. Barton, Oxidative DNA damage through long-range electron transfer. *Nature* **382**, 731–735 (1996).
24. J.J. Storhoff and C.A. Mirkin, Programmed materials synthesis with DNA. *Chem. Rev.* **99**, 1849–1862 (1999).
25. H.W. Fink and C. Schonenberger, Electrical conduction through DNA molecules. *Nature* **398**, 407–410 (1999).
26. Y. Okata, T. Kobayashi, K. Tanaka, and M. Shimomura, Anisotropic electric conductivity in an aligned DNA cast film. *J. Am. Chem. Soc.* **120**, 6165–6166 (1998).

27. C. Mao, W. Sun, Z. Shen, and N.C. Seeman, A nanomechanical device based on the B-Z transition of DNA. *Nature* **397**, 144–146 (1999).
28. L.M. Adlernan, Molecular computation of solutions to combinatorial problems. *Science* **266**, 1021–1024 (1994).
29. N.C. Seeman, Nucleic acid nanostructures and topology. *Angew. Chem., Int. Ed.* **37**, 3220–3238 (1998).
30. T.W. Nilsen, J. Grayzel, and W. Prensky, Dendritic nucleic acid structures. *J. Theor. Biol.* **187**, 273–284 (1997).
31. P.J. Dandliker, R.E. Holmlin, and J.K. Barton, Oxidative thymine dimer repair in the DNA helix. *Science* **275**, 1465–1468 (1997).
32. H.-A. Wagenknecht, Reductive electron transfer and excess electron migration in DNA. *Angew. Chem., Int. Ed.* **42**, 2454–2460 (2003).
33. M.R. Arkin, E.D.A. Stemp, R.E. Holmlin, J.K. Barton, A. Hormann, E.J.C. Olson, and P.F. Barbara, Rates of DNA-mediated electron transfer between metallointercalators. *Science* **273**, 475–480 (1996).
34. A.E.F. Nassar, J.F. Rusling, and N.J. Nakashima, Electron transfer between electrodes and heme proteins in protein-DNA films. *J. Am. Chem. Soc.* **118**, 3043–3044 (1996).
35. Y. Lvov, in *Protein Architecture: Interfacing Molecular Assemblies and Immobilization Biotechnology* (Y. Lvov and H. Mohwald, eds), p. 125. Marcel Dekker, New York (2000).
36. J.F. Rusling, in *Protein Architecture: Interfacing Molecular Assemblies and Immobilization Biotechnology* (Y. Lvov and H. Mohwald, eds), p. 337. Marcel Dekker, New York (2000).
37. Y.M. Lvov, Z.Q. Lu, J.B. Schenkman, and J.F. Rusling, Direct electrochemistry of myoglobin and cytochrome P450cam in alternate polyion layer-by-layer films. *J. Am. Chem. Soc.* **120**, 4073–4080 (1998).
38. P.L. He, N.F. Hu, and G. Zhou, Assembly of electroactive layer-by-layer films of haemoglobin and polycationic poly(diallyldimethylammonium). *Biomacromolecules* **3**, 139–146 (2002).
39. Y.D. Jin, Y. Shao, and S.J. Dong, Direct electrochemistry and surface plasmon resonance characterization of alternate layer-by-layer self-assembled DNA-myoglobin thin films on chemically modified gold surfaces. *Langmuir* **19**, 4771–4777 (2003).
40. N. Kimizuka, M. Tanaka, and T. Kunitake, Spatially controlled synthesis of protein/inorganic nano-assembly: alternate molecular layers of Cyt c and TiO_2 nanoparticles. *Chem. Lett.* **12**, 1333–1334 (1999).
41. K.J. McKenzie and F. Marken, Accumulation and reactivity of the redox protein cytochrome c in mesoporous films of TiO_2 phytate. *Langmuir* **19**, 4327–4331 (2003).
42. Y. Lvov and F. Caruso, Biocolloids with ordered urease multilayer shells as enzymatic reactors. *Anal. Chem.* **73**, 4212–4217 (2001).
43. F. Patolsky, T. Gabriel, and I. Willner, Controlled electrocatalysis by microperoxidase-11 and Au-nanoparticle superstructures on conductive supports. *J. Electroanal. Chem.* **479**, 69–73 (1999).
44. Y. Lvov, B. Munge, O. Giraldo, I. Ichinose, S.L. Suib, and J.F. Rusling, Films of manganese oxide nanoparticles with polycations or myoglobin from alternate-layer adsorption. *Langmuir* **16**, 8850–8857 (2000).
45. Y.L. Zhou, Z. Li, N.F. Hu, Y.H. Zeng, and J.F. Rusling, Layer-by-layer assembly of ultrathin films of haemoglobin and clay nanoparticles with electrochemical and catalytic activity. *Langmuir* **18**, 8573–8579 (2002).
46. F.A. Armstrong, H.A.O. Hill, and N.J. Walton, Direct electrochemistry of redox proteins. *Acc. Chem. Res.* **21**, 407–413 (1998).
47. K.L. Egodage, B.S. de Silva, and G.S. Wilson, Probing the conformation and orientation of adsorbed protein using monoclonal antibodies: cytochrome c3 films on a mercury electrode. *J. Am. Chem. Soc.* **119**, 5295–5301 (1997).
48. D.E. Reed and F.M. Hawkridge, Direct electron transfer reactions of cytochrome c at silver electrodes. *Anal. Chem.* **59**, 2334–2339 (1987).
49. M.J. Eddowes and H.A.O. Hill, Novel method for the investigation of the electrochemistry of metalloproteins: cytochrome c. *J. Chem. Soc. Chem. Commun.* **21**, 771–772 (1977).
50. P. Yeh and T. Kuwana, Reversible electrode reaction of cytochrome c. *Chem. Lett.* 1145–1148 (1977).
51. H. Allen, O. Hill, N.I. Hunt, and A.M. Bond, The transient nature of the diffusion controlled component of the electrochemistry of cytochrome c at "bare" gold electrodes: an explanation based on a self-blocking mechanism. *J. Electroanal. Chem.* **436**, 17–25 (1997).

52. S.M. Chen and S.V. Chen, The bioelectrocatalytic properties of cytochrome c by direct electrochemistry on DNA film modified electrode. *Electrochim. Acta* **48**, 513–529 (2003).
53. Y. Sato and F. Mizutani, Electrochemical responses of cytochrome c on gold electrodes modified with nucleic acid base derivatives – electrochemical and quartz crystal microbalance studies. *Electrochim. Acta* **45**, 2869–2875 (2000).
54. Z.Q. Feng, S. Imabayashi, T. Kakiuchi, and K. Niki, Electroreflectance spectroscopic study of the electron transfer rate of cytochrome c electrostatically immobilized on the ω-carboxyl alkanethiol monolayer modified gold electrode. *J. Electroanal. Chem.* **394**, 149–154 (1995).
55. Y.H. Wu and S.S. Hu, The fabrication of a colloidal gold–carbon nanotubes composite film on a gold electrode and its application for the determination of cytochrome c. *Colloid Surface B* **41**, 299–304 (2005).
56. G. Wang, J.J. Xu, and H.Y. Chen, Interfacing cytochrome c to electrodes with a DNA-carbon nanotube composite film. *Electrochem. Commun.* **4**, 506–509 (2002).
57. C.X. Cai, H.X. Ju, and H.Y. Chen, The effects of temperature and electrolyte on the redox potential of cytochrome c at a chemically modified microband gold electrode. *Electrochim. Acta* **40**, 1109–1112 (1995).
58. J.M. Sevilla, T. Pineda, A.J. Roman, R. Madueno, and M. Blazquez, The direct electrochemistry of cytochrome c at a hanging mercury drop electrode modified with 6-mercaptopurine. *J. Electroanal. Chem.* **451**, 89–93 (1998).
59. F. Lisdat, B. Ge, and F.W. Scheller, Oligonucleotide-modified electrodes for fast electron transfer to cytochrome c. *Electrochem. Commun.* **1**, 65–68 (1999).
60. F.A. Armstrong, A.M. Bond, H.A.O. Hill, B.N. Oliver, and I.S.M. Psalti, Electrochemistry of cytochrome c, plastocyanin, and ferredoxin at edge- and basal-plane graphite electrodes interpreted via a model based on electron transfer at electroactive sites of microscopic dimensions in size. *J. Am. Chem. Soc.* **111**, 91859189 (1989).
61. Y.H. Wu and S.S. Hu, Voltammetric investigation of cytochrome c on gold coated with a self-assembled glutathione monolayer. *Bioelectrochemistry* **68**, 105–112 (2006).
62. C.H. Lei, F. Lisdat, U. Wollenberger, and F.W. Scheller, Cytochrome c/clay-modified electrode. *Electroanalysis* **11**, 274–276 (1999).
63. C.H. Lei and J.Q. Deng, Hydrogen peroxide sensor based on coimmobilized methylene green and horseradish peroxidase in the same montonorillonite-modified bovine serum albumin-glutaraldehyde matrix on a glassy carbon electrode surface. *Anal. Chem.* **68**, 3344–3349 (1996).
64. Y. Sallez, P. Bianco, and E. Lojou, Electrochemical behavior of c-type cytochromes at clay-modified carbon electrodes: a model for the interaction between proteins and soils. *J. Electroanal. Chem.* **493**, 37–49 (2000).
65. O. Ikeda, M. Ohtani, T. Yamaguchi, and A. Komura, Direct electrochemistry of cytochrome c at a glassy carbon electrode covered with a microporous alumina membrane. *Electrochim. Acta* **43**, 833–839 (1998).
66. J.H. Yu and H.X. Ju, Preparation of porous titania sol-gel matrix for immobilization of horseradish peroxidase by a vapor deposition method. *Anal. Chem.* **74**, 3579–3583 (2002).
67. D. Rolison, Zeolite-modified electrodes and electrode-modified zeolites. *Chem. Rev.* **90**, 867–878 (1990).
68. B.H. Liu, R.Q. Hu, and J.Q. Deng, Characterization of immobilization of an enzyme in a modified Y zeolite matrix and its application to an amperometric glucose biosensor. *Anal. Chem.* **69**, 2343–2348 (1997).
69. J. Wang and A. Walcarius, Zeolite containing oxidase-based carbon paste biosensors. *J. Electroanal. Chem.* **404**, 237–242 (1996).
70. B.H. Liu, F. Yan, J.L. Kong, and J.Q. Deng, A reagentless amperometric biosensor based on the coimmobilization of horseradish peroxidase and methylene green in a modified zeolite matrix. *Anal. Chim. Acta* **386**, 31–39 (1999).
71. Z.H. Dai, S.Q. Liu, and H.X. Ju, Direct electron transfer of cytochrome c immobilized on a NaY zeolite matrix and its application in biosensing. *Electrochim. Acta* **49**, 2139–2144 (2004).

72. C. Hinnen, R. Parsons, and K. Niki, Electrochemical and spectroreflectance studies of the adsorbed horse heart cytochrome c and cytochrome c3 from *D. vulgaris*, miyazaki strain, at gold electrode. *J. Electroanal. Chem.* **147**, 329–337 (1983).
73. M.A. Hayat, in *Colloidal Gold: Principles, Methods, and Applications*, vol. 1. Academic Press, San Diego (1989).
74. M. Bendayan, in *Colloidal Gold: Principles, Methods, and Applications* (M.A. Hayat, ed.), vol. 2. Academic Press, San Diego (1989).
75. A.L. Crumbliss, J.Z. Stonehuerner, R.W. Henkens, J. Zhao, and J.P. O'Daly, A carrageenan hydrogel stabilized colloidal gold multi-enzyme biosensor electrode utilizing immobilized horseradish peroxidase and cholesterol oxidase/cholesterol esterase to detect cholesterol in serum and whole blood. *Biosens. Bioelectron.* **8**, 331–337 (1993).
76. J. Zhao, R.W. Henkens, J. Stonehuerner, J.P. O'Daly, and A.L. Crumbliss, Direct electron-transfer at horseradish-peroxidase colloidal gold modified electrodes. *J. Electroanal. Chem.* **327**, 109–119 (1992).
77. A.N. Shipway, E. Katz, and I. Willner, Nanoparticle arrays on surfaces for electronic, optical and sensoric applications. *Chemphyschem* **1**, 18–52 (2000).
78. A. Doron, E. Katz, and I. Willner, Organization of Au colloids as monolayer films onto ITO glass surfaces: application of the metal colloid films as base interfaces to construct redox-active monolayers. *Langmuir* **11**, 1313–1317 (1995).
79. L.A. Lyon, D.J. Pena, and M.J. Natan, Surface plasmon resonance of Au colloid-modified Au films: particle size dependence. *J. Phys. Chem. B* **103**, 5826–5831 (1999).
80. K.R. Brown, L.A. Lyon, A.P. Fox, B.D. Reiss, and M.J. Natan, Hydroxylamine seeding of colloidal Au nanoparticles 3. Controlled formation of conductive Au films. *Chem. Mater.* **12**, 314–323 (2000).
81. Y.D. Jin, X.F. Kang, Y.H. Song, B.L. Zhang, G.J. Cheng, and S.J. Dong, Controlled nucleation and growth of surface-confined gold nanoparticles on a (3-aminopropyl)trimethoxysilane-modified glass slide: a strategy for SPR substrates. *Anal. Chem.* **73**, 2843–2849 (2001).
82. K.R. Brown, A.P. Fox, and M.J. Natan, Morphology-dependent electrochemistry of cytochrome c at Au colloid-modified SnO_2 electrodes. *J. Am. Chem. Soc.* **118**, 1154–1157 (1996).
83. F. Patolsky, T. Gabriel, and I. Willner, Controlled electrocatalysis by microperoxidase-11 and Au-nanoparticle superstructures on conductive supports. *J. Electroanal. Chem.* **479**, 69–73 (1999).
84. C.J. Brinker and G.W. Scherer, in *Sol-Gel Science: The Physics and Chemistry of Sol-Gel Processing*. Academic Press, New York (1990).
85. L.L. Hench and J. West, The sol-gel process. *Chem. Rev.* **90**, 33–72 (1990).
86. B.Q. Wang, B. Li, Z.X. Wang, G.B. Xu, Q. Wang, and S.J. Dong, Sol-gel thin-film immobilized soybean peroxidase biosensor for the amperometric determination of hydrogen peroxide in acid medium. *Anal. Chem.* **71**, 1935–1939 (1999).
87. O.P. Matyshevska, A.Y. Karlash, Y.V. Shtogun, A. Benilov, Y. Kirgizov, K.O. Gorchinskyy, E.V. Buzaneva, Y.I. Prylutskyy, and P. Scharff, Self-organizing DNA/carbon nanotube molecular films. *Mat. Sci. Eng. C-Bio. S.* **15**, 249–252 (2001).
88. E. Buzaneva, A. Karlash, K. Yakovkin, Y. Shtogun, S. Putselyk, D. Zherebetskiy, A. Gorchinskiy, G. Popova, S. Prilutska, O. Matyshevska, Y. Prilutskyy, P. Lytvyn, P. Scharff, and P. Eklund, DNA nanotechnology of carbon nanotube cells: physico-chemical models of self-organization and properties. *Mat. Sci. Eng. C-Bio. S.* **19**, 41–45 (2002).
89. R. Saito, G. Dresselhaus, and M.S. Dresselhaus, Tunneling conductance of connected carbon nanotubes. *Phys. Rev. B* **53**, 2044–2050 (1996).
90. D. Porath, A. Bezryadin, S. de Vries, and C. Dekker, Direct measurements of electrical transport through DNA molecules. *Nature* **403**, 635–637 (2000).
91. S.J. Trans, M.H. Devoret, H. Dai, A. Thess, R.E. Smalley, L.J. Geerligs, and C. Dekker, Individual single-wall carbon nanotubes as quantum wires. *Nature* **386**, 474–476 (1997).
92. G. Wang, J.J. Xu, and H.Y. Chen, Interfacing cytochrome c to electrodes with a DNA-carbon nanotube composite film. *Electrochem. Commun.* **4**, 506–509 (2002).
93. E.F. Bowden, F.M. Hawkridge, and P.M. Blount, in *Bioelectrochemistry* (S. Srinivasan, Y.A. Chizmadzhev, J. Bockris, B.E. Conway, and E. Yeager, eds), p. 297. Plenum Press, New York (1985).

94. J.F. Stargardt, F.M. Hawkridge, and H.L. Landrum, Reversible heterogeneous reduction and oxidation of sperm whale myoglobin at a surface modified gold minigrid electrode. *Anal. Chem.* **50**, 930–932 (1978).
95. I. Taniguchi, K. Watanabe, M. Tominaga, and F.M. Hawkridge, Direct electron transfer of horse heart myoglobin at an indium oxide electrode. *J. Electroanal. Chem.* **333**, 331–338 (1992).
96. M. Tominaga, T. Kumagai, S. Takita, and I. Taniguchi, Effect of surface hydrophilicity of an indium oxide electrode on direct electron transfer of myoglobins. *Chem. Lett.* **10**, 1771–1774 (1993).
97. K. Chattopadhyay and S. Mazumdar, Direct electrochemistry of heme proteins: effect of electrode surface modification by neutral surfactants. *Bioelectrochemistry* **53**, 17–24 (2000).
98. B.C. King and F.M. Hawkridge, A study of the electron transfer and oxygen binding reactions of myoglobin. *J. Electroanal. Chem.* **237**, 81–92 (1987).
99. Q. Lu, X.X. Chen, Y.H. Wu, and S.S. Hu, Studies on direct electron transfer and biocatalytic properties of heme proteins in lecithin film. *Biophys. Chem.* **117**, 55–63 (2005).
100. A.E.F. Nassar, W.S. Willis, and J.F. Rusling, Electron transfer from electrodes to myoglobin: facilitated in surfactant films and blocked by adsorbed biomacromolecules. *Anal. Chem.* **67**, 2386–2392 (1995).
101. D. Mimica, J.H. Zagal, and F. Bedioui, Electroreduction of nitrite by hemin, myoglobin and haemoglobin in surfactant films. *J. Electroanal. Chem.* **497**, 106–113 (2001).
102. A.E.F. Nassar, Z. Zhang, N.F. Hu, J.F. Rusling, and T. Kumosinski, Protein-coupled electron transfer from electrodes to myoglobin in ordered biomembrane-like films. *J. Phys. Chem. B* **101**, 2224–2231 (1997).
103. Y.H. Wu, Q.C. Shen, and S.S. Hu, Direct electrochemistry and electrocatalysis of heme-proteins in regenerated silk fibroin film. *Anal. Chim. Acta* **558**, 179–186 (2006).
104. W.J. Macknight, E.A. Ponomarenko, and D.A. Tirrell, Self-assembled polyelectrolyte-surfactant complexes in nonaqueous solvents and in the solid state. *Acc. Chem. Res.* **31**, 781–788 (1998).
105. Y. Okahata, G. Enna, K. Taguchi, and T. Seki, Electrochemical permeability control through a redox bilayer film. *J. Am. Chem. Soc.* **107**, 5300–5301 (1985).
106. Y.J. Hu, N.F. Hu, and Y.H. Zeng, Electrochemistry and electrocatalysis with myoglobin in biomembrane-like surfactant-polymer $2C_{12}N^+PA^-$ composite films. *Talanta* **50**, 1183–1195 (2000).
107. H. Sun, H.Y. Ma, and N.F. Hu, Electroactive haemoglobin-surfactant-polymer biomembrane-like films. *Bioelectrochem. Bioenerg.* **49**, 1–10 (1999).
108. L.W. Wang and N.F. Hu, Electrochemistry and electrocatalysis with myoglobin in biomembrane-like DHP-PDDA polyelectrolyte-surfactant complex films. *J. Colloid Interface Sci.* **236**, 166–172 (2001).
109. Z.H. Dai, X.X. Xu, and H.X. Ju, Direct electrochemistry and electrocatalysis of myoglobin immobilized on a hexagonal mesoporous silica matrix. *Anal. Biochem.* **332**, 23–31 (2004).
110. J. Ye and R.P. Baldwin, Catalytic reduction of myoglobin and haemoglobin at chemically modified electrodes containing methylene blue. *Anal. Chem.* **60**, 2263–2268 (1988).
111. C.H. Fan, H.Y. Wang, S. Sun, D.X. Zhu, G. Wagner, and G.X. Li, Electron-transfer reactivity and enzymatic activity of haemoglobin in a SP sephadex membrane. *Anal. Chem.* **73**, 2850–2854 (2001).
112. Q. Huang, Z. Lu, and J.F. Rusling, Composite films of surfactants, Nafion, and proteins with electrochemical and enzyme activity. *Langmuir* **12**, 5472–5480 (1996).
113. Q. Lu and S.S. Hu, Studies on direct electron transfer and biocatalytic properties of haemoglobin in polytetrafluoroethylene film. *Chem. Phys. Lett.* **424**, 167–171 (2006).
114. X.J. Han, W.M. Huang, J.B. Jia, S.J. Dong, and E.K. Wang, Direct electrochemistry of haemoglobin in egg–phosphatidylcholine films and its catalysis to H_2O_2. *Biosens. Bioelectron.* **17**, 741–746 (2002).
115. C.H. Fan, G. Wagner, and G.X. Li, Effect of dimethyl sulfoxide on the electron transfer reactivity of haemoglobin. *Bioelectrochemistry* **54**, 49–51 (2001).
116. Y.X. Xu, F. Wang, X.X. Chen, and S.S. Hu, Direct electrochemistry and electrocatalysis of heme-protein based on N,N-dimethylformamide film electrode. *Talanta* **70**, 651–655 (2006).
117. K.M. Clark and C.E. Glatz, A binding model for the precipitation of proteins by carboxymethyl cellulose. *Chem. Eng. Sci.* **47**, 215–224 (1992).
118. H. Zaleska, P. Tomasik, and C.-Y. Lii, Formation of carboxymethyl cellulose–casein complexes by electrosynthesis. *Food Hydrocolloids* **16**, 215–224 (2002).

119. H. Huang, P.L. He, N.F. Hu, and Y.H. Zeng, Electrochemical and electrocatalytic properties of myoglobin and haemoglobin incorporated in carboxymethyl cellulose films. *Bioelectrochemistry* **61**, 29–38 (2003).
120. Q. Lu, J.H. Xu, and S.S. Hu, Studies on the direct electrochemistry of haemoglobin immobilized by yeast cells. *Chem. Commun.* **27**, 2860–2862 (2006).
121. Y.Z. Guo and A.R. Guadalupe, Direct electrochemistry of horseradish peroxidase adsorbed on glassy carbon electrode from organic solutions. *Chem. Commun.* **15**, 1437–1438 (1997).
122. R.A. Marcus and N. Sutin, Electron transfers in chemistry and biology. *BBA-Bioenergetics* **811**, 265–322 (1985).
123. H.B. Gray and B.G. Malmstrom, Long-range electron transfer in multisite metalloproteins. *Biochemistry* **28**, 7499–7505 (1989).
124. C. Bourdillon, C. Demaille, J. Moiroux, and J.M. Saveant, From homogeneous electroenzymatic kinetics to antigen-antibody construction and characterization of spatially ordered catalytic enzyme assemblies on electrodes. *Acc. Chem. Res.* **29**, 529–535 (1996).
125. H.B. Dunford, in *Peroxidases in Chemistry and Biology* (J. Everse, K.E. Everse, and M.B. Grisham, eds), p. 1. CRC Press, Boca Raton, FL (1991).
126. T. Ruzgas, E. Csoregi, J. Emneus, L. Gorton, and G. Marko-Varga, Peroxidase-modified electrodes: fundamentals and application. *Anal. Chim. Acta* **230**, 123–138 (1996).
127. T. Ferri, A. Poscia, and R. Santucci, Direct electrochemistry of membrane entrapped horseradish peroxidase. Part I: A voltammetric and spectroscopic study. *Bioelectrochem. Bioenerg.* **44**, 177–181 (1998).
128. T. Ferri, A. Poscia, and R. Santucci, Direct electrochemisty of a membrane entrapped horseradish peroxidase. Part II: Flowing amperometric detection of H_2O_2 and other analytes. *Bioelectrochem. Bioenerg.* **45**, 221–226 (1998).
129. X.H. Chen, C.M. Ruan, J.L. Kong, and J.Q. Deng, Characterization of the direct electron transfer and bioelectrocatalysis of horseradish peroxidase in DNA film at pyrolytic graphite electrode. *Anal. Chim. Acta* **412**, 89–98 (2000).
130. X.H. Chen, X.S. Peng, J.L. Kong, and J.Q. Deng, Facilitated electron transfer from an electrode to horseradish peroxidase in a biomembrane-like surfactant film. *J. Electroanal. Chem.* **480**, 26–33 (2000).
131. J. Wang and T. Golden, Permselectivity and ion-exchange properties of Eastman-AQ polymers on glassy carbon electrades. *Anal. Chem.* **61**, 1397–1400 (1989).
132. J. Wang, D. Leech, M. Ozsoz, S. Martinez, and M.R. Smyth, One step fabrication of glucose sensors with enzyme entrapment within poly(ester-sulfonic acid). coatings. *Anal. Chim. Acta* **245**, 139–143 (1991).
133. P. Bianco, A. Taye, and J. Haladjian, Incorporation of cytochrome c and cytochrome c3 within poly(ester-sulfonic acid) films cast on pyrolytic graphite electrodes. *J. Electroanal. Chem.* **377**, 299–303 (1994).
134. N.F. Hu and J.F. Rusling, Electrochemistry and catalysis with myoglobin in hydrated poly(ester sulfonic acid) ionomer films. *Langmuir* **13**, 4119–4125 (1997).
135. J. Yang, N.F. Hu, and J.F. Rusling, Enhanced electron transfer for haemoglobin in poly(ester sulfonic acid) films on pyrolytic graphite electrodes. *J. Electroanal. Chem.* **463**, 53–62 (1999).
136. R. Huang and N.F. Hu, Direct electrochemistry and electrocatalysis with horseradish peroxidase in Eastman AQ film. *Bioelectrochemistry* **54**, 75–81 (2001).
137. B.N. Oliver, J.O. Egekeze, and R.W. Murray, Solid-state voltammetry of a protein in a polymer solvent. *J. Am. Chem. Soc.* **110**, 2321–2322 (1988).
138. H. Sun, N.F. Hu, and H.Y. Ma, Direct electrochemistry of haemoglobin in polyacrylamide hydrogel films on pyrolytic graphite electrodes. *Electroanalysis* **12**, 1064–1070 (2000).
139. L. Shen, R. Huang, and N.F. Hu, Myoglobin in polyacrylamide hydrogel films: direct electrochemistry and electrochemical catalysis. *Talanta* **56**, 1131–1139 (2002).
140. R. Huang and N.F. Hu, Direct voltammetry and electrochemical catalysis with horseradish peroxidase in polyacrylamide hydrogel films. *Biophys. Chem.* **104**, 199–208 (2003).

141. P. Aymard, D.R. Martin, K. Plucknett, T.J. Foster, A.H. Clark, and I.T. Norton, Influence of thermal history on the structural and mechanical properties of agarose gels. *Biopolymers* **59**, 131–144 (2001).
142. M. Hara, S. Iazvovskaia, H. Ohkawa, Y. Asada, and J. Miyake, Immobilization of P450 monooxygenase and chloroplast for use in light-driven bioreactors. *J. Biosci. Bioeng.* **87**, 793–797 (1999).
143. H.H. Liu, Z.Q. Tian, Z.X. Lu, Z.L. Zhang, M. Zhang, and D.W. Pang, Direct electrochemistry and electrocatalysis of heme-proteins entrapped in agarose hydrogel films. *Biosens. Bioelectron.* **20**, 294–304 (2004).
144. E. Topoglidis, T. Lutz, R.L. Willis, C.J. Barnett, A.E.G. Cass, and J.R. Durrant, Protein adsorption on nanoporous TiO_2 films: a novel approach to studying photoinduced protein/electrode transfer reactions. *Faraday Discuss.* **116**, 35–46 (2000).
145. E. Topoglidis, C.J. Campbell, A.E.G. Cass, and J.R. Durrant, Factors that affect protein adsorption on nanostructured titania films. A novel spectroelectrochemical application to sensing. *Langmuir* **17**, 7899–7906 (2001).
146. Q.W. Li, G.A. Luo, and J. Feng, Direct electron transfer for heme proteins assembled on nanocrystalline TiO_2 film. *Electroanalysis* **13**, 359–363 (2001).
147. K.J. McKenzie and F. Marken, Accumulation and reactivity of the redox protein cytochrome c in mesoporous films of TiO_2 phytate. *Langmuir* **19**, 4327–4331 (2003).
148. X. Xu, J.Q. Zhao, D.C. Jiang, J.L. Kong, B.H. Liu, and J.Q. Deng, TiO_2 sol-gel derived amperometric biosensor for H_2O_2 on the electropolymerized phenazine methosulfate modified electrode. *Anal. Bioanal. Chem.* **374**, 1261–1266 (2002).
149. Y. Zhang, P.L. He, and N.F. Hu, Horseradish peroxidase immobilized in TiO_2 nanoparticle films on pyrolytic graphite electrodes: direct electrochemistry and bioelectrocatalysis. *Electrochim. Acta* **49**, 1981–1988 (2004).
150. P. Nicholls and G.R. Schonbaum, in *The Enzymes* (P.D. Boyer, H. Lardy, and K. Myrback, eds), p. 158. Academic Press, Orlando (1963).
151. M.R.N. Murthy, T.J. Reid, A. Sicignano, N. Tanaka, and M.G. Rossmann, Structure of beef liver catalase. *J. Mol. Biol.* **152**, 465–499 (1981).
152. D. Voer and J.G. Voet, in *Biochemistry*, 2nd ed. Wiley, New York (1995).
153. B. Halliwell and J.M. Gutteridge, in *Free Radicals in Biologyand Medicine*, 3rd ed., p. 134. New York, Oxford University Press (1999).
154. X.H. Chen, H. Xie, J.L. Kong, and J.Q. Deng, Characterization for didodecy, l-dimethylammonium bromide liquid crystal film entrapping catalase with enhanced direct electron transfer rate. *Biosens. Bioelectron.* **16**, 115–120 (2001).
155. Z. Zhang, S. Chouchane, R.S. Magliozzo, and J.F. Rusling, Direct voltammetry and catalysis with Mycobacterium tuberculosis catalase-peroxidase, peroxidases, and catalase in lipid films. *Anal. Chem.* **74**, 163–170 (2002).
156. H.Y. Lu, Z. Li, and N.F. Hu, Direct voltammetry and electrocatalytic properties of catalase incorporated in polyacrylamide hydrogel films. *Biophys. Chem.* **104**, 623–632 (2003).
157. M.E. Lai and A. Bergel, Direct electrochemistry of catalase on glassy carbon electrodes. *Bioelectrochemistry* **55**, 157–160 (2002).
158. E. Horozova, Z. Jordanowa, and V. Bogdanovskaya, Enzymatic and electrochemical reactions of catalase immobilized on carbon materials. *Z. Naturforsch.* **50**, 499–504 (1995).
159. E. Horozova, N. Dimcheva, and Z. Jordanova, Adsorption, catalytic and electrochemical activity of catalase immobilized on carbon materials. *Z. Naturforsch.* **52**, 639–644 (1997).
160. B.F. Erlanger, B.X. Chen, M. Zhu, and L. Brus, Binding of an anti-fullerene IgG monoclonal antibody to single wall carbon nanotubes. *Nano Lett.* **1**, 465–467 (2001).
161. J.J. Davis, M.L.H. Green, H.A.O. Hill, Y.C. Leung, P.J. Sadler, J. Solan, A.V. Xavier, and S.C. Tsang, The immobilisation of proteins in carbon nanotubes. *Inorg. Chim. Acta* **272**, 261–266 (1998).
162. M. Shim, N.W.S. Kam, R.J. Chen, Y. Li, and H. Dai, Functionalization of carbon nanotubes for biocompatibility and bio-molecular recognition. *Nano. Lett.* **2**, 285–288 (2002).
163. R.J. Chen, Y.G. Zhang, D.W. Wang, and H.J. Dai, Noncovalent sidewall functionalization of single-walled carbon nanotubes for protein immobilization. *J. Am. Chem. Soc.* **123**, 3838–3839 (2001).

164. L. Wang, J.X. Wang, and F.M. Zhou, Direct electrochemistry of catalase at a gold electrode modified with single-wall carbon nanotubes. *Electroanalysis* **16**, 627–632 (2004).
165. R.M. Ianniello, T.J. Lindsay, and A.M. Yacynych, Differential pulse voltammetric study of direct electron transfer in glucose oxidase chemically modified graphite electrodes. *Anal. Chem.* **54**, 1098–1101 (1982).
166. L. Jiang, C.J. McNeil, and J.M. Cooper, Direct electron transfer reaction of glucose oxidase immobilized at a self-assembled monolayer. *J. Chem. Soc. Chem. Commun.* 1293–1295 (1995).
167. Y.D. Zhao, W.D. Zhang, H. Chen, and Q.M. Luo, Direct electron transfer of glucose oxidase molecules adsorbed onto carbon nanotube powder microelectrode. *Anal. Sci.* **18**, 939–941 (2002).
168. A. Guiseppi-Elie, C.H. Lei, and R.H. Baughman, Direct electron transfer of glucose oxidase on carbon nanotubes. *Nanotechnology* **13**, 559–564 (2002).
169. C.X. Cai and J. Chen, Direct electron transfer of glucose oxidase promoted by carbon nanotubes. *Anal. Biochem.* **332**, 75–83 (2004).
170. L. Wang and Z.B. Yuan, Direct electrochemistry of glucose oxidase at a gold electrode modified with single-wall carbon nanotubes. *Sensors* **3**, 544–554 (2003).
171. M. Horisberger, Colloidal gold as a tool in molecular biology. *Trends Biochem. Sci.* **8**, 395–397 (1983).
172. J. Zhao, R. Henkens, J. Stonehuemer, J.P. O'Daly and A.L. Crumbliss, Direct electron transfer at horseradish peroxidase-colloidal gold modified electrodes. *J. Electroanal. Chem.* **327**, 109–119 (1992).
173. K.R. Brown, A.P. Fox, and M.J. Natan, Morphology-dependent electrochemistry of cytochrome c at Au colloid-modified SnO_2 electrodes. *J. Am. Chem. Soc.* **118**, 1154–1157 (1996).
174. S.Q. Liu and H.X. Ju, Reagentless glucose biosensor based on direct electron transfer of glucose oxidase immobilized on colloidal gold modified carbon paste electrode. *Biosens. Bioelectron.* **19**, 177–183 (2003).
175. M. Ghosh, C. Anthony, K. Harlos, M.G. Goodwin, and C. Blake, The refined structure of the quinoprotein methanol dehydrogenase from Methylobacterium extorquens at 1.94Å. *Structure* **3**, 177–187 (1995).
176. A.M. Cleton-Jansen, N. Goosen, O. Fayet, and P. van de Putte, Cloning, mapping, and sequencing of the gene encoding Escherichia coli quinoprotein glucose dehydrogenase. *J. Bacteriol.* **172**, 6308–6315 (1990).
177. M. Yamada, M. Elias, K. Matsushita, C.T. Migita, and O. Adachi, Escherichia coli PQQ-containing quinoprotein glucose dehydrogenase: its structure comparison with other quinoproteins. *Biochim. Biophys. Acta: Proteins Proteomics* **1647**, 185–192 (2003).
178. C. Anthony, M. Ghosh, and C.C. Blake, The structure and function of methanol dehydrogenase and related quinoproteins containing pyrrolo-quinoline quinone. *Biochem. J.* **304**, 665–674 (1994).
179. B. Groen, J. Frank Jr, and J.A. Duine, Quinoprotein alcohol dehydrogenase from ethanol-grown Pseudomonas aeruginosa. *Biochem. J.* **223**, 921–924 (1984).
180. A. Oubrie, H.J. Rozeboom, K.H. Kalk, E.G. Huizinga, and B.W. Dijkstra, Crystal structure of quinohemoprotein alcohol dehydrogenase from Comamonas testosteroni. *J. Biol. Chem.* **277**, 3727–3732 (2002).
181. J. Razumiene, M. Niculescu, A. Ramanavicius, V. Laurinavicius, and E. Csoregi, Direct bioelectrocatalysis at carbon electrodes modified with quinohemoprotein alcohol dehydrogenase from Gluconobacter sp. 33. *Electroanalysis* **14**, 43–49 (2002).
182. T. Ikeda, D. Kobayashi, F. Matsushita, T. Sagara, and K. Niki, Bioelectrocatalysis at electrodes coated with alcohol-dehydrogenase, a quinohemoprotein with heme-c serving as a built-in mediator. *J. Electroanal. Chem.* **361**, 221–228 (1993).
183. J. Okuda and K. Sode, PQQ glucose dehydrogenase with novel electron transfer ability. *Biochem. Biophys. Res. Commun.* **314**, 793–797 (2004).
184. C.J. Gray, in *Enzyme-catalysed Reactions*. Van Nostrand Reinhold Co., London (1971).
185. C.G. Rodrigues and A.G. Wedd, Electrochemistry of xanthine oxidase at glassy carbon and mercury electrodes. *J. Electroanal. Chem.* **312**, 131–140 (1991).
186. T. Nishino and K. Okamoto, The role of the [2Fe-2s] cluster centers in xanthine oxidoreductase. *J. Inorg. Biochem.* **82**, 43–49 (2000).

187. R. Hille and R.F. Anderson, Coupled electron/proton transfer in complex flavoproteins – solvent kinetic isotope effect studies of electron transfer in xanthine oxidase and trimethylamine dehydrogenase. *J. Biol. Chem.* **276**, 31 193–31 201 (2001).
188. F.E. Inscore, R. McNaughton, B.L. Westcott, M.E. Helton, R. Jones, I.K. Dhawan, J.H. Enemark, and M.L. Kirk, Spectroscopic evidence for a unique bonding interaction in oxo-molybdenum dithiolate complexes: implications for sigma electron transfer pathways in the pyranopterin dithiolate centers of enzymes. *Inorg. Chem.* **38**, 1401–1410 (1999).
189. J.G. Zhao, J.P. O'Daly, R.W. Henkens, J. Stonehuerner, and A.L. Crumbliss, A xanthine oxidase/colloidal gold enzyme electrode for amperometric biosensor applications. *Biosens. Bioelectron.* **11**, 493–502 (1996).
190. J.H. Pei and X.Y. Li, Xanthine and hypoxanthine sensors based on xanthine oxidase immobilized on a $CuPtCl_6$ chemically modified electrode and liquid chromatography electrochemical detection. *Anal. Chim. Acta* **414**, 205–213 (2000).
191. K.V. Gobi, Y. Sato, and F. Mizutani, Mediatorless superoxide dismutase sensors using cytochrome c-modified electrodes: xanthine oxidase incorporated polyion complex membrane for enhanced activity and in-vivo analysis. *Electroanalysis* **13**, 397–403 (2001).
192. X.J. Liu, W.L. Peng, H. Xiao, and G.X. Li, DNA facilitating electron transfer reaction of xanthine oxidase. *Electrochem. Commun.* **7**, 562–566 (2005).
193. L. Wang and Z.B. Yuan, Direct electrochemistry of xanthine oxidase at a gold electrode modified with single-wall carbon nanotubes. *Anal. Sci.* **20**, 635–638 (2004).
194. D.G. Buerk, in *Biosensors*, Technomic Publishing, Chap. 4. Pennsylvania (1993).
195. G.C. Zhao, Z.Z. Yin, L. Zhang, and X.W. Wei, Direct electrochemistry of cytochrome c on a multiwalled carbon nanotubes modified electrode and its electrocatalytic activity for the reduction of H_2O_2. *Electrochem. Commun.* **7**, 256–260 (2005).
196. J.J. Feng, G. Zhao, J.J. Xu, and H.Y. Chen, Direct electrochemistry and electrocatalysis of heme proteins immobilized on gold nanoparticles stabilized by chitosan. *Anal. Biochem.* **342**, 280–286 (2005).
197. Q.L. Wang, G.X. Lu, and B.J. Yang, Hydrogen peroxide biosensor based on direct electrochemistry of haemoglobin immobilized on carbon paste electrode by a silica sol-gel film. *Sens. Actuators, B: Chem.* **99**, 50–57 (2004).
198. Y. Xiao, H.X. Ju, and H.Y. Chen, Hydrogen peroxide sensor based on horseradish peroxidase-labeled Au colloids immobilized on gold electrode surface by cysteamine monolayer. *Anal. Chim. Acta.* **391**, 73–82 (1999).
199. A. Fantuzzi, F. Fairhead, and G. Gilardi, Direct Electrochemistry of immobilized human cytochrome P450 2E1. *J. Am. Chem. Soc.* **126**, 5040–5041 (2004).
200. S. Varma and B. Mattiasson, Amperometric biosensor for the detection of hydrogen peroxide using catalase modified electrodes in polyacrylamide. *J. Biotech.* **119**, 172–180 (2005).
201. S.X. Zhang, N. Wang, Y.M. Niu, and C.Q. Sun, Immobilization of glucose oxidase on gold nanoparticles modified Au electrode for the construction of biosensor. *Sens. Actuators, B: Chem.* **109**, 367–374 (2005).
202. C.L. Wang and A. Mulchandani, Ferrocene-conjugated polyaniline-modified enzyme electrodes for determination of peroxides in organic media. *Anal. Chem.* **67**, 1109–1114 (1995).
203. C.X. Lei, H. Wang, G.L. Shen, and R.Q. Yu, Immobilization of enzymes on the nano-Au film modified glassy carbon electrode for the determination of hydrogen peroxide and glucose. *Electroanalysis* **16**, 736–740 (2004).
204. J.L. Tang, B.Q. Wang, Z.Y. Wu, X.J. Han, S.J. Dong, and E.K. Wang, Lipid membrane immobilized horseradish peroxidase biosensor for amperometric determination of hydrogen peroxide. *Biosens. Bioelectron.* **18**, 867–872 (2003).
205. Z.H. Dai, S.Q. Liu, and H.X. Ju, Direct electron transfer of cytochrome c immobilized on a NaY zeolite matrix and its application in biosensing. *Electrochim. Acta* **49**, 2139–2144 (2004).
206. E. Cadenas, Biochemistry of oxygen toxicity. *Annu. Rev. Biochem.* **58**, 79–110 (1989).
207. K.H. Krause, Professional phagocytes: predators and prey of microorganisms. *Schweizerische Medizinische Wochenschrift* **130**, 97–100 (2000).

208. I. Fridovich, Superoxide anion radical (O_2) superoxide dismutases and related matters. *J. Biol. Chem.* **272**, 18 515–18 517 (1997).
209. N.S. Chandel, E. Maltepe, E. Goldwasser, C.E. Mathieu, M.C. Simon, and P.T. Schumacker, Mitochondrial reactive oxygenspecies trigger hypoxia-induced transcription. *Proc. Natl. Acad. Sci.* **95**, 11 715–11 720 (1998).
210. S.H.E. Kaufmann, in *Immunity to Intracellular Bacteria: Fundamental Immunology*. New York, Raven Press (1993).
211. A.J. Kowaltowski and A.E. Vercesi, Mitochondrial damage induced by conditions of oxidative stress. *Free Radic. Biol. Med.* **26**, 463–471 (1999).
212. B. Halliwell, Antioxidants in human health and disease. *Annu. Rev. Nutr.* **16**, 33–50 (1996).
213. S. Ignatov, D. Hishniashvili, B. Ge, F.W. Scheller, and F. Lisdat, Amperometric biosensor based on a functionalized gold electrode for the detection of antioxidants. *Biosens. Bioelectron.* **17**, 191–199 (2002).
214. C.J. McNeil and P. Manning, Sensors-based measurements of the role and interactions of free radicals in cellular systems. *Rev. Mol. Biotech.* **82**, 443–455 (2002).
215. C.J. McNeil, K.R. Greenough, P.A. Weeks, C.H. Self, and J.M. Cooper, Electrochemical sensors for direct reagentless measurement of superoxide production by human neutrophils. *Free Rad. Res. Commun.* **17**, 399–406 (1992).
216. R.H. Fabian, D.S. DeWitt, and T.A. Kent, In vivo detection of superoxide anion production by the brain using a cytochrome c electrode. *Cereb. Blood Flow Metab.* **15**, 242–247 (1995).
217. T. Haruyama, S. Shiino, Y. Yanagida, E. Kobatake, and M. Aizawa, Two types of electrochemical nitric oxide (NO) sensing systems with heat-denatured Cyt C and radical scavenger PTIO. *Biosens. Bioelectron.* **13**, 763–769 (1998).
218. N. Bistolas, U. Wollenberger, C. Jung, and F.W. Scheller, Cytochrome P450 biosensors – a review. *Biosens. Bioelectron.* **20**, 2408–2423 (2005).
219. T.L. Poulos, Cytochrome P450. *Curr. Opin. Struc. Biol.* **5**, 767–774 (1995).
220. G. Brunner and H. Loesgen, in *Artificial Organs* (R. Kenedi, J. Courtney, J. Gaylor, and T. Gilchrist, eds), p. 338. Macmillan Press Ltd, London (1977).
221. F. Scheller, R. Renneberg, G. Strand, K. Pommerening, and P. Mohr, Electrochemical aspects of cytochrome P-450 system from liver microsomes. *Bioelectrochem. Bioenerg.* **4**, 500–507 (1977).
222. V.V. Shumyantseva, T.V. Bulko, S.A. Usanov, R.D. Schmid, C. Nicolini, and A.I. Archakov, Construction and characterization of bioelectrocatalytic sensors based on cytochromes P450. *J. Inorg. Biochem.* **87**, 185–190 (2001).
223. E.I. Iwuoha, S. Joseph, Z. Zhang, M.R. Smyth, U. Fuhr, and P.R. Ortiz de Montellano, Drug metabolism biosensors: electrochemical reactivities of cytochrome CYP101 immobilised in synthetic vesicular systems. *J. Pharm. Biom. Anal.* **17**, 1101–1110 (1998).
224. Y.M. Lvov, Z.Q. Lu, J.B. Schenkman, X.L. Zu, and J.F. Rusling, Direct electrochemistry of myoglobin and cytochrome P450cam in alternate layer-by-layer films with DNA and other polyions. *J. Am. Chem. Soc.* **120**, 4073–4080 (1998).
225. B. Munge, C. Estavillo, J.B. Schenkman, and J.F. Rusling, Optimization of electrochemical and peroxide-driven oxidation of styrene with ultrathin polyion films containing cytochrome P450cam and myoglobin. *Chem. Biol. Biol. Chem.* **4**, 82–89 (2003).
226. X. Zu, Z. Lu, Z. Zhang, J.B. Schenkman, and J.F. Rusling, Electroenzyme-catalyzed oxidation of styrene and cis-β-methylstyrene using thin films of cytochrome P450cam and myoglobin. *Langmuir* **15**, 7372–7377 (1999).
227. C. Estavillo, Z. Lu, I. Jansson, J.B. Schenkman, and J.F. Rusling, Epoxidation of styrene by human cyt P-450 1A2 by thin film electrolysis and peroxide activation compared to solution reactions. *Biophys. Chem.* **104**, 291–296 (2003).
228. S. Joseph, J.F. Rusling, Y.M. Lvov, T. Fredberg, and U. Fuhr, An amperometric biosensor with human CYP3A4 as a novel drug screening tool. *Biochem. Pharmacol.* **65**, 1817–1826 (2003).
229. H. Frauenfelder, B.H. McMahon, R.H. Austin, K. Chu, and J.T. Groves, The role of structure, energy landscape, dynamics, and allostery in the enzymatic function of myoglobin. *Proc. Natl. Acad. Sci.* **98**, 2370 (2001).

230. Y.M. Li, H.H. Liu, and D.W. Pang, Direct electrochemistry and catalysis of heme-proteins entrapped in methyl cellulose films. *J. Electroanal. Chem.* **574**, 23–31 (2004).
231. Y.H. Wu and S.S. Hu, Direct electron transfer of ferritin in dihexadecylphosphate on an Au film electrode and its catalytic oxidation toward ascorbic acid. *Anal. Chim. Acta* **527**, 37–43 (2004).
232. H.H. Liu, Z.Q. Tian, Z.X. Lu, Z.L. Zhang, M. Zhang, and D.W. Pang, Direct electrochemistry and electrocatalysis of heme-proteins entrapped in agarose hydrogel films. *Biosens. Bioelectron.* **20**, 294–304 (2004).
233. M. Bayachou, L. Elkbir, and P.J. Farmer, Catalytic two-electron reductions of N_2O and N_3^- by myoglobin in surfactant films. *Inorg. Chem.* **39**, 289–293 (2000).
234. L. Shen and N.F. Hu, Heme protein films with polyamidoamine dendrimer: direct electrochemistry and electrocatalysis. *BBA-Bioenergetics* **1608**, 23–33 (2004).
235. Z.H. Dai, X.X. Xu, and H.X. Ju, Direct electrochemistry and electrocatalysis of myoglobin immobilized on a hexagonal mesoporous silica matrix. *Anal. Biochem.* **332**, 23–31 (2004).
236. Z. Li and N.F. Hu, Direct electrochemistry of heme proteins in their layer-by-layer films with clay nanoparticles. *J. Electroanal. Chem.* **558**, 155–165 (2003).
237. H.Y. Wang, R. Guan, C.H. Fan, D.X. Zhu, and G.X. Li, A hydrogen peroxide biosensor based on the bioelectrocatalysis of haemoglobin incorporated in a kieselgubr film. *Sens. Actuators, B: Chem.* **84**, 214–218 (2002).
238. J.T. Pang, C.H. Fan, X.J. Liu, T. Chen, and G.X. Li, A nitric oxide biosensor based on the multi-assembly of haemoglobin/montmorillonite/polyvinyl alcohol at a pyrolytic graphite electrode. *Biosens. Bioelectron.* **19**, 441–445 (2003).
239. J.D. Zhang and M. Oyama, A hydrogen peroxide sensor based on the peroxidase activity of haemoglobin immobilized on gold nanoparticles-modified ITO electrode. *Electrochim. Acta* **50**, 85–90 (2004).
240. J.J. Feng, G. Zhao, J.J. Xu, and H.Y. Chen, Direct electrochemistry and electrocatalysis of heme proteins immobilized on gold nanoparticles stabilized by chitosan. *Anal. Biochem.* **342**, 280–286 (2005).
241. A. Salimi, E. Sharifi, A. Noorbakhsh, and S. Soltanian. Direct voltammetry and electrocatalytic properties of haemoglobin immobilized on a glassy carbon electrode modified with nickel oxide nanoparticles. *Electrochem. Commun.* **8**, 1499–1508 (2005).
242. X.J. Liu, Y. Xu, X. Ma, and G.X. Li, A third-generation hydrogen peroxide biosensor fabricated with haemoglobin and Triton X-100. *Sens. Actuators, B: Chem.* **106**, 284–288 (2005).
243. X. Ma, X.J. Liu, H. Xiao, and G.X. Li, Direct electrochemistry and electrocatalysis of haemoglobin in poly-3-hydroxybutyrate membrane. *Biosens. Bioelectron.* **20**, 1836–1842 (2005).
244. Q. Lu, T. Zhou, and S.S. Hu, Direct electrochemistry of haemoglobin in PHEA and its catalysis to H_2O_2. *Biosens. Bioelectron.* **22**, 899–904 (YEAR).
245. X.J. Han, W.M. Huang, J.B. Jia, S.J. Dong, and E.K. Wang, Direct electrochemistry of haemoglobin in egg-phosphatidylcholine films and its catalysis to H_2O_2. *Biosens. Bioelectron.* **17**, 741–746 (2002).
246. R. Huang and N.F. Hu, Direct electrochemistry and electrocatalysis with horseradish peroxidase in Eastman AQ films. *Bioelectrochemistry* **54**, 75–81 (2001).
247. T.J. Castilho, M. del P.T. Sotomayor, and L.T. Kubota, Amperometric biosensor based on horseradish peroxidase for biogenic amine determinations in biological samples. *J. Pharmaceut. Biomed. Anal.* **37**, 785–791 (2005).
248. S. Akgöl and E. Dinckaya, A novel biosensor for specific determination of hydrogen peroxide: catalase enzyme electrode based on dissolved oxygen probe. *Talanta* **48**, 363–367 (1999).
249. A. Salimi, A. Noorbakhsh, and M. Ghadermarz, Direct electrochemistry and electrocatalytic activity of catalase incorporated onto multiwall carbon nanotubes-modified glassy carbon electrode. *Anal. Biochem.* **344**, 16–24 (2005).
250. M. Li, P.L. He, Y. Zhang, and N.F. Hu, An electrochemical investigation of haemoglobin and catalase incorporated in collagen films. *BBA-Bioenergetics* **1749**, 43–51 (2005).
251. H.Y. Lu, Z. Li, and N.F. Hu, Direct voltammetry and electrocatalytic properties of catalase incorporated in polyacrylamide hydrogel films. *Biophys. Chem.* **104**, 623–632 (2003).

252. Y.X. Huang, W.J. Zhang, H. Xiao, and G.X. Li, An electrochemical investigation of glucose oxidase at a CdS nanoparticles modified electrode. *Biosens. Bioelectron.* **21**, 817–821 (2005).
253. Y. Liu, M.K. Wang, F. Zhao, Z.A. Xu, and S.J. Dong, The direct electron transfer of glucose oxidase and glucose biosensor based on carbon nanotubes/chitosan matrix, *Biosens. Bioelectron.* **21**, 984–988 (2005).
254. S.Q. Liu and H.X. Ju, Reagentless glucose biosensor based on direct electron transfer of glucose oxidase immobilized on colloidal gold modified carbon paste electrode. *Biosens. Bioelectron.* **19**, 177–183 (2003).
255. S.G. Wang, Q. Zhang, R.L. Wang, S.F. Yoon, J. Ahn, D.J. Yang, J.Z. Tian, J.Q. Li, and Q. Zhou, Multi-walled carbon nanotubes for the immobilization of enzyme in glucose biosensors. *Electrochem. Commun.* **5**, 800–803 (2003).
256. Y.H. Wu and S.S. Hu, Direct electrochemistry of glucose oxidase in a colloid Au–dihexadecylphosphate composite film and its application to develop a glucose biosensor. *Bioelectrochemistry.* Available online 6 May (2006).
257. M. Viticoli, A. Curulli, A. Cusma, S. Kaciulis, S. Nunziante, L. Pandolfi, F. Valentini, and G. Padeletti, Third-generation biosensors based on TiO_2 nanostructured films. *Mater. Sci. Engine. C* **26**, 947–951 (2006).
258. Y.Q. Zhang, W.D. Shen, R.A. Gu, J. Zhu, and R.Y. Xue, Amperometric biosensor for uric acid based on uricase-immobilized silk fibroin membrane. *Anal. Chim. Acta* **369**, 123–128 (1998).
259. F.F. Zhang, X.L. Wang, S.Y. Ai, Z.D. Sun, Q. Wan, Z.Q. Zhu, Y.Z. Xian, L.T. Jin, and K. Yamamoto, Immobilization of uricase on ZnO nanorods for a reagentless uric acid biosensor. *Anal. Chim. Acta* **519**, 155–160 (2004).
260. J.H. Pei and X.Y. Li, Xanthine and hypoxanthine sensors based on xanthine oxidase immobilized on a CuPtCl6 chemically modified electrode and liquid chromatography electrochemical detection. *Anal. Chim. Acta* **414**, 205–213 (2000).
261. Ü.A. Kirgöz, S. Timur, J. Wang, and A. Teleponcu, Xanthine oxidase modified glassy carbon paste electrode. *Electrochem. Commun.* **6**, 913–916 (2004).
262. X.J. Liu, W.L. Peng, H. Xiao, and G.X. Li, DNA facilitating electron transfer reaction of xanthine oxidase. *Electrochem. Commun.* 7, 562–566 (2005).
263. F. Ghamouss, S. Ledru, N. Ruillé, F. Lantier, and M. Boujtita, Bulk-modified modified screen-printing carbon electrodes with both lactate oxidase (LOD) and horseradish peroxide (HRP) for the determination of l-lactate in flow injection analysis mode. *Anal. Chim. Acta* **570**, 158–164 (2006).

INDEX

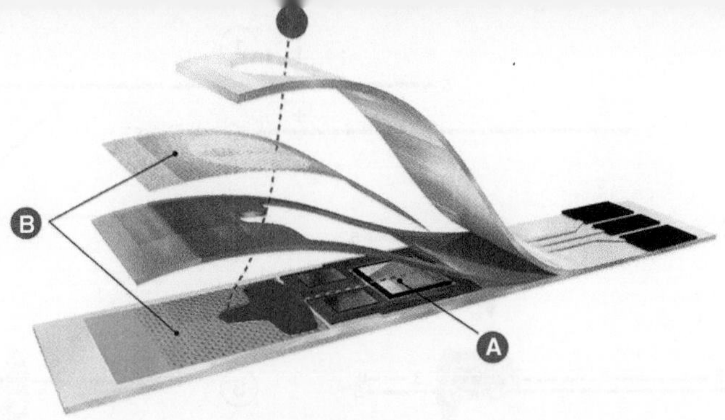

Plate 1 A typical commercial strip for self-testing of blood glucose (based on a biosensor manufactured by Abbott Inc.) (see Figure 3.4).

DNA sample: 5′ A G G T C T C G 3′

AGGA−
AGGT+
AGGC−
AGGG−

GGTA−
GGTT−
GGTC+
GGTG−

GTCA−
GTCT+
GTCC−
GTCG−

TCTA−
TCTT−
TCTC+
TCTG−

CTCA−
CTCT−
CTCC−
CTCG+

AGGT
GGTC
GTCT
TCTC
CTCG
AGGTCTCG

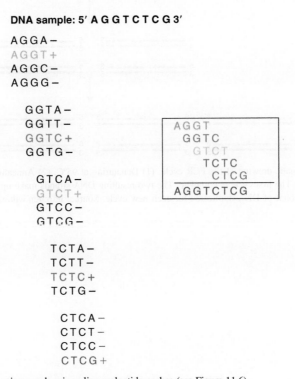

Plate 2 Sequence assembly using overlapping oligonucleotide probes (see Figure 11.6).

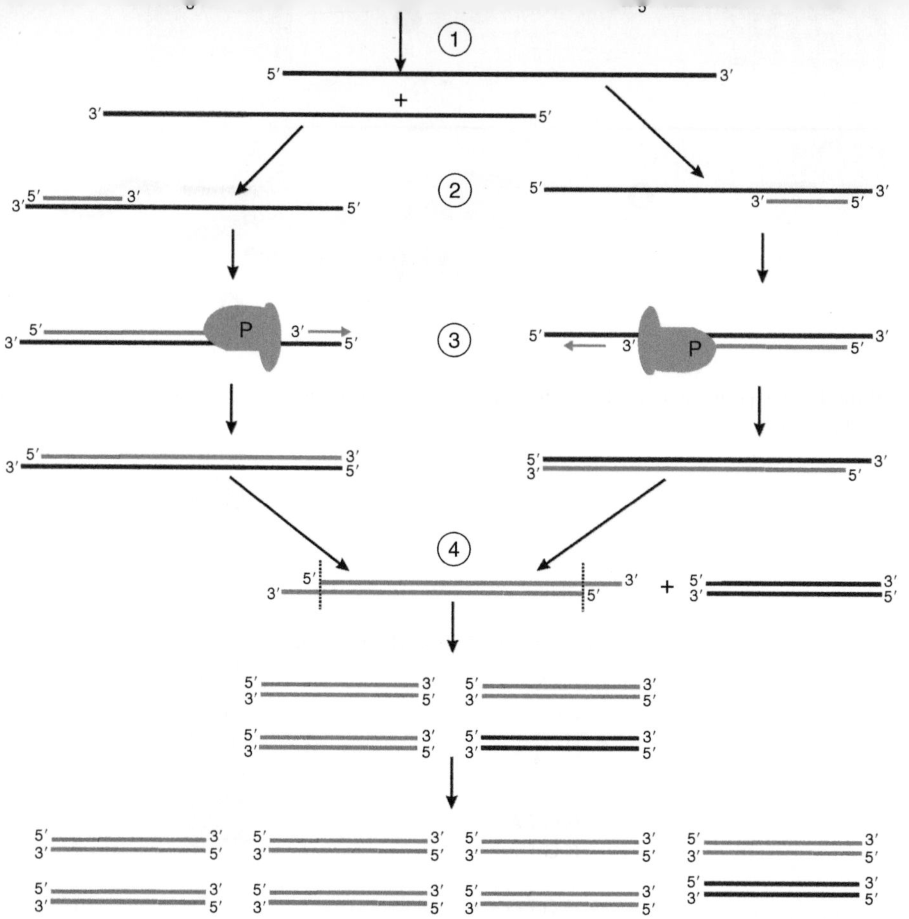

Plate 3 Schematic drawing of the PCR cycle. (1) Denaturing at 95ºC. (2) Annealing at 55ºC. (3) Synthesizing at 72ºC (P = Polymerase). (4) The first cycle is complete. The two resulting DNA strands make up the template DNA for the next cycle, thus doubling the amount of DNA duplicated for each new cycle. Source: http://en.wikipedia.org/wiki/Polymerasechainreaction (see Figure 11.10).

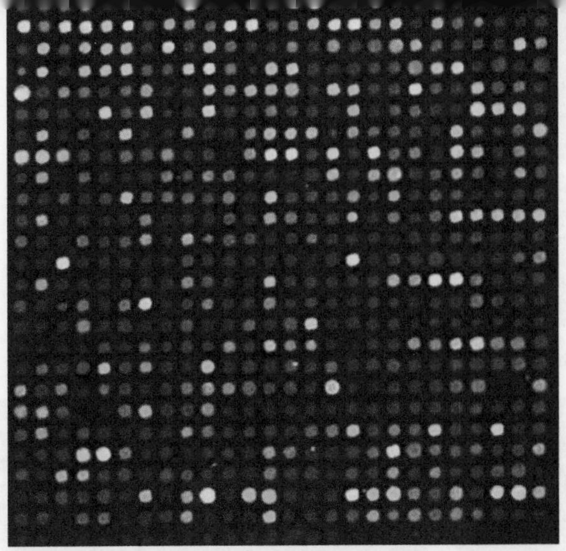

Plate 4 Fluorescence image of a gel-pad array biochip (see Figure 11.14).

Plate 5 Ideal microarray spots (see Figure 11.15).

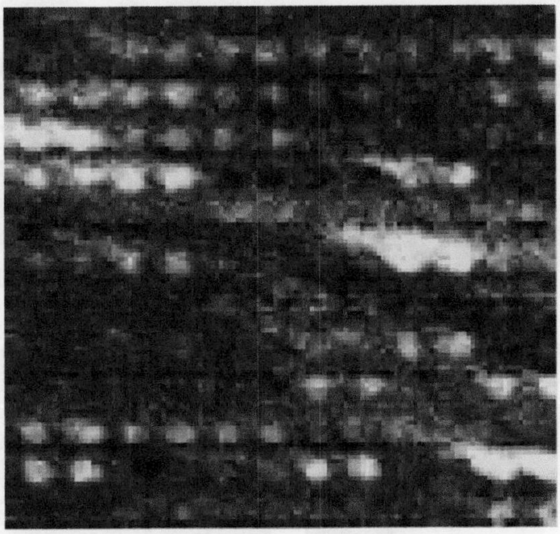

Plate 6 Comet tails (see Figure 11.16).

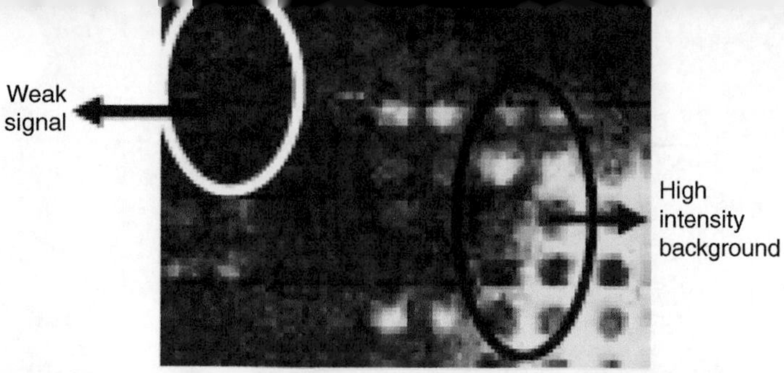

Plate 7 High intensity background and weak signal (see Figure 11.17).

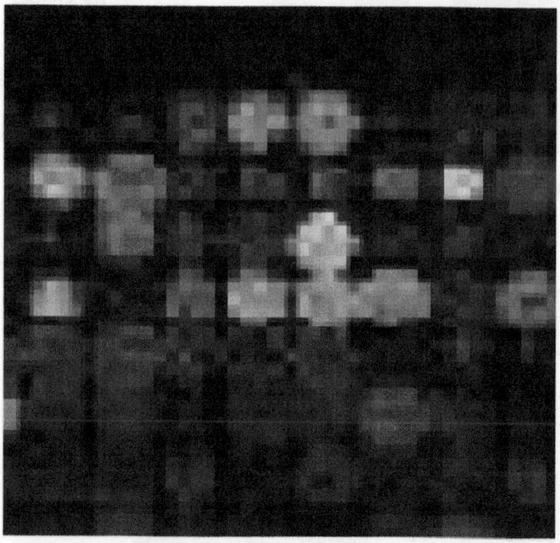

Plate 8 Spot overlap (see Figure 11.18).

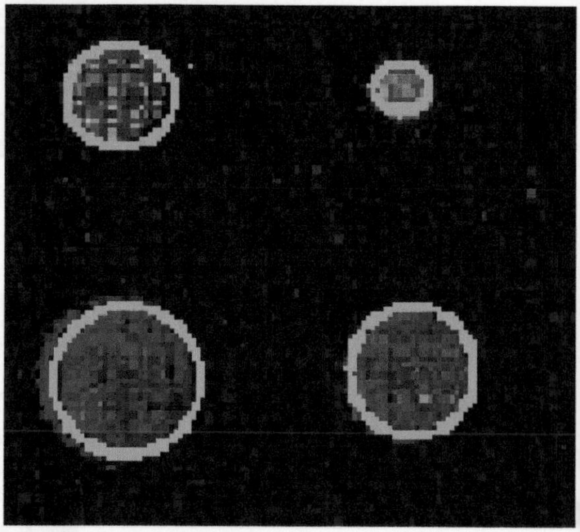

Plate 10 Spots with different sizes (see Figure 11.20).

Plate 9 Fixed circle segmentation (see Figure 11.19).

Plate 11 Silicon wafer-based electronic biochip (10 × 10 arrays) (see Figure 11.28).

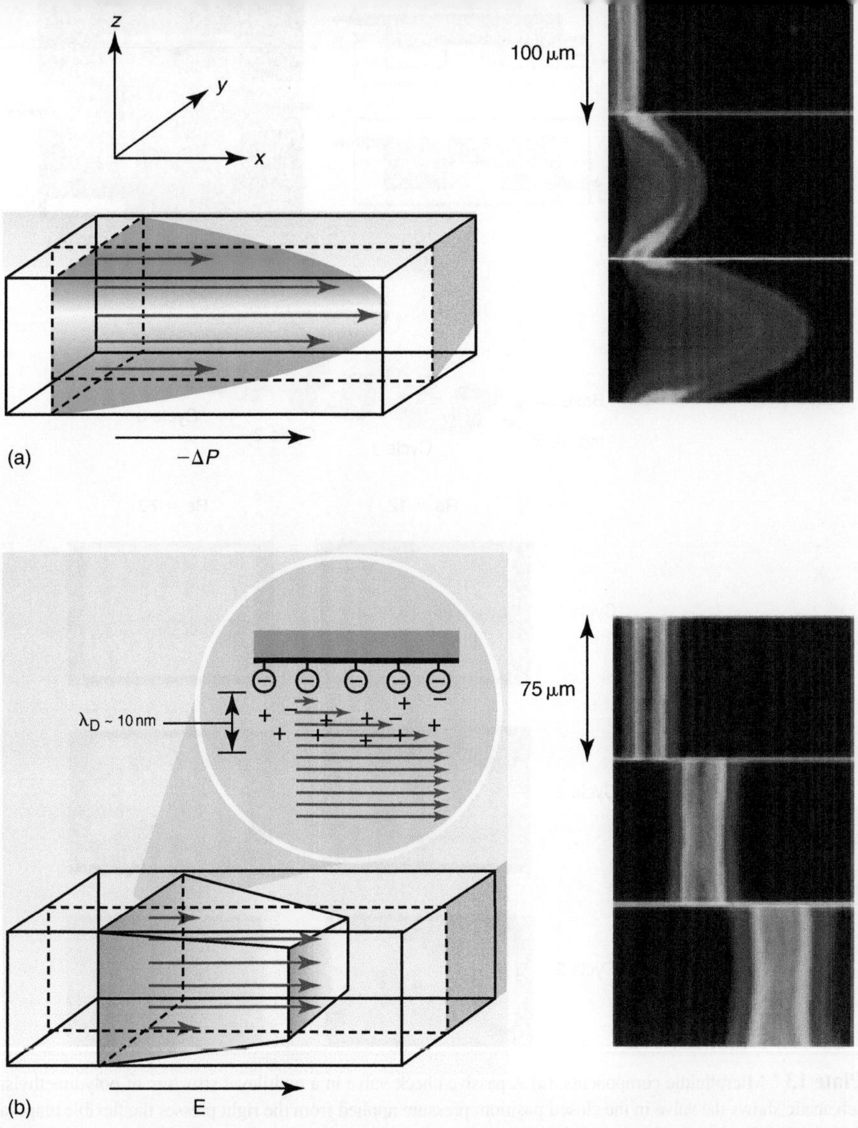

Plate 12 Flow profiles in microchannels. (a) A pressure gradient, $-\Delta P$, along a channel generates a parabolic or Poiseuille flow profile in the channel. The velocity of the flow varies across the entire cross-sectional area of the channel. On the right is an experimental measurement of the distortion of a volume of fluid in a Poiseuille flow. The frames show the state of the volume of fluid 0, 66, and 165 ms after the creation of a fluorescent molecule. (b) In electroosmotic flow in a channel, motion is induced by an applied electric field E. The flow speed only varies within the so-called Debye screening layer, of thickness λ_D. On the right is an experimental measurement of the distortion of a volume of fluid in an electroosmotic flow. The frames show the state of the fluorescent volume of fluid 0, 66, and 165 ms after the creation of a fluorescent molecule [165]. Source: http://www.niherst.gov.tt/scipop/sci-bits/microfluidics.htm (see Figure 11.32).

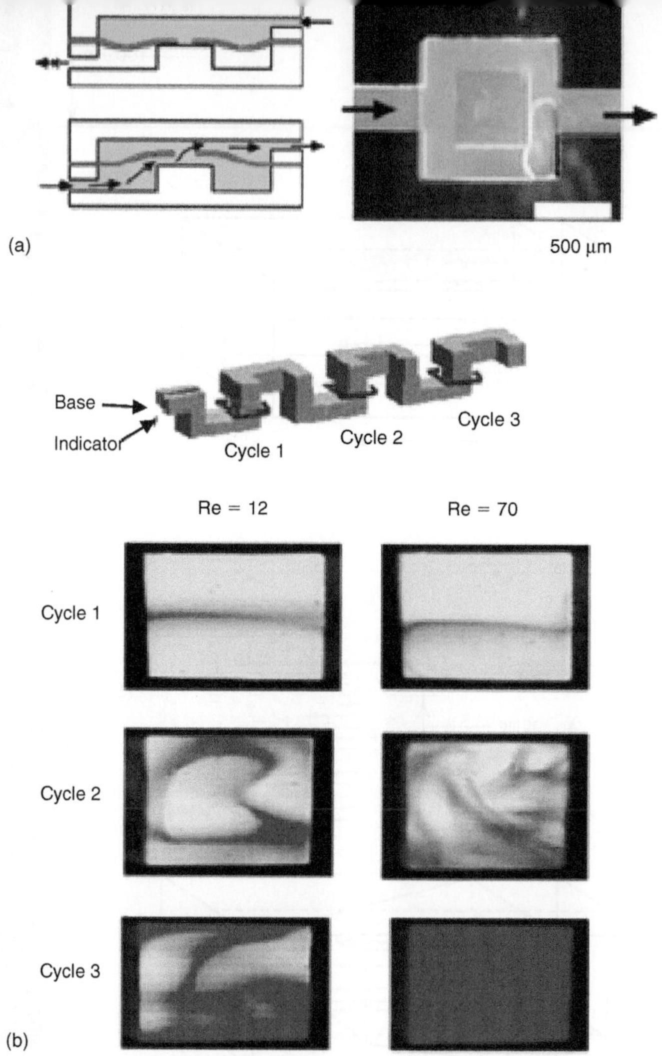

Plate 13 Microfluidic components. (a) A passive check valve in a multilayer structure of polydimethylsiloxane (PDMS). The upper schematic shows the valve in the closed position: pressure applied from the right presses the flexible diaphragm against a post, blocking flow through the hole in the center of the diaphragm. The lower schematic shows the valve in the open position: pressure applied from the left lifts the diaphragm off the post and allows flow through the diaphragm hole. The image on the right shows flow of a fluorescent fluid through the valve. (b) The design of this chaotic advection mixer exploits the character of laminar flow in a three-dimensional serpentine channel to mix adjacent streams in a single flow. A solution of pH indicator runs adjacent to a stream of basic solution and reveals the progress of mixing. As the two solutions mix, the indicator becomes red. Mixing is efficient even at low Reynolds numbers. Although the flow is laminar, there are weak eddies in the corners [163, 167]. Source: http://www.niherst.gov.tt/scipop/sci-bits/microfluidics.htm (see Figure 11.33).

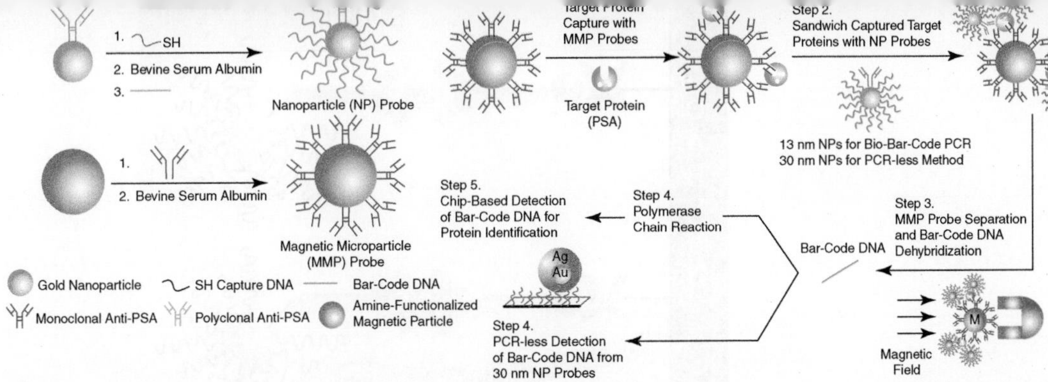

Plate 14 The bio-bar-code assay method. (a) Probe design and preparation. (b) PSA detection and bar-code DNA amplification and identification (Reproduced from [17] with permission.) (see Figure 14.2).

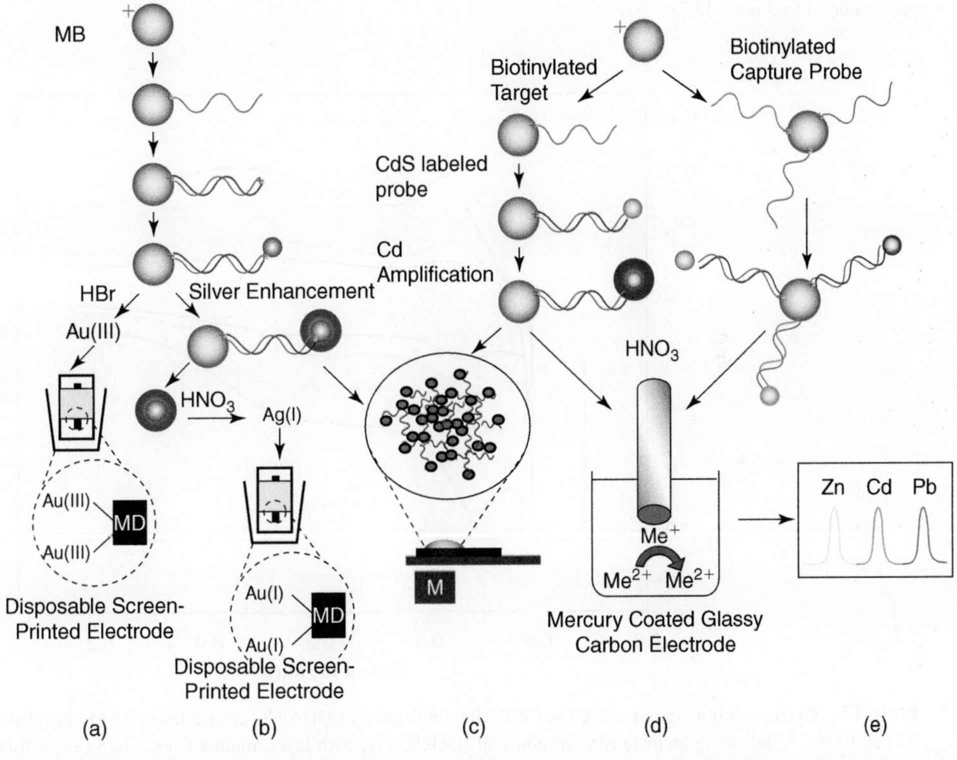

Plate 15 Nanoparticle-based electrochemical detection of DNA. These assays involve the introduction of the probe coated magnetic beads, the addition of the target/hybridization event, magnetic removal of unwanted materials, binding of the metal, and amplified electrochemical detection of the dissolved gold (a), silver (b), and cadmium sulfide (d) nanoparticles. Me: metal tag. Also shown are solid-state stripping (c) and multi-target (e) detection protocols. See text for details (see Figure 14.3).

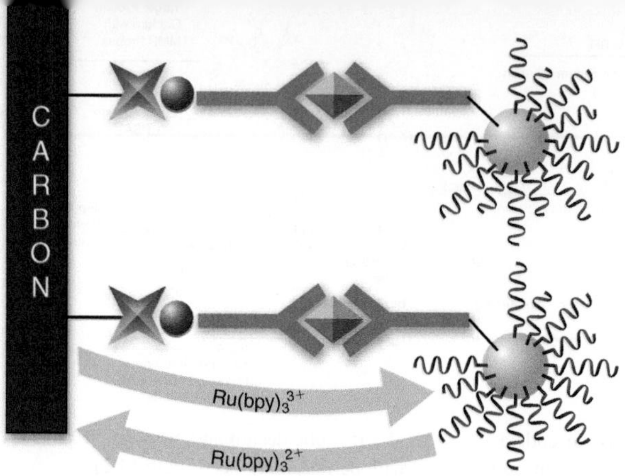

Plate 16 Schematic illustration of an electrochemical immunoassay based on the poly[G]-functionalized silica NPs on carbon electrode. Diamond stands for the mouse IgG; "Y" shape for mouse antibody; red strings for poly[G] (Reproduced from [62] with permission.) (see Figure 14.7)

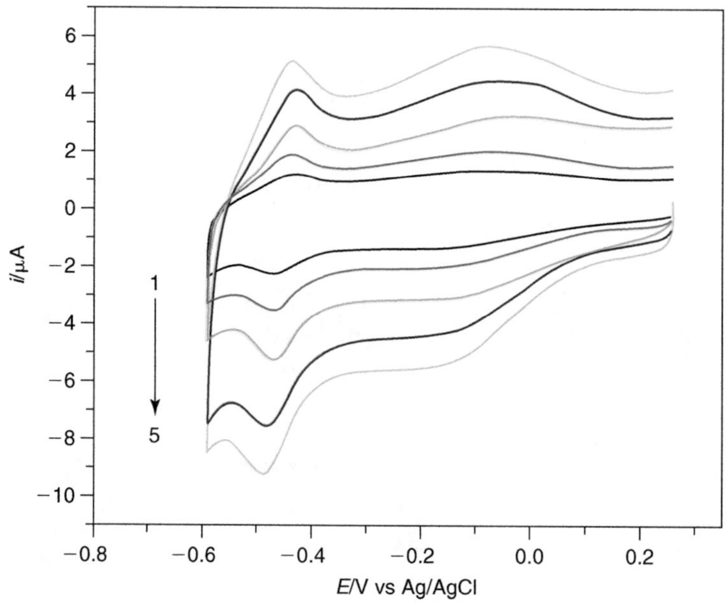

Plate 17 Cyclic voltammograms of $[GOx/CNT]_n$ immobilized on PDDA/GC electrodes in 0.1 M phosphate buffer solution (pH 7.4) at $0.1\,V\,s^{-1}$, indicating an increasing response of $[GOx/CNT]_n$ with layer number from 1 to 5 (unpublished results, J. Zhang, M. Feng, and H. Tachikawa) (see Figure 15.6).

Printed and bound by CPI Group (UK) Ltd, Croydon, CR0 4YY
08/05/2025
01864893-0001